The Largest Suslin Axiom

Developing the theory up to the current state of the art, this book studies the minimal model of the Largest Suslin Axiom (LSA), which is one of the most important determinacy axioms and features prominently in Hugh Woodin's foundational framework known as the Ultimate L.

The authors establish the consistency of LSA relative to large cardinals and develop methods for building models of LSA from other foundational frameworks such as Forcing Axioms. The book significantly advances the Core Model Induction method, which is the most successful method for building canonical inner models from various hypotheses. Also featured is a proof of the Mouse Set Conjecture in the minimal model of the LSA.

It will be indispensable for graduate students as well as researchers in mathematics and philosophy of mathematics who are interested in set theory and, in particular, in descriptive inner model theory.

Grigor Sargsyan is Professor at the Institute of Mathematics of the Polish Academy of Sciences. He holds a PhD in mathematics from the University of California, Berkeley. His thesis, which is the precursor of the work presented in this book, received Association of Symbolic Logic's Sacks Prize. He later received the National Science Foundation's Career Award, which was awarded to support the work carried out in this book.

Nam Trang is Assistant Professor of Mathematics at the University of North Texas. He holds a PhD in mathematics from the University of California, Berkeley. He is the recipient of the National Science Foundation's two regular grants and a Career Award, part of which supports the writing of this book. He also received an Outstanding Research Award from the UNT Department of Mathematics.

LECTURE NOTES IN LOGIC

A Publication of The Association for Symbolic Logic

This series serves researchers, teachers, and students in the field of symbolic logic, broadly interpreted. The aim of the series is to bring publications to the logic community with the least possible delay and to provide rapid dissemination of the latest research. Scientific quality is the overriding criterion by which submissions are evaluated.

More information, including a list of the books in the series, can be found at http://aslonline.org/books/lecture-notes-in-logic/

LECTURE NOTES IN LOGIC 56

The Largest Suslin Axiom

GRIGOR SARGSYAN
Polish Academy of Sciences

NAM TRANG
University of North Texas

ASSOCIATION FOR SYMBOLIC LOGIC

CAMBRIDGE
UNIVERSITY PRESS

CAMBRIDGE
UNIVERSITY PRESS

Shaftesbury Road, Cambridge CB2 8EA, United Kingdom

One Liberty Plaza, 20th Floor, New York, NY 10006, USA

477 Williamstown Road, Port Melbourne, VIC 3207, Australia

314–321, 3rd Floor, Plot 3, Splendor Forum, Jasola District Centre,
New Delhi – 110025, India

103 Penang Road, #05–06/07, Visioncrest Commercial, Singapore 238467

Cambridge University Press is part of Cambridge University Press & Assessment, a department of the University of Cambridge.

We share the University's mission to contribute to society through the pursuit of education, learning and research at the highest international levels of excellence.

www.cambridge.org
Information on this title: www.cambridge.org/9781009520713

DOI: 10.1017/9781009520683

Association for Symbolic Logic
Richard A. Shore, Publisher
Department of Mathematics, Cornell University, Ithaca, NY 14853
http://aslonline.org

When citing this work, please include a reference to the DOI 10.1017/9781009520683

First published 2024

A catalogue record for this publication is available from the British Library.

A Cataloging-in-Publication data record for this book is available from the Library of Congress

ISBN 978-1-009-52071-3 Hardback

The first author dedicates the book to his father, mother, wife, son and daughter:
Edvard Sargsyan, Gayane Shahinyan, Aleksandra Sawa, Tigran Sargsyan and
Nairi Sargsyan.

The second author dedicates the book to his father, mother, wife and son:
Phil Trang, Tram Pham, Tran Tran and Shawn Trang.

CONTENTS

Chapter 1

INTRODUCTION

This book is a contribution to *descriptive inner model theory*, which is the area of set theory that lies between descriptive set theory as developed in [24] and inner model theory. The main goal of this book is to advance the descriptive inner model theoretic methods to the level of the Largest Suslin Axiom (LSA), which is a strong determinacy axiom asserting that there is a largest Suslin cardinal and that the largest Suslin cardinal is a member of the Solovay sequence. In more concrete terms, our goal is twofold: Firstly develop methods for analyzing the minimal model of LSA, and secondly, develop methods for building the minimal model of LSA under various hypotheses such as the Proper Forcing Axiom or Large Cardinals. Since the introduction of Steel's recent book [63], the expository paper [28] and the introduction of [36] contain all the introductory information we need, here we will not introduce the subject matter of this book and instead will hope that the reader has consulted these sources.

The first problem is an instance of the problem Steel mentions on page xii of [63] where he writes: "The most important of the remaining open problems is whether, assuming determinacy, there actually are mouse pairs at every appropriate level of logical complexity". Theorem 10.1.2 shows that the aformentioned problem has a positive solution in the minimal model of LSA. As explained in any of the sources cited above, the goal for doing this is to show that letting Θ be the least ordinal that is not a surjective image of the reals, V_Θ^{HOD} as computed inside a determinacy model is a hod premouse. The above sources explain the importance of having a hod premouse representation of V_Θ^{HOD}.

The second problem amounts to advancing the Core Model Induction to the level of LSA. Corollary 12.0.22 constructs the minimal model of LSA assuming PFA. More dramatically, the paper [36], which extends the methods of this book, demonstrates that the Core Model Induction, in its current form, cannot be used to go much further than LSA.

Corollary 12.0.22 also builds the minimal model of LSA directly from large cardinals, namely strongly compact cardinals. However, Theorem 10.3.1 shows that LSA is weaker than a Woodin cardinal that is a limit of Woodin cardinals, and so strong compactness seems to be much more than needed. Nevertheless, while it is widely believed that strongly compact cardinals are consistency wise stronger than a Woodin cardinal that is a limit of Woodin cardinals, this is not yet known. Still we strongly believe that the methods developed in this book, the methods of [1] and the main theorem of [26] can be used to show that assuming the existence of a Woodin cardinal that is a limit of Woodin cardinals, the minimal model of LSA exists (cf. Definition 1.1.4).

Historically, LSA was introduced by Woodin in [68, Remark 9.28], and it features prominently in Woodin's Ultimate L framework (see [69, Definition 7.14] and Axiom I and Axiom II on page 97 of [69][1]). Theorem 10.3.1 is historically the first proof of the consistency of LSA relative to large cardinals. Cramer and Woodin established the consistency of LSA from large cardinals in the region of I_0 (see [5, Theorem 65]).

1.1. The technical content of the book

1.1.1. The Largest Suslin Axiom. LSA is a determinacy theory whose underlying theory is Woodin's AD^+. Chapter 9.1 of [68] provides a quick overview of AD^+, and Larson's recent manuscript [20] provides more details. Perhaps the most important consequence of AD^+ is the fact that assuming $V = L(\wp(\mathbb{R}))$, the fragment of V coded by the Suslin, co-Suslin sets of reals is Σ_1 elementary in V (see Theorem 9.7 of [68]).

We will need the following concepts to introduce LSA. A cardinal κ is OD-inaccessible if for every $\alpha < \kappa$ there is no surjection $f : \wp(\alpha) \to \kappa$ that is definable from ordinal parameters. A set of reals $A \subseteq \mathbb{R}$ is *κ-Suslin* if for some tree T on κ, $A = p[T]$[2]. A set A is *Suslin* if it is κ-Suslin for some κ; A is *co-Suslin* if its complement $\mathbb{R}\backslash A$ is Suslin. A set A is *Suslin, co-Suslin* if both A and its complement are Suslin. A cardinal κ is a *Suslin cardinal* if there is a set of reals A such that A is κ-Suslin but A is not λ-Suslin for any $\lambda < \kappa$. Suslin cardinals play an important role in the study of models of determinacy as can be seen by just flipping through the Cabal Seminar Volumes ([14, 15, 16, 17]). LSA is then the following theory.

DEFINITION 1.1.1. The Largest Suslin Axiom, abbreviated as LSA, is the conjunction of the following statements:

1. $\mathsf{ZF} + \mathsf{AD}^+$.
2. There is a largest Suslin cardinal.
3. The largest Suslin cardinal is OD-inaccessible.

LSA can also be defined in terms of the *Solovay sequence*.

---

[1]The requirement in these axioms that there is a strong cardinal which is a limit of Woodin cardinals is only possible if $L(A, \mathbb{R}) \vDash \mathsf{LSA}$.

[2]Given a cardinal κ, we say $T \subseteq \bigcup_{n<\omega} \omega^n \times \kappa^n$ is a *tree* on κ if T is closed under initial segments. Given a tree T on κ, we let $[T]$ be the set of its branches, i.e., $b \in [T]$ if $b \in \omega^\omega \times \kappa^\omega$ and letting $b = (b_0, b_1)$, for each $n \in \omega$, $(b_0 \restriction n, b_1 \restriction n) \in T$. We then let $p[T] = \{x \in \mathbb{R} : \exists f((x, f) \in [T])\}$.

DEFINITION 1.1.2. The Solovay sequence is a sequence $(\vartheta_\alpha : \alpha \leq \Omega)$ such that

1. $\vartheta_0 = \sup\{\beta : \exists f : \wp(\omega) \to \beta (f \text{ is an } OD \text{ surjection})\}$,
2. if $\vartheta_\alpha < \Theta$ then $\vartheta_{\alpha+1} = \sup\{\beta : \exists f : \wp(\vartheta_\alpha) \to \beta (f \text{ is an } OD \text{ surjection})\}$,
3. for limit $\lambda \leq \Omega$, $\vartheta_\lambda = \sup_{\alpha<\lambda} \vartheta_\alpha$.
4. $\vartheta_\Omega = \Theta$.

REMARK 1.1.3. LSA is then equivalent to the conjunction of the following axioms:

1. $\mathsf{ZF} + \mathsf{AD}^+$.
2. For some ordinal α, $\Theta = \vartheta_{\alpha+1}$ and ϑ_α is the largest Suslin cardinal.

The above equivalence can be shown using the material of Chapter 9.1 of [68]. We note that it follows from [68, Theorem 9.12] that LSA implies $\neg\mathsf{AD}_{\mathbb{R}}$.

1.1.2. The minimal model of LSA. Suppose V is a model of LSA. Let κ be the largest Suslin cardinal and suppose $A \subseteq \mathbb{R}$ has Wadge rank κ. It then follows that $L(A, \mathbb{R}) \vDash \mathsf{LSA}$. Keeping this fact in mind, we make the following definition.

DEFINITION 1.1.4. Suppose T is a first order theory extending AD^+. We say that M is *a minimal model* of T if

1. M is transitive and $M \vDash T$,
2. $\mathbb{R}, Ord \subseteq M$, and
3. for every N that is a (definable) class of M and contains all the reals and ordinals, either $N = M$ or $N \vDash \neg T$.

It follows that all minimal models of LSA have the form $L(A, \mathbb{R})$. A natural question is whether there is a unique minimal model of LSA. We will show (see the proof of Theorem 10.3.1) that in fact there is a unique minimal model of LSA which is naturally *the* minimal model of LSA. Woodin's proof of the existence of divergent models of AD^+ also shows that not all extensions of AD^+ have a unique minimal model (see [7, Theorem 6.1]).

The minimal model of LSA may not actually be big. For example, if N is a transitive model of AD^+ that contains the minimal model M of LSA and has a Suslin cardinal $> \Theta^M$ then $\Theta^M < \vartheta_0^N$. In particular, every set of reals in M is ordinal definable from a real in N. Motivated by this fact, we make the following definition.

DEFINITION 1.1.5. Suppose M is a transitive model containing all the reals and ordinals and such that $M \vDash \mathsf{AD}^+ + V = L(\wp(\mathbb{R}))$. We say M is *full* if for all transitive N such that

1. $M \subseteq N$ and
2. $N \vDash$ "$\mathsf{AD}^+ + V = L(\wp(\mathbb{R}))$",

Θ^M is a member of the Solovay sequence of N.

The following interesting problem seems central to our understanding of the models of AD^+ that we build from large cardinals or from other hypotheses.

PROBLEM 1.1.6. Do large cardinals or forcing axioms such as PFA imply that there is a full model of LSA?

In particular, whether the models of determinacy obtained as derived models of V contain full models of LSA *or not is a major open problem of the area.* Here we make the following conjecture which is motivated by Woodin's *Sealing Theorem* (see [19]). Below uB stands for the set of universally Baire sets and for a generic g, $\mathsf{uB}_g = (\mathsf{uB})^{V[g]}$ and $\mathbb{R}_g = \mathbb{R}^{V[g]}$.

CONJECTURE 1.1.7. *Suppose κ is a supercompact cardinal and there is a proper class of Woodin cardinals. Let $g \subseteq Coll(\omega, 2^{2^\kappa})$ be generic. Then in $L(\mathsf{uB}_g, \mathbb{R}_g)$, for each $\xi < \Theta$ there is α such that $\vartheta_\alpha \in (\xi, \Theta)$ and ϑ_α is the largest Suslin cardinal below $\vartheta_{\alpha+1}$.*

Thus, in the set up of the conjecture, $L(\mathsf{uB}_g, \mathbb{R}_g)$ has full models of LSA that are cofinal in its Wadge hierarchy. The following is what is known on Conjecture 1.1.7. Woodin (unpublished) has shown that $L(\mathsf{uB}_g, \mathbb{R}_g) \vDash$ "$\mathsf{AD}_{\mathbb{R}} + \Theta$ is a regular cardinal". Sandra Müller and the first author recently showed that $L(\mathsf{uB}_g, \mathbb{R}_g)$ can be represented as a derived model of some iterate of V. They also found a stationary-tower-free proof of Woodin's Sealing Theorem. These results are unpublished. [56] presents a stationary-tower-free proof of the derived model theorem.

The question of whether the Cramer-Woodin model of LSA from [5, Theorem 65] is a full model of LSA or not seems not only interesting but also important for understanding the relationship between large cardinals and models of AD^+.

1.1.3. The content of this book. In this book, we establish three kinds of results that can be stated without mentioning the technology developed to prove them. The first set of results deals with the minimal model of LSA. Assume V is the minimal model of LSA. Then the following hold.

(A) (Theorem 7.2.2) HOD $\vDash$ GCH.
(B) (Theorem 10.2.1) The Mouse Set Conjecture holds.

The second set of results contains a single result which shows the consistency of LSA relative to large cardinals. We will show the following.

(C) (Corollary 10.3.1) Suppose the theory ZFC + "there is a Woodin cardinal that is a limit of Woodin cardinals" is consistent. Then so is LSA.

The third type of result establishes the existence of the minimal model of LSA assuming combinatorial principles or forcing axioms. The following belongs to this group.

(D) (Corollary 12.0.22) Assume PFA. Then the minimal model of LSA exists.

The precursors of these results already exist in print. The first author demonstrated versions of (A), (B), and (C) for the theory $\mathsf{AD}_{\mathbb{R}}$ + "Θ is a regular cardinal". The second author proved the version of (D) for the same theory. The interested reader may consult [29], [31] and [65]. The reason to prove such results is to demonstrate that the underlying technical theory is robust and can be used in a wide range of situations.

Recently the authors of [2] showed that the theory $\mathsf{CH}+$"there is an ω_1-dense ideal on ω_1" implies that the minimal model of $\mathsf{AD}_{\mathbb{R}}+$"$\Theta$ is a regular cardinal" exists. This, along with an earlier result of Woodin, show that these two theories are equiconsistent. This solved part of Problem 12 of [68]. Whether there is a natural hypothesis asserting the existence of an ideal on a small cardinal that is equiconsistent with LSA is an interesting problem. In particular, letting M' be the minimal model of LSA, κ be the largest Suslin cardinal of M' and $M = L(\Gamma,\mathbb{R})$ where $\Gamma = \{A \in \wp(\mathbb{R}) \cap M' : w(A) < \kappa\}$[3], the model $M[G*H]$ where $G*H \subseteq Coll(\omega_1,\mathbb{R}) * Add(1,\omega_2)$ is M-generic has not be studied at all. The model $M[G*H]$ where $G*H \subseteq \mathbb{P}_{max} * Add(1,\omega_3)$ has been investigated in [4], but not much is known beyond [4][4].

1.1.4. The necessity of the short-tree-strategy mice. We do not know how to prove (B)-(D) using the methods of [63], and whether this is possible or not is a very interesting question[5]. The main issue seems to be the absence of an analysis of the LSA stages of the Solovay sequence using the least-branch hierarchy. The main technical concept we use to analyze such levels is the notion of a *short-tree-strategy mice*, which is developed in Chapter 3. Thus, the question is whether it is necessary to develop this theory in order to prove results like (B)-(D).

The main issue is the following. Assume AD^+. Suppose $\vartheta_{\alpha+1} < \Theta$ and ϑ_α is the largest Suslin cardinal below $\vartheta_{\alpha+1}$. Then if $(\mathcal{P},\Sigma)$ is the hod pair generating the pointclass $\Gamma_1 = \{A \subseteq \mathbb{R} : w(A) < \vartheta_{\alpha+1}\}$ then letting δ be the largest Woodin cardinal of $\mathcal{P}$, $((\mathcal{P}|\delta)^\#, \Sigma^{stc})$ is the pair generating the pointclass $\Gamma_0 = \{A \subseteq \mathbb{R} : w(A) < \vartheta_\alpha\}$. If one's goal is to show that assuming $\mathsf{AD}_{\mathbb{R}} + \mathsf{DC} + V = L(\wp(\mathbb{R}))$, $\mathsf{HOD} \vDash \mathsf{GCH}$ then it maybe possible to *skip* Γ_0 and build the generator of Γ_1. The problem with skipping Γ_0 and moving to Γ_1 is exactly the fact that it is then unclear how to prove theorems like (A)-(D). What one would have liked is some sort of hybrid method that does not skip Γ_0 but also incorporates ideas from [63] to avoid the theory of short-tree-strategy mice. It seems to us that this may not be possible.

Suppose then we decide not to skip over Γ_0, and suppose we have succeeded in building a generator $((\mathcal{P}|\delta)^\#, \Sigma^{stc})$ for Γ_0. At this stage, we do not know what $(\mathcal{P},\Sigma)$ must be and can only see $((\mathcal{P}|\delta)^\#, \Sigma^{stc})$. Set then $\mathcal{Q} = (\mathcal{P}|\delta)^\#$ and $\Lambda = \Sigma^{stc}$. What we need to show next is that we can extend $\mathcal{Q}$ to $\mathcal{P}$ in such a way that the following hold[6]:

1. δ is the largest cardinal of $\mathcal{P}$ and $H^{\mathcal{P}}_\delta = \mathcal{Q}|\delta$,
2. for all $A \subseteq \delta$, $A \in \mathcal{P}$ if and only if A is ordinal definable from $(\mathcal{Q},\Lambda)$,
3. $\mathcal{P} \vDash$ "δ is a Woodin cardinal".

[3] $w(A)$ is the Wadge rank of A.

[4] But see also [21].

[5] [63] does show that $\mathcal{H} \vDash \mathsf{GCH}$ but only assuming HPC.

[6] Below $H^{\mathcal{P}}_\delta$ is the set of all $X \in \mathcal{P}$ whose hereditary cardinality is $< \delta$.

The main issue seems to be with proving clause 2. It is a version of MSC for Λ, and the only way we know how to prove it is by building a Λ-mouse over $\mathcal{Q}$ whose derived model contains the set $\{(x,y) \in \mathbb{R}^2 : x$ is ordinal definable from y and $(\mathcal{Q},\Lambda)\}$. This requires a certain level of uniformity: $\mathcal{Q}$ and what we build on the top of $\mathcal{Q}$ have to be the same kind of objects, as otherwise the construction over $\mathcal{Q}$ can project across δ violating clause 3 above.

1.1.5. Some historical remarks on the large cardinal structure of hod mice. The large cardinal structure of hod mice has been somewhat of a mystery. While originally it seemed hod mice must have very limited large cardinal structure, nowadays the prevailing belief is that they in fact can have any large cardinal whatsoever[7]. First we make the following definition.

DEFINITION 1.1.8. Θ_{reg} is the theory $\mathsf{ZF} + \mathsf{AD}_{\mathbb{R}} +$ "Θ is a regular cardinal".

Prior to [29], the theory Θ_{reg} was believed to be beyond the short extender region and was believed to be at the complexity level of supercompact cardinals. Because Woodin was able to force strong combinatorial statements over a model of Θ_{reg} that would normally require large cardinals at the level of supercompact cardinals or beyond[8], the aforementioned belief seemed to be very reasonable.

The main goal of [29], which is based on the first author's PhD thesis, was to analyze HOD of the minimal model of Θ_{reg}[9]. While any model of determinacy has a rich large cardinal structure below its Θ[10], $V_{\Theta}^{\mathrm{HOD}}$ of the minimal model of Θ_{reg} is very simple in the following sense (see Theorem 1.1.9).

Suppose $V \models \mathsf{AD}^+$. The Solovay pointclasses are exactly the stages of the Wadge hierarchy where a "new" *non-definable from below* set appears. For α such that $\vartheta_\alpha \leq \Theta$ let $\mathsf{SP}_\alpha = \{B \subseteq \mathbb{R} : w(B) < \vartheta_\alpha\}$. If $\vartheta_\alpha < \Theta$ and $A \subseteq \mathbb{R}$ has Wadge rank ϑ_α then A is not ordinal definable from any set of reals $B \in \mathsf{SP}_\alpha$ and moreover, every set in $\mathsf{SP}_{\alpha+1}$ is ordinal definable from A and a real. Thus, in a sense, once we perceive a set of reals of Wadge rank ϑ_α, we know everything about $\mathsf{SP}_{\alpha+1}$. Putting it differently,

(†) in the Wadge hierarchy, nothing of any interest happens among sets whose Wadge rank belongs to the interval $(\vartheta_\alpha, \vartheta_{\alpha+1})$.

In general, † is not true. All sorts of structures: Suslin cardinals, large cardinals with complicated partition properties etc, exist in that Wadge interval. However, the hod mice that are below the theory Θ_{reg} cannot have regular limits of Woodin cardinals, and moreover, the Woodin cardinals and their limits of such a hod mouse exactly correspond to the Solovay sequence[11] in the following sense.

---

[7]At least in the short-extender region.

[8]E.g., $\mathsf{MM}^{++}(c)$ (see [68]) and $\mathsf{CH}+$ "there is an ω_1-dense ideal on ω_1" (see [2]).

[9]Prior [29], it was not know that there is a unique minimal model of Θ_{reg}.

[10]E.g., Θ is a limit of strong partition cardinals, see [13].

[11]By a theorem of Woodin, each $\vartheta_{\alpha+1}$ is a Woodin cardinal of HOD. See [18].

THEOREM 1.1.9 ([29] and Theorem 7.2.2). *In the minimal model of Θ_{reg}, and in fact of* LSA, *δ is a Woodin cardinal of* HOD *or a limit of Woodin cardinals of* HOD *if and only if δ is a member of the Solovay sequence.*

Theorem 1.1.9 implies that HOD of the minimal model of Θ_{reg} has no Woodin cardinals in the interval $(\vartheta_\alpha, \vartheta_{\alpha+1})$, and in this sense, † is true below Θ_{reg}[12]. Therefore, to represent V_Θ^{HOD} of the minimal model of Θ_{reg} as a hod mouse, we do not need to understand exactly what happens between $(\vartheta_\alpha, \vartheta_{\alpha+1})$ in V as none of what happens there makes HOD look complicated in that interval[13].

The world of determinacy might have been a simpler place if † was always true, but [29] shows that the theory Θ_{reg} is much weaker than a Woodin cardinal that is a limit of Woodin cardinals. LSA, the main topic of this manuscript, is the next natural determinacy theory that is consistency wise stronger than Θ_{reg}, and while the hod mice of this manuscript do have inaccessible limit of Woodin cardinals, Theorem 1.1.9 is still true. This once again implies that the large cardinal structure of hod mice at the level of LSA is limited and in fact, in such hod mice

(‡) there is no Woodin cardinal δ and a $\kappa < \delta$ such that κ is δ-strong.

Moreover, prior to the current work, it was believed that ‡ and Theorem 1.1.9 are just consequences of AD^+. This belief was based on various arguments due to Woodin that showed that if δ is a member of the Solovay sequence then there cannot be $\kappa < \delta$ whose Mitchell order was much bigger than δ. However, Theorem 10.3.1 shows that LSA is weaker than a Woodin cardinal that is a limit of Woodin cardinals, and further unpublished work of the first author showed that the large cardinal structure of hod mice, at least in the short extender region, may not be limited. In particular, neither † nor Theorem 1.1.9 are consequences of AD^+. The first author then made the following conjecture.

CONJECTURE 1.1.10. *Assume $\mathsf{AD}^+ + V = L(\wp(\mathbb{R}))$. Define the sequence $(\eta_\alpha : \alpha \leq \Omega)$ as follows:*

1. $\eta_0 = \vartheta_0$.
2. *Assuming $\eta_\alpha < \Theta$ and setting $\kappa = (\eta_\alpha^+)^{\mathrm{HOD}}$, $\eta_{\alpha+1}$ is the supremum of all β such that there is an ordinal definable surjection $f : \wp_{\omega_1}(\kappa) \to \beta$.*[14]
3. *For a limit ordinal ξ, $\eta_\xi = \sup_{\alpha<\xi} \eta_\alpha$.*

Then δ is a Woodin cardinal or a limit of Woodin cardinals of HOD *if and only if $\delta = \eta_\alpha$ for some α.*

Using the methods of [63], Steel verified Conjecture 1.1.10 assuming HPC + NLE (see [63, Theorem 11.5.7]). More recently, the first author, using ideas from [63], constructed a hod mouse that has a Woodin cardinal that is a limit of Woodin

[12]It is a well-known fact from inner model theory dating back to [22] that iteration strategies of mice or hod mice acquire complexity only because of Woodin cardinals.

[13]This was the original motivation of the so-called "layering" used both in [29] and in this book.

[14]Recall that $\wp_{\omega_1}(\kappa)$ is the set of countable subsets of κ.

cardinals. This result confirms the belief that hod mice may have a complicated large cardinal structure.

1.1.6. Organization. Chapters 2-8 develop the basic theory of hod mice for AD^+ models up to the minimal model of LSA; a consequence of this analysis is (A). The last four chapters focus on applications. Chapter 11 proves that $\square_{\kappa,2}$ holds in HOD of AD^+ models up to the minimal model of LSA for all HOD-cardinals κ. Our main use of this chapter is Chapter 12, where a proof of (D) is given. Chapter 9 develops the basic theory of *condensing sets*, which is needed in constructions of hod mice in various situations. Chapter 10 uses the material developed in the previous chapters to prove (B) and (C). The last chapter (Chapter 12) proves (D) by constructing a hybrid version of K^c. This chapter uses methods developed in the previous chapters, [36], and [65].

Acknowledgments. We are indebted to Dominik Adolf for a long list of very useful corrections. We are also indebted to the referees for a very thorough list of corrections. In particular, the referee for the earlier chapters and Takehiko Gappo have done a tremendous amount of work correcting numerous typos and identifying many hidden mistakes. We thank Paul Larson for the incredible work he has done as the managing editor. We thank Thomas Piecha for his thorough work typesetting the manuscript.

Both authors would like to thank the National Science Foundation for providing financial assistance through Career Award DMS-1352034, and also to thank the Narodowe Centrum Nauki for providing financial assistance through NCN Grant WEAVE-UNISONO, Id: 567137. The second author would like to thank the National Science Foundation for partial support through NSF Award DMS-1565808 and Career Award DMS-1945592. The first author would like to thank the Oberwolfach Research Institute for Mathematics, Germany for hosting him during the winter of 2012 for seven weeks as a Leibniz Fellow. The first draft of this manuscript was written there. Both authors would like to thank the Oberwolfach Research Institute for Mathematics, Germany for hosting them for a week in the winter of 2012. Several of their joint projects go back to this one week. Both authors would like to thank the Newton Institute at Cambridge, UK for hosting them during the Fall of 2015. Several chapters of this manuscript were written during this period.

Finally, as it must be clear to anyone flipping through the pages of this manuscript, the inspiration behind this work comes from seminal contributions to descriptive inner model theory made by John Steel and Hugh Woodin. We thank them for the monumental work they have done during the past four decades.

Chapter 2

HYBRID $\mathcal{J}$-STRUCTURES

The main goal of this chapter is to prepare some terminology to be used in the rest of this manuscript. One important notion introduced in this chapter is that of the *un-dropping game* (see Definition 2.10.6). We will use it to prove a comparison theorem for hod mice (see Corollary 4.13.5). None of the results stated in this chapter are originally due to the authors, though some of them do not appear in literature in exactly the same form that we state here.

Throughout this book, the reader is assumed to know the basics of inner model theory. Starting from the beginning would have added many more pages to this book, and moreover, the basics of the theory have been developed in several places. The reader is encouraged to review the basic fine structural terminology as presented for example in [3], [41] or in [59].

2.1. $\mathcal{J}$-structures

We say $M = (\lfloor M \rfloor, Q, \in, \ldots)$ is a transitive structure if $\lfloor M \rfloor$ a transitive set. If M is just a set[1] then we let $\lfloor M \rfloor = M$[2]. In what follows, given a transitive set or a structure M we set $\mathrm{ord}(M) = Ord \cap \lfloor M \rfloor$. Also, given a set X, we let $trc(X)$ be the transitive closure of X. We also let $trc^X = (trc(X \cup \{X\}), \in)$.

Recall the inductive definition of $\mathcal{J}^A_{\omega\alpha}(X)$ (for example see [41, Definition 1.6]). In this manuscript, we will also use the round bracket notation while [41, Definition 1.6] only introduces the square bracket notation. We give the definition below, which uses the rud_A function defined in [41, Definition 1.1].

DEFINITION 2.1.1. Suppose $\vec{A} = (A_0, A_1, \ldots, A_n)$ is a finite sequence such that for each $i \leq n$, A_i is a partial set or class function, and suppose X is a set or a transitive structure. Then set

$$\mathcal{J}^{\vec{A}}_0(X) = trc(\{X\}) \text{ if } X \text{ is a set,}$$

[1] All mathematical objects are sets; here we just mean that M does not have any extra structure defined on it.

[2] It seems that this notation is due to Farmer Schlutzenberg.

$$\begin{aligned}
\mathcal{J}_0^{\vec{A}}(X) &= trc(\{\lfloor X \rfloor, Y_0, \dots, Y_n\}) \text{ if } X = (\lfloor X \rfloor, Y_0, Y_1, \dots, Y_n) \text{ is a structure,}\\
\mathcal{J}_{\omega\alpha+\omega}^{\vec{A}}(X) &= rud_{\vec{A}}(\mathcal{J}_{\omega\alpha}^{\vec{A}}(X) \cup \{\mathcal{J}_{\omega\alpha}^{\vec{A}}(X)\}),\\
\mathcal{J}_{\omega\lambda}^{\vec{A}}(X) &= \bigcup_{\alpha<\lambda} \mathcal{J}_{\omega\alpha}^{\vec{A}}(X) \text{ for limit ordinals } \lambda,\\
\mathcal{J}^{\vec{A}}(X) &= \bigcup_{\alpha \in Ord} \mathcal{J}_{\omega\alpha}^{\vec{A}}(X).
\end{aligned}$$

Recall that a transitive structure $\mathcal{M} = (M, A_1, \dots, A_k, \in)$ is called *amenable* if for every $X \in M$ and $1 \leq i \leq n$, $A_i \cap X \in M$. Following [70], we say $\mathcal{M}$ is a *$\mathcal{J}$-structure over* X if $\mathcal{M}$ is an amenable structure, and

$$\mathcal{M} = (\lfloor \mathcal{J}_{\omega\alpha}^{\vec{A}}(X) \rfloor, \vec{A} \cap \mathcal{J}_{\omega\alpha}^{\vec{A}}(X), B_0, \dots, B_m, X, \in)$$

where for any set M, $\vec{A} \cap M = (A_0 \cap M, \dots, A_n \cap M)$.

We think of $\vec{B} = (B_0, \dots, B_m)$ as a sequence of predicates. We will usually just need three such predicates, one for the last extender, one for the last branch and one for the set of layers to be defined later. At most one of B_0 and B_1 will be non-empty. X and its predicates (if there are any) are treated as constants. Thus, the language of $\mathcal{J}$-structures is the language of set theory augmented by infinitely many relation symbols and infinitely many constant symbols[3]. As we said above, most cases that will come up in this book will only have three predicates. X usually will itself be a $\mathcal{J}$-structure.

It is often convenient to think of A_i as a partial function $A_i : \lfloor \mathcal{M} \rfloor \to \lfloor \mathcal{M} \rfloor$ rather than some larger external function. Notice that for any $\vec{A}$, $\mathcal{J}_{\omega\alpha}^{\vec{A}}(X) = \mathcal{J}_{\omega\alpha}^{\vec{A} \cap \mathcal{J}_{\omega\alpha}^{\vec{A}}(X)}(X)$.

DEFINITION 2.1.2. Suppose $\mathcal{M} = (\mathcal{J}_{\omega\alpha}^{\vec{A}}(X), \vec{A}, B_0, \dots, B_n, X, \in)$ is a $\mathcal{J}$-structure with $\vec{A} = (A_0, \dots, A_n)$. We say $\mathcal{M}$ is *hierarchical* if the following clauses hold:

1. $\mathcal{M}$ is amenable.
2. For every $i \leq n$, $\text{dom}(A_i) \subseteq \{\omega\beta : \beta < \alpha\}$.
3. For every $i \leq n$ and for every $\beta < \alpha$ such that $\omega\beta \in \text{dom}(A_i)$, $A_i(\omega\beta) \subseteq \lfloor \mathcal{J}_{\omega\beta}^{\vec{A}}(X) \rfloor$.
4. The structure $(\lfloor \mathcal{J}_{\omega\beta}^{\vec{A}}(X) \rfloor, A_0(\omega\beta), \dots, A_n(\omega))$ is amenable.

The intuition behind a hierarchical structure is that the objects indexed at *active stages*[4] are amenable subsets of the model up to that stage. Often hierarchical structures are not represented in this fashion. For example, if $\vec{E}$ is a fine extender sequence (see [59, Definition 2.4]) then intuitively $\mathcal{J}^{\vec{E}}$ is a hierarchical

[3] We do not need infinitely many such symbols but a large finite number of them.

[4] Here we say that $\omega\beta$ is an active stage of for A_i if $\omega\beta \in \text{dom}(A_i)$.

structures in the above sense, but in reality one needs to use the amenable code (see [59, Lemma 2.9]) of each of the extenders in $\vec{E}$ in order to obtain a hierarchical structure. In this book, to avoid making things even more technical than they are, we will simply let the strategy predicates index the iterations and their branches. Thus, if A_i corresponds to the strategy predicate then according to our definition (see Definition 2.3.1) A_i will be represented as a strategy rather than a function whose domain consists of ordinals. However, it is a simple matter to re-design our hybrid structures so that they fit into our definition of hierarchical. In this book, all our $\mathcal{J}$ structures can be easily represented as hierarchical $\mathcal{J}$-structures.

Suppose now that $\mathcal{M} = (\mathcal{J}^{\vec{A}}_{\omega\alpha}(X), \vec{A}, B_0, \ldots, B_n, X, \in)$ is a hierarchical $\mathcal{J}$-structure and $\omega\beta < \mathrm{ord}(\mathcal{M})$. We then set

$$\mathcal{M}|\omega\beta = (\mathcal{J}^{\vec{A}}_{\omega\beta}(X), \vec{A} \cap \mathcal{J}^{\vec{A}}_{\omega\beta}(X), X, \in).$$

and

$$\mathcal{M}||\omega\beta = (\mathcal{J}^{\vec{A}}_{\omega\beta}(X), \vec{A} \cap \mathcal{J}^{\vec{A}}_{\omega\beta}(X), A_0(\omega\beta), A_1(\omega\beta), \ldots, A_n(\omega\beta), X, \in).$$

Thus, $\mathcal{M}||\mathrm{ord}(\mathcal{M}) = \mathcal{M}$, and $\mathcal{M}|\omega\beta$ is $\mathcal{M}$ "up to" $\omega\beta$ and $\mathcal{M}||\omega\beta$ is $\mathcal{M}$ "up to and including" $\omega\beta$. Below we will say that $A_i(\omega\beta)$ is *indexed* at $\omega\beta$.

REMARK 2.1.3. Thus, $\mathcal{M}|\gamma$ and $\mathcal{M}||\gamma$ are defined only when $\gamma = \omega\alpha$ for some α.

We say X is *self-well-ordered* if there is a wellordering of $\lfloor X \rfloor$ in $\mathcal{J}_1(X)$ definable over $\mathcal{J}_0(X)$ using only the predicates of X as parameters. For example, if X is a premouse then $\vec{E}^X$ is allowed to be used. Unless indicated otherwise, all our $\mathcal{J}$-structures will be over self-well-ordered sets. If $\mathcal{M}$ is a $\mathcal{J}$-structure then we let $X^{\mathcal{M}}$ be the X above. It follows that each $\mathcal{J}$-structure has a canonical well-ordering given we fix a recursive enumeration of formulas. And so in what follows, we will assume that such an enumeration has been fixed and hence, every $\mathcal{J}$-structure, unless otherwise indicated, has a canonical well-ordering which we will denote by $<_{\mathcal{M}}$[5]. We then must have that for $\beta < \alpha$, $<_{\mathcal{M}|\omega\beta} = <_{\mathcal{M}|\omega\alpha} \restriction \lfloor \mathcal{M}|\omega\beta \rfloor$.

2.2. Some fine structure

The goal of this section is to review some fine structural ideas. It is not our goal to develop fine structure, but only import some of the standard terminology that is developed in the literature. It is important to note that while new ideas and concepts do appear in the definition of a short-tree-strategy mouse, no new fine structural issues arise. All such fine structural issues have been handled

[5] $<_{\mathcal{M}}$ depends on the well-ordering of X that is definable over $\mathcal{J}_0(X)$, and there can be many such well-orderings. It does not matter for us which of them is chosen, but one could simply take the well-ordering that is definable via the least formula that defines a well-ordering of X over $\mathcal{J}_0(X)$.

elsewhere, and so we will not dwell on them. The reader unfamiliar with fine structural issues is advised to review some of the following sources: [3, 23, 27, 41, 42, 45, 50, 59, 63, 70]. Our fine structural set up will follow [27] and [41].

We say $\mathcal{M}$ is an *acceptable $\mathcal{J}$-structure* if $\mathcal{M}$ is a $\mathcal{J}$-structure and for all τ and for all β such that $\text{ord}(\mathcal{M}|0) \leq \tau$ and $\tau < \omega\beta$, if $\wp(\tau) \cap \mathcal{M}|\omega(\beta+1) \not\subseteq \mathcal{M}|\omega\beta$ then there is a surjection $f : \tau \to \omega\beta$ in $\mathcal{M}|\omega(\beta+1)$[6].

REMARK 2.2.1. From now on all $\mathcal{J}$-structures we will consider will be assumed to be acceptable and hierarchical.

Suppose $\mathcal{M}$ is a $\mathcal{J}$-structure (over a self-well-ordered set X). We then let $\rho_1(\mathcal{M})$, the Σ_1 projectum of $\mathcal{M}$, be the least $\rho \leq \text{ord}(\mathcal{M})$ such that for some $p \in (\text{ord}(\mathcal{M})^{<\omega})$ and some Σ_1 formula φ[7] the set $A = \{\xi < \rho : \mathcal{M} \vDash \varphi[\xi, p]\}$ is not in $\mathcal{M}$. The Σ_1 standard parameter of $\mathcal{M}$, $p_1(\mathcal{M})$, is the least[8] p as above. The Σ_1-reduct of $\mathcal{M}$ is the J-structure $(\mathcal{M}||\rho, T)$ where T codes the Σ_1 theory of $\mathcal{M}$ with parameters in $\rho_1(\mathcal{M}) \cup \{p_1(\mathcal{M})\}$. The Σ_1 core of $\mathcal{M}$, $\text{core}_1(\mathcal{M})$, is the transitive collapse of the Σ_1 Skolem hull in $\mathcal{M}$ of

$$\rho_1(\mathcal{M}) \cup \{p_1(\mathcal{M})\} \cup X^{\mathcal{M}} \cup \{X^{\mathcal{M}}\}.$$

We say $\mathcal{M}$ is 1-sound if $\text{core}_1(\mathcal{M}) = \mathcal{M}$ and $\mathcal{M}$ is 1-solid. The definition of solidity appears in [59, Definition 2.15] or in [41, Definition 7.5].

DEFINITION 2.2.2. We say $\mathcal{M}$ is a *fine structural $\mathcal{J}$-structure* (f.s. $\mathcal{J}$-structure) if $\mathcal{M} = (\mathcal{M}', k)$ where $\mathcal{M}'$ is a $\mathcal{J}$-structure, $k \leq \omega$ and letting $(\mathcal{M}_i : i \leq k)$ be given by

1. $\mathcal{M}_0 = \mathcal{M}'$ and
2. for $i+1 \leq k$, $\mathcal{M}_{i+1}$ is the Σ_1 reduct of $\mathcal{M}_i$,

then for all $i < k$, $\mathcal{M}_i$ is 1-sound. We say that $(\mathcal{M}_i : i \leq k)$ is the reduct sequence (r-sequence) of $\mathcal{M}$, and set

1. $\rho_0(\mathcal{M}) = \text{ord}(\mathcal{M})$ and $p_0(\mathcal{M}) = \emptyset$,
2. for $i \leq k$, $\rho_{i+1}(\mathcal{M}) = \rho_1(\mathcal{M}_i)$ and $p_{i+1}(\mathcal{M}) = p_i(\mathcal{M})^\frown p_1(\mathcal{M}_i)$,
3. $\rho(\mathcal{M}) = \rho_{k+1}(\mathcal{M})$ and $p(\mathcal{M}) = p_{k+1}(\mathcal{M})$.

We also say that $\mathcal{M}' =_{\text{def}} j(\mathcal{M})$ is the $\mathcal{J}$-component of $\mathcal{M}$ and $k =_{\text{def}} k(\mathcal{M})$ is the f.s.-component of $\mathcal{M}$, and set $l(\mathcal{M}) = (\text{ord}(\mathcal{M}), k)$. Finally, we say $\mathcal{M}$ is sound if $\mathcal{M}_k$ is 1-sound.

We also say $\mathcal{M} = (\mathcal{M}', \omega)$ is an f.s. $\mathcal{J}$-structure in case $(\mathcal{M}', k)$ is an f.s. $\mathcal{J}$-structure for all $k < \omega$. In this case, $\rho(\mathcal{M})$ is the eventual value of $\text{ord}(\mathcal{M}_i)$ for $i < \omega$.

---

[6]For now, we will need this concept only for $\mathcal{M}$ with $X^{\mathcal{M}}$ self-well-ordered.

[7]In the language of $\mathcal{J}$-structures.

[8]With respect to the lexicographic order on decreasing sequences of ordinals.

Suppose now that $\mathcal{M} = (\mathcal{M}', k)$ is an f.s. $\mathcal{J}$-structure and E is an $\mathcal{M}$ extender such that $\text{crit}(E) < \rho_k(\mathcal{M})$. Let $(\mathcal{M}_i : i \leq k)$ be the r-sequence of $\mathcal{M}$. We then let $\text{Ult}(\mathcal{M}, E)$ be the f.s. $\mathcal{J}$-structure whose $\mathcal{J}$-component is obtained by decoding $\text{Ult}_0(\mathcal{M}_k, E)$. We also have a map $\pi_E : \mathcal{M} \to \text{Ult}(\mathcal{M}, E)$ which is a k-embedding. The reader can review the relevant notions by consulting [3, Chapter 2], [41, Chapters 3 and 4], [63, Definition 2.4.1] and [63, Section 2.4].

Suppose $\mathcal{M} = (\mathcal{M}', k)$ is an f.s. $\mathcal{J}$-structure and $(\omega\alpha, m) \leq l(\mathcal{M})$ (here $\leq$ is the lexicographical order). We then let

$$\mathcal{M}|(\omega\alpha, m) = (\mathcal{M}'|\omega\alpha, m) \quad \text{and} \quad \mathcal{M}||(\omega\alpha, m) = (\mathcal{M}'||\omega\alpha, m).$$

Also, we write $\mathcal{N} \trianglelefteq \mathcal{M}$ if for some $(\omega\alpha, m) \leq l(\mathcal{M})$, $\mathcal{N} = \mathcal{M}|(\omega\alpha, m)$ or $\mathcal{N} = \mathcal{M}||(\omega\alpha, m)$. We will often write $\mathcal{M}|\gamma$ or $\mathcal{M}||\gamma$ for $\mathcal{M}'|\gamma$ and $\mathcal{M}'||\gamma$.[9]

The next definition defines the core of a $\mathcal{J}$-structure. One way of defining it is by doing what is described after [41, Definition 7.13]. Here is an outline of essentially that same construction.

DEFINITION 2.2.3. Suppose $\mathcal{M}$ is a $\mathcal{J}$-structure. We define $(\text{core}_k(\mathcal{M}) : k < \omega)$, $(\rho_k(\mathcal{M}) : k < \omega)$ and $(p_k(\mathcal{M}) : k < \omega)$ by induction as follows.

1. Set $\text{core}_0(\mathcal{M}) = \mathcal{M}$.
2. If $\mathcal{M}$ is not 1-solid then let $\text{core}_k(\mathcal{M})$ for $k \geq 1$ be undefined. Otherwise, $\text{core}_1(\mathcal{M})$ is defined as above.
3. Suppose $\text{core}_k(\mathcal{M})$ has been defined and that $\mathcal{N} = (\text{core}_k(\mathcal{M}), k)$ is an f.s. $\mathcal{J}$-structure[10]. Let $(\mathcal{N}_j : j \leq k)$ be the r-sequence of $\mathcal{N}$. If $\mathcal{N}_k$ is not 1-solid then let $\text{core}_i(\mathcal{M})$ for $i \geq k+1$ be undefined. Otherwise, letting $\pi : \text{core}_1(\mathcal{N}_k) \to \mathcal{N}_k$ be the core map, we let $\text{core}_{k+1}(\mathcal{M})$ be the decoding of $\text{core}_1(\mathcal{N}_k)$[11].

If $\text{core}_k(\mathcal{M})$ is defined for all $k < \omega$, then let $\text{core}(\mathcal{M})$ be the eventual value of $\text{core}_k(\mathcal{M})$.

Suppose for some $k < \omega$, $\text{core}_k(\mathcal{M})$ and $p_k(\mathcal{M})$ have been defined. Then letting $(\mathcal{N}_j : j \leq k)$ be the r-sequence of $\text{core}_k(\mathcal{M})$, set $\rho_{k+1}(\mathcal{M}) = \rho_1(\mathcal{N}_k)$ and $p_{k+1}(\mathcal{M}) = p_k(\mathcal{M})^\frown p_1(\mathcal{N}_k)$. Let $\rho(\mathcal{M})$ be the eventual value of the sequence $(\rho_k(\mathcal{M}) : k < \omega)$ and let $\text{ep}(\mathcal{M})$ be the least k such that for all $i \geq k+1$, $\rho_i(\mathcal{M}) = \rho_{k+1}(\mathcal{M})$.

Thinking of $\mathcal{J}$ structures as f.s. $\mathcal{J}$-structures is useful in introducing iteration trees and in the proof of convergence of K^c-constructions (for example see [3]).

[9] Notice that our definitions do not guarantee that $\mathcal{M}|\gamma$ or $\mathcal{M}||\gamma$ are f.s $\mathcal{J}$-structures. However, the structures that we will eventually consider will have this property.

[10] Notice that this condition is simply part of the induction. Above, we have that $(\text{core}_1(\mathcal{M}), 1)$ is an f.s. $\mathcal{J}$-structure.

[11] The decoding process is similar to the Downward Extension of Embeddings Lemma (see [41, Lemma 3.3]). The decoding gives a Σ_1-map $\pi' : \text{core}_{k+1}(\mathcal{M}) \to \text{core}_k(\mathcal{M})$ extending π.

2.3. Layered hybrid $\mathcal{J}$-structures

We say w is a *sequential structure* if $w = (\mathcal{J}_\omega(s), s, \in)$ where s is a sequence $(u_\alpha : \alpha < \gamma^w)$.

DEFINITION 2.3.1 (Definition 1.1 of [29]). Given a function f, we say f is *amenable* if for all $w \in \mathrm{dom}(f)$

1. w is a sequential structure,
2. $f(w) \subseteq \mathrm{ord}(w)$,
3. $\sup f(w) = \gamma^w$ and $0 \in f(w)$,
4. whenever $\eta < \gamma^w$, $f(w) \cap \eta \in w$.

We say f is a *shift of an amenable function* or a *shifted amenable function* if there is an amenable function g with $\mathrm{dom}(g) = \mathrm{dom}(f)$ and such that for all $w \in \mathrm{dom}(f)$,

1. $f(w) \subseteq Ord$,
2. $f(w) \subseteq [\min(f(w)), \min(f(w)) + \gamma^w)$, and
3. $f(w) = \{\min(f(w)) + \omega\gamma : \gamma \in g(w)\}$.

Notice that if f is a shift of an amenable function then it uniquely determines g. We say that g is the *amenable component* of f.

Jumping ahead, we remark that iteration strategies and mouse operators provide an ample source of amenable functions. For instance, let $\mathcal{M} = \mathcal{M}_1^\#$ and let Σ be its canonical iteration strategy. We define f as follows. Let first $\mathrm{dom}(f)$ be the set of structures of the form $w = (\mathcal{J}_\omega(\mathcal{T}^w), \mathcal{T}^w, \in)$ where $\mathcal{T}^w$[12] is a normal iteration tree on $\mathcal{M}$ of limit length and is according to Σ. Next, define $f(w) = b$ where $b = \Sigma(\mathcal{T}^w)$. Then f is amenable. We will refer to such an f as *an amenable function given by an iteration strategy*.

The definitions that follow explain how our indexing schemes work. We first isolate those iterations whose branches will be indexed. The reader may think of the formula φ appearing in Notation 2.3.2 as the formula that defines the set of iterations whose branches need to be indexed. However, φ alone does not define such iterations as we need to add clause 2 for technical reasons. The ordinal β essentially identifies the location where the branch of the iteration tree defined by φ should be indexed.

In general, to develop a reasonable theory of hybrid $\mathcal{J}$-structures, we need to use indexing schemes to index branches of stacks. The reason for this is that if no particular coherent method of indexing is used to organize such structures then one cannot in general hope to develop a comparison theory for the resulting structures. Indeed, if $\mathcal{M}$ and $\mathcal{N}$ are unindexed hybrid $\mathcal{J}$-structures then it is possible that some b is indexed at α in $\mathcal{M}$ but nothing is indexed at α in $\mathcal{N}$, causing a fatal breakdown of the comparison argument. Nevertheless, in Section 3.8, it will be convenient to work with unindexed hybrid $\mathcal{J}$-structures (as defined in Definition 2.5.3).

[12]One could think of $\mathcal{T}$ as a sequence $(\mathcal{T} \restriction \alpha : \alpha < \mathrm{lh}(\mathcal{T}))$.

Another important remark is that it might be convenient to think of the indexing scheme as a parameter of the hybrid $\mathcal{J}$-structures, in the same way we internalize the fine structural parameter. Thus, instead of $\mathcal{M}$ we could consider $(\mathcal{M}, \varphi)$ where φ is the indexing scheme. However, in most cases, isolating φ will not matter so much, and so we will not take this path. We suspect that $\mathcal{M}$ may even recognize many different φs as its own indexing scheme[13]. Thus we may have two φ-indexed $\mathcal{M}$ and $\mathcal{M}'$ and an indexing scheme φ' such that $\mathcal{M}'$ is φ'-indexed while $\mathcal{M}$ is not. However, such issues will not come up in the sequel.

NOTATION 2.3.2. Suppose now that $\mathcal{M} = \mathcal{J}_\iota^{A_0,A_1,\dots,A_n}(X)$ is a $\mathcal{J}$-structure or an f.s. $\mathcal{J}$-structure and $\varphi(\vec{x}, u)$ is a formula in the language of $\mathcal{J}$-structures such that it implies that "u is a sequential structure". Suppose $\vec{s} \in \lfloor \mathcal{M} \rfloor^{<\omega}$. Let $S^{\mathcal{M}}_{\vec{s},\varphi}$ be the set of pairs (β, w) such that

1. $\omega\beta + \omega\gamma^w \leq \mathrm{ord}(\mathcal{M})$,
2. $\mathcal{M}|\omega\beta \models$ "$\mathrm{cf}(\gamma^w)$ is not a measurable cardinal as witnessed by extenders in A_0" (see Remark 2.3.4), and
3. $\mathcal{M}|\omega\beta \models \mathsf{ZFC} + \varphi[\vec{s}, w]$.

Let $\mathsf{nmc}(\alpha)$ be the statement "$\mathrm{cf}(\alpha)$ is not a measurable cardinal as witnessed by the extenders in A_0".

DEFINITION 2.3.3. Suppose that $(\mathcal{M}, \vec{s}, \varphi)$ are as in Notation 2.3.2. Suppose further that f is a shifted amenable function with amenable component g such that $\mathrm{dom}(f) \subseteq \lfloor M \rfloor$ and for all $w \in \mathrm{dom}(f)$, $\min(f(w)) + \gamma^w \leq \mathrm{ord}(\mathcal{M})$[14]. We say w is *weakly $(f, \vec{s}, \varphi)$-minimal* if there is β such that

1. $(\beta, w) \in S^{\mathcal{M}}_{\vec{s},\varphi}$ (in particular, because $\mathcal{M}|\omega\beta \models \mathsf{ZFC}$, $\omega\beta = \beta$),
2. $w \notin \mathrm{dom}(f \cap \lfloor \mathcal{M}|\beta \rfloor)$,
3. $\{u \in \lfloor \mathcal{M}|\beta \rfloor : u <_{\mathcal{M}|\beta} w$ and there is $\xi < \beta$ such that $(\xi, u) \in S^{\mathcal{M}}_{\vec{s},\varphi}\} \subseteq \mathrm{dom}(f \cap \lfloor \mathcal{M}|\beta \rfloor)$.

We say w is *$(f, \vec{s}, \varphi)$-minimal* if there is β witnessing that w is weakly $(f, \vec{s}, \varphi)$-*minimal* and such that w is the $<_{\mathcal{M}|\omega\beta}$-minimal w' which is weakly $(f, \vec{s}, \varphi)$-*minimal* as witnessed by β.

If w is *$(f, \vec{s}, \varphi)$-minimal* then we let $\beta_w^{\mathcal{M}, f, \vec{s}, \varphi}$ be the least β witnessing that w is *$(f, \vec{s}, \varphi)$-minimal*. In many cases, $(\mathcal{M}, f, \varphi)$ will be clear from context and so we will drop it from our notation. If $\vec{s} = \emptyset$ then we drop it from our notation.

REMARK 2.3.4 (The measurable cofinality issue). The reader unfamiliar with strategic mice may find clause 2 of Definition 2.3.8 somewhat odd. This clause has to do with an issue known to experts and was first discovered in earlier versions

[13]Of course, this can be achieved trivially; φ and $0 = 0 \wedge \varphi$ are equivalent. But there could be two different indexing schemes φ and φ' such that ZFC or any natural extension of it, does not prove $\varphi \iff \varphi'$ yet there is $\mathcal{M}$ which is both φ indexed and φ'-indexed.

[14]Recall our convention that $X^{\mathcal{M}}$ is self-well-ordered.

of [40][15]. The problem was fully treated in [50], and the discussion appears in [50, Remark 2.47]. Without getting too much into the technical details, the issue is simply that if $\mathcal{M} = \mathcal{J}^{A,f}$ is a $\mathcal{J}$-structure such that the f predicate codes a strategy for some $\mathcal{N} \in \mathcal{M}$, $w =_{\text{def}} (\mathcal{J}_\omega(\mathcal{T}), \mathcal{T}, \in) \in \text{dom}(f)$, $\kappa =_{\text{def}} \text{cf}^{\mathcal{M}}(\text{lh}(\mathcal{T}))$ is a measurable cardinal in $\mathcal{M}$, $f(w)$ is indexed at λ and $E \in \vec{E}^{\mathcal{M}}$ is an extender with $\text{crit}(E) = \kappa$ then

$$\sup(\pi_E^{\mathcal{M}||(\lambda,0)}[\text{lh}(\mathcal{T})]) < \pi_E(\text{lh}(\mathcal{T})) \text{ while } \lambda = \text{ord}(\text{Ult}(\mathcal{M}||(\lambda,0),E)).$$

The issue is hiding in the fact that in most of the natural attempts to organize strategic mice, $\text{cf}^{\mathcal{M}}(\lambda) = \kappa$, while in the above situation this fails in $\text{Ult}(\mathcal{M}||(\lambda,0),E)$ for $\pi_E^{\mathcal{M}||(\lambda,0)}(\mathcal{T})$.

In general, there may not be a unique w which is *$(f,\vec{s},\varphi)$-minimal*. However, the following holds.

LEMMA 2.3.5. *Suppose $(f,\vec{s},\varphi)$ and $\mathcal{M}$ are as in Definition 2.3.3. Suppose $w \neq w'$ are two $(f,\vec{s},\varphi)$-minimal sets. Set $\beta = \beta_w^{\mathcal{M},f,\vec{s},\varphi}$ and $\beta' = \beta_{w'}^{\mathcal{M},f,\vec{s},\varphi}$, and suppose that $\beta \leq \beta'$. Then $\beta < \beta'$.*

REMARK 2.3.6. The f of Definition 2.3.3 is designed to code an iteration strategy, and it will be the strategy predicate of a hybrid $\mathcal{J}$-structures which indexes an iteration strategy. The iterations that will get indexed are exactly the $(f,\vec{s},\varphi)$-minimal ones, and Lemma 2.3.5 implies that $(f,\vec{s},\varphi)$-minimal w's are well-ordered. We will then use the function $w \mapsto \beta_w$ to index the branch of w at $\beta_w + \omega\gamma^w$.

REMARK 2.3.7. It is perhaps illuminating to figure out the least iteration tree whose branch will be indexed by the predicate f of Definition 2.3.3. We assume $\vec{s} = \emptyset$. First we pick the least β such that for some $\mathcal{T}$ of length ω, $\mathcal{M}|\omega\beta \models \mathsf{ZFC} + \varphi[\mathcal{T}]$. Then we take the $<_{\mathcal{M}|\omega\beta}$-least $\mathcal{T}$ as above and index its branch at $\omega\beta + \omega^2$.

We are now in a position to introduce the *passive hybrid $\mathcal{J}$-structures.*

DEFINITION 2.3.8 (Passive Hybrid $\mathcal{J}$-structures). We say $\mathcal{M}$ is a *passive hybrid $\mathcal{J}$-structure* over a self-well-ordered set X with indexing scheme $\varphi(x)$[16] if $\mathcal{M} = (\mathcal{M}', k)$ is an f.s. $\mathcal{J}$-structure such that the following conditions hold.

1. For some α, $A \subseteq \lfloor \mathcal{M}' \rfloor$ and $f \subseteq \lfloor \mathcal{M}' \rfloor$,
$$\mathcal{M}' = (\mathcal{J}_{\omega\alpha}^{A,f}(X), A, f, X, \in),^{17}$$
2. f is a shift of an amenable function.

[15] The authors were unable to locate the discussion involving the measurable cofinality issue in [40].

[16] φ is in the language of $\mathcal{J}$-structures.

[17] We would like to emphasize that $\mathcal{M}'$ has only the displayed predicates. Also, below $(\mathcal{M}', f, \varphi)$ are omitted from β_w notation.

3. For all $w \in \lfloor \mathcal{M}' \rfloor$, $w \in \mathrm{dom}(f)$ if and only if w is (f, φ)-minimal and $\beta_w + \omega\gamma^w < \mathrm{ord}(\mathcal{M})$[18].
4. For all $w \in \mathrm{dom}(f)$,
   (a) $\beta_w = \min(f(w))$,
   (b) $\lfloor \mathcal{M}' | (\beta_w + \omega\gamma^w) \rfloor = \mathcal{J}_{\beta_w + \omega\gamma^w}(\mathcal{M}' || \omega\beta_w)$ and $A \cap \lfloor \mathcal{M}' | (\beta_w + \omega\gamma^w) \rfloor = A \cap \lfloor \mathcal{M}' | \omega\beta_w \rfloor$[19].

REMARK 2.3.9. Definition 2.3.8 leaves open one important question. Does it follow that $S_\varphi^{\mathcal{M}} = \mathrm{dom}(f^{\mathcal{M}})$? The answer is of course that none of the conditions we have imposed on $f^{\mathcal{M}}$ guarantees that $S_\varphi^{\mathcal{M}} = \mathrm{dom}(f^{\mathcal{M}})$. It could be that some $w \in S_\varphi^{\mathcal{M}}$ but it is not (f, φ)-minimal. However, if $f^{\mathcal{M}}$ is supposed to code an iteration strategy Σ of some P then the fact that $w \in S_\varphi^{\mathcal{M}}$ implies that w is an iteration according to Σ and that we must have that $w \in \mathrm{dom}(f^{\mathcal{M}})$. What will in fact happen, in intuitive terms, is that while w may not be in $\mathrm{dom}(f^{\mathcal{M}})$, it will be in $\mathrm{dom}(f^{\mathcal{N}})$ for some $\mathcal{N}$ extending $\mathcal{M}$. This may not be possible to arrange if for example $\mathcal{N} =_{\mathrm{def}} \mathcal{J}_\omega(\mathcal{M})$ projects in a way that say w is no longer in $\mathsf{core}_1(\mathcal{N})$, but if $\mathcal{M}$ is the final model of some reasonable fully backgrounded construction that produces hybrid premouse then we will indeed have that $S_\varphi^{\mathcal{M}} = \mathrm{dom}(f^{\mathcal{M}})$. This is because it can be shown that any $w \in S_\varphi^{\mathcal{M}}$ is (f, φ)-minimal.

To see this in intuitive terms, suppose towards a contradiction that some $w \in S_\varphi^{\mathcal{M}}$ does not belong to $\mathrm{dom}(f^{\mathcal{M}})$. We can assume w is $<_{\mathcal{M}}$-minimal. Now, and this depends on our choice of φ, the indexing scheme φ will be Σ_1 and hence, it will have the following upward absoluteness property: if for some ν, $\mathcal{M}|\nu \vDash \varphi[w]$ then for all $\nu' \geq \nu$, $\mathcal{M}|\nu' \vDash \varphi[w]$. Let $\xi < \mathrm{ord}(\mathcal{M})$ be such that whenever $u \in S_\varphi^{\mathcal{M}}$ and $u <_{\mathcal{M}} w$ then $\sup(f(u)) < \xi$. Such ξ will exist because, in concrete applications, $\mathrm{ord}(\mathcal{M})$ will be a Woodin cardinal of the universe. It follows that to show that $w \in \mathrm{dom}(f^{\mathcal{M}})$ it is enough to show that there is $\beta \in (\xi, \mathrm{ord}(\mathcal{M}))$ such that $\mathcal{M}|\beta \vDash$ ZFC. Since $\mathrm{ord}(\mathcal{M})$ is a Woodin cardinal of the background universe, there are plenty of such β.

DEFINITION 2.3.10 (Hybrid $\mathcal{J}$-structures). We say $\mathcal{M}$ is a *hybrid $\mathcal{J}$-structure* over a self-well-ordered set X with indexing scheme $\varphi(x)$ if $\mathcal{M} = (\mathcal{M}', k)$ is an f.s. $\mathcal{J}$-structure such that

1. for some α, $A \subseteq \lfloor \mathcal{M}' \rfloor$ and $f \subseteq \lfloor \mathcal{M}' \rfloor$,
$$\mathcal{M}' = (\mathcal{J}_{\omega\alpha}^{A,f}(X), A, f, B, F, X, \in),^{20}$$
2. $(\mathcal{J}_{\omega\alpha}^{A,f}(X), A, f, X, \in)$ is a passive hybrid $\mathcal{J}$-structure,
3. at most one of B and F is not empty,
4. if $F \neq \emptyset$ then F is an ordered pair (w, b) such that if $\beta = min(b)$ then setting $f' = f \cup \{(w, b)\}$,

[18] Here β_w is defined in Definition 2.3.3.
[19] It also follows that $f \cap \lfloor \mathcal{M}' | (\beta_w + \gamma^w) \rfloor = f \cap \lfloor \mathcal{M}' | \beta_w \rfloor$.
[20] Below $(\mathcal{M}', f, \varphi)$ are omitted from β_w notation.

(a) f' is a shift of an amenable function[21],
(b) w is (f',φ)-minimal with $\beta_w^{\mathcal{M},f',\varphi}=\beta$ (in particular, $\omega\beta=\beta$, see Definition 2.3.3),
(c) $\omega\alpha=\beta+\omega\gamma^w$,[22]
(d) $\lfloor\mathcal{M}'\rfloor=\mathcal{J}_{\beta+\omega\gamma^w}(\mathcal{M}'||\beta)$ and $A\cap|\mathcal{M}'|=A\cap|\mathcal{M}'|\beta|$.

For $w\in\mathrm{dom}(f')$, we say that $f'(w)$ is indexed at $\beta_w+\omega\gamma^w$ or that $\beta_w+\omega\gamma^w$ is the index of $f'(w)$.

Suppose $\mathcal{M}$ is a hybrid $\mathcal{J}$-structure with an indexing scheme φ. We will often say that "$\mathcal{M}$ is indexed according to φ" or that "$\mathcal{M}$ is φ-indexed". Notice that only the f predicate is indexed according to φ. In most situations that we will consider A will be an extender sequence. Sometimes, however, we will need to consider cases where there are two or more f predicates.

REMARK 2.3.11. Notice that it follows from Definition 2.3.10 that the function $a\mapsto\min(f(a))$ is injective on $\mathrm{dom}(f)$.

Hod mice are a special blend of *layered hybrid $\mathcal{J}$-structures* introduced below.

DEFINITION 2.3.12 (Passive layered hybrid $\mathcal{J}$-structure). We say $\mathcal{M}$ is a *passive layered hybrid $\mathcal{J}$-structure* over a self-well-ordered set X with indexing scheme $\varphi(x,y)$ if $\mathcal{M}=(\mathcal{M}',k)$ is an f.s. $\mathcal{J}$-structure such that

1. for some α, $A\subseteq\lfloor\mathcal{M}'\rfloor$ and $f\subseteq\lfloor\mathcal{M}'\rfloor$, $\mathcal{M}'=(\mathcal{J}_{\omega\alpha}^{A,f}(X),A,f,Y,X,\in)$,
2. $Y\subseteq\mathcal{J}_{\omega\alpha}^{A,f}(X)$ and $\mathrm{dom}(f)=Y\subseteq\{\mathcal{Q}:\mathcal{Q}\lhd\mathcal{M}\}\cup X$,
3. for all $\mathcal{Q}\in\mathrm{dom}(f)$, $f(\mathcal{Q})=_{\mathrm{def}}f_{\mathcal{Q}}$ is a shift of an amenable function[23],
4. for all $w\in\mathcal{M}'$ and $\mathcal{Q}\in Y$, $w\in\mathrm{dom}(f_{\mathcal{Q}})$ if and only if
   (a) w is $(f_{\mathcal{Q}},\mathcal{Q},\varphi)$-minimal and $\beta_w^{\mathcal{Q}}+\omega\gamma^w<\omega\alpha$, and
   (b) for all $\mathcal{R}\in Y$ such that $\mathcal{R}<_{\mathcal{M}'}\mathcal{Q}$ and for all $(f_{\mathcal{R}},\mathcal{R},\varphi)$-minimal $u\in\mathcal{M}'|\omega\beta_w^{\mathcal{Q}}$, $u\subset\mathrm{dom}(f_{\mathcal{R}}\cap\lfloor\mathcal{M}'|\beta_w^{\mathcal{Q}}\rfloor)$[24],
5. for all $\mathcal{Q}\in Y$ and for all $w\in\mathrm{dom}(f(\mathcal{Q}))$,
   (a) $\beta_w^{\mathcal{Q}}=\min(f(w))$,
   (b) $\lfloor\mathcal{M}'|(\beta_w^{\mathcal{Q}}+\omega\gamma^w)\rfloor=\mathcal{J}_{\beta_w^{\mathcal{Q}}+\omega\gamma^w}(\mathcal{M}'||\beta_w^{\mathcal{Q}})$ and $A\cap\lfloor\mathcal{M}'|(\beta_w^{\mathcal{Q}}+\omega\gamma^w)\rfloor=A\cap\lfloor\mathcal{M}'|\omega\beta_w^{\mathcal{Q}}\rfloor$[25].

DEFINITION 2.3.13 (Layered hybrid $\mathcal{J}$-structure). We say $\mathcal{M}$ is a *layered hybrid $\mathcal{J}$-structure* over self-well-ordered set X with indexing scheme $\varphi(x,y)$ if $\mathcal{M}=(\mathcal{M}',k)$ is an f.s. $\mathcal{J}$-structure such that

---

[21]This implies that w is a sequential structure.

[22]It follows from clause 5 of Definition 2.3.3 that $\mathcal{M}'\models$ "cf(γ) is not a measurable cardinal as witnessed by extenders in A".

[23]Below we will drop $(\mathcal{M}',f_{\mathcal{Q}},\varphi)$ from the $\beta_w^{\mathcal{M}',f_{\mathcal{Q}},\mathcal{Q},\varphi}$ notation.

[24]Implying that $\beta_u^{\mathcal{R}}<\beta_w^{\mathcal{Q}}$.

[25]It also follows that $f\cap\lfloor\mathcal{M}'|(\beta_w^{\mathcal{Q}}+\gamma^w)\rfloor=f\cap\lfloor\mathcal{M}'|\omega\beta_w^{\mathcal{Q}}\rfloor$.

1. for some $A \subseteq \lfloor \mathcal{M}' \rfloor$ and $f \subseteq \lfloor \mathcal{M}' \rfloor$,
$$\mathcal{M}' = (\mathcal{J}_{\omega\alpha}^{A,f}, A, f, Y, B, F, X, \in),$$
2. $\mathcal{M}' = (\mathcal{J}_{\omega\alpha}^{A,f}, A, f, Y, X, \in)$ is a passive layered hybrid $\mathcal{J}$-structure over X,
3. only one of B and F is non-empty,
4. if $F \neq \emptyset$ then F is an ordered pair $(\mathcal{Q}, (w, b))$ such that $\mathcal{Q} \in Y$, $b \subseteq \mathrm{ord}(\mathcal{M})$, and if $\beta = min(b)$ then setting $f' = f \cup \{(\mathcal{Q}, (w, b))\}$,
   (a) $f'(\mathcal{Q})$ is a shift of an amenable function,
   (b) w is $(f', \mathcal{Q}, \varphi)$-minimal with $\beta_w^{\mathcal{M}, f', \mathcal{Q}, \varphi} = \beta$,
   (c) $\alpha = \beta + \omega\gamma^w$,
   (d) $\lfloor \mathcal{M}' \rfloor = \mathcal{J}_{\beta+\omega\gamma^w}(\mathcal{M}'||\beta)$ and $A \cap \lfloor \mathcal{M}' \rfloor = A \cap \lfloor \mathcal{M}'|\beta \rfloor$,
   (e) for all $\mathcal{R} \in Y$ such that $\mathcal{R} <_{\mathcal{M}'} \mathcal{Q}$ and for all $(f'_{\mathcal{R}}, \mathcal{R}, \varphi)$-minimal $u \in \mathcal{M}'|\beta$, $u \in \mathrm{dom}(f_{\mathcal{R}} \cap \lfloor \mathcal{M}'|\beta \rfloor)$[26].

Suppose $\mathcal{M}$ is a layered hybrid $\mathcal{J}$-structure with an indexing scheme φ. We will often say that "$\mathcal{M}$ is indexed according to φ" or that "$\mathcal{M}$ is φ-indexed".

We will often omit φ when discussing a particular layered hybrid $\mathcal{J}$-structure. If $\mathcal{M}$ is a layered hybrid $\mathcal{J}$-structure then we let $f^{\mathcal{M}}$ and $Y^{\mathcal{M}}$ be as in Definition 2.3.12. We again have that for each $\mathcal{Q} \in Y^{\mathcal{M}}$, the function $a \mapsto \min(f^{\mathcal{M}}(\mathcal{Q})(a))$ is injective on $\mathrm{dom}(f(\mathcal{Q}))$.

Notice that hybrid $\mathcal{J}$-structures can be viewed as a special case of layered hybrid $\mathcal{J}$-structures. Because of this, in the sequel we will only establish terminology for layered hybrid $\mathcal{J}$-structures though we might use the same terminology for hybrid $\mathcal{J}$-structures.

Typically, when discussing hybrid $\mathcal{J}$-structures, X will be an *iterable* structure and f will be the predicate coding its strategy.[27]

As mentioned above, hod mice are a special type of layered hybrid $\mathcal{J}$-structures: the f predicate of a hod mouse codes a strategy for its layers. When the A predicate of a layered hybrid $\mathcal{J}$-structure is a coherent sequence of extenders then the resulting model is called a *hybrid layered premouse*.

Results of this manuscript are independent of particular extender-indexing schemes, but for technical reasons we will use a mixture of Jensen indexing as developed in [3, Definition 2.4], [11], [27], [70] and Mitchell-Steel indexing as developed in [23] and [59, Definition 2.4]. Suppose $\mathcal{M} = \mathcal{J}^{\vec{E},f}(X)$ is a φ-indexed layered hybrid $\mathcal{J}$-structure over a self-well-ordered set X and $\vec{E}$ is a sequence of extenders. We say η is a cutpoint of $\mathcal{M}$ if there is no $\alpha \in \mathrm{dom}(\vec{E}^{\mathcal{M}})$ such that $\mathrm{crit}(\vec{E}^{\mathcal{M}}(\alpha)) < \eta \leq \alpha$. We say $\vec{E}$ is a *mixed indexed extender sequence* if the following clauses hold:

---

[26] Implying that $\beta_u^{\mathcal{R}} < \beta$.

[27] In this case, the γ defined in Definition 2.3.10 is the length of a tree $\mathcal{T}$ according to f. The condition "$\mathcal{M} \models \mathrm{cof}(\gamma)$ is not measurable" in Definition 2.3.10 ensures that fine structure is preserved under iterations.

1. (j-like indexing) If $\kappa = \text{crit}(E)$ is a limit of Woodin cardinals of $\mathcal{M}$ and is a cutpoint of $\mathcal{M}$ then letting $\mathcal{N} = \pi_E^{\mathcal{M}}(\mathcal{M}|(\kappa^+)^{\mathcal{M}})$, E is indexed at $(\eta^+)^{\mathcal{N}}$ where $\eta = \sup\{\alpha \in \text{dom}(\vec{E}^{\mathcal{N}}) : \text{crit}(\vec{E}^{\mathcal{N}}(\alpha)) = \kappa\}$.
2. (ms-indexing) All other extenders are indexed according to the ms-indexing.

The initial segment condition for E is clause 3 of [59, Definition 2.4]. There are many papers in the literature that connect the two indexing schemes. For example, the reader may consult [8], [9], and [47]. Our goal in this book is to present the theory of the minimal model of the Largest Suslin Axiom in as short a space as we can, and because of this we will avoid fine structural issues that have been well-treated in the literature.

DEFINITION 2.3.14 (Layered hybrid e-structure). Suppose $\mathcal{M} = \mathcal{J}^{\vec{E},f}(X)$ is a φ-indexed layered hybrid $\mathcal{J}$-structure over a self-well-ordered set X. $\mathcal{M}$ is called a *φ-indexed layered hybrid potential e-structure* (lhpes) if $\vec{E}$ is a mixed indexed extender sequence. We write $\vec{E}^{\mathcal{M}}$ for $\vec{E}$ etc.

If $\mathcal{M}$ is an lhpes and $E = \vec{E}^{\mathcal{M}}(\gamma)$ then we let $\text{ind}^{\mathcal{M}}(E) = \gamma$.

We say that $\mathcal{M}$ is a φ-indexed layered hybrid e-structure (lhes) if $\mathcal{M} = (\mathcal{M}', k)$ is an f.s. $\mathcal{J}$-structure such that $\mathcal{M}'$ is a φ-indexed lhpes and for every $(\omega\beta, m) < l(\mathcal{M})$, $\mathcal{M}||(\omega\beta, m)$ is sound.

Mixed indexing smoothens implementation of certain technical arguments. The most crucial property for us is the following. Suppose E is an extender on the extender sequence of an lhes $\mathcal{M}$ such that $\mathcal{M} \vDash$ "crit(E) is a cutpoint and a limit of Woodin cardinals" (i.e., there is no $F \in \vec{E}^{\mathcal{M}}$ such that $\text{crit}(F) < \text{crit}(E) \leq \text{ind}^{\mathcal{M}}(F)$). So E has j-like indexing. Let $\gamma = \sup\{\xi : \xi < \text{ind}^{\mathcal{M}}(E), \xi \in \text{dom}(\vec{E}^{\mathcal{M}})$ and $\text{crit}(E_\xi) = \text{crit}(E)\}$. Then $\gamma = \sup\{\xi : \xi \in \text{dom}(\vec{E}^{\text{Ult}(\mathcal{M},E)})$ and $\text{crit}(E_\xi) = \text{crit}(E)\}$. The advantage of mixed indexing over other indexing schemes can be seen in Definition 2.8.

For an lhes $\mathcal{M}$ with just one layer (that is, $|Y^{\mathcal{M}}| = 1$), we say $\mathcal{M}$ is a *hybrid e-structure* (hes). Next we introduce lhes that are internally closed under sharps. We will use such a closure to introduce short tree strategy premice (see Definition 2.5.2 and Definition 3.8.17).

DEFINITION 2.3.15 (Closed lhes). Suppose $\mathcal{M}$ is an lhes and $\alpha \leq \text{ord}(\mathcal{M})$. Then we say $\mathcal{M}$ is *closed* below α if for all $\beta < \alpha$ there is $\gamma \in \text{dom}(\vec{E}^{\mathcal{M}})$ such that $\gamma < \alpha$ and $\text{crit}(E_\gamma^{\mathcal{M}}) > \beta$. We say $\mathcal{M}$ is closed if $\mathcal{M}$ is closed below $\text{ord}(\mathcal{M})$.

2.4. Iteration trees and stacks

Below we review iteration trees. Our notation is mostly in line with most of the references we have quoted above in Section 2.2. The only difference is that we incorporated the concepts of a stack of iteration trees into an iteration tree.

Suppose $\mathcal{M}$ is an lhes (or hes). Thus, $\mathcal{M}$ is an f.s. $\mathcal{J}$-structure that has a designated extender sequence $\vec{E}^{\mathcal{M}}$. For a limit ordinal η, we let $\eta - 1 = \eta$.

DEFINITION 2.4.1. We say $\mathcal{T}$ is a *putative iteration tree* on $\mathcal{M}$ if

$$\mathcal{T} = ((\mathcal{M}_\alpha)_{\alpha<\eta}, (E_\alpha)_{\alpha<\eta-1}, D, R, (\beta_\alpha, m_\alpha)_{\alpha\in R}, T)$$

and the following conditions hold.

1. T is a tree order on η.

Let $\mathcal{T}(\alpha+1)$ be the T-predecessor of $\alpha+1$ and $(\alpha,\beta)_{\mathcal{T}}$ be the T-interval (α,β).[28]

2. For all α such that $\alpha+1<\eta$, $\mathcal{M}_\alpha$ is a well-founded lhes (or hes).
3. $R \subseteq \eta - 1$, $0 \in R$ and for all $\alpha \in R$ and for all $\beta \geq \alpha$, $\mathcal{T}(\beta+1) \geq \alpha$.
4. For all $\alpha \in R$, $(\omega\beta_\alpha, m_\alpha) \leq l(\mathcal{M}_\alpha)$.

$$\text{Set } \mathcal{M}'_\alpha = \begin{cases} \mathcal{M}_\alpha & \text{if } \alpha \notin R \vee (\alpha \in R \wedge \omega\beta_\alpha = \operatorname{ord}(\mathcal{M}_\alpha)) \\ \mathcal{M}_\alpha || (\omega\beta_\alpha, m_\alpha) & \text{if } \alpha \in R \wedge \omega\beta_\alpha < \operatorname{ord}(\mathcal{M}_\alpha). \end{cases}$$

5. $\mathcal{M}_0 = \mathcal{M}$.
6. For all $\alpha+1<\eta$, $E_\alpha \in \vec{E}^{\mathcal{M}'_\alpha}$.
7. For all $\alpha+1<\eta$, setting $\beta = \mathcal{T}(\alpha+1)$ and $\kappa_\alpha = \operatorname{crit}(E_\alpha)$,

$$\mathcal{M}_\alpha | (\kappa_\alpha^+)^{\mathcal{M}_\alpha | \operatorname{ind}^{\mathcal{M}_\alpha}(E_\alpha)} \trianglelefteq \mathcal{M}'_\beta.$$

8. for all $\alpha+1<\eta$,

$$\mathcal{M}_{\alpha+1} = \operatorname{Ult}(\mathcal{M}'_\beta || (\omega\xi_\alpha, k_\alpha), E_\alpha)$$

   where
   (a) $\beta = \mathcal{T}(\alpha+1)$,
   (b) $\omega\xi_\alpha \leq \operatorname{ord}(\mathcal{M}'_\beta)$ is the largest such that

$$(\kappa_\alpha^+)^{\mathcal{M}_\alpha | \operatorname{ind}^{\mathcal{M}_\alpha}(E_\alpha)} = (\kappa_\alpha^+)^{\mathcal{M}'_\beta | \omega\xi_\alpha},$$

   (c) k_α is the largest such that $(\omega\xi_\alpha, k_\alpha) \leq l(\mathcal{M}'_\beta)$ and

$$\operatorname{crit}(E_\alpha) < \rho_{k_\alpha}(\mathcal{M}'_\beta || (\omega\xi_\alpha, k_\alpha)).$$

9. $D = \{\alpha+1<\eta : \text{letting } \beta = \mathcal{T}(\alpha+1), (\omega\xi_\alpha, k_\alpha) < l(\mathcal{M}_\beta)\}$.
   Let

$$\pi^{\mathcal{T}}_{\beta,\alpha+1} = \pi_{E_\alpha}^{\mathcal{M}'_\beta || (\omega\xi_\alpha, k_\alpha)} : \mathcal{M}'_\beta || (\omega\xi_\alpha, k_\alpha) \to \mathcal{M}_{\alpha+1}$$

   be the ultrapower map and for $\alpha <_T \gamma < \eta$ let $\pi^{\mathcal{T}}_{\alpha,\gamma} : \mathcal{M}_\alpha \to \mathcal{M}_\gamma$ be the embedding obtained by compositions.[29]

---

[28] Similarly define all other combinations of $(\alpha,\beta)_{\mathcal{T}}$, like $[\alpha,\beta)_{\mathcal{T}}$ and etc.

[29] Assuming these embeddings can be composed. $\pi^{\mathcal{T}}_{\alpha,\gamma}$ is defined if and only if $D \cap (\alpha,\gamma]_{\mathcal{T}} = \emptyset$.

10. For limit $\lambda < \eta$, $D \cap (0,\lambda)_\mathcal{T}$ is finite and letting $\beta \in [0,\lambda)_\mathcal{T}$ be the least such that $D \cap (\beta,\lambda)_\mathcal{T} = \emptyset$, $\mathcal{M}_\lambda$ is the direct limit of the system $(\mathcal{M}_\gamma, \pi^\mathcal{T}_{\gamma,\gamma'} : \gamma < \gamma', \gamma, \gamma' \in [\beta,\lambda)_\mathcal{T})$ and for $\gamma \in [\beta,\lambda)$, $\pi^\mathcal{T}_{\gamma,\lambda} : \mathcal{M}_\gamma \to \mathcal{M}_\lambda$ is the direct limit embedding.

    More precisely, $j(\mathcal{M}_\lambda)$ is the direct limit of $(j(\mathcal{M}_\gamma), \pi^\mathcal{T}_{\gamma,\gamma'} ; \gamma < \gamma', \gamma, \gamma' \in [\beta,\lambda)_\mathcal{T})$ and $k(\mathcal{M}_\lambda) = k(\mathcal{M}_\beta)$ (recall Definition 2.2.2 which defined $j(\mathcal{M})$ and $k(\mathcal{M})$).

If $\alpha + 1 \in D$ then we say that there is a drop at $\alpha + 1$. Suppose $\alpha + 1 \in D$ and $\beta = T(\alpha + 1)$. If $\omega\xi_\alpha < \mathrm{ord}(\mathcal{M}_\beta)$ then we say that there is a drop in model at $\alpha + 1$ and otherwise we say there is a drop in degree.

We set $\mathcal{M}^\mathcal{T}_\alpha = \mathcal{M}_\alpha$, $E^\mathcal{T}_\alpha = E_\alpha$, $\mathrm{ind}^\mathcal{T}_\alpha = \mathrm{ind}^{\mathcal{M}_\alpha}(E_\alpha)$, $\mathrm{lh}(\mathcal{T}) = \eta$, $\kappa^\mathcal{T}_\alpha = \mathrm{crit}(E^\mathcal{T}_\alpha)$ and $\nu^\mathcal{T}_\alpha = \nu(E^\mathcal{T}_\alpha)$[30]. We will drop superscript $\mathcal{T}$ when it is clear from context.

DEFINITION 2.4.2. We say that $\mathcal{T}$ is an *iteration tree* if it is a putative iteration tree such that for every $\alpha < \mathrm{lh}(\mathcal{T})$, $\mathcal{M}^\mathcal{T}_\alpha$ is well-founded.

DEFINITION 2.4.3. Given a putative iteration $\mathcal{T}$ on $\mathcal{M}$ we say that $\mathcal{T}$ is *normal* if

1. $R^\mathcal{T} = \{0\}$,
2. for all $\alpha < \beta < \mathrm{lh}(\mathcal{T})$, $\mathrm{ind}^\mathcal{T}_\alpha < \mathrm{ind}^\mathcal{T}_\beta$, and
3. for all α such that $\alpha + 1 < \mathrm{lh}(\mathcal{T})$, $\beta = \mathcal{T}(\alpha + 1)$ is the least β' such that $(\kappa^+_\alpha)^{\mathcal{M}_\alpha | \mathrm{ind}_\alpha} < \nu_{\beta'}$.

NOTATION 2.4.4. Given a putative iteration tree

$$\mathcal{T} = ((\mathcal{M}_\alpha)_{\alpha<\eta}, (E_\alpha)_{\alpha<\eta-1}, D, R, (\beta_\alpha, m_\alpha)_{\alpha\in R}, T)$$

and ordinals $\gamma < \zeta \leq \eta$, we will use the following notations:

1. $\mathcal{T}_{<\gamma} = ((\mathcal{M}_\alpha)_{\alpha\leq\gamma}, (E_\alpha)_{\alpha<\gamma}, (\xi_\alpha)_{\alpha<\gamma}, D \cap (0,\gamma], R \cap [0,\gamma), (\beta_\alpha, m_\alpha)_{\alpha\in R\cap[0,\gamma)}, T \cap (\gamma+1)^2)$,
2. $\mathcal{T}_{[\gamma,\zeta]} = ((\mathcal{M}_\alpha)_{\alpha\in[\gamma,\zeta]}, (E_\alpha)_{\alpha\in[\gamma,\zeta)}, D \cap (\gamma,\zeta], R \cap [\gamma,\zeta), (\beta_\alpha, m_\alpha)_{\alpha\in R\cap[\gamma,\zeta)}, T \cap [\gamma,\zeta]^2)$.
3. Other notations such as $\mathcal{T}_{<\gamma}$, $\mathcal{T}_{\geq\gamma}$, $\mathcal{T}_{(\gamma,\xi)}$ and etc are defined in the obvious manner.
4. Given $\alpha \in R^\mathcal{T}$,

$$\mathsf{next}^\mathcal{T}(\alpha) = \begin{cases} \min(\mathcal{R}^\mathcal{T} - (\alpha+1)) & \text{if } \mathcal{R}^\mathcal{T} - (\alpha+1) \neq \emptyset, \\ \mathrm{lh}(\mathcal{T}) & \text{otherwise.} \end{cases}$$

5. Given $\alpha \in R^\mathcal{T}$, $\mathsf{nc}^\mathcal{T}_\alpha = \mathcal{T}_{[\alpha,\alpha']}$ where $\alpha' = \mathsf{next}^\mathcal{T}(\alpha)$[31].
6. We say that $\mathcal{U}$ is a *normal component* of $\mathcal{T}$ if for some $\alpha \in R^\mathcal{T}$, $\mathcal{U} = \mathsf{nc}^\mathcal{T}_\alpha$.

[30] Here $\nu(E)$ is the natural length of E. See [59, Definition 2.2].

[31] "nc" stands for "normal component".

7. If $\mathcal{U}$ is a normal component of $\mathcal{T}$ then we let $\alpha^{\mathcal{T}}(\mathcal{U}) = \alpha$ where α is as above. We say $\mathcal{U}$ is the last normal component of $\mathcal{T}$ if $\alpha^{\mathcal{T}}(\mathcal{U}) = \max(R^{\mathcal{T}})$.

DEFINITION 2.4.5. We say that a putative iteration tree $\mathcal{T}$ is a *putative stack* if

1. $R^{\mathcal{T}}$ is closed,
2. for all $\alpha \in R^{\mathcal{T}}$, $\mathsf{nc}^{\mathcal{T}}_\alpha$, after obvious re-enumeration of its members, is a normal iteration tree on $\mathcal{M}^{\mathcal{T}}_\alpha||(\beta^{\mathcal{T}}_\alpha, m^{\mathcal{T}}_\alpha)$.

We say $\mathcal{T}$ is a *stack* if $\mathcal{T}$ is a putative stack that is an iteration tree.

DEFINITION 2.4.6. Suppose $\mathcal{P}$ is an lhes. We say that $\mathcal{T}$ is a *semi-smooth*[32] stack on $\mathcal{P}$ if for all $\alpha \in R^{\mathcal{T}}$,

1. $\mathsf{nc}^{\mathcal{T}}_\alpha$ is normal[33]
2. $\mathcal{M}^{\mathcal{T}}_\alpha||(\omega\beta^{\mathcal{T}}_\alpha, m^{\mathcal{T}}_\alpha)$ is a layer of $\mathcal{M}^{\mathcal{T}}_\alpha$.

REMARK 2.4.7 (Semi-smooth convention). Because we will mostly work with semi-smooth stacks on lhes, we make the convention that all stacks are semi-smooth.

In the sequel, we will often say that $\mathcal{R}$ is a node of $\mathcal{T}$ to mean that $\mathcal{R} = \mathcal{M}^{\mathcal{T}}_\alpha$ for some $\alpha < \mathrm{lh}(\mathcal{T})$.

Branches of iterations. Suppose $\mathcal{M}$ is an lhes and $\mathcal{T}$ is an iteration tree on $\mathcal{M}$ such that $\mathrm{lh}(\mathcal{T})$ is a limit ordinal. We say $b \subseteq \mathrm{lh}(\mathcal{T})$ is a putative cofinal branch of $\mathcal{T}$ if b is cofinal in $\mathrm{lh}(\mathcal{T})$ and for every $\alpha \in b$, $[0,\alpha]_{\mathcal{T}} \subseteq b$. Given a putative cofinal branch b of $\mathcal{T}$, we say b is a cofinal branch of $\mathcal{T}$ if $D^{\mathcal{T}} \cap b$ is finite. Given a cofinal branch b of $\mathcal{T}$ we let $\mathcal{M}^{\mathcal{T}}_b$ be the direct limit of the directed system $((\mathcal{M}_\alpha)_{\alpha \in b'}, (\pi^{\mathcal{T}}_{\alpha,\beta})_{\alpha<\beta, \alpha,\beta \in b'})$ where $b' = b - \max(D^{\mathcal{T}} \cap b)$[34]. We say b is a well-founded cofinal branch if it is a cofinal branch such that $\mathcal{M}^{\mathcal{T}}_b$ is well-founded. If b is a putative cofinal branch of $\mathcal{T}$ then we let $\mathcal{T}^\frown\{b\}$ be the unique putative iteration tree $\mathcal{U}$ such that

1. $\mathrm{lh}(\mathcal{U}) = \mathrm{lh}(\mathcal{T}) + 1$,
2. $T^{\mathcal{T}} = T^{\mathcal{U}} \restriction (\mathrm{lh}(\mathcal{T}) \times \mathrm{lh}(\mathcal{T}))$
3. for all $\alpha < \mathrm{lh}(\mathcal{T})$, $\mathcal{M}^{\mathcal{U}}_\alpha = \mathcal{M}^{\mathcal{T}}_\alpha$ and $E^{\mathcal{T}}_\alpha = E^{\mathcal{U}}_\alpha$,
4. $D^{\mathcal{T}} = D^{\mathcal{U}}$ and $R^{\mathcal{T}} = R^{\mathcal{U}}$,
5. $\mathcal{M}^{\mathcal{U}}_{\mathrm{lh}(\mathcal{T})} = \mathcal{M}^{\mathcal{T}}_b$,
6. $[0, \mathrm{lh}(\mathcal{T}))_{\mathcal{U}} = b$.

We also let for $\alpha \in b$, $\pi^{\mathcal{T}}_{\alpha,b} = \pi^{\mathcal{T}^\frown\{b\}}_{\alpha,\mathrm{lh}(\mathcal{T})}$ given the later embedding is defined, and if $\pi^{\mathcal{T}}_{0,b}$ is defined then we let $\pi^{\mathcal{T}}_b = \pi^{\mathcal{T}}_{0,b}$.

---

[32]"Smooth" is inspired by Jensen's terminology who uses "smooth stack" for stacks that do not allow drops in model or degree at the begining of the rounds. See [11].

[33]After trivial re-organization.

[34]Here the direct limit is defined analogously to clause 7 of Definition 2.4.1.

Strategies. Given a stack $\mathcal{T}$ on $\mathcal{M}$ with last model $\mathcal{N}$ and a stack $\mathcal{U}$ on $\mathcal{N}$ we can form $\mathcal{T}$-followed-$\mathcal{U}$ stack $\mathcal{T}^\frown\mathcal{U}$. More formally, $\mathcal{T}^\frown\mathcal{U}$ is the unique stack $\mathcal{W}$ on $\mathcal{M}$ such that $\mathrm{lh}(\mathcal{W}) = \mathrm{lh}(\mathcal{T}) + \mathrm{lh}(\mathcal{U})$, $\mathcal{W}_{<\mathrm{lh}(\mathcal{T})} = \mathcal{T}$,

$$R^{\mathcal{W}} = R^{\mathcal{T}} \cup \{\mathrm{lh}(\mathcal{T}) + \alpha : \alpha \in R^{\mathcal{U}}\}$$

and $\mathcal{W}_{\geq \mathrm{lh}(\mathcal{T})} = \mathcal{U}$[35]. We may often say that $\mathcal{T}^\frown\mathcal{U}$ is a normal iteration tree if after straightforward re-enumeration of its members it becomes a normal iteration tree.[36]

Suppose $\mathcal{M}$ is an lhes and Σ is a function. We say Σ is an iteration strategy for $\mathcal{M}$ if whenever $\mathcal{T} \in \mathrm{dom}(\Sigma)$ then $\mathcal{T}$ is a stack on $\mathcal{M}$, $\mathrm{lh}(\mathcal{T})$ is a limit ordinal and $\Sigma(\mathcal{T})$ is a cofinal well-founded branch of $\mathcal{T}$.

A putative iteration tree $\mathcal{T}$ on $\mathcal{M}$ is according to Σ if for all limit ordinals $\alpha < \mathrm{lh}(\mathcal{T})$, $\mathcal{T}_{<\alpha} \in \mathrm{dom}(\Sigma)$ and $\Sigma(\mathcal{T}_{<\alpha}) = [0,\alpha)_{\mathcal{T}}$. We say Σ is a κ-strategy for $\mathcal{M}$ if

1. Σ is an iteration strategy for $\mathcal{M}$ such that if $\mathcal{T} \in \mathrm{dom}(\Sigma)$ then $\mathcal{T}$ is a normal iteration tree on $\mathcal{M}$ of length $< \kappa$,
2. if $\mathcal{T}$ is a normal iteration tree on $\mathcal{M}$ such that $\mathrm{lh}(\mathcal{T}) < \kappa$, $\mathrm{lh}(\mathcal{T})$ is a limit ordinal and $\mathcal{T}$ is according to Σ then $\mathcal{T} \in \mathrm{dom}(\Sigma)$,
3. if $\mathcal{T} \in \mathrm{dom}(\Sigma)$, $b = \Sigma(\mathcal{T})$ and $\mathcal{U}$ is a normal finite putative iteration tree on $\mathcal{M}_b^{\mathcal{T}}$ such that $\mathrm{lh}(\mathcal{T}) + \mathrm{lh}(\mathcal{U}) < \kappa$ and $\mathcal{T}^\frown\mathcal{U}$ is a normal putative iteration tree on $\mathcal{M}$ (after obvious re-enumeration of it) then $\mathcal{U}$ is an iteration tree.

Alternatively, Σ is a κ-strategy if it is a winning strategy for II in the iteration game $\mathcal{G}(\mathcal{M},\kappa)$. This game is defined immediately after [59, Definition 3.3]. The subscript k that appears in this game is just $k(\mathcal{M})$.

Similarly we can define (κ,λ)-strategy for $\mathcal{M}$ that acts on stacks. Here κ bounds the number of normal components and λ bounds the length of the normal components of the stacks. The relevant iteration game, $\mathcal{G}(\mathcal{M},\kappa,\lambda)$ appears soon after [59, Remark 4.3].

The common part. Given a normal iteration tree $\mathcal{T}$ we let the common part of $\mathcal{T}$ be

$$\mathrm{m}(\mathcal{T}) = \cup_{\alpha<\mathrm{lh}(\mathcal{T})} \mathcal{M}_\alpha^{\mathcal{T}} | \mathrm{ind}_\alpha^{\mathcal{T}}.$$

Usually in literature, for example in [59], $\mathrm{m}(\mathcal{T})$ is denoted by $\mathcal{M}(\mathcal{T})$. However, inner model theorists make $\mathcal{M}$ tired, and so in this book we give it less weight to carry than its usual heavy load. Following [59] we let $\delta(\mathcal{T}) = \mathrm{ord}(\mathrm{m}(\mathcal{T})) = \sup\{\mathrm{ind}_\alpha^{\mathcal{T}} : \alpha < \mathrm{lh}(\mathcal{T})\}$.

Restrictions.

TERMINOLOGY 2.4.8. Suppose $\mathcal{M}$ is an lses and $\mathcal{T}$ is a stack on $\mathcal{M}$.

---

[35] After obvious re-enumeration of its members.

[36] In this re-enumeration we must set $R^{\mathcal{T}^\frown\mathcal{U}} = \{0\}$.

1. Given $\eta \leq \mathrm{ord}(\mathcal{M})$, we say $\mathcal{T}$ is *above* η if for all $\alpha+1 < \mathrm{lh}(\mathcal{T})$, $\mathrm{crit}(E^{\mathcal{T}}_{\alpha}) \geq \eta$.
2. We say that $\mathcal{T}$ is *below* η if for all $\alpha+1 < \mathrm{lh}(\mathcal{T})$, either $\pi^{\mathcal{T}}_{0,\alpha}$ is undefined or $\mathrm{ind}^{\mathcal{T}}_{\alpha} < \pi^{\mathcal{T}}_{0,\alpha}(\eta)$.
3. If $\mathcal{N} \trianglelefteq \mathcal{M}$ then we say $\mathcal{T}$ is *based on* $\mathcal{N}$ if $\mathcal{T}$ is below $\mathrm{ord}(\mathcal{N})$.

DEFINITION 2.4.9. Suppose $\mathcal{M}$ is an lhes and $\mathcal{N} \trianglelefteq \mathcal{M}$ is such that $ord(\mathcal{N})$ is a regular cardinal of $\mathcal{M}$ such that $\rho(\mathcal{M}) > \mathrm{ord}(\mathcal{N})$. Suppose further that $\mathcal{T}$ is a stack on $\mathcal{M}$. We then let $\downarrow(\mathcal{T},\mathcal{N})$ be the portion of $\mathcal{T}$ that is based on $\mathcal{N}$. More precisely, if

$$\mathcal{T} = ((\mathcal{M}_{\alpha})_{\alpha<\eta}, (E_{\alpha})_{\alpha<\eta-1}, D, R, (\beta_{\alpha}, m_{\alpha})_{\alpha\in R}, T).$$

then

$$\downarrow(\mathcal{T},\mathcal{N}) = ((\mathcal{M}'_{\alpha})_{\alpha<\eta'}, (E'_{\alpha})_{\alpha<\nu-1}, D', R', (\beta'_{\alpha}, m'_{\alpha})_{\alpha\in R'}, T').$$

is such that there is an order preserving map $\sigma : \eta' \to \eta$ such that

1. for all $\alpha < \alpha' < \eta'$, $(\alpha, \alpha') \in T' \iff (\sigma(\alpha), \sigma(\alpha')) \in T$,
2. for all $\alpha < \eta'$, $\alpha \in R' \iff \sigma(\alpha) \in R$,
3. for all $\alpha < \eta'$, $\mathcal{M}'_{\alpha} \trianglelefteq \mathcal{M}_{\sigma(\alpha)}$ and $E'_{\alpha} = E_{\sigma(\alpha)}$,
4. for all $\alpha < \eta'$, $\beta'_{\alpha} = \beta_{\sigma(\alpha)}$ and $m'_{\alpha} = m_{\sigma(\alpha)}$,
5. for all $\alpha+1 < \eta'$, $\sigma(\alpha)$ is the least $\beta \in (\sup_{\gamma<\alpha} \sigma(\gamma), \eta)$ such that $\mathrm{ind}^{\mathcal{T}}_{\beta} \leq \mathrm{ord}(\mathcal{M}'_{\alpha+1})$,
6. if $\alpha+1 = \eta'$ and α is a limit ordinal then $\sigma(\alpha) = \sup_{\gamma<\alpha} \sigma(\gamma)$,
7. if $\alpha+1 = \eta'$ and $\alpha = \beta+1$ then $\sigma(\alpha) = \sigma(\beta)+1$,
8. for all $\alpha < \eta'$, if there is $\beta < \eta$ such that $\mathrm{ind}^{\mathcal{T}}_{\beta} \leq \mathrm{ord}(\mathcal{M}'_{\alpha})$ then $\alpha+1 < \eta'$.

DEFINITION 2.4.10. Suppose $\mathcal{M}$ is an lhes and $\mathcal{N} \trianglelefteq \mathcal{M}$ is such that $\mathrm{ord}(\mathcal{N})$ is a regular cardinal of $\mathcal{M}$ such that $\rho(\mathcal{M}) > \mathrm{ord}(\mathcal{N})$. Suppose further that $\mathcal{T}$ is a stack on $\mathcal{N}$. We then let $\uparrow(\mathcal{T},\mathcal{M})$ be the result of "applying" $\mathcal{T}$ to $\mathcal{M}$. More precisely, if

$$\mathcal{T} = ((\mathcal{N}_{\alpha})_{\alpha<\eta}, (E_{\alpha})_{\alpha<\eta-1}, D, R, (\beta_{\alpha}, m_{\alpha})_{\alpha\in R}, T).$$

then

$$\uparrow(\mathcal{T},\mathcal{M}) = ((\mathcal{M}_{\alpha})_{\alpha<\eta}, (E_{\alpha})_{\alpha<\eta-1}, D, R, (\beta_{\alpha}, m_{\alpha})_{\alpha\in R}, T).$$

with $\mathcal{M}_0 = \mathcal{M}$.

Suppose for some $\eta \leq \mathrm{ord}(\mathcal{Q})$, η is a regular cardinal of $\mathcal{P}$, $\rho(\mathcal{P}) > \eta$ and $\mathcal{T}$ is below η. We then have that $\mathcal{T} \restriction \mathcal{Q}$ is the unique stack $\mathcal{U}$ on $\mathcal{Q}$ such that the copy of $\mathcal{U}$ onto $\mathcal{P}$ via *id* is $\mathcal{T}$.

2.5. Layered strategy e-structures

In this manuscript, we are mostly concerned with lhes whose f predicate codes a strategy. The goal of this section is to introduce the language used to describe such structures.

Suppose that $\mathcal{M}$ is an lhes. We then say that a shifted amenable function f *codes a partial strategy function* for $\mathcal{M}$ if letting g be the amenable component of f, the following conditions hold:

1. $\mathrm{dom}(f) \subseteq \{(\mathcal{J}_\omega(\mathcal{T}),\mathcal{T},\in) : \mathcal{T}$ is a stack on $\mathcal{M}$ without a last model$\}$.
2. Whenever $\mathcal{T}$ is a stack on $\mathcal{M}$ such that $(\mathcal{J}_\omega(\mathcal{T}),\mathcal{T},\in) \in \mathrm{dom}(f)$ and whenever $\mathcal{U}$ is an initial segment of $\mathcal{T}$ without a last model, $(\mathcal{J}_\omega(\mathcal{U}),\mathcal{U},\in) \in \mathrm{dom}(f)$ and
$$g((\mathcal{J}_\omega(\mathcal{U}),\mathcal{U},\in)) = [0,\mathrm{lh}(\mathcal{U}))_{\mathcal{T}}.$$
3. For all $(\mathcal{J}_\omega(\mathcal{T}),\mathcal{T},\in) \in \mathrm{dom}(f)$, $g((\mathcal{J}_\omega(\mathcal{T}),\mathcal{T},\in))$ is a cofinal branch of $\mathcal{T}$.

Notice that we do not require that

(a) $g((\mathcal{J}_\omega(\mathcal{T}),\mathcal{T},\in))$ is a well-founded branch of $\mathcal{T}$,
(b) if $(\mathcal{J}_\omega(\mathcal{T}),\mathcal{T},\in) \in \mathrm{dom}(f)$ and $b = g((\mathcal{J}_\omega(\mathcal{T}),\mathcal{T},\in))$ is a cofinal well-founded branch of $\mathcal{T}$ then any reasonable finite extension of $\mathcal{T}^\frown\{b\}$ has well-founded models.

Conditions (a) and (b) will be part of a more restrictive notion.

When defining short tree strategy mice, we will encounter hybrid structures whose f predicate does not necessarily code a strategy but a partial strategy. We make this notion more precise. First we make a useful definition.

DEFINITION 2.5.1. Suppose $\mathcal{M}$ is an lhes. We then say that Σ is a *semi-strategy* for $\mathcal{M}$ if the domain of Σ consists of quadruples $(\mathcal{M}_0,\mathcal{T}_0,\mathcal{M}_1,\mathcal{U})$ such that

1. $\mathcal{M}_0 = \mathcal{M}$,
2. $\mathcal{T}_0$ is a normal tree on $\mathcal{M}_0$,
3. $\mathcal{M}_1$ is either the last model of $\mathcal{T}_0$ or $\mathcal{T}_0$ does not have a last model and $\mathcal{M}_1 = (\mathrm{m}(\mathcal{T}_0))^{\#}$[37], and
4. $\mathcal{U}$ is a stack on $\mathcal{M}_1$ below $\delta(\mathcal{T}_0)$[38].

We can then consider amenable functions that code partial semi-iteration strategies. We will abuse our terminology and will treat semi-iteration strategies as if they were just strategies.

Suppose then a shifted amenable function f codes a partial strategy function for $\mathcal{M}$. We then let Σ^f be the partial strategy function coded by f. More precisely, letting g be the amenable component of f,

[37]This is the true, ω_1-iterable, sharp of $\mathrm{m}(\mathcal{T}_0)$.

[38]This means that all extenders used in $\mathcal{U}$ have lengths below the image of $\delta(\mathcal{T}_0)$. That is, for each $\alpha < \mathrm{lh}(\mathcal{T}_0)$ either $[0,\alpha)_{\mathcal{T}} \cap D^{\mathcal{T}} \neq$ or $\mathrm{ind}^{\mathcal{T}}_\alpha < \pi^{\mathcal{T}}_{0,\alpha}(\delta(\mathcal{T}_0))$.

1. $\mathrm{dom}(\Sigma^f) = \{\mathcal{T} : (\mathcal{J}_\omega(\mathcal{T}), \mathcal{T}, \in) \in \mathrm{dom}(f)\}$ and
2. for all $\mathcal{T} \in \mathrm{dom}(\Sigma^f)$, $\Sigma^f(\mathcal{T}) = g((\mathcal{J}_\omega(\mathcal{T}), \mathcal{T}, \in))$.

We say f *codes a partial strategy* if Σ^f chooses cofinal and well-founded branches. We say f *codes a total A-strategy* if $\Sigma^f(\mathcal{T})$ is defined whenever $\mathcal{T} \in A$ is of limit length and is according to Σ. If A is clear from context then we will drop it from our notation.

Following [29], if $\mathcal{M}$ is an lhes, $\mathcal{N} \trianglelefteq \mathcal{M}$ and Σ is an iteration strategy for $\mathcal{M}$ then $\Sigma_{\mathcal{N}}$ is the strategy of $\mathcal{N}$ we get by the copy construction. More precisely, $\Sigma_{\mathcal{N}}$ is the *id*-pullback of Σ. Like in [29], if a transitive structure P has a distinguished sequence of extenders then when discussing iterability of P we will always mean iterability with respect to that extender sequence.

DEFINITION 2.5.2 (Strategic e-structure, ses). Suppose $\mathcal{P}$ is a transitive structure, X is a self-well-ordered set such that $\mathcal{P} \in X$ and $\mathcal{M}$ is a φ-indexed hes. We say $\mathcal{M}$ is a *φ-indexed strategic e-structure* (ses) over X based on $\mathcal{P}$ if $f^{\mathcal{M}}$ codes a partial iteration strategy for $\mathcal{P}$ and for any $w \in \mathrm{dom}(f^{\mathcal{M}})$ if $\beta = \min(f^{\mathcal{M}}(w))$ then $\mathcal{M}|\beta$ is closed[39].

We say $\mathcal{M}$ is based on $\mathcal{P}$ if $\mathcal{M}$ is over $\mathcal{J}_\omega[\mathcal{P}]$ and is based on $\mathcal{P}$.

In Section 3.8, we will also need unindexed ses[40].

DEFINITION 2.5.3 (Unindexed ses). Suppose $\mathcal{P}$ is a transitive structure, X is a self-well-ordered set such that $\mathcal{P} \in X$ and $\mathcal{M} = \mathcal{J}^{\vec{E},f}(X)$ is a hybrid $\mathcal{J}$-structure over X. We say $\mathcal{M}$ is an *unindexed strategic e-structure* (unindexed ses) over X based on $\mathcal{P}$ if the following clauses hold.

1. $f^{\mathcal{M}}$ codes a partial iteration strategy for $\mathcal{P}$ such that for any $w \in \mathrm{dom}(f^{\mathcal{M}})$ if $\beta = min(f^{\mathcal{M}}(w))$ then $\mathcal{M}|\beta$ is closed[41].
2. $\vec{E}$ is a mixed indexed extender sequence.
3. If $\mathcal{M} = (\mathcal{M}', k)$[42] then for every $(\omega\beta, m) < l(\mathcal{M})$, $\mathcal{M}||(\omega\beta, m)$ is sound.

We say $\mathcal{M}$ is based on $\mathcal{P}$ if $\mathcal{M}$ is over $\mathcal{J}_\omega[\mathcal{P}]$ and is based on $\mathcal{P}$.

DEFINITION 2.5.4 (Layered strategic e-structure, lses). Suppose $\mathcal{M}$ is a φ-indexed lhes. We say $\mathcal{M}$ is a *φ-indexed layered strategic e-structure* (lses) if for all $\mathcal{Q} \in Y^{\mathcal{M}}$, in $\mathcal{M}$,

1. $f^{\mathcal{M}}(\mathcal{Q})$ codes a partial iteration strategy for $\mathcal{Q}$ such that for every $w \in \mathrm{dom}(f^{\mathcal{M}}(\mathcal{Q}))$, if $\beta = \min(f^{\mathcal{M}}(\mathcal{Q})(w))$ then $\mathcal{M}|\beta$ is closed, and

[39] See Definition 2.3.15. Also, recall that for such β we have $\omega\beta = \beta$.

[40] Notice that in Definition 2.5.3, *unindexed* simply means that no indexing is specified. It is possible that a given unindexed ses $\mathcal{M}$ is in fact φ-indexed for some φ.

[41] See Definition 2.3.15. Also, recall that for such β we have $\omega\beta = \beta$.

[42] See Definition 2.2.2.

2. if $\mathcal{Q}_0, \mathcal{Q}_1 \in Y^{\mathcal{M}} - (X^{\mathcal{M}} \cup \{X^{\mathcal{M}}\})$[43] are such that $\mathcal{Q}_0 \trianglelefteq \mathcal{Q}_1$ then letting, for $i \in 2$, Σ_i be the partial iteration strategy coded by $f^{\mathcal{M}}(\mathcal{Q}_i)$ and Λ be the id-pullback of Σ_1, then $\Lambda \subseteq \Sigma_0$[44].

If $\mathcal{Q} \in Y^{\mathcal{M}}$ then we let $\Sigma^{\mathcal{M}}_{\mathcal{Q}}$ be the partial strategy function coded by $f^{\mathcal{M}}(\mathcal{Q})$ and let $\Sigma^{\mathcal{M}}$ be the function with domain $Y^{\mathcal{M}}$ such that $\Sigma^{\mathcal{M}}(\mathcal{Q}) = \Sigma^{\mathcal{M}}_{\mathcal{Q}}$. The next definition isolates the language of lses and ses.

DEFINITION 2.5.5. We let $\mathcal{L}_{\mathsf{ses}}$ be the language of ses intended for lightface ses, where we say $\mathcal{M}$ is a lightface ses if for some $\mathcal{P}$, $\mathcal{M}$ is an ses over $\mathcal{J}_\omega[\mathcal{P}]$ based on $\mathcal{P}$. Thus, $\mathcal{L}_{\mathsf{ses}}$ augments the ordinary language for premice as introduced in [59, Definition 2.10] by adding one constant symbol $\dot{\mathcal{P}}$ for $\mathcal{P}$ and a predicate symbol $\dot{f}$ for f. $\mathcal{L}_{\mathsf{ses}}$ can be further augmented by a constant symbol for X (see Definition 2.5.2), and this language can be used for boldface ses.

We let $\mathcal{L}_{\mathsf{lses}}$ be the language of lses over $\emptyset$ (those are the lses whose X predicate is the $\emptyset$). Thus, $\mathcal{L}_{\mathsf{lses}}$ is the language of premice augmented by symbols $\{\dot{B}, \dot{f}, \dot{Y}\}$.

In some cases, it is convenient to use the symbol $\dot{\mathcal{V}}$ to denote the universe of lses or ses, and also the symbol $\dot{\Sigma}$ to indicate the strategy function coded by $\dot{f}$. Moreover, if $\mathcal{Q} \in \mathrm{dom}(\dot{f})$ then we will use $\dot{\Sigma}_{\mathcal{Q}}$ to denote the strategy function given by $\dot{f}(\mathcal{Q})$.

In most applications, lses have a very canonical indexing scheme which is originally due to Woodin. At each stage the stack whose branch is being indexed by f is the least stack whose branch has not yet been indexed. We call this the *standard indexing scheme* (see Section 3.9).

REMARK 2.5.6. Unless indicated otherwise, we will always tacitly assume that the extenders used to witness the existence of large cardinals in lses belong to the extender sequence of the lses. Thus, when we say "κ is a measurable cardinal in $\mathcal{M}$" we mean that there is an extender $E \in \vec{E}^{\mathcal{M}}$ such that E witnesses that κ is a measurable cardinal in $\mathcal{M}$. In [44], Schlutzenberg extensively studied the problem of whether in pure extender models all large cardinal properties are witnessed by extenders that are indexed on the extender sequence. In particular, he showed that measurability and Woodinness are witnessed by extenders that are on the extender sequence.

REMARK 2.5.7. Suppose $\mathcal{M}$ is an lses and $\beta < \mathrm{ord}(\mathcal{M})$. The notations $\mathcal{M}|\beta$ and $\mathcal{M}||\beta$ were introduced just before Remark 2.1.3. In this remark, we would like to clarify the meaning of $Y^{\mathcal{M}|\beta}$. It is not hard to re-formulate Definition 2.3.13 in a way that lses become hierarchical $\mathcal{J}$-structures (see Definition 2.1.2) with the property that $Y^{\mathcal{M}|\beta} = X^{\mathcal{M}} \cup \{\mathcal{Q} \trianglelefteq \mathcal{M} : \mathcal{Q} \in Y^{\mathcal{M}} \wedge \mathrm{ord}(\mathcal{Q}) < \omega\beta\}$[45].

[43] Recall $X^{\mathcal{M}}$ is the set or structure over which $\mathcal{M}$ is defined.

[44] Here, we cannot demand equality as there maybe $\mathcal{T} \in \mathrm{dom}(\Sigma_0)$ such that if $\mathcal{U}$ is the od-copy of $\mathcal{T}$ on $\mathcal{Q}_1, \mathcal{U} \notin \mathrm{dom}(\Lambda)$.

[45] One could for example index every $\mathcal{M}||\omega\beta \in Y^{\mathcal{M}}$ at $\omega\beta + \omega$.

Suppose $\mathcal{M}$ is an lses and Σ is a (κ,ϑ)-iteration strategy for $\mathcal{Q}$ for some $\mathcal{Q}\in Y^{\mathcal{M}}$. Then it can be the case that $\Sigma_{\mathcal{Q}}^{\mathcal{M}}\subseteq\Sigma$. When this happens we get structures relative to Σ.

DEFINITION 2.5.8 ((Σ,φ)-premouse). Suppose X is a transitive self-well-ordered structure and $\mathcal{P}\in X$ is an ses or lses or just a transitive self-well-ordered set. Suppose further that Σ is a (κ,ϑ)-iteration strategy for $\mathcal{P}$ and $\mathcal{M}$ is a φ-indexed ses over X based on $\mathcal{P}$. Then $\mathcal{M}$ is called a (Σ,φ)*-premouse* over X based on $\mathcal{P}$ if $\Sigma^{\mathcal{M}}\subseteq\Sigma\restriction\mathcal{M}$.

Similarly, if $\mathcal{M}$ is a (Σ,φ)-premouse over X based on $\mathcal{P}$, $X=(\mathcal{J}_\omega[\mathcal{P}],\mathcal{P},\in)$ and Σ is a (κ,ϑ)-iteration strategy for $\mathcal{P}$ then $\mathcal{M}$ is called a (Σ,φ)*-premouse* over X.

We then say $\mathcal{M}$ is a (Σ,φ)-premouse if one of the cases in Definition 2.5.8 holds.

DEFINITION 2.5.9 ((Σ,φ)-mouse). Keeping the notation of Definition 2.5.8, we say $\mathcal{M}$ is a (Σ,φ)*-mouse* if $\mathcal{M}$ has an ω_1+1-iteration strategy Λ such that whenever $\mathcal{N}$ is a Λ-iterate of $\mathcal{M}$ then $\mathcal{N}$ is a (Σ,φ)-premouse.

We warn the reader that we will often omit φ from our notation and say "$\mathcal{M}$ is a Σ-mouse" instead of "$\mathcal{M}$ is a (Σ,φ)-mouse" if φ is clear from the context.

2.6. Iterations of (Σ,φ)-mice

Suppose X is a transitive self-well-ordered structure such as ses or lses or just a transitive self-well-ordered set. Suppose further that Σ is an (ω_1,ω_1)-iteration strategy for some $\mathcal{P}\in X$ (which is also ses or lses or some transitive set) and φ is an indexing scheme. Given two (Σ,φ)-mice, we can compare them using the usual comparison argument.

THEOREM 2.6.1 (Theorem 3.11 of [59]). *Suppose $\mathcal{M}$ and $\mathcal{N}$ are two countable (Σ,φ)-mice with (ω_1+1)-iteration strategies Λ and Γ respectively. Then there are iteration trees $\mathcal{T}$ and $\mathcal{U}$ on $\mathcal{M}$ and $\mathcal{N}$ respectively according to Λ and Γ respectively, having last models $\mathcal{M}_\alpha^{\mathcal{T}}$ and $\mathcal{N}_\eta^{\mathcal{N}}$ such that either*

1. *the iteration embedding $\pi_{0,\alpha}^{\mathcal{T}}$-exists and $\mathcal{M}_\alpha^{\mathcal{T}}$ is an initial segment of $\mathcal{M}_\eta^{\mathcal{U}}$, or*
2. *the iteration embedding $\pi_{0,\eta}^{\mathcal{U}}$-exists, and $\mathcal{M}_\eta^{\mathcal{U}}$ is an initial segment of $\mathcal{M}_\alpha^{\mathcal{T}}$.*

Comparison for lses is more involved and we do not know how to do it in general. Below we recall our primary method of identifying the good branches of iteration trees. Recall that the strategy for a sound mouse projecting to ω is determined by $\mathcal{Q}$*-structures*. For $\mathcal{T}$ normal, let $\Phi(\mathcal{T})$ be the phalanx of $\mathcal{T}$ (see Definition 6.6 of [52]).

DEFINITION 2.6.2. Suppose $\mathcal{M}$ is an lses (or ses). Let $\mathcal{T}$ be a normal tree of limit length on $\mathcal{M}$ and let b be a cofinal branch of $\mathcal{T}$. Then $\mathcal{Q}(b,\mathcal{T})$ is the shortest initial segment $\mathcal{Q}$ of $\mathcal{M}_b^{\mathcal{T}}$, if one exists, such that $\mathcal{Q}$ projects strictly across $\delta(\mathcal{T})$ (i.e., $\rho(\mathcal{Q}) < \delta(\mathcal{T})$) or defines a function witnessing $\delta(\mathcal{T})$ is not a Woodin cardinal as witnessed by the extenders on the sequence of $\mathrm{m}(\mathcal{T})$. Equivalently, $\mathcal{Q}(b,\mathcal{T}) = \mathcal{M}_b^{\mathcal{T}}||\omega\xi$ such that ξ is the largest ξ' with the property that $\mathcal{M}_b^{\mathcal{T}}||\omega\xi' \models$ "$\delta(\mathcal{T})$ is a Woodin cardinal".

Next we would like to state a general result stating that branches identified by $\mathcal{Q}$-structures are unique.

DEFINITION 2.6.3. Suppose that $\mathcal{M}$ is an lses and Σ is a strategy for $\mathcal{M}$. If $\mathcal{N}$ is a Σ-iterate of $\mathcal{M}$ via $\mathcal{T}$ then we let $\Sigma_{\mathcal{N},\mathcal{T}}$ be the strategy of $\mathcal{N}$ given by $\Sigma_{\mathcal{N},\mathcal{T}}(\mathcal{U}) = \Sigma(\mathcal{T}^\frown\mathcal{U})$. If then $\mathcal{Q} \trianglelefteq \mathcal{N}$ then we let $\Sigma_{\mathcal{Q},\mathcal{T}}$ be the *id*-pullback of $\Sigma_{\mathcal{N},\mathcal{T}}$.

DEFINITION 2.6.4. Suppose $\mathcal{M}$ is a φ-indexed lses (perhaps over some set X and based on some $\mathcal{P} \in X$) and Σ is an iteration strategy for $\mathcal{M}$. We say $(\mathcal{M},\Sigma)$ is a *layered strategy φ-mouse* (φ-lsm) pair if Σ has hull condensation (see Definition 1.30 of [29]) and whenever $\mathcal{N}$ is a Σ-iterate of $\mathcal{M}$ via $\mathcal{T}$ then $\mathcal{N}$ is a φ-indexed lses and for any $\mathcal{Q} \in Y^{\mathcal{N}} - X$, $\Sigma_{\mathcal{Q}}^{\mathcal{N}} \subseteq \Sigma_{\mathcal{Q},\mathcal{T}}$. We say $(\mathcal{M},\Sigma)$ is sound if $\mathcal{M}$ is sound.

Similarly we can define φ-sm. We will say that $\mathcal{M}$ is a (Σ,φ)-lsm or (Σ,φ)-sm if $(\mathcal{M},\Sigma)$ is respectively a φ-lsm or φ-sm.

TERMINOLOGY 2.6.5. Suppose $\mathcal{M}$ is an lses.

1. We say γ is a *cutpoint* of $\mathcal{M}$ if there is no extender $E \in \vec{E}^{\mathcal{M}}$ such that $\mathrm{crit}(E) < \gamma \leq \mathrm{ind}^{\mathcal{M}}(E)$.
2. We say γ is a *strong cutpoint* of $\mathcal{M}$ if there is no extender $E \in \vec{E}^{\mathcal{M}}$ such that $\mathrm{crit}(E) \leq \gamma \leq \mathrm{ind}^{\mathcal{M}}(E)$.
3. An extender $E \in \vec{E}^{\mathcal{M}}$ *overlaps* κ if $\mathrm{crit}(E) < \kappa \leq \mathrm{lh}(E)$, and *weakly overlaps* κ if $\mathrm{crit}(E) \leq \kappa \leq \mathrm{lh}(E)$.
4. $\mathrm{ord}(Y^{\mathcal{M}}) = \sup\{\mathrm{ord}(\mathcal{Q}) : \mathcal{Q} \in Y^{\mathcal{M}}\}$.

THEOREM 2.6.6. *Suppose $(\mathcal{M},\Sigma)$ is a sound φ-lsm pair, and suppose $\gamma < \mathrm{ord}(\mathcal{M})$ is a strong cutpoint of $\mathcal{M}$ such that*

$$\mathrm{ord}(Y^{\mathcal{M}}) \leq \gamma \quad \textit{and} \quad \rho(\mathcal{M}) \leq \gamma.$$

Then $\mathcal{M}$ has at most one (ω_1+1)-iteration strategy Λ that acts on iteration trees that are strictly above γ and whenever $\mathcal{N}$ is a Λ-iterate of $\mathcal{M}$ then $\mathcal{N}$ is a φ-indexed lses *and $\Sigma^{\mathcal{N}} \subseteq \Sigma \restriction \mathcal{N}$.*

Moreover, any such strategy Λ is determined by: for countable length normal iteration trees $\mathcal{T}$, $\Lambda(\mathcal{T})$ is the unique cofinal wellfounded b such that the phalanx

$$\Phi(\mathcal{T})^\frown(\delta(\mathcal{T}),\mathcal{Q}(b,\mathcal{T}))$$

is ω_1+1-iterable (as a (Σ,φ)-phalanx, see Definition 2.6.2 for the meaning of $\mathcal{Q}(b,\mathcal{T})$).[46]

[46]The meaning of this is left to the reader, but see [52, Definition 6.7] or [3, Definition 2.22].

In some cases, however, it is enough to assume that $\mathcal{Q}(b, \mathcal{T})$ is countably iterable. This happens, for instance, when $\mathcal{M}$ has no local Woodin cardinals with extenders overlapping it. While the lses we will consider may have initial segments that have Woodin cardinals that are not cutpoints, no such cardinal will be Woodin in the entire model. This simplifies our situation somewhat, and below we describe exactly how this will be used.

DEFINITION 2.6.7 (Definition 2.1 of [53]). Let $(\mathcal{M}, \Sigma)$ be a sound φ-lsm pair and let $\gamma < \mathrm{ord}(\mathcal{M})$ be such that $\tau = \mathrm{ord}(Y^{\mathcal{M}}) \le \gamma$. Suppose $\mathcal{T}$ is a normal iteration tree on $\mathcal{M}$ that is above $\gamma+1$; then $\mathcal{Q}(\mathcal{T})$, if exists, is the unique $\mathcal{Q}$ that has the following properties.

1. $\mathcal{Q}$ is a $(\Sigma_{\mathcal{M}||\tau}, \varphi)$-sm over $\mathrm{m}(\mathcal{T})$ based on $\mathcal{M}||\tau$ (in particular, $\delta(\mathcal{T})$ is a strong cutpoint of $\mathcal{Q}$).
2. $\mathcal{J}_\omega(\mathcal{Q}) \models$ "$\delta(\mathcal{T})$ is not a Woodin cardinal",
3. $k(\mathcal{Q})$ is the least k such that
   (a) $\rho_k(\mathcal{Q}) < \delta(\mathcal{T})$ or
   (b) $\rho_k(\mathcal{Q}) = \delta(\mathcal{T})$ and there is $r\Sigma_k^{\mathcal{Q}}$-definable function $f : \delta(\mathcal{T}) \to \delta(\mathcal{T})$ witnessing that $\delta(\mathcal{T})$ is not a Woodin cardinal as witnessed by the extenders of $\mathrm{m}(\mathcal{T})$.

Countable iterability is usually enough to guarantee there is at most one lses with the properties of $\mathcal{Q}(\mathcal{T})$. If it exists, $\mathcal{Q}(\mathcal{T})$ might identify the good branch of $\mathcal{T}$, the one any sufficiently powerful iteration strategy must choose. This is the content of the next lemma which can be proved by analyzing the proof of Theorem 6.12 of [59]. To state it we need to introduce fatal drops.

DEFINITION 2.6.8 (Fatal drop). Suppose $\mathcal{M}$ is a φ-indexed lses and $\mathcal{T}$ is an iteration tree on $\mathcal{M}$. We say $\mathcal{T}$ has a *fatal drop* if for some $\alpha < \mathrm{lh}(\mathcal{T})$ and $\eta < \mathrm{ord}(\mathcal{M}_\alpha^{\mathcal{T}})$,

1. η is a cutpoint of $\mathcal{M}_\alpha^{\mathcal{T}}||\omega\xi_\alpha^{\mathcal{T}}$,
2. $\sup\{\mathrm{ind}_\beta^{\mathcal{T}} : \beta < \alpha\} \le \eta$,
3. $\rho(\mathcal{M}_\alpha^{\mathcal{T}}||(\omega\xi_\alpha^{\mathcal{T}}, k_\alpha^{\mathcal{T}})) \le \eta$[47],
4. $\mathcal{T}_{\ge\alpha}$ is a normal iteration tree on $\mathcal{M}_\alpha^{\mathcal{T}}||(\omega\xi_\alpha^{\mathcal{T}}, k_\alpha^{\mathcal{T}})$ that is above η.

We then say $\mathcal{T}$ has a fatal drop at (α, η) if the pair is the lexicographically least satisfying the above condition.

The following is the lemma mentioned above.

LEMMA 2.6.9. *Let $(\mathcal{M}, \Sigma)$ be a φ-lsm pair such that $\mathrm{ord}(Y^{\mathcal{M}})$ is a strong cutpoint of $\mathcal{M}$[48] and let $\gamma < \mathrm{ord}(\mathcal{M})$ be such that $\mathrm{ord}(Y^{\mathcal{M}}) \le \gamma$. Suppose $\mathcal{T}$ is a normal iteration tree on $\mathcal{M}$ that is above $\gamma+1$ and has limit length.*

---

[47] $\omega\xi_\alpha^{\mathcal{T}}$ and $k_\alpha^{\mathcal{T}}$ are defined in clause 8 of Definition 2.4.1.

[48] The hod mice considered in the manuscript satisfy this condition.

1. *Suppose $\mathcal{Q}(\mathcal{T})$ exists. Then there is at most one cofinal branch b of $\mathcal{T}$ such that either $\mathcal{Q}(\mathcal{T}) = \mathcal{M}_b^{\mathcal{T}}$ or $\mathcal{Q}(\mathcal{T}) = \mathcal{M}_b^{\mathcal{T}}||\omega\xi$ for some ξ in the wellfounded part of $\mathcal{M}_b^{\mathcal{T}}$.*
2. *Suppose further no measurable cardinal of $\mathcal{M}$ which is $\geq \gamma$ is a limit of Woodin cardinals. Suppose further that $\mathcal{T}$ is according to Σ, $\mathcal{T}$ does not have a fatal drop and if $b = \Sigma(\mathcal{T})$ then $\mathcal{Q}(b,\mathcal{T})$-exists. Then $\mathcal{Q}(b,\mathcal{T}) = \mathcal{Q}(\mathcal{T})$.*

$\mathcal{Q}(\mathcal{T})$ identifies b because it determines a canonical cofinal subset of $\mathrm{rng}(\pi_{\alpha,b}^{\mathcal{T}} \cap \delta(\mathcal{T}))$, for some $\alpha \in b$, to which we can apply Lemma 1.13 of [29] (which is an immediate consequence of the zipper argument from [22]).

REMARK 2.6.10. Suppose $(\mathcal{M},\Sigma)$ is a φ-lsm pair and $\mathcal{Q} \in Y^{\mathcal{M}} - X^{\mathcal{M}}$. Let $\mathcal{R} = \mathcal{M}$ if $\mathcal{Q}$ is the largest initial segment of $\mathcal{M}$ in $Y^{\mathcal{M}}$ and otherwise, let $\mathcal{R}$ be the least member of $Y^{\mathcal{M}}$ properly extending $\mathcal{Q}$. Suppose $\mathcal{T}$ is a tree on $\mathcal{M}$ which is above $\mathrm{ord}(\mathcal{Q})+1$ and is based on $\mathcal{R}$. Notice that in this case we can define $\mathcal{Q}(\mathcal{T})$ just as in Definition 2.6.7 by using $\mathcal{R}$ instead of $\mathcal{M}$.

We end this section by introducing the $\mathcal{O}$-stack. Suppose $\mathcal{P}$ is an lses, $\alpha,\eta < \mathrm{ord}(\mathcal{P})$ and $\mathcal{Q} \trianglelefteq \mathcal{P}||\eta$. Let $e_{\eta,\alpha}^{\mathcal{P}}$ be the least ordinal $\beta > \eta$, if it exists, such that $\beta \in \mathrm{dom}(\vec{E}^{\mathcal{P}})$, and letting $E = \vec{E}^{\mathcal{P}}(\beta)$, $\mathrm{crit}(E) \in (\alpha,\eta)$. Thus, $e_{\eta,\alpha}^{\mathcal{P}}$ is the index of the first extender that overlaps $\eta+1$ and has a critical point $> \alpha$. Otherwise, if there is no such extender then set $e_{\eta,\alpha}^{\mathcal{P}} = \mathrm{ord}(\mathcal{P})$.

Let $s_{\eta,\mathcal{Q}}^{\mathcal{P}}$ be the least ordinal $\beta > \eta$, if it exists, such that for some $\mathcal{R} \in Y^{\mathcal{P}} - X^{P}$ with $\mathcal{Q} \triangleleft \mathcal{R}$ letting F be the set indexed at β in $\mathcal{P}$, F is a pair of the form $(\mathcal{R},a)$. Thus, $s_{\eta,\mathcal{Q}}^{\mathcal{P}}$ is the first place above η where a branch of some iteration tree $\mathcal{T}$ that is based on a strictly longer layer than $\mathcal{Q}$ is added. If there is no such $\mathcal{R}$ then let $s_{\eta,\mathcal{Q}}^{\mathcal{P}} = \mathrm{ord}(\mathcal{P})$. Let $\eta' = (\eta^+)^{\mathcal{P}}$ if $(\eta^+)^{\mathcal{P}}$ exists and otherwise let $\eta' = \mathrm{ord}(\mathcal{P})$. Set $\alpha_{\eta,\mathcal{Q},\alpha}^{\mathcal{P}} = \min\{e_{\eta,\alpha}^{\mathcal{P}}, s_{\eta,\mathcal{Q}}^{\mathcal{P}}, \eta'\}$.

Suppose $\mathcal{M}$ is f.s. $\mathcal{J}$-structure and $\eta < \mathrm{ord}(\mathcal{M})$ is the largest cardinal of $\mathcal{M}$. We then let $\mathcal{M}|(\eta^+)^{\mathcal{M}} = \mathcal{M}|\mathrm{ord}(\mathcal{M})$ and $\mathcal{M}||(\eta^+)^{\mathcal{M}} = \mathcal{M}$.

DEFINITION 2.6.11 ($\mathcal{O}^{\mathcal{P}}$-stack). Suppose $\mathcal{P}$ is an lses, $\alpha,\eta < \mathrm{ord}(\mathcal{P})$ and $\mathcal{Q} \trianglelefteq \mathcal{P}||\eta$.

We now set

$$\mathcal{O}_{\eta,\mathcal{Q},\alpha}^{\mathcal{P}} = \mathcal{P}|(\eta^+)^{\mathcal{P}|\alpha_{\eta,\mathcal{Q},\alpha}^{\mathcal{P}}}.$$

Next we define the stack $(\mathcal{O}_{\eta,\mathcal{Q},\alpha}^{\mathcal{P},\xi} : \xi \leq \Omega_{\eta,\mathcal{Q},\alpha}^{\mathcal{P}})$ according to the following recursion:

1. $\mathcal{O}_{\eta,\mathcal{Q},\alpha}^{\mathcal{P},0} = \mathcal{O}_{\eta,\mathcal{Q},\alpha}^{\mathcal{P}}$,
2. for $\xi+1 \leq \Omega_{\eta,\mathcal{Q},\alpha}^{\mathcal{P}}$, $\mathcal{O}_{\eta,\mathcal{Q},\alpha}^{\mathcal{P},\xi+1} = \mathcal{O}_{\mathrm{ord}(\mathcal{O}_{\eta,\mathcal{Q},\alpha}^{\mathcal{P},\xi}),\mathcal{Q},\alpha}^{\mathcal{P}}$,
3. for limit $\lambda \leq \Omega_{\eta,\mathcal{Q},\alpha}^{\mathcal{P}}$, $\mathcal{O}_{\eta,\mathcal{Q},\alpha}^{\mathcal{P},\lambda} = \bigcup_{\xi<\lambda} \mathcal{O}_{\eta,\mathcal{Q},\alpha}^{\mathcal{P},\xi}$, and
4. $\Omega_{\eta,\mathcal{Q},\alpha}^{\mathcal{P}}$ is the least ν such that $\mathrm{ord}(\mathcal{O}_{\eta,\mathcal{Q},\alpha}^{\mathcal{P},\nu}) = \alpha_{\eta,\mathcal{Q},\alpha}^{\mathcal{P}}$.

If $\mathcal{Q} = \mathcal{P}||\kappa$, then we write $\mathcal{O}^{\mathcal{P}}_{\eta,\kappa,\alpha}$ for $\mathcal{O}^{\mathcal{P}}_{\eta,\mathcal{Q},\alpha}$; if $\alpha = 0$, we also write $\mathcal{O}^{\mathcal{P}}_{\eta,\kappa}$ for $\mathcal{O}^{\mathcal{P}}_{\eta,\mathcal{Q},\alpha}$. For $\xi \leq \Omega^{\mathcal{P}}_{\eta,\mathcal{P}||\eta,\alpha}$, we let $\mathcal{O}^{\mathcal{P},\xi}_{\eta} = \mathcal{O}^{\mathcal{P},\xi}_{\eta,\mathcal{P}||\eta,0}$ with $\mathcal{O}^{\mathcal{P}}_{\eta} = \mathcal{O}^{\mathcal{P}}_{\eta,\mathcal{P}||\eta,0}$.

2.7. Hod-like layered hybrid premice

The difference between the lses considered here and those considered in [29] is that here we will have lses whose predicate codes the *short tree strategy* of its initial segments. The hod mice we will consider in this paper are all *layered*, and we start by introducing these objects.

If $\mathcal{M}$ is an lses and κ is an $\mathcal{M}$-cardinal then we set $E^{\mathcal{M}}_{\xi} = \vec{E}^{\mathcal{M}}(\xi)$ and

$$X^{\mathcal{M}}_{\kappa} = \{\xi : E^{\mathcal{M}}_{\xi} \neq \emptyset \text{ and } \mathrm{crit}(E^{\mathcal{M}}_{\xi}) = \kappa\}.$$

We also let

$$o^{\mathcal{M}}(\kappa) = \max(\sup X^{\mathcal{M}}_{\kappa}, (\kappa^+)^{\mathcal{M}}).$$

Suppose M is a transitive structure and η is an ordinal. Then we let $(\eta^{+\alpha})^M$ be the αth-cardinal successor of η in M if it exists and otherwise, we let it be $\mathrm{ord}(M)$.

DEFINITION 2.7.1 (Pre-hod-like). Suppose $\mathcal{P}$ is an lses. We say $\mathcal{P}$ is *pre-hod-like* if one of the following holds:

1. (Meek) There is δ such that
   (a) $\mathcal{P} \models$ "δ is a Woodin cardinal or a limit of Woodin cardinals",
   (b) δ is a cutpoint of $\mathcal{P}$[49],
   (c) if $\kappa < \mathrm{ord}(\mathcal{P})$ is a limit of Woodin cardinals of $\mathcal{P}$ then $o^{\mathcal{P}}(\kappa) < \delta$,
   (d) $\mathcal{P} \models \mathsf{ZFC} - \mathsf{Replacement}$ and
   (e) if δ is a Woodin cardinal of $\mathcal{P}$ then $\mathcal{P} = \bigcup_{n<\omega} \mathcal{P}|(\delta^{+n})^{\mathcal{P}}$, and if δ is a limit of Woodin cardinals of $\mathcal{P}$ then δ is the largest cardinal of $\mathcal{P}$.
2. (Non-meek) There is $\delta \leq \mathrm{ord}(\mathcal{P})$ such that
   (a) there is $\kappa < \delta$ such that $\delta \leq o^{\mathcal{P}}(\kappa)$,
   (b) if κ is the least $\eta < \delta$ such that $\delta \leq o^{\mathcal{P}}(\eta)$ then $o^{\mathcal{P}}(\kappa) = \delta$ and $\mathcal{P} \models$ "κ is a limit of Woodin cardinals",
   (c) letting $\kappa < \delta$ be the least such that $o^{\mathcal{P}}(\kappa) = \delta$, $\rho(\mathcal{P}) \in (\kappa, \delta]$ or $\mathrm{ord}(\mathcal{P})$ is a limit of ordinals ξ such that $\rho(\mathcal{P}||(\xi, \omega)) \in (\kappa, \delta]$[50].
   (d) $\mathcal{P}$ is δ-sound,
   (e) if $\mathrm{dom}(\vec{E}^{\mathcal{P}}) \cap (\delta^{\mathcal{P}}, \mathrm{ord}(\mathcal{P})] = \emptyset$ then $\mathcal{J}_{\omega}[\mathcal{P}] \models$ "$\delta^{\mathcal{P}}$ is not a Woodin cardinal".
3. (Gentle) $\delta =_{\mathrm{def}} \mathrm{ord}(\mathcal{P})$ is a limit of Woodin cardinals of $\mathcal{P}$ and $\mathcal{P} \models \mathsf{ZFC} - \mathsf{Replacement}$.

We let $\delta^{\mathcal{P}}$ be the δ above.

[49] This condition follows from the other conditions, but we would like to isolate it.

[50] Here, we implicitly assuming that $\xi = \omega\beta$ for some β. See Remark 2.1.3.

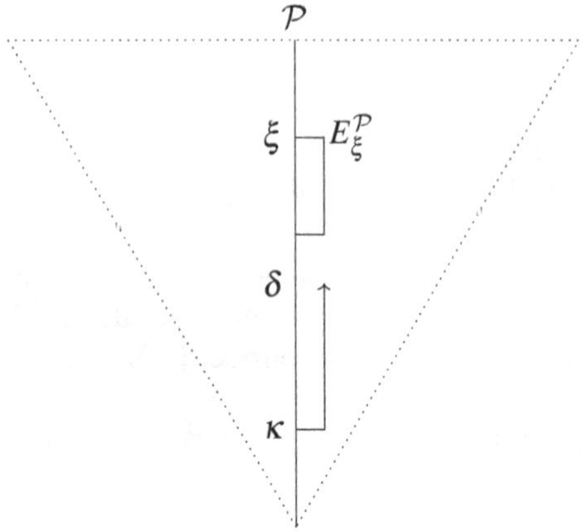

FIGURE 1. Lsa type lses. Here, $\mathcal{P}$ is an lsa type lses. κ is a limit of Woodin cardinals in $\mathcal{P}$, $\delta = \delta^{\mathcal{P}}$ is Woodin in $\mathcal{P}$, and $o^{\mathcal{P}}(\kappa) = \delta$. $\mathcal{P}|\xi$ is the least active level of $\mathcal{P}$ above δ.

The next definition is somewhat technical. The meaning of it is that we will wait until we see the sharp of a layer before we will activate the strategy.

DEFINITION 2.7.2 (Properly non-meek). Suppose $\mathcal{P}$ is a non-meek pre-hod-like lses. We say $\mathcal{P}$ is *properly non-meek* if there is $\xi \in \mathrm{dom}(\vec{E}^{\mathcal{P}})$ (ξ may be $o(\mathcal{P})$) such that $\mathrm{crit}(E_\xi^{\mathcal{P}}) > \delta^{\mathcal{P}}$ and $\mathcal{P}|\xi = \mathcal{J}_\xi[\mathcal{P}|\delta^{\mathcal{P}}]$.

The next definition isolates the type of hod premice that give rise to pointclasses satisfying the Largest Suslin Axiom.

DEFINITION 2.7.3 (Lsa type, Figure 1). Suppose $\mathcal{P}$ is a pre-hod-like lses. We say $\mathcal{P}$ is of *lsa type* if

1. $\mathcal{P}$ is properly non-meek,
2. $\mathcal{P} \vDash$ "$\delta^{\mathcal{P}}$ is a Woodin cardinal".

Suppose $\mathcal{P}$ is a pre-hod-like lses of lsa type. We let $\mathcal{P}_{\mathrm{ex}} \trianglelefteq \mathcal{P}$ be the longest initial segment $\mathcal{P}'$ of $\mathcal{P}$ such that $\mathcal{P}'$ is of lsa type, $\delta^{\mathcal{P}} = \delta^{\mathcal{P}'}$ and letting $k = k(\mathcal{P}')$, for every $\kappa < \delta^{\mathcal{P}}$ there is no cofinal $f : \kappa \to \delta^{\mathcal{P}}$ that is $r\Sigma_k^{\mathcal{P}'}$-definable over $\mathcal{P}'$[51]. We then say that $\mathcal{P}$ is *exact* if $\mathcal{P} = \mathcal{P}_{\mathrm{ex}}$.

Continuing with $\mathcal{P}$, let $\alpha = \min(\mathrm{dom}(\vec{E}^{\mathcal{P}}) - \delta^{\mathcal{P}})$ and set $\mathcal{P}_\# = \mathcal{P}||\alpha$. We then say that $\mathcal{P}$ is of *#-lsa type* if $\mathcal{P}_\# = \mathcal{P}$ and $\mathcal{J}_\omega[\mathcal{P}] \vDash$ "$\delta^{\mathcal{P}}$ is a Woodin cardinal".

If Σ is a strategy of $\mathcal{P}$ then we let Σ_{ex} be the strategy of $\mathcal{P}_{\mathrm{ex}}$ with the property that $\Sigma_{\mathrm{ex}} = $ (id-pullback of Σ).

In this paper we will consider hod mice that are lsa small.

[51] "ex" stands for "exact".

DEFINITION 2.7.4 (Lsa small). Suppose $\mathcal{P}$ is a pre-hod-like lses. We say $\mathcal{P}$ is *lsa small* if for all $\mathcal{P}$-cardinals κ such that $o^{\mathcal{P}}(\kappa) < \delta^{\mathcal{P}}$ and $\mathcal{P} \models$ "κ is a limit of Woodin cardinals", $\mathcal{P} \models$ "$o^{\mathcal{P}}(\kappa)$ is not a Woodin cardinal".

REMARK 2.7.5. From now on we tacitly assume that all lses considered in this paper are lsa-small. We will, from time to time, remind the reader of this.

We can now isolate the layers of pre-hod-like lses.

REMARK 2.7.6. Before we give the definition we make the following intuitive remarks. Suppose $\mathcal{P}$ is a hod premouse, which are the objects that we eventually want to define (see Definition 3.10.2).

1. The philosophy behind "layering" is the desire to make maximal complexity jumps in the Wadge hierarchy. Ordinary mice and premice are designed to reach large cardinals by using the least amount of information large cardinals give us, namely the extenders that induce those embeddings that we use to define the large cardinal in question. For example, to reach a measurable cardinal in a mouse we only use ultrafilters. However, measurability tells us much more than just that there is a nice ultrafilter on some cardinal. For example, if κ is measurable then every Π^1_1 set is κ-homogenously Suslin, and in trying to build mice with measurable cardinals we ignore this extra information. We justify our ignorance by claiming that our algorithms that produce mice (e.g., fully backgrounded constructions, K^c constructions and etc) using extenders as oracles output structures that do inherit all the important properties of large cardinals. That this indeed happens has been verified by Neeman for large cardinals in the region of Woodin cardinal that is a limit of Woodin cardinals (see [26]). However, a priori, this dream-like solution may have been wrong, and more of the information given to us by large cardinals might have been required to reach them in canonical structures, and perhaps the fact that we cannot do significantly better than a Woodin cardinal that is a limit of Woodin cardinals is a sign that only extenders will not do.

Hod mice have an entirely different purpose. Instead of large cardinals the aim is to reach or rather "capture" the Wadge hierarchy inside canonical structures. This is parallel to Shoenfield's Absoluteness, namely that L is Σ^1_2-correct. Each layer of a hod mouse corresponds to a new level of the Wadge hierarchy. However, what the philosophy of layering claims to be possible is that we can reach all levels of the Wadge hierarchy by simply jumping to the most significant levels of it, and here the significant levels of the Wadge hierarchy are defined to be *the Solovay pointclass*.

DEFINITION 2.7.7. Assume $\mathsf{ZF}+\mathsf{AD}^+$. We say Γ is a *Solovay pointclass* if there is κ such that κ is a member of the Solovy sequence and $\Gamma = \{A \subseteq \mathbb{R} : w(A) < \kappa\}$[52].

[52]See Definition 1.1.2.

The strategy of each layer of a hod mouse generates a Solovay pointclass in the sense that the named strategy has Wadge rank ϑ_α for some α. The dream of the "layering" philosophy is that by only generating the Solovay pointclasses we will reach all levels of the Wadge hierarchy. Internalizing this idea would help the reader with a knowledge of AD^+ theory to understand why layers are defined the way they are defined: every initial segment whose strategy corresponds to a Solovay pointclass is a layer.

Of course, at this stage the idea is vague. If $\mathcal{P}$ is our hod mouse, Σ is an iteration strategy for $\mathcal{P}$, $\mathcal{Q}_0 \lhd \mathcal{Q}_1 \unlhd \mathcal{P}$ then the reader should expect that $\Sigma_{\mathcal{Q}_0}$ is not more complex then $\Sigma_{\mathcal{Q}_1}$, and one can easily build many situations where in fact $\Sigma_{\mathcal{Q}_0}$ is Wadge reducible to $\Sigma_{\mathcal{Q}_1}$ but not vice a versa. However, it may be the case that neither $\mathcal{Q}_0$ nor $\mathcal{Q}_1$ are layers of $\mathcal{P}$. To make the idea work we need to *anticipate* the initial segments of $\mathcal{P}$ whose strategies generate the Solovay pointclasses. Below we spell out what initial segments should be layers in the minimal model of LSA.

2. The basic phenomenon that guides us in our definition of layers is the following: Suppose $(\mathcal{P},\Sigma)$ is a pair such that $\mathcal{P}$ is a hod mouse and Σ is its iteration strategy[53]. Let $\mathcal{M}_\infty(\mathcal{P},\Sigma)$ be the direct limit of all countable Σ-iterates of $\mathcal{P}$ and $\pi_{\mathcal{P},\infty} : \mathcal{P} \to \mathcal{M}_\infty(\mathcal{P},\Sigma)$ be the direct limit embedding.

> *Key Phenomenon:* For $\delta \leq \delta^{\mathcal{P}}$, $\pi_{\mathcal{P},\infty}(\delta)$ is a member of the Solovay sequence if and only if δ is either a cutpoint Woodin cardinal of $\mathcal{P}$ or a cutpoint limit of Woodin cardinals of $\mathcal{P}$.

The way we use the Key Phenomenon is as follows. Suppose we have declared $\mathcal{Q}$ a layer of $\mathcal{P}$.

(a) If $\delta^{\mathcal{Q}}$ is a cutpoint Woodin cardinal of $\mathcal{P}$ or a cutpoint limit of Woodin cardinals of $\mathcal{P}$ then $\mathcal{Q}$ is the unique layer $\mathcal{Q}'$ of $\mathcal{P}$ such that $\delta^{\mathcal{Q}'} = \delta^{\mathcal{Q}}$.

(b) If, however, $\delta^{\mathcal{Q}} = o^{\mathcal{Q}}(\kappa)$ for some κ and $\mathcal{Q}'$ is such that $\rho(\mathcal{Q}') \leq \delta^{\mathcal{Q}}$ then $\pi_{\mathcal{Q},\infty}(\kappa) \leq \pi_{\mathcal{Q}',\infty}(\kappa)$ and the strict inequality cannot be ruled out. Thus, we declare $\mathcal{Q}'$ a layer as it can generate a new Solovay pointclass.

(c) Also, suppose $\kappa < \delta^{\mathcal{P}}$ is a limit of cutpoint Woodin cardinals of $\mathcal{P}$ and suppose that α is such that either $\alpha = \mathrm{ind}^{\mathcal{P}}(E)$ for some $E \in \vec{E}^{\mathcal{P}}$ with $\mathrm{crit}(E) = \kappa$ or α is a limit of such points. Notice now that for every $\beta < \alpha$, $\pi_{\mathcal{P}||(\omega\beta,\omega),\infty}(\kappa) < \pi_{\mathcal{P}||(\omega\alpha,\omega),\infty}(\kappa)$. This is because we must have that $\pi_{\mathcal{P}||(\omega\beta,\omega),\infty} = \pi_{\mathrm{Ult}(\mathcal{P},E)||(\omega\beta,\omega),\infty}$. Therefore, $\mathcal{P}||(\omega\alpha,\omega)$ must be a layer of $\mathcal{P}$.

3. Layers of $\mathcal{P}$ are those proper initial segments of $\mathcal{P}$ whose strategy is being indexed on the strategy predicate of $\mathcal{P}$, with the exception that $\mathcal{P}$ is also considered to be a layer of itself.

4. Meek layers of our hod mice were already studied in [29].

5. The key new ingredient of our hod mice is the way we treat the lsa type layers. Given an lsa type layer $\mathcal{Q}$ of $\mathcal{P}$, say $\mathcal{Q}$ is minimal in $\mathcal{P}$ if there is no layer $\mathcal{Q}' \lhd \mathcal{Q}$ such that $\delta^{\mathcal{Q}} = \delta^{\mathcal{Q}'}$. Given a minimal lsa type layer $\mathcal{Q}$, we start

[53]For explanatory reasons, we are being somewhat vague.

indexing the short-tree-strategy of $\mathcal{Q}$ into the strategy predicate, and this leads to our notion of short-tree-strategy (sts) premouse (see Definition 3.8.17). There are two possibilities here. Either (a) we reach a level $\mathcal{Q}'$ with the property that $\mathcal{Q}'$ is an sts premouse over $\mathcal{Q}$ such that $\delta^{\mathcal{Q}}$ is not a Woodin cardinal definably over $\mathcal{Q}'$ or (b) $\delta^{\mathcal{P}} = \delta^{\mathcal{Q}}$, $\delta^{\mathcal{P}}$ is Woodin in $\mathcal{P}$ and $\mathcal{P}$ above $\delta^{\mathcal{Q}}$ is an sts premouse. If we do reach such a $\mathcal{Q}'$ then $\mathcal{Q}'$ becomes a layer and we start adding the strategy of $\mathcal{Q}'$. Notice that $\mathcal{Q}'$ itself is of lsa type.

6. The following conditions essentially characterize all proper layers of $\mathcal{P}$, but the conditions below do not spell out the actual definition and are given for explanatory purposes.

(a) (Woodin cardinals) If $\eta < \delta^{\mathcal{P}}$ is a Woodin cardinal of $\mathcal{P}$ then there is a layer $\mathcal{Q}$ of $\mathcal{P}$ such that $\delta^{\mathcal{Q}} = \eta$.

(b) (Limit of Woodin cardinals) If $\kappa < \delta^{\mathcal{P}}$ is a limit of Woodin cardinals of $\mathcal{P}$ then $\mathcal{Q} =_{\text{def}} \mathcal{P}|(\kappa^+)^{\mathcal{P}}$ is a layer of $\mathcal{P}$ such that $\delta^{\mathcal{Q}} = \kappa$ and $\mathcal{Q}$ is the unique layer $\mathcal{Q}'$ of $\mathcal{P}$ with $\delta^{\mathcal{Q}'} = \kappa$. Moreover, κ is a strong cutpoint in $\mathcal{Q}$ (in particular, if $E \in \vec{E}^{\mathcal{P}}$ is such that $\text{crit}(E) = \kappa$ then E is total.)

(c) (Active layers) If $\kappa < \delta^{\mathcal{P}}$ is a limit of Woodin cardinals and $E \in \vec{E}^{\mathcal{P}}$ is such that $\text{crit}(E) = \kappa$ then there is a layer $\mathcal{Q}$ of $\mathcal{P}$ such that $\delta^{\mathcal{Q}} = \text{ind}^{\mathcal{P}}(E)$.

(d) (Limits of layers) If $\nu < \delta^{\mathcal{P}}$ is a limit of ordinals of the form $\delta^{\mathcal{Q}}$ where $\mathcal{Q}$ is a layer of $\mathcal{P}$ then there is a layer of $\mathcal{P}$ such that $\delta^{\mathcal{Q}} = \nu$.

(e) If $\mathcal{Q}$ is a layer of $\mathcal{P}$ with $\delta^{\mathcal{Q}} < \delta^{\mathcal{P}}$ and $\mathcal{Q}' \trianglelefteq \mathcal{O}^{\mathcal{P}}_{\text{ord}(\mathcal{Q})}$ is such that $\rho(\mathcal{Q}') \leq \delta^{\mathcal{Q}}$ then $\mathcal{Q}'$ is a layer of $\mathcal{P}$ with $\delta^{\mathcal{Q}'} = \delta^{\mathcal{Q}}$.

7. The layers of a hod-like lses are defined in a way that all non-meek layers are properly non-meek. There is no deep reason for doing this. The theory can be developed without this condition, but having more room above $\delta^{\mathcal{P}}$ is a convenience.

DEFINITION 2.7.8 (Layers of lses). Suppose $\mathcal{P}$ is an lsa small pre-hod-like lses. We define the *layers* $(\mathcal{P}_{\xi,\xi'} : \xi \leq \eta \wedge \xi' \leq \nu_\xi)$ of $\mathcal{P}$ as follows. As part of the definition, we will also define a sequence $(\delta_\xi, \iota_{\xi,\xi'} : \xi \leq \eta \wedge \xi' \leq \nu_\xi)$. The sequences are subject to the following requirements.

The Condition Defining the Sequence $(\delta_\xi : \xi \leq \eta)$*:*

R0 : The sequence $(\delta_\xi : \xi \leq \eta)$ enumerates in increasing order the set consisting of the following ordinals.

1. Woodin cardinals of $\mathcal{P}$ that are $\leq \delta^{\mathcal{P}}$[54].
2. The limits of Woodin cardinals of $\mathcal{P}$ that are $\leq \delta^{\mathcal{P}}$.
3. Ordinals ν with the property that $\nu \in \text{dom}(\vec{E}^{\mathcal{P}})$ and $\text{crit}(\vec{E}^{\mathcal{P}}(\nu)) < \delta^{\mathcal{P}}$ is a limit of Woodin cardinals of $\mathcal{P}$.
4. Ordinals ν which are limits of ordinals as in clause 3 above.

[54]Examining Definition 2.7.1, one could see that clause 2c leaves open the possibility of $\mathcal{P}$ having Woodin cardinals $> \delta^{\mathcal{P}}$.

The Conditions Defining the Sequence $(\iota_{\xi,0} : \xi < \eta)$*:*

R1: Suppose $\xi < \eta$ and δ_ξ is a Woodin cardinal of $\mathcal{P}$. Then

$$\iota_{\xi,0} = \text{ord}(\mathcal{O}^{\mathcal{P},\omega}_{\delta_\xi,\mathcal{P}|\delta_{\xi-1}}).^{55}$$

R2: Suppose $\xi < \eta$ and δ_ξ is a limit of Woodin cardinals of $\mathcal{P}$. Then $\iota_{\xi,0} = \delta_\xi$.
R3: Suppose $\xi < \eta$ and δ_ξ is neither a Woodin cardinal of $\mathcal{P}$ nor a limit of Woodin cardinals of $\mathcal{P}$. Then

$$\iota_{\xi,0} = \min(\text{dom}(\vec{E}^{\mathcal{P}}) - (\delta_\xi + 1)).$$

The Conditions Defining the Sequence $(\iota_{\xi,1} : \xi < \eta)$ *for* ξ *as in* R3*:*

In R4-R5, suppose $\xi < \eta$ and δ_ξ is neither a Woodin cardinal of $\mathcal{P}$ nor a limit of Woodin cardinals of $\mathcal{P}$.

R4: Suppose $rud(\mathcal{P}|\iota_{\xi,0}) \vDash$ "δ_ξ is not a Woodin cardinal". Then $\iota_{\xi,1}$ is the least ordinal $\beta > \iota_{\xi,0}$ such that $\rho(\mathcal{P}||(\beta,\omega)) \leq \delta_\xi$.

R5: Suppose $rud(\mathcal{P}|\iota_{\xi,0}) \vDash$ "δ_ξ is a Woodin cardinal". Then $\iota_{\xi,1}$ is the largest ordinal $\beta > \iota_{\xi,0}$ such that $\mathcal{P}|\beta \vDash$ "δ_ξ is a Woodin cardinal".

The Conditions Defining the Sequence $(\iota_{\xi,\xi'} : \xi' \leq \nu_\xi)$ *for* $\xi < \eta$*:*

R6: Suppose δ_ξ is a Woodin cardinal of $\mathcal{P}$ or a limit of Woodin cardinals of $\mathcal{P}$. Then $\iota_{\xi,0}$ is defined as in R1 and R2. If δ_ξ is a Woodin cardinal then $\nu_\xi = 0$. If δ_ξ is a limit of Woodin cardinals then $\nu_\xi = 1$ and

$$\iota_{\xi,1} = \text{ord}(\mathcal{O}^{\mathcal{P}}_{\delta_\xi+1,\mathcal{P}|\delta_\xi+1}).$$

R7: Suppose δ_ξ is neither a Woodin cardinal of $\mathcal{P}$ nor a limit of Woodin cardinals of $\mathcal{P}$. Then $\iota_{\xi,0}$ is defined as in R3, $\iota_{\xi,1}$ is defined as in R4-R5, and the sequence $(\iota_{\xi,\xi'} : \xi' \in (1,\nu_\xi])$ enumerates in increasing order the closure of the set

$$\{\alpha < \delta_{\xi+1} : \rho(\mathcal{P}||(\alpha,\omega)) \leq \delta_\xi\}.$$

When $\xi = \eta$*:*

R8: Suppose $\mathcal{P}$ is meek. If $\delta^{\mathcal{P}}$ is a Woodin cardinal then $\nu_\eta = 0$ and $\iota_{\eta,0} = \text{ord}(\mathcal{P})$. If $\delta^{\mathcal{P}}$ is a limit of Woodin cardinals then $\nu_\eta = 1$, $\iota_{\eta,0} = \delta_\eta$ and $\iota_{\eta,1} = \text{ord}(\mathcal{P})$.

R9: Suppose $\mathcal{P}$ is non-meek and $\text{dom}(\vec{E}^{\mathcal{P}}) - (\delta_\eta + 1) = \emptyset$[56]. Then $\iota_{\eta,0} = \text{ord}(\mathcal{P})$ and $\nu_\eta = 0$ (in this case, we have that $\mathcal{P} = \mathcal{J}_{\iota_{\eta,0}}[\mathcal{P}|\delta_\eta]$).

R10: Suppose $\mathcal{P}$ is non-meek and $\text{dom}(\vec{E}^{\mathcal{P}}) - (\delta_\eta + 1) \neq \emptyset$. Then

$$\iota_{\eta,0} = \min(\text{dom}(\vec{E}^{\mathcal{P}}) - (\delta_\eta + 1))$$

and one of the following conditions holds:

---

[55] Since we do not have a Woodin limit of Woodin cardinals in our $\mathcal{P}$, $\xi - 1$ makes sense. For $\xi = 0$, we let $\delta_{-1} = 0$.

[56] If $\text{ord}(\mathcal{P}) = \delta_\eta$ then this condition is satisfied.

1. If $rud(\mathcal{P}||\iota_{\eta,0}) \vDash$ "δ_η is not a Woodin cardinal" then $(\iota_{\eta,\xi} : \xi \in [1, \nu_\eta])$ enumerates in increasing order the closure of the set
$$\{\alpha \leq \text{ord}(\mathcal{P}) : \rho(\mathcal{P}||(\alpha, \omega)) \leq \delta_\eta\}.$$
2. If $rud(\mathcal{P}||\iota_{\eta,0}) \vDash$ "δ_η is a Woodin cardinal" but $\mathcal{P} \vDash$ "δ_η is not a Woodin cardinal" then $\iota_{\xi,1}$ is the largest ordinal β such that $\mathcal{P}|\beta \vDash$ "δ_η is a Woodin cardinal" and the sequence $(\iota_{\eta,\xi} : \xi \in [2, \nu_\eta])$ enumerates in increasing order the closure of the set
$$\{\alpha \leq \text{ord}(\mathcal{P}) : \rho(\mathcal{P}||(\alpha, \omega)) \leq \delta_\eta\}.$$
3. If $\mathcal{P}| \vDash$ "δ_η is a Woodin cardinal" then $\iota_{\eta,1} = \text{ord}(\mathcal{P})$ and $\nu_\eta = 1$.

R11: Suppose $\mathcal{P}$ is gentle. Then $\nu_\eta = 0$ and $\iota_{\eta,0} = \delta^{\mathcal{P}}$.

The Definition of $(\mathcal{P}_{\xi,\xi'} : \xi \leq \eta, \xi' \leq \nu_\xi)$*:*

R12: $\mathcal{P}_{\xi,\xi'} = \mathcal{P}||\iota_{\xi,\xi'}$.

We say $\mathcal{Q}$ is a layer of $\mathcal{P}$ if for some $\xi \leq \eta$ and $\xi' \leq \nu_\xi$,
$$\mathcal{Q} = \mathcal{P}||\iota_{\xi,\xi'}.$$
We say $\mathcal{Q}$ is a proper layer of $\mathcal{P}$ if $\mathcal{Q}$ is a layer of $\mathcal{P}$ and $\mathcal{Q} \neq \mathcal{P}$. We write $\mathcal{Q} \trianglelefteq_{hod} \mathcal{P}$ if and only if $\mathcal{Q}$ is a layer of $\mathcal{P}$, and we write $\mathcal{Q} \triangleleft_{hod} \mathcal{P}$ if and only if $\mathcal{Q}$ is a proper layer of $\mathcal{P}$.

REMARK 2.7.9. Suppose $\mathcal{P}$ is an active lses such that

1. if $\alpha = \text{ord}(\mathcal{P})$ then $\mathcal{P}|\alpha$ is a hod-like lses,
2. if $E = \vec{E}^{\mathcal{P}}(\alpha)$ then $o^{\mathcal{P}|\alpha}(\text{crit}(E)) = \delta^{\mathcal{P}|\alpha}$ and
3. $\rho(\mathcal{P}) > \text{crit}(E)$.

Then $\mathcal{P}$ itself is hod-like. It falls under clause 2c of Definition 2.7.1. Notice that α is enumerated in the δ-sequence of $\mathcal{P}$.

Next we introduce hod-like lses. These will eventually turn into hod premice. To do this we need to impose conditions on the layers of lses, which are just the members of $Y^{\mathcal{P}}$ where $\mathcal{P}$ is an lses.

DEFINITION 2.7.10 (Hod-like lses). Suppose $\mathcal{P}$ is a pre-hod-like lses. We say $\mathcal{P}$ is *hod-like* if the following conditions hold.

1. $\{\mathcal{Q} : \mathcal{Q} \text{ is a proper layer of } \mathcal{P}\} = (Y^{\mathcal{P}} - X^{\mathcal{P}})$.
2. For all layers $\mathcal{Q}$ of $\mathcal{P}$ such that $\delta^{\mathcal{Q}}$ is a limit of Woodin cardinals of $\mathcal{P}$, $\text{ord}(\mathcal{Q})$ is a cardinal of $\mathcal{P}$.

REMARK 2.7.11. Perhaps clause 2 of Definition 2.7.10 needs some more explanation. According to Definition 2.7.8 if ξ is such that $\mathcal{Q} = \mathcal{Q}_{\xi,0}$ then
$$\text{ord}(\mathcal{Q}) = \text{ord}(\mathcal{O}^{\mathcal{P}}_{\delta_\xi+1,\mathcal{P}|\delta_\xi})$$
which is the longest initial segment of $\mathcal{P}$ whose strategy predicate codes a strategy for $\mathcal{P}|\delta_\xi = \mathcal{Q}|\delta^{\mathcal{Q}}$. A priori there is no reason for $\text{ord}(\mathcal{Q})$ to be a cardinal. Clause 3,

following [29], makes this demand. It is a fullness condition that we will have to verify every time we build a hod premouse.

REMARK 2.7.12. Continuing with the set up of clause 2 of Definition 2.7.10, it follows that if $E \in \vec{E}^{\mathcal{P}}$ is such that $\text{crit}(E) = \delta^{\mathcal{Q}}$ then E is total. For if E is such that $\text{crit}(E) = \delta^{\mathcal{Q}}$ then

$$\mathcal{O}^{\mathcal{P}}_{\delta^{\mathcal{Q}}+1,\mathcal{P}|\delta^{\mathcal{Q}}} \trianglelefteq \mathcal{P}|\text{ind}^{\mathcal{P}}(E).$$

But if E is not total we must also have that $\mathcal{P}|\text{ind}^{\mathcal{P}}(E) \trianglelefteq \mathcal{P}|((\delta^{\mathcal{Q}})^+)^{\mathcal{P}}$.

REMARK 2.7.13. Each lses comes with its own Y predicate and the role of Definition 2.7.8 and Definition 2.7.10 is to impose conditions that the Y predicate of a hod like lses must have. One important point is that conditions like R1 and R6 depend on external factors. For example, in R6 we demand that $\iota_{\xi,1} = \text{ord}(\mathcal{O}^{\mathcal{P}}_{\delta_\xi+1,\mathcal{P}|\delta_\xi+1})$ while there can be many ordinals $\alpha \in (\delta_\xi, \iota_{\xi,1})$ with the property that $\mathcal{P}|\alpha$ is hod-like, yet none of them determine a layer. When designing $\mathcal{P}$ via hod pair constructions (see Definition 4.3.3), we will need to choose $\iota_{\xi,1}$, and its choice depends on the pointclass Γ that we attempt to generate via the hod pair construction. Intuitively, $\iota_{\xi,1}$ is defined to be the ordinal height of the stack of all sound ses over $\mathcal{P}|\delta_\xi + 1$ that are based on $\mathcal{P}|\delta_\xi$, have a projectum $\leq \delta_\xi$ and have an iteration strategy in Γ.

NOTATION 2.7.14. Suppose $\mathcal{P}$ is a hod-like lses. Let

$$L^{\mathcal{P}} = \{\delta : \exists \mathcal{Q} \in Y^{\mathcal{P}} - X^{\mathcal{P}}(\delta^{\mathcal{Q}} = \delta)\} \cup \{\delta^{\mathcal{P}}\}.$$

Let $\lambda^{\mathcal{P}} + 1$ be the order type of $L^{\mathcal{P}}$. We let $(\delta^{\mathcal{P}}_\alpha : \alpha \leq \lambda^{\mathcal{P}})$ be the increasing enumeration of $L^{\mathcal{P}}$. Also for $\xi \leq \lambda^{\mathcal{P}}$, set

$$\mathcal{P}(\xi) = \cup\{\mathcal{Q} \in Y^{\mathcal{P}} - X^{\mathcal{P}} : \delta_\xi = \delta^{\mathcal{Q}}\}.$$

We say $w = (\eta^w, \delta^w)$ is a window of $\mathcal{P}$ if

1. η^w is the least η such that $\delta^w = o^{\mathcal{P}}(\eta^w)$ and
2. there is a layer $\mathcal{Q}$ of $\mathcal{P}$ such that $\delta^w = \delta^{\mathcal{Q}}$.

We say w is the top window of $\mathcal{P}$ if $\delta^w = \delta^{\mathcal{P}}$. Given a hod-like lses $\mathcal{P}$, we set $\text{ml}(\mathcal{P}) = \cup(Y^{\mathcal{P}})$[57].

We say that $\mathcal{Q}$ is a *complete* layer of $\mathcal{P}$ if $\mathcal{Q}$ is a layer of $\mathcal{P}$ such that if $\mathcal{Q}$ is non-meek then there is no layer of $\mathcal{R}$ of $\mathcal{P}$ with the property that $\mathcal{Q} \triangleleft_{hod} \mathcal{R}$ and $\mathcal{R}^b = \mathcal{Q}^b$.

If $\mathcal{Q}$ is a layer of $\mathcal{P}$ of successor type then letting ξ be such that $\delta^{\mathcal{Q}} = \delta^{\mathcal{P}}_{\xi+1}$, $\mathcal{Q}^- =_{\text{def}} \mathcal{Q}(\xi)$. Thus, $\mathcal{Q}^-$ is the longest complete layer that is in an initial segment of $\mathcal{Q}$.

DEFINITION 2.7.15 (Germane lses). Suppose $\mathcal{M}$ is an lses. We say $\mathcal{M}$ is *germane* if letting $\alpha = \sup\{\text{ord}(\mathcal{Q}) : \mathcal{Q} \in Y^{\mathcal{M}} - X^{\mathcal{M}}\}$, the following conditions hold:

---

[57]"ml" stands for "maximal layer".

1. If $\mathcal{Q} \in Y^{\mathcal{M}} - X^{\mathcal{M}}$ then $\mathcal{Q}$ is a hod-like lses[58].
2. If $\mathcal{Q} \in Y^{\mathcal{M}} - X^{\mathcal{M}}$ and $\mathcal{Q}$ is meek then for all $\omega\beta \in [\mathrm{ord}(\mathcal{Q}), \mathrm{ord}(\mathcal{M}))$, $\rho(\mathcal{M}||(\omega\beta, \omega)) > \delta^{\mathcal{Q}}$.
3. $\alpha + 1$ is a cutpoint of $\mathcal{M}$.
4. If $\mathcal{M}$ is pre-hod-like then it is hod-like.
5. One of the following conditions holds:
   (a) $\mathcal{M}$ is pre-hod-like.
   (b) $\mathcal{M}$ is not pre-hod like and one of the following holds:
      (i) $\alpha = \mathrm{ord}(\mathcal{M})$ and α is a limit of Woodin cardinals of $\mathcal{M}$.
      (ii) $\alpha < \mathrm{ord}(\mathcal{M})$, $\mathcal{M}||\alpha \in Y^{\mathcal{M}}$ and α is a cardinal of $\mathcal{M}$ (see Remark 2.7.16).

If $\mathcal{M}$ is a germane lses then we let

$$\mathsf{hl}(\mathcal{M}) = \begin{cases} \mathcal{M} & \text{if clause 5(a) holds,} \\ \mathcal{M}||\alpha & \text{otherwise,} \end{cases}$$

where $\alpha = \sup\{\mathrm{ord}(\mathcal{Q}) : \mathcal{Q} \in Y^{\mathcal{M}} - X^{\mathcal{M}}\}$[59].

REMARK 2.7.16. Continuing with the set up of Definition 2.7.15, suppose $\mathcal{M}$ is germane but not hod-like. Clause 5(b) then says that either α is a limit of Woodin cardinals of $\mathcal{M}$ or $\mathcal{M}||\alpha$ is the longest hod-like initial segment of $\mathcal{M}$ and, moreover, it is declared to be a layer of $\mathcal{M}$[60].

It is not hard to create examples of germane lses that are not hod-like. For example, if $\mathcal{P}$ is hod-like and Σ is its strategy then Σ-premie over $\mathcal{P}$ will be germane. This comment is not literally true as such premice can project in ways not allowed by Definition 2.7.15, but also such premice need to be re-organized into lses.

TERMINOLOGY 2.7.17. Suppose $\mathcal{P}$ is a hod-like lses.

1. *(Successor type)* We say $\mathcal{P}$ has a *successor type* if $\mathcal{P}$ has a top window (η, δ) and η is not a limit of Woodin cardinals of $\mathcal{P}$.
2. *(Limit type)* We say $\mathcal{P}$ has a *limit type* if either $\mathcal{P}$ does not have a top window or if (η, δ) is the top window of $\mathcal{P}$ then η is a limit of Woodin cardinals of $\mathcal{P}$.

If $\mathcal{M}$ is germane then we say $\mathcal{M}$ is of successor type if $\mathsf{hl}(\mathcal{M})$ is of successor type hod-like lses. Otherwise we say that $\mathcal{M}$ is of limit type. We say $\mathcal{M}$ is of *b-type*[61] if $\mathcal{M}$ is of limit type and letting $\alpha = \sup\{\mathrm{ord}(\mathcal{Q}) : \mathcal{Q} \in Y^{\mathcal{M}} - X^{\mathcal{M}}\}$, α is not a limit of Woodin cardinals of $\mathcal{M}$[62].

---

[58]Recall that if $\mathcal{Q} \in Y^{\mathcal{M}} - X^{\mathcal{M}}$ then $Y^{\mathcal{Q}} = \{\mathcal{R} \in Y^{\mathcal{M}} : \mathrm{ord}(\mathcal{R}) < \mathrm{ord}(\mathcal{Q})\}$.

[59]"hl" stands for "hod-like".

[60]See R9 and R10 of Definition 2.7.8. We demand that α be a cardinal of $\mathcal{M}$ because otherwise $\mathcal{M}$ would be pre-hod-like and hence, hod-like.

[61]"b" stands for "bottom", see below.

[62]This means that $\mathcal{M}||\alpha$ is hod-like and is of limit type.

Next, we isolate the bottom part of b-type germane lses. For non-meek hod-like lses, this is essentially the part of $\mathcal{P}$ that is below the largest measurable limit of cutpoint Woodin cardinals.

DEFINITION 2.7.18 (The bottom part of lses). Given a limit type hod-like lses $\mathcal{P}$ we let $\mathcal{P}^b = \mathcal{P}$ if $\mathcal{P}$ does not have a top window and otherwise, letting (η, δ) be the top window of $\mathcal{P}$, we let

$$\mathcal{P}^b = \mathcal{P}|(\eta^+)^{\mathcal{P}}$$

where "b" stands for "bottom". We say that $\mathcal{P}^b$ is the *bottom* part of $\mathcal{P}$. It follows that $\mathcal{P}^b$ is a hod-like meek lses of limit type.

Similarly, if $\mathcal{M}$ is germane of b-type then $\mathcal{M}^b = (\mathsf{hl}(\mathcal{M}))^b$.

DEFINITION 2.7.19. Suppose $\mathcal{M}$ is germane. We say $\mathcal{M}$ is *projecting well* if letting $k = k(\mathcal{M})$[63] one of the following clauses holds:

1. $\mathcal{M}$ is of successor type and setting $\delta = \delta^{\mathsf{hl}(\mathcal{M})}$, δ is Woodin with respect to all $f : \delta \to \delta$ which are $r\Sigma_{k+1}^{\mathcal{M}}$-definable as witnessed by the extender sequence $\vec{E}^{\mathcal{M}|\delta}$.
2. $\mathcal{M}$ is of b-type and $\rho_{k+1}(\mathcal{M}) > \delta^{\mathcal{M}^b}$.
3. $\mathcal{M}$ is of limit type but not of b-type and $\rho_{k+1}(\mathcal{M}) > \mathrm{ord}(\mathsf{hl}(\mathcal{M}))$.

Otherwise we say that $\mathcal{M}$ *projects badly*. We say $\mathcal{M}$ *projects precisely* if $\mathcal{M}$ projects well and if there is n such that $(\mathcal{M}, n)$ projects badly then letting $k = k(\mathcal{M})$, $\mathcal{M}' = (\mathcal{M}, k+1)$ projects badly.

Clearly if $\mathcal{M}$ projects badly then there is always an initial segment $\mathcal{M}'$ of $\mathcal{M}$ such that $\mathrm{ord}(\mathcal{M}') = \mathrm{ord}(\mathcal{M})$ and $\mathcal{M}'$ projects precisely.

REMARK 2.7.20. We are interested in germane $\mathcal{M}$ that project precisely because we would like to apply stacks that are based on $\mathsf{hl}(\mathcal{M})$ to $\mathcal{M}$ without changing the stack.

For example, assume $\mathsf{hl}(\mathcal{M}) =_{\mathrm{def}} \mathcal{P}$ and $\mathcal{P}$ is a meek hod-like lses of limit type. Suppose $\mathcal{M}$ projects badly. If now $E \in \vec{E}^{\mathcal{P}|\delta^{\mathcal{P}}}$ then $\mathrm{Ult}(\mathcal{M}, E)$ may have more layers than $\mathrm{Ult}(\mathcal{P}, E)$, and $\mathcal{P}$'s strategy does not act on these new layers. On the other hand if $\mathcal{M}$ projects precisely then this is no longer the case as the functions used to compute $\pi_E^{\mathcal{M}}(\delta^{\mathcal{P}})$ and $\pi_E^{\mathcal{P}}(\delta^{\mathcal{P}})$ are the same, and they all are in $\mathcal{P}$.

We will use this sort of arguments later, when we need to show that if $\mathcal{P}$ is full, Σ is its strategy, $\mathcal{M}$ is germane such that $\mathsf{hl}(\mathcal{M}) = \mathcal{P}$ and $\mathcal{M}$ is a Σ-mouse over $\mathcal{P}$ then $\mathcal{M}$ does not project badly.

Notice that our comment above concerns only to germane $\mathcal{M}$ which are not themselves hod-like. If $\mathcal{M}$ is of b-type and projects across $\mathrm{ord}(\mathsf{hl}(\mathcal{M}))$ but it does not project badly then $\mathcal{M}$ itself is hod-like.

Summarizing, if $\mathcal{M}$ projects precisely and $\mathcal{T}$ is a stack on $\mathsf{hl}(\mathcal{M})$ then we define $\uparrow(\mathcal{T}, \mathcal{M})$ just like we did in Definition 2.4.10.

[63]See Section 2.2.

DEFINITION 2.7.21 (Almost non-dropping stacks). Suppose $\mathcal{M}$ is germane of b-type and projects precisely. Suppose further that $\mathcal{T}$ is a stack on $\mathcal{M}$ that is based on $\mathsf{hl}(\mathcal{M})$. We say that $\mathcal{T}$ is *almost non-dropping* if one of the following holds:

1. There is $\alpha \in R^{\mathcal{T}}$ such that $\pi^{\mathcal{T}_{\leq \alpha}}$ exists and $\mathcal{T}_{\geq \alpha}$ is above $\operatorname{ord}(\mathcal{M}^b_\alpha)$.
2. $\mathcal{T}$ has a last model and $\pi^{\mathcal{T}}$ exists.

If $\mathcal{T}$ is almost non-dropping and the first clause holds then let $\alpha(\mathcal{T})$ witness it. If $\mathcal{T}$ is almost non-dropping then we set

$$\pi^{\mathcal{T},b} = \begin{cases} \pi^{\mathcal{T}} \restriction \mathcal{M}^b & \text{if } \pi^{\mathcal{T}} \text{ exists,} \\ \pi^{\mathcal{T}_{\leq \alpha(\mathcal{T})}} \restriction \mathcal{M}^b & \text{otherwise.} \end{cases}$$

Suppose Σ is an iteration strategy for $\mathcal{M}$[64]. We then let

$$I(\mathcal{M},\Sigma) = \{(\mathcal{T},\mathcal{R}) : \mathcal{T} \text{ is according to } \Sigma,\ \mathcal{T} \text{ is based on } \mathsf{hl}(\mathcal{M}), \\ \mathcal{R} \text{ is the last model of } \mathcal{T} \text{ and } \pi^{\mathcal{T}} \text{ is defined}\}.$$

$$I^b(\mathcal{M},\Sigma) = \{(\mathcal{T},\mathcal{R}) : \mathcal{T} \text{ is according to } \Sigma,\ \mathcal{T} \text{ is based on } \mathsf{hl}(\mathcal{M}), \\ \mathcal{R} \text{ is the last model of } \mathcal{T} \text{ and } \pi^{\mathcal{T},b} \text{ is defined}\}.$$

REMARK 2.7.22. Notice that if $\mathcal{T}$ is almost non-dropping then it may only have drops in some image of the top window of $\mathcal{P}$.

DEFINITION 2.7.23. Suppose $\mathcal{P}$ is an lses and $\alpha < \operatorname{lh}(\mathcal{T})$. We say α is a *cutpoint* of $\mathcal{T}$ if $\mathcal{T}_{\geq \alpha}$ is a stack (after trivial re-enumeration) on $\mathcal{M}^{\mathcal{T}}_\alpha$, or equivalently, if for every $\beta + 1 \in (\alpha, \operatorname{lh}(\mathcal{T})))$, $\mathcal{T}(\beta+1) \geq \alpha$.

The reader may benefit from reviewing Notation 2.4.4.

DEFINITION 2.7.24. Suppose $\mathcal{M}$ is germane lses and

$$\mathcal{T} = ((\mathcal{M}_\alpha)_{\alpha<\eta}, (E_\alpha)_{\alpha<\eta-1}, D, R, (\beta_\alpha, m_\alpha)_{\alpha \in R}, T)$$

is a stack on $\mathcal{M}$ that is based on $\mathcal{P} =_{\text{def}} \mathsf{hl}(\mathcal{M})$. We say $\mathcal{T}$ is a *proper stack* if the following conditions hold:

1. $\mathcal{T}$ is semi-smooth[65].
2. $R = \{\alpha : \alpha \text{ is a cutpoint of } \mathcal{T}\}$.
3. For all $\alpha \in R$ such that $\alpha \neq \max(R)$, if $\mathsf{nc}^{\mathcal{T}}_\alpha$ has a fatal drop then $\mathcal{T}_{\geq \alpha}$ is a normal stack.
4. If $\omega\beta_\alpha < \operatorname{ord}(\mathcal{M}_\alpha)$ then $\mathcal{M}_\alpha || (\omega\beta_\alpha, m_\alpha)$ is a non-meek hod-like lses.
5. For all $\alpha \in R$, if $\mathcal{M}_\alpha$ is of b-type[66] and $\operatorname{ind}_\alpha < \delta^{\mathcal{M}^b_\alpha}$ then letting γ be the least such that

[64] It is worth remembering that this entails that Σ-iterates of $\mathcal{M}$ have the same indexing scheme as $\mathcal{M}$.

[65] This condition is already built into our definition of stack. See Remark 2.4.7. We are only making it explicit here.

[66] See Notation 2.7.17.

(a) $\text{ind}_\alpha < \text{ord}(\mathcal{M}_\alpha(\gamma))$[67] and
(b) $\text{ord}(\mathcal{M}_\alpha(\gamma))$ is a cutpoint of $\mathcal{M}_\alpha$,

if $\mathcal{M}_\alpha(\gamma)$ is of successor type and $\text{next}^{\mathcal{T}}(\alpha) \in R$ then $\pi^{\text{nc}_\alpha^{\mathcal{T}}}$ exists.

The following lemma summarizes the properties of proper stacks.

LEMMA 2.7.25. *Suppose*

$$\mathcal{T} = ((\mathcal{M}_\alpha)_{\alpha<\eta}, (E_\alpha)_{\alpha<\eta-1}, D, R, (\beta_\alpha, m_\alpha)_{\alpha\in R}, T)$$

is a proper stack on a germane $\mathcal{M}$. Then the following conditions hold.

1. *For all $\alpha \in R$, if $\mathcal{M}_\alpha$ is of b-type and $\text{ind}_\alpha < \delta^{\mathcal{M}_\alpha^b}$ then letting γ be the least such that*
   (a) $\text{ind}_\alpha < \text{ord}(\mathcal{M}_\alpha(\gamma))$ and
   (b) $\text{ord}(\mathcal{M}_\alpha(\gamma))$ is a cutpoint of $\mathcal{M}_\alpha$,
   then the following conditions hold:
   (a) $\text{nc}_\alpha^{\mathcal{T}}$ is based on $\mathcal{M}_\alpha(\gamma)$.
   (b) If $\mathcal{M}_\alpha(\gamma)$ is of successor type and $\text{next}^{\mathcal{T}}(\alpha) \in R$ then $\pi^{\text{nc}_\alpha^{\mathcal{T}}}$ exists and $\text{nc}_\alpha^{\mathcal{T}}$ is above $\text{ord}(\mathcal{M}_\alpha(\gamma-1))$[68].
   (c) If $\mathcal{M}_\alpha(\gamma)$ is of limit type and $\text{next}^{\mathcal{T}}(\alpha) \in R$ then $\pi^{\text{nc}_\alpha^{\mathcal{T}},b}$ exists and $\text{nc}_\alpha^{\mathcal{T}}$ is above $\delta^{\mathcal{M}_\alpha(\gamma)^b}$[69].
2. *For all $\alpha \in R$, if $\mathcal{M}_\alpha$ is of b-type and $\text{ind}_\alpha > \delta^{\mathcal{M}_\alpha^b}$ then*
   (a) $\text{nc}_\alpha^{\mathcal{T}}$ is above $\delta^{\mathcal{M}_\alpha^b}$[70] and
   (b) if $\text{next}^{\mathcal{T}}(\alpha) \in R$ then $\pi^{\text{nc}_\alpha^{\mathcal{T}},b}$ exists[71].

NOTATION 2.7.26. Suppose

$$\mathcal{T} = ((\mathcal{M}_\alpha)_{\alpha<\eta}, (E_\alpha)_{\alpha<\eta-1}, D, R, (\beta_\alpha, m_\alpha)_{\alpha\in R}, T)$$

is a proper stack on a germane $\mathcal{M}$. For $\alpha \in R$, we define $\text{layer}_\alpha^{\mathcal{T}}$ to be the least complete[72] layer $\mathcal{N}$ of $\mathcal{M}_\alpha$ such that $\text{ind}_\alpha^{\mathcal{T}} \in \mathcal{N}$. We also let $\text{rnc}_\alpha^{\mathcal{T}} = \downarrow(\text{nc}_\alpha^{\mathcal{T}}, \text{layer}_\alpha^{\mathcal{T}})$[73]. Often, we will represent $\mathcal{T}$ as

$$\mathcal{T} = ((\mathcal{M}_\alpha)_{\alpha<\eta}, (E_\alpha)_{\alpha<\eta-1}, D, R, (\text{rnc}_\alpha, \text{layer}_\alpha)_{\alpha\in R}, (\beta_\alpha, m_\alpha), T).$$

Suppose $\alpha \in R$ is such that $\mathcal{M}_\alpha$ is of b-type. We say α is of *limit type* if either $\text{ind}_\alpha > \delta^{\mathcal{M}_\alpha^b}$ or $\text{layer}_\alpha^{\mathcal{T}}$ is of limit type. Otherwise we say that α is of *successor type*. We say α is of *bottom* type if $\text{ind}_\alpha < \delta^{\mathcal{M}_\alpha^b}$.

---

[67] See Notation 2.7.14 for the definition of $\mathcal{P}(\beta)$. The definition obviously carries over to germane lses.

[68] This condition follows from the requirement that all cutpoints of $\mathcal{T}$ are in R. Similarly the last portion of the next clause.

[69] Here, $\pi^{\text{nc}_\alpha^{\mathcal{T}},b}$ is defined provided $\text{nc}_\alpha^{\mathcal{T}}$ is on $\mathcal{M}_\alpha(\gamma)$ which may not be the case. The meaning of this here and in the sequel is that letting $\mathcal{U} = \downarrow(\text{nc}_\alpha^{\mathcal{T}}, \mathcal{M}_\alpha(\gamma))$, $\pi^{\mathcal{U},b}$ is defined.

[70] This condition follows from clause 2 of Definition 2.7.24.

[71] This condition can be deduced from clause 3 of Definition 2.7.24.

[72] See Notation 2.7.14.

[73] See Definition 2.4.9.

REMARK 2.7.27 (Proper-stacks Convention). In this book all stacks on hod-like lses are proper stacks.

2.8. The iteration embedding $\pi^{\mathcal{T},b}$

Recall our convention regarding stacks (see Remark 2.7.27). In this section, we define the embedding $\pi^{\mathcal{T},b}$ via an inductive process. The reader may skip this section. The one important point that will come up later is the following. Suppose $\alpha \in R^{\mathcal{T}}$ and $\beta < \mathrm{lh}(\mathcal{T})$. Then if $\pi^{\mathcal{T},b}_{\alpha,\beta}$ is defined then its domain is $(\mathcal{M}_\alpha||(\omega\beta_\alpha, m_\alpha))^b$ which in general may not be the same as $\mathcal{M}^b_\alpha$.

Assume that $\mathcal{P}$ is a limit type hod-like lses which is not meek and suppose $\mathcal{T}$ is an iteration tree on $\mathcal{P}$. Again, we will not be concerned with the particular indexing scheme that $\mathcal{P}$ has. In some cases, regardless of whether $\mathcal{T}$ has a last model or not, it is possible to extract an embedding out of the iteration embeddings given by $\mathcal{T}$ that acts on $\mathcal{P}^b$. We describe this embedding below. First we define it by assuming that $\mathcal{T}$ is a normal iteration tree and then extend the definition to stacks. Recall that our lses are lsa-small (see Definition 2.7.4).

DEFINITION 2.8.1. Suppose $\mathcal{P}$ is a non-meek hod-like lses[74]. Suppose

$$\mathcal{T} = ((\mathcal{M}_\alpha)_{\alpha<\eta}, (E_\alpha)_{\alpha<\eta-1}, D, R, (\beta_\alpha, m_\alpha), T)$$

is a normal iteration tree on $\mathcal{P}$. We define $(\pi^{\mathcal{T},b}_{\alpha,\alpha'} : \alpha < \alpha' < \eta \wedge (\alpha,\alpha') \in T)$ by induction maintaining that if $\pi^{\mathcal{T},b}_{\alpha,\alpha'} \neq \emptyset$ then

(a) if $\alpha \notin R$ then $\pi^{\mathcal{T},b}_{\alpha,\alpha'} : \mathcal{M}^b_\alpha \to \mathcal{M}^b_{\alpha'}$ is an elementary embedding, and
(b) if $\alpha \in R$ then $\pi^{\mathcal{T},b}_{\alpha,\alpha'} : (\mathcal{M}_\alpha||(\omega\beta_\alpha, m_\alpha))^b \to \mathcal{M}^b_{\alpha'}$ is an elementary embedding.

The successor case. Suppose $\beta + 1 < \eta$ and we have defined $(\pi^{\mathcal{T},b}_{\alpha,\alpha'} : \alpha < \alpha' \leq \beta \wedge (\alpha,\alpha') \in T)$. Let $\gamma' = T(\beta+1)$. For $\gamma < \beta + 1$ such that $(\gamma, \beta+1) \in T$, we define $\pi^{\mathcal{T},b}_{\gamma,\beta+1}$ as follows.

1. If $\pi^{\mathcal{T},b}_{0,\gamma'} = \emptyset$ then set $\pi^{\mathcal{T},b}_{\gamma,\beta+1} = \emptyset$.

In the next three clauses we assume that $\pi^{\mathcal{T},b}_{0,\gamma'} \neq \emptyset$ and that $\gamma' \notin R$.

2. Suppose $\mathrm{crit}(E_\beta) > \mathrm{ord}(\mathcal{M}^b_{\gamma'})$. Then set $\pi^{\mathcal{T},b}_{\gamma,\beta+1} = \pi^{\mathcal{T},b}_{\gamma,\gamma'}$.
3. Suppose $\mathrm{crit}(E_\beta) \leq \mathrm{ord}(\mathcal{M}^b_{\gamma'})$ and $\beta + 1 \in D$. Then $\pi^{\mathcal{T},b}_{\gamma,\beta+1} = \emptyset$.
4. Suppose $\mathrm{crit}(E_\beta) \leq \mathrm{ord}(\mathcal{M}^b_{\gamma'})$ and $\beta + 1 \notin D$. Then $\pi^{\mathcal{T},b}_{\gamma,\beta+1} = (\pi^{\mathcal{T}}_{\gamma',\beta+1} \restriction \mathcal{M}^b_{\gamma'}) \circ \pi^{\mathcal{T},b}_{\gamma,\gamma'}$.

[74]In particular, $\mathcal{P} \neq \mathcal{P}^b$.

In the next three clauses we assume that $\pi_{0,\gamma'}^{\mathcal{T},b} \neq \emptyset$ and that $\gamma' \in R$.

5. Suppose $\mathrm{crit}(E_\beta) > \mathrm{ord}((\mathcal{M}_{\gamma'}||(\omega\beta_{\gamma'}, m'_\gamma))^b)$. Then set $\pi_{\gamma,\beta+1}^{\mathcal{T},b} = \pi_{\gamma,\gamma'}^{\mathcal{T},b}$.
6. Suppose $\mathrm{crit}(E_\beta) \leq \mathrm{ord}((\mathcal{M}_{\gamma'}||(\omega\beta_{\gamma'}, m'_\gamma))^b)$ and $\beta+1 \in D$. Then $\pi_{\gamma,\beta+1}^{\mathcal{T},b} = \emptyset$.
7. Suppose $\mathrm{crit}(E_\beta) \leq \mathrm{ord}((\mathcal{M}_{\gamma'}||(\omega\beta_{\gamma'}, m'_\gamma))^b)$ and $\beta+1 \notin D$. Then $\pi_{\gamma,\beta+1}^{\mathcal{T},b} = (\pi_{\gamma',\beta+1}^{\mathcal{T}} \restriction \mathcal{M}_{\gamma'}^b) \circ \pi_{\gamma,\gamma'}^{\mathcal{T},b}$.

The limit case. Suppose next that $\beta < \eta$ is a limit ordinal and we have defined $(\pi_{\alpha,\alpha'}^{\mathcal{T},b} : \alpha < \alpha' < \beta \wedge (\alpha, \alpha') \in T)$. Then we define $\pi_{\gamma,\beta}^{\mathcal{T},b}$ for $\gamma \in [0,\beta)_T$ according to the following cases:

1. If $\gamma \in [0,\beta)_T$ is such that there is $\gamma' \in [0,\beta)_T$ with the property that $\pi_{\gamma,\gamma'}^{\mathcal{T},b} = \emptyset$ then $\pi_{\gamma,\beta}^{\mathcal{T},b} = \emptyset$.
2. Suppose $\gamma \in [0,\beta)_T$ is such that for all $\gamma' \in [0,\beta)_T$, $\pi_{\gamma,\gamma'}^{\mathcal{T},b}$ is defined. Then letting $\nu+1 \in b$ be such that $T(\nu+1) = \gamma$, $\pi_{\gamma,\beta}^{\mathcal{T},b} = \pi_{\nu+1,\beta}^{\mathcal{T},b} \circ \pi_{\gamma,\nu+1}^{\mathcal{T}}$ where $\pi_{\nu+1,\beta}^{\mathcal{T},b}$ is the direct limit embedding given by the directed system $(\mathcal{M}_\xi^b, \pi_{\xi,\xi'}^{\mathcal{T},b} : \nu+1 \leq \xi < \xi' \wedge (\xi,\xi') \in c^2)$ where $c = [0,\beta)_T$.

The iteration embedding $\pi^{\mathcal{T},b}$. Continuing with the $\mathcal{T}$ above, we let $\pi^{\mathcal{T},b}$ be defined according to the following clauses.

1. Suppose $\mathrm{lh}(\mathcal{T}) = \gamma+1$. Then set $\pi^{\mathcal{T},b} = \pi_{0,\gamma}^{\mathcal{T},b}$.

In all the clauses below we assume that $\mathrm{lh}(\mathcal{T})$ is a limit ordinal.

2. Suppose there is $\gamma < \mathrm{lh}(\mathcal{T})$ such that $\pi_{0,\gamma}^{\mathcal{T},b} = \emptyset$ and $\mathcal{T}_{\geq\gamma}$ is a normal stack on $\mathcal{M}_\gamma$. Then set $\pi^{\mathcal{T},b} = \emptyset$.

In all the clauses below we assume that if $\gamma < \mathrm{lh}(\mathcal{T})$ is such that $\mathcal{T}_{\geq\gamma}$ is a normal stack on $\mathcal{M}_\gamma$ then $\pi_{0,\gamma}^{\mathcal{T},b} \neq \emptyset$.

3. Suppose there is $\gamma < \mathrm{lh}(\mathcal{T})$ such that $\mathcal{T}_{\geq\gamma}$ is a normal stack on $\mathcal{M}_\gamma$ based on $\mathcal{M}_\gamma^b$. Then set $\pi^{\mathcal{T},b} = \emptyset$.
4. Suppose there is $\gamma < \mathrm{lh}(\mathcal{T})$ such that $\mathcal{T}_{\geq\gamma}$ is a normal stack on $\mathcal{M}_\gamma$ above $\mathrm{ord}(\mathcal{M}_\gamma^b)$. Then set $\pi^{\mathcal{T},b} = \pi_{0,\gamma}^{\mathcal{T},b}$.
5. Suppose there is a cofinal $c \subseteq \mathrm{lh}(\mathcal{T})$ such that $\{\gamma < \mathrm{lh}(\mathcal{T}) : \exists\gamma' \in c((\gamma,\gamma') \in T)\}$ is a well-founded branch of $\mathcal{T}$ and for all $\gamma < \gamma'$ with $(\gamma,\gamma') \in c^2$, $\pi_{0,\gamma}^{\mathcal{T},b} \neq \pi_{0,\gamma'}^{\mathcal{T},b}$. Then set $\pi^{\mathcal{T},b} = \pi_c^{\mathcal{T}} \restriction \mathcal{P}^b$.

Given $(\alpha,\alpha') \in T$, we say $\pi_{\alpha,\alpha'}^{\mathcal{T},b}$ is *defined* or *exists* if $\pi_{\alpha,\alpha'}^{\mathcal{T},b} \neq \emptyset$. Similarly we say $\pi^{\mathcal{T},b}$ is *defined* or *exists* if $\pi^{\mathcal{T},b} \neq \emptyset$.

REMARK 2.8.2. Suppose $\mathcal{P}$ is a non-meek hod-like lses and $\mathcal{T}$ is a stack on $\mathcal{P}$. For all $\alpha < \alpha'$ such that $(\alpha,\alpha') \in T$, if $\pi_{\alpha,\alpha'}^{\mathcal{T},b}$ exists then it is essentially the iteration embedding. However, given how $\pi^{\mathcal{T}}_{\alpha,\alpha'}$ is defined, it is possible that $\pi^{\mathcal{T},b}_{\alpha,\alpha'}$ exists yet $\pi^{\mathcal{T}}_{\alpha,\alpha'}$ is undefined.

In general, we have that $\pi^{\mathcal{T},b}_{\alpha,\alpha'}$ is defined if and only if for all γ such that $\gamma+1 \in [\alpha,\alpha')_{\mathcal{T}} \cap D^{\mathcal{T}}$, $\mathrm{crit}(E_\gamma) > \mathrm{ord}(\mathcal{M}^b_\gamma)$.

Notice that in Definition 2.8.1 we are not assuming that the stack has a last model. The fragment of the eventual iteration embedding $\pi^{\mathcal{T}}$ restricted to $\mathcal{P}^b$ can be seen without actually having the last branch.

2.9. Canonical singularizing sequence

The following notion will be used throughout this paper.

DEFINITION 2.9.1 (Canonical singularizing sequences). Suppose $\mathcal{P}$ is a germane lses of b-type that projects precisely and $\mathcal{T}$ is an almost non-dropping stack on $\mathcal{P}$. Let $\mathcal{Q} = \pi^{\mathcal{T},b}(\mathcal{P}^b)$. Then $\mathcal{Q}$ is a hod-like lses. If $w = (\eta,\delta)$ is a window of $\mathcal{Q}$ then we let

$$s(\mathcal{T},w) = \{\alpha : \exists a \in \eta^{<\omega} \exists f \in \mathcal{P}^b(\alpha = \pi^{\mathcal{T},b}(f)(a))\} \cap \delta$$

The following is an easy lemma, which is a consequence of our assumption that all hod-like lses are lsa small. It traces back to the fact that if $\mathcal{P}$ is a hod-like lses and $E \in \vec{E}^{\mathcal{P}}$ is an extender such that $\mathrm{crit}(E) = \delta^{\mathcal{Q}}$ for some layer $\mathcal{Q}$ of $\mathcal{P}$ and $\nu \in [\mathrm{crit}(E), \mathrm{ind}^{\mathcal{P}}(E)]$ then $\mathrm{Ult}(\mathcal{P},E) \models$ "ν is not a Woodin cardinal".

The following definition will be used in the next few lemmas.

DEFINITION 2.9.2. Suppose $\mathcal{T}$ is an almost non-dropping stack on (an lsa small) germane, b-type lses $\mathcal{P}$ that projects precisely, $\mathrm{lh}(\mathcal{T}) = \alpha+1$ and $\mathcal{Q} = \pi^{\mathcal{T},b}(\mathcal{P}^b)$[75]. Suppose ξ is a cardinal of $\mathcal{Q}$. Let $\iota \leq \alpha$ be the least such that $\pi^{\mathcal{T}_{\leq \iota+1},b}$ is defined, $\iota+1 \in [0,\alpha]_{\mathcal{T}}$ and for some $\xi' \in \mathcal{M}^b_{\iota+1}$, $\pi_{\iota+1,\alpha}(\xi') = \xi$[76]. We say ξ is $\mathcal{T}$-critical if $\xi' = \mathrm{crit}(E^{\mathcal{T}}_\iota)$.

Given $A \subseteq X \times Y$ and $x \in X$, we set $A_x = \{y : (x,y) \in A\}$.

LEMMA 2.9.3. *Suppose $\mathcal{T}$ is an almost non-dropping stack on (an lsa small) germane, b-type* lses *$\mathcal{P}$ that projects precisely, $\mathrm{lh}(\mathcal{T}) = \alpha+1$ and $\mathcal{Q} = \pi^{\mathcal{T},b}(\mathcal{P}^b)$. Suppose ξ is $\mathcal{T}$-critical. Then there is a finite sequence $(\gamma_i, \gamma'_i, \xi_i : i \leq n+1)$ such that*

1. *$(\gamma_i : i \leq n)$ is increasing and for each $i \leq n$, $\gamma_i \in [0,\alpha]_{\mathcal{T}}$,*
2. *$\gamma_{n+1} = \gamma'_{n+1} = \alpha$ and $\xi_{n+1} = \xi$,*
3. *for every $i \leq n$, $\gamma_i = \mathcal{T}(\gamma'_i + 1)$,*

[75]We drop $\mathcal{T}$ from our notation.

[76]This embedding may not be defined, but because both $\pi^{\mathcal{T},b}$ and $\pi^{\mathcal{T}_{\leq \iota+1},b}$ are defined, $\pi_{\iota+1,\alpha} \restriction \mathcal{M}^b_{\iota+1}$ is meaningful.

4. *ξ_0 is not $\mathcal{T}_{\leq\gamma_0}$-critical,*
5. *for every $i \leq n$, $\pi^{\mathcal{T}_{\leq\gamma_i},b}$ is defined,*
6. *for every $i \leq n$, $\xi_i = \mathrm{crit}(E_{\gamma'_i})$,*
7. *for every i such that $i+1 \leq n$, $\xi_{i+1} = \pi_{\gamma'_i+1,\gamma_{i+1}}(\xi_i)$,*
8. *$\xi = \pi_{\gamma'_n+1,\alpha}(\xi_n)$,*
9. *for every $i \leq n$,*

$$\wp(\xi_{i+1}) \cap \mathcal{M}^{\mathcal{T}}_{\gamma_{i+1}} = \{\pi^{\mathcal{T}}_{\gamma'_i+1,\gamma_{i+1}}(g)(t) : g \in \mathcal{M}^{\mathcal{T}}_{\gamma'_i+1}|(\xi_i^+)^{\mathcal{M}^{\mathcal{T}}_{\gamma'_i+1}} \wedge t \in [\xi_{i+1}]^{<\omega}\}.$$

10. *for every $i \leq n+1$, for every $(m_0, m_1, \ldots, m_k) \in \mathbb{N}^{<\omega}$ and for every $A \in \mathcal{M}^{\mathcal{T}}_{\gamma_i}$ such that*

$$A \subseteq [\xi_i]^{m_0} \times [\xi_i]^{m_1} \times \cdots \times [\xi_i]^{m_k}$$

there is $B \in \mathcal{M}^{\mathcal{T}}_{\gamma_0}|(\xi_0^+)^{\mathcal{M}^{\mathcal{T}}_{\gamma_0}}$ and $t \in [\xi_i]^{<\omega}$ such that

$$A = \pi^{\mathcal{T}}_{\gamma_0,\gamma_i}(B)_t \cap ([\xi_i]^{m_0} \times [\xi_i]^{m_1} \times \cdots \times [\xi_i]^{m_k}).$$

PROOF. We first get a finite sequence satisfying clauses 1-8, and then show that any such sequence also satisfies clauses 9 and 10. Because ξ is $\mathcal{T}$-critical, we have some (ι, ξ') satisfying the clauses of Definition 2.9.2. Let $\gamma = \mathcal{T}(\iota+1)$. The claim now can be proven by induction. Assuming our claim is true for $\mathcal{T}_{\leq\gamma}$ we have two cases. Suppose first that ξ' is not $\mathcal{T}_{\leq\gamma}$-critical. Set then $n = 1$, $\gamma_0 = \gamma$, $\gamma'_0 = \iota$ and $\xi_0 = \xi'$. Otherwise let $(\gamma_i, \gamma'_i, \xi_i : i \leq m)$ witness the claim for the pair $(\xi', \mathcal{T}_{\leq\gamma})$. Then set $n = m+2$, $\gamma_{m+1} = \gamma$, $\xi_{m+1} = \xi'$ and $\gamma'_{m+1} = \iota$. This finishes the proof that there is a finite sequence satisfying clauses 1-8.

We now want to show that any sequence that satisfies clauses 1-8 also satisfies clauses 9 and 10. Let then $(\gamma_i, \gamma'_i, \xi_i : i \leq n+1)$ be a sequence satisfying clauses 1-8. Clause 9 is easy to show as the generators of $\pi^{\mathcal{T}}_{\gamma'_i+1,\gamma_{i+1}} \restriction \mathcal{M}^{\mathcal{T}}_{\gamma'_i+1}|(\xi_i^+)^{\mathcal{M}^{\mathcal{T}}_{\gamma'_i+1}}$ are contained in $\xi_{i+1} = \pi^{\mathcal{T}}_{\gamma'_i+1,\gamma_{i+1}}(\xi_i)$.

We now show clause 10. Fix $i+1 \leq n+1$. Without loss of generality we can assume that clause 10 holds for all $j \leq i$. We then want to prove it for $i+1$. Below we drop $\mathcal{T}$ from superscripts. Fix $(m_0, m_1, \ldots, m_k) \in \mathbb{N}^{<\omega}$ and $A \in \mathcal{M}_{\gamma_{i+1}}$ such that

$$A \subseteq [\xi_{i+1}]^{m_0} \times [\xi_{i+1}]^{m_1} \times \cdots \times [\xi_{i+1}]^{m_k}.$$

We want to find a $B \in \mathcal{M}_{\gamma_0}|(\xi_0^+)^{\mathcal{M}_{\gamma_0}}$ and $t \in [\xi_{i+1}]^{<\omega}$ such that

$$A = \pi_{\gamma_0,\gamma_{i+1}}(B)_t \cap ([\xi_{i+1}]^{m_0} \times [\xi_{i+1}]^{m_1} \times \cdots \times [\xi_{i+1}]^{m_k}).$$

It follows from clause (9) that for some $u \in [\xi_{i+1}]^{<\omega}$, $A = \pi_{\gamma'_i+1,\gamma_{i+1}}(g)(u)$. Let $p = |u|$. Notice that

$$g : [\xi_i]^p \to \wp([\xi_i]^{m_0} \times [\xi_i]^{m_1} \times \cdots \times [\xi_i]^{m_k}).$$

Let then $G \subseteq [\xi_i]^p \times [\xi_i]^{m_0} \times [\xi_i]^{m_1} \times \cdots \times [\xi_i]^{m_k}$ be given by

$$(x,y) \in G \iff y \in g(x).$$

We thus have that

(1) for all $(x,y) \in [\xi_i]^p \times [\xi_i]^{m_0} \times [\xi_i]^{m_1} \times \cdots \times [\xi_i]^{m_k}$,

$$y \in g(x) \iff (x,y) \in \pi_{\gamma_i,\gamma_i'+1}(G),$$

implying that

(2) for all $x \in [\xi_i]^p$,

$$g(x) = \pi_{\gamma_i,\gamma_i'+1}(G)_x \cap [\xi_i]^{m_0} \times [\xi_i]^{m_1} \times \cdots \times [\xi_i]^{m_k}.$$

Because clause 10 holds for i, we get some $H \in \mathcal{M}_{\gamma_0}|(\xi_0^+)^{\mathcal{M}_{\gamma_0}}$ and $s \in [\xi_i]^{<\omega}$ such that

$$G = \pi_{\gamma_0,\gamma_i}(H)_s \cap [\xi_i]^p \times [\xi_i]^{m_0} \times [\xi_i]^{m_1} \times \cdots \times [\xi_i]^{m_k}.$$

Combining the above with (2) we get that

(3) for all $x \in [\xi_i]^p$,

$$g(x) = (\pi_{\gamma_0,\gamma_i'+1}(H)_s)_x \cap [\xi_i]^{m_0} \times [\xi_i]^{m_1} \times \cdots \times [\xi_i]^{m_k}.$$

Applying $\pi_{\gamma_i'+1,\gamma_{i+1}}$ to the equation above and recalling that $A = \pi_{\gamma_i'+1,\gamma_{i+1}}(g)(u)$ for some $u \in [\xi_{i+1}]^{<\omega}$, we get that

$$A = (\pi_{\gamma_0,\gamma_{i+1}}(H)_s)_u \cap ([\xi_{i+1}]^{m_0} \times [\xi_{i+1}]^{m_1} \times \cdots \times [\xi_{i+1}]^{m_k}).$$

Because both $s,u \in [\xi_{i+1}]^{<\omega}$, we now can find some B and t such that

$$A = \pi_{\gamma_0,\gamma_{i+1}}(B)_t \cap ([\xi_{i+1}]^{m_0} \times [\xi_{i+1}]^{m_1} \times \cdots \times [\xi_{i+1}]^{m_k})$$

with $t \in [\xi_{i+1}]^{<\omega}$. ⊣

LEMMA 2.9.4. *Suppose $\mathcal{T}$ is an almost non-dropping stack on (an lsa small) germane, b-type* lses *$\mathcal{P}$ that projects precisely and $\mathcal{Q} = \pi^{\mathcal{T},b}(\mathcal{P}^b)$. Suppose ξ is a limit of Woodin cardinals of $\mathcal{Q}$ and $A \in \mathcal{Q}$. Then the following holds.*

1. *Suppose $A \in \wp(\xi)$ and ξ is not $\mathcal{T}$-critical. Then $A = \pi^{\mathcal{T},b}(B)_t$ where $B \in \mathcal{P}^b$ and $t \in [\xi]^{<\omega}$.*
2. *Suppose for some $(m_0, m_1, \ldots, m_k) \in \mathbb{N}^{<\omega}$,*

$$A \subseteq [\xi]^{m_0} \times [\xi]^{m_1} \times \cdots \times [\xi]^{m_k}.$$

and ξ is $\mathcal{T}$-critical. Then there is $B \in \mathcal{P}^b$ and $t \in [\xi]^{<\omega}$ such that

$$A = \pi^{\mathcal{T},b}(B)_t \cap ([\xi]^{m_0} \times [\xi]^{m_1} \times \cdots \times [\xi]^{m_k}).$$

PROOF. We drop $\mathcal{T}$ from superscripts. Let α be the least such that $\mathcal{Q} = \mathcal{M}_\alpha^b$. Notice that because $\pi^{\mathcal{T},b}$ is defined, $\pi_{0,\alpha} \upharpoonright \mathcal{P}^b$ makes sense and is equal to $\pi^{\mathcal{T},b}$. Therefore, we will simply use $\pi_{\iota,\iota'}$ as if it is defined on all of $\mathcal{M}_\iota$. To prove our claim, we may just as well assume, without losing generality, that $\alpha + 1 = \mathrm{lh}(\mathcal{T})$.

Towards a contradiction suppose our claim is false. Without loss of generality we may assume that

(∗) for every ι such that $\iota+1<\mathrm{lh}(\mathcal{T})$ and $\pi^{\mathcal{T}_{\leq\iota},b}$ is defined, for every ν which is a limit of Woodin cardinals of $\mathcal{Q}'=_{\mathrm{def}}\pi^{\mathcal{T}_{\leq\iota},b}(\mathcal{P}^b)$, and for every $C\in\mathcal{Q}'$ the following holds:

1. Suppose $C\in\wp(\nu)$ and ν is a not $\mathcal{T}$-critical. Then $C=\pi^{\mathcal{T}_{\leq\iota},b}(D)_t$ where $D\in\mathcal{P}^b$ and $t\in[\nu]^{<\omega}$.
2. Suppose $(n_0,n_1,\ldots,n_k)\in\mathbb{N}^{<\omega}$ is such that
$$C\subseteq[\nu]^{n_0}\times[\nu]^{n_1}\times\cdots\times[\nu]^{n_k}.$$
and ν is $\mathcal{T}$-critical. Then there is $D\in\mathcal{P}^b$ and $t\in[\xi]^{<\omega}$ such that
$$C=\pi^{\mathcal{T}_{\leq\iota},b}(D)_t\cap([\nu]^{n_0}\times[\nu]^{n_1}\times\cdots\times[\nu]^{n_k}).$$

A direct limit argument then shows that $\alpha=\beta+1$. Let $\kappa=\mathrm{crit}(E_\beta)$ and let $\gamma=\mathcal{T}(\alpha)$. Notice that if $\xi<\kappa$ then our claim follows from (∗)[77]. Thus, we assume that $\xi\geq\kappa$.

Assume first that $\xi\leq\mathrm{ind}_\beta$. Because ξ is a limit of Woodin cardinals of $\mathcal{Q}$ and there are no Woodin cardinals of $\mathcal{Q}$ in the interval $(\kappa,\mathrm{ind}_\beta]$[78], we have that in fact $\xi=\kappa$. Because $\kappa=\mathrm{crit}(E_\beta)$, we have that ξ is $\mathcal{T}$-critical. Applying Lemma 2.9.3 to $(\xi,\mathcal{T})$ we get a finite sequence $(\gamma_i,\gamma_i',\xi_i:i\leq n)$ satisfying the clauses of Lemma 2.9.3. In particular, $\xi=\xi_{n+1}$, $\xi_n=\kappa$, $\gamma_n=\gamma$, $\gamma_n'=\beta$ and $\gamma_{n+1}=\alpha$. Clause 10 of Lemma 2.9.3 implies that there is $B'\in\mathcal{M}_{\gamma_0}|(\xi_0^+)^{\mathcal{M}_{\gamma_0}}$ and $s\in[\xi]^{<\omega}$ such that

(1) $A=\pi_{\gamma_0,\alpha}(B')_s\cap([\xi]^{m_0}\times[\xi]^{m_1}\times\cdots\times[\xi]^{m_k})$.

Because ξ_0 is not $\mathcal{T}_{\leq\gamma_0}$-critical and because ξ_0 is a limit of Woodin cardinals of $\mathcal{M}_{\gamma_0}$, applying (∗) to $(\xi_0,\mathcal{T}_{\leq\gamma_0})$ we get some $B''\in\mathcal{P}^b$ and $s'\in[\xi_0]^{<\omega}$ such that

(2) $\pi_{0,\gamma_0}(B'')_{s'}=B'$.

Putting (1) and (2) together and rearranging B'' we get some $B\in\mathcal{P}^b$ and $t\in[\xi]^{<\omega}$ such that
$$A=\pi^{\mathcal{T},b}(B)_t\cap([\xi]^{m_0}\times[\xi]^{m_1}\times\cdots\times[\xi]^{m_k}).$$

Assume now that $\xi>\mathrm{ind}_\beta$. Let λ be the least such that $\pi_{\gamma,\alpha}(\lambda)\geq\xi$. Because ξ is a limit of Woodin cardinals of $\mathcal{Q}$, we have that

(1) $\mathcal{M}_\gamma\models$ "λ is a limit of Woodin cardinals".

We now have some $g\in\mathcal{M}_\gamma$, $g:\kappa\to\lambda$ such that $\pi_{\gamma,\alpha}(g)(s)=A$ where $s\in[\mathrm{ind}_\beta]^{<\omega}\subset[\xi]^{<\omega}$.

Suppose first that λ is not $\mathcal{T}$-critical. Since λ is not $\mathcal{T}$-critical, applying (∗) to $(\lambda,\mathcal{T}_{\leq\gamma})$, we get some $f\in\mathcal{P}^b$ and some $t\in[\lambda]^\omega$ such that $g=\pi_{0,\gamma}(f)(t)$.

[77] Notice that we must have that $\pi^{\mathcal{T}_{\leq\gamma},b}$ is defined as otherwise $\pi^{\mathcal{T},b}$ cannot be defined.

[78] This is consequence of the fact that $\mathcal{P}$ is lsa small.

Therefore, $A = \pi_{0,\alpha}(f)(u)(s)$ where $u = \pi_{\gamma,\alpha}(t)$. But because $u \in [\xi]^{<\omega}$, we can find some $f^* \in \mathcal{P}^b$ and some $u^* \in [\xi]^{<\omega}$ such that $A = \pi_{0,\alpha}(f^*)(u^*)$. Rearranging f^* we get some $B \in \mathcal{P}^b$ and $t \in [\xi]^{<\omega}$ such that $A = \pi_{0,\alpha}(B)_t$.

Finally, suppose that λ is $\mathcal{T}$-critical. In this case, λ is a regular cardinal of $\mathcal{M}_\gamma$ and so we have two cases. Either $\lambda = \kappa$ or $\pi_{\gamma,\alpha}(\lambda) = \xi$. In both cases, we have some $B' \in \mathcal{M}_\gamma|(\lambda^+)^{\mathcal{M}_\gamma}$ such that $A = \pi_{\gamma,\alpha}(B')_t$ for $t \in [\xi]^{<\omega}$.

We now have two cases. Suppose first that $\pi_{\gamma,\alpha}(\lambda) = \xi$. Applying $(*)$ to $(\lambda, \mathcal{T}_{\leq\gamma})$ we get some $B'' \in \mathcal{P}^b$ and some $t' \in [\lambda]^{<\omega}$ such that

$$B' = \pi_{0,\gamma}(B'')_{t'} \cap ([\lambda]^{|t|} \times [\lambda]^{m_0} \times [\lambda]^{m_1} \times \cdots \times [\lambda]^{m_k}).$$

and so rearranging B'' we get some $B \in \mathcal{P}^b$ and $s \in [\xi]^{<\omega}$ such that

$$A = \pi_{0,\alpha}(B)_s \cap ([\xi]^{m_0} \times [\xi]^{m_1} \times \cdots \times [\xi]^{m_k}).$$

Suppose next that $\pi_{\gamma,\alpha}(\lambda) > \xi$. As λ is a regular cardinal of $\mathcal{M}_\gamma$ this is only possible if $\lambda = \kappa$. Since κ is a $\mathcal{T}$-critical point we have that there is some $B'' \in \mathcal{P}^b$ and some $t' \in [\kappa]^{<\omega}$

$$B' = \pi_{0,\gamma}(B'')_{t'} \cap ([\kappa]^{|t|} \times [\kappa]^{m_0} \times [\kappa]^{m_1} \times \cdots \times [\kappa]^{m_k}).$$

Therefore, we get that

$$A = (\pi_{0,\alpha}(B'')_{t'} \cap ([\pi_{\gamma,\alpha}(\kappa)]^{|t|} \times [\pi_{\gamma,\alpha}(\kappa)]^{m_0} \times \cdots \times [\pi_{\gamma,\alpha}(\kappa)]^{m_k})_t.$$

Therefore, for some $B''' \in \mathcal{P}^b$ and $t'' \in [\xi]^{<\omega}$,

$$A = \pi_{0,\alpha}(B''')_{t''} \cap ([\xi]^{m_0} \times [\xi]^{m_1} \times \cdots \times [\xi]^{m_k}).$$

Since $\xi \in (\mathrm{ind}_\beta, \pi_{\gamma,\alpha}(\kappa))$, we have that $\xi = \pi_{\gamma,\alpha}(g)(u)$ for some $u \in [\xi]^{<\omega}$. Therefore, rearranging B''' we get some $B \in \mathcal{P}^b$ and $s \in [\xi]^{<\omega}$ such that $A = \pi_{0,\alpha}(B)_s$. ⊣

LEMMA 2.9.5. *Suppose $\mathcal{P}$ is a germane, b-type* lses *that projects precisely and $\mathcal{T}$ is an almost non-dropping stack on $\mathcal{P}$. Let $\mathcal{Q} = \pi^{\mathcal{T},b}(\mathcal{P}^b)$. Then for any window $w = (\eta, \delta)$ of $\mathcal{Q}$ (see Notation 2.7.14) such that $\mathcal{Q} \vDash$ "δ is a Woodin cardinal",*

$$\sup(s(\mathcal{T}, w)) = \delta.$$

PROOF. We drop $\mathcal{T}$ from superscripts. Let α^* be the least such that $\mathcal{Q} = \mathcal{M}^b_{\alpha^*}$. To prove our claim, we may just as well assume, without losing generality, that $\alpha^* + 1 = \mathrm{lh}(\mathcal{T})$.

Suppose to the contrary that $w = (\eta, \delta)$ is a window of $\mathcal{Q}$ such that $\mathcal{Q} \vDash$ "δ is a Woodin cardinal" but $\sup(s(\mathcal{T}, w)) < \delta$. Let then α be the least α' such that $\delta < \delta^{(\mathcal{M}_{\alpha'})^b}$ and $\mathcal{Q}|\delta = \mathcal{M}_{\alpha'}|\delta$. As $\pi_{\alpha,\alpha^*} \restriction \delta + 1 = id$[79], we can now assume without loss of generality that $\alpha = \alpha^*$.

We can also assume, without loss of generality, that

[79]Notice that while π_{α,α^*} may not exist, $\pi_{\alpha,\alpha^*} \restriction \mathcal{M}^b_\alpha$ must be defined, and so the use of π_{α,α^*} is justified.

(∗) for any $\beta \in [0,\alpha)_{\mathcal{T}}$, $\delta \notin \mathrm{rge}(\pi_{\beta,\alpha})$.[80]

This is because for any such β, letting δ' be such that $\pi_{\beta,\alpha}(\delta') = \delta$, we must have that $\sup(\pi_{\beta,\alpha}[\delta']) = \delta$[81].

Because δ has no pre-image in any $\mathcal{M}_\beta$, it must be the case that $\alpha = \beta + 1$ for some β. Let $\gamma = \mathcal{T}(\beta+1)$. We thus have that $\mathcal{M}_\alpha = \mathrm{Ult}(\mathcal{M}_\gamma, E_\beta)$[82] and that

(1) $\delta \notin \mathrm{rge}(\pi_{\gamma,\alpha})$ and hence, $\delta > \mathrm{crit}(E_\beta)$.[83]

Furthermore, notice that

(2) $\delta > \mathrm{ind}_\beta$

as otherwise in the case that $\delta = \mathrm{ind}_\beta$ we have that δ is a successor cardinal in $\mathcal{M}_\alpha$ and hence not a Woodin cardinal, or in the case that $\delta \in (\mathrm{crit}(E_\beta), \mathrm{ind}_\beta)$ we have that δ is not a Woodin cardinal in $\mathcal{M}_\alpha$ as it is not a Woodin cardinal in $\mathrm{Ult}(\mathcal{M}_\beta, E_\beta)$ and $\mathcal{M}_\alpha \cap \wp(\delta) = \mathrm{Ult}(\mathcal{M}_\beta, E_\beta) \cap \wp(\delta)$.

Notice next that (2), and more relevantly the argument used to establish (2), also implies that

(3) $\eta > \mathrm{ind}_\beta$.

It then follows that if ξ is least such that $\pi_{\gamma,\alpha}(\xi) > \eta$ then

$$\delta = \sup(\{\pi_{\gamma,\alpha}(f)(s) : f \in \mathcal{M}_\gamma,\ f : \mathrm{crit}(E_\beta)^{|s|} \to \xi \text{ and } s \in [\mathrm{ind}_\beta]^{<\omega}\} \cap \delta)$$

and therefore, it follows from (3) that

4 $\delta = \sup(\{\pi_{\gamma,\alpha}(f)(s) : f \in \mathcal{M}_\gamma,\ f : \mathrm{crit}(E_\beta)^{|s|} \to \xi \text{ and } s \in [\eta]^{<\omega}\} \cap \delta)$.

Notice that it follows from our choice of ξ that $\mathcal{M}_\gamma \models$ "ξ is a cutpoint limit of Woodin cardinals". (4) and Lemma 2.9.4 now give a contradiction[84], as we get that $\sup(s(\mathcal{T}, w)) = \delta$.

Below we calculate the details in the case ξ is $\mathcal{T}$-critical. In this case, if $f \in \mathcal{M}_\gamma$ is such that $f : \mathrm{crit}(E_\beta)^{|s|} \to \xi$ then letting F be the graph of f, we can find $G \in \mathcal{P}^b$ and $t \in [\xi]^{<\omega}$ such that $F = \pi_{0,\gamma}(G)_t \cap [\xi]^{|s|} \times [\xi]$. But then for every $u \in [\mathrm{ind}_\beta]^{<\omega}$, if $\pi_{\gamma,\alpha}(f)(u)$ is defined then it is the unique x such that

$$(s,x) \in \pi_{0,\alpha}(G)_t \cap [\pi_{\gamma,\alpha}(\xi)]^{|s|} \times [\pi_{\gamma,\alpha}(\xi)].$$

Setting then $g(v) = G_v$, we get that for every $u \in [\mathrm{ind}_\beta]^{<\omega}$, if $\pi_{\gamma,\alpha}(f)(u)$ is defined then it is equal to $\pi_{0,\alpha}(g)(t)(u)$. It then follows that there is $h \in \mathcal{P}^b$ such that for

[80]Notice that while $\pi_{\beta,\alpha}$ may not be defined, it nevertheless is defined on $\mathcal{M}_\beta^b$, and so here and in the sequel we will ignore the fact that $\pi_{\beta,\alpha}$ may not be defined.

[81]This is a consequence of the fact that because we are only considering lsa small hod-like lses, δ' is not a critical point of any $E \in \vec{E}^{\mathcal{M}_\beta}$.

[82]Notice that E_β cannot cause a drop as we are assuming that $\pi^{\mathcal{T},b}$ exists.

[83]$\delta = \mathrm{crit}(E_\beta)$ is not possible because of lsa smallness.

[84]We apply Lemma 2.9.4 to functions f used in (4).

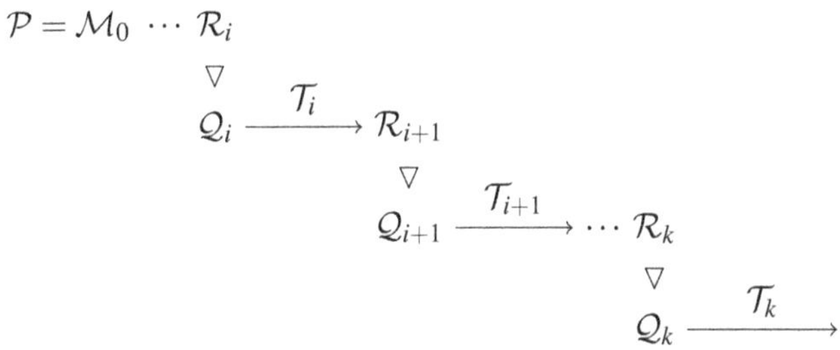

FIGURE 2. A stack with neat drops.

every $u \in [\text{ind}_\beta]^{<\omega}$ there is $u' \in [\text{ind}_\beta]^{<\omega}$ such that if $\pi_{\gamma,\alpha}(f)(u)$ is defined then $\pi_{\gamma,\alpha}(f)(u) = \pi_{0,\alpha}(h)(u')$. It then follows from (4) that $\sup(s(\mathcal{T},w)) = \delta$. $\dashv$

2.10. The un-dropping game

Recall our convention regarding proper stacks (see Remark 2.7.27). Before we proceed, we explain the meaning of the un-dropping game. Suppose we are comparing the strategies of two lsa type hod-like lses $\mathcal{P}$ and $\mathcal{Q}$. Let Σ be the strategy of $\mathcal{P}$ and Λ be the strategy of $\mathcal{Q}$. Let us assume that the pointclasses generated by $(\mathcal{P},\Sigma)$ and $(\mathcal{Q},\Lambda)$ are the same. We are then searching for $\mathcal{R}$ which is an iterate of $\mathcal{P}$ and $\mathcal{Q}$ and $\Sigma_{\mathcal{R}} = \Lambda_{\mathcal{R}}$. In this comparison we might be forced to consider iteration trees $\mathcal{T}$ and $\mathcal{U}$ with last models $\mathcal{M}$ and $\mathcal{N}$ such that $\pi^{\mathcal{T}}$ and $\pi^{\mathcal{U}}$ do not exist and for some $\mathcal{K} \trianglelefteq_{hod} \mathcal{M}$ and $\mathcal{K} \trianglelefteq_{hod} \mathcal{N}$, $\Sigma_{\mathcal{K}} \neq \Lambda_{\mathcal{K}}$. We can continue the comparison by comparing $(\mathcal{M},\Sigma_{\mathcal{M}})$ and $(\mathcal{N},\Lambda_{\mathcal{N}})$ and producing $(\mathcal{S},\Phi)$ which is a common tail of $(\mathcal{M},\Sigma_{\mathcal{M}})$ and $(\mathcal{N},\Lambda_{\mathcal{N}})$. However, $(\mathcal{S},\Phi)$ cannot be thought of as a last model of a successful comparison of $(\mathcal{P},\Sigma)$ and $(\mathcal{Q},\Lambda)$ simply because $\pi^{\mathcal{T}}$ and $\pi^{\mathcal{U}}$ do not exist. What we need to do is to compare $(\mathcal{M},\Sigma_{\mathcal{M}})$ and $(\mathcal{N},\Lambda_{\mathcal{N}})$ and then somehow get back to $\mathcal{P}$ and $\mathcal{Q}$. This is what the un-dropping game achieves.

DEFINITION 2.10.1 (The main drops of a stack, Figure 2). Suppose $\mathcal{P}$ is a germane, b-type lses that projects precisely and

$$\mathcal{T} = ((\mathcal{M}_\alpha)_{\alpha<\eta}, (E_\alpha)_{\alpha<\eta-1}, D, R, (\mathsf{rnc}_\alpha, \mathsf{layer}_\alpha)_{\alpha\in R}, (\beta_\alpha, m_\alpha), T)$$

is a (proper) stack on $\mathcal{P}$ based on $\mathsf{hl}(\mathcal{P})$.

We say that $\alpha \in R$ is a *main drop* if

1. α is of limit type and of bottom type,
2. $\text{lh}(\mathcal{T}_\alpha)$ is a successor ordinal,
3. $\pi^{\mathcal{T}_\alpha}$ is undefined (see Definition 2.7.24),
4. $\pi^{\mathcal{T}_\alpha,b}$ is defined.

We say T has a *main drop* if there is $\alpha \in R$ which is a *main drop*.

Suppose $\mathcal{T}$ has main drops and let $(\alpha_i : i \in [1,k]) \subseteq R$ be the sequence of *main drops* of $\mathcal{T}$ enumerated in increasing order. We then set

1. $\alpha_0 = 0$ and $\alpha_{k+1} = \mathrm{lh}(\mathcal{T}) - 1$,
2. for $i \leq k+1$, $\mathcal{R}_i = \mathcal{M}_{\alpha_i}$ and for $i \leq k$, $\mathcal{Q}_i = \mathsf{layer}_{\alpha_i}$[85],
3. for $i \leq k$, $\mathcal{T}_i = \mathcal{T}_{\lfloor \alpha_i, \alpha_{i+1} \rfloor}$,
4. $\mathcal{T}_{k+1}$ and $\mathcal{Q}_{k+1}$ are undefined,
5. $md^{\mathcal{T}} = (\alpha_i, \mathcal{R}_i, \mathcal{T}_i, \mathcal{Q}_i : i \leq k+1)$.

We then say that $md^{\mathcal{T}} = (\alpha_i, \mathcal{R}_i, \mathcal{T}_i, \mathcal{Q}_i : i \leq k+1)$ is the *md*-sequence of $\mathcal{T}$.

Next we define the un-dropping extender of a stack. This is essentially the extender given by dovetailing the embeddings $\pi^{\mathcal{T}_i,b}$. The un-dropping extender allows us to get back to the original model, and hence it "undrops" the main drops of $\mathcal{T}$. First notice that the following is true.

DEFINITION 2.10.2. Suppose

1. $\mathcal{P}$ is a germane, b-type lses that projects precisely and
2. $\mathcal{T}$ is a stack on $\mathcal{P}$ such that $\mathcal{T}$ has a last model and it is based on $\mathsf{hl}(\mathcal{P})$.

Let $\nu + 1 = \mathrm{lh}(\mathcal{T})$. We say $\mathcal{T}$ has a *one point extension* if letting

$$\mathcal{T} = ((\mathcal{M}_\alpha)_{\alpha \leq \nu}, (E_\alpha)_{\alpha < \nu}, D, R, T),$$

and

$$\mathcal{T}^{ope} = ((\mathcal{M}_\alpha)_{\alpha \leq \nu}, (E_\alpha)_{\alpha < \nu}, D, R \cup \{\nu\}, T),$$

is a proper stack (or according to our convention Remark 2.7.27 just a stack)[86].

The following can now be demonstrated by examining Definition 2.7.24 and the definition of $\pi^{\mathcal{T},b}$.

LEMMA 2.10.3. *Suppose*

1. *$\mathcal{P}$ is a germane, b-type* lses *that projects precisely and*
2. *$\mathcal{T}$ is a stack on $\mathcal{P}$ that has a one point extension and $\pi^{\mathcal{T},b}$ is undefined.*

Then $\mathcal{T}$ has a main drop and letting $md^{\mathcal{T}} = (\alpha_i, \mathcal{R}_i, \mathcal{T}_i, \mathcal{Q}_i : i \leq k+1)$ be the md-sequence of $\mathcal{T}$, $\pi^{\mathcal{T}_k \restriction \mathcal{Q}_k, b}$ exists.

We make the following convention.

TERMINOLOGY 2.10.4. Suppose $j : M \to N$ is a map between two transitive sets or classes M and N, and suppose (κ, λ) is such that $j(\kappa) \geq \lambda$ and $j \restriction \kappa = 1$ $\not{d}$. We then say that E is the (κ, λ)-*extender* derived from j if

$$E = \{(a, A) : A \in \wp([\kappa]^{|a|}) \cap M \wedge a \in [\lambda]^{<\omega} \wedge a \in j(A)\}.$$

[85] See Notation 2.7.26.

[86] See Definition 2.7.24.

We say E is a *short* extender if $\mathrm{crit}(j) = \kappa$ and otherwise we say E is *long*. All extenders used to build extender sequences that we consider in this book are short extenders. In particular, when discussing fully backgrounded constructions (e.g., Definition 4.3.3) we tacitly assume that all extenders are short. However, we may from time to time derive an extender from a given embedding and not specify whether it is short or long. For example, see the definition of $E^{\mathcal{T}}_{\mathcal{Q}}$ below.

DEFINITION 2.10.5 (The un-dropping extender of a proper stack). Suppose

1. $\mathcal{P}$ is a germane, b-type lses that projects precisely and
2. $\mathcal{T}$ is a stack on $\mathcal{P}$ such that $\mathcal{T}$ is based on $\mathsf{hl}(\mathcal{P})$ and $\mathcal{T}$ has a one point extension.

When $\pi^{\mathcal{T},b}$ is undefined. Let $md^{\mathcal{T}} = (\alpha_i, \mathcal{R}_i, \mathcal{T}_i, \mathcal{Q}_i : i \leq k+1)$ be the *md*-sequence of $\mathcal{T}$. For $i \leq k+1$, set $\kappa_i = \delta^{\mathcal{R}_i^b}$ and for $i \leq k$, let

$$\sigma_i^{\mathcal{T}} : (\wp(\kappa_i))^{\mathcal{R}_i} \to (\wp(\kappa_{i+1}))^{\mathcal{R}_{i+1}}$$

be given by

$$\sigma_i^{\mathcal{T}}(A) = \pi^{\mathcal{T}_i \restriction \mathcal{Q}_i, b}(A) \cap \kappa_{i+1}.$$

Set $\sigma^{\mathcal{T}} = \sigma_k^{\mathcal{T}} \circ \sigma_{k-1}^{\mathcal{T}} \cdots \circ \sigma_0^{\mathcal{T}}$.

Suppose $\mathcal{Q} \trianglelefteq_{hod} \mathcal{R}^b_{k+1}$ is meek. We then let $E^{\mathcal{T}}_{\mathcal{Q}}$ be the $(\kappa_0, \delta^{\mathcal{Q}})$-extender derived from $\sigma^{\mathcal{T}}$. More precisely,

$$E^{\mathcal{T}}_{\mathcal{Q}} = \{(a, A) : a \text{ is a finite subset of } \delta^{\mathcal{Q}},\ A \in (\wp([\kappa_0]^{|a|}))^{\mathcal{P}}, \text{ and } a \in \sigma^{\mathcal{T}}(A)\}.$$

When $\pi^{\mathcal{T},b}$ is defined. Suppose $\mathcal{Q} \trianglelefteq_{hod} \pi^{\mathcal{T},b}(\mathcal{P}^b)$ is a complete layer[87] of $\pi^{\mathcal{T},b}(\mathcal{P}^b)$. We then let $E^{\mathcal{T}}_{\mathcal{Q}}$ be the $(\delta^{\mathcal{P}^b}, \delta^{\mathcal{Q}})$-extender derived from $\sigma^{\mathcal{T}}$. More precisely,

$$E^{\mathcal{T}}_{\mathcal{Q}} = \{(a, A) : a \text{ is a finite subset of } \delta^{\mathcal{Q}},\ A \in (\wp([\delta^{\mathcal{P}^b}]^{|a|}))^{\mathcal{P}}, \text{ and } a \in \pi^{\mathcal{T},b}(A)\}.$$

We then say that $E^{\mathcal{T}}_{\mathcal{Q}}$ is the $\mathcal{Q}$-un-dropping extender of $\mathcal{T}$. We also say that E is the *main un-dropping* extender of $\mathcal{T}$ if $E = E^{\mathcal{T}}_{\mathcal{R}^b_{k+1}}$ or if $E = E^{\mathcal{T}}_{\pi^{\mathcal{T},b}(\mathcal{P}^b)}$.

When comparing hod premice we need to consider iterations in which at certain stages I is allowed to use the un-dropping extender of the resulting stack. The game producing such iterations is defined below.

DEFINITION 2.10.6 (The un-dropping iteration game).
Suppose $\mathcal{P}$ is a germane, b-type lses that projects precisely. The un-dropping iteration game on $\mathcal{P}$, $\mathcal{G}^u(\mathcal{P}, \kappa, \lambda, \alpha)$, is an iteration game satisfying the following conditions:

[87] See Notation 2.7.14.

1. In $\mathcal{G}^u(\mathcal{P},\kappa,\lambda,\alpha)$, player I and II collaborate to produce a sequence

$$p = (\mathcal{M}_\beta, \mathcal{T}_\beta, \mathcal{Q}_\beta, E_\beta : \beta < \gamma)$$

such that

(a) $\gamma \leq \kappa$,

(b) $\mathcal{M}_0 = \mathcal{P}$,

(c) for all $\beta < \gamma$, $\mathcal{T}_\beta$ is a stack on $\mathcal{M}_\beta$ (and is produced via the rules of $\mathcal{G}(\mathcal{M}_\beta,\lambda, \alpha)$[88]),

(d) for each β such that $\beta+1<\gamma$, the iteration embedding $\pi^p_{0,\beta} : \mathcal{M}_0 \to \mathcal{M}_\beta$ is defined,

(e) for each β such that $\beta+1<\gamma$, either

  (i) E_β is the $\mathcal{Q}_\beta$- un-dropping extender of $\mathcal{T}_\beta$ and $\mathcal{M}_{\beta+1} = \mathrm{Ult}(\mathcal{M}_\beta, E_\beta)$, or

  (ii) $E_\beta = \mathcal{Q}_\beta = \emptyset$, $\pi^{\mathcal{T}_\beta}$ exists and $\mathcal{M}_{\beta+1}$ is the last model of $\mathcal{T}_\beta$,

(f) for a limit ordinal $\beta<\gamma$, $\mathcal{M}_\beta$ is the direct limit of $(\mathcal{M}_\xi, \pi^p_{\xi,\zeta} : \xi < \zeta < \beta)$ where $\pi^p_{\xi,\zeta} : \mathcal{M}_\xi \to \mathcal{M}_\zeta$ is the iteration embedding,

(g) player I is the player that chooses extenders while playing $\mathcal{G}(\mathcal{M}_\beta,\lambda,\alpha)$ to produce $\mathcal{T}_\beta$,

(h) player I is the player that chooses to stop the run of $\mathcal{G}(\mathcal{M}_\beta,\lambda,\alpha)$ by either playing the $\mathcal{Q}_\beta$-un-dropping extender E_β or by letting $\mathcal{M}_{\beta+1}$ be the last model of $\mathcal{T}_\beta$ (in which case $\pi^{\mathcal{T}_\beta}$ must be defined),

(i) player II chooses branches while playing $\mathcal{G}(\mathcal{M}_\beta,\lambda,\alpha)$.

2. Player II loses a run p of $\mathcal{G}^u(\mathcal{P},\kappa,\lambda,\alpha)$ if one of the models appearing in p is ill-founded.

We say Σ is a (κ,λ,α)-*strategy* for $\mathcal{P}$ if it is a strategy for II in $\mathcal{G}^u(\mathcal{P},\kappa,\lambda,\alpha)$ such that any run of $\mathcal{G}^u(\mathcal{P},\kappa,\lambda,\alpha)$ in which player II plays according to Σ is not a loss for II.

We say $p = (\mathcal{M}_\beta,\mathcal{T}_\beta,E_\beta : \beta<\gamma)$ is a *generalized stack* on $\mathcal{P}$ if it is produced by a run of $\mathcal{G}^u(\mathcal{P},\kappa,\lambda,\alpha)$ and p is not a loss for II. Since $\mathrm{lh}(E_\beta) = \delta^{\mathcal{Q}_\beta}$ there is no ambiguity in omitting $\mathcal{Q}_\beta$s.

REMARK 2.10.7. Suppose $\mathcal{P}$ is a germane, b-type lses that projects precisely and Σ is a (κ,λ,α)-strategy-strategy. Suppose $\mathcal{Q}'$ is a Σ-iterate of $\mathcal{P}$ via $p = (\mathcal{M}_\beta,\mathcal{T}_\beta,E_\beta : \beta<\gamma)$[89] such that $\pi^{p,b}$ is defined (see Definition 2.10.13) and either $\mathcal{Q} = \mathcal{Q}'$ or $\mathcal{Q} \in Y^{\mathcal{Q}'}$ is such that $\mathcal{Q}^b = (\mathcal{Q}')^b$. Then $\Sigma_{\mathcal{Q},p}$ is the (κ',λ,α)-strategy of $\mathcal{Q}$ given by $\Sigma_{\mathcal{Q},p}(q) = \Sigma(p^\frown q)$. Here, $\gamma+\kappa' = \kappa$.

Suppose next that $\mathcal{Q}'$ is a Σ-iterate of $\mathcal{P}$ via $p = (\mathcal{M}_\beta,\mathcal{T}_\beta,E_\beta : \beta<\gamma)$ and $\mathcal{Q} \trianglelefteq \mathcal{Q}'$ is such that at least one of the following holds:

[88]This is the game defined in [59, Chapter 4].

[89]Thus, $\mathcal{Q}$ is the last model of p.

1. $\pi^{p,b}$ does not exist.[90]
2. $\pi^{p,b}$ exists and $\mathcal{Q} \trianglelefteq \pi^{p,b}(\mathcal{P}^b)$.

Then $\Sigma_{\mathcal{Q},p}$ is defined like in the previous case but only for stacks produced by $\mathcal{G}(\mathcal{Q},\lambda,\alpha)$.

Just like with ordinary strategies, it also possible to pullback (κ,λ,α)-strategies. The proof of the fallowing theorem is just like the proof of the same theorem for ordinary strategies.

THEOREM 2.10.8. *Suppose $\mathcal{P}$ and $\mathcal{Q}$ are germane, b-type* lses *which project precisely, $\sigma : \mathcal{Q} \to \mathcal{P}$ is a weak embedding*[91] *and Σ is a (κ,λ,α)-strategy. Then $\mathcal{Q}$ has a (κ,λ,α)-iteration strategy, Λ, with the following property. For all generalized stack $q = (\mathcal{Q}_\beta, \mathcal{U}_\beta, F_\beta : \beta < \gamma)$ on $\mathcal{Q}$, q is according to Λ if and only if there is a generalized stack $p = (\mathcal{P}_\beta, \mathcal{T}_\beta, E_\beta : \beta < \gamma)$ on $\mathcal{P}$ and sequences $(\sigma_\beta : \beta < \gamma)$ and $(\tau_{\beta,\iota} : \beta < \gamma \wedge \iota < \mathrm{lh}(\mathcal{U}_\beta))$ such that the following clauses hold:*

1. *$\sigma_0 = \sigma$ and for all $\beta < \gamma$, $\sigma_\beta : \mathcal{Q}_\beta \to \mathcal{P}_\beta$ is a weak embedding.*
2. *For all $\beta < \gamma$, $\mathcal{T}_\beta = \sigma_\beta \mathcal{U}_\beta$, i.e., $\mathcal{T}_\beta$ is obtained from $\mathcal{U}_\beta$ via the σ_β-copying construction (see [59, Chapter 4.1]).*
3. *For all $\beta < \gamma$, $(\tau_{\beta,\iota} : \iota < \mathrm{lh}(\mathcal{U}_\beta))$ is the sequence of copy maps produced during the construction of $\mathcal{T}_\beta$.*
4. *For each $\beta < \gamma$, F_β is the undropping extender of $\mathcal{U}_\beta$ if and only if E_β is the undropping extender of $\mathcal{T}_\beta$.*
5. *For each $\beta < \gamma$, $F_\beta = \emptyset$ and $\mathcal{Q}_\beta$ is the last model of $\mathcal{U}_\beta$ if and only if $E_\beta = \emptyset$ and $\mathcal{Q}_\beta$ is the last model of $\mathcal{U}_\beta$.*
6. *For each β such that $\beta + 1 < \gamma$ and F_β is the undropping extender of $\mathcal{U}_\beta$, letting $\nu = \mathrm{lh}(\mathcal{U}_\beta)$, for all $a \in \mathrm{lh}(F_\beta)^{<\omega}$ and $A \in \mathcal{M}_\beta$, $(a,A) \in F_\beta$ if and only if $(\tau_{\beta,\nu}(a), \sigma_\beta(A)) \in E_\beta$.*
7. *For each β such that $\beta + 1 < \gamma$ and F_β is the undropping extender of $\mathcal{U}_\beta$, letting $\nu = \mathrm{lh}(\mathcal{U}_\beta)$, $\sigma_{\beta+1} : Ult(\mathcal{Q}_\beta, F_\beta) \to \mathrm{Ult}(\mathcal{P}_\beta, E_\beta)$ is such that $\sigma_{\beta+1}([a,f]_{F_\beta}) = [\sigma_{\beta,\nu}(a), \sigma_\beta(f)]_{E_\beta}$.*
8. *For each β such that $\beta + 1 < \gamma$ and F_β is the undropping extender of $\mathcal{U}_\beta$, letting $\nu = \mathrm{lh}(\mathcal{U}_\beta)$, $\sigma_{\beta+1} = \sigma_{\beta,\nu}$.*

NOTATION 2.10.9. Suppose $\mathcal{T} = (\mathcal{M}_\alpha, \mathcal{T}_\alpha, E_\alpha : \alpha < \gamma)$ is a generalized stack.

1. For $\alpha < \gamma$ and $\alpha' < \mathrm{lh}(\mathcal{T}_\alpha)$, we let $\mathcal{M}^{\mathcal{T}}_{\alpha,\alpha'} = \mathcal{M}^{\mathcal{T}_\alpha}_{\alpha'}$.
2. For $\alpha < \gamma$ and $\iota_0 \leq \iota_1 < \mathrm{lh}(\mathcal{T}_\alpha)$ such that $\iota_0 \in [0,\iota_1)_{\mathcal{T}}$, $\pi^{\mathcal{T},\alpha}_{\iota_0,\iota_1} : \mathcal{M}^{\mathcal{T}}_{\alpha,\iota_0} \to \mathcal{M}^{\mathcal{T}}_{\alpha,\iota_1}$ is the iteration embedding $\pi^{\mathcal{T}_\alpha}_{\iota_0,\iota_1}$ provided it is defined.
3. Suppose next that $\alpha_0 \leq \alpha_1 < \gamma$ and $\iota < \mathrm{lh}(\mathcal{T}_{\alpha_1})$. We then let $\pi^{\mathcal{T}}_{\alpha_0,\alpha_1} : \mathcal{M}_{\alpha_0} \to \mathcal{M}_{\alpha_1}$ be the iteration embedding and $\pi^{\mathcal{T}}_{\alpha_0,(\alpha_1,\iota)} = \pi^{\mathcal{T}_{\alpha_1}}_{\alpha_1,\iota} \circ \pi^{\mathcal{T}}_{\alpha_0,\alpha_1}$ given that $\pi^{\mathcal{T}_{\alpha_1}}_{\alpha_1,\iota}$ is defined.

---

[90] This means that if $\beta + 1 = \gamma$ then $\pi^{\mathcal{T}_\beta}$ is undefined.

[91] In the sense of [3, Fact 2.13]. See the paragraph after [3, Fact 2.13].

4. We let $\mathcal{T}^{\text{ue}}$ be the *un-dropping extension* of $\mathcal{T}$. More precisely, $\mathcal{T}^{\text{ue}}$ is defined assuming $\eta = \beta + 1$ and $\mathcal{T}_\beta$ has a one point extension[92], in which case $\mathcal{T}^{\text{ue}}$ is obtained by letting E_β be the un-dropping extender of $\mathcal{T}_\beta$, $\mathcal{M}_{\beta+1} = \text{Ult}(\mathcal{M}_\beta, E_\beta)$ and $\mathcal{T}_{\beta+1} = \emptyset$.
5. We can also define $\mathcal{T}^{\text{ue}}_{\mathcal{Q}}$ assuming $\mathcal{Q} \trianglelefteq \mathcal{R}^b$ where $\mathcal{R}$ is the last model of $\mathcal{T}_\beta$. Here, we let F_β be the $\mathcal{Q}$ un-dropping extender of $\mathcal{T}_\beta$.
6. Again assuming $\text{lh}(\mathcal{T}) = \beta + 1$ and $\mathcal{T}$ has a one point extension, letting $\mathcal{R}$ the last model of $\mathcal{T}_\beta$ and $\mathcal{Q} \trianglelefteq \mathcal{R}^b$ be a complete layer of $\mathcal{R}$, we can define the $\mathcal{Q}$-un-dropping extender of $\mathcal{T}$ by setting:
$$E^{\mathcal{T}}_{\mathcal{Q}} = \{(a,A) : a \in [\delta^{\mathcal{Q}}]^{<\omega} \wedge A \in \wp(\delta^{\mathcal{P}^b}) \cap \mathcal{P} \wedge a \in \sigma^{\mathcal{T}_\beta}(\pi^{\mathcal{T}}_{0,\beta}(A))\},$$
where $\sigma^{\mathcal{X}}$ is defined in Definition 2.10.5. We then set
$$\sigma^{\mathcal{T}} = \sigma^{\mathcal{T}_b} \circ \pi^{\mathcal{T}}_{0,\beta}.$$
Alternatively, $E^{\mathcal{T}}_{\mathcal{Q}}$ is the $(\delta^{\mathcal{P}^b}, \delta^{\mathcal{Q}})$-extender derived from $\pi^{\mathcal{T}^{\text{ue}}}$. We say $E^{\mathcal{T}}$ is the un-dropping extender of $\mathcal{T}$ if $\mathcal{T}$ is the $\mathcal{R}^b$-un-dropping extender of $\mathcal{T}$.
7. As ordinary stacks are instances of generalized stacks, $\mathcal{T}^{\text{ue}}$ and $\mathcal{T}^{\text{ue}}_{\mathcal{Q}}$ can also be used for ordinary stacks.

Often, when $\mathcal{T}$ is clear from the context, we will omit it from our notation.

The next definition introduces *self-cohering* iteration strategies. The idea is as follows. Suppose $\mathcal{P}$ is a non-meek hod-like lses and suppose $\mathcal{T} = (\mathcal{M}_\alpha, \mathcal{T}_\alpha, F_\alpha : \alpha < \eta)$ is a generalized stack on $\mathcal{P}$ according to some iteration strategy Σ. Let $\mathcal{R}$ be the last model of $\mathcal{T}_0$. Then $\mathcal{R}^b \trianglelefteq \mathcal{M}^b_1$. But it is not clear that $\Sigma_{\mathcal{R}^b, \mathcal{T}_0} = \Sigma_{\mathcal{R}^b, \mathcal{T}_0^\frown\{F_0\}}$. Self-cohering strategies have this property. We will use this property in our diamond comparison argument (see Definition 4.14).

DEFINITION 2.10.10. Suppose $\mathcal{T}$ is a stack and $\mathcal{R} = \mathcal{M}^{\mathcal{T}}_\alpha$ for some $\alpha < \text{lh}(\mathcal{T})$. We then say that $\mathcal{R}$ is a node of $\mathcal{T}$ and write $\mathcal{T}_{\leq \mathcal{R}}$ for $\mathcal{T}_{\leq \alpha}$. Similarly if $\mathcal{R}' = \mathcal{M}^{\mathcal{T}}_\beta$ for $\beta > \alpha$ then we can define $\mathcal{T}_{<\mathcal{R}}$, $\mathcal{T}_{\mathcal{R},\mathcal{R}'}$ and $\mathcal{T}_{\geq \mathcal{R}}$. Similar notation can be introduced for generalized stacks in the obvious way.

DEFINITION 2.10.11. Suppose $(\mathcal{P}, \Sigma)$ is a hod-like lses pair (see Definition 2.10.12). We say that Σ is *self-cohering* if whenever

1. $\mathcal{T} = (\mathcal{M}_\alpha, \mathcal{T}_\alpha, F_\alpha : \alpha < \eta)$ is a generalized stack according to Σ,
2. $\alpha_0, \alpha_1 < \eta$,
3. $\xi_0 < \text{lh}(\mathcal{T}_{\alpha_0})$ and $\xi_1 < \text{lh}(\mathcal{T}_{\alpha_1})$,
4. $\mathcal{R} \triangleleft_{hod} \mathcal{M}^{\mathcal{T}_{\alpha_0}}_{\xi_0} =_{\text{def}} \mathcal{S}_0$ and $\mathcal{R} \triangleleft_{hod} \mathcal{M}^{\mathcal{T}_{\alpha_1}}_{\xi_1} =_{\text{def}} \mathcal{S}_1$,

$$\Sigma_{\mathcal{R}, \mathcal{T}_{\leq \mathcal{S}_0}} = \Sigma_{\mathcal{R}, \mathcal{T}_{\leq \mathcal{S}_1}},$$ [93]

where the equality is between the (ω_1, ω_1) portions of both strategies.

[92] See Definition 2.10.2.

[93] See Definition 2.6.3.

Self-cohering is a desired property and we will have to establish that our constructions produce strategies that are self-cohering. However, it is more convenient not to make it part of our definitions.

DEFINITION 2.10.12 (Hod-like lses pair). We say $(\mathcal{P},\Sigma)$ is a *hod-like* lses *pair* (with an indexing scheme φ) if

1. $\mathcal{P}$ is a hod-like lses (with an indexing scheme φ),
2. if $\mathcal{P}$ is non-meek then Σ is a (κ,λ,ν)-strategy,
3. if $\mathcal{P}$ is meek or gentle then Σ is a (λ,ν)-strategy,
4. if $\mathcal{Q}$ is a Σ-iterate of $\mathcal{P}$ via $\mathcal{T}$ and $\mathcal{R} \trianglelefteq_{hod} \mathcal{Q}$ then $\Sigma^{\mathcal{R}} \subseteq \Sigma_{\mathcal{R},\mathcal{T}}$.[94]

We say $(\mathcal{P},\Sigma)$ is a *simple hod-like* lses *pair* if $\mathcal{P}$ is a hod-like lses , Σ is a (λ,ν)-iteration strategy and clause 4 above holds.

In the context of AD^+, unless otherwise specified, the strategy of a hod-like lses pair or a simple hod-like lses pair is an $(\omega_1,\omega_1,\omega_1)$-strategy or an (ω_1,ω_1)-strategy.

Finally we finish this section by stating the version of Lemma 2.9.5 for generalized stacks. Its proof is just like the proof of Lemma 2.9.5. First we generalize $\pi^{\mathcal{T},b}$ and Definition 2.9.1 to generalized stacks.

DEFINITION 2.10.13 (Almost non-dropping generalized stacks). Suppose $\mathcal{M}$ is germane of b-type and projects precisely. Suppose further that

$$\mathcal{T} = (\mathcal{M}_\alpha, \mathcal{T}_\alpha, E_\alpha : \alpha < \gamma)$$

is a generalized stack on $\mathcal{M}$ that is based on $\mathsf{hl}(\mathcal{M})$. We say that $\mathcal{T}$ is *almost non-dropping* if either γ is a limit ordinal or $\gamma = \alpha + 1$ and $\pi^{\mathcal{T}_\alpha,b}$ exists. Assuming $\mathcal{T}$ is almost non-dropping we set

$$\pi^{\mathcal{T},b} = \begin{cases} \pi^{\mathcal{T}}_c & \text{if } \gamma \text{ is a limit ordinal and} \\ & c \text{ is the unique branch of } \mathcal{T}, \\ \pi^{\mathcal{T}_\alpha,b} \circ \pi^{\mathcal{T}}_{0,\alpha} \restriction \mathcal{M}^b & \text{otherwise.} \end{cases}$$

Suppose Σ is a (κ,λ,α)-iteration strategy for $\mathcal{M}$[95]. We then let

$$\begin{aligned} I(\mathcal{M},\Sigma) = \{(\mathcal{T},\mathcal{R}) : {} & \mathcal{T} \text{ is according to } \Sigma,\ \mathcal{T} \text{ is based on } \mathsf{hl}(\mathcal{M}), \\ & \mathcal{R} \text{ is the last model of } \mathcal{T} \text{ and } \pi^{\mathcal{T}} \text{ is defined}\}; \\ I^b(\mathcal{M},\Sigma) = \{(\mathcal{T},\mathcal{R}) : {} & \mathcal{T} \text{ is according to } \Sigma,\ \mathcal{T} \text{ is based on } \mathsf{hl}(\mathcal{M}), \\ & \mathcal{R} \text{ is the last model of } \mathcal{T} \text{ and } \pi^{\mathcal{T},b} \text{ is defined}\}; \\ I^{ope}(\mathcal{M},\Sigma) = \{(\mathcal{T},\mathcal{R}) : {} & \mathcal{T} \text{ is according to } \Sigma,\ \mathcal{T} \text{ is based on } \mathsf{hl}(\mathcal{M}), \\ & \mathcal{T} \text{ has a one point extension and } \mathcal{R} \text{ is the last model of } \mathcal{T}\}, \end{aligned}$$[96]

[94]This clause is asserting that the internal strategy of $\mathcal{R}$ agrees with $\Sigma_{\mathcal{R},\mathcal{T}}$.

[95]It is worth remembering that this entails that Σ-iterates of $\mathcal{M}$ have the same indexing scheme as $\mathcal{M}$.

$B^{ope}(\mathcal{M},\Sigma) = \{(\mathcal{T},\mathcal{R}) : \text{there is } (\mathcal{T},\mathcal{R}') \in I^{ope} \text{ and } \mathcal{R} \text{ is a layer of } \mathcal{R}'\}$.

We remark that we will use $I^{ope}(\mathcal{M},\Sigma)$ and $B^{ope}(\mathcal{M},\Sigma)$ even when Σ is an iteration strategy acting on stacks.

DEFINITION 2.10.14 (Canonical singularizing sequences). Suppose $\mathcal{P}$ is a germane lses of b type that projects precisely and $\mathcal{T}$ is an almost non-dropping generalized stack on $\mathcal{P}$. Let $\mathcal{Q} = \pi^{\mathcal{T},b}(\mathcal{P}^b)$. Then $\mathcal{Q}$ is a hod-like lses. If $w = (\eta,\delta)$ is a window of $\mathcal{Q}$ then we let

$$s(\mathcal{T},w) = \{\alpha : \exists a \in \eta^{<\omega} \exists f \in \mathcal{P}^b(\alpha = \pi^{\mathcal{T},b}(f)(a))\} \cap \delta$$

LEMMA 2.10.15. *Suppose $\mathcal{P}$ is germane, b-type* lses *that projects precisely and*

$$\mathcal{T} = (\mathcal{M}_\alpha, \mathcal{T}_\alpha, E_\alpha : \alpha < \gamma)$$

is a generalized stack $\mathcal{P}$ such that $\pi^{\mathcal{T},b}$ exists. Let $\mathcal{Q} = \pi^{\mathcal{T},b}(\mathcal{P}^b)$. Then for any window $w = (\eta,\delta)$ of $\mathcal{Q}$ (see Notation 2.7.14) such that $\mathcal{Q} \vDash$ "δ is a Woodin cardinal", $\sup(s(\mathcal{T},w)) = \delta$.

[96]See Definition 2.10.2. Here one point extension of a generalized stack $\mathcal{T} = (\mathcal{M}_\alpha, \mathcal{T}_\alpha, E_\alpha : \alpha < \eta)$ is $\mathcal{T}$ unless $\eta = \beta + 1$, in which case we let $\mathcal{T}^{ope} = (\mathcal{M}_\alpha, \mathcal{T}_\alpha, E_\alpha : \alpha < \beta)^\frown(\mathcal{M}_\beta, \mathcal{T}_\beta^{ope})$.

Chapter 3

SHORT TREE STRATEGY MICE

The main purpose of this chapter is to isolate the definition of short tree strategy mice. As was mentioned many times before, the main problem with defining this concept is the fact that it is possible that maximal iteration trees (which should not have branches indexed in the strategy predicate) may *core down* to short iteration trees (which must have branches indexed in the strategy predicate), thus causing indexing issues. To solve this issue we will design an authentication procedure which will carefully choose iteration trees and index their branches. Thus, if some iteration tree does not have a branch indexed in the strategy predicate then it is because the authentication procedure has not yet found an authenticated branch, and therefore, such iteration trees cannot core down to an iteration tree whose branch is authenticated.

The following is a rough roadmap of the chapter. Section 3.1 introduces the short tree component of an iteration strategy, while Section 3.2 introduces the short tree strategy as an abstract object. This is an important step as the strategy predicate of a short tree strategy mouse codes a short tree strategy in the sense of Definition 3.2.4. The next important step is the isolation of two different kinds of iterations, those that are universally short (see Definition 3.3.2), i.e., short with respect to any strategy, and those that are ambiguous. As there is no ambiguity involved in determining whether a universally short iteration trees are short or not and moreover, since universal shortness is preserved under Mostowski collapses, we will simply add the branches of such iterations to the strategy predicate without authenticating them first. The branches of ambiguous iterations will be authenticated before being added to the strategy predicate.

A key tool in the authentication procedure is the fully backgrounded constructions that produce iterates of hod like lses (see Definition 3.5.1). Such constructions are used to find a branch of an iteration tree with the property that the branch model itself iterates to the same construction. Definition 3.6.4 and Definition 3.8.7 introduce the particular ways branches will be indexed. Our authentication procedure appears as Definition 3.7.3, and Definition 3.8.9 defines indexing scheme. Then Definition 3.8.17 introduces the short tree strategy mice and Remark 3.8.20 explains exactly how branches get indexed. Definition 3.10.2 finally introduces the concept of a hod premouse.

REMARK 3.0.16. All the notions introduced in this chapter can be routinely carried over to germane lses that project precisely. Thus, when discussing germane lses, we will freely use the language developed in the sections of this chapter.

3.1. The short tree component of a strategy

Suppose $(\mathcal{P},\Sigma)$ is a hod-like lses pair or a simple hod-like lses pair such that $\mathcal{P}$ is of lsa type (see Definition 2.10.12). Since the particular indexing scheme will not matter for what follows, we suppress the indexing scheme that the pair $(\mathcal{P},\Sigma)$ has. The next definition isolates the *short tree component* of Σ denoted by Σ^{stc}. Let $\kappa = \delta^{\mathcal{P}^b}$ and $\delta = \delta^{\mathcal{P}}$. Recall that all our stacks are proper stacks (see Remark 2.7.27). The next few concepts will be introduced for generalized stacks, and as stacks are instances of generalized stacks, they can be used in connection with stacks.

REMARK 3.1.1. The short tree component of Σ is a strategy that acts on $\mathcal{P}_{\#}$[1]. Thus, the short tree component does not in general produce stacks that can be applied to $\mathcal{P}$ without dropping in degree. Such dropping can happen, for example, when $\rho_{k(\mathcal{P})}(\mathcal{P}) < \delta^{\mathcal{P}}$.

DEFINITION 3.1.2. Suppose $(\mathcal{P},\Sigma)$ is a hod-like lses pair or a simple hod-like lses pair such that $\mathcal{P}$ is of lsa type. Suppose $\mathcal{T} = (\mathcal{M}_\beta, \mathcal{T}_\beta, E_\beta : \beta < \gamma)$ is a generalized stack on $\mathcal{P}_{\mathrm{ex}}$ according to Σ_{ex}[2]. We say $\mathcal{T}$ is Σ-*short* if $\mathcal{T} \in \mathrm{dom}(\Sigma_{\mathrm{ex}})$ and letting $b = \Sigma_{\mathrm{ex}}(\mathcal{T})$ one of the following conditions holds:

1. $\pi_b^{\mathcal{T}}$ is undefined.
2. γ is a limit ordinal.
3. γ is a successor ordinal and $\mathrm{lh}(\mathcal{T}_{\gamma-1})$ is a limit ordinal.
4. γ is a successor ordinal, $\mathrm{lh}(\mathcal{T}_{\gamma-1})$ is a successor ordinal and letting
$$\alpha = \max(R^{\mathcal{T}_{\gamma-1}}),$$
$\pi_b^{\mathcal{T}}(\delta) > \delta(\mathcal{T}_{\geq\alpha})$.

We then say that $\mathcal{T}$ is Σ-*maximal* if it is not Σ-short.

REMARK 3.1.3. Recall that according to our convention Remark 2.7.27, if $\mathcal{T}$ is a stack and α is a cutpoint of $\mathcal{T}$ then $\alpha \in R^{\mathcal{T}}$. Hence, if for some $\alpha < \mathrm{lh}(\mathcal{T})$, $\mathcal{T}_{<\alpha}$ is Σ-maximal then $\alpha \in R^{\mathcal{T}}$.

Notice that if $\mathcal{T}$ is Σ-short then it does not follow that initial segments of $\mathcal{T}$ are also Σ-short. If $\mathcal{T}$ is a generalized stack or just a stack then we let $\mathcal{T}^-$ be $\mathcal{T}$ without its last model if it exists and $\mathcal{T}$ otherwise. The next definitions describe exactly when a stack is according to the short tree strategy component of

[1] See Definition 2.7.3.
[2] See Definition 2.7.3.

Σ. Definition 3.1.6 introduces the domain of Σ^{stc} restricted to ordinary indexable stacks and Definition 3.1.7 introduces the domain of Σ^{stc}.

DEFINITION 3.1.4. Suppose $\mathcal{T}$ is a normal iteration tree of limit length. We then let $\mathrm{m}^{+}(\mathcal{T}) = (\mathrm{m}(\mathcal{T}))^{\#}$.

Definition 3.1.6 needs a slight modification of the concept of a tree order.

DEFINITION 3.1.5. Suppose $(\iota_\tau : \tau < \nu)$ is an increasing sequence of ordinals and for all $\tau < \nu$ such that $\tau + 1 < \nu$, I_τ is either the interval $[\iota_\tau, \iota_{\tau+1})$ or the interval $[\iota_\tau, \iota_{\tau+1}]$. We say I_τ is right-open if $I_\tau = [\iota_\tau, \iota_{\tau+1})$ and otherwise we say I_τ is right-closed. Let $\iota = \sup\{\iota_\tau + 1 : \tau < \nu\}$. We then say that U is a *tree order* on $\prod_{\tau<\nu} I_\tau$ if $U \subseteq \iota^2$ such that the following clauses hold.

1. U is a partial order preserving the usual order on ordinals.
2. If $(\alpha, \beta) \in U$ then for some $\tau < \nu$, $(\alpha, \beta) \in I_\tau \times I_\tau$.
3. For all limit ordinals $\lambda < \iota$, either
   (a) for some $\tau < \nu$, $\lambda = \iota_{\tau+1}$ and I_τ is right-open, or
   (b) $\{\alpha < \lambda : (\alpha, \lambda) \in U\}$ is a closed unbounded subset of λ.

We will freely adopt the usual notation used for ordinary tree orders. For example, $<_U$ is the order given by U, $U(\alpha+1)$ is the U-predecessor of $\alpha+1$ and $[\alpha, \beta]_U = \{\gamma : \alpha \leq_U \gamma \leq_U \beta\}$.

Suppose T is a tree order on ι. We then let $T \restriction \prod_{\tau<\nu} I_\tau$ be the unique tree order U on $\prod_{\tau<\nu} I_\tau$ with the property that for all $\tau < \nu$ and for all $(\alpha, \beta) \in I_\tau \times I_\tau$, $(\alpha, \beta) \in U \iff (\alpha, \beta) \in T$.

DEFINITION 3.1.6 (The domain of the short tree component of a strategy I). Suppose $(\mathcal{P}, \Sigma)$ is a hod-like lses pair or a simple hod-like lses pair such that $\mathcal{P}$ is of lsa type. We let

$$\mathcal{U} = ((\mathcal{M}_\alpha)_{\alpha<\eta}, (E_\alpha)_{\alpha<\eta-1}, D, R, (\beta_\alpha, m_\alpha)_{\alpha\in R}, \mathsf{short}, \mathsf{max}, U) \in \mathrm{dom}(\Sigma^{stc})$$

if there is a stack

$$\mathcal{T} = ((\mathcal{M}'_\alpha)_{\alpha<\eta}, (E'_\alpha)_{\alpha<\eta-1}, D', R', (\beta'_\alpha, m'_\alpha)_{\alpha\in R'}, T) \in \mathrm{dom}(\Sigma_{\mathsf{ex}})^3$$

such that $\mathcal{U}$ is the same as $\mathcal{T}$ except it does not have the maximal branches of $\mathcal{T}$; more precisely, the following conditions hold.

1. $\mathcal{M}_0 = \mathcal{P}_\#$ and $\mathcal{T}$ is below $\delta^{\mathcal{P}}$.[4]
2. $D = D'$, $R' = R = \mathsf{short} \cup \mathsf{max}$, $\mathsf{short} \cap \mathsf{max} = \emptyset$ and max is finite.
3. For all $\alpha < \eta$, $E_\alpha = E'_\alpha$, $\beta_\alpha = \beta'_\alpha$ and $m_\alpha = m'_\alpha$.
4. For all successor $\alpha < \eta$, $\mathcal{M}_\alpha = \mathcal{M}'_\alpha$.
5. For all limit $\alpha < \eta$ such that $\mathcal{T}_{<\alpha}$ is Σ-short, $\mathcal{M}_\alpha = \mathcal{M}'_\alpha$.

[3]See Definition 2.7.3.

[4]That is, for every $\alpha < \eta$ if $\pi^{\mathcal{T}}_{0,\alpha}$ is defined then $\mathrm{ind}^{\mathcal{T}}_\alpha < \pi^{\mathcal{T}}_{0,\alpha}(\delta^{\mathcal{P}})$.

6. For all limit $\alpha < \eta$ such that $\mathcal{T}_{<\alpha}$ is Σ-maximal, letting $\mathcal{X}$ be the last normal component of $\mathcal{T}_{<\alpha}$, $\mathcal{M}_\alpha = \mathrm{m}^+(\mathcal{X})$.[5]

Let $\nu = o.t.(R)$ and $(\iota_\tau : \tau < \nu)$ be the increasing enumeration of R. If $\tau + 1 = \nu$ then set

$$\iota_{\iota+1} = \begin{cases} \eta & \text{if } \eta \text{ is a limit ordinal,} \\ \eta - 1 & \text{otherwise.} \end{cases}$$

We say $\tau+1$ is irrelevant if $\tau+1 = \nu$ and $\iota_{\tau+1} = \eta$, and if $\tau+1$ is not irrelevant then we say τ is relevant. For $\tau < \nu$ such that $\iota_\tau \in \mathsf{short}$ let $I_\tau = [\iota_\tau, \iota_{\tau+1}]$ and otherwise set $I_\tau = [\iota_\tau, \iota_{\tau+1})$.

7. $U = T \restriction \prod_{\tau<\nu} I_\tau$.
8. For every $\tau \leq \nu$ if ι_τ is defined[6] then $\iota_\tau \in \mathsf{max}$ if and only if τ is relevant and $\mathcal{T}_{<\iota_{\tau+1}}$ is Σ-maximal.

If $\mathcal{T}$ and $\mathcal{U}$ are as above then we write $\mathcal{U} = \mathcal{T}^{sc}$ and say that $\mathcal{U}$ is the short component of $\mathcal{T}$.

DEFINITION 3.1.7 (The short tree component of a strategy II). Suppose $(\mathcal{P}, \Sigma)$ is a hod-like lses pair or a simple hod-like lses pair such that $\mathcal{P}$ is of lsa type. We set

$$\mathcal{U} = (\mathcal{N}_\alpha, \mathcal{U}_\alpha, E_\alpha : \alpha < \eta) \in \mathrm{dom}(\Sigma^{stc})$$

if there is a generalized stack $\mathcal{T} = (\mathcal{M}_\alpha, \mathcal{T}_\alpha, F_\alpha : \alpha < \eta) \in \mathrm{dom}(\Sigma_{\mathsf{ex}})$ (see Definition 2.10.6) such that $\mathcal{U}$ is the same as $\mathcal{T}$ except it does not have the maximal branches of $\mathcal{T}$; more precisely,

1. $\mathcal{M}_0 = \mathcal{P}||\alpha_0$,
2. for every $\alpha < \eta$, $\mathcal{N}_\alpha = \mathcal{M}_\alpha$ and $E_\alpha = F_\alpha$,
3. for every $\alpha < \eta$, $\mathcal{U}_\alpha = \mathcal{T}_\alpha^{sc}$,
4. there are at most finitely many α such that $\mathcal{U}_\alpha \neq \mathcal{T}_\alpha$, and
5. either η is a limit ordinal or the last normal component of $\mathcal{T}_{\eta-1}$ has a limit length (this condition is redundant as $\mathcal{T} \in \mathrm{dom}(\Sigma)$).

If $\mathcal{T}$ and $\mathcal{U}$ are as above then we write $\mathcal{U} = \mathcal{T}^{sc}$ and say that $\mathcal{U}$ is the short component of $\mathcal{T}$.

Conditions (3)-(4) in Definition 3.1.7 ensure that if the relevant stacks are of limit length, we can take the direct limit. We will not be concerned with quasi-limits (cf. [40]) here. The next definition defines the short tree component of Σ.

DEFINITION 3.1.8 (The short tree component of a strategy). Suppose $(\mathcal{P}, \Sigma)$ is a hod-like lses pair or a simple hod-like lses pair such that $\mathcal{P}$ is of lsa type and is

[5] If $\mathcal{T}_{<\alpha}$ is Σ-maximal then because of clause 1 its last normal component cannot be normally continued implying that $\alpha \in R$.

[6] if ν is limit then ι_τ is not defined.

exact. Suppose

$$\mathcal{T} = (\mathcal{M}_\alpha, \mathcal{T}_\alpha, F_\alpha : \alpha < \eta)$$

is a generalized stack on $\mathcal{P}$ such that $\mathcal{T} \in \mathrm{dom}(\Sigma_{\mathrm{ex}})$ and $\mathcal{T}^{sc} \in \mathrm{dom}(\Sigma^{stc})$. Let $b = \Sigma(\mathcal{T})$. We then set $\Sigma^{stc}(\mathcal{U}) = x$ where x is defined as follows:

1. If η is a successor ordinal, $\mathcal{T}_{\eta-1}$ has a last normal component[7] and letting $\mathcal{X}$ be the last normal component of $\mathcal{T}_{\eta-1}$, $\pi_b^{\mathcal{T}}$ is defined and $\pi_b^{\mathcal{T}}(\delta) = \delta(\mathcal{X})$ then $x = \mathrm{m}^+(\mathcal{X})$.
2. Otherwise $x = b$[8].

Thus, $\Sigma^{stc}(\mathcal{T})$ either returns the value of $\Sigma_{\mathrm{ex}}(\mathcal{T})$ or $\mathrm{m}^+(\mathcal{X})$ where $b = \Sigma_{\mathrm{ex}}(\mathcal{T})$. From now on, we will use this notation even when Σ is a partial iteration strategy.

Notice the similarity with the short tree iterability for suitable mice in the context of core model induction or in the context of HOD analysis and Σ^{stc}. If $\mathcal{P}$ is a Σ_1^2-suitable premouse and Σ is fullness preserving iteration strategy for $\mathcal{P}$,[9] Σ^{stc} is just the short tree iterability strategy of $\mathcal{P}$.

3.2. Short tree stacks and short tree strategies

In order to define the *short tree strategy mice*, we will need to introduce the concept of *short tree strategy* that is independent of a particular strategy. We start by defining short-tree-stacks, or just st-stacks. Recall our convention that all stacks are proper (see Remark 2.7.27). We will not take the usual route of first defining putative st-stacks and then defining st-stacks, and leave such matters to the reader. Our goal is to concentrate on the important new property that st-stacks have.

DEFINITION 3.2.1. Suppose $\mathcal{P}$ is a hod-like #-lsa type lses[10]. Set $\delta = \delta^{\mathcal{P}}$. We say that $\mathcal{T}$ is an *st-stack* on $\mathcal{P}$ if

$$\mathcal{T} = ((\mathcal{M}_\alpha)_{\alpha<\eta}, (E_\alpha)_{\alpha<\eta-1}, D, R, (\beta_\alpha, m_\alpha)_{\alpha\in R}, \mathsf{short}, \mathsf{max}, T)$$

and the following conditions hold.

1. R is a closed subset of η and $0 \in R$.
2. $R = \mathsf{short} \cup \mathsf{max}$, $\mathsf{short} \cap \mathsf{max} = \emptyset$ and max is finite.
   Let $\nu = o.t.(R)$ and $(\iota_\tau : \tau < \nu)$ be the increasing enumeration of R. For each $\alpha < \eta$, let τ_α be the largest $\tau < \nu$ such that $\iota_\tau \leq \alpha$ and set $\iota^\alpha = \iota_{\tau_\alpha}$. If

[7] See Definition 2.4.4.

[8] For the reader's convenience we spell out the exact clauses of "Otherwise".

(a) η is a limit ordinal.
(b) η is a successor ordinal and $\mathcal{T}_{\eta-1}$ does not have a last normal component or $\pi_b^{\mathcal{T}}$ is undefined.
(c) η is a successor ordinal, $\mathcal{T}_{\eta-1}$ has a last normal component $\mathcal{X}$, $\pi_b^{\mathcal{T}}$ is defined and $\pi_b^{\mathcal{T}}(\delta) > \delta(\mathcal{X})$.

[9] Here Σ_1^2 and fullness preservation are relative to an AD^+-model.

[10] See Definition 2.7.3.

$\tau+1=\nu$ then set

$$\iota_{\tau+1}=\begin{cases}\eta & \text{if } \eta \text{ is a limit ordinal,}\\ \eta-1 & \text{otherwise.}\end{cases}$$

We say $\tau+1$ is irrelevant if $\tau+1=\nu$ and $\iota_{\tau+1}=\eta$, and if $\tau+1$ is not irrelevant then we say τ is relevant. For $\tau<\nu$ such that $\iota_\tau\in$ short let $I_\tau=[\iota_\tau,\iota_{\tau+1}]$ and otherwise set $I_\tau=[\iota_\tau,\iota_{\tau+1})$. We say that $(\iota_\tau:\tau\le\nu)$ is the ι-sequence of $\mathcal{T}$

3. T is a tree order on $\prod_{\tau<\nu}I_\tau$. Set

$$\mathcal{M}'_\alpha=\begin{cases}\mathcal{M}_\alpha & \text{if } \alpha\notin R\vee(\alpha\in R\wedge\omega\beta_\alpha=\operatorname{ord}(\mathcal{M}_\alpha)),\\ \mathcal{M}_\alpha||(\omega\beta_\alpha,\omega) & \text{if } \alpha\in R\wedge\omega\beta_\alpha<\operatorname{ord}(\mathcal{M}_\alpha).\end{cases}$$

4. $\mathcal{M}_0=\mathcal{P}$.
5. For all $\alpha+1<\eta$, $E_\alpha\in\vec{E}^{\mathcal{M}'_\alpha}$.
6. Normality conditions hold. More precisely, the following conditions hold.
   (a) For all $\alpha+1<\eta$, letting $\beta=T(\alpha+1)$ and $\kappa_\alpha=\operatorname{crit}(E_\alpha)$, then β is the least ordinal $\gamma\ge\tau_\alpha$ such that

$$(\kappa_\alpha^+)^{\mathcal{M}_\alpha|\operatorname{ind}^{\mathcal{M}_\alpha}(E_\alpha)}<\nu(E_\gamma).$$

   (b) For all $\alpha<\beta$ such that $\beta+1<\eta$ and $\iota^\alpha=\iota^\beta$, $\operatorname{ind}^{\mathcal{M}_\alpha}(E_\alpha)<\operatorname{ind}^{\mathcal{M}_\beta}(E_\beta)$.
7. For all $\alpha+1<\eta$,

$$\mathcal{M}_{\alpha+1}=\operatorname{Ult}(\mathcal{M}'_\beta||(\omega\xi_\alpha,k_\alpha),E_\alpha)$$

   where
   (a) $\beta=T(\alpha+1)$,
   (b) $\omega\xi_\alpha\le\operatorname{ord}(\mathcal{M}'_\beta)$ is the largest such that

$$(\kappa_\alpha^+)^{\mathcal{M}_\alpha|\operatorname{ind}^{\mathcal{M}_\alpha}(E_\alpha)}=(\kappa_\alpha^+)^{\mathcal{M}'_\beta|\omega\xi_\alpha},$$

   (c) k_α is the largest such that

$$(\omega\xi_\alpha,k_\alpha)\le l(\mathcal{M}'_\beta)\quad\text{and}\quad\operatorname{crit}(E_\alpha)<\rho_{k_\alpha}(\mathcal{M}'_\beta||(\omega\xi_\alpha,k_\alpha)).$$

8. $D=\{\alpha+1<\eta:\text{letting }\beta=T(\alpha+1),\ (\omega\xi_\alpha,k_\alpha)<l(\mathcal{M}_\beta)\}$.
   Let

$$\pi^{\mathcal{T}}_{\beta,\alpha+1}=\pi_{E_\alpha}^{\mathcal{M}'_\beta||(\omega\xi_\alpha,k_\alpha)}:\mathcal{M}'_\beta||(\omega\xi_\alpha,k_\alpha)\to\mathcal{M}_{\alpha+1}$$

   be the ultrapower map and for $\alpha<\gamma<\eta$ such that $\tau_\alpha=\tau_\gamma$ and $\alpha<_T\gamma<\eta$ let $\pi^{\mathcal{T}}_{\alpha,\gamma}:\mathcal{M}_\alpha\to\mathcal{M}_\gamma$ be the embedding obtained by compositions.[11]
9. Suppose $\lambda<\eta$ is a limit ordinal. Then the following clauses hold.
   (a) Suppose $\lambda\notin R$. Then $D\cap(\iota^\lambda,\lambda)_T$ is finite and letting $\beta\in[\iota^\lambda,\lambda)_T$ be the least such that $D\cap(\beta,\lambda)_T=\emptyset$, $\mathcal{M}_\lambda$ is the direct limit of the system $(\mathcal{M}_\gamma,\pi^{\mathcal{T}}_{\gamma,\gamma'}:\gamma<\gamma',\gamma,\gamma'\in[\beta,\lambda)_T)$ and for $\gamma\in[\beta,\lambda)$, $\pi^{\mathcal{T}}_{\gamma,\lambda}:\mathcal{M}_\gamma\to\mathcal{M}_\lambda$ is the direct limit embedding.

[11] Assuming these embeddings can be composed. $\pi^{\mathcal{T}}_{\alpha,\gamma}$ is defined if and only if $D\cap(\alpha,\gamma]_T=\emptyset$.

(b) Suppose $\lambda \in \mathsf{short}$. Then $\sup(\mathsf{max} \cap \lambda) < \lambda$[12] and setting $\lambda_0 = \sup(\mathsf{max} \cap \lambda)$ and $\lambda_1 = \sup(D \cap \lambda)$, $\mathcal{M}_\lambda$ is the direct limit of

$$(\mathcal{M}_\alpha, \pi_{\alpha,\beta} : \max \lambda_0, \lambda_1 < \alpha < \beta, (\alpha, \beta) \in \mathsf{short}^2 \cap \lambda^2).$$

For each $\tau < \nu$ such $\iota_\tau \in \mathsf{short}$, let $\mathcal{T}^\tau$ be the re-organization of $\mathcal{T}_{[\iota_\tau, \iota_{\tau+1}]}$[13] as a normal iteration tree on $\mathcal{M}_{\iota_\tau}$ and for each $\tau < \nu$ such that $\iota_\tau \in \mathsf{max}$, let $\mathcal{T}^\tau$ be the re-organization of $\mathcal{T}_{[\iota_\tau, \iota_{\tau+1})}$ as a normal iteration tree on $\mathcal{M}_{\iota_\tau}$.

10. For each $\tau < \nu$ such that $\iota_\tau \in \mathsf{max}$, $\mathcal{M}_{\iota_{\tau+1}} = m^+(\mathcal{T}^\tau)$, $\delta^{\mathcal{M}_{\iota_{\tau+1}}} = \delta(\mathcal{T}^\tau)$ and

$$\mathcal{J}_\omega[\mathcal{M}_{\iota_{\tau+1}}] \vDash \text{“}\delta(\mathcal{T}^\tau) \text{ is a Woodin cardinal”}.$$

11. For each $\tau < \nu$ such that $\iota_\tau \in \mathsf{short}$, $\tau + 1$ is relevant and $\pi_{\iota_\tau, \iota_{\tau+1}}$ is defined, $\pi_{\iota_\tau, \iota_{\tau+1}}(\delta^{\mathcal{M}_{\iota_\tau}}) > \delta(\mathcal{T}^\tau)$.
12. For all $\tau < \nu$ such that $\iota_\tau \in \mathsf{max}$, $\pi^{\mathcal{T}^\tau, b}$ is defined[14].
13. For every $\alpha < \eta$, if $\pi_{0,\alpha}$ is defined then $\mathrm{ind}^{\mathcal{M}_\alpha}(E_\alpha) < \pi_{0,\alpha}(\delta)$.
14. If $\tau < \nu$ is such that ι_τ is the least member of max then π_{0,ι_τ} is defined.
15. If $\tau_0 < \tau_1$ are such that $\iota_{\tau_0} \in \mathsf{max}$, $\iota_{\tau_1} \in \mathsf{max}$ and $[\iota_{\tau_0+1}, \iota_{\tau_1}) \cap \mathsf{max} = \emptyset$ then (provided $\tau_0 + 1 < \tau_1$) $\pi_{\iota_{\tau_0+1}, \iota_{\tau_1}}$ is defined and for every $\alpha \in [\iota_{\tau_0+1}, \iota_{\tau_1+1})$, if $\pi_{\iota_{\tau_0+1}, \alpha}$ is defined then $\mathrm{ind}^{\mathcal{M}_\alpha}(E_\alpha) < \pi_{\iota_{\tau_0+1}, \alpha}(\delta^{\mathcal{M}_{\iota_{\tau_0+1}}})$.
16. If $\tau < \nu$ is such that ι_τ is the largest member of max then for every $\alpha \in [\iota_\tau, \eta)$ if $\pi_{\iota_\tau, \alpha}$ is defined then $\mathrm{ind}^{\mathcal{M}_\alpha}(E_\alpha) < \pi_{\iota_\tau, \alpha}(\delta^{\mathcal{M}_{\iota_\tau}})$.

We say $\mathcal{T}$ is (an ordinary) normal st-stack if $R^{\mathcal{T}} = \{0\}$ and $(\beta_0^{\mathcal{T}}, m_0^{\mathcal{T}}) = l(\mathcal{M}_0^{\mathcal{T}})$.

We adopt our proper stacks convention, Remark 2.7.27, and in particular demand that all cutpoints of $\mathcal{T}$ are in $R^{\mathcal{T}}$.

REMARK 3.2.2. $\pi^{\mathcal{T}, b}$ can also be defined for st-stacks. See Definition 2.8.1.

REMARK 3.2.3 (Proper st-stack convention). We again make the convention that st-stacks are proper stacks. Adopting the definition of proper stack to st-stacks is a straightforward matter which we leave to the reader.

We will use superscript $\mathcal{T}$ to denote the objects introduced in Definition 3.2.1 (e.g., $\mathsf{max}^{\mathcal{T}}$ or $\iota_\tau^{\mathcal{T}}$). Also, we write $\mathrm{lh}(\mathcal{T})$ for the ordinal η.

It is now straightforward to define the concept of *generalized st-stacks* on $\mathcal{P}$ following the definition of Definition 2.10.6. These have the form $(\mathcal{M}_\beta, \mathcal{T}_\beta, E_\beta : \beta < \gamma)$ where $\mathcal{T}_\beta$ is an st-stack on $\mathcal{M}_\beta$ and E_β is the un-dropping extender. We leave the details of the definition to the reader. Next we define st-strategy and leave it to the reader to define *generalized st-strategies*.

DEFINITION 3.2.4 (St-strategy). Suppose $\mathcal{P}$ is a hod-like #-lsa type lses[15]. We say that Λ is an *st-strategy* for $\mathcal{P}$ if Λ is a function with the following properties.

[12] This is a consequence of the fact that max is finite.
[13] If $\iota_{\tau+1} = \eta$, we let $\mathcal{T}_{[\iota_\tau, \iota_{\tau+1}]} = \mathcal{T}_{[\iota_\tau, \iota_{\tau+1})}$. Also, see the discussion after Definition 2.4.1.
[14] See Definition 2.8.1.
[15] See Definition 2.7.3.

1. If $x \in \text{dom}(\Lambda)$ then x is an st-stack on $\mathcal{P}$ such that if

$$x =_{\text{def}} \mathcal{T} = ((\mathcal{M}_\alpha)_{\alpha<\eta}, (E_\alpha)_{\alpha<\eta-1}, D, R, (\beta_\alpha)_{\alpha\in R}, \mathsf{short}, \mathsf{max}, T)$$

then η is a limit ordinal.

2. If $\mathcal{T} \in \text{dom}(\Lambda)$,

$$\mathcal{T} = ((\mathcal{M}_\alpha)_{\alpha<\eta}, (F_\alpha)_{\alpha<\eta-1}, D, R, (\beta_\alpha)_{\alpha\in R}, \mathsf{short}, \mathsf{max}, T)$$

and $\Lambda(\mathcal{T}) = x$ then $\mathcal{T}^\frown\{x\}$ is an st-stack on $\mathcal{P}$. More precisely the following conditions hold.
   (a) If $o.t.(R)$ is a limit ordinal then letting $\alpha \in R$ be such that $\mathsf{max} \bigcup D \subseteq \alpha$, x is the direct limit of $(\mathcal{M}_\beta, \pi_{\beta,\gamma} : \beta < \gamma, (\beta, \gamma) \in (R-\alpha)^2)$.
   (b) If $\tau + 1 = o.t.(R)$ then x is either a branch of $\mathcal{T}_{\iota_\tau}$ or $x = \mathrm{m}^+(\mathcal{T}_{\iota_\tau})$[16].
3. If $\mathcal{T} \in \text{dom}(\Lambda)$ then $\mathcal{T}$ is according to Λ, i.e., for every limit ordinal $\eta' < \eta$, $\mathcal{T}_{<\eta'} \in \text{dom}(\Lambda)$ and $\mathcal{T}_{\leq\eta'} = \mathcal{T}^{\eta'\frown}\{x\}$ where $x = \Lambda(\mathcal{T}_{<\eta'})$.

We say that $\mathcal{T}$ is a (κ, λ)-st-stack on $\mathcal{P}$ if $\mathcal{T}$ is an st-stack on $\mathcal{P}$ such that $o.t.(R^\mathcal{T}) < \kappa$ and for every $\tau < o.t.(R^\mathcal{T})$, $\text{lh}(\mathcal{T}^\tau) < \lambda$. As we said above, we could define the concept of putative st-stack similarly to Definition 2.4.1. As doing this is straightforward, we leave it to the reader. Putative essentially means that all models in the stack except possibly the last one are well-founded.

DEFINITION 3.2.5. Suppose $\mathcal{P}$ is a hod-like lsa type φ-indexed lses. We say Λ is a (κ, λ)-*st-strategy* for $\mathcal{P}$ if the following clauses hold.

1. Λ is an st-strategy.
2. If $\mathcal{T}$ is a putative (κ, λ)-st-stack that is according to Λ then $\mathcal{T}$ is a (κ, λ)-stack.
3. If $\mathcal{T}$ is a (κ, λ)-st-stack that is according to Λ such that
   (a) $\text{lh}(\mathcal{T})$ is a limit ordinal and
   (b) if $o.t.(R^\mathcal{T}) = \tau + 1$ then $\text{lh}(\mathcal{T}^\tau) + 1 < \lambda$,
   then $\mathcal{T} \in \text{dom}(\Lambda)$.

As we said above, we can then define generalized (κ, λ, ν)-st-strategy which acts on generalized st-stacks. The definition of this notion is rather straightforward.

Suppose now $\mathcal{P}$ and Λ are as in Definition 3.2.5. We let $b(\Lambda)$ be the set of all $\mathcal{T} \in \text{dom}(\Lambda)$ such that $\mathcal{T}$ has a last normal component of limit length and $\Lambda(\mathcal{T})$ is a cofinal wellfounded branch of $\mathcal{T}$. Let $m(\Lambda) = \text{dom}(\Lambda) - b(\Lambda)$. We call $m(\Lambda)$ the *model component of* Λ. Given $\mathcal{U} \in \text{dom}(\Lambda)$ such that the last component of $\mathcal{U}$ has a limit length, we let

$$\mathcal{M}(\Lambda, \mathcal{U}) = \begin{cases} \mathcal{M}_b^\mathcal{U} & \text{if } \Lambda(\mathcal{U}) = b, \\ \Lambda(\mathcal{U}) & \text{otherwise.} \end{cases}$$

If Λ is an st-strategy for $\mathcal{P}$ and $\mathcal{T}$ is a stack on $\mathcal{P}$ according to Λ with last model $\mathcal{N}$ then we let $\Lambda_{\mathcal{N},\mathcal{T}}$ be the short tree strategy of $\mathcal{N}$ induced by Λ, i.e., for every $\mathcal{U}$

[16] See Definition 3.1.4.

on $\mathcal{N}$, $\Lambda_{\mathcal{N},\mathcal{T}}(\mathcal{U}) = \Lambda(\mathcal{T}^\frown\mathcal{U})$. Here $\mathcal{T}^\frown\mathcal{U}$ is an st-stack defined in a natural way so that the normal components of $\mathcal{T}$ and $\mathcal{U}$ are the normal components of $\mathcal{T}^\frown\mathcal{U}$.

REMARK 3.2.6. In many situations, it is expected that finding (κ, λ)-st-strategies must be easy. For example, whenever $\mathcal{T}$ is normal iteration tree of length ω such that $\mathcal{J}_\omega(\mathrm{m}^+(\mathcal{T})) \vDash$ "$\delta(\mathcal{T})$ is a Woodin", we can set $\Lambda(\mathcal{T}) = \mathrm{m}^+(\mathcal{T})$. Thus, instead of working hard to define the correct branch, we declare success by setting $\Lambda(\mathcal{T}) = \mathrm{m}^+(\mathcal{T})$. However, we will be interested in st-strategies that have certain fullness preservation properties. For instance, suppose $\mathcal{M}$ is just a suitable mouse in the sense of $L(\mathbb{R})$. If we now demand that Λ must have the property that whenever $\mathcal{T} \in \mathrm{dom}(\Lambda)$ is such that $\mathcal{Q}(\mathcal{T})$ exists then $\Lambda(\mathcal{T})$ must be a branch b with the property that $\mathcal{Q}(b, \mathcal{T}) = \mathcal{Q}(\mathcal{T})$ then Λ would be a rather complex object. We will have that $\Lambda(\mathcal{T})$ is a model only in the case when $\mathcal{T}$ is a maximal iteration tree. In this case, Λ is in fact a "short tree iterability strategy" in the sense of $L(\mathbb{R})$. Such strategies are difficult to construct, and in our current situation, we will be interested in a notion of fullness preservation with respect to a much more complicated pointclass than $(\Sigma^2_1)^{L(\mathbb{R})}$.

3.3. Hull and branch condensation for short tree strategy

The goal of this section is to introduce *hull condensation* for st-strategies. Hull condensation for iteration strategies was introduced in Definition 1.31 of [29]. It is an important property that is used to show that when doing hod pair constructions no discrepancies arise due to the coring down process. Thus if $\mathcal{T}$ is according to a strategy with hull condensation and $\mathcal{U}$ is a *hull* of $\mathcal{T}$ (cf. Definition 3.3.4) then it is according to the strategy.

The difference between strategies and st-strategies is that st-strategies have a model component, and this difference causes some complications when trying to outright generalize hull condensation: such a direct generalization leads to a very strong property. Our definition is based on our indexing scheme Definition 3.6.4. In short tree strategy mice, we only index branches of a certain kinds of iterations, and we need to apply hull condensation to these types of iterations. We start by introducing these iterations.

First we define the *universally short normal trees* which are essentially those normal iteration trees that are short with respect to any iteration strategy.

DEFINITION 3.3.1. We say that $\mathcal{T}$ is a *normal stack* on M if letting

$$\mathcal{T} = ((\mathcal{M}_\alpha)_{\alpha<\eta}, (E_\alpha)_{\alpha<\eta-1}, D, R, (\beta_\alpha, m_\alpha)_{\alpha\in R}, T),$$

for all $\alpha < \eta$, $\beta_\alpha = \mathrm{ord}(\mathcal{M}_\alpha)$, $m_\alpha = k(\mathcal{M}_\alpha)$ and setting

$$\mathcal{U} = ((\mathcal{M}_\alpha)_{\alpha<\eta}, (E_\alpha)_{\alpha<\eta-1}, D, T),$$

$\mathcal{U}$ is a normal iteration tree[17]. Given an st-stack

$$\mathcal{T} = ((\mathcal{M}_\alpha)_{\alpha<\eta}, (E_\alpha)_{\alpha<\eta-1}, D, R, (\beta_\alpha, m_\alpha)_{\alpha\in R}, \mathsf{short}, \mathsf{max}, T),$$

we say $\mathcal{T}$ is a *normal stack* if

1. $\mathsf{max} = \emptyset$ and letting

$$\mathcal{U} = ((\mathcal{M}_\alpha)_{\alpha<\eta}, (E_\alpha)_{\alpha<\eta-1}, D, T),$$

   $\mathcal{U}$ is a normal iteration tree, or
2. $|\mathsf{max}| = 1$ and letting α be the unique element of max, $\mathsf{next}_\alpha^{\mathcal{T}} = \mathrm{lh}(\mathcal{T})$ and

$$\mathcal{U} = ((\mathcal{M}_\alpha)_{\alpha<\eta-1}, (E_\alpha)_{\alpha<\eta-1}, D, T),$$

   $\mathcal{U}$ is a normal iteration tree.

DEFINITION 3.3.2 (Universally short stacks). Suppose $\mathcal{P}$ is a hod-like #-lsa type lses and $\mathcal{T}$ is a normal stack on $\mathcal{P}$ (see Definition 3.3.1) such that $\mathrm{lh}(\mathcal{T})$ is a limit ordinal. We say

$$\mathcal{T} = ((\mathcal{M}_\alpha)_{\alpha<\eta}, (E_\alpha)_{\alpha<\eta-1}, D, R, (\beta_\alpha, m_\alpha)_{\alpha\in R}, T),$$

is *universally short* (uvs) if one of the following holds:

1. $\pi^{\mathcal{T},b}$ is undefined.

Suppose next $\pi^{\mathcal{T},b}$ is defined and let $\alpha < \mathrm{lh}(\mathcal{T})$ be the least such that $\pi_{0,\alpha}$ is defined and $\mathcal{M}_\alpha^b = \pi^{\mathcal{T},b}(\mathcal{P}^b)$. It then follows that $\mathcal{T}_{\geq\alpha}$ is a stack on $\mathcal{M}_\alpha$ that is above $\mathrm{ord}(\mathcal{M}_\alpha^b)$[18].

2. R is cofinal in $\mathrm{lh}(\mathcal{T})$.
3. $\mathcal{T}$ has a fatal drop (see Definition 2.6.8).
4. For some $\beta \in R - (\alpha+1)$, $D \cap (\alpha, \beta]_T \neq \emptyset$.
5. For some $\beta \in R - (\alpha+1)$ and some $\eta \in (\delta^{\mathcal{S}^b}, \delta^{\mathcal{M}_\beta})$, $\mathcal{T}_{\geq\beta}$ is a normal stack on $\mathcal{M}_\beta$ that is below η.
6. There is $\mathcal{Q} \trianglelefteq \mathrm{m}(\mathcal{T})^\#$ such that $\mathcal{Q} \vDash$ "$\delta(\mathcal{T})$ is a Woodin cardinal" and $\mathcal{J}_\omega(\mathcal{Q}) \vDash$ "$\delta(\mathcal{T})$ is not a Woodin cardinal".

If $\mathcal{T}$ is not uvs then we say that $\mathcal{T}$ is *non-universally short* (nuvs).

DEFINITION 3.3.3 (Indexable stack). Suppose $\mathcal{P}$ is a hod-like #-lsa type lses[19]. We say that an st-stack[20]

$$\mathcal{T} = ((\mathcal{M}_\alpha)_{\alpha<\eta}, (E_\alpha)_{\alpha<\eta-1}, D, R, (\beta_\alpha, m_\alpha)_{\alpha\in R}, \mathsf{short}, \mathsf{max}, T)$$

is an *indexable stack* on $\mathcal{P}$ if one of the following clauses hold:

1. $\mathsf{max} = \emptyset$ and there is $\alpha \in R^{\mathcal{T}}$ such that $\pi^{\mathcal{T}_{\leq\alpha},b}$ is defined and $\mathcal{T}_{\geq\alpha}$ is based on $\pi^{\mathcal{T}_{\leq\alpha},b}(\mathcal{P}^b)$.

---

[17] Recall our general convention that all cutpoints of a stack a $\mathcal{W}$ belong to $R^{\mathcal{W}}$.

[18] The condition that $\pi_{0,\alpha}$ is defined follows from the equality $\mathcal{S}^b = \pi^{\mathcal{T},b}(\mathcal{P}^b)$.

[19] See Definition 2.7.3.

[20] See Definition 3.2.1.

2. $|\mathsf{max}| = 1$, $\mathcal{T}$ is a normal stack[21] and if α is the unique element of max then $\pi^{\mathcal{T}}_{0,\alpha}$ is defined and $\mathsf{next}^{\mathcal{T}}(\alpha) = \mathrm{lh}(\mathcal{T})$[22].

Below and elsewhere we will use the notation $\mathcal{T} = (\mathcal{P}_0, \mathcal{T}_0, \mathcal{P}_1, \mathcal{T}_1)$ to denote indexable stacks. Here $\mathcal{T}_0 = \mathcal{T}_{\leq\alpha}$ where α is either as in clause 1 or 2 and $\mathcal{T}_1 = \mathcal{T}_{\geq\alpha}$. We will say that the indexable stack is *ordinary* if $\mathsf{max}^{\mathcal{T}} = \emptyset$.

The iterations that we will index in short tree strategy mice are finite st-stacks of length 2. We define hull condensation for such stacks.

DEFINITION 3.3.4 (Hull of a stack). Suppose $\mathcal{M}$ and $\mathcal{M}'$ are hod-like lses and $\mathcal{T}$ and $\mathcal{T}'$ are stacks on $\mathcal{M}$ and $\mathcal{M}'$ respectively. Set

$$\mathcal{T} = ((\mathcal{M}_\alpha)_{\alpha<\eta}, (E_\alpha)_{\alpha<\eta-1}, D, R, (\beta_\alpha, m_\alpha)_{\alpha\in R}, T),$$
$$\mathcal{T}' = ((\mathcal{M}'_\alpha)_{\alpha<\eta}, (E'_\alpha)_{\alpha<\eta-1}, D', R', (\beta'_\alpha, m_\alpha)_{\alpha\in R'}, T').$$

Let $(\iota_\beta : \beta \leq o.t.(R))$ and $(\iota'_\gamma : \gamma \leq o.t.(R'))$ be the ι-sequences of $\mathcal{T}$ and $\mathcal{T}'$ respectively (see Definition 3.2.1). Let $i_{\alpha,\beta} = \pi^{\mathcal{T}}_{\alpha,\beta}$ and $i'_{\alpha,\beta} = \pi^{\mathcal{T}'}_{\alpha,\beta}$ provided the aforementioned embeddings exist.

We say $(\mathcal{M}', \mathcal{T}')$ is a *hull* of $(\mathcal{M}, \mathcal{T})$ if there is a tuple

$$(\sigma, (\tau_\alpha)_{\alpha<\mathrm{lh}(\mathcal{T}')})$$

such that the following clauses hold.

1. $\sigma : \mathrm{lh}(\mathcal{T}') \to \mathrm{lh}(\mathcal{T})$ is an injective map that preserves the tree order and is such that $\sigma[R'] \subseteq R$ and $\sigma(0) = 0$.
2. For all α, β such that $\alpha + \beta < \mathrm{lh}(\mathcal{T}')$, $\sigma(\alpha+\beta) = \sigma(\alpha) + \sigma(\beta)$.
3. For every $\beta < o.t.(R')$, $\sigma(\iota'_{\beta+1}) = \iota_{\sigma(\beta)+1}$.
4. For every $\alpha < \mathrm{lh}(\mathcal{T}')$, $\tau_\alpha : \mathcal{M}'_\alpha \to_{\Sigma_1} \mathcal{M}_{\sigma(\alpha)}$ and $E_{\sigma(\alpha)} = \tau_\alpha(E'_\alpha)$.
5. For all $\alpha < \beta < \mathrm{lh}(\mathcal{T}')$, $[\alpha, \beta]_{T'} \cap D' = \emptyset \iff [\sigma(\alpha), \sigma(\beta)]_T \cap D = \emptyset$.
6. For every $\alpha < \beta < \mathrm{lh}(\mathcal{T}')$, if $\sup(R' \cap (\alpha+1)) = \sup(R' \cap \beta)$ then
$$\tau_\alpha \restriction \mathrm{lh}(E'_\alpha) + 1 = \tau_\beta \restriction \mathrm{lh}(E'_\alpha) + 1.$$
7. For every $\alpha < \beta < \mathrm{lh}(\mathcal{T}')$ such that $\alpha \leq_{T'} \beta$ and $(\alpha, \beta]_{T'} \cap R' = \emptyset$,
$$\tau_\beta \circ i'_{\alpha,\beta} = i_{\sigma(\alpha),\sigma(\beta)} \circ \tau_\alpha.$$
8. For every $\alpha + 1 < \mathrm{lh}(\mathcal{T}')$, if $\beta = T'(\alpha+1)$ then $\sigma(\beta) = T(\sigma(\alpha)+1)$ and
$$\tau_{\alpha+1}([a, f]_{E'_\alpha}) = [\tau_\alpha(a), \tau_\beta(f)]_{E_{\sigma(\alpha)}}.$$

We say $(\sigma, (\tau_\alpha)_{\alpha<\mathrm{lh}(\mathcal{T}')})$ *witnesses* that $(\mathcal{M}', \mathcal{T}')$ is a hull of $(\mathcal{M}, \mathcal{T})$.

If $\mathcal{M} = \mathcal{M}'$ then we say that $(\mathcal{M}, \mathcal{T}')$ is a *hull* of $(\mathcal{M}, \mathcal{T})$ if there is a tuple $(\sigma, (\tau_\alpha)_{\alpha<\mathrm{lh}(\mathcal{T}')})$ witnessing that $(\mathcal{M}, \mathcal{T}')$ is a hull of $(\mathcal{M}, \mathcal{T})$ and such that $\tau_0 = id$.

---

[21] See Definition 3.3.1.

[22] It follows that $\mathcal{T}_{\geq\alpha}$ is above $\pi^{\mathcal{T}}_{0,\alpha}(\delta^{\mathcal{P}^b})$. See also Notation 2.4.4.

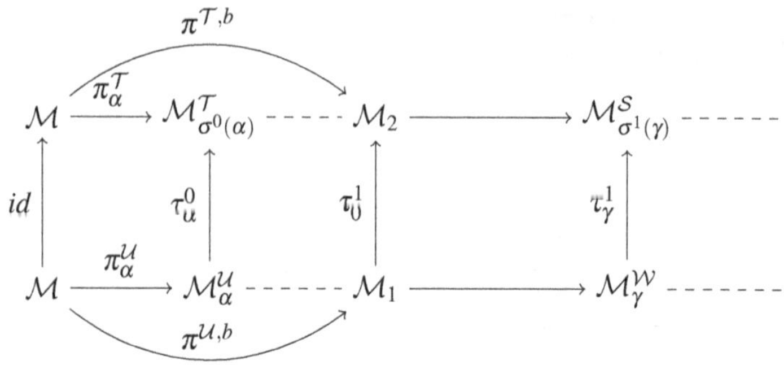

FIGURE 3. Hull of a stack of length 2. $(\mathcal{M},\mathcal{U},\mathcal{M}_1,\mathcal{W})$ is a hull of $(\mathcal{M},\mathcal{T},\mathcal{M}_2,\mathcal{S})$.

Both in the case $\mathcal{M}=\mathcal{M}'$ and $\mathcal{M}\neq\mathcal{M}'$, it is not ambiguous to simply say that $\mathcal{T}'$ is a hull of $\mathcal{T}$ to mean that $(\mathcal{M}',\mathcal{T}')$ is a hull of $(\mathcal{M},\mathcal{T})$, and so we will use this terminology[23].

DEFINITION 3.3.5 (Hull of an indexable stack). (See Figure 3.) Suppose $\mathcal{M}$ is a hod-like #-lsa type lses and

$$u=(\mathcal{M},\mathcal{U},\mathcal{M}_1,\mathcal{W}),$$
$$t=(\mathcal{M},\mathcal{T},\mathcal{M}_2,\mathcal{S})$$

are two indexable stacks. We say $(\mathcal{M},u)$ is a *hull* of $(\mathcal{M},t)$ if either

1. both u and t are ordinary (see Definition 3.3.3) and $(\mathcal{M},u)$ is a hull of $(\mathcal{M},t)$ (in the sense of Definition 3.3.4) or
2. both u and t are not ordinary, and there are two tuples $(\sigma^0,(\tau^0_{\alpha})_{\alpha<\mathrm{lh}(\mathcal{U})})$ and $(\sigma^1,(\tau^1_{\alpha})_{\alpha<\mathrm{lh}(\mathcal{W})})$ such that the following holds.
   (a) $(\sigma^0,(\tau^0_{\alpha})_{\alpha<\mathrm{lh}(\mathcal{U})})$ witnesses that $(\mathcal{M},\mathcal{U})$ is a hull of $(\mathcal{M},\mathcal{T})$.
   (b) $(\sigma^1,(\tau^1_{\alpha})_{\alpha<\mathrm{lh}(\mathcal{W})})$ witnesses that $(\mathcal{M}_1,\mathcal{W})$ is a hull of $(\mathcal{M}_2,\mathcal{S})$.
   (c) $\tau^1_0\restriction(\mathcal{M}^b_1)\circ\pi^{\mathcal{U},b}=\pi^{\mathcal{T},b}$.

To finally define hull condensation for short tree strategy, we need to introduce a few more definitions. Suppose $(\mathcal{P},\Sigma)$ such that $\mathcal{P}$ is a hod-like #-lsa type lses and Σ is a st-strategy for $\mathcal{P}$. First we introduce two sorts of iterates of $(\mathcal{P},\Sigma)$, $I^b(\mathcal{P},\Sigma)$ and $I(\mathcal{P},\Sigma)$.

NOTATION 3.3.6. Suppose $\mathcal{P}$ is a hod-like #-lsa type lses[24] and Σ is a st-strategy[25] for $\mathcal{P}$. We let $\max(\mathcal{P},\Sigma)$ be the set of Σ-maximal iterations. More precisely, $\max(\mathcal{P},\Sigma)$ consists of pairs $(\mathcal{T},\mathcal{Q})$ such that $\mathcal{T}\in m(\Sigma)$ and $\mathcal{Q}=\mathrm{m}^+(\mathcal{T})$.

[23]Notice that in the case $\mathcal{M}=\mathcal{M}'$, we must have that $\tau_0=id$.
[24]See Definition 2.7.3.
[25]See Definition 3.2.5.

In the following definition, we recycle the notations used in Definition 2.7.21. The difference here is that Σ is the short-tree strategy.

DEFINITION 3.3.7 ($I^b(\mathcal{P},\Sigma)$ and $I(\mathcal{P},\Sigma)$). Suppose $(\mathcal{P},\Sigma)$ is a pair such that $\mathcal{P}$ is a hod-like #-lsa type lses and Σ is an st-strategy for $\mathcal{P}$. We then let

$$I^b(\mathcal{P},\Sigma) = \{(\mathcal{T},\mathcal{Q}) : \mathcal{T} \text{ is according to } \Sigma,\ \mathcal{Q} \text{ is the last model of } \mathcal{T} \text{ and } \pi^{\mathcal{T},b} \text{ exists}\},$$

$$I(\mathcal{P},\Sigma) = \{(\mathcal{T},\mathcal{Q}) : \text{either } (\mathcal{T},\mathcal{Q}) \in \max(\mathcal{P},\Sigma) \text{ or for some } \beta \in \max(\mathcal{T}),\ \pi^{\mathcal{T}_{\geq\beta}} \text{ exists}\}.$$

NOTATION 3.3.8. We let HC be the set of all hereditarily countable sets. In Definition 4.1.1, we fix a coding of elements of HC by reals. This coding then induces a coding of elements of $\bigcup_{n\in\omega}\wp(\mathsf{HC}^n)$ by sets of reals. Let Code be the coding function introduced in Definition 4.1.1. Thus for $A \subseteq \mathsf{HC}^n$, $\mathsf{Code}(A)$ is the set of reals that codes A.

DEFINITION 3.3.9. Suppose $(\mathcal{P},\Sigma)$ is a pair such that $\mathcal{P}$ is a hod-like #-lsa type lses and Σ is an st-strategy for $\mathcal{P}$. We then let

$$B(\mathcal{P},\Sigma) = \{(\mathcal{T},\mathcal{Q}) : \exists\mathcal{R}((\mathcal{T},\mathcal{R}) \in I^b(\mathcal{P},\Sigma) \wedge \mathcal{Q} \trianglelefteq_{hod} \mathcal{R}^b)\},$$

and

$$\Gamma^b(\mathcal{P},\Sigma) = \{A \subseteq \mathbb{R} : \exists(\mathcal{T},\mathcal{Q}) \in B(\mathcal{P},\Sigma)(A \leq_w \mathsf{Code}(\Sigma_{\mathcal{Q},\mathcal{T}}))\}.$$

DEFINITION 3.3.10 (Hull condensation). Suppose $\mathcal{P}$ is a hod-like #-lsa type lses and Σ is a st-strategy for $\mathcal{P}$. We say Σ has *hull condensation* if the following clauses hold.

1. For all $(\mathcal{T},\mathcal{Q}) \in B(\mathcal{P},\Sigma)$, $\Sigma_{\mathcal{Q},\mathcal{T}}$ has hull condensation, and
2. Whenever $(\mathcal{T},\mathcal{Q}) \in I(\mathcal{P},\Sigma)$, $u = (\mathcal{Q},\mathcal{U},\mathcal{Q}_1,\mathcal{W})$ and $t = (\mathcal{Q},\mathcal{T}',\mathcal{Q}_2,\mathcal{W}')$ are two indexable stacks on $\mathcal{Q}$ such that t is according to $\Sigma_{\mathcal{Q},\mathcal{T}}$ and $(\mathcal{Q},u)$ is a hull of $(\mathcal{Q},t)$ then u is according to $\Sigma_{\mathcal{Q},\mathcal{T}}$.

Next we introduce branch condensation for short tree strategies. We will need this notion in the definition of hod mice (see Definition 3.10.2).

DEFINITION 3.3.11 (Branch condensation for st-strategies; see Figure 4). Suppose $(\mathcal{P},\Sigma)$ is such that $\mathcal{P}$ is a hod-like #-lsa type lses and Σ is a st-strategy for $\mathcal{P}$. We say Σ has *branch condensation* if whenever $(\mathcal{T},\mathcal{Q},\mathcal{U},\mathcal{R},\tau,\mathcal{S},c,\alpha,\beta)$ is such that

1. $(\mathcal{T},\mathcal{Q}),(\mathcal{U},\mathcal{R}) \in I^b(\mathcal{P},\Sigma)$,
2. $\alpha < \lambda^{\mathcal{R}^b}$ and $\delta^{\mathcal{R}(\alpha+1)}$ is a Woodin cardinal of $\mathcal{R}$[26],
3. $\mathcal{S}$ is a normal iteration tree of limit length according to $\Sigma_{\mathcal{R}^b,\mathcal{U}}$ that is based on $\mathcal{R}(\alpha+1)$ and is above $\delta^{\mathcal{R}}_{\alpha}$,

[26] See Notation 2.7.14 for the definition of $\mathcal{R}(\tau)$.

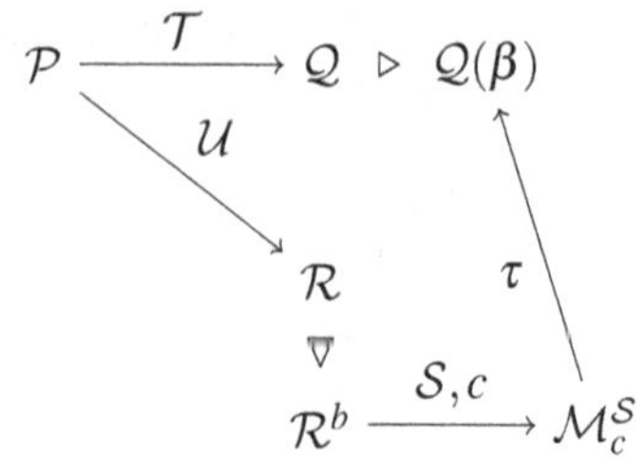

FIGURE 4. Branch condensation for short tree strategies. Notations are as in Definition 3.3.11. In the above, $\pi^{\mathcal{T},b} = \pi \circ \pi_c^{\mathcal{S}} \circ \pi^{\mathcal{U},b}$.

4. c is a branch of $\mathcal{S}$ such that $\pi_c^{\mathcal{S}}$ exists, and
5. $\tau : \mathcal{M}_c^{\mathcal{S}} \to \mathcal{Q}(\beta)$ and $\pi^{\mathcal{T},b} = \tau \circ \pi_c^{\mathcal{S}} \circ \pi^{\mathcal{U},b}$

then $c = \Sigma_{\mathcal{R},\mathcal{U}}(\mathcal{S})$.

3.4. St-type pairs

Suppose $\mathcal{P}$ is a hod-like #-lsa type lses[27] and suppose Λ is an st-strategy for $\mathcal{P}$. We would like to introduce the notion of short tree premice and in particular, Λ-premice. The main technical problem is that we do not have a reasonable notion of condensation for st-strategies. In particular, if $\Lambda = \Sigma^{stc}$ for some strategy Σ, then it may well be that there is a Σ-maximal iteration tree $\mathcal{T}$ on $\mathcal{P}$ such there is a Σ-short hull $\mathcal{U}$ of $\mathcal{T}$.

The above scenario is the main difficulty with defining short tree strategy mice. We have to find a particular indexing of short tree strategies, or rather carefully skip over "bad trees", in a way that when $\mathcal{T}$ above is "cored down" to $\mathcal{U}$ above then our indexing is still preserved. In particular, the branch of $\mathcal{T}$ cannot be added too early. The idea is to wait until the branch of $\mathcal{T}$ or rather its correct $\mathcal{Q}$-structure is *certified*. Before we define short tree hybrids, however, we have to make a few definitions that will be useful to us in the future.

We will only consider st-strategies Λ with the property that whenever $\mathcal{T} \in \mathrm{dom}(\Lambda)$ is uvs then $\Lambda(\mathcal{T})$ is a branch.

DEFINITION 3.4.1 (Faithful short tree strategy). Suppose $\mathcal{P}$ is a hod-like #-lsa type lses and Λ is a (κ, λ, η)-st-strategy for $\mathcal{P}$. We say Λ is a *faithful* (κ, λ, η)*-st-strategy* if whenever $\mathcal{T} \in \mathrm{dom}(\Lambda)$ is uvs, $\mathcal{T} \in b(\Lambda)$.

DEFINITION 3.4.2 (St-type pair). We say $(\mathcal{P}, \Lambda)$ is a *hod-like st-type pair* if

1. $\mathcal{P}$ is a hod-like #-lsa type lses,

[27] See Definition 2.7.3.

2. Λ is a faithful $(\omega_1,\omega_1,\omega_1)$-st-strategy,
3. if $\mathcal{Q}$ is a Λ-iterate of $\mathcal{P}$ via $\mathcal{T}$ and $\mathcal{R}\in Y^{\mathcal{Q}}$ then $\Sigma^{\mathcal{R}}\subseteq\Lambda_{\mathcal{R},\mathcal{T}}$,[28]
4. Λ has hull condensation[29].

Similarly we can define *simple* hod-like st type pairs by demanding that Λ is a faithful (ω_1,ω_1)-strategy and that clause 3 above holds.

3.5. $(\mathcal{P},\Sigma)$-hod pair construction

Suppose that $(\mathcal{P},\Sigma)$ is a hod-like st-type pair. Below we describe a fully backgrounded construction that, if successful, constructs a Σ-iterate of $\mathcal{P}$. To learn more about such backgrounded constructions the reader may consult [23, Chapter 11] and also various papers of Schlutzenberg and Schindler-Steel-Zeman that deal with certain fine structural issues present in [23] (for example, [59, Chapter 2.2, Definition 2.4] and the discussion after it, and also [45] and [43]). We say a (κ,λ)-extender E coheres Σ if $\mathcal{P}\in V_\kappa$, $V_\lambda\subseteq\mathrm{Ult}(V,E)$ and $\pi_E(\Sigma)\restriction V_\lambda=\Sigma\restriction V_\lambda$.

In this manuscript, our goal is to deal with novel issues arising from the theory of short tree strategy mice, such as developing an indexing scheme for short tree strategies, proving a comparison theorem for lsa small hod pairs and obtain core model induction applications at the level of LSA, to list a few. We don not have space to also carefully develop the theory of fully backgrounded constructions, but all issues that arise have been handled in literature. For example, to deal with issues arising from our mixing indexing we refer the reader to Schlutzenberg's [47] and to deal with issues regarding inheriting large cardinals we refer the reader to [23, Chapters 9-12] and to [46].

Unlike in [3] and [23], and other similar places in literature where the convergence of the backgrounded constructions is established, here we will not be concerned with iterability issues of the backgrounded constructions and just simply assume that such constructions converge provided the background universe is iterable. Our assumption is justified by the results of [23, Chapter 12]. The consequence of our assumption is that in clause (3) below we simply take the core rather than the dropdown sequence. See Definition 2.2.3 for the definition of core.

DEFINITION 3.5.1 ($(\mathcal{P},\Sigma)$-coherent fully backgrounded constructions). Suppose κ is an inaccessible cardinal and $(\mathcal{P},\Sigma)$ is a hod-like st-type pair such that Σ is a (κ,κ,κ)-st-strategy. Then for $\eta\leq\kappa$, we say $((\mathcal{M}_\gamma,\mathcal{N}_\gamma:\gamma\leq\eta),(F_\gamma:\gamma<\eta),(\mathcal{T}_\gamma:\gamma\leq\eta))$ is the output of the $(\mathcal{P},\Sigma)$-*coherent fully backgrounded construction* of V_κ if the following holds.

1. $\mathcal{M}_0=\emptyset$.

[28] This clause asserts that the internal strategy of $\mathcal{R}$ agrees with $\Lambda_{\mathcal{R},\mathcal{T}}$.

[29] See Definition 3.3.10.

2. $\mathcal{M}_\gamma$ is a hod-like lses such that there is a tree $\mathcal{T}_\gamma$[30] on $\mathcal{P}$ according to Σ such that either
   (a) $\mathcal{T}_\gamma$ has a last model $\mathcal{M}$ such that if $\xi = \mathrm{ord}(\mathcal{M}_\gamma)$ then $\mathcal{M}_\gamma|\xi = \mathcal{M}|\xi$ or
   (b) $\mathcal{M}_\gamma = \mathrm{m}(\mathcal{T}_\gamma)$.
3. Suppose $\gamma \leq \eta$ is such that either $\mathcal{T}_\gamma$ has a last model or $\mathcal{T}_\gamma \in b(\Sigma)$. Let $\mathcal{M}$ be the last model of $\mathcal{T}_\gamma$ if it exists and otherwise, setting $b = \Sigma(\mathcal{T}_\gamma)$, let $\mathcal{M} = \mathcal{M}_b^{\mathcal{T}_\gamma}$. Let $\xi = \mathrm{ord}(\mathcal{M}_\gamma)$ and suppose $\mathcal{M}_\gamma = \mathcal{J}_\xi^{\vec{E},f}$. Then the following statements hold.
   (a) If $\mathcal{M}_\gamma = \mathcal{M}$ then $\gamma = \eta$.
   (b) If $\mathcal{M}_\gamma$ is active and $\mathcal{M}_\gamma \neq \mathcal{M}||\xi$ then $\gamma = \eta$.
   (c) If $\mathcal{M}_\gamma$ is active and $\mathcal{M}_\gamma = \mathcal{M}||\xi$ then $\mathcal{N}_{\gamma+1} = \mathcal{J}_\omega[\mathcal{M}_\gamma]$ and $\mathcal{M}_{\gamma+1} = \mathsf{core}(\mathcal{N}_\gamma)$.
   (d) Suppose $\mathcal{M}_\gamma$ is passive and $\mathcal{M}_\gamma \triangleleft \mathcal{M}$. Suppose there is no pair (F^*, F) and an ordinal $\zeta < \xi$ such that $F^* \in V_\kappa$ is an extender over V that coheres Σ, F is an extender over $\mathcal{M}_\gamma$, $V_{\zeta+\omega} \subseteq \mathrm{Ult}(V, F^*)$ and
   $$F = F^* \cap ([\zeta]^\omega \times \mathcal{J}_\xi^{\vec{E},f})$$
   such that $(\mathcal{J}_\xi^{\vec{E},f}, \in, \vec{E}, f, \tilde{F})$ is a hod-like lses[31]. Then $\mathcal{N}_\gamma = \mathcal{J}_\omega(\mathcal{M}_\gamma)$ and $\mathcal{M}_{\gamma+1} = \mathsf{core}(\mathcal{N}_\gamma)$.
   (e) Again suppose $\mathcal{M}_\gamma$ is passive and $\mathcal{M}_\gamma \triangleleft \mathcal{M}$ but there is a pair (F^*, F) and an ordinal ζ satisfying the above conditions. Then if $F \in \vec{E}^{\mathcal{M}}$ then we let
   $$\mathcal{N}_\gamma = (\mathcal{J}_\xi^{\vec{E},f}, \in, \vec{E}, f, \tilde{F}) \quad \text{and} \quad \mathcal{M}_{\gamma+1} = \mathsf{core}(\mathcal{N}_\gamma).$$
   (f) Again suppose $\mathcal{M}_\gamma$ is passive, $\mathcal{M}_\gamma \triangleleft \mathcal{M}$ and that $\mathcal{M}||\xi$ is an active $\mathcal{J}$-structure such that its last predicate codes a set A that is not an extender. Let then $\mathcal{N}_\gamma = (\mathcal{M}_\gamma, A, \in)$[32] and $\mathcal{M}_{\gamma+1} = \mathsf{core}(\mathcal{N}_\gamma)$.
4. Suppose $\gamma \leq \eta$ is such that $\mathcal{T}_\gamma$ is of limit length and $\mathcal{T}_\gamma \notin b(\Sigma)$. Then $\gamma = \eta$.

REMARK 3.5.2. Notice that the constructions introduced in Definition 3.5.1 can be carried out even when Σ is a partial strategy. Thus, for example, we may say that "$((\mathcal{M}_\gamma, \mathcal{N}_\gamma : \gamma \leq \eta), (F_\gamma : \gamma < \eta), (\mathcal{T}_\gamma : \gamma \leq \eta))$ is the output of the $(\mathcal{P}, \Sigma)$-coherent fully backgrounded construction of N" the meaning of which should be self-evident with one wrinkle. It may be that for some $\gamma \leq \eta$, $\Sigma(\mathcal{T}_\gamma)$ is undefined. In this case, we have that $\gamma = \eta$ and we stop the construction.

If the background universe has a distinguished extender sequence then we tacitly assume that the extenders appearing in the $(\mathcal{P}, \Sigma)$-coherent fully background construction come from this distinguished extender sequence.

[30] Notice that if there is such a $\mathcal{T}_\gamma$ then it is unique.

[31] Here $\tilde{F}$ is the amenable code of F, see the discussion after [59, Lemma 2.9].

[32] Here we mean that A is being indexed in the strategy predicate of $\mathcal{N}_\gamma$.

3.6. A short tree strategy indexing scheme

Our goal here is to introduce the notion of a *short tree strategy premouse* (*sts premouse*). As we mentioned in the previous section, the difficulty with doing this lies in the fact that maximal trees might "core down" to short trees and thus, creating indexing issues. The idea behind the solution presented here is to add a branch for a tree as soon as we see a certificate of shortness, which in our case will be a $\mathcal{Q}$-structure. As the $\mathcal{Q}$-structures that we will be looking for are themselves sts premice, this inevitably leads to an induction.

Technically speaking $\mathcal{M}$ in Definition 3.6.2 should not be ses (see Definition 2.5.2) as $f^{\mathcal{N}}$ does not quite code an iteration strategy. Its domain consists of indexable stacks (see Definition 3.3.3). But recall the abuse of terminology proposed after Definition 2.5.1. Also, recall the definition of $\mathrm{m}^+(\mathcal{T}) = \mathrm{m}(\mathcal{T})^\#$ (see Definition 3.1.4).

The language of unindexed ses[33] includes constant symbols for $\vec{E}$, f, X and $\mathcal{P}$. We denote these symbols by $\dot{E}$, $\dot{f}$, $\dot{X}$ and $\dot{\mathcal{P}}$. Also, we let $\dot{<}$ be the symbol denoting the constructibility order and $\dot{\Sigma}$ be the partial strategy coded by $\dot{f}$. $\dot{<}$ and $\dot{\Sigma}$ are not symbols in the language but they can be easily defined from the other symbols.

DEFINITION 3.6.1 (φ^*-formula). We let $\varphi^*(x)$ be the conjunction of the following statements in the language of ses.

1. x is a sequential structure of the form $(\mathcal{J}_\omega(t), t, \in)$ where $t = (\mathcal{P}_0, \mathcal{T}_0, \mathcal{P}_1, \mathcal{T}_1)$ is an indexable stack on $\dot{\mathcal{P}}$,
2. t is according to $\dot{\Sigma}$ where $\dot{\Sigma}$ is the partial strategy coded by $\dot{f}$, and
3. $\mathrm{cf}(\mathrm{lh}(\mathcal{T}_0))$ and $\mathrm{cf}(\mathrm{lh}(\mathcal{T}_1))$ are not measurable cardinals.
4. there is (ν, ξ) such that letting $((\mathcal{M}_\gamma, \mathcal{N}_\gamma : \gamma \leq \eta), (F_\gamma : \gamma < \eta), (\mathcal{W}_\gamma : \gamma < \eta))$ be the output of the $(\dot{\mathcal{P}}, \dot{\Sigma})$-coherent fully backgrounded construction of the universe[34] in which extenders used have critical points $> \nu$[35], $\mathcal{W}_\xi = \mathcal{T}_0$.

DEFINITION 3.6.2 (Unambiguous ses). Suppose $\mathcal{M}$ is an unindexed ses[36] over some self-well-ordered set X based on a hod-like #-lsa type lses $\mathcal{P}$. We say $\mathcal{M}$ is *unambiguous* if $\mathcal{M}$ is closed[37] and whenever w is a sequential structure of the form $(\mathcal{J}_\omega(t), t, \in)$ where $t = (\mathcal{P}_0, \mathcal{T}_0, \mathcal{P}_1, \mathcal{T}_1) \in \mathcal{M}$ is an indexable stack according to $\Sigma^{\mathcal{M}}$ and such that

1. $\mathcal{M} \models \varphi^*[w]$ and
2. either
   (a) $\mathcal{T}_1 = \emptyset$ and $\mathcal{M} \models$ "$\mathcal{T}_0$ is a uvs[38] of limit length" or

[33] See Definition 2.5.3.
[34] See Remark 3.5.2.
[35] See Definition 3.5.1.
[36] See Definition 2.5.3.
[37] See Definition 2.3.15.
[38] See Definition 3.3.2.

(b) $\mathcal{T}_1$ is a nonempty stack of limit length

then $t \in \mathrm{dom}(\Sigma^{\mathcal{M}})$. We say $\mathcal{M}$ is ambiguous if it is not unambiguous.

Notice that ambiguity is a first order property of unindexed ses. The next definition introduces an indexing scheme that we will use to define short tree premice. The indexing scheme only defines the strategy on certain carefully chosen stacks. It turns out that this much information is enough to extend the strategy to all stacks (see Chapter 6).

REMARK 3.6.3. The reader may find the following remark helpful. Definition 3.3.2 introduced the uvs stacks, which are stacks that are short with respect to all reasonable strategies. Definition 3.6.2 introduces unambiguous ses, which are the ses whose internal strategy predicate is total on all indexable uvs stacks that satisfy the formula φ^* (see Definition 3.6.1). Negating this, we have that if $\mathcal{N}$ is ambiguous ses then in $\mathcal{N}$ there is a uvs $\mathcal{T}$ of limit length that is according to the internal strategy of $\mathcal{N}$ yet no branch of $\mathcal{T}$ is indexed in the strategy predicate of $\mathcal{N}$.

DEFINITION 3.6.4 (ψ-sts indexing scheme). Suppose $\varphi(x)$ and $\psi(x,y)$ are two formulas in the language of ses. We say φ is a *ψ-sts indexing scheme* if $\varphi(w)$ is the conjunction of the following clauses:

1. For all ordinals γ there is $\xi > \gamma$ such that $\dot{E}(\xi)$ is defined[39].
2. $\dot{\Sigma}$ is a partial faithful st-strategy such that $m(\dot{\Sigma}) = \emptyset$[40].
3. $\varphi^*(w)$.
4. Either
   (a) The universe is ambiguous and w is the $\dot{<}$-least sequential structure w' witnessing ambiguity of the universe. Or
   (b) The universe is unambiguous and w is the $\dot{<}$-least sequential structure w' of the form $(\mathcal{J}_\omega(t), t, \in)$ with the property that $t = (\mathcal{P}_0, \mathcal{T}_0, \mathcal{P}_1, \mathcal{T}_1)$ is an indexable stack on $\dot{\mathcal{P}}$, $\varphi^*(w')$ holds, $\dot{\Sigma}(\mathcal{T}_0)$ is undefined and there is a unique cofinal well-founded branch b of $\mathcal{T}_0$ such that $\psi(\mathcal{T}_0, b)$ holds.

REMARK 3.6.5. The reader may find it useful to compare Definition 3.6.4 with Definition 2.3.3, Definition 2.3.8 and Definition 2.3.10. The model over which we intend to evaluate φ in Definition 3.6.4 corresponds to $\mathcal{M}|\omega\beta$ in Definition 2.3.3. More precisely, if φ is as in Definition 3.6.4 and w is a sequential structure, then to decide whether we need to index a branch of w or not we need to look for β such that $\mathcal{M}|\omega\beta \vDash \varphi[w]$.

The meaning of clause 4(b) is that ψ is the certification of b as the correct branch, but Definition 3.6.4 does not say anything about a particular certification procedure that we will use. The exact certification method is presented in Definition 3.8.9.

---

[39]That is, the universe is closed in the sense of Definition 2.3.15.

[40]Notice that clause 4 below guarantees that $\dot{\Sigma}$ is really a partial strategy rather than an st-strategy. We emphasize the fact that $\dot{\Sigma}$ is an st-strategy to point out the fact that there is no iteration according to $\dot{\Sigma}$ that is $\dot{\Sigma}$-maximal.

Notice that φ is uniquely determined by ψ. The next definition uses ideas from Definition 2.3.3 and Definition 2.3.8, and it may be useful to review those definitions (in particular clause 4(a) of Definition 2.3.8).

DEFINITION 3.6.6 (Sts ψ-premouse). Suppose X is a self-well-ordered set, $\mathcal{P} \in X$ is a hod-like #-lsa type lses and $\psi(x,y)$ is a formula in the language of unindexed ses. Let φ be the ψ-sts indexing scheme. Then $\mathcal{M}$ is an *sts ψ-premouse* over X based on $\mathcal{P}$ if $\mathcal{M}$ is a φ-indexed ses over X based on $\mathcal{P}$ and if $w \in \text{dom}(f^{\mathcal{M}})$ is such that clause 4(b) of Definition 3.6.4 applies to $w =_{\text{def}} (\mathcal{J}_\omega(t), t, \in)$ where $t =_{\text{def}} (\mathcal{P}_0, \mathcal{T}_0, \mathcal{P}_1, \mathcal{T}_1)$ then letting $\beta = \min(f^{\mathcal{M}}(w))$,

$$f^{\mathcal{M}}(w) = \{\beta + \omega\gamma : \gamma \in b\}$$

where $b \in \mathcal{M}|\beta$ is the unique branch of $\mathcal{T}_0$ such that $\mathcal{M}|\beta \vDash \psi[\mathcal{T}_0, b]$.

If $\psi(x,y) =$ "$0 = 1$" then we say $\mathcal{M}$ has a trivial indexing scheme and also say that $\mathcal{M}$ is a trivial sts premouse. Notice that in a trivial sts premouse indexable nuvs stacks do not have branches indexed in the strategy predicate.

$\mathcal{Q}$-structures are sts ψ-premouse. Suppose $\mathcal{P}$ is an #-lsa type hod like lses and $\mathcal{T}$ is a normal nuvs tree on $\mathcal{P}$. Suppose b is a well-founded branch of $\mathcal{T}$ such that $\mathcal{Q}(b, \mathcal{T})$ exists. Does it follow that $\mathcal{Q}(b, \mathcal{T})$ is an sts ψ-premouse in some reasonable sense? The following lemma gives the answer we need.

DEFINITION 3.6.7. Suppose $\mathcal{P}$ is a hod-like #-lsa type lses and $\psi(x,y)$ is a formula in the language of unindexed ses. We say $\mathcal{P}$ is *uniformly ψ-organized* if for each #-lsa type layer $\mathcal{Q}$ of $\mathcal{P}$ such that $\mathcal{Q}^b = \mathcal{P}^b$ and $\delta^{\mathcal{Q}} < \delta^{\mathcal{P}}$, if ν is the largest such that $\mathcal{P}||\nu \vDash$ "$\delta^{\mathcal{Q}}$ is a Woodin cardinal" then $\mathcal{P}||\nu$ is an sts ψ-premouse over $\mathcal{Q}$.

LEMMA 3.6.8. *Suppose $\mathcal{P}$ is a uniformly ψ-organized hod-like #-lsa type* lses. *Suppose $\mathcal{T}$ is a normal iteration tree on $\mathcal{P}$. Suppose $\alpha < \text{lh}(\mathcal{T})$ and $\mathcal{R} \trianglelefteq_{hod} \mathcal{M}_\alpha^{\mathcal{T}}$ is such that letting $(\mathcal{P}_{\xi,\xi'} : \xi \leq \eta, \xi' \leq \nu_\xi)$ be the layers of $\mathcal{M}_\alpha^{\mathcal{T}}$, for some $\xi \leq \eta$, $\mathcal{R} = \mathcal{P}_{\xi,0}$ and $\mathcal{P}_{\xi,1}$ is defined either according to condition* R5 *of Definition 2.7.8 or clause 2 of* R10 *of Definition 2.7.8*[41]. *Then $\mathcal{P}_{\xi,1}$ is an sts ψ-premouse over $\mathcal{R}$.*

PROOF. We prove the claim by induction. Suppose first $\alpha = \beta + 1$ and the claim is true for β (i.e., the claim is true for all $\zeta \leq \beta$ and hod initial segments of $\mathcal{M}_\zeta^{\mathcal{T}}$). In this case, we have that $\mathcal{M}_\alpha^{\mathcal{T}} = \text{Ult}(\mathcal{N}, E_\beta^{\mathcal{T}})$ where $\gamma = \mathcal{T}(\alpha)$ and $\mathcal{N} \trianglelefteq \mathcal{M}_\gamma^{\mathcal{T}}$ is the appropriate initial segment of $\mathcal{M}_\gamma^{\mathcal{T}}$. If now $\delta^{\mathcal{R}} < \delta^{\mathcal{M}_\alpha^{\mathcal{T}}}$ then the claim follows from elementarity of $\pi_{\gamma,\alpha}^{\mathcal{T}}$ restricted to strict initial segments of $\mathcal{N}$.

Assume then that $\delta^{\mathcal{R}} = \delta^{\mathcal{M}_\alpha^{\mathcal{T}}}$. If $\mathcal{N} \vDash$ "$\delta^{\mathcal{N}}$ is not a Woodin cardinal" then once again elementarity implies the claim (as $\mathcal{P}_{\xi,1} \in \text{rge}(\pi_{\gamma,\alpha}^{\mathcal{T}})$). Assume then $\mathcal{N} \vDash$ "$\delta^{\mathcal{N}}$ is a Woodin cardinal". In this case, we have that $\mathcal{P}_{\xi,1} = \text{Ult}(\mathcal{N}, E_\beta^{\mathcal{T}})$ and $\mathcal{N}$ is an sts ψ-premouse over $\mathcal{N}|\tau$ where $\tau = \min(\vec{E}^{\mathcal{N}} - \delta^{\mathcal{N}})$.

[41]This simply means that $\mathcal{R}$ is a #-lsa type layer of $\mathcal{M}_\alpha^{\mathcal{T}}$.

Set now $\mathcal{Q} =_{\text{def}} \mathcal{P}_{\xi,1}$, $E = E^{\mathcal{T}}_{\beta}$ and $j = \pi^{\mathcal{T}}_{\gamma,\alpha}$. Let f be the strategy predicate of $\mathcal{Q}$ and suppose that $\mathcal{Q}$ is not an sts ψ-premouse over $\mathcal{R}$. Notice that because j is Σ_1-elementary, we must have that for every $\omega\zeta < \text{ord}(\mathcal{Q})$, $\mathcal{Q}||\omega\zeta$ is an sts ψ-premouse over $\mathcal{R}$. This is because otherwise $\mathcal{N}$ would satisfy that there is some $\omega\zeta' + \omega < \text{ord}(\mathcal{N})$ such that $\mathcal{N}|\omega\zeta' + \omega \vDash$ "$\mathcal{N}||\omega\zeta'$ is not an sts ψ-premouse".

Thus, it is enough to show that if $\mathcal{Q}$ is active and its last predicate is a pair $(\mathcal{T}, b) \in f$ then $(\mathcal{T}, b)$ conforms the rules of sts ψ-premice. Set then $w = (\mathcal{J}_\omega(\mathcal{T}), \mathcal{T}, \in)$ and $\nu = \min(f(w))$. Let $\nu' = j^{-1}(\nu)$ and $\mathcal{T}' = j^{-1}(\mathcal{T})$. Because $j \restriction \mathcal{N}|\nu'$ is fully elementary, we have that

(1) $\mathcal{T}'$ is chosen in $\mathcal{N}|\nu'$ according to clause 4(a) of Definition 3.6.4 if and only if $\mathcal{T}$ is chosen in $\mathcal{Q}|\nu$ according to clause 4(a) of Definition 3.6.4.
(2) $\mathcal{T}'$ is chosen in $\mathcal{N}|\nu'$ according to clause 4(b) of Definition 3.6.4 if and only if $\mathcal{T}$ is chosen in $\mathcal{Q}|\nu$ according to clause 4(b) of Definition 3.6.4.

Thus, to finish, we need to verify that letting $b^* = j^{-1}[b]$ and b' be the closure of b^* in $\mathcal{T}'$ then

(*) b' is as in clause 4(b) of Definition 3.6.4 if and only if b is as in clause 4(b) of Definition 3.6.4.

(*) is straightforward because if b' is as in clause 4(b) then $b = j(b')$, and if b is as in clause 4(b) then $b' = j^{-1}(b)$[42].

The case when α is a limit ordinal is very similar, and we leave it to the reader. ⊣

3.7. Authentic indexable stacks

Suppose $(\mathcal{P}, \Sigma)$ is a hod-like limit type pair. Suppose $\mathcal{T}$ is a tree on $\mathcal{P}$ according to Σ such that $\pi^{\mathcal{T},b}$ exists and $\text{m}^+(\mathcal{T}) \vDash$ "$\delta(\mathcal{T})$ is a Woodin cardinal" (see Definition 3.1.4). When defining short tree strategy mice, we will be faced with the following question. How can we guess the correct branches of iteration trees that are on $\text{m}^+(\mathcal{T})$ and are according to $\Sigma_{\text{m}^+(\mathcal{T}),\mathcal{T}}$? In this section, we present an authentication process that allows us to guess the correct branches of such iterations.

The main technical object used in our authentication process is $s(\mathcal{T}, w)$ introduced in Definition 2.9.1. In the light of [63], we could use *strong hull condensation* instead of $s(\mathcal{T}, w)$, similar to the way optimal Suslin representations are obtained in [62, Chapter 2]. We, however, do not know if the core model induction applications of this book could be done using the ideas of [62].

We start by recalling $s(\mathcal{T}, w)$ (and slightly modifying it). Suppose $\mathcal{P}$ is a non-meek hod-like lses and $\mathcal{T}$ is a stack on $\mathcal{P}$ such that $\pi^{\mathcal{T},b}$ exists. Let $\mathcal{S} = \pi^{\mathcal{T},b}(\mathcal{P}^b)$,

[42]The equivalence follows from the fact that because $\text{cf}^{\mathcal{N}}(\text{lh}(\mathcal{T}'))$ is not a measurable cardinal in $\mathcal{N}$, $j \restriction \text{lh}(\mathcal{T}')$ is cofinal in $\text{lh}(\mathcal{T})$.

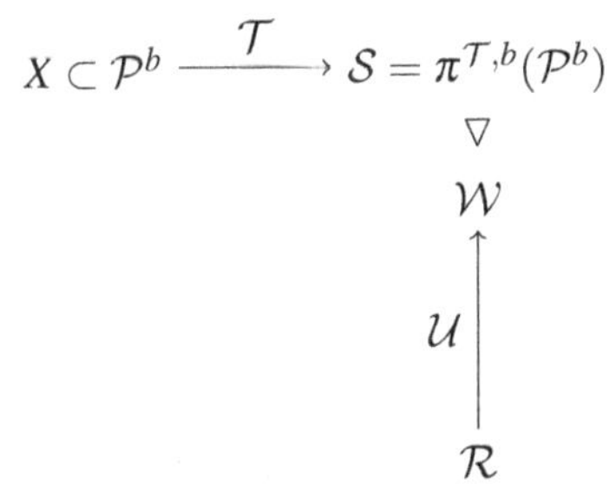

FIGURE 5. $(\mathcal{T},X)$ authenticates $\mathcal{R}$. The objects $\xi,\mathcal{U}$ etc. are as in Definition 3.7.3.

$w=(\eta,\delta)$ be a window[43] of $\mathcal{S}$ and $X\subseteq\mathcal{P}^b$. We then set

$$s(\mathcal{T},X,w)=\{\alpha:\exists a\in\eta^{<\omega}\exists f\in X(\alpha=\pi^{\mathcal{T},b}(f)(a))\}\cap\delta.$$

When $X=\mathcal{P}^b$ then we just write $s(\mathcal{T},w)$.

DEFINITION 3.7.1. Suppose $\mathcal{P}$ is a hod-like limit type lses and $X\subseteq\mathcal{P}^b$. We then say that X is *useful* if whenever $\mathcal{T}$ is a stack on $\mathcal{P}$ such that $\pi^{\mathcal{T},b}$ is defined, δ is a Woodin cardinal of $\mathcal{S}=_{\text{def}}\pi^{\mathcal{T},b}(\mathcal{P}^b)$ and w is a window of $\mathcal{S}$ such that $\delta^w=\delta$ then $s(\mathcal{T},X,w)$ is cofinal in δ.

Recall that Lemma 2.9.5 shows that $X=\mathcal{P}^b$ is useful.

NOTATION 3.7.2. Here and elsewhere in the manuscript , given a collection of formulas Γ, by $cHull_\Gamma^{\mathcal{M}}(Y)$ we mean the transitive collapse of X where $a\in X$ if and only if there is a formula $\varphi\in\Gamma$ and $s\in Y^{<\omega}$ such that $a\in\mathcal{M}$ is the unique b with the property that $\mathcal{M}\vDash\varphi[b,s]$. If Γ contains all formulas then we omit it from the notation. If Γ is the set of all Σ_n formulas then we just write $cHull_n^{\mathcal{M}}(Y)$. If Γ is the set of all formulas then we just write $cHull^{\mathcal{M}}(Y)$.

DEFINITION 3.7.3 (Authentic hod-like lses; see Figure 5). Suppose $(\mathcal{P},\Sigma)$ is a hod-like st-type pair, $\mathcal{T}$ is a normal iteration tree on $\mathcal{P}$ according to Σ such that $\pi^{\mathcal{T},b}$ exists and $X\subseteq\mathcal{P}^b$ is useful. Let $\mathcal{S}=\pi^{\mathcal{T},b}(\mathcal{P}^b)$ and suppose $\mathcal{R}$ is a hod-like lses. We say $(\mathcal{T},X)$ *authenticates* $\mathcal{R}$ if there is a normal iteration tree $\mathcal{U}$ on $\mathcal{R}$ such that the following clauses hold.

1. $\mathcal{U}$ has a last model $\mathcal{W}$ such that $\pi^{\mathcal{U}}$ is defined and $\mathcal{W}\trianglelefteq_{hod}\mathcal{S}$.
2. If $\gamma<\text{lh}(\mathcal{U})$ is a limit ordinal such that $\mathcal{S}\vDash$ "$\delta(\mathcal{U}\restriction\gamma)$ is a Woodin cardinal"[44], letting w be the unique window of $\mathcal{S}$ such that $\delta(\mathcal{U}\restriction\gamma)=\delta^w$ and setting $b=[0,\gamma)_{\mathcal{U}}$, for some $\tau\in b$,
$$s(\mathcal{T},X,w)\subseteq\text{rge}(\pi^{\mathcal{U}}_{\tau,b}).$$

[43]See Definition 2.7.14.

[44]This condition then implies that for some window $w=(\eta,\delta)$, $\mathcal{S}\vDash$ "δ is a Woodin cardinal" and $\text{m}(\mathcal{U}\restriction\gamma)=\mathcal{S}|\delta$. See Definition 2.7.8.

3. If $\mathcal{R}$ is of limit type then

$$\mathcal{W}^b = cHull^{\mathcal{S}^b}(\pi^{\mathcal{T},b}[X] \cup \delta^{\mathcal{W}^b})$$

and if $\sigma : \mathcal{W}^b \to \mathcal{S}^b$ is the uncollapse map then

$$\sigma^{-1}[\pi^{\mathcal{T},b}[X]] \subset \text{rge}(\pi^{\mathcal{U},b})$$

We say $\mathcal{R}$ is $(\mathcal{P},\Sigma,X)$-authentic if there is $\mathcal{T}$ on $\mathcal{P}$ according to Σ such that $(\mathcal{T},X)$ authenticates $\mathcal{R}$. We also say that $\mathcal{R}$ is $(\mathcal{P},\Sigma,X,\mathcal{T})$-authentic.

Notice that there is only one iteration tree $\mathcal{U}$ with the above properties. We then say that $\mathcal{U}$ is the $(\mathcal{T},X)$-authentication tree on $\mathcal{R}$. When $X = \mathcal{P}^b$ we simply omit it from terminology.

Clearly the tree $\mathcal{U}$ in Definition 3.7.3 is a tree built via a comparison process in which $\mathcal{S}$ does not move. A typical $\mathcal{R}$ that we would like to authenticate will be an iterate of $\mathcal{P}$. If Σ has nice properties, such as *strong branch condensation* (see Definition 4.9.2) then clauses 2 and 3 of Definition 3.7.3 hold for the iterates of $\mathcal{P}$. Next, we would like to define *authentic iterations*.

DEFINITION 3.7.4 (Authentic iterations). Suppose $(\mathcal{P},\Sigma)$ is a hod-like st-type pair, $\mathcal{T}$ is a normal tree on $\mathcal{P}$ according to Σ such that $\pi^{\mathcal{T},b}$ exists and $X \subseteq \mathcal{P}^b$ is useful. Let $\mathcal{S} = \pi^{\mathcal{T},b}(\mathcal{P}^b)$. Suppose $\mathcal{R}$ is an lses and $\mathcal{X}$ is a stack on $\mathcal{R}$. We say $(\mathcal{T},X)$ *authenticates* $(\mathcal{R},\mathcal{X})$ if $(\mathcal{T},X)$ authenticates $\mathcal{R}$ and, letting $\mathcal{U}$ be the $(\mathcal{T},X)$-authentication tree on $\mathcal{R}$ and $\mathcal{W}$ be the last model of $\mathcal{U}$, $\mathcal{X}$ is according to $\pi^{\mathcal{U}}$-pullback of $\Sigma_{\mathcal{W},\mathcal{T}}$.

Again we omit X when $X = \mathcal{P}^b$. We say $(\mathcal{R},\mathcal{X})$ is a $(\mathcal{P},\Sigma,X)$-*authenticated iteration* if there is a tree $\mathcal{T}$ on $\mathcal{P}$ according to Σ such that $(\mathcal{T},X)$ authenticates $(\mathcal{R},\mathcal{X})$. We also say that $(\mathcal{R},\mathcal{X})$ is $(\mathcal{P},\Sigma,X,\mathcal{T})$-*authentic*. When $X = \mathcal{P}^b$ we simply omit it from our terminology.

Next we define *authentic indexable stacks*. These are stacks that will be important in our definition of short tree strategy mice (see Definition 3.8.9). It maybe helpful to review the notation introduced in Notation 2.4.4.

DEFINITION 3.7.5 (Authentic indexable stacks). Suppose $(\mathcal{P},\Sigma)$ is a hod-like st-type pair, $X \subseteq \mathcal{P}^b$ is useful and $\mathcal{R}$ is a hod-like #-lsa type lses. Suppose

$$t = (\mathcal{R}_0,\mathcal{U},\mathcal{R}_1,\mathcal{W})$$

is an indexable stack on $\mathcal{R} = \mathcal{R}_0$. We say t is $(\mathcal{P},\Sigma,X)$-*authenticated* if the following conditions hold.

1. Suppose $\alpha \in R^{\mathcal{U}}$ is such that $\pi^{\mathcal{U}_{\leq\alpha},b}$ exists. Then for all

$$\alpha' \in (R^{\mathcal{U}} - (\alpha+1)) \cup \{\text{lh}(\mathcal{U})\}$$

such that $\mathcal{K} =_{\text{def}} \mathcal{U}_{[\alpha,\alpha']}$ is based on $\mathcal{S} =_{\text{def}} \mathcal{M}^b_\alpha$, $(\mathcal{S},\mathcal{K})$ is $(\mathcal{P},\Sigma,X)$-authenticated iteration.

2. Suppose $\alpha \in R^{\mathcal{U}}$ is such that $\pi^{\mathcal{U}_{\leq \alpha}, b}$ exists. Then for all
$$\alpha' \in (R^{\mathcal{U}} - (\alpha+1)) \cup \{\mathrm{lh}(\mathcal{U})\}$$
such that $\mathcal{K} =_{\mathrm{def}} \mathcal{U}_{[\alpha,\alpha']}$ is above $\mathrm{ord}(\mathcal{S})$ where $\mathcal{S} =_{\mathrm{def}} \mathcal{M}^b_\alpha$, the following conditions hold.
   (a) Suppose $\mathcal{K}$ does not have any fatal drops[45]. Then for any limit $\alpha < \mathrm{lh}(\mathcal{K})$, if b is the branch of $\mathcal{K} \restriction \alpha$ then $\mathcal{Q}(b, \mathcal{K} \restriction \alpha)$ exists and is $(\mathcal{P}, \Sigma, X)$-authentic.
   (b) Suppose $\mathcal{K}$ has a fatal drop at (α, η). Let $\mathcal{Q} = \mathcal{M}^{\mathcal{K}}_\tau || \omega\xi^{\mathcal{K}}_\tau$. Then $(\mathcal{Q}, \mathcal{K}_{\geq \mathcal{Q}})$ is a $(\mathcal{P}, \Sigma, X)$-authenticated iteration.
3. $(\mathcal{R}^b_1, \mathcal{W})$ is a $(\mathcal{P}, \Sigma, X)$-authenticated iteration.

When $X = \mathcal{P}^b$ we simply omit it from our terminology.

It is of course desirable that $(\mathcal{P}, \Sigma, X)$-authenticated stacks are according to Σ. In the next section, we will use our authentication idea to define certified stacks.

3.8. Short-tree-strategy mice

We now have developed enough terminology and tools to define sts premice. We use the following notation below. Suppose $\mathcal{M}$ is a transitive model of some fragment of set theory and λ is a limit of Woodin cardinals. Let $g \subseteq Coll(\omega, <\lambda)$ be $\mathcal{M}$-generic. For $\alpha < \lambda$, let $g_\alpha = g \cap Coll(\omega, <\alpha)$. We let $D(\mathcal{M}, \lambda, g)$ stand for the derived model of $\mathcal{M}$ at λ computed using g. More precisely, letting $\mathbb{R}^* = \bigcup_{\alpha<\lambda} \mathbb{R}^{\mathcal{M}[g_\alpha]}$, $D(\mathcal{M}, \lambda, g)$ is defined in $\mathcal{M}(\mathbb{R}^*)$ by letting

1. Γ be the set of A such that for some $\alpha < \lambda$ and some $B \in \mathcal{M}[g_\alpha]$ such that $\mathcal{M}[g_\alpha] \models$ "B is $<\lambda$-universally Baire", A is the interpretation of B in $\mathcal{M}(\mathbb{R}^*)$[46].
2. $D(\mathcal{M}, \lambda, g) = L(\Gamma, \mathbb{R}^*)$.

Woodin's derived model theorem says that $D(\mathcal{M}, \lambda, g) \models \mathsf{AD}^+$ (see [58]). We will use this theorem throughout this book.

Before we introduce the notion of short tree strategy premouse, we take a moment to describe the intuition behind the definition. Suppose $\mathcal{P}$ is a hod-like #-lsa type lses and $\mathcal{T}$ is a normal nuvs tree on $\mathcal{P}$. We would like to find the correct $\mathcal{Q}$-structure for $\mathcal{T}$. We first attempt to find this $\mathcal{Q}$-structure among ses that have the trivial indexing scheme ψ_0, i.e., no indexable nuvs stack has an indexed branch. However, there may never be such an ses that can be used as $\mathcal{Q}$-structure. Assume then that this is the case. We then immediately encounter two problems.

---

[45] See Definition 2.6.8.

[46] The meaning of this is the following. For each $\mathcal{M}$-cardinal $\beta \in (\alpha, \lambda)$, let $(T_\beta, S_\beta) \in \mathcal{M}[g \cap Coll(\omega, <\alpha)]$ be β-absolutely complementing trees such that $p[T_\beta] = B$. We then have that $A = \bigcup_{\beta<\lambda}(p[T_\beta])^{\mathcal{M}(\mathbb{R}^*)}$. It is customary to set $\Gamma = Hom^*$. See [58].

The first problem has to do with determining the exact stage of the constructibility order where we must stop looking for a $\mathcal{Q}$-structure among the ses that have the trivial indexing scheme. We will do this as soon as we reach a sufficiently closed stage. To know that we have reached such a level, we need to address the second problem.

The second problem is to describe the next type of gadgets that can be used as $\mathcal{Q}$-structures. A natural choice is the collection of ses over $\mathrm{m}(\mathcal{T})$ in which all nuvs trees have $\mathcal{Q}$-structures with the trivial indexing scheme. This is our second indexing scheme. Let us call it ψ_1. One wrinkle is that we need a certification method for the $\mathcal{Q}$-structures that are used in a ψ_1-sts premouse. This is done by using the ideas from Definition 3.7.5.

The way we put the two ideas together is as follows. We first search for a $\mathcal{Q}$-structure among ses with the trivial indexing scheme ψ_0. If we reach a level $\mathcal{M}_0$ that has a ψ_0-sts $\mathcal{Q}_1 \in \mathcal{M}_0$ that can be used as a $\mathcal{Q}$-structure then we stop and see if $\mathcal{M}_0$ certifies $\mathcal{Q}_1$ (see Definition 3.8.9). If yes, then we declare success. If no, then we continue with the trivial indexing. This naturally leads to an induction, in which we define more and more complex indexing schemes which themselves are indexed by ordinals. One issue is that the most straightforward approach to the problem of defining the indexing schemes involves extending the language of ses to have names for ordinals, and this creates several unpleasant issues. Instead, we will first introduce ses whose indexing scheme may not be first order definable, the externally-φ-ses. Afterwards, it will be straightforward to verify that being a short-tree-strategy premouse is in fact first order.

Another issue is to show that if there is a $\mathcal{Q}$-structure for some tree $\mathcal{T}$ then we will indeed reach this $\mathcal{Q}$-structure inside our short-tree-strategy mice. For this, we will use an appropriate notion of fullness. Finally, the reader may find Remark 3.8.20 useful. What follows is parallel to Section 2.3. The reader may want to review Definition 2.1.1, Definition 2.3.1, Notation 2.3.2, Definition 2.3.3, Definition 2.3.8, Definition 2.3.10, Definition 2.3.14 and Definition 2.5.3.

Externally-indexed hes. The main difference between Definition 3.8.1 and Definition 2.3.2 is clause 4 below.

NOTATION 3.8.1. Suppose that $\mathcal{M} = \mathcal{J}_\iota^{A_0,\dots,A_n,f}(X)$ is a $\mathcal{J}$-structure or an f.s. $\mathcal{J}$-structure, $P \in X$ and Φ is a set of triples (x,y,z) such that if $(x,y,z) \in \Phi$ then x is a sequential structure. Let $S^{\mathcal{M}}_{P,\Phi}$ be the set of pairs (β, w) such that

1. $\omega\beta + \omega\gamma^w \leq \mathrm{ord}(\mathcal{M})$,
2. $\mathcal{M}|\omega\beta \models$ "$\mathrm{cf}(\gamma^w)$ is not a measurable cardinal as witnessed by extenders in A_0"[47], and
3. $\mathcal{M}|\omega\beta \models \mathsf{ZFC}$, and
4. $(w, \mathcal{M}|\omega\beta, P) \in \Phi$.

[47] See Remark 2.3.4.

DEFINITION 3.8.2. Suppose that $(\mathcal{M}, P, \Phi)$ are as in Notation 3.8.1. Suppose further that f is a shifted amenable function with amenable component g such that $\text{dom}(f) \subseteq \lfloor M \rfloor$ and for all $w \in \text{dom}(f)$, $\min(f(w)) + \gamma^w \leq \text{ord}(\mathcal{M})$[48]. We say w is *weakly (f,P,Φ)-minimal* if there is β such that

1. $(\beta, w) \in S^{\mathcal{M}}_{P,\Phi}$ (in particular, because $\mathcal{M}|\omega\beta \models \mathsf{ZFC}$, $\omega\beta = \beta$),
2. $w \notin \text{dom}(f \cap \lfloor \mathcal{M}|\beta \rfloor)$,
3. $\{u \in \lfloor \mathcal{M}|\beta \rfloor : u <_{\mathcal{M}|\beta} w$ and there is $\xi < \beta$ such that $(\xi, u) \in S^{\mathcal{M}}_{P,\Phi}\} \subseteq \text{dom}(f \cap \lfloor \mathcal{M}|\beta \rfloor)$.

We say w is *(f,P,Φ)-minimal* if there is β witnessing that w is weakly (f,P,Φ)-*minimal* and such that w is the $<_{\mathcal{M}|\beta}$-minimal w' which is weakly (f,P,Φ)-*minimal* as witnessed by β.

If w is *(f,P,Φ)-minimal* then we let $\beta_w^{\mathcal{M},f,P,\Phi}$ be the least β witnessing that w is *(f,P,Φ)-minimal.* In many cases, $(\mathcal{M}, f, P, \Phi)$ will be clear from context and so we will drop it from our notation.

We are now in a position to introduce the *externally-Φ-indexed passive hybrid $\mathcal{J}$-structures*, or just $e\Phi$-indexed passive hybrid $\mathcal{J}$-structures.

DEFINITION 3.8.3 ($e\Phi$-indexed Passive Hybrid $\mathcal{J}$-structures). We say $\mathcal{M}$ is an *$e\Phi$-indexed passive hybrid $\mathcal{J}$-structure* over a self-well-ordered set X based on P if $\mathcal{M} = (\mathcal{M}', k)$ is an f.s. $\mathcal{J}$-structure such that the following conditions hold.

1. For some α, $A \subseteq \lfloor \mathcal{M}' \rfloor$ and $f \subseteq \lfloor \mathcal{M}' \rfloor$,
$$\mathcal{M}' = (\mathcal{J}^{A,f}_{\omega\alpha}(X), A, f, X, \in),$$[49]
2. f is a shift of an amenable function.
3. For all $w \in \lfloor \mathcal{M}' \rfloor$, $w \in \text{dom}(f)$ if and only if w is (f,P,Φ)-minimal.
4. For all $w \in \text{dom}(f)$,
   (a) $\beta_w = \min(f(w))$ and $\beta_w + \omega\gamma^w < \text{ord}(\mathcal{M})$[50],
   (b) $\lfloor \mathcal{M}'|(\beta_w + \omega\gamma^w) \rfloor = \mathcal{J}_{\beta_w+\omega\gamma^w}(\mathcal{M}'||\omega\beta_w)$ and $A \cap \lfloor \mathcal{M}'|(\beta_w + \omega\gamma^w) \rfloor = A \cap \lfloor \mathcal{M}'|\omega\beta_w \rfloor$[51].

DEFINITION 3.8.4 ($e\Phi$-indexed Hybrid $\mathcal{J}$-structures). We say $\mathcal{M}$ is an *$e\Phi$-indexed hybrid $\mathcal{J}$-structure* over a self-well-ordered set X based on P if $\mathcal{M} = (\mathcal{M}', k)$ is an f.s. $\mathcal{J}$-structure such that

1. for some α, $A \subseteq \lfloor \mathcal{M}' \rfloor$ and $f \subseteq \lfloor \mathcal{M}' \rfloor$,
$$\mathcal{M}' = (\mathcal{J}^{A,f}_{\omega\alpha}(X), A, f, B, F, X, \in),$$[52]
2. $(\mathcal{J}^{A,f}_{\omega\alpha}(X), A, f, X, \in)$ is an $e\Phi$-indexed passive hybrid $\mathcal{J}$-structure,

[48]Recall our convention that $X^{\mathcal{M}}$ is self-well-ordered.

[49]We would like to emphasize that $\mathcal{M}'$ has only the displayed predicates. Also, below $(\mathcal{M}', f, \varphi)$ are omitted from β_w notation.

[50]Here β_w is defined in Definition 3.8.2.

[51]It also follows that $f \cap \lfloor \mathcal{M}'|(\beta_w + \gamma^w) \rfloor = f \cap \lfloor \mathcal{M}'|\beta_w \rfloor$.

[52]Below $(\mathcal{M}', f, \mathcal{P}, \Phi)$ are omitted from β_w notation.

3. at most one of B and F is not empty,
4. if $F \neq \emptyset$ then F is an ordered pair (w,b) such that if $\beta = min(b)$ then setting $f' = f \cup \{(w,b)\}$,
   (a) f' is a shift of an amenable function[53],
   (b) w is (f',P,Φ)-minimal with $\beta_w^{\mathcal{M},f',P,\Phi} = \beta$ (in particular, $\omega\beta = \beta$, see Definition 3.8.2),
   (c) $\omega\alpha = \beta + \omega\gamma^w$,[54]
   (d) $\lfloor \mathcal{M}' \rfloor = \mathcal{J}_{\beta+\omega\gamma^w}(\mathcal{M}'||\beta)$ and $A \cap \lfloor \mathcal{M}' \rfloor = A \cap \lfloor \mathcal{M}'|\beta \rfloor$.

For $w \in \text{dom}(f')$, we say that $f'(w)$ is indexed at $\beta_w + \omega\gamma^w$ or that $\beta_w + \omega\gamma^w$ is the index of $f'(w)$.

DEFINITION 3.8.5 ($e\Phi$-indexed Strategic e-structure, $e\Phi -$ ses). Suppose $\mathcal{P}$ is a transitive structure, X is a self-well-ordered set such that $\mathcal{P} \in X$ and $\mathcal{M} = \mathcal{J}^{\vec{E},f}(X)$ is an $e\Phi$-indexed hybrid $\mathcal{J}$-structure over X based on P. We say $\mathcal{M}$ is an *$e\Phi$-indexed strategic e-structure* ($e\Phi$-ses) over X based on $\mathcal{P}$ if the following clauses hold.

1. $f^{\mathcal{M}}$ codes a partial iteration strategy for $\mathcal{P}$ such that for any $w \in \text{dom}(f^{\mathcal{M}})$ if $\beta = min(f^{\mathcal{M}}(w))$ then $\mathcal{M}|\beta$ is closed[55].
2. $\vec{E}$ is a mixed indexed extender sequence.
3. If $\mathcal{M} = (\mathcal{M}',k)$[56] then for every $(\omega\beta,m) < l(\mathcal{M})$, $\mathcal{M}||(\omega\beta,m)$ is sound.

We say $\mathcal{M}$ is based on $\mathcal{P}$ if $\mathcal{M}$ is over $\mathcal{J}_\omega[\mathcal{P}]$ and is based on $\mathcal{P}$.

External Ψ-sts indexing scheme. The reader may want to review Definition 2.5.3.

DEFINITION 3.8.6 ($e\Psi$-sts indexing scheme). Suppose Φ and Ψ are two sets. We say Φ is an *external Ψ-sts indexing scheme* (or just an $e\Psi$-sts indexing scheme) if for all triples $(w,\mathcal{N},\mathcal{P})$, $(w,\mathcal{N},\mathcal{P}) \in \Phi$ if and only if the following clauses hold.

1. X is a self-well-ordered set, $\mathcal{P} \in X$ is a hod-like #-lsa type lses and $w \in \mathcal{N}$.
2. $\mathcal{N}$ is an unindexed ses over X based on $\mathcal{P}$.
3. $\mathcal{N}$ is closed[57].
4. $\mathcal{N} \models$ "$\Sigma^{\mathcal{N}}$ is a partial faithful st-strategy with $m(\Sigma^{\mathcal{N}}) = \emptyset$"[58],
5. $\mathcal{N} \models \mathsf{ZFC} + \varphi^*(w)$[59].
6. Either

[53]This implies that w is a sequential structure.

[54]It follows from clause 5 of Definition 3.8.2 that $\mathcal{M}' \models$ "cf(γ) is not a measurable cardinal as witnessed by extenders in A".

[55]See Definition 2.3.15. Also, recall that for such β we have $\omega\beta = \beta$.

[56]See Definition 2.2.2.

[57]See Definition 2.3.15.

[58]This comment was made before as well, but we remind the reader. Notice that clause 4 below guarantees that $\Sigma^{\mathcal{N}}$ is really a partial strategy rather than st-strategy. We emphasize the fact that $\Sigma^{\mathcal{N}}$ is an st-strategy to point out the fact that there is no iteration according to $\Sigma^{\mathcal{N}}$ that is $\Sigma^{\mathcal{N}}$-maximal.

[59]See Definition 3.6.1.

(a) $\mathcal{N}$ is ambiguous and w is the $<_{\mathcal{N}}$-least sequential structure witnessing the ambiguity of $\mathcal{N}$. Or

(b) $\mathcal{N}$ is unambiguous and w is the $<_{\mathcal{N}}$-least sequential structure $w' \in \mathcal{N}$ of the form $w' = (\mathcal{J}_\omega(t), t, \in)$ where $t = (\mathcal{P}, \mathcal{T}_0, \mathcal{P}_1, \mathcal{T}_1)$ such that $\mathcal{N} \models \varphi^*[w']$, t is an indexable nuvs such that $\dot{\Sigma}(\mathcal{T}_0)$ is undefined and there is a cofinal well-founded branch b of $\mathcal{T}_0$ such that $b \in \mathcal{N}$ and $(\mathcal{T}_0, \mathcal{N}, b) \in \Psi$.

DEFINITION 3.8.7 (Sts Ψ-premouse). Suppose X is a self-well-ordered set, $\mathcal{P} \in X$ is a hod-like #-lsa type lses and Ψ is a set. Let Φ be the $e\Psi$-sts indexing scheme. Then $\mathcal{M}$ is an *sts Ψ-premouse* over X based on $\mathcal{P}$ if $\mathcal{M}$ is a $e\Phi$-ses over X based on $\mathcal{P}$ and if $w \in \text{dom}(f^{\mathcal{M}})$ is such that clause 6(b) of Definition 3.8.6 applies to $w =_{\text{def}} (\mathcal{J}_\omega(t), t, \in)$ where $t =_{\text{def}} (\mathcal{P}_0, \mathcal{T}_0, \mathcal{P}_1, \mathcal{T}_1)$ then letting $\beta = \min(f^{\mathcal{M}}(w))$,

$$f^{\mathcal{M}}(w) = \{\beta + \omega\gamma : \gamma \in b\}$$

where $b \in \mathcal{M}|\beta$ is the unique branch of $\mathcal{T}_0$ such that $(\mathcal{T}_0, \mathcal{M}|\beta, b) \in \Psi$.

Notice that in Definition 3.8.6, Φ is uniquely determined by Ψ. We now by induction define a sequence of sets $(\Psi_\beta : \beta \in Ord)$ and for $\beta \in Ord$, we let Φ_β be the eΨ_β-sts indexing scheme. To start we let $\Psi_0 = \emptyset$. Thus, if $\mathcal{M}$ is an $e\Phi_0 -$ ses then $\mathcal{M}$ does not have branches for nuvs stacks. We will use the following concept.

DEFINITION 3.8.8 (Terminal tree). Suppose X is a self-well-ordered set, $\mathcal{P} \in X$ is a hod-like #-lsa type lses, $\mathcal{N}$ is an ses over X based on $\mathcal{P}$. Given $\mathcal{T} \in \mathcal{N}$ on $\mathcal{P}$, we say $\mathcal{T}$ is *$\mathcal{N}$-terminal* if $\mathcal{T}$ is nuvs, $\mathcal{T}$ is according to $\Sigma^{\mathcal{N}}$ and $\mathcal{T} \notin \text{dom}(\Sigma^{\mathcal{N}})$.

DEFINITION 3.8.9 (Sts indexing scheme). Let $\Psi_0 = \emptyset$ and suppose $(\Psi_\xi : \xi < \alpha)$ have been defined. For $\xi < \alpha$ let Φ_ξ be the eΨ_ξ-sts indexing scheme (see Definition 3.8.6). We let Ψ_α be the set of triples $(\mathcal{T}, \mathcal{M}, b)$ such that $\mathcal{M}$ is an ses over X based on $\mathcal{P}$ and $(\mathcal{T}, b)$ is the $\mathcal{M}$-lexicographically least[60] pair such that $\mathcal{T}$ is a normal iteration tree on $\mathcal{P}$, $\mathcal{T}$ is $\mathcal{M}$-terminal, and b is a cofinal branch through $\mathcal{T}$ such that for some pair $(\gamma, \xi) \in \text{ord}(\mathcal{M}) \times \alpha$ the following clauses hold:

1. $\mathcal{M}|\gamma$ is unambiguous[61] and $\mathcal{M}|\gamma \models \mathsf{ZFC}+$ "there are infinitely many Woodin cardinals $> \delta(\mathcal{T})$".
2. $\mathcal{M}|\gamma \models$ "$\text{lh}(\mathcal{T})$ is not of measurable cofinality".
3. $b \in \mathcal{M}|\gamma$ and $\mathcal{M}|\gamma \models$ "b is a well-founded branch".
4. $\mathcal{M}|\gamma \models$ "$\mathcal{Q}(b, \mathcal{T})$ exists" and $\mathcal{Q}(b, \mathcal{T})$ is an $e\Phi_\xi -$ ses over $\text{m}^+(\mathcal{T})$[62].
5. Letting $(\delta_i : i < \omega)$ be the first ω Woodin cardinals of $\mathcal{M}|\gamma$ that are strictly greater than $\delta(\mathcal{T})$, the following holds in $\mathcal{M}|\gamma$: $\mathcal{Q}(b, \mathcal{T})$ is $< Ord$-iterable above $\delta(\mathcal{T})$ via a strategy Λ such that letting $\lambda = \sup_{i<\omega} \delta_i$, for every generic $g \subseteq Coll(\omega, < \lambda)$, Λ has an extension $\Lambda^+ \in D(\mathcal{M}|\gamma, \lambda, g)$ such that
   (a) $D(\mathcal{M}, \lambda, g) \models$ "Λ^+ is an ω_1-iteration strategy" and

[60]This is just the order defined by: first order the first coordinate by $<_{\mathcal{M}}$, the canonical well-order of $\mathcal{M}$, then order the second coordinate by $<_{\mathcal{M}}$.

[61]See Definition 3.6.2.

[62]This last statement about $\mathcal{Q}(b, \mathcal{T})$ may not be first order over $\mathcal{M}|\gamma$.

(b) whenever $\mathcal{R} \in D(\mathcal{M}|\gamma, \lambda, g)$ is a Λ^+-iterate of $\mathcal{Q}(b, \mathcal{T})$ above $\delta(\mathcal{T})$ and $t \in \mathcal{R}$ is an indexable stack on $\mathrm{m}^+(\mathcal{T})$ according to $\Sigma^{\mathcal{R}}$,

$$\mathcal{M}|\gamma[g] \models \text{“}t \text{ is } (\mathcal{P}, \Sigma^{\mathcal{M}|\gamma})\text{-authenticated”}.$$ [63]

The lexicographically least pair (γ, ξ) satisfying the above conditions is called the least $(\mathcal{M}, \Psi_\alpha)$-shortness witness for $(\mathcal{T}, b)$. We also say that (γ, ξ, b) is an $\mathcal{M}$-minimal shortness witness for $\mathcal{T}$. We also say that $\mathcal{T}$ has an $\mathcal{M}$-shortness witness.

We say $\mathcal{M}$ is a *potential* sts premous if $\mathcal{M}$ is an sts Ψ_α-premouse for some α.

Notice that because we minimized b, $\mathcal{M}$ has at most one $\mathcal{M}$-shortness witness for $\mathcal{T}$. The next lemma can now be established via an induction on ordinals.

LEMMA 3.8.10. *Suppose M is a transitive model of* ZFC *and $\mathcal{R} \in M$. Then*

1. *For every $\alpha < \mathrm{ord}(\mathcal{R})$, $M \models$ “$\mathcal{R}$ is an* sts *Ψ_α-premouse” if and only if $\mathcal{R}$ is an* sts *Ψ_α-premouse.*
2. *For every $\alpha < \mathrm{ord}(\mathcal{M})$, $M \models$ “$\mathcal{R}$ is an $e\Phi_\alpha -$* ses*” if and only if $\mathcal{R}$ is an $e\Phi_\alpha -$* ses.
3. *For every $\alpha < \mathrm{ord}(\mathcal{R})$, Definition 3.8.6 and Definition 3.8.9 define the sequences $(\Psi_\beta \cap M : \beta \leq \alpha)$ and $(\Phi_\beta \cap M : \beta \leq \alpha)$ in M.*

PROOF. The claim is obvious for $\alpha = 0$. In this case, $\Psi_0 = \emptyset$, and since clause 6b of Definition 3.8.6 is not applicable, the statement “$(w, \mathcal{N}, \mathcal{P}) \in \Phi_0$” is a first order (over $\mathcal{N}$) property of $\mathcal{N}$. Thus, the three clauses above follow.

Suppose now that for some $\alpha < \mathrm{ord}(M)$, the three clauses have been verified for all $\beta < \alpha$. We want to verify it for α.

We start with clause 3 of Lemma 3.8.10. Notice that all clauses of Definition 3.8.9 except the second half of clause 4 are internal properties of $\mathcal{M}$, where $\mathcal{M}$ is as in Definition 3.8.9. But our induction hypothesis implies that for every $\xi < \alpha$, being $e\Phi_\xi -$ ses is absolute between M and V, implying that the second half of clause 4 of Definition 3.8.9 is absolute between M and V (notice that in Definition 3.8.9 the branch b is in $\mathcal{M}$). This means that $\Psi_\alpha^M = \Psi_\alpha \cap M$.

Next, fix $(w, \mathcal{N}, P) \in M$. Notice that if $(w, \mathcal{N}, P) \in \Phi_\alpha^M$ then

1. if clause 6(a) of Definition 3.8.6 applies then $(w, \mathcal{N}, P) \in \Phi_\alpha$ and
2. if clause 6(b) of Definition 3.8.6 applies then $(\mathcal{T}_0, \mathcal{N}, b) \in \Psi_\alpha^M$, where $(\mathcal{T}_0, b)$ are as in clause 6b (and in particular, $b \in \mathcal{N}$).

The first statement above holds as all clauses of Definition 3.8.6 except 6(b) are internal properties of $\mathcal{N}$ and as such are absolute between M and V. Notice next that because we already have that $\Psi_\alpha^M = \Psi_\alpha \cap M$, the second statement above implies that $(\mathcal{T}_0, \mathcal{N}, b) \in \Psi_\alpha$ and hence, $(w, \mathcal{N}, P) \in \Phi_\alpha$. Thus, $\Phi_\alpha^M = \Phi_\alpha \cap M$.

The proof of the remaining clauses are very similar, and can be easily established by examining Definition 3.8.7. ⊣

[63] The witness for t being $(\mathcal{P}, \Sigma^{\mathcal{M}|\gamma})$-authenticated is in $\mathcal{M}|\gamma$.

COROLLARY 3.8.11. *Suppose M is a transitive model of* ZFC *and* $\alpha < \text{ord}(M)$. *Then* $M \vDash$ "$\mathcal{M}$ *is an* $e\Phi_\alpha -$ ses" *if and only if* $\mathcal{M}$ *is an* $e\Phi_\alpha -$ ses. *Also,* $M \vDash$ "$\mathcal{M}$ *is an* sts Ψ_α*-premouse*" *if and only if* $\mathcal{M}$ *is an* sts Ψ_α*-premouse,* $\alpha < \text{ord}(M)$ *and* $\mathcal{M} \in M$.

DEFINITION 3.8.12. Suppose $\mathcal{M}$ is a potential sts premouse. We say α is the *shortness degree* of $\mathcal{M}$ if α is the least for which $\mathcal{M}$ is a $e\Phi_\alpha -$ ses. We let $sd(\mathcal{M})$ be the shortness degree of $\mathcal{M}$.

The shortness degree of a potential sts premouse. Suppose $\mathcal{M}$ is a potential sts. We now describe a well-founded tree $U(\mathcal{M})$ whose rank bounds $sd(\mathcal{M})$. The nodes in $U(\mathcal{M})$ consist of finite sequences of the form $(x_0, x_1, \ldots, x_n)$ such that the following conditions hold:

1. For each $i \leq n$, $x_i = (t_i, b_i, \mathcal{M}_i)$ where $t_i = (\mathcal{M}_i, \mathcal{T}_i)$ is an indexable stack on $\mathcal{M}_i$, $\mathcal{T}_i$ is a normal tree on $\mathcal{M}_i$, b_i is a branch of $\mathcal{T}_i$ and for $i+1 \leq n$, $\mathcal{M}_{i+1} = \mathcal{Q}(b_i, \mathcal{T}_i)$.
2. $\mathcal{M}_0 = \mathcal{M}$.
3. For each $i \leq n$, $t_i \in \text{dom}(\Sigma^{\mathcal{M}_i})$[64] and $b_i = \Sigma^{\mathcal{M}_i}(\mathcal{T}_i)$.
4. For each $i \leq n$, $\mathcal{J}_\omega[\text{m}^+(\mathcal{T}_i)] \vDash$ "$\delta(\mathcal{T}_i)$ is a Woodin cardinal".

Notice that if $(x_0, \ldots, x_n) \in U(\mathcal{M})$ then for each $i < n$, $\mathcal{M}_{i+1} \in \mathcal{M}_i$[65]. Therefore, $U(\mathcal{M})$ is well-founded. If $p \in U(\mathcal{M})$ then we use superscript p to denote the objects that appear in p. For example $\mathcal{M}_n^p$, x_i^p or $\mathcal{T}_i^p$. Given a well-founded tree S we let $rank(S)$ be its rank.

LEMMA 3.8.13. $sd(\mathcal{M}) = rank(U(\mathcal{M}))$.

PROOF. The proof is by induction on $sd(\mathcal{M})$. Suppose $sd(\mathcal{M}) = 0$. In this case, $U(\mathcal{M}) = \emptyset$ and hence, its rank is also 0. On the other hand, if $U(\mathcal{M}) = \emptyset$ then clearly $\mathcal{M}$'s strategy predicate does not index any branch for an nuvs indexable stack, and therefore, $sd(\mathcal{M}) = 0$. For β an ordinal, let $I(\beta)$ be the conjunction of the following two statements.

(1) For all $\mathcal{M}'$, if $sd(\mathcal{M}') = \beta$ then $rank(U(\mathcal{M}')) = \beta$.
(2) For all $\mathcal{M}'$, if $rank(U(\mathcal{M}')) = \beta$ then $sd(\mathcal{M}') = \beta$.

We want to prove that for all β, $I(\beta)$ is true. Assume then that for some $\alpha \geq 1$, for all $\beta < \alpha$, $I(\beta)$ holds. We want to prove $I(\alpha)$, which amounts to proving that the following two statements hold.

(A) For any $\mathcal{M}$ such that $sd(\mathcal{M}) = \alpha$, $rank(U(\mathcal{M})) = \alpha$.
(B) For any $\mathcal{M}$ such that $rank(U(\mathcal{M})) = \alpha$, $sd(\mathcal{M}) = \alpha$.

We now prove (A). Fix $\mathcal{M}$ such that $sd(\mathcal{M}) = \alpha$. We want to see that

$$rank(U(\mathcal{M})) = \alpha.$$

[64] In particular, $t_i \in \mathcal{M}_i$.

[65] Notice that because $\delta(\mathcal{T}_i)$ is a Woodin cardinal, t_i is an nuvs and therefore, $b_i, \mathcal{M}_{i+1} \in \mathcal{M}_i$. See clause 3 of Definition 3.8.9.

Suppose first that $p = (x_0, \dots, x_n) \in U(\mathcal{M})$. Then we have that

(∗) $U(\mathcal{M}^p_{n+1}) = \{q : p^\frown q \in U(\mathcal{M})\}$.

We now show that the rank of $U(\mathcal{M})$ is at least as big as α. To see this, it is enough to show that for each $\beta < sd(\mathcal{M})$ there is a node $p = (x_0, \dots, x_n) \in U(\mathcal{M})$ such that letting $\mathcal{M}_{n+1} = \mathcal{Q}(b^p_n, \mathcal{T}^p_n)$, $sd(\mathcal{M}_{n+1}) > \beta$. (∗) then will imply that in fact $\beta \leq rank(U(\mathcal{M}))$. But because $\beta < sd(\mathcal{M})$, we must have a pair $(\mathcal{P}, \mathcal{T}) \in \mathcal{M}$ such that $\mathcal{T} \in \mathrm{dom}(\Sigma^{\mathcal{M}})$ and if $b = \Sigma^{\mathcal{M}}(\mathcal{T})$ and $t = (\mathcal{P}, \mathcal{T})$ then $(t, b, \mathcal{M}) \in U(\mathcal{M})$ and $sd(\mathcal{Q}(b, \mathcal{T})) \geq \beta$. It then follows that $p = (t, b, \mathcal{M})$ is as desired.

We now show that $rank(U(\mathcal{M})) \leq \alpha$. Indeed, let $p = (x_0) \in U(\mathcal{M})$ and set $\mathcal{M}_1 = \mathcal{Q}(b^p_0, \mathcal{T}^p_0)$. Because $sd(\mathcal{M}) = \alpha$, it follows that $sd(\mathcal{M}_1) < \alpha$. Therefore, it follows from (∗) and $\forall \beta < \alpha I(\beta)$ that $rank(U(\mathcal{M}_1)) < \alpha$. As this is true for any node of $U(\mathcal{M})$ of length 1, we have that $rank(U(\mathcal{M})) \leq \alpha$.

The proof of (B) is very similar. Indeed, if $\mathcal{M}$ is such that $rank(U(\mathcal{M})) = \alpha$ then (A) implies that $sd(\mathcal{M}) \geq \alpha$. Suppose then $sd(\mathcal{M}) > \alpha$. We then claim that there is $p \in U(\mathcal{M})$ such that p has length $n+1$ and $sd(\mathcal{M}^p_{n+1}) = \alpha$. Suppose otherwise. Thus the following is true:

(∗∗) whenever $p \in U(\mathcal{M})$ is of length $n+1$, either $sd(\mathcal{M}^p_{n+1}) > \alpha$ or $sd(\mathcal{M}^p_{n+1}) < \alpha$.

We now inductively define $(p_i : i < \omega)$ such that for all $i < \omega$,

1. $p_i \in U(\mathcal{M})$,
2. p_{i+1} extends p_i,
3. p_i has length $i+1$,
4. $sd(\mathcal{M}^{p_i}_{i+1}) > \alpha$.

Let $p_0 \in U(\mathcal{M})$ be of length 1 and such that $sd(\mathcal{M}^{p_0}_1) \geq \alpha$. There is indeed such a p_0 as all $\mathcal{Q}$-structures used in $\mathcal{M}$ would have shortness degree $< \alpha$ implying that $sd(\mathcal{M}) \leq \alpha$. (∗∗) now implies that in fact $sd(\mathcal{M}^{p_0}_1) > \alpha$. Repeating this construction ω times produces our desired sequence. It now follows that $U(\mathcal{M})$ is not well-founded, which is a contradiction and hence, (B) holds. ⊣

LEMMA 3.8.14. *Suppose M is a transitive model of* ZFC, *$\mathcal{M} \in M$ and $\mathcal{M}$ is a potential* sts *premouse. Then $M \vDash$ "$\mathcal{M}$ is potential* sts *premouse".*

PROOF. It follows from Lemma 3.8.11 that it is enough to establish that $sd(\mathcal{M}) \in M$.

But this follows from the fact that $U(\mathcal{M}) \in M$ (and hence, $rank(U(\mathcal{M})) \in M$) and $sd(\mathcal{M}) = rank(U(\mathcal{M}))$. ⊣

Authenticated potential sts premouse.

DEFINITION 3.8.15. Suppose $\mathcal{M}$ is an unindexed ses over X based on $\mathcal{P}$ and $\mathcal{Q} \in \mathcal{M}$ is a potential sts premouse over some #-lsa type hod-like lses $\mathcal{S}$. We say $\mathcal{Q}$ is *$\mathcal{M}$-authenticated* if the following clauses hold:

1. $\mathcal{M}$ has at least ω many Woodin cardinals $> \mathrm{ord}(\mathcal{Q})$.

2. Letting $(\delta_i : i < \omega)$ be the first ω Woodin cardinals of $\mathcal{M}$ that are strictly greater than $\mathrm{ord}(\mathcal{Q})$, the following holds in $\mathcal{M}$: $\mathcal{Q}$ is $< Ord$-iterable above $\delta^{\mathcal{S}}$ via a strategy Λ such that letting $\lambda = \sup_{i<\omega} \delta_i$, for every generic $g \subseteq Coll(\omega, <\lambda)$, Λ has an extension $\Lambda^+ \in D(\mathcal{M}, \lambda, g)$ such that
   (a) $D(\mathcal{M}, \lambda, g) \vDash$ "Λ^+ is an ω_1-iteration strategy" and
   (b) whenever $\mathcal{R} \in D(\mathcal{M}, \lambda, g)$ is a Λ^+-iterate of $\mathcal{Q}$ above $\delta^{\mathcal{S}}$ and $t \in \mathcal{R}$ is an indexable stack on $\mathrm{m}^+(\mathcal{T})$ according to $\Sigma^{\mathcal{S}}$,

$$\mathcal{M}[g] \vDash \text{"}t \text{ is } (\mathcal{P}, \Sigma^{\mathcal{M}})\text{-authenticated"}.[66]$$

Being $\mathcal{M}$-authenticated is a first order property of $\mathcal{M}$, and so we write $\mathcal{M} \vDash$ "$\mathcal{Q}$ is authenticated" for the statement $\mathcal{Q}$ is $\mathcal{M}$-authenticated.

The definition of short tree strategy premouse. Let $U_0(x,y)$ be the formula in the language of unindexed ses expressing the statement that "x is an ordinal and y is the universe up to ωx". Thus, $U_0(x,y)$ defines the function $\gamma \mapsto \mathcal{M}|\omega\gamma$ over any ses $\mathcal{M}$.

DEFINITION 3.8.16. We let $\mathsf{sts}_0(x,y)$ be the formula in the language of ses expressing the following: there is an ordinal γ such that letting M be such that $U_0(\gamma, M)$, the following clauses hold:

1. $M \vDash \mathsf{ZFC}$.
2. x is a normal iteration tree of limit length and y is a cofinal well-founded branch of x.
3. $M \vDash$ "$\mathcal{Q}(y,x)$ exists and is an authenticated potential sts premouse".
4. For any well founded branch y' of x and an ordinal γ', letting M' be such that $U(\gamma', M')$, if
   (a) $y \neq y'$,
   (b) $M' \vDash \mathsf{ZFC}$, and
   (c) $M' \vDash$ "$\mathcal{Q}(b', \mathcal{T})$ exists and is an authenticated potential sts premouse",
   then letting $\mathcal{Q} = \mathcal{Q}(y,x)$ and $\mathcal{Q}' = \mathcal{Q}(y',x)$, either $(\gamma, sd(\mathcal{Q})) <_{lex} (\gamma', sd(\mathcal{Q}'))$ or $(\gamma, sd(\mathcal{Q})) = (\gamma', sd(\mathcal{Q}'))$ and $b <_M b'$.

DEFINITION 3.8.17 (Sts-indexed ses, Sts mouse). Suppose X is a self-well-ordered set and $\mathcal{P} \in X$ is a hod-like #-lsa type lses. Let sts be the sts_0-sts indexing scheme[67]. We say $\mathcal{M}$ is an *sts premouse* over X based on $\mathcal{P}$ if $\mathcal{M}$ is an sts-indexed ses over X based on $\mathcal{P}$. If additionally $\mathcal{M}$ is $\omega_1 + 1$-iterable the we say that $\mathcal{M}$ is an *sts mouse*.[68]

The following is an easy lemma.

LEMMA 3.8.18. *Suppose M is a transitive set and $\mathcal{M} \in M$. Then $\mathcal{M}$ is an* sts *premouse if and only if $M \vDash$ "$\mathcal{M}$ is an* sts *premosue".*

---

[66]The witness for t being $(\mathcal{P}, \Sigma^{\mathcal{M}})$-authenticated is in $\mathcal{M}$.

[67]See Definition 3.6.4.

[68]Here implicit in this is the demand that iterates of $\mathcal{P}$ according to the strategy are sts premice.

The following is a corollary to Lemma 3.6.8. It implies that the certified $\mathcal{Q}$-structures themselves are sts premice.

LEMMA 3.8.19. *Suppose $\mathcal{P}$ is a uniformly*[69] sts-*organized #-lsa type hod like* lses. *Suppose $\mathcal{T}$ is a normal* nuvs *tree on $\mathcal{P}$ and b is well-founded branch of $\mathcal{T}$ such that $\mathcal{Q}(b,\mathcal{T})$ exists. Then $\mathcal{Q}(b,\mathcal{T})$ is an* sts *premouse based on $m^+(\mathcal{T})$.*

REMARK 3.8.20 (On how branches get indexed). The first key point is that $\mathcal{M}|\gamma$ in Definition 3.8.9 is not the analogue of $\mathcal{M}|\beta$ in Definition 2.3.3. The analogue of $\mathcal{M}|\beta$ in the sense of Definition 2.3.3 is $\mathcal{M}$ itself. Recall that the indexing scheme is not Ψ_α but rather Φ_α, and so the relevant definitions for determining the analogue of $\mathcal{M}|\beta$ in the sense of Definition 2.3.3 are Definition 3.8.16 and Definition 3.8.17.

Definition 2.3.10 introduced layered hybrid $\mathcal{J}$-structures, and a key aspect of that definition is the indexing of branches. The indexing scheme φ (in the sense of Definition 2.3.10) is only picking the iteration trees that we would like to index, where the branches are indexed is then uniquely determined by the procedure described in Definition 2.3.10. Definition 3.8.16 and Definition 3.8.17 are relevant definitions, and explain what the φ in Definition 2.3.10 should be.

The reader may wonder why we have concentrated so much on nuvs iterations. The point is that clause 4(b) of Definition 3.6.4 requires that we add the branches of uvs iterations, and these branches are not branches that we intend to certify. These branches are told to the model by consulting an outside strategy. It is only the branches of nuvs iterations, the ones that appear in clause 4(b) of Definition 3.6.4, are being certified. The schemes introduced in Definition 3.8.9 determine our certification procedures.

DEFINITION 3.8.21 (Λ-sts premouse). Suppose X is a self-well-ordered set, $\mathcal{P} \in X$ is a hod-like #-lsa type lses, Λ is an st-strategy for $\mathcal{P}$ and $\mathcal{M}$ is an sts premouse over X based on $\mathcal{P}$. Then we say $\mathcal{M}$ is a Λ-*sts premouse* over X based on $\mathcal{P}$ if $\Sigma^{\mathcal{M}} \subseteq \Lambda \restriction \mathcal{M}$.

DEFINITION 3.8.22 (Λ-sts mouse). Suppose X is a self-well-ordered set, $\mathcal{P} \in X$ is a hod-like #-lsa type lses, Λ is an st-strategy for $\mathcal{P}$ and $\mathcal{M}$ is a Λ-sts premouse over X based on $\mathcal{P}$. Then we say $\mathcal{M}$ is a Λ-*sts mouse* over X based on $\mathcal{P}$ if $\mathcal{M}$ has an ω_1+1-iteration strategy Σ such that whenever $\mathcal{N}$ is a Σ-iterate of $\mathcal{M}$ via Σ, $\mathcal{N}$ is a Λ-sts premouse over X based on $\mathcal{P}$.

We say $\mathcal{M}$ is a Λ-sts (pre)mouse over $\mathcal{P}$ if $\mathcal{M}$ is a Λ-sts (pre)mouse over $\mathcal{J}_\omega[\mathcal{P}]$ based on $\mathcal{P}$.

[69] See Definition 3.6.7.

3.9. The hod premouse indexing scheme

The goal of this short section is to introduce *the hod premouse indexing scheme* (hp indexing scheme). This scheme combines the standard indexing scheme with the sts indexing scheme The standard indexing scheme which is used in [29] is due to Woodin. According to this scheme we must pick the least iteration whose branch has not yet been indexed in the strategy predicate and index the branch of this iteration in the strategy predicate. Below we give a formal definition of the hp indexing scheme.

The reader may find it helpful to review Definition 2.3.3 and Definition 2.5.5. In particular, the reader should keep in mind that the intended universes where indexing schemes are evaluated are the models of the form $\mathcal{M}|\omega\beta$ of Notation 2.3.2. Thus, these universes themselves are not hod like lses (see Definition 2.7.10). But each such $\mathcal{M}|\omega\beta$ has its own predicate $Y^{\mathcal{M}|\omega\beta}$ which is what we will use below to describe the hp-indexing scheme. Perhaps reviewing Remark 2.5.7 may clarify some of the questions that the reader might have.

DEFINITION 3.9.1. We say that lses $\mathcal{M}$ is *strategy-ready* if letting $\iota = \text{ord}(Y^{\mathcal{M}})$, $\omega\iota + \omega^2 < \text{ord}(\mathcal{M})$.

DEFINITION 3.9.2 (Hod premouse indexing scheme, hp indexing scheme). We say $\varphi(x,y)$ is the *hod premouse indexing scheme* (hp indexing scheme) if φ is the conjunction of the following clauses.

1. The universe is closed (see Definition 2.3.15).
2. The universe is strategy-ready (see Definition 3.9.1).
3. $x = \cup Y^{\dot{\mathcal{V}}}$.
4. If x is lsa like then
   (a) $\dot{\mathcal{V}}$ is an sts premouse over $\dot{\mathcal{V}}|\iota+\omega$ based on x (see Definition 3.8.17),
   (b) $\text{sts}[y]$.
5. If x is not lsa like then y is the $<_{\dot{\mathcal{V}}}$-least sequential structure of the form $(\mathcal{J}_\omega(\mathcal{T}^y), \mathcal{T}^y, \in)$ where $\mathcal{T}^y$ is a stack on x that is according to $\dot{\Sigma}_x$ and does not have a last model.

We let hp denote the hp-indexing scheme.

REMARK 3.9.3. The determination of the Y predicate of the models appearing in the hod pair constructions (see Definition 4.3.3) is an important step in such constructions.

The next definition isolates the standard indexing scheme. It is defined in the language of ses which has a constant symbol for a structure whose strategy is indexed on the sequence of ses. We let $\dot{\mathcal{P}}$ be this constant.

DEFINITION 3.9.4. We say $\varphi(y)$ is the *standard indexing scheme* (sis-indexing scheme) if φ expresses the following statement: y is then $<_{\dot{\mathcal{V}}}$-least sequential structure of the form $(\mathcal{J}_\omega(\mathcal{T}^y), \mathcal{T}^y, \in)$ where $\mathcal{T}^y$ is a stack on $\dot{\mathcal{P}}$ that is according to $\dot{\Sigma}$ and $\mathrm{lh}(\mathcal{T}^y)$ is a limit ordinal. We let sis denote the sis indexing scheme.

REMARK 3.9.5. Woodin's method of feeding the branch information into the model (as described in clause 4 of Definition 3.9.2) is easy to comprehend and allows us to develop the basic theory of hod mice in this manuscript; however, it does not seem to allow for the proof of $\square$ to generalize easily. An alternative method to feeding in branch information that does allow for the $\square$ proof to generalize is to use the $\mathfrak{B}$-operator (see [50]). This method is summarized in Section 11.1; we also describe where the $\square$ proof seems to break down if Woodin's method was used. Nevertheless, a hod mouse constructed using Woodin's method constructs the same sets as the one using the $\mathfrak{B}$-operator (given that everything else is the same). Woodin's method is used from now on to the end of Chapter 10 because of its simplicity.

3.10. Hod mice

The main goal of this section is to introduce *lsa small hod premice*. The reader might find it helpful to review Section 2.7. In particular, we will use Definition 2.6.11, Definition 2.7.1, Definition 2.7.2, Definition 2.7.3, Definition 2.7.8, Definition 2.7.10, Notation 2.7.14, and Terminology 2.7.17. Also recall our convention introduced in Remark 2.7.5. According to this convention all our hod-like lses are lsa small.

We start by isolating the types of points in $Y^{\mathcal{P}}$ where $\mathcal{P}$ is hod-like lses.

NOTATION 3.10.1 (Meek and lsa points). Suppose $\mathcal{P}$ is a hod-like lses.

1. $meek(\mathcal{P}) = \{\mathcal{Q} \in Y^{\mathcal{P}} : \mathcal{Q}$ is meek[70]$\}$.
2. $lsa(\mathcal{P}) = \{\mathcal{Q} \in Y^{\mathcal{P}} : \mathcal{Q}$ is of #-lsa type[71]$\}$.
3. $\mathrm{ml}(\mathcal{P}) = \bigcup Y^{\mathcal{P}}$[72].

Definition 2.7.1 and Definition 2.7.10 do most of the job that we need to do to define hod premice. Essentially what is missing from Definition 2.7.10 is the exact nature of premice at lsa layers. In the next definition, we will not repeat what has already been introduced in Definition 2.7.1 and Definition 2.7.10.

DEFINITION 3.10.2 (Hod premouse). Suppose $\mathcal{P}$ is a (lsa small) hod-like lses[73]. Let $(\mathcal{P}_{\xi,\xi'} : \xi \leq \eta \wedge \xi' \leq \nu_\xi)$ be the sequence of layers of $\mathcal{P}$ and $(\delta_\xi, \iota_{\xi,\xi'} : \xi \leq$

[70] See Definition 2.7.1.

[71] See Definition 2.7.3.

[72] This object was introduced in Definition 2.7.14.

[73] See Definition 2.7.1 and Definition 2.7.10.

$\eta \wedge \xi' \leq \nu_\xi)$ be the sequence of ordinal parameters associated with it (see Definition 2.7.8). We say $\mathcal{P}$ is an *lsa small hod premouse* or just a *hod premouse* if $\mathcal{P}$ is hp-indexed[74] hod-like lses that has the following properties:

1. Suppose ν is a cutpoint of $\mathcal{P}$. Then the following holds.
   (a) If $\mathcal{P}$ is meek and $\nu < \delta^{\mathcal{P}}$ then $\mathcal{P} \models$ "$\mathcal{O}^{\mathcal{P}}_{\nu,\nu}$ has an *Ord*-strategy (sts strategy respectively) acting on iteration trees that are above[75] ν".
   (b) If $\mathcal{P}$ is non-meek and $\nu < \delta^{\mathcal{P}}$ then $\mathcal{P}|\delta^{\mathcal{P}} \models$ "$\mathcal{O}^{\mathcal{P}}_{\nu,\nu}$ has a $\delta^{\mathcal{P}}$-strategy acting on trees that are above ν".
2. If $\mathcal{P}$ is of successor type[76], $\xi + 1 = \eta$ and $\mathcal{Q} = \mathcal{P}_{\xi,\iota_\xi}$ then for any $\eta \in (\delta^{\mathcal{Q}}, \delta^{\mathcal{P}})$,

   $\mathcal{P} \models$ "$\mathcal{P}|\eta^+$ is (Ord, Ord)-iterable for stacks that are above $\mathrm{ord}(\mathcal{Q})$".
3. If $\mathcal{P}$ is of lsa type and $\eta \in (\mathrm{ord}(\mathcal{P}^b), \delta^{\mathcal{P}})$ then $\mathcal{P}|\delta^{\mathcal{P}} \models$ "$\mathcal{P}|\eta^+$ is (Ord, Ord)-iterable for stacks that are above $\mathrm{ord}(\mathcal{P}^b)$"

Next we define hod pairs.

DEFINITION 3.10.3 (Hod pairs). We say $(\mathcal{P}, \Sigma)$ is a (simple) hod pair if $(\mathcal{P}, \Sigma)$ is a (simple) hod-like lses pair[77], $\mathcal{P}$ is a hod premouse and Σ has hull condensation.

Next we introduce the collection of sets generated by hod pairs.

DEFINITION 3.10.4 ($\Gamma(\mathcal{P}, \Sigma)$ and $B(\mathcal{P}, \Sigma)$). Suppose $(\mathcal{P}, \Sigma)$ is a hod pair of limit type. We then let

$$B(\mathcal{P}, \Sigma) = \{(\mathcal{T}, \mathcal{Q}) : \exists \mathcal{R}((\mathcal{T}, \mathcal{R}) \in I(\mathcal{P}, \Sigma) \wedge \mathcal{Q} \trianglelefteq_{hod} \mathcal{R}^b)\}, \text{ and}$$
$$\Gamma(\mathcal{P}, \Sigma) = \{A \subseteq \mathbb{R} : \exists (\mathcal{T}, \mathcal{Q}) \in B(\mathcal{P}, \Sigma)(A \leq_w \mathsf{Code}(\Sigma_{\mathcal{Q},\mathcal{T}})\}.$$

DEFINITION 3.10.5 (Pre-sts hod pairs). We say $(\mathcal{P}, \Sigma)$ is a *pre-sts hod pair* if $(\mathcal{P}, \Sigma)$ is a hod-like st-type pair[78] and Σ is a (κ, λ, ν)-st-strategy for $\mathcal{P}$ with hull condensation.

We say $(\mathcal{P}, \Sigma)$ is a *simple pre-sts hod pair* if $(\mathcal{P}, \Sigma)$ is a hod-like st-type pair and Σ is a (λ, ν)-st-strategy for $\mathcal{P}$ with hull condensation.

To define sts hod pairs, we will make use of the notation introduced in Definition 3.3.9. Recall that in Definition 3.3.9, we introduced $\Gamma^b(\mathcal{P}, \Sigma)$ but not $\Gamma(\mathcal{P}, \Sigma)$. We will define $\Gamma(\mathcal{P}, \Sigma)$ for sts hod pairs in Section 8.1.

Suppose now that X is a self-well-ordered set, $(\mathcal{P}, \Sigma)$ is a pre-sts pair such that $\mathcal{P} \in X$ and $\mathcal{Q}$ is a Σ-sts mouse over X based on $\mathcal{P}$. Let Λ be the strategy of $\mathcal{Q}$. We then let $\Gamma(\mathcal{Q}, \Lambda)$ be the collection of all sets of reals A such that for some Λ-iterate $\mathcal{R}$ of $\mathcal{Q}$, there is $(\mathcal{T}, \mathcal{S}) \in B(\mathcal{P}, \Sigma^{\mathcal{R}})$ such that $A \leq_w \Sigma_{\mathcal{S},\mathcal{T}}$.

[74] See Definition 3.9.2.
[75] See Terminology 2.4.8.
[76] See Terminology 2.7.17.
[77] See Definition 2.10.12.
[78] See Definition 3.4.2.

DEFINITION 3.10.6 (Sts hod pairs). We say $(\mathcal{P},\Sigma)$ is an *sts hod pair* if $(\mathcal{P},\Sigma)$ is a pre-sts pair such that whenever $(\mathcal{T},\mathcal{R},\tau)$ is such that letting $(\mathcal{R}_{\xi,\xi'} : \xi \leq \eta \wedge \xi' \leq \nu_\xi)$ be the sequence of layers of $\mathcal{R}$ and $(\delta_\xi, \iota_{\xi,\xi'} : \xi \leq \eta \wedge \xi' \leq \nu_\xi)$ be the sequence of ordinal parameters associated with it (see Definition 2.7.8),

1. $(\mathcal{T},\mathcal{R}) \in I(\mathcal{P},\Sigma)$[79] and
2. $\mathcal{R}_{\tau,0} \in lsa(\mathcal{R})$ and $\delta_\iota < \delta^{\mathcal{R}}$,

then $\mathcal{R}_{\tau,1}$ has an iteration strategy $\Phi \in \Gamma^b(\mathcal{P},\Sigma)$ witnessing that $\mathcal{R}_{\tau,1}$ is a $\Sigma_{\mathcal{R}_{\tau,0},\mathcal{T}}$-sts mouse based on $\mathcal{R}_{\tau,0}$[80] and such that $\Gamma(\mathcal{R}_{\tau,1},\Phi) \subset \Gamma^b(\mathcal{P},\Sigma)$.

Similarly we can define *simple* sts hod pairs.

DEFINITION 3.10.7. We say $(\mathcal{P},\Sigma)$ is an *allowable pair* if it is one of the hod pairs introduced above. More precisely, one of the following holds:

1. $(\mathcal{P},\Sigma)$ is a hod pair.
2. $(\mathcal{P},\Sigma)$ is a simple hod pair.
3. $(\mathcal{P},\Sigma)$ is an sts hod pair.
4. $(\mathcal{P},\Sigma)$ is a simple sts hod pair.

In the context of AD^+, unless otherwise specified, the strategy component of any of the above pairs will always be $(\omega_1,\omega_1,\omega_1)$ or (ω_1,ω_1) strategy or st-strategies.

Definition 3.10.6 imposes conditions on sts hod pairs that may seem unnatural. However, these conditions are needed to prove that sts hod pairs behave nicely. These clauses will be used in Chapter 6.

---

[79] See Definition 3.3.7.

[80] Thus, all the iterates of $\mathcal{R}_{\tau,1}$ via Φ are above $\mathrm{ord}(\mathcal{R}_{\tau,0}) = \iota_{\tau,0}$.

Chapter 4

A COMPARISON THEORY OF HOD MICE

This section is devoted to proving a comparison theorem for hod pairs. We will have two comparison theorems, Corollary 4.13.4 and Corollary 4.14.4. Corollary 4.13.4 is useful in determinacy context while Corollary 4.14.4 is useful in Core Model Induction applications. The following is a key hypothesis used in many of the theorems of this chapter.

DEFINITION 4.0.8. We let NsesS stand for the statement "there is no ω_1-iterable ses with a superstrong cardinal".

4.1. Backgrounds and Suslin capturing

The goal of this section is to introduce *backgrounds* and the concept of *Suslin, co-Suslin capturing*. We will use these notions to build hod pairs with desired properties, such as *fullness preservation* and *branch condensation*. Before we do this, we fix a coding of hereditarily countable sets by reals. We will use this coding throughout this book.

DEFINITION 4.1.1. Given a real $x \in \mathbb{R}$, we let $E_x = \{(m,n) : x(2^m 3^n) = 1\}$. We let $x \in \mathsf{Code}$ if $m_x =_{\text{def}} (\omega, E_x)$ is a well-founded model satisfying the Axiom of Extensionality. If $x \in \mathsf{Code}$ then we let $\pi_x : m_x \to M_x$ be the transitive collapse of m_x and let $c_x = \pi_x(0) = \{\pi_x(m) : x(2^m) = 1\}$. We then say that *$x$ codes c_x*.

Recall that HC is the set of hereditarily countable sets (see Definition 3.3.8). Given $n \in \omega$ and $A \subseteq \mathsf{HC}^n$ we let $\mathsf{Code}(A) = \{x \in \mathsf{Code} : c_x \in A\}$. Notice that

$$\mathsf{Code} : \cup_{n\in\omega}\wp(\mathsf{HC}^n) \to \wp(\mathbb{R})$$

is an injective function.

DEFINITION 4.1.2. Let $(p_i : i < \omega)$ be the sequence of prime numbers. Let

$$\mathsf{merge} : \mathbb{R}^{\leq\omega} \to \mathbb{R}$$

be given by $\text{merge}(q) = y$ if letting $q = (y_i)_{i<n}$,

$$y(j) = \begin{cases} y_{j_0}(j_1) & \text{if } j = p_{j_0}^{j_1}, \\ 0 & \text{otherwise.} \end{cases}$$

NOTATION 4.1.3. If M is a transitive set and $\alpha \leq \text{ord}(M)$ then we let $M|\alpha = V_\alpha^M$.

DEFINITION 4.1.4 (Background). We say

$$\mathbb{M} = (M, \delta, \vec{G})$$

is a *background* if

1. $M \vDash \mathsf{ZFC} +$ "δ is a Woodin cardinal",
2. $\vec{G} : \delta \to V_\delta^M$ is a partial function such that for each $\alpha \in \text{dom}(\vec{G})$, for some (κ, λ), $M \vDash$ "$\vec{G}(\alpha)$ is a (κ, λ)-extender such that $M|\lambda \subseteq \text{Ult}(M, \vec{G}(\alpha))$ and λ is inaccessible",
3. $\mathcal{M} \vDash$ "$\vec{G}$ witnesses that δ is a Woodin cardinal",[1]

We say

$$\mathbb{M} = (M, \delta, \vec{G}, \Sigma)$$

is an *internally iterable background* if in addition to the three clauses above, the following clauses hold:

1. $\Sigma \in M$ and $M \vDash$ "Σ is a winning strategy for II in the version of the iteration game $\mathcal{G}(M, \delta, \delta + 1)$ in which player I is required to choose extenders whose (natural) lengths are inaccessible cardinals in the model they are chosen from and are also below the image of δ".
2. Σ has hull condensation,
3. $\text{dom}(\Sigma) \subseteq \mathcal{J}_\omega(V_\delta^M)$.

We say $\mathbb{M} = (M, \delta, \vec{G}, \Sigma)$ is an *externally iterable background* if $(M, \delta, \vec{G})$ is a background and Σ is a winning strategy for II in the version of $\mathcal{G}(M, \omega_1, \omega_1)$ mention in clause 1 above. We say $\mathbb{M} = (M, \delta, \vec{G}, \Sigma)$ is an *iterable background* if it is either internally or externally iterable background.

We say that an externally iterable background $(M, \delta, \vec{G}, \Sigma)$ is *self-knowledgable* if $(M, \delta, \vec{G}, \Sigma \restriction \mathcal{J}_\omega(M|\delta))$ is an internally iterable background.

Suppose $(M, \delta, \vec{G}, \Sigma)$ is a background and N is a Σ-iterate of M. Let $i : M \to N$ be the iteration embedding. We set

$$\mathbb{M}_N = (N, i(\delta), i(\vec{G}), \Sigma_N).$$

In most cases considered in this book, Σ_N will not depend on the iteration producing N.

[1] That is, $M \vDash$ "for every $A \subseteq \delta$ there is κ such that for every $\lambda < \delta$ there is an α with the property that letting $E = \vec{G}(\alpha)$, $\text{crit}(E) = \kappa$, $\text{lh}(E) \geq \lambda$ and $A \cap \text{lh}(E) = \pi_E(A) \cap \text{lh}(E)$".

Suppose $\mathbb{M} = (M, \delta, \vec{G}, \Sigma)$ is an externally iterable background and $A \subseteq \mathbb{R}$. We review the standard capturing notions (for example see [59] or [57] and references presented in those papers). We say $\mathbb{M}$ Suslin captures A at an M-cardinal η if there is a tree $T \in M$ such that whenever N is a Σ-iterate of M with $i : M \to N$ the iteration embedding and whenever g is $< i(\eta)$-generic over N, $(p[i(T)])^{N[g]} = A \cap N[g]$. We say $\mathbb{M}$ Suslin, co-Suslin captures A at η if it Suslin captures both A and A^c at η. We say $\mathbb{M}$ Suslin captures A if $\mathbb{M}$ Suslin captures A at $(\delta^+)^M$, and similarly $\mathbb{M}$ Suslin, co-Suslin captures A if $\mathbb{M}$ Suslin, co-Suslin captures A at $(\delta^+)^M$.

Finally we recall the notion of *self-capturing background* (Definition 2.24 of [29]).

DEFINITION 4.1.5. Suppose $\mathbb{M} = (M, \delta, \vec{G}, \Sigma)$ is a self-knowledgable, externally iterable background. We say $\mathbb{M}$ is *self-capturing* if Σ is positional[2] and for every M-inaccessible cardinal $\lambda < \delta$ there is a name $\dot{X} \in M^{Coll(\omega, M|\lambda)}$ such that for any M-generic $g \subseteq Coll(\omega, M|\lambda)$, $(M[g], \delta, \vec{G}, \Sigma)$[3] Suslin, co-Suslin captures $\mathsf{Code}(\Sigma_{M|\lambda})$ at $(\delta^+)^M$ as witnessed by $\dot{X}_g = (T, S)$.

Theorem 4.1.12 is the main method for producing self-capturing backgrounds.

4.1.1. Capturing pointclasses. We recall the definition of a good pointclass (see [57, Definition 9.12]). Unlike [57, Definition 9.12] we include scale property into the definition of good pointclass.

DEFINITION 4.1.6. We say Γ is a good pointclass if Γ is closed under recursive substitutions, is closed under quantification over ω, is closed under existential quantification over $\mathbb{R}$, is ω-parametrized[4] and has the scale property.

Suppose Γ is a good pointclass. For $x \in \mathbb{R}$, we let $C_\Gamma(x)$ be the largest countable $\Gamma(x)$-set of reals. For transitive $a \in \mathsf{HC}$[5] and surjection $g : \omega \to a$, we let a_g be the real coding $(a, \in)$ via g. More precisely,

$$a_g(k) = \begin{cases} 1 & \text{if } k = 2^m 3^n \text{ and } g(m) \in g(n), \\ 0 & \text{otherwise.} \end{cases}$$

Clearly $M_{a_g} = (a, \in)$. If $b \subseteq a$, then we let $b_g = \{m : g(m) \in b\}$. We then let $C_\Gamma(a) = \{b \subseteq a : \text{for comeager many } g : \omega \to a,\ b_g \in C_\Gamma(a_g)\}$.

Continuing with Γ, we say P is a Γ-Woodin if there is a P-cardinal δ_P such that

1. P is countable,
2. $P = C_\Gamma(C_\Gamma(V^P_{\delta_P}))$,
3. $P \vDash$ "δ_P is the only Woodin cardinal" and
4. for every $\eta < \delta_P$, $C_\Gamma(V_\eta) \vDash$ "η is not a Woodin cardinal".

---

[2]That is, whenever N is a Σ-iterate of M via $\mathcal{X}$, $\Sigma_{N,\mathcal{X}}$ is independent of $\mathcal{X}$. See [29, Definition 2.35].

[3]Here we abuse the notation a bit. In reality we should use Σ' which is the portion of Σ that acts on stacks above $\lambda + 1$.

[4]This means that there is $U \subseteq \omega \times \mathbb{R}$ such that $U \in \Gamma$ and $\{A \subseteq \mathbb{R} : A \in \Gamma\} = \{U_e : e \in \omega\}$.

[5]HC is the set of hereditarily countable sets.

We say (P,Ψ) is a Γ-*Woodin pair* if

1. Ψ is an ω_1-iteration strategy for P and
2. for every Ψ-iterate Q of P, Q is a Γ-Woodin[6].

Woodin, assuming AD^+, showed that if Γ is a good pointclass not closed under $\forall^{\mathbb{R}}$ then there are Γ-Woodin pairs (see [57, Theorem 10.3]). To learn more on Woodin's work one may consult [32].

Suppose Γ is a good pointclass and (P,Ψ) is a Γ-Woodin pair. Let $\mathcal{L}_\Psi$ be the extension of the language of set theory obtained by adding one predicate symbol $\dot{\Psi}$ and one constant symbol e. The intended interpretation of $\dot{\Psi}$ is $\mathsf{Code}(\Psi)$. e wlll denote a real number. Given $u \in \mathbb{R}$, we define $T'_n(\Psi,u)$ to be the set of $(\varphi,\vec{x})$ such that φ is a Σ_n-formula in $\mathcal{L}_\Psi$, $\vec{x} \in \mathbb{R}^m$ where m is the number of free variables of φ and

$$(\mathsf{HC},\mathsf{Code}(\Psi),u,\in) \models \varphi[\vec{x}].$$

We let $T'_n(\Psi) = T'_n(\Psi,0)$.

Next we code $T'_n(\Psi,u)$ by a set of reals as follows. First let G_Ψ be the set of natural numbers that are Gödel numbers for $\mathcal{L}_\Psi$-formulae. We say $y \in \mathbb{R}$ is Ψ-appropriate if $y(0)$ is a Gödel number of an $\mathcal{L}_\Psi$ formula. If y is Ψ-appropriate then we let φ_y be the formula that $y(0)$ codes and l_y be the number of free variables of φ_y. Let $(p_i : i < \omega)$ be the sequence of prime numbers in increasing order. For $i \leq l_y$, let $y_i \in \mathbb{R}$ be such that for all $k \in \omega$, $y_i(k) = y(p_i^{k+1})$. If y is Ψ-appropriate then we say y is neat if for all k' such that $k' \neq 0$ and $k' \notin \{p_i^k : i < l_y \wedge k \in \omega\}$, $y(k') = 0$. Let then $T_n(\Psi,u)$ be the set of Ψ-appropriate neat $y \in \mathbb{R}$ such that

$$(\varphi_y, \mathsf{merge}(y_i : i < l_y)) \in T'_n(\Psi,u).$$

Again, set $T_n(\Psi) = T_n(\Psi,0)$.

Suppose $z \in \mathbb{R}$, φ is an $\mathcal{L}_\Psi$-formula with $l+1$ free variables and $(x_i : 2 \leq i \leq l) \in \mathbb{R}^m$. Let $y_0 \in \mathbb{R}$ be such that $y_0(0)$ is the Gödel number of φ and for $i > 0$, $y_0(i) = 0$. Let $y_1 = z$ and for $2 \leq i \leq l$, $y_i = x_i$. Set $a(\varphi,z,\vec{x}) = \mathsf{merge}((y_i : i \leq l))$. Notice that $(\varphi,z,\vec{x})$ is uniquely determined by $a(\varphi,z,\vec{x})$. In fact, the function $(\varphi,z,\vec{x}) \mapsto a(\varphi,z,\vec{x})$ is a Π^0_1 injection.

Assuming AD, if $A \subseteq \mathbb{R}$ then $w(A)$ is its Wadge rank, and if Γ is a pointclass then $w(\Gamma) = \sup\{w(A) : A \in \Gamma\}$.

NOTATION 4.1.7. Suppose Γ is a pointclass closed under continous preimages and $A \subseteq \mathbb{R}$. We say A is a least upper bound for Γ if $\Gamma = \{B \subseteq \mathbb{R} : w(B) < w(A)\}$. Set then $lub(\Gamma) = \{A \subseteq \mathbb{R} : A$ is a least upper bound for $\Gamma\}$.

DEFINITION 4.1.8. Suppose Γ is any pointclass closed under the continuous preimages. We say that the tuple $(\mathbb{M},(\mathcal{P},\Psi),\Gamma^*,A)$ Suslin, co-Suslin captures Γ if the following conditions hold:

---

[6] P is a coarse structure, there is no notion of dropping for iterations of P, so P-to-Q embedding always exists.

1. $A \in lub(\Gamma)$,
2. Γ^* is the least good pointclass such that $\Gamma \subseteq \underset{\sim}{\Delta}_{\Gamma^*}$.
3. (P,Ψ) is a Γ^*-Woodin pair.
4. (P,δ_P,Ψ) Suslin, co-Suslin captures A.
5. $\mathbb{M}$ is a self-capturing background.
6. $\mathbb{M}$ Suslin, co-Suslin captures the sequence $(T_n(\Psi) : n < \omega)$.

NOTATION 4.1.9. Suppose Γ is a pointclass closed under the continuous preimages, $\mathsf{C} = (\mathbb{M},(\mathcal{P},\Psi),\Gamma^*,A)$ Suslin, co-Suslin captures Γ and $\mathbb{M} = (M,\delta,\vec{G},\Sigma)$. If N is a Σ-iterate of M then we set $\mathsf{C}_N = (\mathbb{M}_N,(\mathcal{P},\Psi),\Gamma^*,A)$.

The following is an important yet straightforward lemma that we will use throughout this book.

TERMINOLOGY 4.1.10. Below and throughout this book we say that "g is $<\eta$-generic" to mean that the poset for which g is generic has size $<\eta$. Similarly we say that "g is $\leq\eta$ generic" to mean that the poset for which g is generic has size $\leq\eta$.

LEMMA 4.1.11 (Correctness of backgrounds). *Suppose* $(\mathbb{M},(\mathcal{P},\Psi),\Gamma^*,A)$ *Suslin, co-Suslin captures* Γ *and set* $\mathbb{M} = (M,\delta,\vec{G},\Sigma)$. *Suppose* $x \in \mathbb{R}\cap M$. *Let* $(S_n,U_n : n<\omega) \in M$ *be the sequence of trees on* $\omega\times(\delta^+)^M$ *such that* (S_n,U_n) *Suslin, co-Suslin captures* $T_n(\Psi)$. *Let g be $<\delta$-generic over M. Then for any real* $u \in M[g]$,

$$(\mathsf{HC}^{M[g]},\mathsf{Code}(\Psi)\cap M[g],u,\in) \prec (\mathsf{HC},\mathsf{Code}(\Psi),u,\in).$$

PROOF. It is enough to verify that if φ is a formula, $m+1$ is the number of its free variables and $\vec{x}\in\mathbb{R}^m\cap M[g]$ then if $(\mathsf{HC},\mathsf{Code}(\Psi),\in)\vDash\exists v\varphi[v,\vec{x}]$ then there is $v\in M[g]\cap\mathbb{R}$ such that $(\mathsf{HC},\mathsf{Code}(\Psi),\in)\vDash\varphi[v,\vec{x}]$. Let n be such that φ is Σ_n. Then there is v such that $a(\varphi,v,\vec{x})\in T_n(\Psi)$.

Working in $M[g]$, let $S' = \{(s,h)\in S_n : s(0) \text{ is the Gödel number of } \varphi\}$, and let S be the tree on $\omega\times\delta$ whose branches are pairs $(y',f)\in\mathbb{R}\times\delta^\omega$ with the property that if $y = a(\varphi,y',\vec{x})$ then $(y,f)\in[S']$.

We now have that whenever h is any $Coll(\omega,\delta)$-generic extension of $M[g]$, in $M[g][h]$, $p[S]$ is the set of $y'\in\mathbb{R}^{M[g][h]}$, such that if $y = a(\varphi,y',\vec{x})$ then $y\in p[S']$. Because S_n Suslin captures $T_n(\Psi)$ we have that $(p[S])^{M[g]}\neq\emptyset$[7]. Notice next that if $v\in p[S]\cap M[g]$ then $(\mathsf{HC},\mathsf{Code}(\Psi),\in)\vDash\varphi[v,\vec{x}]$. ⊣

Self-capturing backgrounds are very useful for building hod pairs and proving comparison. The following theorem of Woodin shows that under AD^+, self-capturing backgrounds are abundant. [32] has an outline of the proof of Theorem 4.1.12.

[7]This follows from genericity iterations. One can iterate $M[g]$ via Σ to obtain $i : M[g]\to N$ such that $(p[i(S)])^N\neq\emptyset$. For example if v is generic over N for the extender algebra at $i(\delta)$ then $v\in(p[i(S)])^N$.

THEOREM 4.1.12 (Woodin, Theorem 10.3 of [57]). *Assume* AD^+. *Suppose* Γ *is a good pointclass and there is a good pointclass* Γ^* *such that* $\Gamma \subseteq \Delta_{\Gamma^*}$. *Suppose* (N,Ψ) *is* Γ^**-Woodin which Suslin, co-Suslin captures some* $A \in lub(\Gamma)$. *There is then a function* F *defined on* $\mathbb{R}$ *such that for a Turing cone of* x, $F(x) = (\mathcal{N}_x^*, \mathcal{M}_x, \delta_x, \Sigma_x)$ *is such that*

1. $N \in L_1[x]$,
2. $\mathcal{N}_x^* | \delta_x = \mathcal{M}_x | \delta_x$,
3. $\mathcal{M}_x$ *is a* Ψ*-mouse over* x*: in fact,* $\mathcal{M}_x = \mathcal{M}_1^{\Psi,\#}(x) | \kappa_x$ *where* κ_x *is the least inaccessible cardinal of* $\mathcal{M}_1^{\Psi,\#}(x)$ *that is* $> \delta_x$,
4. $\mathcal{N}_x^* \models$ "δ_x *is the only Woodin cardinal*",
5. Σ_x *is the unique iteration strategy of* $\mathcal{M}_x$,
6. $\mathcal{N}_x^* = L(\mathcal{M}_x, \Lambda)$ *where* $\Lambda = \Sigma_x \restriction \mathrm{dom}(\Lambda)$ *and*

$$\mathrm{dom}(\Lambda) = \{\mathcal{T} \in \mathcal{M}_x : \mathcal{T} \textit{ is a normal iteration tree on } \mathcal{M}_x,$$
$$\mathrm{lh}(\mathcal{T}) \textit{ is a limit ordinal and } \mathcal{T} \textit{ is below } \delta_x\},$$

7. *setting* $\vec{G} = \{(\alpha, \vec{E}^{\mathcal{N}_x^*}(\alpha)) : \mathcal{N}_x^* \models \text{"}\mathrm{lh}(\vec{E}^{\mathcal{N}_x^*}(\alpha))$ *is an inaccessible cardinal* $< \delta_x$"$\}$ *and* $\mathbb{M}_x = (\mathcal{N}_x^*, \delta_x, \vec{G}, \Sigma_x)$, $(\mathbb{M}_x, (N,\Psi), \Gamma^*, A)$ *Suslin, co-Suslin captures* Γ[8].

4.1.2. The meaning of Lp^Γ, HP^Γ **and** Mice^Γ**.** The reader may find it helpful to review Definition 3.9.4 and Definition 3.10.7. Recall that we say X is *self-well-ordered* if there is a wellordering of $\lfloor X \rfloor$ in $\mathcal{J}_1(X)$ definable over $\mathcal{J}_0(X)$.

DEFINITION 4.1.13 (The Lp function). Suppose Γ is a pointclass and $(\mathcal{P}, \Sigma)$ is an allowable pair[9] such that $\mathsf{Code}(\Sigma) \in \Gamma$. Suppose X is a self-well-ordered set such that $\mathcal{P} \in X$.

1. If Σ is an iteration strategy then $\mathsf{Lp}^{\Gamma,\Sigma}(X)$ is the stack of all sound (Σ, sis)-mice $\mathcal{M}$ over X based on $\mathcal{P}$[10] such that $\rho(\mathcal{M}) = \mathrm{ord}(trc(X))$[11] and $\mathcal{M}$ has a strategy in Γ.
2. If Σ is a st-strategy $\mathsf{Lp}^{\Gamma,\Sigma}(X)$ is the stack of all sound Σ-sts mice $\mathcal{M}$ over X based on $\mathcal{P}$ such that $\rho(\mathcal{M}) = \mathrm{ord}(trc(X))$ and $\mathcal{M}$ has a strategy in Γ[12].

We set $\mathsf{Lp}^{\Gamma,\Sigma}(\mathcal{P}) = \mathsf{Lp}^{\Gamma,\Sigma}(\mathcal{J}_\omega[\mathcal{P}])$.

Below if Ψ is an iteration strategy or an st-strategy then we let M_Ψ be the structure that Ψ is iterating.

---

[8] Hence, $(\mathcal{N}_x^*, \delta_x, \vec{G}, \Sigma_x)$ is a self-capturing background.

[9] See Definition 3.10.7.

[10] See Definition 2.5.2, Definition 2.5.8 and Definition 3.9.4.

[11] Our fine structural notation was introduced in Definition 2.2.3.

[12] From here on, "Lp" means "g-organized Lp" as defined in [50] unless explicitly stated otherwise. We will occasionally remind the reader of this convention. The reason we need to use g-organization is so that S-constructions go through.

NOTATION 4.1.14. Suppose Γ is a pointclass. Following Section 2.5 of [29] we let

$$\mathsf{Hp}^\Gamma = \{(\mathcal{P},\Sigma) : (\mathcal{P},\Sigma) \text{ is an allowable pair such that } \mathsf{Code}(\Sigma) \in \Gamma\},$$
$$\mathsf{Mice}^\Gamma = \{(a,\Sigma,\mathcal{M}) : a \in \mathsf{HC} \wedge a \text{ is a self-well-ordered} \wedge (M_\Sigma,\Sigma) \in \mathsf{Hp}^\Gamma \wedge \mathcal{M}_\Sigma \in a \wedge \mathcal{M} \trianglelefteq \mathsf{Lp}^{\Gamma,\Sigma}(a) \wedge \rho(\mathcal{M}) = \mathrm{ord}(trc(a))\}$$

and given $(\mathcal{P},\Sigma) \in \mathsf{Hp}^\Gamma$,

$$\mathsf{Mice}^\Gamma_\Sigma = \{(a,\mathcal{M}) : a \in \mathsf{HC} \wedge a \text{ is a self-well-ordered} \wedge \mathcal{P} \in a \wedge \mathcal{M} \trianglelefteq \mathsf{Lp}^{\Gamma,\Sigma}(a) \wedge \rho(\mathcal{M}) = \mathrm{ord}(trc(a))\}.$$

When $\Gamma = \wp(\mathbb{R})$, we omit it from our notation.

Suppose $A \subseteq \mathbb{R}$ with $w(\Gamma) \leq w(A)$. We say $\sigma \in \mathbb{R}$ is an A-code if $\sigma(0)$ is a Gödel number for some formula φ, and if B is the set of reals definable over $(\mathsf{HC},A,\sigma,\in)$ via φ[13] then $B \in \mathrm{rge}(\mathsf{Code})$. We then let $C_\sigma = \mathsf{Code}^{-1}(B)$ and $A\mathsf{Code}$ be the set of A-codes.

Given a set $A \subseteq \mathbb{R}$ with $w(\Gamma) \leq w(A)$, we let $\mathsf{Code}(\mathsf{Hp}^\Gamma,A)$ be the set of $\sigma \in A\mathsf{Code}$ such that $C_\sigma \in \mathsf{Hp}^\Gamma$. If $\sigma \in \mathsf{Code}(\mathsf{Hp}^\Gamma,A)$ then we let $(\mathcal{P}_\sigma,\Sigma_\sigma)$ be the pair determined by σ.

Given a set $A \subseteq \mathbb{R}$ with $w(\Gamma) \leq w(A)$, we let $\mathsf{Code}(\mathsf{Mice}^\Gamma,A)$ be the set of (σ_0,σ_1) such that $\sigma_0 \in A\mathsf{Code}$, $\sigma_1 \in A\mathsf{Code}$ and $C_{\sigma_1} = \mathsf{Mice}^\Gamma_{\Sigma_{\sigma_0}}$.

Given a set $A \subseteq \mathbb{R}$ with $w(\Gamma) \leq w(A)$, we let A_Γ be the set of triples $(\sigma_0,\sigma_1,\sigma_2)$ such that

1. For each $i < 3$, $\sigma_i \in A\mathsf{Code}$,
2. $\sigma_0 \in \mathsf{HP}^\Gamma$,
3. $C_{\sigma_1} = (a,\Sigma_{\sigma_0},\mathcal{M}) \in \mathsf{Mice}^\Gamma$,
4. C_{σ_2} is the unique ω_1-iteration strategy of $\mathcal{M}$.

The following is an easy consequence of Lemma 4.1.11. It follows from the fact that each of

$$\mathsf{Code}(A)_\Gamma,\ \mathsf{Code}(\mathsf{Hp}^\Gamma,A) \quad \text{and} \quad \mathsf{Code}(\mathsf{Mice}^\Gamma,A).$$

is definable over $(\mathsf{HC},\mathsf{Code}(\Psi),\in)$, where (P,Ψ) is as below.

COROLLARY 4.1.15. *Suppose $\mathbb{M} = (M,\delta,\vec{G},\Sigma)$ and $(\mathbb{M},(P,\Psi),\Gamma^*,A)$ Suslin, co-Suslin captures Γ. Then $\mathbb{M}$ Suslin, co-Suslin captures*

$$\mathsf{Code}(A)_\Gamma,\ \mathsf{Code}(\mathsf{Hp}^\Gamma,A) \quad \textit{and} \quad \mathsf{Code}(\mathsf{Mice}^\Gamma,A).$$

[13]That is, $u \in B \iff (\mathsf{HC},A,\sigma,\in) \vDash \varphi[u]$.

4.1.3. Internalizing HP^Γ**.** Suppose next that Γ is a pointclass and $(\mathbb{M},(P,\Psi),\Gamma^*,A)$ Suslin, co-Suslin captures Γ. In Section 4.3, we will describe the Γ-hod pair construction of $\mathbb{M}$ that produces hod pairs. When describing this construction, we will use the following concepts and simple observations.

DEFINITION 4.1.16. Suppose M is a transitive model of ZFC, $X \in M$ and φ is a formula. We say (X,φ) is (M,η)*-generically absolute* if for some $\vartheta \geq \eta$ such that $X \in V_\vartheta^M$, for all $Y \prec (V_\vartheta^M, X, \eta \in)$ such that $Y \in M$ and $M \models$ "Y is countable", letting N_Y be the transitive collapse of Y and $\pi_Y : Y \to N_Y$ be the collapse map, whenever $g \in M$ is $\leq \pi(\eta)$-generic over N_Y and $x \in N_Y[g] \cap \mathbb{R}$,

$$N_Y[g] \models \varphi[\pi_Y(X), x] \iff M \models \varphi[X,x].$$

DEFINITION 4.1.17. Suppose M is a transitive model of ZFC, $X \in M$ and φ is a formula. We say (X,φ) is (M,η,α)*-generically absolute* if $\alpha < \eta$ and whenever $g \subseteq Coll(\omega, M|\alpha)$ is M-generic, $((X,g),\varphi)$ is $(M[g],\eta)$*-generically absolute.*

The following theorem can be proven by using the *Tree Production Lemma* (see [58, Lemma 4.1]).

LEMMA 4.1.18. *Suppose M is a transitive model of* ZFC *and (X,φ) is (M,η)-generically absolute. There is then a pair $(T,S) \in OD_X^M$ such that (T,S) is $\leq \eta$-absolutely complementing and whenever g is $\leq \eta$-generic*

$$(p[T])^{M[g]} = \{x : M[g] \models \varphi[X,x]\}.$$

DEFINITION 4.1.19. Suppose Γ is a pointclass, $(\mathbb{M},(P,\Psi),\Gamma^*,A)$ Suslin, co-Suslin captures Γ, $\mathbb{M} = (M,\delta,\vec{G},\Sigma)$ and (X,φ) is (M,δ,α)-generically absolute. We then write

$$M \models (X,\varphi) \in \mathsf{Hp}^\Gamma$$

to mean that the following holds.

Whenever $g \subseteq Coll(\omega, M|\alpha)$ is M-generic, there is a real

$$\sigma \in M[g] \cap \mathsf{Code}(\mathsf{Hp}^\Gamma, A)$$

such that letting τ be the formula coded by $\sigma(0)$, whenever h is $\leq \delta$-generic over $M[g]$, in $M[g][h]$,

$$\{x \in \mathbb{R} : \varphi[(X,g),x]\} = \{x : (\mathsf{HC}^{M[g][h]}, A \cap M[g][h], \sigma, \in) \models \tau[c_x]\}.$$

where c_x is the set coded by x.

The following lemma is a straightforward consequence of genericity iterations.

LEMMA 4.1.20. *Suppose Γ is a pointclass, $(\mathbb{M},(P,\Psi),\Gamma^*,A)$ Suslin, co-Suslin captures Γ, $\mathbb{M} = (M,\delta,\vec{G},\Sigma)$, (X,φ) is (M,δ,α)-generically absolute and $M \models (X,\varphi) \in \mathsf{Hp}^\Gamma$. Suppose $g \subseteq Coll(\omega,M|\alpha)$ is M-generic and $\sigma_0,\sigma_1 \in M[g] \cap \mathsf{Code}(\mathsf{Hp}^\Gamma, A)$ are two reals witnessing that $M \models (X,\varphi) \in \mathsf{Hp}^\Gamma$. Then $(\mathcal{P}_{\sigma_0},\Sigma_{\sigma_0}) = (\mathcal{P}_{\sigma_1},\Sigma_{\sigma_1})$.*

The following now is not hard to show. It follows from Lemma 4.1.11, which implies that

$$(\mathsf{HC}^{M[g][h]}, A \cap M[g][h], \sigma, \in) \prec (\mathsf{HC}, A, \sigma, \in),$$

and also from Lemma 4.1.18.

LEMMA 4.1.21. *Suppose Γ is a pointclass, $(\mathbb{M}, (P, \Psi), \Gamma^*, A)$ Suslin, co-Suslin captures Γ, $\mathbb{M} = (M, \delta, \vec{G}, \Sigma)$ and (X, φ) is (M, δ, α)-generically absolute. Suppose further that $g \subseteq Coll(\omega, M|\alpha)$ is M-generic, $\sigma \in M[g] \cap \mathsf{Code}(\mathsf{Hp}^\Gamma, A)$ witnesses that $M \vDash (X, \varphi) \in \mathsf{Hp}^\Gamma$ and τ is the formula coded by $\sigma(0)$. Set $u \in C$ if and only if*

> *there is an iteration $i : M \to N$ according to Σ such that* $\mathrm{crit}(i) > \alpha$ *and for some $N[g]$-generic $h \subseteq Coll(\omega, i(\delta))$, $u \in N[g][h]$ and $N[g][h] \vDash \tau[c_x]$.*

Then $\mathsf{Code}^{-1}(C) = (\mathcal{P}_\sigma, \Sigma_\sigma) \in \mathsf{HP}^\Gamma$ and C is Suslin, co-Suslin captured by $(M[g], \delta, \Sigma)$.

4.2. Fully backgrounded constructions relative to short tree strategy

Suppose $(M, \delta, \vec{G}, \Sigma)$ is an iterable background and $\mathcal{P} \in V_\delta^M$ is a #-lsa type hod premouse (see Definition 2.7.3). Suppose $\Lambda \in M$ is a (δ, δ, δ) st-strategy for $\mathcal{P}$ and $X \in V_\delta^M$ is a transitive self-well-ordered set such that $\mathcal{P} \in X$. We can then define the model $\mathcal{J}^{\vec{E}, \Lambda}(X)$ exactly like in the case Λ is an iteration strategy. The construction will ensure that the model $\mathcal{J}^{\vec{E}, \Lambda}(X)$ is an sts premouse over X based on $\mathcal{P}$. Here is the precise definition.

Recall that if $(\mathcal{M}_\alpha : \alpha < \xi)$ is a sequence of $\mathcal{J}$-structures and ξ is a limit ordinal then $\mathcal{M} = lim_{\alpha \to \xi} \mathcal{M}_\alpha$ is the $\mathcal{J}$-structure with the property that for each β such that $\mathcal{J}_\beta^{\mathcal{M}}$ is defined, there is $\gamma < \xi$ such that for all $\alpha \in (\gamma, \xi)$, $\mathcal{J}_\beta^{\mathcal{M}_\alpha} = \mathcal{J}_\beta^{\mathcal{M}}$.

Suppose $(M, \delta, \vec{G}, \Sigma)$ is an internally or externally iterable background, $A \subseteq V_\delta^M$ and $E \in V_\delta^M$ is an extender. Then we say E *coheres* or *reflects* A if $\nu(E)$ is an inaccessible cardinal of M, $V_{\nu(E)}^M \subseteq \mathrm{Ult}(\mathcal{M}, E)$ and $A \cap V_{\nu(E)}^M = \pi_E(A) \cap V_{\nu(E)}^M$. Recall that an lses $\mathcal{M}$ is reliable if for all k, $\mathsf{core}_k(\mathcal{M})$ exists and $(\mathsf{core}_k(\mathcal{M}), k)$ is $\omega_1 + 1$-iterable (see Definition 2.2.3 and [23, Chapter 11]). Finally recall our notation $\lfloor M \rfloor$ denoting the universe of M. This notation was introduced in Section 2.1. Finally recall that sts premice are sts-indexed (see Definition 3.8.16 and Definition 3.8.17).

As was stated many times, in this book we are mostly concerned with new issues that arise from dealing with sts mice. Reproving all the well-established facts will add 1000s of more pages to this book without adding any new ideas. In particular, our exposition of the fully backgrounded constructions heavily relies on

[23] and [48, Chapter 5]. The later proves the uniqueness of the next extender in full generality.

DEFINITION 4.2.1. Suppose $(M,\delta,\vec{G})$ is a background and $(\mathcal{P},\Lambda)\in M$ is an sts hod pair[14], Λ is (δ,δ,δ) st-strategy for $\mathcal{P}$ and $X\in V_\delta^M$ is a transitive self-well-ordered set such that $\mathcal{P}\in X$. Suppose further that Λ has hull condensation. Then

$$\mathsf{Le}((\mathcal{P},\Lambda),X)=(\mathcal{M}_\gamma,\mathcal{N}_\gamma,F_\gamma^+,F_\gamma,b_\gamma:\gamma\leq\delta)$$

is the output of *the fully backgrounded construction of* $(M,\delta,\vec{G})$ *relative to* Λ *done over* X *using the coherence condition* if the following conditions hold.

1. $\mathcal{M}_0=\mathcal{J}_\omega(X)$, and for all $\gamma<\delta$, each of $\mathcal{M}_\gamma$ and $\mathcal{N}_\gamma$ is either undefined or is a Λ-sts premouse[15].

2. If for some $\xi\leq\eta$, $\mathcal{N}_\xi$ is defined but is not a reliable sts premouse over X based on $\mathcal{P}$ then all other objects with index $\geq\xi$ are undefined.

3. Suppose for some $\xi<\delta$, for all $\gamma\leq\xi$, both $\mathcal{M}_\gamma$ and $\mathcal{N}_\gamma$ are defined. Then $\mathcal{M}_{\xi+1}$, $\mathcal{N}_{\xi+1}$, F_ξ^+, F_ξ and b_ξ are determined as follows.

(a) Suppose $\mathcal{M}_\xi=(\mathcal{J}_{\omega\alpha}^{\vec{E},f},\in,\vec{E},f)$ is a passive ses[16] and there is an extender $F^*\in\vec{G}$, an extender F over $\mathcal{M}_\xi$, and an ordinal $\nu<\omega\alpha$ such that
  (i) $\nu<\nu(F^*)$,
  (ii) $F=F^*\cap([\nu]^\omega\times\lfloor\mathcal{M}_\xi\rfloor)$, and
  (iii) setting

$$\mathcal{N}_{\xi+1}=(\mathcal{J}_{\omega\alpha}^{\vec{E},f},\in,\vec{E},f,\tilde{F})$$

  where $\tilde{F}$ is the amenable code of F[17], clause 2 fails for $\xi+1$.

  Then $\mathcal{M}_{\xi+1}=\mathsf{core}(\mathcal{N}_{\xi+1})$[18], $F_\xi^+=\vec{G}(\xi)$ where ξ is the least such that $F^*=\vec{G}(\xi)$ has the above properties, $F_\xi=F^+\cap([\nu]^\omega\times\lfloor\mathcal{M}_\xi\rfloor)$ where ν is chosen so that the above clauses hold and $b_\xi=\emptyset$.

(b) Suppose $\mathcal{M}_\xi=(\mathcal{J}_{\omega\alpha}^{\vec{E},f},\in,\vec{E},f)$ is a passive ses, $\alpha=\beta+\gamma$ and there is $t=(\mathcal{P}_0,\mathcal{T},\mathcal{P}_1,\mathcal{U})\in\lfloor\mathcal{M}_\xi|\omega\beta\rfloor\cap\mathrm{dom}(\Lambda)$ such that setting $w=(\mathcal{J}_\omega(t),t,\in)$, w is (f,sts)-minimal as witnessed by β[19] and $\gamma=\mathrm{lh}(t)$. Set $b=\Lambda(t)$ and

$$\mathcal{N}_{\xi+1}=(\mathcal{J}_{\omega\beta+\omega\gamma}^{\vec{E},f^+},\in,\vec{E},f,\tilde{b})$$

---

[14] In particular, Λ has hull condensation. An easy Skolem hull and a realizability argument implies that if E is a countably complete total extender in M then $\pi_E(\Lambda)=\Lambda\restriction\mathrm{Ult}(M,E)$.

[15] See Definition 3.8.21.

[16] That is, with no last predicate.

[17] For the definition of the "amenable code" see the last paragraph on page 14 of [59].

[18] Recall that $\mathsf{core}(\mathcal{M})$ is the core of $\mathcal{M}$.

[19] See Definition 2.3.3. In particular, this means that we have to index the branch of t at $\omega\alpha$.

where $\tilde{b} \subseteq \omega\beta + \omega\gamma$ is defined by $\omega\beta + \omega\nu \in \tilde{b} \iff \nu \in b$. Assuming clause 2 fails for $\xi + 1$, $\mathcal{M}_{\xi+1} = \mathsf{core}(\mathcal{N}_{\xi+1})$, $F_\xi^+ = F_\xi = \emptyset$ and $b_\xi = \tilde{b}$.

Important Anomaly: Suppose t is nuvs and suppose $e \in \mathcal{M}_\xi|\omega\beta$ is such that $\mathcal{M}_\xi|\omega\beta \models \mathsf{sts}_0(t,e)$[20]. If $e \neq b$ then $\mathcal{N}_{\xi+1}$ is not an sts premouse over X based on $\mathcal{P}$, and so clause 2 holds.

(c) If $\mathcal{M}_\xi$ does not satisfy clause 2(a) or 2(b) then set $\mathcal{N}_{\xi+1} = \mathcal{J}_\omega[\mathcal{M}_\xi]$. Assuming clause 2 fails for $\xi + 1$, $\mathcal{M}_{\xi+1} = \mathsf{core}(\mathcal{N}_{\xi+1})$, $F_\xi^+ = F_\xi = b_\xi = \emptyset$.

4. Suppose $\xi \leq \delta$ is a limit ordinal and for all $\gamma < \xi$, both $\mathcal{M}_\gamma$ and $\mathcal{N}_\gamma$ are defined. Then $\mathcal{M}_\xi$ and $\mathcal{N}_\xi$ are determined as follows[21]. Set $\mathcal{N}_\xi = lim_{\alpha\to\xi}\mathcal{M}_\alpha$. Assuming clause 2 fails for ξ, $\mathcal{M}_\xi = \mathsf{core}(\mathcal{N}_\xi)$.

5. $\mathcal{M}_\delta = \mathcal{N}_\delta$ and $F_\delta^+ = F_\delta = b_\delta = \emptyset$.

We say that $\mathsf{Le}((\mathcal{P},\Lambda),X)$ is *successful* if for all $\xi < \delta$ clause 2 above fails. Given $\kappa < \delta$, we can also define $\mathsf{Le}((\mathcal{P},\Lambda),X)_{\geq\kappa}$ by requiring that in clause 3(a), $\mathrm{crit}(F) \geq \kappa$.

We will use the following terminology. We say $\mathcal{Q}$ is an $\mathcal{N}$-*model* of

$$\mathsf{Le}((\mathcal{P},\Lambda),X)_{\geq\kappa}$$

if for some $\gamma \leq \delta$, $\mathcal{Q} = \mathcal{N}_\gamma$. Similarly we define $\mathcal{M}$-model and other such expressions. We say $\mathcal{Q}$ is *the last model* of $\mathsf{Le}((\mathcal{P},\Lambda),X)_{\geq\kappa}$ if $\mathcal{Q} = \mathcal{N}_\delta$.

The fully backgrounded constructions of both [23] and [48, Chapter 5] do not use the coherence condition. In most cases considered in this book, we also do not need the coherence condition. The following theorem is essentially a corollary to [23, Chapter 12].

THEOREM 4.2.2. *Suppose $(M,\delta,\vec{G},\Sigma)$ is an iterable background and $(\mathcal{P},\Lambda) \in M$ is an sts hod pair, Λ is (δ,δ,δ) st-strategy for $\mathcal{P}$ and $X \in V_\delta^M$ is a transitive self-well-ordered set such that $\mathcal{P} \in X$. Then for any $\kappa < \delta$, $\mathsf{Le}((\mathcal{P},\Lambda),X)_{\geq\kappa}$ is not successful if and only if for some $\xi < \delta$, the Anomaly stated in clause 3(b) of Definition 4.2.1 holds.*

REMARK 4.2.3. Assuming that $(\mathcal{P},\Lambda)$ is a pair with the property that $\mathcal{P}$ is an lses and Λ is an iteration strategy for $\mathcal{P}$ with hull condensation, we could define $\mathsf{Le}((\mathcal{P},\Lambda),X)_{\geq\kappa}$ just like above except in clause 3(b) we require t to be a stack on $\mathcal{P}$ according to Λ, $b = \Lambda(t)$ and w be (f,sis)-minimal (see Definition 3.9.4). $\mathsf{Le}((\mathcal{P},\Lambda),X)_{\geq\kappa}$ can be defined for various types of strategies; in particular, it can be defined for ω_1-strategies and for (ω_1,ω_1)-strategies. We also let Le be the construction relative the $\emptyset$. Thus, the models of Le are simply ordinary premice. We leave the details of the above mentioned constructions to the reader who may want to consult [29, Definition 2.3].

[20] See Definition 3.8.16. This means that e is the branch of t we must choose.

[21] F_ξ, b_ξ will be defined at the next stage of the induction as in clause 2.

The important comment in clause 3(b) is a non-trivial matter. Recall that according to our sts indexing scheme (see Definition 3.8.9), the branch we have to index at stage ξ in clause 3(b) is e not b. However, if $e \neq b$ then the resulting structure cannot be a Λ-sts mouse. Thus, if $e \neq b$ then we have to halt the construction. When Λ has nice properties such as *strong branch condensation* (see Definition 4.9.2) then such anomaly will never arise. See Remark 4.12.6 for an in-depth discussion of this issue.

4.3. Hod pair constructions

In this section we introduce the Γ-hod pair constructions. The goal of such a construction is to produce a hod pair $(\mathcal{P},\Sigma)$ such that $w(\mathsf{Code}(\Sigma)) \geq w(\Gamma)$ but for any hod initial segment $\mathcal{Q} \lhd \mathcal{P}$, $w(\mathsf{Code}(\Sigma_{\mathcal{Q}})) < w(\Gamma)$ (or equivalently $(\mathcal{Q},\Sigma_{\mathcal{Q}}) \in \mathsf{Hp}^{\Gamma}$).

The reader may benefit from reviewing the concept of fully backgrounded constructions as presented in [23, Chapter 11 and Chapter 12]. Such constructions inherit a strategy from the background model[22] via the procedure described in [23, Chapter 12]. Other forms of such constructions also have appeared in [47] and [46].

Suppose Γ is a good pointclass and $\mathbb{M} = (M, \delta, \vec{G}, \Sigma)$ is a background Suslin, co-Suslin capturing Γ (see Definition 4.1.8). We will work with $\mathbb{M}$ and Γ, but we will omit both from our notations.

All concepts introduced here depend on $\mathbb{M}$. For instance, E below should really be $\mathsf{E}^{\mathbb{M}}$. Also, all fully backgrounded constructions that we will use are fully backgrounded constructions in the sense of V_{δ}^{M}, and if M is equipped with a distinguished extender sequence then we tacitly assume that all the backgounded constructions use extenders from this particular extender sequence.

The reader may find it helpful to review Definition 2.7.8, Definition 2.7.14, Terminology 2.7.17, Definition 2.7.18 and Definition 3.10.7. We start by introducing those hod premice that can be used as layers in the Γ-hod pair construction.

DEFINITION 4.3.1. We say that an allowable pair $(\mathcal{R},\Lambda)$ *can be* Γ*-layered* or is just $\Gamma - \mathsf{cbl}$ if one of the following conditions hold:

1. $\mathcal{R}$ is a hod premouse of successor type and $\mathcal{R} = \mathsf{Lp}_{\omega}^{\Gamma,\Lambda_{\mathcal{R}^-}}(\mathcal{R}|\delta^{\mathcal{R}})$.
2. $\mathcal{R}$ is a properly non-meek[23] hod premouse of limit type and letting $\kappa = \delta^{\mathcal{R}^b}$,
$$\mathcal{R}^b = \mathsf{Lp}^{\Gamma,\Lambda_{\mathcal{R}|\kappa}}(\mathcal{R}^b|\kappa).$$
3. $\mathcal{R}$ is a gentle hod premouse such that if $\mathcal{Q} \in Y^{\mathcal{R}}$ then $(\mathcal{Q},\Lambda_{\mathcal{Q}})$ is $\Gamma - \mathsf{cbl}$.

[22] The model where the construction is being done.

[23] See Definition 2.7.2.

Recall that an lses $\mathcal{M}$ over $\emptyset$ set has a predicate, $Y^{\mathcal{M}}$, whose members are the layers of the lses (see Definition 2.3.13). Thus, below, when describing $\mathcal{M}_\gamma$, we must also simultaneously define $Y^{\mathcal{M}_\gamma}$.

Φ_γ below will be the iteration strategy induced by Σ essentially via the resurrection procedure describe in [23, Chapter 12]. The procedure described in [23, Chapter 12] only induces (ω_1, ω_1)-iteration strategies, but it is not hard to modify it to obtain an $(\omega_1, \omega_1, \omega_1)$-strategy (see Definition 2.10.6). We will give an outline of how to do this after Definition 4.3.3.

TERMINOLOGY 4.3.2. Suppose $\mathcal{M}$ is an lses and $\alpha \leq \text{ord}(\mathcal{M})$. We say that a stack $\mathcal{T}$ on $\mathcal{M}$ is *below* α if for every $\gamma < \text{lh}(\mathcal{T})$ such that $[0,\gamma)_{\mathcal{T}} \cap D^{\mathcal{T}} = \emptyset$, $\text{ind}_\gamma^{\mathcal{T}} < \pi_{0,\gamma}^{\mathcal{T}}(\alpha)$. Similarly we define the meaning of "below α" for generalized stacks.

Suppose $\mathcal{M}$ is an lses and $\mathcal{Q} \trianglelefteq \mathcal{M}$. We say Σ is a strategy of $\mathcal{M}$ *based* on $\mathcal{Q}$ if whenever $\mathcal{T}$ is according to Σ, $\mathcal{T}$ is below $\text{ord}(\mathcal{Q})$.

As was mentioned before, our exposition of the fully backgrounded constructions heavily relies on [23] and [48, Chapter 5]. As was mentioned before, the later reference proves the uniqueness of the next extender in full generality.

DEFINITION 4.3.3. Suppose Γ is a pointclass, $\mathsf{C} = (\mathbb{M}, (P, \Psi), \Gamma^*, A)$ Suslin, co-Suslin captures Γ and $\mathbb{M} = (M, \delta, \vec{G}, \Sigma)$. Then

$$\mathsf{hpc} = (\mathcal{M}_\gamma, \mathcal{N}_\gamma, Y_\gamma, \Phi_\gamma, F_\gamma^+, F_\gamma, b_\gamma : \gamma \leq \delta)$$

is the output of the Γ-*hod pair construction* ($\Gamma - \mathsf{hpc}$) of $\mathbb{M}$ if the following conditions hold (the construction is over $\emptyset$).

1. $\mathcal{M}_0 = \mathcal{J}_\omega$, and for all $\gamma \leq \delta$, each of $\mathcal{M}_\gamma$ and $\mathcal{N}_\gamma$ is either undefined or is an hp-indexed lses (see Definition 3.9.2).
2. For all $\gamma \leq \delta$, if $\mathcal{M}_\gamma$ is defined then $Y_\gamma = Y^{\mathcal{M}_\gamma}$ (see Definition 2.3.13).
3. For all $\gamma \leq \delta$, if $\mathcal{M}_\gamma$ is defined then Φ_γ is the strategy defined in Definition 4.3.8[24].
4. For all $\gamma \leq \delta$, if $\mathcal{N}_\gamma$ is defined and either

(a) $\mathcal{N}_\gamma$ is not a reliable hp-indexed lses[25] or
(b) $\mathcal{N}_\gamma$ is a reliable hp-indexed lses but for some $\mathcal{Q} \in Y^{\mathcal{N}_\gamma}$ such that $\mathcal{Q}$ is meek or gentle[26] and for some $n < \omega$, $\rho_n(\mathcal{N}_\gamma) \leq \delta^{\mathcal{Q}}$,

then all remaining objects with index $\geq \gamma$ are undefined.

For all $\gamma \leq \eta$ for which clause 4 (the above statement) fails, $\pi_\gamma : \mathsf{core}(\mathcal{N}_\gamma) \to \mathcal{N}_\gamma$ is the uncollapse map.

5. Suppose for some $\xi < \delta$, for all $\gamma \leq \xi$, both $\mathcal{M}_\gamma, \mathcal{N}_\gamma$ are defined. Then $\mathcal{M}_{\xi+1}, \mathcal{N}_{\xi+1}, Y_{\xi+1}, \Phi_{\xi+1}, F_\xi^+, F_\xi$ and b_ξ are determined as follows.

---

[24] This strategy is induced by Σ essentially via the resurrection procedure of [23, Chapter 12].

[25] Recall clause 2 of Definition 2.5.4. To verify that $\mathcal{N}_\gamma$ is lses, we need to verify that clause 2 of Definition 2.5.4 holds.

[26] See Definition 2.7.1.

(a) Suppose $\mathcal{M}_\xi = (\mathcal{J}_{\omega\alpha}^{\vec{E},f}, \in, \vec{E}, f, Y_\xi, \in)$ is a passive hp-indexed lses[27], there is an extender $H^* \in \vec{G}$ an extender H over $\mathcal{M}_\xi$, and an ordinal $\nu < \omega\alpha$ such that $\nu < \mathrm{lh}(H^*)$ and setting

$$H = H^* \cap ([\nu]^\omega \times \lfloor \mathcal{M}_\xi \rfloor), \text{ and}$$
$$\mathcal{N}_{\xi+1} = (\mathcal{J}_{\omega\alpha}^{\vec{E},f}, \in, \vec{E}, f, Y_\xi, \tilde{H}, \in)$$

where $\tilde{H}$ is the amenable code of H, clause 4(a) fails for $\xi + 1$. Then letting $\iota \in \mathrm{dom}(\vec{G})$ be the least such that $H^* =_{\mathrm{def}} \vec{G}(\iota)$ has the above properties,

$$\mathcal{N}_{\xi+1} = (\mathcal{J}_{\omega\alpha}^{\vec{E},f}, \in, \vec{E}, f, Y_\xi, \tilde{H}, \in)$$

where $\tilde{H}$ is the amenable code of H[28]. Assuming clause 4 fails for $\xi + 1$, the remaining objects are defined as follows.
  (i) $\mathcal{M}_{\xi+1} = \mathsf{core}(\mathcal{N}_{\xi+1})$[29],
  (ii) $F_\xi^+ = H^*$ and $F_\xi = H$,
  (iii) $b_\xi = \emptyset$ and
  (iv) $Y_{\xi+1} = \pi_{\xi+1}^{-1}(Y_\xi)$.

(b) Suppose $\mathcal{M}_\xi = (\mathcal{J}_{\omega\alpha}^{\vec{E},f}, \in, \vec{E}, f, Y_\xi, \in)$ is a passive hp-indexed lses, $\mathcal{M}_\xi$ is strategy-ready[30], $\alpha = \beta + \gamma$ and there is $t \in \lfloor \mathcal{M}_\xi | \omega\beta \rfloor$ such that setting $w = (\mathcal{J}_\omega(t), t, \in)$, w is (f, hp)-minimal as witnessed by β[31] and $\gamma = \mathrm{lh}(t)$. Set $b = \Phi_\xi(t)$ and

$$\mathcal{N}_{\xi+1} = (\mathcal{J}_{\omega\beta+\omega\gamma}^{\vec{E},f^+}, \in, \vec{E}, f, Y_\xi, \tilde{b}, \in)$$

where $\tilde{b} \subseteq \omega\beta + \omega\gamma$ is defined by $\omega\beta + \omega\nu \in \tilde{b} \iff \nu \in b$. Assuming clause 4 fails for $\xi + 1$, the remaining objects are defined as follows.
  (i) $\mathcal{M}_{\xi+1} = \mathsf{core}(\mathcal{N}_{\xi+1})$,
  (ii) $F_\xi = F_\xi^+ = \emptyset$,
  (iii) $b_\xi = \tilde{b}$ and
  (iv) $Y_{\xi+1} = \pi_{\xi+1}^{-1}(Y_\xi)$.

Important Anomaly: Suppose $\cup Y_\xi$ is #-lsa type[32] and t is nuvs. Suppose $e \in \mathcal{M}_\xi | \omega\beta$ is such that $\mathcal{M}_\xi | \omega\beta \models \mathsf{sts}_0(t, e)$[33]. If $e \neq b$ then $\mathcal{N}_{\xi+1}$ is not an sts premouse over $\mathcal{J}_\omega(\cup Y_\xi)$ based on $\cup Y_\xi$, and so the construction must stop.

---

[27] That is, with no last predicate.

[28] Here H is what is determined by H^*. For the definition of the "amenable code" see the last paragraph on page 14 of [59].

[29] Recall that $\mathsf{core}(\mathcal{M})$ is the core of $\mathcal{M}$.

[30] See Definition 3.9.1.

[31] See Definition 2.3.3. In particular, this means that we have to index the branch of t at $\omega\alpha$.

[32] See Definition 2.7.3.

[33] See Definition 3.8.16. This means that e is the branch of t we must choose.

(c) If $\mathcal{M}_\xi$ does not satisfy clause 5(a) or 5(b) then set $\mathcal{N}_{\xi+1} = \mathcal{J}_\omega[\mathcal{M}_\xi]$ (this presupposes that $Y^{\mathcal{N}_{\xi+1}} = Y_\xi$). Assuming clause 4 fails for $\xi+1$, the remaining objects are defined as follows.

(I) $\mathcal{M}_{\xi+1} = \mathsf{core}(\mathcal{N}_{\xi+1})$[34],

(II) $F_\xi = F_\xi^+ = \emptyset$,

(III) $b_\xi = \emptyset$,

and $Y_{\xi+1}$ is defined as follows.

(i) If $M \vDash$ "$(\mathcal{M}_\xi, \Phi_\xi)$ is Γ-$\mathsf{cbl}+(X_\xi, \varphi_\xi) \in \mathsf{Hp}^\Gamma$"[35] then $Y_{\xi+1} = \pi_{\xi+1}^{-1}(Y_\xi) \cup \{\pi_{\xi+1}^{-1}(\mathcal{M}_\xi)\}$.

(ii) If $M \vDash$ "$(\mathcal{M}_\xi, \Phi_\xi)$ is Γ-$\mathsf{cbl}+(X_\xi, \varphi_\xi) \notin \mathsf{Hp}^\Gamma$" then all remaining objects with index $\geq \xi$ are undefined.

(iii) If both 5(c)(i) and 5(c)(ii) fail then $Y_{\xi+1} = \pi_{\xi+1}^{-1}(Y_\xi)$.

6. Suppose $\xi \leq \delta$ is a limit ordinal and for all $\gamma < \xi$, both $\mathcal{M}_\gamma$ and $\mathcal{N}_\gamma$ are defined. Then $\mathcal{M}_\xi$ and $\mathcal{N}_\xi$ are determined as follows[36]. Set $\mathcal{N}_\xi = lim_{\alpha\to\xi}\mathcal{M}_\alpha$. Assuming clause 4 fails for $\xi+1$, the remaining objects are defined as follows.

(a) $\mathcal{M}_\xi = \mathsf{core}(\mathcal{N}_\xi)$ and

(b) $Y_\xi = \pi_\xi^{-1}(Y^{\mathcal{N}_\xi})$[37].

7. $\mathcal{M}_\delta = \mathcal{N}_\delta$ and $Y_\delta, \Phi_\delta, F_\delta^+, F_\delta$, and b_δ are undefined.

Let

$$\mathsf{hpc} = (\mathcal{M}_\gamma, \mathcal{N}_\gamma, Y_\gamma, \Phi_\gamma, F_\gamma^+, F_\gamma, b_\gamma : \gamma \leq \delta)$$

be the output of the $\Gamma - \mathsf{hpc}$ of $\mathbb{M}$. We say that the $\Gamma - \mathsf{hpc}$ of $\mathbb{M}$ is *successful* if clause 4 fails for all $\gamma < \delta$. We say that the $\Gamma - \mathsf{hpc}$ of $\mathbb{M}$ *reaches its goal* if the $\Gamma - \mathsf{hpc}$ of $\mathbb{M}$ is successful and for some $\xi < \delta$, clause 5(c)(ii) holds.

For each $\gamma \leq \delta$, we let Φ_γ^+ be the extension of Φ_γ defined in Section 4.3.1. We then set

$$\mathsf{hpc}^+ = (\mathcal{M}_\gamma, \mathcal{N}_\gamma, Y_\gamma, \Phi_\gamma^+, F_\gamma^+, F_\gamma, b_\gamma : \gamma \leq \delta).$$

Notice that $\mathsf{hpc} \in M$ while $\mathsf{hpc}^+ \notin M$.

Also, given $\xi \leq \delta$ and $\alpha \leq \delta$, we set

$$\mathsf{hpc} \restriction (\xi, \alpha) = (\mathcal{M}_\gamma, \mathcal{N}_\gamma, Y_\gamma, \Phi_\gamma \restriction M|\alpha, F_\gamma^+, F_\gamma, b_\gamma : \gamma \leq \xi)$$

and finally we let

$$\mathsf{hpc}^- = (\mathcal{M}_\gamma, \mathcal{N}_\gamma, Y_\gamma, F_\gamma^+, F_\gamma, b_\gamma : \gamma \leq \delta)$$

---

[34]Recall that $\mathsf{core}(\mathcal{M})$ is the core of $\mathcal{M}$.

[35](X_ξ, φ_ξ) is defined in Definition 4.3.13. The meaning of $M \vDash (X_\xi, \varphi_\xi) \in \mathsf{Hp}^\Gamma$ is essentially that $M \vDash (\mathcal{M}_\xi, \Phi_\xi) \in \mathsf{Hp}^\Gamma$.

[36]The rest of the objects will be defined at the next stage of the induction as in clause 4.

[37]F_ξ and b_ξ are defined at step $\xi+1$.

We will often use (C,Γ) as a subscript to emphasize the dependence on (C,Γ). Thus, we will write $\mathsf{hpc}_{\mathsf{C},\Gamma}$ and etc. Also, to emphasize the dependence on C, we may also say that $\Gamma-\mathsf{hpc}$ of C is successful or reaches its goal.

We say that $\mathcal{Q}$ is an $\mathcal{N}$*-model* of hpc if for some $\gamma \leq \delta$, $\mathcal{Q} = \mathcal{N}_\delta$. We define other such expressions (e.g., $\mathcal{M}$-model and etc) in a similar fashion. We say $\mathcal{W}$ is *the last model* of hpc if $\mathcal{Q} = \mathcal{N}_\gamma$, the last defined $\mathcal{N}$-model of hpc.

REMARK 4.3.4. Section 4.1.3 defines the meaning of $M \vDash (X_\xi, \varphi_\xi) \in \mathsf{Hp}^\Gamma$, which in reality formalizes $M \vDash (\mathcal{M}_\xi, \Phi_\xi) \in \mathsf{Hp}^\Gamma$. Using very similar ideas, one can also easily formalize the meaning of $M \vDash$ "$(\mathcal{M}_\xi, \Phi_\xi)$ is Γ-cbl". Such a formalism will refer to some set Z_ξ and a formula ψ_ξ. The definition of these will be similar to the definitions of X_ξ and φ_ξ (see Definition 4.3.13). We leave the details to the reader.

REMARK 4.3.5. Notice that each $\mathcal{M}_\gamma$ and $\mathcal{N}_\gamma$ are germane (see Definition 2.7.15), and so we can use the concepts introduced in Section 2.10.

4.3.1. The construction of Φ^+_γ. We are continuing with the objects defined in Section 4.3 and in particular, in Definition 4.3.3. Recall that Φ^+_γ must be an $(\omega_1,\omega_1,\omega_1)$-iteration strategy. Its (ω_1,ω_1) component can be defined using the procedure of [23, Chapter 12]. Also Φ^+_γ is the strategy of $\mathcal{M}_\gamma$ that is based on $\cup Y^{\mathcal{M}_\gamma}$. Thus, to define Φ^+_γ we may just as well assume that $\cup Y^{\mathcal{M}_\gamma}$ is a limit type hod premouse, as this is when an $(\omega_1,\omega_1,\omega_1)$-iteration strategy is used. As the process is a straightforward adaptation of [23, Chapter 12], we will only give a short outline.

The procedure of [23, Chapter 12] gives an (ω_1,ω_1)-strategy Φ_γ for $\mathcal{M}_\gamma$. Set $\mathcal{P}^+ = \mathcal{M}_\gamma$ and $\mathcal{P} = \cup Y^{\mathcal{M}_\gamma}$. Suppose N is a Σ-iterate of M via $\mathcal{X}$ and $i: M \to N$ is the iteration embedding. Recall that we had

$$\mathsf{hpc}_{\mathsf{C},\Gamma} = (\mathcal{M}_\gamma, \mathcal{N}_\gamma, Y_\gamma, \Phi_\gamma, F^+_\gamma, F_\gamma, b_\gamma : \gamma < \delta)$$

and our background is $\mathbb{M} = (M, \delta, \vec{G}, \Sigma)$. Suppose $\alpha < \mathrm{lh}(\mathcal{X})$ and $\beta \leq \pi^{\mathcal{X}}_{0,\alpha}(\delta)$. We then let $\mathcal{R}^{\mathcal{X}}_{\alpha,\beta}$ be the β-th $\mathcal{N}$-model of $\pi^{\mathcal{X}}_{0,\alpha}(\mathsf{hpc}_{\mathsf{C},\Gamma})$, and also we let $\Phi^N_{\alpha,\beta}$ be the (ω_1,ω_1)-iteration strategy of $\mathcal{R}^{\mathcal{X}}_{\alpha,\beta}$ induced by Σ_N. We let Φ^N be the strategy of $i(\mathcal{P}^+)$ induced by Σ_N.

Given N as above and a stack $\mathcal{T}$ on $i(\mathcal{P}^+)$ that is based on $i(\mathcal{P})$ and is according to Φ^N, we let $r\mathcal{T}$ be the *resurrection of* $\mathcal{T}$. The reader may wish to review properties H1-H7 on pages 113–115 of [23], which outline the construction of $r\mathcal{T}$. Below we outline the description of Φ^M and leave Φ^N to the reader. Assuming

$$\mathcal{T} = ((\mathcal{M}_\alpha)_{\alpha<\eta}, (E_\alpha)_{\alpha<\eta-1}, D, R, (\beta_\alpha, m_\alpha)_{\alpha\in R}, T)^{38}$$

and

[38] Recall that our stacks are proper, see Definition 2.7.27.

$$r\mathcal{T} = ((r\mathcal{M}_\alpha)_{\alpha<\eta}, (rE_\alpha)_{\alpha<\eta-1}, rD, rR, (r\beta_\alpha, rm_\alpha)_{\alpha\in rR}, rT)^{39}$$

there are sequences $\vec{\sigma} = (\sigma_\alpha : \alpha < \eta)$ and $\vec{v} = (v_\alpha : \alpha < \eta)$ satisfying the following conditions:

1. $rD = \emptyset$, $T = rT$ and $R = rR$.
2. For each $\alpha < \eta$, $v_\alpha \leq \pi^{r\mathcal{T}}_{0,\alpha}(\gamma)$ and $\sigma_\alpha : \mathcal{M}_\alpha \to \mathcal{R}^{r\mathcal{T}}_{\alpha,v_\alpha}$ is a weak embedding[40].
3. If $[0,\alpha)_\mathcal{T} \cap D = \emptyset$ then $v_\alpha = \pi^{r\mathcal{T}}_{0,\alpha}(\gamma)$ and $\mathcal{R}^{r\mathcal{T}}_{\alpha,v_\alpha} = \pi^{r\mathcal{T}}_{0,\alpha}(\mathcal{P})$.
4. For each $\alpha < \alpha'$ such that $(\alpha,\alpha') \cap R = \emptyset$, $\sigma_\alpha \restriction \mathrm{ind}^\mathcal{T}_\alpha = \sigma_{\alpha'} \restriction \mathrm{ind}^\mathcal{T}_{\alpha'}$.
5. For each $\alpha < \alpha'$ such that $\alpha \mathcal{T} \alpha'$ and $\pi^\mathcal{T}_{\alpha,\alpha'}$ is defined, $\pi^{r\mathcal{T}}_{\alpha,\alpha'} \circ \sigma_\alpha = \sigma_{\alpha'} \circ \pi^\mathcal{T}_{\alpha,\alpha'}$.
6. Moreover, $(\sigma_\alpha : \alpha < \eta)$ and $(v_\alpha : \alpha < \eta)$ are uniquely determined via the procedure described on pages 113-115 of [23].

We then say that $(\vec{\sigma}, \vec{v})$ are the $r\mathcal{T}$-sequences.

DEFINITION 4.3.6. Suppose now that $p = (\mathcal{P}_\beta, \mathcal{T}_\beta, E_\beta : \beta < \gamma)$ is a generalized stack on $\mathcal{P}^+$ that is based on $\mathcal{P}$. We say p is *correct* if there is a stack $q = ((\mathcal{Q}_\alpha)_{\alpha<\eta}, (F_\alpha)_{\alpha<\eta-1}, D, R, (\beta_\alpha, m_\alpha)_{\alpha\in R}, Q)$ according to Σ and a sequence of embeddings $(\sigma_\beta : \beta < \gamma)$ such that the following conditions hold:

1. $\eta = \Sigma_{\beta<\gamma} \mathrm{lh}(\mathcal{T}_\beta)$ and $\eta_\beta =_{\mathrm{def}} \Sigma_{\beta'<\beta} \mathrm{lh}(\mathcal{T}_{\beta'})$.
2. For all $\beta < \gamma$, $\sigma_\beta : \mathcal{P}_\beta \to \pi^q_{0,\eta_\beta}(\mathcal{P}^+)$ is a weak embedding.
3. $\sigma_0 = id$.
4. For all $\beta < \gamma$, $\eta_\beta \in R$.
5. For all $\beta < \gamma$, $q_{[\eta_\beta,\eta_{\beta+1})} = r(\sigma_\beta \mathcal{T}_\beta)$.
6. For all $\beta < \gamma$ such that $\beta + 1 < \gamma$ and E_β is an un-dropping extender[41], letting
   (a) $md^{\mathcal{T}_\beta} = (\alpha_i, \mathcal{R}_i, \mathcal{W}_i, \mathcal{S}_i : i \leq k+1)$ be the main drops of $\mathcal{T}_\beta$,
   (b) for $i \leq k+1$, $\kappa_i = \delta^{\mathcal{R}^b_i}$[42],
   (c) $\xi + 1 = \mathrm{lh}(\mathcal{T}_\beta)$,
   (d) $m : \mathcal{M}^{\mathcal{T}_\beta}_\xi \to \mathcal{M}^{\sigma_\beta \mathcal{T}_\beta}_\xi$ is the map obtained via the copying process,
   (e) $(\vec{n}, \vec{v})$ are the $r(\sigma_\beta \mathcal{T}_\beta)$-sequences,
   (f) $k = \sigma^\mathcal{T}$,
   the embedding
   $$\sigma_{\beta+1} : \mathcal{P}_{\beta+1} \to \pi^q_{0,\eta_{\beta+1}}(\mathcal{P}^+)$$
   is given by
   $$\sigma_{\beta+1}(\pi_{E_\beta}(f)(a)) = \pi^q_{\eta_\beta,\eta_{\beta+1}}(\sigma_\beta(f))(n_\xi(m(a))).$$

[39] Here we only use (ω_1, ω_1)-portion of Λ_N.

[40] For example, see the discussion after Fact 2.13 of [3].

[41] The case when $\pi^{\mathcal{T}_\beta,b}$ is defined is easier and very similar, and we leave it to the reader.

[42] See Definition 2.10.5.

The definition of $\sigma_{\beta+1}$ works because we have that $(a,A) \in E_\beta$ if and only if $n_\xi \circ m(a) \in \pi^q_{\eta_\beta,\eta_{\beta+1}}(\sigma_\beta(A))$.

Notice that both q and the embeddings $\vec{\sigma} = (\sigma_\beta : \beta < \gamma)$ are uniquely determined. We then set $q = \mathsf{res}(p)$ and $\vec{\sigma} = \mathsf{emb}(p)$.

The following is an easy lemma. It uses the objects introduced above.

LEMMA 4.3.7. $\sigma_{\beta+1} \restriction (\mathcal{P}_{\beta+1}|\mathrm{lh}(E_\beta)) = n_\xi \circ m \restriction (\mathcal{P}_{\beta+1}|\mathrm{lh}(E_\beta))$.

It is now straightforward to show, using the resurrection process of [23, Chapter 12], that if p is a correct generalized stack on $\mathcal{P}^+$ based on $\mathcal{P}$ of limit length then there is a unique branch b of p such that $p^\frown\{b\}$ is also correct. We then let Φ^+_γ be the unique $(\omega_1,\omega_1,\omega_1)$-strategy of $\mathcal{P}^+$ with the property that p is according to Φ^+_γ if and only if p is a correct generalized stack on $\mathcal{P}^+$ based on $\mathcal{P}$. Notice finally that the definition of Φ^+_γ can be done locally inside M.

DEFINITION 4.3.8. If $\cup Y^{\mathcal{M}_\gamma}$ is not of #-lsa type then $\Phi_\gamma = \Phi^+_\gamma \restriction M|(\delta^+)^M$. If $\cup Y^{\mathcal{M}_\gamma}$ is of #-lsa type then $\Phi_\gamma = (\Phi^+_\gamma)^{stc} \restriction M|(\delta^+)^M$.

The following lemma summarizes Definition 4.3.6.

LEMMA 4.3.9. *Suppose Γ is a pointclass, $\mathsf{C} = (\mathbb{M},(P,\Psi),\Gamma^*,A)$ Suslin, co-Suslin captures Γ and $\mathbb{M} = (M,\delta,\vec{G},\Sigma)$. Set*

$$\mathsf{hpc}^+_{\mathsf{C},\Gamma} = (\mathcal{M}_\gamma,\mathcal{N}_\gamma,Y_\gamma,\Phi^+_\gamma,F^+_\gamma,F_\gamma,b_\gamma : \gamma \leq \delta).$$

Suppose $\gamma \leq \delta$ is such that Y_γ is defined. Set $\mathcal{P} = \cup Y_\gamma$ and suppose $\mathcal{M}_\gamma$ is of b-type. Suppose $\mathcal{T}$ is a generalized stack according to Φ^+_γ with last model $\mathcal{Q}$. There is then a Σ-iterate N of M such that letting $i : M \to N$ be the iteration embedding and

$$\mathsf{hpc}^+_{\mathsf{C}_N,\Gamma} = (\mathcal{R}_\gamma,\mathcal{S}_\gamma,Z_\gamma,\Psi^+_\gamma,E^+_\gamma,E_\gamma,c_\gamma : \gamma \leq i(\delta)),$$

there is $\nu \leq i(\gamma)$ and a weak embedding $\sigma : \mathcal{Q} \to \mathcal{S}_\nu$ such that the following holds.

1. *If $\pi^{\mathcal{T}}$ is defined then $\nu = i(\gamma)$ and $\sigma \circ \pi^{\mathcal{T}} = i \restriction \mathcal{P}$.*
2. *If $\pi^{\mathcal{T},b}$ is defined then $i(\mathcal{P}^b) = \mathcal{S}^b_\nu$ and $\sigma \circ \pi^{\mathcal{T},b} = i \restriction \mathcal{P}^b$.*
3. *$\Phi^+_{\mathcal{Q},\mathcal{T}}$ is the σ-pullback of Ψ^+_ν.*

We remark that a similar result holds for all γ. We now have the following lemma connecting different strategies to each other.

LEMMA 4.3.10. *Suppose Γ is a pointclass, $\mathsf{C} = (\mathbb{M},(P,\Psi),\Gamma^*,A)$ Suslin, co-Suslin captures Γ and $\mathbb{M} = (M,\delta,\vec{G},\Sigma)$. Set*

$$\mathsf{hpc}^+_{\mathsf{C},\Gamma} = (\mathcal{M}_\gamma,\mathcal{N}_\gamma,Y_\gamma,\Phi^+_\gamma,F^+_\gamma,F_\gamma,b_\gamma : \gamma \leq \delta).$$

Suppose $\alpha < \beta \leq \delta$ are such that $\mathcal{N}_\alpha$ and $\mathcal{N}_\beta$ are defined. Let $\mathcal{Q} \in \mathcal{Y}_\alpha$ be a meek hod premouse. Set $\Psi^0 = \Phi^+_{\alpha+1}$, $\mathcal{P}_1 = \mathcal{M}_{\alpha+1}$ and define Ψ^1 as follows:

1. *If $\rho(\mathcal{N}_\beta) \leq \delta^{\mathcal{Q}}$ then let n be the largest such that for every $\kappa < \delta^{\mathcal{Q}}$, any $r\Sigma_n^{\mathcal{N}_\beta}$-definable $f : \kappa \to \delta^{\mathcal{Q}}$ is in $\mathcal{Q}$ and let Ψ^1 be the strategy of $\mathcal{P}_1 =_{def} \mathsf{core}_n(\mathcal{N}_\beta)$ defined via the resurrection procedure described above.*
2. *If $\rho(\mathcal{N}_\beta) > \delta^{\mathcal{Q}}$ then let Ψ^1 be the strategy of $\mathcal{P}_1 =_{def} \mathsf{core}(\mathcal{N}_\beta)$ defined via the resurrection procedure described above.*

Then $\Psi^0_{\mathcal{Q}} = \Psi^1_{\mathcal{Q}}$.

The proof of the lemma is straightforward. Let γ be such that $\mathcal{M}_\gamma = \mathcal{Q}$ and let $\zeta = \sup\{\mathrm{lh}(F_\iota^+) : \iota < \gamma\}$. Observe now that because we assume that $\mathcal{Q}$ is meek, if $\mathcal{T}$ is a stack on $\mathcal{Q}$ then the *id*-copy of $\mathcal{T}$ onto $\mathcal{P}_0$ and onto $\mathcal{P}_1$ is simply $\mathcal{T}_0 =_{\mathrm{def}} \uparrow(\mathcal{T}, \mathcal{P}_0)$ and $\mathcal{T}_1 =_{\mathrm{def}} \uparrow(\mathcal{T}, \mathcal{P}_1)$ respectively, and these stacks use exactly the same extenders as $\mathcal{T}$. Therefore the resurrection procedure resurrects both $\mathcal{T}_0$ and $\mathcal{T}_1$ to stacks based on $M|\zeta$. Hence both $\Psi^0_{\mathcal{Q}}$ and $\Psi^1_{\mathcal{Q}}$ are determined by $\Sigma_{M|\zeta}$.

Lastly we state the following consequence of Lemma 4.3.7.

DEFINITION 4.3.11. Suppose $(\mathcal{P}, \Sigma)$ is a hod pair or an sts pair. We say Σ is *weakly self-cohering* if the following clauses hold:

1. Whenever $\mathcal{T}$ is a generalized stack on $\mathcal{P}$ according to Σ with last model $\mathcal{S}$ and $\mathcal{Q}$ is a complete layer of $\mathcal{S}^b$ such that $\mathcal{T}^{\mathsf{ue}}_{\mathcal{Q}}$ is defined[43], $\Sigma_{\mathcal{Q},\mathcal{T}} = \Sigma_{\mathcal{Q},\mathcal{T}^{\mathsf{ue}}}$.
2. Whenever $\mathcal{T}$ is a generalized stack on $\mathcal{P}$ according to Σ with last model $\mathcal{S}$ and such that $\mathcal{T}$ has a one point extension[44], $\mathcal{Q} \trianglelefteq_{hod} \mathcal{S}^b$ is of limit type, and $\mathcal{U}$ is a stack on $\mathcal{Q}$ according to $\Sigma_{\mathcal{Q},\mathcal{T}}$ such that $\mathcal{U}$ has a one point extension then letting E be the un-dropping extender of $\mathcal{U}$, $\mathrm{Ult}(\mathcal{S}, E)$ is well-founded.

Suppose next that $(\mathcal{P}, \Sigma)$ is a simple hod pair or an sts hod pair. Then we say that Σ is *weakly self cohering* if the following clauses hold:

1. Whenever $\mathcal{T}$ is a stack on $\mathcal{P}$ according to Σ with last model $\mathcal{S}$ and $\mathcal{Q}$ is a complete layer of $\mathcal{S}^b$ such that $\mathcal{T}^{\mathsf{ue}}_{\mathcal{Q}}$ is defined, the last model of $\mathcal{T}^{\mathsf{ue}}_{\mathcal{Q}}$ is well-founded.
2. Whenever $\mathcal{T}$ is a stack on $\mathcal{P}$ according to Σ with last model $\mathcal{S}$ and such that $\mathcal{T}$ has a one point extension, $\mathcal{Q} \trianglelefteq_{hod} \mathcal{S}^b$ is of limit type, and $\mathcal{U}$ is a stack on $\mathcal{Q}$ according to $\Sigma_{\mathcal{Q},\mathcal{T}}$ such that $\mathcal{U}$ has a one point extension then letting E be the un-dropping extender of $\mathcal{U}$, $\mathrm{Ult}(\mathcal{S}, E)$ is well-founded.

The following now is an easy consequence of Lemma 4.3.7.

LEMMA 4.3.12. *Suppose Γ is a pointclass, $\mathsf{C} = (\mathbb{M}, (P, \Psi), \Gamma^*, A)$ Suslin, co-Suslin captures Γ and $\mathbb{M} = (M, \delta, \vec{G}, \Sigma)$. Set*

$$\mathsf{hpc}^+_{\mathsf{C},\Gamma} = (\mathcal{M}_\gamma, \mathcal{N}_\gamma, Y_\gamma, \Phi^+_\gamma, F^+_\gamma, F_\gamma, b_\gamma : \gamma \leq \delta).$$

Suppose $\gamma \leq \delta$ is such that Y_γ is defined. Set $\mathcal{P} = \cup Y_\gamma$ and suppose $\mathcal{M}_\gamma$ is of b-type. Then $(\Phi_\gamma)_{\mathcal{P}}$ is weakly self-cohering.

[43] See Notation 2.10.9.

[44] See Definition 2.10.2.

4.3.2. The definition of $(X_\gamma, \varphi_\gamma)$. Notice that the map $\mathcal{T} \mapsto \mathsf{res}(\mathcal{T})$ can be defined without any reference to any strategy for $\mathcal{P}$ or M. In this view, $\mathsf{res}(\mathcal{T})$ may not have well-founded models. Moreover, the construction of $\mathsf{res}(\mathcal{T})$ only depends on $\mathsf{hpc} \restriction \gamma+1$.

DEFINITION 4.3.13. Let $\xi_\gamma < \delta$ be the least inaccessible cardinal of M such that

$$\mathsf{hpc}^- \restriction \gamma+1 \in M|\xi_\gamma.$$

Let $\dot{X}_\gamma \in M^{Coll(\omega, M|\xi_\gamma)}$ witness that M is self-capturing for $M|\xi_\gamma$ (see Definition 4.1.5) and set

$$X_\gamma = (\dot{X}_\gamma, \mathcal{M}_\gamma, \mathsf{hpc} \restriction (\gamma+1, \xi_\gamma), M|\xi_\gamma).$$

Let $\psi(x,y,z,w)$ be a formula such that

$$\psi[\mathcal{M}_\gamma, \mathsf{hpc} \restriction (\gamma+1, \xi_\gamma), M|\xi_\gamma]$$

expresses all the clauses of Definition 4.3.3 except the portion of clause 5(c) that defines $Y_{\xi+1}$. Let $\varphi_\gamma(u,v,w)$ be the conjunction of the following formulas.

1. $u = (Y,g)$ such that $Y = (Z,Q,h,f,N)$, Z is $Coll(\omega,N)$-name and $g \subseteq Coll(\omega,N)$ is a filter,
2. $\psi(Q,h,f,N)$,
3. w is a stack on Q, and
4. letting $Z_g = (U,W)$, $\mathsf{res}(w) \in p[U]$.

4.4. On backgrounded constructions

The following sequence of lemmas will be used in the proof of Theorem 4.5.6.

DEFINITION 4.4.1. We say $(M, \delta, \vec{G}, \Sigma, \mathcal{P})$ has the property $(*)$ if

1. $(M, \delta, \vec{G})$ is a background[45],
2. $\Sigma \in M$ is a (δ, δ)-iteration or δ-iteration strategy for $\mathcal{P}$ with hull condensation[46], or $(\mathcal{P}, \Sigma)$ is an sts hod pair and Σ is a (δ, δ, δ) st-strategy.
3. $\mathsf{Le}(\Sigma, \mathcal{J}_\omega[\mathcal{P}])^{(M,\delta,\vec{G})}$ is successful.

We say that $(M, \delta, \vec{G}, \Sigma, \mathcal{P})$ has the property $(*+)$ if in addition to the above clauses $M \models$ "Σ is a (δ^+, δ^+)-iteration strategy, δ^+-iteration strategy or $(\delta^+, \delta^+, \delta^+)$ st-strategy". If $Q \subseteq M$ then we let $\Sigma^Q = \Sigma \restriction (Q|\delta)$.

LEMMA 4.4.2. *Assume $(M, \delta, \vec{G}, \Sigma, \mathcal{P})$ has the property $(*)$. Set $P = \mathcal{J}_\omega[\mathcal{P}]$. Suppose $\lambda < \delta$ is such that $P \in M|\lambda$ and $\mathcal{N}$ is the last model of $\mathsf{Le}(\Sigma^M, P)^{(M,\delta,\vec{G})}$. Suppose $F^* \in \vec{G}$ is such that*

[45] See Definition 4.1.4.

[46] The exact nature of $\mathcal{P}$ is irrelevant.

1. $\mathrm{lh}(F^*) = \eta$ *is an inaccessible cardinal of* M,
2. $\pi_{F^*}(\mathcal{N})|\eta = \mathcal{N}|\eta$.

Set $\kappa = \mathrm{crit}(F^*)$ *and let* F' *be the* (κ, η) *extender derived from*

$$\pi_{F^*} \restriction \mathcal{N} : \mathcal{N} \to \pi_{F^*}(\mathcal{N}).$$

Then for any $\rho \in [(\kappa^+)^{\mathcal{N}}, \eta)$ *such that* ρ *is the natural length of* $F' \restriction \rho$, *letting* F *be the trivial completion of* $F' \restriction \rho$, *one of the following conditions hold:*

1. $\mathrm{lh}(F) \in \mathrm{dom}(\vec{E}^{\mathcal{N}})$ *and* $F = \vec{E}^{\mathcal{N}}(\mathrm{lh}(F))$ *or*
2. $\mathrm{lh}(F) \notin \mathrm{dom}(\vec{E}^{\mathcal{N}})$, ρ *is a limit ordinal* $> (\kappa^+)^{\mathcal{N}}$, ρ *is a generator of* F, $\rho \in \mathrm{dom}(\vec{E}^{\mathcal{N}})$ *and letting* $E = \vec{E}^{\mathcal{N}}(\rho)$, $F = \pi_E^{\mathcal{N}|\rho}(\vec{E}^{\mathcal{N}|\rho})(\mathrm{lh}(F))$.

Suppose $(M, \delta, \vec{G})$ is a background. We write $(M, \vec{G}) \models$ "κ reflects A" to mean that κ reflects A using extenders in $\vec{G}$[47]. Working in M, let $(A_i^M : i < \omega)$ be defined by the following induction:

1. $A_0^M \subseteq \delta$ is the set of $< \delta$-strong cardinals κ such that $(M, \vec{G}) \models$ "κ reflects $\vec{G}$.
2. $A_{i+1}^M \subseteq \delta$ is the set of $< \delta$-strong cardinals κ such that $(M, \vec{G}) \models$ "κ reflects A_i^M".

Clearly, A_i^M depends on both δ and $\vec{G}$, but in all of the lemmas below $(\delta, \vec{G})$ is clear from the context. Sometimes, when $(\delta, \vec{G})$ is not clear from the context, we will write $A_i^{M,\delta,\vec{G}}$. We have the following straightforward lemma.

LEMMA 4.4.3. *Suppose* $(M, \delta, \vec{G})$ *is a background. Then the following holds in* M.

1. Suppose $A \subseteq \delta$ *and* $\kappa_0 < \kappa_1 < \delta$ *are such that* $(M, \vec{G}) \models$ "κ_1 *reflects* $(A, \vec{G})$" *and* $(M|\kappa_1, \vec{G} \restriction \kappa) \models$ "κ_0 *reflects* $(A \cap \kappa_1, \vec{G} \restriction \kappa_1)$". *Then* $(M, \vec{G}) \models$ "κ_0 *reflects* $(A, \vec{G})$".
2. *For each* $i < \omega$, *if* $\lambda \in A_{i+1}^M$ *or is a limit point of* A_{i+1}^M *then* λ *is a limit point of* A_i^M.
3. *For all* $i < \omega$ *and for every* λ, *which is a member or a limit point of* A_i^M, $A_i^M \cap \lambda = A_i^{M|\lambda, \vec{G} \restriction \lambda}$.
4. *For all* $i \in \omega$, $A_{i+1}^M \subseteq A_i^M$.
5. $\kappa \in \cap_{i \in \omega} A_i^M$ *if and only if for each* $i \in \omega$, $(M, \vec{G}) \models$ "κ *reflects* A_i^M". *Hence,* $\cap_{i \in \omega} A_i^M \neq \emptyset$.
6. *If* $(M, \vec{G}) \models$ "$\kappa < \delta$ *reflects* $(A_i : i \in \omega)$" *then* $\kappa \in \cap_{i < \omega} A_i^M$.

LEMMA 4.4.4. *Assume* $(M, \delta, \vec{G}, \Sigma, \mathcal{P})$ *has the property* $(*)$. *Set* $P = \mathcal{J}_\omega[\mathcal{P}]$. *Suppose* $\lambda < \delta$ *is such that* $P \in M|\lambda$, $\mathcal{N}'$ *is the last model of* $\mathsf{Le}(\Sigma^M, P)^{(M,\delta,\vec{G})}$ *and*

[47]That is, the set of κ such that for every $\lambda < \delta$ there is $F \in \vec{G}$ such that $\pi_F(\vec{G}) \restriction \lambda = \vec{G} \restriction \lambda$.

$\mathcal{N} = L_{\mathrm{ord}(M)}[\mathcal{N}']$[48]. *Let* $\vec{H} = \{E \in \vec{E}^{\mathcal{N}} : \mathcal{N} \vDash \text{“}\nu(E) \text{ is inaccessible”}\}$. *Then*

$$\cap_{i<\omega} A_i^M = \cap_{i<\omega} A_i^{\mathcal{N},\vec{H}}.$$

PROOF. We will use $A_n^{\mathcal{N}}$ for $A_n^{\mathcal{N},\vec{H}}$. It is enough to show that $i < \omega$, $A_{i+1}^M \subseteq A_i^{\mathcal{N}} \subseteq A_i^M$. Notice first that

(1) in M, if $\kappa \in A_0^M - (\lambda+1)$ and $\mathcal{Q}$ is an $\mathcal{N}$-model of $\mathsf{Le}(\Sigma^M, P)^{(M,\delta,\vec{G})}$ such that $\mathcal{Q} \in M|\kappa$ then $\mathcal{Q}$ is an $\mathcal{N}$-model of $\mathsf{Le}(\Sigma^M, P)^{(M,\kappa,\vec{G}\restriction\kappa)}$.

(1) then easily implies that

(2) if $\kappa \in A_0^M - (\lambda+1)$ then $\mathcal{N}|\kappa$ is the last model of $\mathsf{Le}(\Sigma^M, P)^{(M,\kappa,\vec{G}\restriction\kappa)}$[49] and the κth $\mathcal{N}$-model of $\mathsf{Le}(\Sigma^M, P)^{(M,\delta,\vec{G})}$.

$A_0^{\mathcal{N}} \subseteq A_0^M$ follows from the fact that the backgrounding extenders used in

$$\mathsf{Le}(\Sigma^M, P)^{(M,\delta,\vec{G})}$$

are all total M-extenders.

Suppose now that $\kappa \in A_1^M$. We want to see that $\kappa \in A_0^{\mathcal{N}}$. Let $\eta_0 < \eta_1$ be two members of A_0^M such that $\kappa < \eta_0$. Let $F^* \in \vec{G}$ be an extender such that $\pi_{F^*}(A_0^M) \cap \eta_1 + 1 = A_0^M \cap \eta_1 + 1$ and $\pi_{F^*}(\vec{G}) \restriction \eta_1 = \vec{G} \restriction \eta_1$. (2) then implies that

(3) $\pi_{F^*}(\mathcal{N})|\eta_1 = \mathcal{N}|\eta_1$.

Indeed, it follows from (2) that $\pi_{F^*}(\mathcal{N})|\eta_1$ is the last model of

$$(\mathsf{Le}(\Sigma^{\mathrm{Ult}(M,F^*)}, P)_{>\lambda})^{(\mathrm{Ult}(M,F^*)|\eta_1,\eta_1,\pi_{F^*}(\vec{G})\restriction\eta_1)},$$

and since $\mathrm{Ult}(M,F^*)|\eta_1 = M|\eta_1$, we have that $\pi_{F^*}(\mathcal{N})|\eta_1$ is the last model of $\mathsf{Le}(\Sigma^M, P)^{(M|\eta_1,\eta_1,\vec{G}\restriction\eta_1)}$, which according to (2) is just $\mathcal{N}|\eta_1$.

Let now F be the (κ, η_1) extender derived from $\pi_{F^*} \restriction \mathcal{N}$. Since η_0 is a regular cardinal of $\mathcal{N}$ and hence, $\eta_0 \notin \mathrm{dom}(\vec{E}^{\mathcal{N}})$, it follows from Lemma 4.4.2 that the trivial completion of $F \restriction \eta_0$ is on $\vec{E}^{\mathcal{N}}$. As η_0 was arbitrary, we have that $\delta = \sup\{\mathrm{lh}(E) : E \in \vec{E}^{\mathcal{N}} \wedge \mathrm{crit}(E) = \kappa\}$, implying that $\kappa \in A_0^{\mathcal{N}}$.

Assume now that $A_{n+1}^M \subseteq A_n^{\mathcal{N}} \subseteq A_n^M$. We want to see that

(a) $A_{n+1}^{\mathcal{N}} \subseteq A_{n+1}^M$,
(b) $A_{n+2}^M \subseteq A_{n+1}^{\mathcal{N}}$.

---

[48] δ is a Woodin cardinal of $\mathcal{N}$ and all bounded subsets of δ in $\mathcal{N}$ are in $\mathcal{N}'$. The first claim can be shown by the results of [23, Chapter 11], and the second follows from the fact that δ is a regular cardinal, which allows us to take Skolem hulls of M that are transitive below δ.

[49] This is a mild abuse of our notation as κ may not be a Woodin cardinal of M. But Le construction do not depend on the Woodinness of δ.

First suppose $\kappa \in A_{n+1}^{\mathcal{N}}$. To show that $\kappa \in A_{n+1}^{M}$, we need to show that in M, κ reflects A_n^M. Let $\eta \in A_0^{\mathcal{N}}$ be a limit point of $A_n^{\mathcal{N}}$. Let $E \in \vec{H}$ be a (κ,η)-extender that reflects $A_n^{\mathcal{N}}$[50]. Thus, $\pi_E(A_n^{\mathcal{N}}) \cap \eta = A_n^{\mathcal{N}} \cap \eta$. Let $E^* \in \vec{G}$ be the resurrection of E. We then have that $E = E^* \cap (\eta^{<\omega} \times \mathcal{N})$ and an embedding $\sigma : \mathrm{Ult}(\mathcal{N},E) \to \pi_{E^*}(\mathcal{N})$ such that $\mathrm{crit}(\sigma) \geq \eta$. Because $A_n^{\mathcal{N}} \subseteq A_n^M$ we have that

(4) $A_n^{\pi_{E^*}(\mathcal{N})} \subseteq A_n^{\mathrm{Ult}(M,E^*)}$,
(5) $\sigma(A_n^{\mathcal{N}} \cap \eta) = \sigma(A_n^{\mathrm{Ult}(\mathcal{N},E)} \cap \eta) = A_n^{\pi_{E^*}(\mathcal{N})} \cap \sigma(\eta) \subseteq A_n^{\mathrm{Ult}(M,E^*)}$.

Since $\sigma[A_n^{\mathcal{N}} \cap \eta] = A_n^{\mathcal{N}} \cap \eta$, it follows that

(6) $\pi_{E^*}(A_n^M) \cap \eta$ is cofinal in η.

It then follows that $\pi_E^*(A_n^M) \cap \eta = A_n \cap \eta$ (see Lemma 4.4.3). Thus, $\kappa \in A_{n+1}^M$.

Finally suppose that $\kappa \in A_{n+2}^M$. We want to see that $\kappa \in A_{n+1}^{\mathcal{N}}$. Let $\eta_0 < \eta_1$ be two members of A_{n+1}^M such that η_0 is a limit of A_{n+1}^M and $\kappa < \eta_0$. Let F^* be such that $\pi_{F^*}(A_{n+1}^M) \cap \eta_1 + 1 = A_{n+1}^M \cap \eta_1 + 1$. Like in the $n = 0$ case, we have that if F' is the (κ, η_1)-extender derived from $\pi_{F^*} \restriction \mathcal{N}$ and F is the trivial completion of $F' \restriction \eta_0$ then $F \in \vec{E}^{\mathcal{N}}$. Let now $\sigma : Ult(\mathcal{N},F) \to \pi_{F^*}(\mathcal{N})$ be the canonical factor map. We have that $\mathrm{crit}(\sigma) \geq \eta_0$. We also have that $A_{n+1}^M \cap \eta_1 \subseteq A_n^{\mathcal{N}} \cap \eta_1$. Arguing as above, we get that, in $\mathcal{N}$, F reflects $A_n^{\mathcal{N}}$. ⊣

LEMMA 4.4.5. *Assume $(M,\delta,\vec{G},\Sigma,\mathcal{P})$ has the property (∗). Set $P = \mathcal{J}_\omega[\mathcal{P}]$. Suppose $\lambda < \delta$ is such that $P \in M|\lambda$, $\mathcal{N}'$ is the last model of* $\mathsf{Le}(\Sigma^M,P)^{(M,\delta,\vec{G})}$ *and $\mathcal{N} = L_{\mathrm{ord}(M)}[\mathcal{N}']$. Suppose $F^* \in \vec{G}$ is such that*

1. $\mathrm{lh}(F^*) = \eta \in A_1^M$,
2. $\pi_{F^*}(\mathcal{N})|\eta = \mathcal{N}|\eta$.

Set $\kappa = \mathrm{crit}(F^*)$ and let F' be the (κ,η)-extender derived from $\pi_{F^*} \restriction \mathcal{N} : \mathcal{N} \to \pi_{F^*}(\mathcal{N})$. Let F be the trivial completion of $F' \restriction \eta$. Then $F \in \vec{E}^{\mathcal{N}}$.

PROOF. Let $\eta' \in A_1^M - (\eta+1)$ and let $H \in \vec{G}$ be an extender such that $\mathrm{crit}(H) = \eta$, $\mathrm{lh}(H) > \eta'$, and $\pi_H(A_0^M) \cap (\eta'+1) = A_0^M \cap \eta' + 1$. We have that

(1) $\mathcal{N}|\eta'$ is the last model of both

$$(\mathsf{Le}(\Sigma^M,P)_{>\lambda})^{(M|\eta',\eta',\vec{G}\restriction\eta')} \quad \text{and} \quad (\mathsf{Le}(\Sigma^{\mathrm{Ult}(M,H)},P)_{>\lambda})^{(\mathrm{Ult}(M,H)|\eta',\eta',\vec{G}\restriction\eta')},$$

(2) $\pi_H(\mathcal{N})|\eta' = \mathcal{N}|\eta'$.

It follows from Lemma 4.4.2 that all initial segments of F are on the sequence of $\mathcal{N}$ or an ultrapower away. Thus, in $\mathrm{Ult}(M,H)$, we have that all initial segments of $\pi_H(F)$ are on the extender sequence of $\pi_H(\mathcal{N})$ or an ultrapower away. As $F \restriction \eta = \pi_H(F) \restriction \eta$, we have that in $\mathrm{Ult}(M,H)$, the trivial completion of $F \restriction \eta$ is on the sequence of $\pi_H(\mathcal{N})$. But the trivial completion of $F \restriction \eta$ both in M and

[50]By this we mean an extender whose natural length is η. As η is a regular cardinal of $\mathcal{N}$, there are no (κ,η)-extenders on the sequence of $\mathcal{N}$.

in Ult(M,H) is F, as it only depends on $\pi_{F|\eta}(\mathcal{N}|(\kappa^+)^{\mathcal{N}})$ which is computed the same way in both models. Thus, F is on the extender sequence of $\pi_H(\mathcal{N})$. Since $\pi_H(\mathcal{N})|(\eta^+)^{\pi_H(\mathcal{N})} = \mathcal{N}|(\eta^+)^{\mathcal{N}}$, we have that F is on the extender sequence of $\mathcal{N}$. ⊣

DEFINITION 4.4.6. Suppose $(M,\delta,\vec{G})$ is a background. Let $\vec{K}^M$ consist of all extenders $E \in \vec{G}$ such that

1. $\nu(E) \in \cap_{i<\omega} A_i^M$ and is a limit point of $\cap_{i<\omega} A_i^M$,
2. E reflects $(A_i^M : i < \omega)$.

We say $\mathcal{S}$ is the *fully backgrounded λ-core* of $(M,\delta,\vec{G})$ if $\mathcal{S}$ is the last model of $\mathsf{Le}_{>\lambda}^{(M,\delta,\vec{K}^M)}$. We let $\mathsf{LeCore}_{>\lambda}^{(M,\delta,\vec{G})}$ be the fully backgrounded λ-core of $(M,\delta,\vec{G})$.

Clearly $(M,\delta,\vec{K}^M)$ is a background.

LEMMA 4.4.7. *Assume $(M,\delta,\vec{G},\Sigma,\mathcal{P})$ has the property $(*)$. Set $P = \mathcal{J}_\omega[\mathcal{P}]$. Suppose $\lambda < \delta$ is such that $P \in M|\lambda$, $\mathcal{R}'$ is the last model of $\mathsf{Le}(\Sigma^M,P)_{>\lambda}^{(M,\delta,\vec{G})}$ and $\mathcal{R} = L_{\mathrm{ord}(M)}[\mathcal{R}']$. Then $\mathsf{LeCore}^{(M,\delta,\vec{G})}$ is the last model of $\mathsf{Le}_{>\lambda}^{(\mathcal{R},\delta,\vec{K}^{\mathcal{R}})}$ where $\vec{K}^{\mathcal{R}}$ is computed relative to $\vec{H}^{\mathcal{R}} = \{E \in \vec{E}^{\mathcal{R}} : \mathcal{R} \vDash$ "$\nu(E)$ is an inaccessible cardinal"$\}$.*

PROOF. It is enough to show that if $\mathcal{Q}$ is an $\mathcal{M}$-model of both $\mathsf{Le}_{>\lambda}^{(M,\delta,\vec{K}^M)}$ and $\mathsf{Le}_{>\lambda}^{(\mathcal{R},\delta,\vec{K}^{\mathcal{R}})}$ then the $\mathcal{N}$-models of $\mathsf{Le}_{>\lambda}^{(M,\delta,\vec{K}^M)}$ and $\mathsf{Le}_{>\lambda}^{(\mathcal{R},\delta,\vec{K}^{\mathcal{R}})}$ constructed immediately after $\mathcal{Q}$ coincide. Assume then the $\mathcal{N}$-model of $\mathsf{Le}_{>\lambda}^{(M,\delta,\vec{K}^M)}$ constructed immediately after $\mathcal{Q}$ is $\mathcal{Q}'$. The only non-trivial case is when $\mathcal{Q}'$ is obtained by adding an extender to $\mathcal{Q}$. Thus, assume $\mathcal{Q}' = (\mathcal{Q},F)$. We need to see that $(\mathcal{Q},F)$ is the $\mathcal{N}$-model of $\mathsf{Le}_{>\lambda}^{(\mathcal{N},\delta,\vec{K}^{\mathcal{R}})}$ constructed immediately after $\mathcal{Q}$. Let F^* be the background extender of F. It follows that

1. $\nu(F) < \nu(F^*)$ and $F^* \in \vec{K}^M$,
2. $\nu(F^*) \in \cap A_i^M$ and is a limit point of $\cap_{i<\omega} A_i^M$,
3. F^* reflects $(A_i^M : i < \omega)$.

Set $\eta = \mathrm{lh}(F^*)$ and let F' be the (κ,η)-extender derived from $\pi_{F^*} \restriction \mathcal{R}$. Let E be the trivial completion of $F'|\eta$. It follows from Lemma 4.4.5 that in fact $E \in \vec{E}^{\mathcal{R}}$ and it also follows from Lemma 4.4.4 that $E \in \vec{K}^{\mathcal{R}}$. Since $F = F^* \cap (\nu(F)^{<\omega} \times \lfloor \mathcal{Q} \rfloor)$, we have that $\mathcal{Q} \subseteq \mathcal{R}$ and in $\mathcal{R}$, E is a background certificate of F. It then follows from the uniqueness of the next extender (see [23, Chapter 9] and [48, Theorem 5.1][51]) that in fact that $\mathcal{Q}'$ is the $\mathcal{N}$-model of $\mathsf{Le}_{>\lambda}^{(\mathcal{R},\delta,\vec{K}^{\mathcal{R}})}$ constructed immediately after $\mathcal{Q}$.

Conversely, suppose the $\mathcal{N}$-model of $\mathsf{Le}_{>\lambda}^{(\mathcal{R},\delta,\vec{K}^{\mathcal{R}})}$ constructed immediately after $\mathcal{Q}$ is $(\mathcal{Q},F)$ and let $F^* \in \vec{K}^{\mathcal{R}}$ be the background extender of F. We then have that

[51]This reference contains the proof of non-existence of mixed bicephali, completing [23, Chapter 9].

1. $\nu(F) < \nu(F^*)$,
2. $\nu(F^*) \in \bigcap_{i<\omega} A_i^{\mathcal{R}}$ and is a limit point of $\bigcap_{i<\omega} A_i^{\mathcal{R}}$,
3. F^* reflects $(A_i^{\mathcal{R}} : i < \omega)$.

It then follows from Lemma 4.4.4 that letting F^{**} be the background extender of F^* and $E = F^{**} | \mathrm{lh}(F^*)$, $E \in \vec{K}^M$ and E backgrounds F. It then follows from the uniqueness of the next extender (see the above references) that $(\mathcal{Q}, F)$ is indeed the $\mathcal{N}$-model of $\mathsf{Le}(\Sigma^M, P)_{>\lambda}^{(M,\delta,\vec{G})}$ constructed immediately after $\mathcal{Q}$. ⊣

COROLLARY 4.4.8. *Suppose $(M, \delta, \vec{G})$ is a background and $\lambda < \delta$. Then for any $(\mathcal{P}, \Sigma)$ such that $(M, \delta, \vec{G}, \Sigma, \mathcal{P})$ has the property $(*)$ and $\mathcal{J}_\omega[\mathcal{P}] \in M|\lambda$, letting $\mathcal{R}$ be the last model of* $\mathsf{Le}(\Sigma^M, P)_{>\lambda}^{(M,\delta,\vec{G})}$, $\mathsf{LeCore}_{>\lambda}^{(M,\delta,\vec{G})}$ *is a definable class of $\mathcal{R}$.*

4.5. On the existence of thick sets

[29, Chapter 5.1] develops a methodology for proving branch condensation and various uniqueness results for iteration strategies. The basic idea, due to Jensen[52] and Steel[53], is that the *stack* over a fully backgrounded construction has covering properties. However, both [29] and our current exposition needs, in addition, that *thick sets* exist. While [29] uses their existence, it seems that [29] does not establish their existence. In this section, we take a moment to fill this gap.

First we import one important definition from [29, Chapter 5.1]. Recall that if M is a transitive set then we let $M|\alpha$ be V_α^M.

DEFINITION 4.5.1. Suppose κ is a regular cardinal, Σ is a κ^+-iteration strategy[54] and $\mathcal{M}$ is a Σ-premouse (possibly over some set X) such that $\mathcal{M} \subseteq H_\kappa$. We let $\mathsf{stack}(\mathcal{M}, \Sigma)$ be the union of all sound countably iterable Σ-premice $\mathcal{S}$ such that $\mathcal{M} \trianglelefteq \mathcal{S}$ and $\rho(\mathcal{S}) = \kappa$.

If the stack is computed inside an inner model M then to emphasize the dependence on M, we will write $\mathsf{stack}^M(\mathcal{M}, \Sigma)$.

DEFINITION 4.5.2. Suppose κ is a regular cardinal, Σ is a κ^+-iteration strategy and $\mathcal{M}$ is a Σ-premouse such that $\mathcal{M} \subseteq H_\kappa$. We say $\mathcal{M}$ is κ-*fat* if $\kappa = \mathrm{ord}(\mathcal{M})$ and letting $\mathcal{M}' = \mathsf{stack}(\mathcal{M}, \Sigma)$, $\mathrm{cf}(\mathrm{ord}(\mathcal{M}')) \geq \kappa$. To emphasize the dependence on Σ, we say that $\mathcal{M}$ is (κ, Σ)-fat.

We say $\mathcal{M}$ has *thick sets* (or κ-*thick sets* or (κ, Σ)-*thick sets*) if $\mathcal{M}$ is κ-fat and $\mathcal{M}' =_{\mathrm{def}} \mathsf{stack}(\mathcal{M}, \Sigma)$ has a $(\kappa, \kappa+1)$-iteration strategy Λ (as a Σ-premouse) such that whenever $\mathcal{X}$ is a stack on $\mathcal{M}'$ according to Λ such that $\mathcal{X}$ is below κ, $\pi^{\mathcal{X}}$ exists and $\pi^{\mathcal{X}}(\kappa) = \kappa$,

[52] Jensen developed similar ideas for the K^c constructions, see [12].

[53] The first author learnt about the main idea behind [29, Chapter 5.1] from Steel sometime between 2004-2006. To the author's best knowledge [29, Chapter 5.1] is the first written account of this material.

[54] The nature of the structure that Σ is a strategy of is not important.

1. $\pi^{\mathcal{X}}(\mathcal{M}') = \mathsf{stack}(\pi^{\mathcal{X}}(\mathcal{M}), \Sigma)$, and
2. for some club $C \subseteq \kappa$, whenever $\tau \in C$ is a non-measurable inaccessible cardinal, $\pi^{\mathcal{X}}[\mathrm{ord}(\mathcal{M}')]$ contains a τ-club.

If Λ is as above then we say that $(\mathcal{M}, \Lambda)$ has *thick sets.*

The following lemma is due to Steel. Its proof can be found in [29, Lemma 5.2]. Below H_λ is the set of hereditarily size $< \lambda$ sets.

LEMMA 4.5.3. *Assume* NsesS *and suppose* $(M, \delta, \vec{G}, \Sigma, \mathcal{P})$ *has the property* $(*+)$[55]. *Let* $\lambda < \delta$ *and* $\mathcal{M}$ *be the last model of* $(\mathsf{Le}(\Sigma^M, \mathcal{J}_\omega[\mathcal{P}])_{>\lambda})^{M|\delta}$. *Then* $M \vDash$ *"$\mathcal{M}$ is δ-fat".*

DEFINITION 4.5.4. Suppose $\mathcal{M}$ is a Σ-premouse and κ is a cardinal such that $\mathcal{M} \subseteq H_\kappa$. We say $\mathcal{M}$ is (κ, Σ)-*universal* if $\mathcal{M}$ has a $(\kappa, \kappa+1)$-iteration strategy Λ (as a Σ-mouse) such that for all $(\mathcal{N}, \Phi, \mathcal{Q}, \mathcal{T})$ with the property that

1. $\mathcal{N} \subseteq H_\kappa$ is a Σ-premouse,
2. Φ is a $\kappa+1$-iteration strategy for $\mathcal{N}$ (as a Σ-premouse),
3. $\mathcal{X}$ is an iteration of $\mathcal{M}$ according to Λ such that $\pi^{\mathcal{X}}$ is defined and $\pi^{\mathcal{X}}(\kappa) = \kappa$,
4. $\mathcal{Q}$ is the last model of $\mathcal{X}$,

$(\mathcal{Q}, \Lambda_{\mathcal{Q},\mathcal{X}})$ wins the coiteration with $(\mathcal{N}, \Phi)$. More precisely, if $(\mathcal{T}, \mathcal{U})$ are the normal stacks on $\mathcal{Q}$ and $\mathcal{N}$ respectively that are produced according to the ordinary comparison procedure by using $\Lambda_{\mathcal{Q},\mathcal{X}}$ on the $\mathcal{Q}$ side and Φ on the $\mathcal{N}$ side, then letting $\mathcal{Q}'$ and $\mathcal{N}'$ be the last models of $\mathcal{T}$ and $\mathcal{U}$ respectively, $\mathcal{N}' \trianglelefteq \mathcal{Q}$.

If Λ is as above then we say that $(\mathcal{M}, \Lambda)$ is (κ, Σ)-*universal.*

The following simple lemma will be used in the proof of Theorem 4.5.6.

LEMMA 4.5.5. *Suppose* $\lambda < \delta$ *are cardinals,* δ *is a regular cardinal,* $M \subseteq \delta$, $E \in V_\lambda$ *is an (possible long)* M*-extender and* $N = \mathrm{Ult}(M, E)$[56]. *Then the following holds:*

1. *Suppose* $\kappa \in (\lambda, \delta)$ *and* $\mathrm{cf}^M(\kappa) \geq \lambda$. *Then* $\sup(\pi_E[\kappa]) = \pi_E(\kappa)$.
2. *If* $\kappa \in (\lambda, \delta)$ *is an inaccessible cardinal then* $\pi_E(\kappa) = \kappa$.
3. *Suppose* $\kappa > \lambda$ *is an inaccessible cardinal and* $\eta \in (\kappa, \delta)$ *is a measurable cardinal of* N *such that* $\mathrm{cf}(\eta) < \eta$. *Then there is* $\eta' > \kappa$ *such that* $M \vDash$ *"η' is a measurable cardinal" and* $\mathrm{cf}(\eta') < \eta'$.

PROOF. As clause 1 and 2 are straightforward, we only prove clause 3. Let $f \in M$ be such that for some $a \in \mathrm{lh}(E)^{<\omega}$, $\eta = \pi_E(f)(a)$. Let $(f_i : i < \kappa) \subseteq M$ and $(a_i : i < \kappa) \subseteq \mathrm{lh}(E)^{<\omega}$ be such that $(\pi_E(f_i)(a_i) : i < \kappa)$ is increasing and cofinal in $\pi_E(f)(a)$. Let $\tau \leq \lambda$ be the least such that $\pi_E(\tau) \geq \mathrm{lh}(E)$ and set $h_i(s) = \sup\{f_i(t) : t \in \tau^{<\omega} \wedge f_i(t) < f(s)\}$. We then have that $(\pi_E(h_i)(a) : i < \kappa)$ is cofinal in $\pi_E(f)(a)$. It follows that for E_a measure one many s, $(h_i(s) : i < \kappa)$ is

[55] See Definition 4.4.1.

[56] This is the ultrapower that is constructed using functions in M.

cofinal in $f(s)$, as otherwise if $h(s) = \sup\{h_i(s)+1 : i < \kappa\}$ then we would have $\pi_E(h)(a) < \pi_E(f)(a)$ and for each i, $\pi_E(h_i)(a) < \pi_E(h)(a)$. Fix one such s with the property that $f(s) > \kappa$ and $f(s)$ is a measurable cardinal of M (E_a-measure one many s have this property). Because $(h_i(s) : i < \kappa)$ is cofinal in $f(s)$, we have that $\mathrm{cf}(f(s)) < f(s)$. Hence, $\eta' = f(s)$ is as desired. ⊣

Theorem 4.5.6 is the main theorem on thick sets that we will use throughout this book.

THEOREM 4.5.6. *Assume* NsesS. *Suppose*

$$(M,\delta,\vec{G},\mathcal{P},\Sigma,\mathcal{P}',\mathcal{P}^+,\mathcal{Q},\Lambda,\Lambda',E,\mathcal{R},\Phi)$$

has the following properties:

1. *$(M,\delta,\vec{G})$ is a background.*
2. *$\lambda < \delta$, $\mathbb{P} \in M|\lambda$ is a poset and $g \subseteq \mathbb{P}$ is M-generic.*
3. *$(\mathcal{P},\Sigma)$ and $(\mathcal{Q},\Lambda)$ are allowable pairs with the property that*
   (a) *$\mathcal{P} \in M|\lambda$ and $(M,\delta,\vec{G},\Sigma,\mathcal{P})$ has the property $(*+)$,*
   (b) *$\mathcal{Q} \in M|\lambda[g]$ is a successor type and $M[g] \models$ "Λ is a (δ^+,δ^+)-strategy".*
4. *$\mathcal{P}'$ is the last model of $\mathsf{Le}((\mathcal{P},\Sigma),\mathcal{J}_\omega[\mathcal{P}])_{>\lambda}^{(M,\delta,\vec{G})}$ and $\mathcal{P}^+ = L_{\mathrm{ord}(M)}[\mathcal{P}']$.*
5. *$E \in M|\lambda[g]$ is a $\mathcal{P}^+$-extender such that*
   (a) *$\mathcal{P}_E^+ =_{def} \mathrm{Ult}(\mathcal{P}^+,E)$ is well-founded,*
   (b) *$\Lambda' =_{def} \Lambda \restriction \mathcal{P}_E^+ \in \mathcal{P}_E^+$,*
   (c) *letting $\vec{H} = \{E' \in \vec{E}^{\mathcal{P}_E^+} : \mathrm{crit}(E') > \pi_E(\mathrm{ord}(\mathcal{P}))$ and $\mathcal{P}_E^+ \models$ "$\nu(E')$ is an inaccessible cardinal"$\}$, $(\mathcal{P}_E^+,\delta,\vec{H},\Lambda',\mathcal{Q})$ has the property $(*)$,*
   (d) *letting $\mathsf{Le}((\mathcal{Q}^-,\Lambda_{\mathcal{Q}^-}),\mathcal{J}_\omega[\mathcal{Q}])_{>\lambda}^{(\mathcal{P}_E^+,\delta,\vec{H})} = (\mathcal{Q}_\gamma,\mathcal{Q}'_\gamma,F_\gamma^+,F_\gamma,b'_\gamma : \gamma < \delta)$,*
      (i) *for all $\gamma < \delta$, $\rho(\mathcal{Q}'_\gamma) > \delta^{\mathcal{Q}}$,*
      (ii) *$\mathcal{R} = \mathcal{Q}'_\delta$.*
6. *$\Phi \in M[g]$ is a $(\delta,\delta+1)$-iteration strategy of $\mathcal{R}$.*

Suppose that $\Phi_{\mathcal{Q}^-} = \Lambda_{\mathcal{Q}^-}$. Then for every stack $\mathcal{X}$ according to Φ such that $\mathrm{lh}(\mathcal{X}) < \delta$ and $\pi^{\mathcal{X}}$ exists, letting $\mathcal{R}_1$ be the last model of $\mathcal{X}$,

1. *$(\mathcal{R}_1,\Phi_{\mathcal{R}_1,\mathcal{X}})$ has $(\delta,\Lambda_{\pi^{\mathcal{X}}(\mathcal{Q}^-),\mathcal{X}})$-thick sets, and*
2. *(consequently) $(\mathcal{R}_1,\Phi_{\mathcal{R}_1,\mathcal{X}})$ is $(\delta,\Lambda_{\pi^{\mathcal{X}}(\mathcal{Q}^-),\mathcal{X}})$-universal*[57]*.*

Furthermore, in $M[g]$, $\Phi_{\mathcal{Q}}$ is the unique $(\delta,\delta+1)$-strategy Ψ_0 of $\mathcal{Q}$ such that for some $\mathcal{S}$ and a $(\delta,\delta+1)$-iteration strategy Ψ of $\mathcal{S}$,

1. *$\mathcal{Q} = \mathcal{S}|\delta^{\mathcal{Q}}$ and $\delta^{\mathcal{Q}}$ is a regular cardinal of $\mathcal{S}$,*
2. *$\Psi_{\mathcal{Q}} = \Psi_0$,*
3. *$\Psi_{\mathcal{Q}^-} = \Lambda_{\mathcal{Q}^-}$,*
4. *for every stack $\mathcal{X}$ according to Ψ such that $\mathrm{lh}(\mathcal{X}) < \delta$ and $\pi^{\mathcal{X}}$ exists, letting $\mathcal{S}_1$ be the last model of $\mathcal{X}$, $(\mathcal{S}_1,\Psi_{\mathcal{S}_1,\mathcal{X}})$ has $(\delta,\Lambda_{\pi^{\mathcal{X}}(\mathcal{Q}^-),\mathcal{X}})$-thick sets.*

[57] Universality follows from the existence of thick sets, for example see the proof of [29, Lemma 5.4].

PROOF. The proofs of all of the claims made above are essentially contained in [29]. We first prove the statements made before the "furthermore" clause. The following is the first important step. Set $\Lambda_{\mathcal{X}} = \Lambda_{\pi^{\mathcal{X}}(\mathcal{Q}^-),\mathcal{X}}$.

SUBLEMMA 4.5.7. *In $M[g]$, $\mathcal{R}_1$ is $(\delta,\Lambda_{\mathcal{X}})$-fat.*

PROOF. Towards a contradiction assume not. Let $(Z_\alpha : \alpha < \delta)$ be a continuous chain of submodels of $H_{\delta^+}[g]$ of size $< \delta$ such that for a club of α, letting N_α be the transitive collapse of Z_α and $\tau_\alpha : N_\alpha \to Z_\alpha$ be the inverse of the collapse, $\alpha = \mathrm{crit}(\tau_\alpha)$ and $\wp(\alpha)^{\mathcal{R}_1} \subseteq N_\alpha$. Such a sequence can be constructed following the construction given in the proof of [29, Lemma 5.2].

Let $\mathcal{W} = \mathsf{LeCore}^{(\mathcal{P}^+,\delta,\vec{H})}_{>\lambda}$ where $\vec{H}$ consists of those extenders of $\vec{E}^{\mathcal{P}^+}$ whose natural length is an inaccessible cardinal of $\mathcal{P}^+$. It follows from Lemma 4.4.7 that $\pi_E(\mathcal{W})$ is a class of $\mathcal{R}$ and therefore, $\mathcal{W}_{\mathcal{X}} =_{\mathrm{def}} \pi^{\mathcal{X}}(\pi_E(\mathcal{W}))$ is a class of $\mathcal{R}_1$. Hence, for a club of $\alpha < \delta$ the following conditions are true:

(1) $E \in Z_\alpha$, $\pi^{\mathcal{X}} \circ \pi_E(\alpha) = \alpha$, $\mathrm{crit}(\tau_\alpha) = \alpha$ and $\wp(\alpha)^{\mathcal{W}_{\mathcal{X}}} \subseteq N_\alpha$.

If α is as in (1) then we in fact have that $\wp(\alpha)^{\mathcal{W}} \subseteq N_\alpha$. However, as in the proof of [29, Lemma 5.2], we can find an extender $F^* \in \vec{G}$ such that for some ν, the trivial completion of $F^* \cap (\nu^{<\omega} \times \lfloor \mathcal{W} \rfloor)$ is on $\vec{E}^{\mathcal{W}}$ and witnesses that $\mathrm{crit}(F)$ is a superstrong cardinal in $\mathcal{W}$, contradicting NsesS. ⊣

Set $\mathcal{R}_1^+ = \mathsf{stack}(\mathcal{R}_1, \Lambda_{\mathcal{X}})$ and let $\mathcal{X}^+ = \uparrow(\mathcal{X}, \mathcal{R}_1^+)$. Applying the proof of [29, Lemma 5.3] we get the following.

SUBLEMMA 4.5.8. *Suppose $\mathcal{Y}$ is an iteration of $\mathcal{R}$ according to Φ such that $\pi^{\mathcal{Y}}$ is defined and $\pi^{\mathcal{Y}}(\delta) = \delta$. Then all models of $\mathcal{Y}^+ =_{def} \uparrow(\mathcal{Y}, \mathcal{R}^+)$ are well-founded and if $\mathcal{S}$ is the last model of $\mathcal{Y}^+$ then $\mathcal{S} = \mathsf{stack}(\mathcal{S}|\delta, \Phi_{\pi^{\mathcal{X}^\frown\mathcal{Y}}(\mathcal{Q}^-),\mathcal{X}^\frown\mathcal{Y}})$.*

We now have that $(\mathcal{R}_1, \Phi_{\mathcal{R}_1,\mathcal{X}})$ is $(\delta, \Lambda_{\mathcal{Q}^-})$-universal (e.g., see the proof of [29, Lemma 5.4]). Next we show that $(\mathcal{R}_1, \Phi_{\mathcal{R}_1,\mathcal{X}})$ has $(\delta, \Lambda_{\mathcal{X}})$-thick sets. Let $\mathcal{U}$ be a stack on $\mathcal{R}_1$ according to $\Phi_{\mathcal{R}_1,\mathcal{X}}$ and let $\mathcal{U}^+ = \uparrow(\mathcal{U}, \mathcal{R}^+)$. We are assuming that $\pi^{\mathcal{U}^+}(\delta) = \delta$[58] and want to show that

(a) for some club $C \subseteq \delta$, whenever $\kappa \in C$ is a non-measurable inaccessible cardinal, $\pi^{\mathcal{U}^+}[\mathrm{ord}(\mathcal{R}_1^+)]$ contains a κ-club.

Set $\sigma = (\pi^{\mathcal{U}^+} \restriction \delta+1) \circ (\pi^{\mathcal{X}^+} \restriction \delta+1) \circ (\pi_E \restriction \delta+1)$ and $\nu = \mathrm{ord}(\mathcal{R}_1^+)$. Notice that since $\sigma(\delta) = \delta$, we have a club $C \subseteq \delta$ such that for each $\alpha \in C$, $\sigma[\alpha] \subseteq \alpha$. Let $\lambda' < \delta$ be such that $\max(\lambda, \mathrm{ord}(\mathcal{P})) < \lambda'$ and $\mathcal{X} \in M|\lambda'[g]$. We want to show that $C - (\lambda'+1)$ witnesses (a). Suppose then $\kappa \in (\lambda', \delta)$ is an inaccessible cardinal of M which is not measurable in M and $\kappa \in C$. It then follows that $\sigma(\kappa) = \kappa$. Indeed, because $E, \mathcal{X} \in M|\lambda'[g]$ we have that $\pi^{\mathcal{X}}(\pi_E(\kappa)) = \kappa$. Notice now that because κ is not measurable in M, κ is not measurable in $\mathcal{P}^+$ and therefore, in $\mathcal{P}_E^+$ and consequently in $\mathcal{R}$ and $\mathcal{R}_1$. Hence, it follows from $\pi^{\mathcal{U}}[\kappa] \subseteq \kappa$ that $\pi^{\mathcal{U}}(\kappa) = \kappa$.

[58]We in fact should also assume that $\mathcal{U}$ is above $\pi^{\mathcal{X}}(\mathrm{ord}(\mathcal{Q}^-))$ but this is irrelevant to the proof.

Suppose now that $\alpha \in [\delta, \nu)$ and $\mathrm{cf}^M(\alpha) = \kappa$. We claim that $\sup(\pi^{\mathcal{U}}[\alpha]) = \pi^{\mathcal{U}}(\alpha)$. The claim is clear if $\mathrm{cf}^{\mathcal{R}_1^+}(\alpha) = \kappa$. Suppose then that $\eta =_{\mathrm{def}} \mathrm{cf}^{\mathcal{R}_1^+}(\alpha) > \kappa$. Notice that we have that $\mathrm{cf}^M(\eta) = \kappa$. We claim that

(b) η is not a measurable cardinal of $\mathcal{R}_1$.

Assume η is measurable in $\mathcal{R}_1$. Then it follows from Lemma 4.5.5 that there is $\eta' > \kappa$ such that η' is a measurable cardinal of $\mathcal{R}$ and $\mathrm{cf}^M(\eta') < \eta'$. Because η' is measurable in $\mathcal{R}$, η' is a measurable cardinal of $\mathcal{P}_E^+$. Since $\mathrm{cf}^M(\eta') < \eta'$, Lemma 4.5.5 implies that there is a measurable cardinal η'' of $\mathcal{P}^+$ such that $\eta'' > \kappa$ and $\mathrm{cf}^M(\eta'') < \eta''$. But each measurable cardinal of $\mathcal{P}^+$ that is $> \kappa$ is a measurable cardinal of $M[g]$, contradiction! Thus, (b) holds.

Since η is not a measurable cardinal of $\mathcal{R}_1$ we get that $\sup(\pi^{\mathcal{U}^+}[\eta]) = \eta$. Hence, $\sup(\pi^{\mathcal{U}^+}[\alpha]) = \pi^{\mathcal{U}^+}(\alpha)$. It then follows that $\pi^{\mathcal{U}^+}[\nu]$ is a κ-club.

Next, we prove that in $M[g]$, $\Phi_{\mathcal{Q}}$ is the unique $(\delta, \delta+1)$-strategy Ψ_0 of $\mathcal{Q}$ such that for some $\mathcal{S}$ and a $(\delta, \delta+1)$-iteration strategy Ψ of $\mathcal{S}$,

1. $\mathcal{Q} = \mathcal{S}|\delta^{\mathcal{Q}}$ and $\delta^{\mathcal{Q}}$ is a regular cardinal of $\mathcal{S}$,
2. $\Psi_{\mathcal{Q}} = \Psi_0$,
3. $\Psi_{\mathcal{Q}^-} = \Lambda_{\mathcal{Q}^-}$,
4. for every stack $\mathcal{X}$ according to Ψ such that $\mathrm{lh}(\mathcal{X}) < \delta$ and $\pi^{\mathcal{X}}$ exists, letting $\mathcal{S}_1$ be the last model of $\mathcal{X}$, $(\mathcal{S}_1, \Psi_{\mathcal{S}_1,\mathcal{X}})$ has $(\delta, \Lambda_{\pi^{\mathcal{X}}(\mathcal{Q}^-),\mathcal{X}})$-thick sets.

Fix then $(\mathcal{S}, \Psi)$ that satisfies clause 1, 3, and 4 above. We have that $(\mathcal{R}, \Phi)$ also satisfies those clauses. It is then enough to show that $\Psi_{\mathcal{Q}} = \Phi_{\mathcal{Q}}$. Assume not. Let $\mathcal{U}$ be a stack on $\mathcal{Q}$ such that $\Phi_{\mathcal{Q}}(\mathcal{U}) \neq \Psi_{\mathcal{Q}}(\mathcal{U})$. Let $\mathcal{R}^+ = \mathsf{stack}(\mathcal{R}, \Lambda_{\mathcal{Q}^-})$, $\mathcal{S}^+ = \mathsf{stack}(\mathcal{S}, \Lambda_{\mathcal{Q}^-})$, $\mathcal{U}_0 = \uparrow(\mathcal{U}, \mathcal{R}^+)$ and $\mathcal{U}_1 = \uparrow(\mathcal{U}, \mathcal{S}^+)$. Because $\Phi_{\mathcal{Q}^-} = \Psi_{\mathcal{Q}^-}$, we have some $\alpha < \mathrm{lh}(\mathcal{U})$ such that $\pi^{\mathcal{U}}_{0,\alpha}$ is defined and $\mathcal{U}_{\geq\alpha}$ is a normal stack on $\mathcal{M}^{\mathcal{U}}_{\alpha}$ and is above $\mathrm{ord}(\pi^{\mathcal{U}}_{0,\alpha}(\mathcal{Q}^-))$. Let then $\mathcal{X}_0 = (\mathcal{U}_0)_{\leq\alpha}$, $\mathcal{X}_1 = (\mathcal{U}_1)_{\leq\alpha}$ and $\mathcal{Y} = \mathcal{U}_{\geq\alpha}$. Set $\mathcal{R}_1 = \mathcal{M}^{\mathcal{X}_0}_{\alpha}$, $\mathcal{S}_1 = \mathcal{M}^{\mathcal{X}_1}_{\alpha}$, $\mathcal{Y}_0 = \uparrow(\mathcal{Y}, \mathcal{R}_1)$ and $\mathcal{Y}_1 = \uparrow(\mathcal{Y}, \mathcal{S}_1)$. Finally set $\Phi_{\mathcal{Q}}(\mathcal{U}) = b_0$ and $\Psi_{\mathcal{Q}}(\mathcal{U}) = b_1$.

We claim that $\mathcal{Q}(b_0, \mathcal{U})$ does not exist. Towards a contradiction assume it does exist. Assume first that $\mathcal{Q}(b_1, \mathcal{U})$ does not exist. It follows that $\delta(\mathcal{Y})$ is not a limit of Woodin cardinals of $\mathrm{m}(\mathcal{Y})$, and therefore, $\mathcal{Q}(b_0, \mathcal{U})$ is a $\Lambda_{\pi^{\mathcal{U}}_{0,\alpha}(\mathcal{Q}^-),\mathcal{U}_{\leq\alpha}}$-mouse over $\mathrm{m}(\mathcal{Y})$, and since $\mathcal{M}^{\mathcal{Y}_1}_{b_1}$ is universal, $\mathcal{Q}(b_0, \mathcal{U}) \trianglelefteq \mathcal{M}^{\mathcal{Y}_1}_{b_1}$[59]. Thus, we must have that both $\mathcal{Q}(b_0, \mathcal{U})$ and $\mathcal{Q}(b_1, \mathcal{U})$ exist. A similar argument shows that $\mathcal{U}$ cannot have a fatal drop, implying that $\mathcal{Q}(b_0, \mathcal{U})$ and $\mathcal{Q}(b_1, \mathcal{U})$ are $\Lambda_{\pi^{\mathcal{U}}_{0,\alpha}(\mathcal{Q}^-),\mathcal{U}_{\leq\alpha}}$-mice over $\mathrm{m}(\mathcal{Y})$. Hence, $\mathcal{Q}(b_0, \mathcal{U}) = \mathcal{Q}(b_1, \mathcal{U})$ implying that $b_0 = b_1$, contradiction. Hence, $\mathcal{Q}(b_0, \mathcal{U})$ does not exist. A symmetric argument shows that $\mathcal{Q}(b_1, \mathcal{U})$ also does not exist.

[59] More precisely, setting $\mathcal{S}' = \mathcal{M}^{\mathcal{Y}_1}_{b_1}$, $(\mathcal{S}', \Psi_{\mathcal{S}', \mathcal{U}_1^\frown\{b\}})$ is $(\delta, \Lambda_{\pi^{\mathcal{U}}_{0,\alpha}(\mathcal{Q}^-),\mathcal{U}_{\leq\alpha}})$-universal.

We thus have that for $i \in 2$, $\pi^{\mathcal{Y}_i}_{b_i}$ is defined. Let $\mathcal{R}_2 = \mathcal{M}^{\mathcal{Y}_0}_{b_0}$ and $\mathcal{S}_2 = \mathcal{M}^{\mathcal{Y}_1}_{b_1}$. Both $\mathcal{R}_2$ and $\mathcal{S}_2$ are $\Lambda_{\pi^{\mathcal{U}}_{0,\alpha}(\mathcal{Q}^-),\mathcal{U}_{\leq\alpha}}$-mice. We can then find $\mathcal{W}$ such that

(1) $\mathcal{W}$ is a $\Phi_{\mathcal{R}_2,\mathcal{U}_0^\frown\{b_0\}}$-iterate of $\mathcal{R}_2$ and the iteration embedding $j_0 : \mathcal{R}_2 \to \mathcal{W}$ exists and has the property that $j_0(\delta) = \delta$, and
(2) $\mathcal{W}$ is a $\Psi_{\mathcal{S}_2,\mathcal{U}_1^\frown\{b_1\}}$-iterate of $\mathcal{S}_2$ and the iteration embedding $j_1 : \mathcal{S}_2 \to \mathcal{W}$ exists and has the property that $j_1(\delta) = \delta$.

Because of our assumption on thick sets, we have a club $C_0 \subseteq \delta$ and a club $C_1 \subseteq \delta$ such that for every $\kappa \in C_0 \cap C_1$ that is an inaccessible cardinal of M but not a measurable cardinal of M,

(3) $j_0 \circ \pi^{\mathcal{Y}_0}_{b_0}[\text{ord}(\mathcal{R}_1)]$ and $j_1 \circ \pi^{\mathcal{Y}_1}_{b_1}[\text{ord}(\mathcal{S}_1)]$ contain a κ-club.

(3) then implies that

(4) $(j_0 \circ \pi^{\mathcal{Y}_0}_{b_0}[\text{ord}(\mathcal{R}_1)]) \cap (j_1 \circ \pi^{\mathcal{Y}_1}_{b_1}[\text{ord}(\mathcal{S}_1)])$ contains a κ-club.

Let then $D \subseteq ((j_0 \circ \pi^{\mathcal{Y}_0}_{b_0}[\text{ord}(\mathcal{R}_1)]) \cap (j_1 \circ \pi^{\mathcal{Y}_1}_{b_1}[\text{ord}(\mathcal{S}_1)])$ be a κ-club and set $D_0 = (j_0 \circ \pi^{\mathcal{Y}_0}_{b_0})^{-1}[D]$ and $D_1 = (j_1 \circ \pi^{\mathcal{Y}_1}_{b_1})^{-1}[D]$. Let now $\mathcal{Q}_0 = \pi^{\mathcal{U}}_{0,\alpha}(\mathcal{Q})$. Notice that $\mathcal{Y}$ is a normal stack on $\mathcal{Q}_0$ that is above $\text{ord}(\mathcal{Q}_0^-)$ and below $\delta^{\mathcal{Q}_0}$. We now have that

(5) $\delta^{\mathcal{Q}_0} = \sup(Hull^{\mathcal{R}_1}(D_0 \cup \mathcal{Q}_0^-) \cap \delta^{\mathcal{Q}_0})$,
(6) $\delta^{\mathcal{Q}_0} = \sup(Hull^{\mathcal{S}_1}(D_1 \cup \mathcal{Q}_0^-) \cap \delta^{\mathcal{Q}_0})$,
(7) $\delta(\mathcal{Y}) = \sup(Hull^{\mathcal{R}_2}(\pi^{\mathcal{Y}_0}_{b_0}[D_0] \cup \mathcal{Q}_0^-) \cap \delta(\mathcal{Y}))$,
(8) $\delta(\mathcal{Y}) = \sup(Hull^{\mathcal{S}_2}(\pi^{\mathcal{Y}_1}_{b_1}[D_1] \cup \mathcal{Q}_0^-) \cap \delta(\mathcal{Y}))$.

(5)-(8) are consequences of universality. For example, (5) can be shown as follows. Suppose $\delta^{\mathcal{Q}_0} > \sup(Hull^{\mathcal{R}_1}(D_0 \cup \mathcal{Q}_0^-) \cap \delta^{\mathcal{Q}_0})$ and set $\gamma = \sup(Hull^{\mathcal{R}_1}(D_0 \cup \mathcal{Q}_0^-) \cap \delta^{\mathcal{Q}_0})$. Let $\mathcal{R}' = cHull^{\mathcal{R}_1}(D_0 \cup \gamma)$ and let $\tau : \mathcal{R}' \to \mathcal{R}_1$ be the inverse of the transitive collapse. Then because $\tau(\mathcal{Q}_0^-) = \mathcal{Q}_0^-$, we have that $\mathcal{R}'$ is a $\Lambda_{\mathcal{Q}_0^-,\mathcal{U}_{\leq\alpha}}$-mouse as witnessed by $\Phi' = (\tau\text{-pullback of }\Phi_{\mathcal{R}_1,\mathcal{X}_0})$. Moreover, it follows from [29, Lemma 5.4] that $\mathcal{R}' = \mathsf{stack}(\mathcal{R}'|\delta, \Lambda_{\mathcal{Q}_0^-,\mathcal{U}_{\leq\alpha}})$ and $(\mathcal{R}', \Phi')$ is $(\delta, \Lambda_{\mathcal{Q}_0^-,\mathcal{U}_{\leq\alpha}})$-universal. But because $\mathcal{R}' \models$ "γ is a Woodin cardinal" and $\mathcal{R}_1 \models$ "γ is not a Woodin cardinal", we have a contradiction.

(5)-(8) easily imply that $\text{rge}(\pi^{\mathcal{Y}_0}_{b_0}) \cap \text{rge}(\pi^{\mathcal{Y}_1}_{b_1})$ is cofinal in $\delta(\mathcal{Y})$. Hence, because $\pi^{\mathcal{Y}_0}_{b_0} \restriction \delta^{\mathcal{Q}_0} = \pi^{\mathcal{Y}}_{b_0} \restriction \delta^{\mathcal{Q}_0}$ and $\pi^{\mathcal{Y}_1}_{b_1} \restriction \delta^{\mathcal{Q}_0} = \pi^{\mathcal{Y}}_{b_1} \restriction \delta^{\mathcal{Q}_0}$, we have that $b_0 = b_1$. $\dashv$

The next few chapters are essentially applications of Theorem 4.5.6.

4.6. Fullness preservation

Throughout this section we assume AD^+. Below, we use $\mathcal{R}^*$ to denote the $*$-translation of $\mathcal{R}$ (cf. [39] or [57, Remark 12.7].). Suppose η is a cutpoint cardinal of a hod premouse $\mathcal{R}$. The $*$-translation is used to translate $\mathcal{R}|(\eta^+)^{\mathcal{R}}$ into an lses over $\mathcal{R}|\eta$. More precisely, $\lfloor(\mathcal{R}|(\eta^+)^{\mathcal{R}})^*\rfloor = \lfloor\mathcal{R}|(\eta^+)^{\mathcal{R}}\rfloor$ but η is a strong cutpoint of $(\mathcal{R}|(\eta^+)^{\mathcal{R}})^*$. If in fact η is already a strong cutpoint of $\mathcal{R}$ then $(\mathcal{R}|(\eta^+)^{\mathcal{R}})^* = \mathcal{R}|(\eta^+)^{\mathcal{R}}$. Thus, as far as grasping the main ideas are concerned, the reader will lose little by treating all cutpoint cardinals as strong cutpoint cardinals.

DEFINITION 4.6.1. We say Γ is *projectively closed* if whenever A is a set of reals such that for some $B \in \Gamma$, A is first order definable over $(\mathsf{HC}, B, \in)$ (with parameters), $A \in \Gamma$.

DEFINITION 4.6.2 (Γ-Fullness preservation). Suppose $(\mathcal{P}, \Sigma)$ is a hod pair or an sts hod pair[60] such that $\mathcal{P} \in \mathsf{HC}$ and Γ is a projectively closed pointclass. We say Σ is Γ-*fullness preserving* if the following holds for all $(\mathcal{T}, \mathcal{Q}) \in I(\mathcal{P}, \Sigma)$.

1. For all meek[61] layers $\mathcal{R}$ of $\mathcal{Q}$ such that $\mathcal{R}$ is of successor type[62], letting $\mathcal{S} = \mathcal{R}^{-}$[63], for all $\eta \in (\mathrm{ord}(\mathcal{S}), \mathrm{ord}(\mathcal{R}))$ if η is a cutpoint cardinal of $\mathcal{R}$ then
$$(\mathcal{R}|(\eta^+)^{\mathcal{R}})^* = \mathsf{Lp}^{\Sigma_{\mathcal{S},\mathcal{T}}}(\mathcal{R}|\delta).$$
2. For all meek[64] layers $\mathcal{R}$ of $\mathcal{Q}$ such that $\mathcal{R}$ is of limit type,
$$\mathcal{R} = \mathsf{Lp}^{\Sigma_{\mathcal{R}|\delta^{\mathcal{R}},\mathcal{T}}}(\mathcal{R}|\delta^{\mathcal{R}}).$$
3. If $\mathcal{P}$ is of #-lsa type then $\mathsf{Lp}^{\Gamma,\Sigma^{stc}_{\mathcal{Q},\mathcal{T}}}(\mathcal{Q}) \models$ "$\delta^{\mathcal{Q}}$ is a Woodin cardinal"[65].

If only conditions 1 and 2 hold then we say that Σ is *almost* Γ-*fullness preserving*. We say that Σ is *lower-level* Γ-*fullness preserving* if the above clauses hold for $\mathcal{R} \lhd_{hod} \mathcal{Q}$[66].

Suppose $(\mathcal{P}, \Sigma)$ is a hod pair such that $\mathcal{P}$ is gentle. Then we say that Σ is Γ-*fullness preserving* if for every $\mathcal{Q} \in Y^{\mathcal{P}}$, $\Sigma_{\mathcal{Q}}$ is Γ-fullness preserving.

If Γ is a Solovay pointclass then we will omit it from the terminology.

THEOREM 4.6.3 (Fullness preservation of induced strategies). *Assume* AD^+. *Suppose* Γ *is a pointclass such that for some* α *with* $\vartheta_\alpha < \Theta$, $\Gamma = \{A \subseteq \mathbb{R} : w(A) < \vartheta_\alpha\}$, $\mathsf{C} = (\mathbb{M}, (P, \Psi), \Gamma^*, A)$ *Suslin, co-Suslin captures* Γ *and* $\mathbb{M} = (M, \delta, \vec{G}, \Sigma)$. *Set*
$$\mathsf{hpc}^+_{\mathsf{C},\Gamma} = (\mathcal{M}_\gamma, \mathcal{N}_\gamma, Y_\gamma, \Phi^+_\gamma, F^+_\gamma, F_\gamma, b_\gamma : \gamma \leq \delta).$$

---

[60] Recall that if $(\mathcal{P}, \Sigma)$ is an sts hod pair then $\mathcal{P} = (\mathcal{P}|\delta^{\mathcal{P}})^\#$. See Definition 3.10.5.

[61] See Definition 2.7.1.

[62] See Definition 2.7.17.

[63] This is the longest proper layer of $\mathcal{R}$. See Notation 2.7.14.

[64] See Definition 2.7.1.

[65] Here, if Σ is a short tree strategy then $\Sigma^{stc} = \Sigma$.

[66] We will use this version of fullness preservation when studying anomalous hod pairs (see Section 5.4). For now, the reader may ignore it. The concept will became important in Theorem 10.1.5.

Suppose $\beta < \delta$, $\mathcal{P} \in Y_\beta$ and $M \models$ "$(\mathcal{P}, (\Phi_\beta)_\mathcal{P}) \in \mathsf{HP}^\Gamma$". Then $(\Phi^+_\beta)_\mathcal{P}$ is almost Γ-fullness preserving.

Moreover, assuming that

1. *$\mathcal{P}$ is of #-lsa type,*
2. *letting $\Psi = (\Phi^+_\beta)^{stc}_\mathcal{P}$, $\mathsf{Lp}^{\Gamma,\Psi}(\mathcal{P}) \models$ "$\delta^\mathcal{P}$ is a Woodin cardinal" and*
3. *$\mathsf{L}\bullet((\mathcal{P}, (\Phi^{stc}_\beta)_\mathcal{P}), \mathcal{J}_\omega[\mathcal{P}])$ does not break down because of the anomaly stated in clause 3(b) of Definition 4.2.1,*

Ψ is Γ-fullness preserving. Also, the above clauses hold for ξ as in the definition of $Y_{\xi+1}$ that appears in clause 5(c)(ii) of Definition 4.3.3.

PROOF. Below we will use the universality clause of Theorem 4.5.6. Towards a contradiction, assume $\Lambda =_{\text{def}} (\Phi^+_\beta)_\mathcal{P}$ is not Γ-fullness preserving. We have that Λ is the *id*-pullback of Φ^+_β[67]. It follows by absoluteness[68] that there is a counterexample in $M[g]$ where $g \subseteq Coll(\omega, \nu)$ is M-generic and $\nu < \delta$. All the clauses of Γ-fullness preservation are very similar and follow from the universality of background constructions. Below we derive a contradiction from the failure of clause 2 of Definition 4.6.2 and leave the rest to the reader. We also leave the "moreover" clause to the reader as it is very similar to the other cases. We can then further assume that $\mathcal{P}$ is of limit type as otherwise we would just be re-proving [29, Lemma 5.7].

Fix $(\mathcal{T}, \mathcal{Q}) \in I(\mathcal{P}, \Lambda)$[69] and fix $\mathcal{R}$ which is as in clause 2 of Definition 4.6.2 (so $\mathcal{R}$ is a meek layer of $\mathcal{Q}$ of limit type, implying that $\mathcal{R} = \mathcal{R}^b$). Let $\kappa = \delta^\mathcal{R}$. We need to see that

$$\mathcal{R} = \mathsf{Lp}^{\Gamma, \Lambda_{\mathcal{R}|\kappa, \mathcal{T}}}(\mathcal{R}|\kappa).$$

Using Lemma 4.3.9, we can find a Σ-iterate N of M such that letting $i : M \to N$ be the iteration embedding and

$$\mathsf{hpc}^+_{\mathsf{CN},\Gamma} = (\mathcal{S}'_\gamma, \mathcal{S}_\gamma, Z_\gamma, \Psi^+_\gamma, E^+_\gamma, E_\gamma, c_\gamma : \gamma \leq i(\delta)),$$

there is a $\gamma < i(\delta)$ and a weak embedding $\sigma : \mathcal{Q} \to i(\mathcal{P})$ such that

(1) $\sigma \circ \pi^\mathcal{T} = i \restriction \mathcal{P}$,
(2) $\Lambda_{\mathcal{Q},\mathcal{T}}$ is the σ-pullback of $(\Psi^+_\gamma)_{i(\mathcal{P})}$,
(3) $i(\mathcal{P}) \in Z_\gamma$.

Suppose first that $\mathcal{M} \trianglelefteq \mathcal{R}$ is such that $\mathcal{R}|\kappa \triangleleft \mathcal{M}$. We need to see that

(a) $\mathcal{M}$ as a $\Lambda_{\mathcal{R}|\kappa,\mathcal{T}}$-premouse has an ω_1-iteration strategy in Γ.

Clearly (2) and (3) above easily imply (a)[70].

[67] See Definition 2.6.3.

[68] See Lemma 4.1.11 and Corollary 4.1.15. Here we use the fact that $M \models (\mathcal{P}, (\Phi_\beta)_\mathcal{P}) \in \mathsf{HP}^\Gamma$.

[69] It is irrelevant whether $\mathcal{T}$ is an ordinary stack or a generalized stack.

[70] In fact this also follows from our choice of Γ as since $\mathsf{Code}(\Lambda_{\mathcal{R}|\kappa,\mathcal{T}}) \in \Gamma$, any $\Lambda_{\mathcal{R}|\kappa,\mathcal{T}}$-mouse $\mathcal{M}$ such that $\rho(\mathcal{M}) = \kappa$ has an iteration strategy in Γ.

Fix now $\mathcal{M} \trianglelefteq \mathsf{Lp}^{\Gamma,\Lambda_{\mathcal{R}|\kappa,\mathcal{T}}}(\mathcal{R}|\kappa)$ such that $\rho(\mathcal{M}) = \kappa$. We want to see that

(b) $\mathcal{M} \trianglelefteq \mathcal{R}$.

We let $\pi = \pi^{\mathcal{T}}$, $\tau = \delta^{\mathcal{P}^b}$, $\zeta = \sup\{\mathrm{lh}(F_\gamma^+) : \gamma < \beta\}$, $\vec{G}' = \{F \in \vec{G} : \mathrm{crit}(F) > \max(\zeta, \nu)\}$ and $\mathcal{N}$ be the last model of

$$(\mathsf{Le}((\mathcal{P}|\tau, \Lambda_{\mathcal{P}|\tau}), \mathcal{P}^b)_{>\zeta})^{(M[g],\delta,\vec{G}')}.$$

Notice that because of our choice of Γ (see the footnote above), the fact that $(\mathcal{P}, \Lambda)$ is a $\Gamma-\mathsf{cbl}$ and the $(\delta, \Lambda_{\mathcal{P}|\tau})$-universality of $\mathcal{N}$,

(4) $\mathcal{P} = \mathcal{N}|(\tau^+)^{\mathcal{N}}$.

Notice next that if E is the $(\tau, \delta^{\mathcal{Q}^b})$-extender derived from $\pi^{\mathcal{T}}$ then

(5) $M[g] \vDash$ "$\mathrm{Ult}(\mathcal{N}, E)$ is δ-iterable".

This is because $\sigma : \mathcal{Q} \to i(\mathcal{P})$ can be extended to $\sigma^+ : Ult(\mathcal{N}, E) \to i(\mathcal{N})$.

Let then $\pi^+ = \pi_E^{\mathcal{N}}$, $\vec{H} = \{E \in \vec{E}^{\mathrm{Ult}(\mathcal{N},E)} : \mathrm{crit}(E) > \delta^{\mathcal{Q}}$ and $\nu(E)$ is an inaccessible cardinal of $\mathrm{Ult}(\mathcal{N}, E)\}$, and $\mathcal{N}^*$ be the last model of

$$(\mathsf{Le}((\mathcal{R}|\kappa, \Lambda_{\mathcal{R}|\kappa,\mathcal{T}}), \mathcal{R}|\kappa))^{\mathrm{Ult}(\mathcal{N},E),\delta,\vec{H}}.$$

It then follows from $(\delta, \Lambda_{\mathcal{R}|\kappa,\mathcal{T}})$-universality of $\mathcal{N}^*$ that $\mathcal{M} \trianglelefteq \mathcal{N}^*$. Therefore, $\mathcal{M} \in \mathrm{Ult}(\mathcal{N}, E)$, and since $\mathcal{R} = \mathrm{Ult}(\mathcal{N}, E)|(\kappa^+)^{\mathrm{Ult}(\mathcal{N},E)}$, $\mathcal{M} \in \mathcal{R}$. Since $\mathcal{M}$ is ω_1-iterable, it follows that $\mathcal{M} \trianglelefteq \mathcal{R}$. ⊣

The proof actually gives more.

DEFINITION 4.6.4 (Strongly Γ-fullness preserving). Suppose $(\mathcal{P}, \Sigma)$ is a hod pair or an sts hod pair and Γ is a pointclass. We say Σ is strongly Γ-fullness preserving if Σ is Γ-fullness preserving and whenever

1. $\mathcal{T}$ is a stack according to Σ with last model $\mathcal{S}$ such that if $\mathcal{P}$ is of limit type then $\pi^{\mathcal{T},b}$ exists and otherwise $\pi^{\mathcal{T}}$ exists, and
2. $\mathcal{R}$ is such that there are elementary embedding (σ, τ) with the property that
   (a) if $\mathcal{P}$ is of limit type then $\sigma : \mathcal{P}^b \to \mathcal{R}$, $\tau : \mathcal{R} \to \mathcal{S}^b$ and $\pi^{\mathcal{T},b} = \tau \circ \sigma$, and
   (b) if $\mathcal{P}$ is of successor type then $\sigma : \mathcal{P} \to \mathcal{R}$, $\tau : \mathcal{R} \to \mathcal{S}$ and $\pi^{\mathcal{T}} = \tau \circ \sigma$,

then the τ-pullback strategy of $\Sigma_{\mathcal{S}^b,\mathcal{T}}$ if 2(a) holds and of $\Sigma_{\mathcal{S},\mathcal{T}}$ if 2(b) holds is Γ-fullness preserving. Following Definition 4.6.2 we can also define the meaning of *strongly almost* Γ*-fullness preserving* as well as the meaning of *strongly low-level* Γ*-fullness preserving*.

The following is then a corollary to the proof of Theorem 4.6.3 and we leave it to the reader.

THEOREM 4.6.5 (Strong fullness preservation of induced strategies). *Assume* AD^+ *and suppose* Γ *is a pointclass such that for some* α *with* $\vartheta_\alpha < \Theta$,

$\Gamma = \{A \subseteq \mathbb{R} : w(A) < \vartheta_\alpha\}$, $\mathsf{C} = (\mathbb{M}, (P, \Psi), \Gamma^*, A)$ *Suslin, co-Suslin captures* Γ *and* $\mathbb{M} = (M, \delta, \vec{G}, \Sigma)$. *Set*

$$\mathsf{hpc}^+_{\mathsf{C},\Gamma} = (\mathcal{M}_\gamma, \mathcal{N}_\gamma, Y_\gamma, \Phi^+_\gamma, F^+_\gamma, F_\gamma, b_\gamma : \gamma \leq \delta).$$

Suppose $\beta < \delta$ *and* $\mathcal{P} \in Y_\beta$. *Then* $(\Phi^+_\beta)_\mathcal{P}$ *is almost* Γ*-fullness preserving.*

Moreover, assuming that

1. $\mathcal{P}$ *is of #-lsa type,*
2. *letting* $\Psi = (\Phi^+_\beta)^{stc}_\mathcal{P}$, $\mathsf{Lp}^{\Gamma,\Psi}(\mathcal{P}) \vDash$ "$\delta^\mathcal{P}$ *is a Woodin cardinal" and*
3. *for every* $\zeta < \delta$ *the* $\mathsf{Le}((\mathcal{P}, (\Phi^{stc}_\beta)_\mathcal{P}), \mathcal{J}_\omega[\mathcal{P}])_{>\zeta}$ *does not break down because of the anomaly stated in clause 3(b) of Definition 4.2.1*[71]*,*

Ψ *is* Γ*-fullness preserving.*

The following is an easy yet useful consequence of strong fullness preservation.

LEMMA 4.6.6. *Assume* AD^+ *and suppose* Γ *is a pointclass. Suppose further that* $(\mathcal{P}, \Sigma)$ *is a hod pair or an sts hod pair such that* Σ *is strongly* Γ*-fullness preserving. Let* $\mathcal{T}$ *be a stack on* $\mathcal{P}$ *according to* Σ *with last model* $\mathcal{S}$ *such that if* $\mathcal{P}$ *is of limit type then* $\pi^{\mathcal{T},b}$ *exists and otherwise* $\pi^\mathcal{T}$ *exists. Suppose* $(\mathcal{R}, \sigma, \tau)$ *is such that*

1. *if* $\mathcal{P}$ *is of limit type then* $\sigma : \mathcal{P}^b \to \mathcal{R}$, $\tau : \mathcal{R} \to \mathcal{S}^b$ *and* $\pi^{\mathcal{T},b} = \tau \circ \sigma$, *and*
2. *if* $\mathcal{P}$ *is of successor type then* $\sigma : \mathcal{P} \to \mathcal{R}$, $\tau : \mathcal{R} \to \mathcal{S}$ *and* $\pi^\mathcal{T} = \tau \circ \sigma$.

Let E be such that

1. *if* $\mathcal{P}$ *is of limit type then E is the* $(\delta^{\mathcal{P}^b}, \delta^\mathcal{R})$*-extender derived from* σ*, and*
2. *if* $\mathcal{P}$ *is of successor type then E is the* $(\delta^\mathcal{P}, \delta^\mathcal{R})$*-extender derived from* σ

Then $\mathcal{R} = \mathrm{Ult}(\mathcal{P}, E)$. *In particular,* $\mathcal{R} = \{\pi_E(f)(a) : f \in \mathcal{P}$ *and* $a \in (\delta^\mathcal{R})^{<\omega}\}$.

PROOF. Let $k : Ult(\mathcal{P}, E) \to \mathcal{R}$ be the factor map, i.e., $k(\pi(f)(a)) = \sigma(f)(a)$. Then if $\mathcal{P}$ is of limit type then $\pi^{\mathcal{T},b} = \tau \circ k \circ \pi_E$ and if $\mathcal{P}$ is of successor type then $\pi^\mathcal{T} = \tau \circ k \circ \pi_E$. Notice that $\mathrm{crit}(k) > \delta^\mathcal{R}$. It now follows from strong Γ-fullness preservation of Σ that $\Sigma^{\tau\circ k}_{\mathcal{S},\mathcal{T}}$, the $\tau \circ k$-pullback of $\Sigma_{\mathcal{S},\mathcal{T}}$, is Γ-fullness preserving. But because $k \restriction \delta^\mathcal{R} = id$, we have that for every $\mathcal{R}' \in Y^\mathcal{R}$,

$$(\Sigma^{\tau\circ k}_{\mathcal{S},\mathcal{T}})_{\mathcal{R}'} = (\Sigma^{\tau}_{\mathcal{S},\mathcal{T}})_{\mathcal{R}'}.$$

It then follows that $\mathcal{R} = \mathrm{Ult}(\mathcal{P}, E)$. ⊣

4.7. Tracking disagreements

Here we introduce terminology that we will use to track the disagreements between strategies. The reader may wish to review Notation 2.7.14, Definition 3.10.7 and Terminology 2.7.17.

[71] We only need this condition for $\zeta = \sup\{\mathrm{lh}(F^+)_\gamma : \gamma < \beta\}$.

DEFINITION 4.7.1 (Low level disagreement between strategies). Suppose $(\mathcal{P}, \Sigma)$ and $(\mathcal{P}, \Lambda)$ are two allowable pairs. We say that there is a *low level disagreement* between Σ and Λ if one of the following conditions holds:

1. $\mathcal{P}$ is of successor type and $\Sigma_{\mathcal{P}^-} \neq \Lambda_{\mathcal{P}^-}$.
2. $\mathcal{P}$ is gentle and for some complete proper layer $\mathcal{Q}$ of $\mathcal{P}$, $\Sigma_{\mathcal{Q}} \neq \Lambda_{\mathcal{Q}}$.
3. $\mathcal{P}$ is of limit type, $\mathcal{P}$ is meek and there is $(\mathcal{T}, \mathcal{Q}) \in B(\mathcal{P}, \Sigma) \cap B(\mathcal{P}, \Lambda)$ such that $\Sigma_{\mathcal{Q},\mathcal{T}} \neq \Lambda_{\mathcal{Q},\mathcal{T}}$.
4. $\mathcal{P}$ is of limit type, $(\mathcal{P}, \Sigma)$ and $(\mathcal{P}, \Lambda)$ are hod pairs or sts hod pairs and there is $(\mathcal{T}_1, \mathcal{P}_1) \in I^b(\mathcal{P}, \Sigma)$ and $(\mathcal{T}_2, \mathcal{P}_2) \in I^b(\mathcal{P}, \Lambda)$ such that
   (a) $\mathcal{Q} =_{\text{def}} \mathcal{P}_1^b = \mathcal{P}_2^b$,
   (b) $\pi^{\mathcal{T}_1, b} = \pi^{\mathcal{T}_2, b}$, and
   (c) $\Sigma_{\mathcal{Q},\mathcal{T}_1} \neq \Lambda_{\mathcal{Q},\mathcal{T}_2}$.
5. $\mathcal{P}$ is of limit type, $(\mathcal{P}, \Sigma)$ and $(\mathcal{P}, \Lambda)$ are simple hod pairs or simple sts hod pairs and there is $(\mathcal{T}_1, \mathcal{P}_1)$ and $(\mathcal{T}_2, \mathcal{P}_2)$ such that
   (a) $(\mathcal{T}_1, \mathcal{P}_1) \in I^{ope}(\mathcal{P}, \Sigma)$[72],
   (b) $(\mathcal{T}_2, \mathcal{P}_2) \in I^{ope}(\mathcal{P}, \Lambda)$,
   (c) $\mathcal{Q} =_{\text{def}} \mathcal{P}_1^b = \mathcal{P}_2^b$,
   (d) the $\mathcal{Q}$-un-dropping extenders of $\mathcal{T}_1$ and $\mathcal{T}_2$ are the same,
   (e) $\Sigma_{\mathcal{Q},\mathcal{T}_1} \neq \Lambda_{\mathcal{Q},\mathcal{T}_2}$.

If clause 4 or 5 holds then we say that $(\mathcal{T}_1, \mathcal{P}_1, \mathcal{T}_2, \mathcal{P}_2)$ is a low level disagreement between Σ and Λ. Suppose next that $\mathcal{P}$ is of limit type. We say $(\mathcal{T}_1, \mathcal{P}_1, \mathcal{T}_2, \mathcal{P}_2, \mathcal{Q})$ is a *minimal low level disagreement* if,

1. $(\mathcal{T}_1, \mathcal{P}_1, \mathcal{T}_2, \mathcal{P}_2)$ is a low level disagreement between Σ and Λ,
2. $\mathcal{Q}$ is of successor type and $\mathcal{Q} \trianglelefteq \mathcal{P}_1^b = \mathcal{P}_2^b$,
3. $\Sigma_{\mathcal{Q}^-,\mathcal{T}_1} = \Lambda_{\mathcal{Q}^-,\mathcal{T}_2}$,
4. $\Sigma_{\mathcal{Q},\mathcal{T}_1} \neq \Lambda_{\mathcal{Q},\mathcal{T}_2}$.

Next we show that the existence of a disagreement translates into the existence of a minimal low level disagreement. The reader may wish to review Definition 2.10.10, Definition 2.10.1, Notation 2.10.9, Remark 2.10.7 and Definition 3.10.7.

LEMMA 4.7.2 (Disagreement implies low level disagreement). *Suppose Γ is a projectively closed pointclass, and $(\mathcal{P}, \Sigma)$ and $(\mathcal{P}, \Lambda)$ are allowable pairs such that both Σ and Λ are almost Γ-fullness preserving. Suppose that one of the following conditions holds:*

1. *$\mathcal{P}$ is of limit type but not of lsa type, and $\Sigma \neq \Lambda$.*
2. *$(\mathcal{P}, \Sigma)$ and $(\mathcal{P}, \Lambda)$ are sts pairs or simple sts pairs, $\Sigma \neq \Lambda$ and both Σ and Λ are fullness preserving.*
3. *$(\mathcal{P}, \Sigma)$ is a (simple) sts pair, $(\mathcal{P}, \Lambda)$ is a (simple) hod pair, $\Sigma \neq \Lambda$ and both Σ and Λ are fullness preserving.*

[72]See Definition 2.10.2 and Definition 2.10.13. We mainly use this to conclude that the un-dropping extender exists.

4. $\mathcal{P}$ is of lsa type, $\mathcal{J}_\omega[\mathcal{P}] \models$ "$\delta^{\mathcal{P}}$ is not a Woodin cardinal" and $\Sigma \neq \Lambda$.
Then there is a low level disagreement between Σ and Λ.

PROOF. We give the proof from clause 2, which is the hardest, and leave the rest to the reader. The proof from clause 1 is easier and is similar to [29, Proposition 2.41]). We also assume that $(\mathcal{P},\Sigma)$ and $(\mathcal{P},\Lambda)$ are sts pairs (as apposed to simple sts pairs).

Thus, we assume that $(\mathcal{P},\Sigma)$ and $(\mathcal{P},\Lambda)$ are sts hod pairs and $\Sigma \neq \Lambda$. We then have that $\mathcal{P} = (\mathcal{P}|\delta^{\mathcal{P}})^{\#}$. Assume there is no low level disagreement between Σ and Λ and let $\mathcal{T} = (\mathcal{S}_\alpha, \mathcal{Y}_\alpha, E_\alpha : \alpha < \eta)$ be any disagreement between Σ and Λ. Because $\Sigma(\mathcal{T}) \neq \Lambda(\mathcal{T})$ we must have that

(1) $\eta = \gamma + 1$, $\mathcal{Y}_\gamma \neq \emptyset$, $E_\gamma = \emptyset$ and $\mathrm{lh}(\mathcal{Y}_\gamma)$ is a limit ordinal.

Set $\mathcal{U} = \mathcal{Y}_\gamma$. For $\xi < \mathrm{lh}(\mathcal{U})$ we let $\mathcal{M}_\xi = \mathcal{M}^{\mathcal{U}}_\xi$.

SUBLEMMA 4.7.3. *The following holds.*
1. If $\alpha \in R^{\mathcal{U}}$ is such that $\pi^{\mathcal{U}}_{0,\alpha}$ is defined then $\Sigma_{\mathcal{M}^b_\alpha} = \Lambda_{\mathcal{M}^b_\alpha}$.
2. $\mathcal{U}$ does not have a main drop[73]*.*
3. $R^{\mathcal{U}}$ has a largest element and if $\alpha = \max(\mathcal{R}^{\mathcal{U}})$ then $\mathcal{U}_{\geq\alpha}$ is above $\mathrm{ord}(\mathcal{M}^b_\alpha)$.

PROOF. Clause 1 is an immediate consequence of our assumption that there are no low level disagreements between Σ and Λ. To see that $\mathcal{U}$ does not have a main drop, suppose that it does and let

$$md^{\mathcal{U}} = (\alpha_i, \mathcal{R}_i, \mathcal{W}_i, \mathcal{R}'_i : i \leq k+1)$$

be the *md*-sequence of $\mathcal{U}$. It follows that $(\mathcal{U})_{\geq\alpha_1}$ is based on $\mathcal{R}'_1 \trianglelefteq \mathcal{R}^b_1$ and therefore, $\Sigma_{\mathcal{R}'_1, \mathcal{T}_{\leq \mathcal{R}'_1}} \neq \Lambda_{\mathcal{R}'_1, \mathcal{T}_{\leq \mathcal{R}'_1}}$. Let $\mathcal{T}' = (\mathcal{S}'_\xi, \mathcal{Y}'_\xi, E'_\xi : \xi < \eta)$ be such that

1. for $\xi \leq \gamma$, $\mathcal{S}'_\xi = \mathcal{S}_\xi$ and $E'_\xi = E_\xi$,
2. for $\xi < \gamma$, $\mathcal{Y}'_\xi = \mathcal{Y}_\xi$,
3. $\mathcal{Y}'_\gamma = \mathcal{U}_{\leq\beta}$ where β is the least such that $\mathcal{M}^b_\beta = \mathcal{R}^b_1$[74].

Then $(\mathcal{T}', \mathcal{R}^b_1)$ is a low level disagreement between Σ and Λ.

To see that clause 3 holds, notice that if $R^{\mathcal{U}}$ does not have a maximal element then $\mathcal{U}$ has a unique branch which must be chosen by both Σ and Λ. Suppose now that $\alpha = \max(R^{\mathcal{U}})$. Recall our convention on proper stacks (see Remark 2.7.27). Thus, every cutpoint of $\mathcal{U}$ belongs to $R^{\mathcal{U}}$. Therefore, as $\alpha = \max(R^{\mathcal{U}})$ and as every cutpoint of $\mathcal{U}$ belongs to $R^{\mathcal{U}}$, we have the following four possibilities.

1. $\mathcal{U}_{\geq\alpha}$ is above $\mathrm{ord}(\mathcal{M}^b_\alpha)$.
2. $\mathcal{U}_{\geq\alpha}$ is above $\delta^{\mathcal{M}^b_\alpha}$ but below $\mathrm{ord}(\mathcal{M}^b_\alpha)$.
3. $\mathcal{U}_{\geq\alpha}$ is below $\delta^{\mathcal{M}^b_\alpha}$.

[73] See Definition 2.10.1.

[74] Notice that it follows that $\pi^{\mathcal{U}_{\leq\beta}}$ is defined.

4. $\mathrm{lh}(\mathcal{U}_{\geq\alpha}) = 2$ and $\mathrm{crit}(E_0^{\mathcal{U}_{\geq\alpha}}) = \delta^{\mathcal{M}_\alpha^b}$.

If 1 holds then there is nothing to prove. Clearly 4 fails as $\mathcal{U}$ has a limit length.

We now show that neither 2 nor 3 can hold. Assume 2 holds. Because both Σ and Λ are Γ-fullness preserving, $\Sigma(\mathcal{T}) = \Lambda(\mathcal{T})$.

Assume now that 3 holds. Let $\mathcal{T}' = (\mathcal{S}'_\xi, \mathcal{Y}'_\xi, E'_\xi : \xi < \eta)$ be such that

1. for $\xi \leq \gamma$, $\mathcal{S}'_\xi = \mathcal{S}_\xi$ and $E'_\xi = E_\xi$,
2. for $\xi < \gamma$, $\mathcal{Y}'_\xi = \mathcal{Y}_\xi$,
3. $\mathcal{Y}'_\gamma = \mathcal{U}_{\leq\beta}$ where β is the least such that $\mathcal{M}_\beta^b = \mathcal{M}_\alpha^b$[75].

Then $(\mathcal{T}', \mathcal{M}_\beta^b)$ constitutes a low level disagreement between Σ and Λ. ⊣

Let $\alpha_0 = \max(R^{\mathcal{U}})$ and $\mathcal{X} = \mathcal{U}_{\geq\alpha_0}$. Set $\mathcal{P}_1 = \mathrm{m}^+(\mathcal{X})$.

SUBLEMMA 4.7.4. *There are ordinary stacks*[76] *$\mathcal{T}_1$ and $\mathcal{T}_2$ on $\mathcal{M}_{\alpha_0}^{\mathcal{U}}$ such that*

1. *$(\mathcal{T}_{\leq\mathcal{M}_\alpha^{\mathcal{U}}})^\frown\mathcal{T}_1$ is according to Σ and $(\mathcal{T}_{\leq\mathcal{M}_\alpha^{\mathcal{U}}})^\frown\mathcal{T}_2$ is according to Λ,*
2. *$\mathcal{T}_1$ and $\mathcal{T}_2$ use the same extenders,*
3. *both $\pi^{\mathcal{T}_1,b}$ and $\pi^{\mathcal{T}_2,b}$ exist and $\pi^{\mathcal{T}_1,b} = \pi^{\mathcal{T}_2,b}$,*
4. *$(\mathcal{T}_{\leq\mathcal{M}_\alpha^{\mathcal{U}}})^\frown\mathcal{T}_1 \in b(\Sigma)$ and $(\mathcal{T}_{\leq\mathcal{M}_\alpha^{\mathcal{U}}})^\frown\mathcal{T}_2 \in b(\Lambda)$, that is, $\Sigma((\mathcal{T}_{\leq\mathcal{M}_\alpha^{\mathcal{U}}})^\frown\mathcal{T}_1)$ and $\Lambda((\mathcal{T}_{\leq\mathcal{M}_\alpha^{\mathcal{U}}})^\frown\mathcal{T}_2)$ are branches rather than models,*
5. *$\mathcal{T}_1$ and $\mathcal{T}_2$ have last normal components $\mathcal{X}_1$ and $\mathcal{X}_2$,*
6. *letting $b = \Sigma((\mathcal{T}_{\leq\mathcal{M}_\alpha^{\mathcal{U}}})^\frown\mathcal{T}_1)$ and $c = \Lambda((\mathcal{T}_{\leq\mathcal{M}_\alpha^{\mathcal{U}}})^\frown\mathcal{T}_2)$, $\mathcal{Q}(b, \mathcal{X}_1) \neq \mathcal{Q}(c, \mathcal{X}_2)$*[77].

PROOF. Towards a contradiction, suppose not. As $\mathcal{T}$ is a disagreement between Σ and Λ, we have that $\mathcal{T} \notin b(\Sigma) \cap b(\Lambda)$ as otherwise we could just take $\mathcal{T}_1 = \mathcal{X} = \mathcal{T}_2$. Notice that since $\mathcal{T}$ is a disagreement between Σ and Λ, $\mathcal{T} \notin m(\Sigma) \cap m(\Lambda)$, as otherwise $\Sigma(\mathcal{T}) = \mathcal{P}_1 = \Lambda(\mathcal{T})$. Assume without loss of generality that $\mathcal{T} \in m(\Sigma)$ and $\mathcal{T} \in b(\Lambda)$. Then letting $c = \Lambda(\mathcal{T})$, $\mathcal{Q}(c, \mathcal{X})$ exists. Let Σ_1 be the (ω_1, ω_1)-portion of $\Sigma_{\mathcal{P}_1, \mathcal{T}}$ and Λ_1 be the (ω_1, ω_1)-portion of $(\Lambda_{\mathcal{Q}(c,\mathcal{X}),\mathcal{T}})_{\mathrm{ex}}$[78].

It follows from Γ fullness preservation that $\Sigma_1 \neq \Lambda_1^{stc}$. Indeed, if $\Sigma_1 = \Lambda_1^{stc}$ then $\mathcal{Q}(c, \mathcal{T})$ is a Σ_1-sts mouse over $\mathcal{P}_1$ with an iteration strategy in Γ[79]. Hence, $\mathcal{T} \in b(\Sigma)$ and $\Sigma(\mathcal{T}) = b$.

Notice now that there is no low level disagreement between Σ_1 and Λ_1^{stc} since if $(\mathcal{T}_1, \mathcal{P}_1, \mathcal{T}_2, \mathcal{P}_2)$ is a low level disagreement between Σ_1 and Λ_1^{stc} then letting E be the $\mathcal{P}_1^b = \mathcal{P}_2^b$-un-dropping extender of $\mathcal{T}_1$ and $\mathcal{T}_2$, $\mathcal{T}^\frown\mathcal{T}_1^\frown\{\mathrm{Ult}(\mathcal{S}_\gamma, E), E\}$ and $\mathcal{T}^\frown\mathcal{T}_2^\frown\{\mathrm{Ult}(\mathcal{S}_\gamma, E), E\}$ induce a low level disagreement between Σ and Λ.

Let $\mathcal{U}_1$ be a disagreement between Σ_1 and Λ_1^{stc}. Arguing as we have argued for $\mathcal{T}$, we get that $\pi^{\mathcal{U}_1,b}$ is defined. Let $\mathcal{R}_1$ be the least node of $\mathcal{U}_1$ such that

[75] Notice that it follows that $\pi^{\mathcal{U}_{\leq\beta}}$ is defined.

[76] As apposed to generalized stacks.

[77] Because $\mathcal{T}_1$ and $\mathcal{T}_2$ use the same extenders, we have that $\mathrm{m}^+(\mathcal{X}_1) = \mathrm{m}^+(\mathcal{X}_2)$.

[78] Here and below, if Ψ is an st-strategy then $\Psi^{stc} = \Psi$. Also, if for example $\mathcal{T} \in m(\Sigma)$ then $\Sigma_{\mathcal{P}_1,\mathcal{T}}$ is an $(\omega_1, \omega_1, \omega_1)$-st-strategy. The definition of Ψ_{ex} appeared in Definition 2.7.3.

[79] Recall that our sts indexing scheme indexes branches of (ω_1, ω_1)-iterations and not generalized stacks.

$\mathcal{R}_1^b = \pi^{\mathcal{U}_1,b}(\mathcal{P}_1^b)$. It then follows that $(\mathcal{U}_1)_{\geq \mathcal{R}_1}$ is a stack on $\mathcal{R}_1$ that is above $\mathrm{ord}(\mathcal{R}_1^b)$. Let $\mathcal{X}_1$ be the last normal component of $(\mathcal{U}_1)_{\geq \mathcal{R}_1}$. It follows from Γ-fullness preservation that $\mathcal{X}_1$ does not have a fatal drop and $\mathrm{m}^+(\mathcal{X}_1) \vDash$ "$\delta(\mathcal{X}_1)$ is a Woodin cardinal". Set $\mathcal{P}_2 = \mathrm{m}^+(\mathcal{X}_1)$. We now claim that

CLAIM 4.7.5. $\mathcal{U}_1 \in b(\Lambda_1^{stc})$.

PROOF. Assume that $\mathcal{U}_1 \in m(\Lambda_1^{stc})$. Because $\mathcal{U}_1$ is a disagreement between Σ_1 and Λ_1^{stc}, we must have that $\mathcal{U}_1 \in b(\Sigma_1)$. Let $\mathcal{U}^* \in \mathrm{dom}(\Lambda_1)$ be such that $(\mathcal{U}^*)^{sc} = \mathcal{U}_1$[80]. It then follows that both $\Sigma(\mathcal{T}^\frown\mathcal{U}_1)$ and $\Lambda(\mathcal{T}^\frown\{c\}^\frown\mathcal{U}^*)$ are branches, and therefore, letting $b_1 = \Sigma_1(\mathcal{U}_1)$ and $c_1 = \Lambda_1(\mathcal{U}^*)$[81], we must have that $\mathcal{Q}(b_1, \mathcal{X}_1) = \mathcal{Q}(c_1, \mathcal{X}_1)$. Indeed, if $\mathcal{Q}(b_1, \mathcal{X}_1) \neq \mathcal{Q}(c_1, \mathcal{X}_1)$ then letting $\mathcal{T}_1 = \mathcal{X}^\frown\mathcal{U}_1$ and $\mathcal{T}_2 = \mathcal{X}^\frown\{c\}^\frown\mathcal{U}^*$[82], $\mathcal{T}_1$ and $\mathcal{T}_2$ are as desired. It follows that

(2) $b_1 = c_1$.

We now have that because $\mathcal{U}_1 \in m(\Lambda_1^{stc})$, letting β be the largest member of $\mathsf{max}^{\mathcal{U}_1}$,

(3) $\pi^{\mathcal{U}^*}_{\beta,c_1}$ is defined and $\pi^{\mathcal{U}^*}_{\beta,c_1}(\delta^{\mathcal{M}^{\mathcal{U}_1}_\beta}) = \delta^{\mathcal{P}_2}$.

Because $\mathcal{U}_1 \in b(\Sigma_1)$, we must have that

(4) either $\pi^{\mathcal{U}_1}_{\beta,b_1}$ is undefined or $\pi^{\mathcal{U}_1}_{\beta,b_1}(\delta^{\mathcal{M}^{\mathcal{U}_1}_\beta}) > \delta^{\mathcal{P}_2}$.

But because of (2)

(5) $\pi^{\mathcal{U}_1}_{\beta,b_1}$ is undefined if and only if $\pi^{\mathcal{U}^*}_{\beta,c_1}$ is undefined, and if $\pi^{\mathcal{U}_1}_{\beta,b_1}$ is defined then $\pi^{\mathcal{U}^*}_{\beta,c_1}(\delta^{\mathcal{M}^{\mathcal{U}_1}_\beta}) = \pi^{\mathcal{U}_1}_{\beta,b_1}(\delta^{\mathcal{M}^{\mathcal{U}_1}_\beta})$,

as the calculation of both depends on the functions in $\mathcal{M}^{\mathcal{U}_1}_\beta | \delta^{\mathcal{M}^{\mathcal{U}_1}_\beta}$. Clearly (2), (3), (4) and (5) contradict each other. ⊣

Since $\mathcal{U}_1$ is a disagreement, we have that $\mathcal{U}_1 \in m(\Sigma_1)$. Let then $c_1 = \Lambda_1^{stc}(\mathcal{U}_1)$. Notice that

(6) $\mathcal{Q}(c_1, \mathcal{X}_1)$ exists and if $\mathcal{U}^* \in \mathrm{dom}(\Lambda_1)$ is such that $(\mathcal{U}^*)^{sc} = \mathcal{U}_1$ then either $\pi^{\mathcal{U}^*}_{c_1}$ is undefined or $\pi^{\mathcal{U}^*}_{c_1}(\delta^{\mathcal{P}_1}) > \delta^{\mathcal{P}_2}$.

We now continue in the above manner by letting

$$\Sigma_2 = (\Sigma_1)_{\mathcal{P}_2,\mathcal{U}_1} \quad \text{and} \quad \Lambda_2 = ((\Lambda_1)_{\mathcal{Q}(c_1,\mathcal{X}_1),\mathcal{U}^{*\frown}\{c_1\}})_{\mathsf{ex}}.$$

Notice that Γ-fullness preservation once again implies that $\Sigma_2 \neq \Lambda_2^{stc}$. By repeating in the above manner we obtain sequences $(\mathcal{U}_i^* : i \in [1, \omega))$, $(\Lambda_i : i \in [1, \omega))$ and $(c_i : i \in [1, \omega))$ such that the following conditions are satisfied:

[80] See Definition 3.1.6.

[81] Recall that Λ_1 is a strategy.

[82] In this iteration, player I starts a new round of the iteration after player II plays c. At the beginning of this round, player I drops to $\mathcal{Q}(c, \mathcal{X})_{\mathsf{ex}}$.

1. $\mathcal{U}_1^* = \mathcal{U}^*$ where $\mathcal{U}^*$ is as in (6).
2. For each $i < \omega$, $\mathcal{U}_i^*$ is according to Λ_i and $c_i = \Lambda_i(\mathcal{U}_i^*)$.
3. For each $i < \omega$, $\mathcal{U}_i^*$ has a last normal component $\mathcal{X}_i$ and $\mathcal{Q}(c_i, \mathcal{X}_i)$ exists.
4. For each $i < \omega$, $\Lambda_{i+1} = ((\Lambda_i)_{\mathcal{Q}(c_i,\mathcal{X}_i),\mathcal{U}_i^{*\frown}\{c_i\}})_{\text{ex}}$,
5. For each $i \in \omega$, letting $\delta_{i+1} = \delta(\mathcal{X}_i)$, either $\pi_{c_i}^{\mathcal{U}_i^*}$ is undefined or $\pi_{c_i}^{\mathcal{U}_i^*}(\delta_i) > \delta_{i+1}$.

Concatenating the $\mathcal{U}_i^*$s we get $\mathcal{U}$ according to Λ_1 without a well-founded branch. ⊣

Let $\mathcal{T}_1$ and $\mathcal{T}_2$ be as in Sublemma 4.7.4. Set

$$\mathcal{U}_1 = (\mathcal{T}_{\leq \mathcal{M}_{\alpha_0}^{\mathcal{U}}})^\frown \mathcal{T}_1,$$
$$\mathcal{U}_2 = (\mathcal{T}_{\leq \mathcal{M}_{\alpha_0}^{\mathcal{U}}})^\frown \mathcal{T}_2,$$

$b = \Sigma(\mathcal{U}_1)$ and $c = \Lambda(\mathcal{U}_2)$. Let $\mathcal{X}_1$ and $\mathcal{X}_2$ be the last normal components of $\mathcal{T}_1$ and $\mathcal{T}_2$. It follows that $\mathrm{m}^+(\mathcal{X}_1) = \mathrm{m}^+(\mathcal{X}_2)$, both $\mathcal{Q}(b, \mathcal{X}_1)$ and $\mathcal{Q}(c, \mathcal{X}_2)$ exist and $\mathcal{Q}(b, \mathcal{X}_1) \neq \mathcal{Q}(c, \mathcal{X}_2)$.

Let $\mathcal{P}_2 = \mathrm{m}^+(\mathcal{X})$. Notice that it follows from our smallness assumption on hod mice, namely that hod mice do not have lsa hod initial segments, that $\delta^{\mathcal{P}_2}$ is a strong cutpoint of both $\mathcal{Q}(b, \mathcal{X}_1)$ and $\mathcal{Q}(c, \mathcal{X}_2)$. We then have that $\mathcal{Q}(b, \mathcal{X}_1)$ is a $\Sigma^{stc}_{\mathcal{P}_2,\mathcal{U}_1}$-sts mouse over $\mathcal{P}_2$, $\mathcal{Q}(c, \mathcal{X}_2)$ is a $\Lambda^{stc}_{\mathcal{P}_2,\mathcal{U}_2}$-sts mouse over $\mathcal{P}_2$, and the comparison of $\mathcal{Q}(b, \mathcal{X}_1)$ and $\mathcal{Q}(c, \mathcal{X}_2)$ does not halt (as otherwise we would have $\mathcal{Q}(b, \mathcal{X}_1) = \mathcal{Q}(c, \mathcal{X}_2)$). Set $\nu = \delta^{\mathcal{P}_2}$, $\mathcal{M}_0 = \mathcal{Q}(b, \mathcal{X}_1)$ and $\mathcal{M}_1 = \mathcal{Q}(c, \mathcal{X}_2)$. We now have that

(7) $\mathcal{M}_0 \ntrianglelefteq \mathcal{M}_1$, $\mathcal{M}_1 \ntrianglelefteq \mathcal{M}_0$, $\mathcal{M}_0||\nu = \mathcal{M}_1||\nu$, $\mathcal{M}_0$ and $\mathcal{M}_1$ are ν-sound and project to ν, and ν is a strong cutpoint of both $\mathcal{M}_0$ and $\mathcal{M}_1$.
(8) $\mathcal{M}_0$ is a $\Sigma^{stc}_{\mathcal{P}_2,\mathcal{U}_1}$-sts mouse over $\mathcal{P}_2$ and $\mathcal{M}_1$ is a $\Lambda^{stc}_{\mathcal{P}_2,\mathcal{U}_2}$-sts mouse over $\mathcal{P}_2$.
(9) The comparison of $\mathcal{M}_0$ and $\mathcal{M}_1$ cannot halt.

(9) holds as otherwise its failure implies that either $\mathcal{M}_0 \trianglelefteq \mathcal{M}_1$ or $\mathcal{M}_1 \trianglelefteq \mathcal{M}_0$, both of which are impossible (because of (7)).

It follows that the comparison of $\mathcal{M}_0$ and $\mathcal{M}_1$ encounters disagreements involving strategies, as otherwise the usual comparison argument would imply that the comparison halts. Let Ψ_0 and Ψ_1 be the canonical strategies of $\mathcal{M}_0$ and $\mathcal{M}_1$ respectively. Thus, Ψ_0 witnesses that $\mathcal{M}_0$ is a $\Sigma^{stc}_{\mathcal{P}_2,\mathcal{U}_1}$-sts mouse, and Ψ_1 witnesses that $\mathcal{M}_1$ is a $\Lambda^{stc}_{\mathcal{P}_2,\mathcal{U}_2}$-sts mouse.

We can then find Ψ_0-iterate $\mathcal{K}_0$ of $\mathcal{M}_0$ and Ψ_1-iterate $\mathcal{K}_1$ of $\mathcal{M}_1$ such that $\mathcal{K}_0$ and $\mathcal{K}_1$ are produced via the usual extender comparison procedure (this implies that both iterations are above ν) and for some α,

(10) $\mathcal{K}_0|\alpha = \mathcal{K}_1|\alpha$, $\mathcal{K}_0||\alpha \neq \mathcal{K}_1||\alpha$, $\alpha \notin \mathrm{dom}(\vec{E}^{\mathcal{K}_0}) \cup \mathrm{dom}(\vec{E}^{\mathcal{K}_1})$.

Notice that it follows from our indexing scheme (see Definition 3.6.4) that there must be a branch indexed at α in both $\mathcal{K}_0$ and $\mathcal{K}_1$. Let then $t = (\mathcal{P}_2, \mathcal{W}, \mathcal{P}_3, \mathcal{W}') \in \mathcal{K}_0|\alpha$ be such that its branch is indexed at α in both $\mathcal{K}_0$ and $\mathcal{K}_1$.

We now have to analyze exactly what kind of stack t is. Recall that our indexing scheme is so that we add branches for two kinds of stacks that we now list.

Case 1. $\mathcal{W}$ is a $\mathcal{K}_0|\alpha$-terminal tree[83] and $\mathcal{W}'$ is undefined.
Case 2. $\mathcal{W}'$ is defined and is a stack on $(\mathcal{P}_3)^b$.

We can immediately rule out case 1 above: $\mathcal{K}|\alpha = \mathcal{N}|\alpha$ and the branch of $\mathcal{W}$ just depends on $\mathcal{K}|\alpha$[84]. On the other hand, case 2, just like in the proof of Lemma 4.7.3, leads to a low level disagreement between Σ and Λ, which is contrary to our assumption. This contradiction implies that the comparison of $\mathcal{M}_0$ and $\mathcal{M}_1$ does not encounter strategy disagreement implying that (7) is false. This contradiction also completes our proof of Lemma 4.7.2. ⊣

LEMMA 4.7.6 (Minimal low level disagreement). *Suppose Γ is a pointclass projectively closed pointclass, and $(\mathcal{P},\Sigma)$ and $(\mathcal{P},\Lambda)$ are allowable pairs such that both Σ and Λ are almost Γ-fullness preserving. Suppose that one of the following conditions holds:*

1. *$\mathcal{P}$ is of limit type but not of lsa type, and $\Sigma \neq \Lambda$.*
2. *$(\mathcal{P},\Sigma)$ and $(\mathcal{P},\Lambda)$ are sts pairs or simple sts pairs, $\Sigma \neq \Lambda$ and both Σ and Λ are fullness preserving and are weakly self-cohering.*
3. *$(\mathcal{P},\Sigma)$ is a (simple) sts pair, $(\mathcal{P},\Lambda)$ is a (simple) hod pair, $\Sigma \neq \Lambda$ and both Σ and Λ are fullness preserving and are weakly self-cohering.*
4. *$\mathcal{P}$ is of lsa type, $\mathcal{J}_\omega[\mathcal{P}] \vDash$ "$\delta^{\mathcal{P}}$ is not a Woodin cardinal" and $\Sigma \neq \Lambda$.*

Then there is a minimal low level disagreement between Σ and Λ.

PROOF. Again we give the proof from clause 2. Assume there is no minimal low level disagreement between Σ and Λ. It follows from Lemma 4.7.1 that there is a low level disagreement between Σ and Λ. Let $(\mathcal{T}_1,\mathcal{P}_1) \in I^b(\mathcal{P},\Sigma)$ and $(\mathcal{U}_1,\mathcal{R}_1) \in I^b(\mathcal{P},\Lambda)$ be a low level disagreement. Set $\mathcal{Q} = \mathcal{P}_1^b (= \mathcal{R}_1^b)$. We thus have that $\Sigma_{\mathcal{Q},\mathcal{T}_1} \neq \Lambda_{\mathcal{Q},\mathcal{U}_1}$. Notice that if $\Sigma_{\mathcal{Q}|\delta^{\mathcal{Q}},\mathcal{T}_1} = \Lambda_{\mathcal{Q}|\delta^{\mathcal{Q}},\mathcal{U}_1}$ then Γ-fullness preservation implies that $\Sigma_{\mathcal{Q},\mathcal{T}_1} = \Lambda_{\mathcal{Q},\mathcal{U}_1}$. Thus, there is $\beta < \lambda^{\mathcal{Q}}$ such that[85]

1. $\Sigma_{\mathcal{Q}(\beta),\mathcal{T}_1} \neq \Lambda_{\mathcal{Q}(\beta),\mathcal{U}_1}$,
2. $\mathrm{ord}(\mathcal{Q}(\beta))$ is a cutpoint of $\mathcal{Q}$.

Let β_1 be the least such ordinal and set $\mathcal{Q}_1 = \mathcal{Q}(\beta_1)$. If $\mathcal{Q}_1$ is of successor type then by minimality of β_1 we get that $(\mathcal{T}_1,\mathcal{U}_1,\mathcal{Q}_1)$ is a minimal low level disagreement. Thus, we have that $\mathcal{Q}_1$ is limit type. The minimality of β_1 then implies that

(1) $\mathcal{Q}_1$ is non-meek,
(2) $\Sigma_{\mathcal{Q}_1^b,\mathcal{T}_1} = \Lambda_{\mathcal{Q}_1^b,\mathcal{U}_1}$[86].

---

[83] See Definition 3.8.8.
[84] See Definition 3.8.9.
[85] See Notation 2.7.14.
[86] Notice that these are ordinary strategies not generalized strategies. The reason is that $\Sigma_{\mathcal{P}'}$ for $\mathcal{P}' \lhd \mathcal{P}^b$ is an ordinary strategy.

Applying Lemma 4.7.1, we get $(\mathcal{T}_2, \mathcal{P}_2, \mathcal{U}_2, \mathcal{R}_2)$ that constitute a low level disagreement between $\Sigma_{\mathcal{Q}_1,\mathcal{T}_1}$ and $\Lambda_{\mathcal{Q}_1,\mathcal{U}_1}$. Let β_2 be the least β such that

1. $\mathcal{P}_2(\beta) = \mathcal{R}_2(\beta)$,
2. $\Sigma_{\mathcal{P}_2(\beta),\mathcal{T}_1^\frown\mathcal{T}_2} \neq \Lambda_{\mathcal{R}_2(\beta),\mathcal{U}_1^\frown\mathcal{U}_2}$,
3. $\text{ord}(\mathcal{P}_2(\beta))$ is a cardinal of both $\mathcal{P}_2$ and $\mathcal{R}_2$.

Thus,

(3) $\mathcal{P}_2(\beta_2) \lhd \mathcal{P}_2^b$.

Set $\mathcal{Q}_2 = \mathcal{P}_2(\beta)$. We claim that

CLAIM 4.7.7. *$\mathcal{Q}_2$ is not of successor type.*

PROOF. To see this, suppose that $\mathcal{Q}_2$ is of successor type. Let $\mathcal{T}_1 = (\mathcal{M}_\alpha, \mathcal{X}_\alpha, F_\alpha : \alpha \leq \eta)$ and $\mathcal{U}_1 = (\mathcal{N}_\alpha, \mathcal{Y}_\alpha, G_\alpha : \alpha \leq \eta)$. We have that $\mathcal{M}_\eta, F_\eta, \mathcal{N}_\eta$ and G_η are undefined. Let $\mathcal{X} = \mathcal{X}_\eta^\frown\mathcal{T}_2$ and $\mathcal{Y} = \mathcal{Y}_\eta^\frown\mathcal{U}_2$. In forming $\mathcal{X}$, we let player I start a new round on $\mathcal{P}_1$ by dropping to $\mathcal{Q}_1$. The same happens in $\mathcal{Y}$ as well. Let then F_η be the $\mathcal{Q}_2$-un-dropping extender of $\mathcal{X}$ and G_η be the $\mathcal{Q}_2$-un-dropping extender of $\mathcal{Y}$ and set $\mathcal{X}' = \mathcal{X}^\frown\{\text{Ult}(\mathcal{M}_\eta, F_\eta), F_\eta\}$ and $\mathcal{Y}' = \mathcal{Y}^\frown\{\text{Ult}(\mathcal{N}_\eta, G_\eta), G_\eta\}$. Notice that $F_\eta = G_\eta$ as $\pi^{\mathcal{T}_1,b} = \pi^{\mathcal{U}_1,b}$ and the $\mathcal{Q}_2$-un-dropping extenders of $\mathcal{T}_2$ and $\mathcal{U}_2$ are the same. Because Σ and Λ are weakly self-cohering, we have that $\Sigma_{\mathcal{Q}_2,\mathcal{X}'} = \Sigma_{\mathcal{Q}_2,\mathcal{X}}$ and $\Lambda_{\mathcal{Q}_2,\mathcal{Y}'} = \Lambda_{\mathcal{Q}_2,\mathcal{Y}}$. Thus, $\Sigma_{\mathcal{Q}_2,\mathcal{X}'} \neq \Lambda_{\mathcal{Q}_2,\mathcal{Y}'}$ and hence, $(\mathcal{X}, \mathcal{Y}, \mathcal{Q}_2)$ is a minimal low level disagreement. ⊣

Continuing in this fashion we can now produce a sequence $(\mathcal{P}_i, \mathcal{T}_i, \mathcal{Q}_i : i \in [2, \omega))$ such that the following conditions hold.

1. For all $i \in [2, \omega)$, $\mathcal{Q}_i \lhd_{hod} \mathcal{P}_i^b$.
2. $\mathcal{Q}_i$ is non-meek.
3. $\mathcal{T}_i$ is a stack on $\mathcal{Q}_i$ such that $\pi^{\mathcal{T}_i,b}$ exists and $\mathcal{P}_{i+1}$ is the last model of $\mathcal{T}_i$.

Clearly, the concatenation of $\mathcal{T}_i$'s is an iteration according to $\Sigma_{\mathcal{P}_2,\mathcal{X}}$ without a well-founded branch. ⊣

Next we introduce several definitions that will be useful in the sequel.

DEFINITION 4.7.8 (Comparison stack). Suppose $(\mathcal{P}, \Sigma)$ and $(\mathcal{Q}, \Lambda)$ are two hod pairs or sts hod pairs. Then we say $(\mathcal{T}, \mathcal{R}, \mathcal{U}, \mathcal{S})$ are comparison stacks for

$$((\mathcal{P}, \Sigma), (\mathcal{Q}, \Lambda))$$

with last models $(\mathcal{R}, \mathcal{S})$ if $(\mathcal{T}, \mathcal{R}) \in I(\mathcal{P}, \Sigma)$, $(\mathcal{U}, \mathcal{S}) \in I(\mathcal{Q}, \Lambda)$, and either

1. $\mathcal{S} \in Y^{\mathcal{R}}$ and $\Sigma_{\mathcal{S},\mathcal{T}} = \Lambda_{\mathcal{S},\mathcal{U}}$.
2. $\mathcal{R} \in Y^{\mathcal{S}}$ and $\Sigma_{\mathcal{R},\mathcal{T}} = \Lambda_{\mathcal{R},\mathcal{U}}$.

DEFINITION 4.7.9 (Agreement up to the top). Suppose $\mathcal{P}$ and $\mathcal{Q}$ are two hod premice of limit type. Then we say $\mathcal{P}$ and $\mathcal{Q}$ agree up to the top if $\mathcal{P}^b = \mathcal{Q}^b$. Suppose further that Σ and Λ are such that $(\mathcal{P}, \Sigma)$ and $(\mathcal{Q}, \Lambda)$ are two hod pairs or sts hod pairs. Then we say $(\mathcal{P}, \Sigma)$ and $(\mathcal{Q}, \Lambda)$ agree up to the top if $\mathcal{P}$ and $\mathcal{Q}$ agree up to the top and $\Sigma_{\mathcal{P}^b} = \Lambda_{\mathcal{Q}^b}$.

DEFINITION 4.7.10 (Extender and strategy disagreement). Given two hod premice $\mathcal{P}$ and $\mathcal{Q}$ such that $\mathcal{P} \neq \mathcal{Q}$, we let $\beta(\mathcal{P},\mathcal{Q})$ be the least ordinal γ such that $\mathcal{P}|\gamma = \mathcal{Q}|\gamma$ but $\mathcal{P}||\gamma \neq \mathcal{Q}||\gamma$. We say $\mathcal{P}$ and $\mathcal{Q}$ have an extender disagreement if $\beta(\mathcal{P},\mathcal{Q}) \in \text{dom}(\vec{E}^{\mathcal{R}}) \cup \text{dom}(\vec{E}^{\mathcal{Q}})$. We say $\mathcal{P}$ and $\mathcal{Q}$ have a strategy disagreement if $\beta(\mathcal{P},\mathcal{Q}) \notin \text{dom}(\vec{E}^{\mathcal{R}}) \cup \text{dom}(\vec{E}^{\mathcal{Q}})$. In this case, we let

$$\mathcal{R}_{\mathcal{P},\mathcal{Q}} = \cup Y^{\mathcal{P}|\beta(\mathcal{P},\mathcal{Q})} (= \cup Y^{\mathcal{Q}|\beta(\mathcal{P},\mathcal{Q})})$$

Thus, both $\mathcal{P}$ and $\mathcal{Q}$ have a branch indexed at $\beta(\mathcal{P},\mathcal{Q})$ for some $\mathcal{T}$ on $\mathcal{R}_{\mathcal{P},\mathcal{Q}}$. We say $\mathcal{R}_{\mathcal{P},\mathcal{Q}}$ is the disagreement layer of $\mathcal{P}$ and $\mathcal{Q}$.

DEFINITION 4.7.11 (Extender comparison). Suppose that $(\mathcal{P},\Sigma)$ and $(\mathcal{Q},\Lambda)$ are two allowable pairs which agree up to the top. Then we say $(\mathcal{T},\mathcal{R},\mathcal{U},\mathcal{S})$ are the trees of the extender comparison of $(\mathcal{P},\Sigma)$ and $(\mathcal{Q},\Lambda)$ if

1. $\mathcal{T}$ is according to Σ and $\mathcal{R}$ is its last model,
2. $\mathcal{U}$ is according to Λ and $\mathcal{S}$ is its last model, and
3. $\mathcal{T}$ and $\mathcal{U}$ are obtained by using the usual extender comparison process (i.e., by removing the least extender disagreements) for comparing the top windows of $\mathcal{P}$ and $\mathcal{Q}$ until a strategy disagreement appears.

It follows that if in Definition 4.7.11, $\mathcal{R} \neq \mathcal{S}$ then $\mathcal{R}$ and $\mathcal{S}$ have a strategy disagreement.

4.8. Self-cohering

Here our goal is to show that the strategies appearing in hod pair constructions are self-cohering[87].

THEOREM 4.8.1. *Assume* AD^+. *Suppose* Γ *is a pointclass such that for some* α *with* $\vartheta_\alpha < \Theta$, $\Gamma = \{A \subseteq \mathbb{R} : w(A) < \vartheta_\alpha\}$, $\mathsf{C} = (\mathbb{M},(P,\Psi),\Gamma^*,A)$ *Suslin, co-Suslin captures* Γ *and* $\mathbb{M} = (M,\delta,\vec{G},\Sigma^*)$. *Set*

$$\mathsf{hpc}^+_{\mathsf{C},\Gamma} = (\mathcal{M}_\gamma,\mathcal{N}_\gamma,Y_\gamma,\Phi^+_\gamma,F^+_\gamma,F_\gamma,b_\gamma : \gamma \leq \delta).$$

Suppose $\beta < \delta$, $\mathcal{P} \in Y_\beta$ *and* $M \models$ "$(\mathcal{P},(\Phi_\beta)_{\mathcal{P}}) \in \mathsf{HP}^\Gamma$. *Then* $(\Phi^+_\beta)_{\mathcal{P}}$ *is self-cohering.*

PROOF. Set $\Sigma = (\Phi^+_\beta)_{\mathcal{P}}$. The hardest case is when $\mathcal{P}$ is non-meek and Σ is generalized strategy. Suppose

1. $\mathcal{T} = (\mathcal{M}_\alpha,\mathcal{T}_\alpha,F_\alpha : \alpha < \eta)$ is a generalized stack according to Σ,
2. $\alpha_0,\alpha_1 < \eta$,
3. $\xi_0 < \text{lh}(\mathcal{T}_{\alpha_0})$ and $\xi_1 < \text{lh}(\mathcal{T}_{\alpha_1})$, and
4. $\mathcal{R} \lhd_{hod} \mathcal{M}^{\mathcal{T}_{\alpha_0}}_{\xi_0} =_{\text{def}} \mathcal{S}_0$ and $\mathcal{R} \lhd_{hod} \mathcal{M}^{\mathcal{T}_{\alpha_1}}_{\xi_1} =_{\text{def}} \mathcal{S}_1$.

[87] See Definition 2.10.11.

By absoluteness, we can find such a $\mathcal{T}$ in $M[g]$ where $g \subseteq Coll(\omega, \zeta_0)$ is M-generic and $\zeta_0 < \delta$. We want to see that

$$\Sigma_{\mathcal{R},\mathcal{T}_{\leq S_0}} = \Sigma_{\mathcal{R},\mathcal{T}_{\leq S_1}}.^{88}$$

Assume then that $\Sigma_{\mathcal{R},\mathcal{T}_{\leq S_0}} \neq \Sigma_{\mathcal{R},\mathcal{T}_{\leq S_1}}$. Again, the hard case is when $\mathcal{R}$ is of limit type, and so we assume this.

It follows from Lemma 4.7.6 that there is a minimal low level disagreement $(\mathcal{U}_0, \mathcal{R}_0, \mathcal{U}_1, \mathcal{R}_1, \mathcal{Q})$ between $\Sigma_{\mathcal{R},\mathcal{T}_{\leq S_0}}$ and $\Sigma_{\mathcal{R},\mathcal{T}_{\leq S_1}}$. We thus have that

1. $\mathcal{Q}$ is of successor type,
2. $\Sigma_{\mathcal{Q},(\mathcal{T}_{\leq S_0})^\frown\mathcal{U}_0} \neq \Sigma_{\mathcal{Q},(\mathcal{T}_{\leq S_1})^\frown\mathcal{U}_1}$ and
3. $\Sigma_{\mathcal{Q}^-,(\mathcal{T}_{\leq S_0})^\frown\mathcal{U}_0} = \Sigma_{\mathcal{Q}^-,(\mathcal{T}_{\leq S_1})^\frown\mathcal{U}_1}$.

Let $\mathcal{X}_0 = ((\mathcal{T}_{\alpha_0})_{\leq S_0})^\frown\mathcal{U}_0$ and $\mathcal{X}_1 = ((\mathcal{T}_{\alpha_1})_{\leq S_1})^\frown\mathcal{U}_1$. Let H_0 and H_1 be the $\mathcal{Q}$-un-dropping extenders of $\mathcal{X}_0$ and $\mathcal{X}_1$[89]. Finally, let for $i \in 2$, $\mathcal{Y}_i = (\mathcal{M}^i_\xi, T^i_\xi, F^i_\xi : \xi \leq \alpha_i + 1)$ be the generalized stack that has the following properties:

1. For $\alpha \leq \alpha_i$, $\mathcal{M}^i_\xi = \mathcal{M}_\xi$.
2. For $\alpha < \alpha_i$, $\mathcal{T}_\alpha = T^i_\alpha$ and $F^i_\xi = F_\xi$.
3. $T^i_{\alpha_i} = \mathcal{X}_i$, $\mathcal{M}^i_{\alpha_i+1} = \mathrm{Ult}(\mathcal{M}^i_{\alpha_i}, H_i)$ and $F^i_{\alpha_i} = H_i$.

Let for $i \in 2$, E_i be the $(\delta^{\mathcal{P}^b}, \pi^{Y_i,b}(\delta^{\mathcal{P}^b}))$-extender derived from $\pi^{\mathcal{Y}_i,b}$. Because Σ is a weakly self-cohering[90], we have that

1. $\mathcal{Q}$ is of successor type,
2. $\Sigma_{\mathcal{Q},\mathcal{Y}_0} \neq \Sigma_{\mathcal{Q},\mathcal{Y}_1}$ and
3. $\Sigma_{\mathcal{Q}^-,\mathcal{Y}_0} = \Sigma_{\mathcal{Q}^-,\mathcal{Y}_1}$.

Set $\kappa = \delta^{\mathcal{P}^b}$ and $\zeta_1 = \sup\{\mathrm{lh}(F^+_\gamma) : \gamma \leq \beta\}$. Set $\zeta = \max((\zeta_0^+)^M, (\zeta_1^+)^M)$. Let now $\mathcal{N}$ be the last model of

$$(\mathsf{Le}((\mathcal{P}|\kappa, \Sigma_{\mathcal{P}|\kappa}), \mathcal{J}_\omega[\mathcal{P}^b])_{>\zeta})^{(M,\delta,\vec{G})}.$$

We have that $\mathcal{N}|(\kappa^+)^{\mathcal{N}} = \mathcal{P}^b$[91]. We then set for $i \in 2$, $\mathcal{N}_i = \mathrm{Ult}(\mathcal{N}, E_i)$. Because for $i \in 2$, $\mathcal{Y}_i$ is an iteration of $\mathcal{P}$ according to the strategy induced by Σ^*, we have a Σ^*-iterate M_i of M such that letting $j_i : M \to M_i$ be the iteration embedding the following clauses hold:

(1) For each $i \in 2$, there is an $\mathcal{M}$-model $\mathcal{M}_i$ of $\mathsf{hpc}^{M_i}_\Gamma$ and an elementary embedding $\sigma_i : \mathcal{Q} \to \mathcal{M}_i$.
(2) For each $i \in 2$, $\mathcal{M}_i \lhd_{hod} j_i(\mathcal{P})$.[92]

[88] See Definition 2.6.3.
[89] See Definition 2.10.5.
[90] See Lemma 4.3.12.
[91] See Theorem 4.6.3, which implies that $\mathcal{P}$ is full.
[92] This follows from the fact that we are not allowed to project across $\delta^{\mathcal{M}_i}$.

(3) For each $i \in 2$, there is an $\mathcal{M}$-model $\mathcal{M}'_i$ of $\mathsf{hpc}_\Gamma^{M_i}$ with index $\leq j_i(\beta)$ such that $\mathcal{M}_i \in Y^{\mathcal{M}'_i}$ and $\Sigma_{\mathcal{Q},\mathcal{Y}_i}$ is the σ_i-pullback of Φ_i, where Φ_i is the strategy of $\mathcal{M}'_i$ induced by $\Sigma^*_{M_i}$.

(4) For each $i \in 2$, σ_i extends to $\sigma_i^+ : \mathcal{N}_i \to j_i(\mathcal{N})$ and $j_i = \sigma_i^+ \circ \pi_{E_i}$.

For each $i \in 2$, let Λ_i^* be the strategy of $j_i(\mathcal{N})$ induced by $\Sigma^*_{M_i}$ and let Λ_i be the σ_i^+-pullback of Λ_i^*.

LEMMA 4.8.2. *For each $i \in 2$, $(\Lambda_i)_{\mathcal{Q}} = \Sigma_{\mathcal{Q},\mathcal{Y}_i}$.*

PROOF. It is enough to show that $(\Phi_i)_{\mathcal{M}_i} = (\Lambda^*)_{\mathcal{M}_i}$. This follows easily from the fact that $\delta^{\mathcal{M}_i}$[93] is a regular cardinal both in $\mathcal{M}'_i$ and in $j_i(\mathcal{N})$. Because of this, both $(\Phi_i)_{\mathcal{M}_i}$ and $(\Lambda^*)_{\mathcal{M}_i}$ are the strategy of $\mathcal{M}_i$ induced by $\Sigma^*_{M_i|\tau_i}$ where τ_i is the least such that $\mathcal{M}_i$ is constructed inside $M_i|\tau_i$. ⊣

We now let for $i \in 2$, $\mathcal{W}_i$ be the last model of

$$(\mathsf{Le}((\mathcal{Q}^-, (\Lambda_i)_{\mathcal{Q}^-}), \mathcal{J}_\omega[\mathcal{Q}])_{>\pi_{E_i}(\zeta)})^{(\mathcal{N}'_i, \delta, \vec{K}_i)}$$

where $\mathcal{N}'_i = L_{\mathrm{ord}(M)}[\mathcal{N}_i]$ and $\vec{K}_i = \{K \in \vec{E}^{\mathcal{N}_i} : \nu(K)$ is an inaccessible cardinal of $\mathcal{N}_i\}$. Let Ω_i be the strategy of $\mathcal{W}_i$ induced by Λ_i. We once again have that $(\Omega_i)_{\mathcal{Q}} = (\Lambda_i)_{\mathcal{Q}}$. Applying the "furthermore" clause of Theorem 4.5.6 to $((\Omega_0)_{\mathcal{Q}}, \mathcal{Q})$ and $((\Omega_1)_{\mathcal{Q}}, \mathcal{Q})$, we get that $(\Omega_0)_{\mathcal{Q}} = (\Omega_1)_{\mathcal{Q}}$. However, since $(\Lambda_i)_{\mathcal{Q}} = \Sigma_{\mathcal{Q},\mathcal{Y}_i}$ and $(\Omega_i)_{\mathcal{Q}} = (\Lambda_i)_{\mathcal{Q}}$, we have that $(\Omega_0)_{\mathcal{Q}} \neq (\Omega_1)_{\mathcal{Q}}$. This contradiction completes the proof of Theorem 4.8.1. ⊣

4.9. Branch condensation

In this subsection we prove that the hod pair constructions produce strategies with branch condensation and in fact more. In order, however, to prove that hod pair constructions converge, we will need to establish the solidity and universality of the standard parameter of the models appearing in such constructions. Establishing such fine structural facts was not an issue in [29] as the fine structure for hod mice considered in that paper was a routine generalization of the fine structure theory developed in [23]. Here the matter is somewhat more complicated as the fine structure of non-meek hod mice cannot be viewed as a routine generalization of the fine structure of [23]. Nevertheless, the matter is not too complicated as a simple generalization of branch condensation, *strong branch condensation*, allows us to reduce our case to the one in [23]. In this subsection, we will establish that hod pair constructions produce strategies with strong branch condensation. The next definition will use concepts from Notation 2.7.14, Definition 2.10.2, Definition 2.10.13 and Definition 3.10.7.

[93] Recall that because $(\mathcal{U}, \mathcal{Q})$ is a minimal disagreement, $\mathcal{Q}$ is of successor type. Thus, in fact $\delta^{\mathcal{M}_i}$ is a Woodin cardinal of $j_i(\mathcal{N}_i)$ and $\mathcal{M}'_i$.

DEFINITION 4.9.1. Suppose $\mathcal{P}$ is a non-gentle hod premouse.
Suppose next that either

1. $\mathcal{P}$ is of successor type or
2. $\mathcal{P}$ is of lsa type and $\mathcal{J}_\omega[\mathcal{P}] \models$ "$\delta^{\mathcal{P}}$ is a Woodin cardinal".

Suppose $\sigma : \mathcal{R} \to \mathcal{Q}$ is an elementary embedding. We say that there is a *total* $(\mathcal{Q}, \mathcal{R}, \sigma)$*-b-condensation diagram* on $\mathcal{P}$ if there is (π, τ) such that

1. $\pi : \mathcal{P} \to \mathcal{Q}$ is an elementary embedding,
2. $\tau : \mathcal{P} \to \mathcal{R}$ is an elementary embedding,
3. $\pi = \sigma \circ \tau$,

We then say that (π, τ) *supports* a total $(\mathcal{Q}, \mathcal{R}, \sigma)$-b-condensation diagram on $\mathcal{P}$.

Suppose next that $\mathcal{P}$ is of limit type and $\sigma : \mathcal{R} \to \mathcal{Q}$ is an elementary embedding. We say there is a *bottom-type* $(\mathcal{Q}, \mathcal{R}, \sigma)$*-b-condensation diagram* on $\mathcal{P}$ if there is $((\pi, \mathcal{Q}'), (\tau, \mathcal{R}'), \sigma')$ such that

1. $\pi : \mathcal{P}^b \to \mathcal{Q}'$ is an elementary embedding,
2. $\tau : \mathcal{P}^b \to \mathcal{R}'$ is an elementary embedding,
3. $\sigma' : \mathcal{R}' \to \mathcal{Q}'$ is an elementary embedding,
4. $\pi = \sigma' \circ \tau$,

and either

5. $\mathcal{Q}$ and $\mathcal{R}$ are of successor type, $\mathcal{Q} \trianglelefteq_{hod} \mathcal{Q}'$, $\mathcal{R} \trianglelefteq_{hod} \mathcal{R}'$ and $\sigma' \restriction \mathcal{R} = \sigma$, or
6. $\mathcal{Q}$ and $\mathcal{R}$ are of limit type, $\mathcal{Q}^b \trianglelefteq_{hod} \mathcal{Q}'$, $\mathcal{R}^b \trianglelefteq_{hod} \mathcal{R}'$ and $\sigma' \restriction \mathcal{R}^b = \sigma \restriction \mathcal{R}^b$.

We then say that $((\pi, \mathcal{Q}'), (\tau, \mathcal{R}'), \sigma')$ *supports* a bottom-type $(\mathcal{Q}, \mathcal{R}, \sigma)$-b-condensation diagram on $\mathcal{P}$. We say that $((\pi, \mathcal{Q}'), (\tau, \mathcal{R}'), \sigma')$ supports a *strict* bottom-type $(\mathcal{Q}, \mathcal{R}, \sigma)$-b-condensation diagram on $\mathcal{P}$ if in clause 6, $\mathcal{Q}^b \triangleleft_{hod} \mathcal{Q}'$.

Suppose now that $\mathcal{P}$ is as above and $(\mathcal{P}, \Sigma)$ is an allowable pair. We then say that there is a $(\mathcal{P}, \Sigma)$-supported $(\mathcal{Q}, \mathcal{R}, \sigma)$*-b-condensation diagram* on $\mathcal{P}$ if there is $(\mathcal{T}, \mathcal{Q}^*) \in I^{ope}(\mathcal{P}, \Sigma)$ such that one of the following clauses holds:

1. (a) $(\mathcal{P}, \Sigma)$ is a hod pair,
   (b) $\mathcal{P}$ is either of successor type or of lsa type and such that $\mathcal{J}_\omega[\mathcal{P}] \models$ "$\delta^{\mathcal{P}}$ is a Woodin cardinal",
   (c) $\mathcal{Q}^* = \mathcal{Q}$,
   (d) $(\mathcal{T}, \mathcal{Q}) \in I(\mathcal{P}, \Sigma)$, and
   (e) there is $\tau : \mathcal{P} \to \mathcal{R}$ such that $(\pi^{\mathcal{T}}, \tau)$-supports a total $(\mathcal{Q}, \mathcal{R}, \sigma)$-b-condensation diagram on $\mathcal{P}$.
2. (a) $\mathcal{P}$ is of limit type[94]
   (b) $\mathcal{Q}$ is a complete layer of $\mathcal{Q}^*$ and
   (c) letting

$$E = \begin{cases} E^{\mathcal{T}}_{\mathcal{Q}^b} & \text{if } \mathcal{Q} \text{ is of limit type,} \\ E^{\mathcal{T}}_{\mathcal{Q}} & \text{if } \mathcal{Q} \text{ is of successor type,} \end{cases}$$

[94]This clause also works for simple hod pairs and simple sts hod pairs.

there are $\tau : \mathcal{P}^b \to \mathcal{R}'$ and $\sigma' : \mathcal{R}' \to \mathcal{Q}' =_{\text{def}} \pi_E(\mathcal{P}^b)$ such that $((\pi_E \upharpoonright \mathcal{P}^b, \mathcal{Q}'), (\tau, \mathcal{R}'), \sigma')$ supports a bottom type $(\mathcal{Q}, \mathcal{R}, \sigma)$-b-condensation diagram on $\mathcal{P}$,

(d) (*The sts conditions*)[95] if $(\mathcal{P}, \Sigma)$ is an sts hod pair or a simple sts hod pair then $\mathcal{Q} \neq \mathcal{Q}^*$ provided one of the following holds:
  (i) $\pi^{\mathcal{T}}$ exists.
  (ii) $\pi^{\mathcal{T}}$ does not exist but letting $\mathcal{T} = (\mathcal{P}_\alpha, \mathcal{X}_\alpha, G_\alpha : \alpha \leq \beta)$ and γ be the largest element of $\max^{\mathcal{X}}_\beta$, $\pi^{(\mathcal{X}_\beta)_{\geq\gamma}}$ exists.

DEFINITION 4.9.2 (Strong branch condensation). Suppose $(\mathcal{P}, \Sigma)$ is an allowable pair and $\mathcal{P}$ is not gentle. We say Σ has *strong branch condensation with low-level-agreements* if

1. Σ has branch condensation[96],
2. whenever
   (a) $(\mathcal{T}, \mathcal{Q}), (\mathcal{U}, \mathcal{R}) \in I^{ope}(\mathcal{P}, \Sigma)$,
   (b) $\pi : \mathcal{R}^b \to \mathcal{Q}^b$ is such that $\pi^{\mathcal{T},b} = \pi \circ \pi^{\mathcal{U},b}$,
   (c) $\mathcal{X}$ is a stack on $\mathcal{R}^b$ according to $\Sigma_{\mathcal{R}^b,\mathcal{U}}$,
   (d) c is a branch of $\mathcal{X}$ such that $\pi^{\mathcal{X}}_c$ is defined and there is $\sigma : \mathcal{M}^{\mathcal{X}}_c \to \mathcal{Q}^b$ with the property that $\pi = \sigma \circ \pi^{\mathcal{X}}_c$,

   $c = \Sigma(\mathcal{U}^\frown \mathcal{X})$.
3. whenever $(\mathcal{Q}, \mathcal{R}, \sigma)$, $(\mathcal{T}, \mathcal{Q}^*) \in I^{ope}(\mathcal{P}, \Sigma)$ and
$$(\mathcal{W}, \mathcal{R}) \in B^{ope}(\mathcal{P}, \Sigma) \cup I^{ope}(\mathcal{P}, \Sigma)$$
   are such that
   (a) there is a $(\mathcal{P}, \Sigma)$-supported $(\mathcal{Q}, \mathcal{R}, \sigma)$-b-condensation diagram on $\mathcal{P}$ as witnessed by $(\mathcal{T}, \mathcal{Q}^*)$ and
   (b) letting Λ be the σ-pullback of $\Sigma_{\mathcal{Q},\mathcal{T}}$, there is no low level disagreement between $\Sigma_{\mathcal{R},\mathcal{W}}$ and Λ,

   then one of the following holds:
   (a) If
      (i) $\mathcal{P}$ is of lsa type, $(\mathcal{P}, \Sigma)$ is a hod pair and $\mathcal{J}_\omega[\mathcal{P}] \models$ "$\delta^{\mathcal{P}}$ is a Woodin cardinal",
      (ii) $(\mathcal{T}, \mathcal{Q}^*)$ supports a bottom-type $(\mathcal{Q}, \mathcal{R}, \sigma)$-b-condensation diagram
$$((\pi, \mathcal{Q}'), (\tau, \mathcal{R}'), \sigma')$$
      on $\mathcal{P}$ and
      (iii) $\pi^{\mathcal{T}}$ is defined,

      then $\mathcal{R}$ is of lsa type and $(\Sigma_{\mathcal{R},\mathcal{W}})^{stc} = \Lambda^{stc}$.
   (b) In all other cases, $\Sigma_{\mathcal{R},\mathcal{W}} = \Lambda$.

[95]We need this conditions in order to make sense of σ-pullback of $\Sigma_{\mathcal{Q},\mathcal{T}}$.

[96]See [29, Definition 2.14]. If Σ is an st-strategy then we apply [29, Definition 2.14] to stacks $\mathcal{T}$ and $\mathcal{U}$ such that $\max^{\mathcal{T}} = \max^{\mathcal{U}} = \emptyset$.

We say Σ has *strong branch condensation* if $\Lambda = \Sigma_{\mathcal{R},\mathcal{W}}$ holds without the requirement that there is no low level disagreement between Λ and $\Sigma_{\mathcal{R},\mathcal{W}}$.

REMARK 4.9.3. The proof of Theorem 4.9.5 only establishes clause 3 of strong branch condensation, but the proof can be easily modified to show clause 1 and 2 as well.

The following is an easily provable lemma, which establishes the equivalence between strong branch condensation and strong branch condensation with low-level-agreements. The reader may wish to review Definition 4.3.11.

LEMMA 4.9.4. *Suppose $(\mathcal{P},\Sigma)$ is a hod pair or an sts hod pair, Σ is weakly self-cohering and Γ is a projectively closed pointclass. Suppose that*

1. *Σ has strong branch condensation with low-level-agreements,*
2. *Σ is Γ-strongly fullness preserving,*
3. *if $\mathcal{P}$ is of successor type then $\Sigma_{\mathcal{P}^-}$ has strong branch condensation.*

Then $(\mathcal{P},\Sigma)$ has strong branch condensation.

PROOF. Suppose that $(\mathcal{Q},\mathcal{R},\sigma)$ is such that there is a $(\mathcal{P},\Sigma)$-supported $(\mathcal{Q},\mathcal{R},\sigma)$-b-condensation diagram on $\mathcal{P}$ as witnessed by $(\mathcal{T},\mathcal{Q}^*)$. Let Λ be the σ-pullback of $\Sigma_{\mathcal{Q},\mathcal{T}}$. Fix a pair $(\mathcal{W},\mathcal{R}) \in B^{ope}(\mathcal{P},\Sigma) \cup I^{ope}(\mathcal{P},\Sigma)$. Our goal is to argue that $\Sigma_{\mathcal{R},\mathcal{W}} = \Lambda$. Towards a contradiction assume that $\Sigma_{\mathcal{R},\mathcal{W}} \neq \Lambda$. Thus, we must have that there is a lower level disagreement between $\Sigma_{\mathcal{R},\mathcal{W}}$ and Λ.

Suppose first that $\mathcal{P}$ is of successor type. Because there is a lower level disagreement between $\Sigma_{\mathcal{R},\mathcal{W}}$ and Λ, we must have that $\Sigma_{\mathcal{R}^-,\mathcal{W}} \neq \Lambda_{\mathcal{R}^-}$. However, it is not hard to see that there is a $(\mathcal{P}^-,\Sigma_{\mathcal{P}^-})$-supported $(\mathcal{Q}^-,\mathcal{R}^-,\sigma \restriction \mathcal{R}^-)$-b-condensation diagram on $\mathcal{P}^-$ as witnessed by $(\downarrow(\mathcal{T},\mathcal{P}^-),\mathcal{Q}^-)$. Because $\sigma \restriction \mathcal{R}^-$-pullback of $\Sigma_{\mathcal{Q}^-,\downarrow(\mathcal{T},\mathcal{P}^-)}$ is just $\Lambda_{\mathcal{R}^-}$ and because $\Sigma_{\mathcal{P}^-}$ has strong branch condensation, we have that $\Sigma_{\mathcal{R}^-,\mathcal{W}} = \Lambda_{\mathcal{R}^-}$

We now assume that $\mathcal{P}$ is of limit type. Since all the cases are very similar, we will examine two representative cases, namely:

(A) $\mathcal{P}$ is of lsa type, $\mathcal{J}_\omega[\mathcal{P}] \vDash$ "$\delta^{\mathcal{P}}$ is a Woodin cardinal", $(\mathcal{P},\Sigma)$ is a hod pair and there is a total $(\mathcal{P},\Sigma)$-supported $(\mathcal{Q},\mathcal{R},\sigma)$-b-condensation diagram on $\mathcal{P}$ as witnessed by $(\mathcal{T},\mathcal{Q}^*)$.

(B) $(\mathcal{P},\Sigma)$ is an sts hod pair and $\mathcal{Q}$ is of limit type.

We start with (A). In this case, $\mathcal{Q}^* = \mathcal{Q}$ and $\pi^{\mathcal{T}}$ exists. Let $\tau : \mathcal{P} \to \mathcal{R}$ be such that $\pi^{\mathcal{T}} = \sigma \circ \tau$. Let then $(\mathcal{W}_1,\mathcal{R}_1,\mathcal{W}_1',\mathcal{R}_1',\mathcal{R}_2)$ be a minimal low level disagreement between $(\mathcal{R},\Lambda)$ and $(\mathcal{R},\Sigma_{\mathcal{R},\mathcal{W}})$. Thus,

1. (since $(\mathcal{R},\Lambda)$ and $(\mathcal{R},\Sigma_{\mathcal{R},\mathcal{W}})$ are hod pairs), we have that $\mathcal{W}_1$ and $\mathcal{W}_1'$ are generalized stacks.
2. $\mathcal{R}_2$ is of successor type,
3. $\Lambda_{\mathcal{R}_2^-,\mathcal{W}_1} = \Sigma_{\mathcal{R}_2^-,\mathcal{W}^\frown\mathcal{W}_1'}$ and
4. $\Lambda_{\mathcal{R}_2,\mathcal{W}_1} \neq \Sigma_{\mathcal{R}_2,\mathcal{W}^\frown\mathcal{W}_1'}$.

Let $\mathcal{T}_1 = \sigma\mathcal{W}_1$ and let $\mathcal{Q}_1$ be the last model of $\mathcal{T}_1$. Let $\sigma_1 : \mathcal{R}_1 \to \mathcal{Q}_1$ be the copy map and set $\mathcal{Q}_2 = \sigma_1(\mathcal{R}_2)$. Notice that both $\mathcal{R}_2$ and $\mathcal{Q}_2$ are of successor type.

Let $\mathcal{T} = (\mathcal{P}_\alpha, \mathcal{X}_\alpha, G_\alpha : \alpha < \eta)$ and let $\mathcal{U} = \mathcal{T}^\frown\mathcal{T}_1$ where we construct this by setting $\mathcal{P}_\eta = \mathcal{Q}$ and $\mathcal{X}_{\eta+1} = \mathcal{T}_1$. Thus, $\mathcal{U} \restriction \eta = \mathcal{T}$. Combining $\mathcal{T}$ and $\mathcal{T}_1$ this way is a legal way of producing a generalized stack because $\pi^\mathcal{T}$ is defined[97]. Let then

1. E be the $\mathcal{Q}_2$-un-dropping extender $\mathcal{U}$ and $\mathcal{Q}' = \pi_E(\mathcal{P}^b)$,
2. F' be the $\mathcal{R}_2$-un-dropping extender of $\mathcal{W}_1$,
3. $F = \{(a,A) : (a, \tau(A)) \in F'\}$,
4. $\sigma_2 : \mathcal{R}' \to \mathcal{Q}'$ be the map given by $\sigma_2([a,f]_F) = [\sigma_1(a), f]_E$.

It follows that

(A1) $(a,X) \in F \iff (\sigma_1(a), X) \in E$, and hence σ_2 is an elementary embedding, and

(A2) $\sigma_2 \restriction \mathcal{R}_2 = \sigma_1 \restriction \mathcal{R}_2$ and $\pi_E \restriction \mathcal{P}^b = \sigma_2 \circ \pi_F \restriction \mathcal{P}^b$.

Therefore, $(\mathcal{T}^\frown\mathcal{T}_1, \mathcal{Q}_1)$ and $((\pi_E \restriction \mathcal{P}^b, (\mathcal{Q}_2')^b), ((\pi_F \restriction \mathcal{P}^b, (\mathcal{R}_2')^b), \sigma_2)$ support a bottom-type $(\mathcal{Q}_2, \mathcal{R}_2, \sigma_1 \restriction \mathcal{R}_2)$-b-condensation diagram on $\mathcal{P}$. Therefore, since $\Lambda_{\mathcal{R}_2^-, \mathcal{W}_1} = \Sigma_{\mathcal{R}_2^-, \mathcal{W}^\frown\mathcal{W}_1'}$ (i.e., there is no low level disagreement between $\Lambda_{\mathcal{R}_2, \mathcal{W}_1} = \Sigma_{\mathcal{R}_2, \mathcal{W}^\frown\mathcal{W}_1'}$) and $\Lambda_{\mathcal{R}_2, \mathcal{W}_1}$ is the σ_2-pullback of $\Sigma_{\mathcal{Q}_2, \mathcal{T}^\frown\mathcal{T}_1}$, $\Lambda_{\mathcal{R}_2, \mathcal{W}_1} = \Sigma_{\mathcal{R}_2, \mathcal{W}^\frown\mathcal{W}_1'}$, contradiction!

We now work assuming (B). Most of what we say below is very similar to the above with only minor differences. In this case we have that clause 2 of Definition 4.9.1 holds. Thus,

1. $\mathcal{Q}$ is a complete layer of $\mathcal{Q}^*$,
2. $\mathcal{Q}$ is of limit type and
3. letting $E = E^\mathcal{T}_{\mathcal{Q}^b}$, there are $\tau : \mathcal{P}^b \to \mathcal{R}'$ and $\sigma' : \mathcal{R}' \to \mathcal{Q}' =_{\text{def}} \pi_E(\mathcal{P}^b)$ such that $((\pi_E \restriction \mathcal{P}^b, \mathcal{Q}'), (\tau, \mathcal{R}'), \sigma')$ supports a bottom type $(\mathcal{Q}, \mathcal{R}, \sigma)$-b-condensation diagram on $\mathcal{P}$.

Set $\pi = \pi_F \restriction \mathcal{P}^b$. Let then $(\mathcal{W}_1, \mathcal{R}_1, \mathcal{W}_1', \mathcal{R}_1', \mathcal{R}_2)$ be a minimal low level disagreement between $(\mathcal{R}, \Lambda)$ and $(\mathcal{R}, \Sigma_{\mathcal{R},\mathcal{W}})$. Let $\mathcal{T}_1 = \sigma\mathcal{W}_1$ and let $\mathcal{Q}_1$ be the last model of $\mathcal{T}_1$. Let $\sigma_1 : \mathcal{R}_1 \to \mathcal{Q}_1$ be the copy map and let $\mathcal{Q}_2 = \sigma_1(\mathcal{R}_2)$. Notice that both $\mathcal{R}_2$ and $\mathcal{Q}_2$ are of successor type.

We now define $\mathcal{U}$ as follows. If $\pi^\mathcal{T}$ is defined then we let $\mathcal{U} = \mathcal{T}^\frown\mathcal{T}_1$ be as in (A). Assume then $\pi^\mathcal{T}$ is not defined. In this case, $\mathcal{T} = (\mathcal{P}_\alpha, \mathcal{X}_\alpha, G_\alpha : \alpha \le \beta)$, $\mathcal{Q}^*$ is the last model of $\mathcal{X}_\beta$ and $\pi^{\mathcal{X}_\beta}$ is not defined. Let then $\mathcal{U}$ be the same as $\mathcal{T}$ except that the βth stack used in $\mathcal{U}$ is $\mathcal{X}_\beta^\frown\mathcal{T}_1$[98].

Just like in case (A), we have that if E' is the $\mathcal{Q}_2$-un-dropping extender of $\mathcal{U}$, F' is the $\mathcal{R}_2$-un-dropping extender of $\mathcal{W}_1$, $F = \{(a,A) : (a, \tau(A)) \in F'\}$ and $\sigma_2 : \pi_F(\mathcal{P}^b) \to \pi_{E'}(\mathcal{P}^b)$ is the map given by $\sigma_2([a,f]_F) = [\sigma_1(a), f]_E$ then

[97]If $(\mathcal{P}, \Sigma)$ was a simple hod pair then at this step we would let $\mathcal{T}^\frown\mathcal{T}_1$ be a stack.

[98]Thus, we have that for some $\gamma \in R^{\mathcal{X}_\beta^\frown\mathcal{T}_1}$, $\mathcal{Q}^* = \mathcal{M}_\gamma^{\mathcal{X}_\beta^\frown\mathcal{T}_1}$ and $(\omega\beta_\gamma^{\mathcal{X}_\beta^\frown\mathcal{T}_1}, m_\gamma^{\mathcal{X}_\beta^\frown\mathcal{T}_1}) = (\text{ord}(\mathcal{Q}), \omega)$.

(B1) $(a,X) \in F \iff (\sigma_1(a),X) \in E'$, and hence σ_2 is an elementary embedding, and

(B2) $\sigma_2 \restriction \mathcal{R}_2 = \sigma_1 \restriction \mathcal{R}_2$ and $\pi_{E'} \restriction \mathcal{P}^b = \sigma_2 \circ \pi_F \restriction \mathcal{P}^b$.

Here the situation may seem somewhat more complicated as $\mathcal{W}_1$ is on $\mathcal{R}$ and not on $\mathcal{R}'$. But since $\mathcal{R}^b \trianglelefteq_{hod} \mathcal{R}'$, F' is an $\mathcal{R}'$-extender. Moreover, since $\mathcal{T}_1 = \sigma\mathcal{W}_1$ and $\sigma' \restriction \mathcal{R}^b = \sigma \restriction \mathcal{R}^b$, we have that for each $A \in \wp(\delta^{\mathcal{P}^b}) \cap \mathcal{P}$,

$$\sigma_2(\sigma^{\mathcal{W}_1}(\tau(A))) = \sigma^{\mathcal{T}_1}(\pi(A)).$$[99]

We then once again, just like in (A), have that

$$(\mathcal{T}^\frown\mathcal{T}_1, \mathcal{Q}_1) \quad \text{and} \quad ((\pi_{E'} \restriction \mathcal{P}^b, \pi_{E'}(\mathcal{P}^b)), ((\pi_F \restriction \mathcal{P}^b, \pi_F(\mathcal{P}^b)), \sigma_2)$$

support a bottom-type $(\mathcal{Q}_2, \mathcal{R}_2, \sigma_1 \restriction \mathcal{R}_2)$-b-condensation diagram on $\mathcal{P}$, and which implies, just like in (A), that $\Lambda_{\mathcal{R}_2,\mathcal{W}_1} = \Sigma_{\mathcal{R}_2,\mathcal{W}^\frown\mathcal{W}_1'}$. ⊣

THEOREM 4.9.5. *Assume* $\mathsf{AD}^+ + \mathsf{NsesS}$. *Suppose*

1. *for some* α_0 *such that* $\vartheta_{\alpha_0} < \Theta$, $\Gamma = \{A \subseteq \mathbb{R} : w(A) < \vartheta_{\alpha_0}\}$,
2. $\mathsf{C} = (\mathbb{M}, (P, \Psi), \Gamma^*, A)$ *Suslin, co-Suslin captures* Γ,
3. $\mathbb{M} = (M, \delta, \vec{G}, \Sigma^*)$,
4. $\mathsf{hpc} = (\mathcal{M}_\gamma, \mathcal{N}_\gamma, Y_\gamma, \Phi_\gamma, F_\gamma^+, F_\gamma, b_\gamma : \gamma \leq \delta)$ *is the output of the* $\Gamma-\mathsf{hpc}$ *of* $\mathbb{M}$,
5. $\xi < \delta$ *is such that* $(\mathcal{M}_\xi, \Phi_\xi^+)$ *is a hod pair,* $\mathcal{M}_\xi$ *is not gentle and* $M \vDash (\mathcal{M}_\xi, \Phi_\xi) \in \mathsf{Hp}^\Gamma$.

Then Φ_ξ^+ *has strong branch condensation.*

Also if $\xi < \delta$ *is such that* $(\mathcal{M}_\xi, (\Phi_\xi^+)^{stc})$ *is an sts pair and* $M \vDash (\mathcal{M}_\xi, \Phi_\xi^{stc}) \in \mathsf{Hp}^\Gamma$ *then* $(\Phi_\xi^+)^{stc}$ *has strong branch condensation.*

PROOF. The proof of the second half of the theorem is similar to the first and so we will prove the first and leave the second to the reader. The proof of the branch condensation is very similar to the proof of the second half of strong branch condensation, and so we give the proof of the second half of strong branch condensation.

Towards a contradiction, suppose that for some ξ', $\mathcal{M}_{\xi'}$ is a hod premouse and $\Phi_{\xi'}^+$ does not have strong branch condensation, and let ξ be the least such ξ'. Because of Lemma 4.9.4, it is enough to show that Φ_ξ^+ has strong branch condensation with low-level-agreements.

Just like in the proof of fullness preservation (see Theorem 4.6.3), if Φ_ξ^+ does not have strong branch condensation then for some $\zeta_0 < \delta$ the witness can be found in some $M[g]$ where $g \subseteq Coll(\omega, \zeta_0)$ is M-generic. Let $\zeta_1 = \{\sup(F_\gamma^+) : \gamma < \xi\}$ and set $\zeta = \max((\zeta_0^+)^M, (\zeta_1^+)^M)$.

Let $\mathcal{P} = \mathcal{M}_\xi$ and $\Sigma = \Phi_\xi^+$. The difficult case is when $\mathcal{P}$ is non-meek, and so we assume this. We start working in $M[g]$. What we need to show is that whenever

[99] Here, $\sigma^{\mathcal{X}}$ is defined in Definition 2.10.5 and Notation 2.10.9.

1. $(\mathcal{Q},\mathcal{R},\sigma)$ is such that there is a $(\mathcal{P},\Sigma)$-supported $(\mathcal{Q},\mathcal{R},\sigma)$-b-condensation diagram on $\mathcal{P}$ as witnessed by $(\mathcal{T},\mathcal{Q}^*)$, and
2. $(\mathcal{W},\mathcal{R})\in B^{ope}(\mathcal{P},\Sigma)\cup I^{ope}(\mathcal{P},\Sigma)$ is such that letting Λ be the σ-pullback of $\Sigma_{\mathcal{Q},\mathcal{T}}$, there is no low level disagreement between $\Sigma_{\mathcal{R},\mathcal{W}}$ and Λ,

then $\Sigma_{\mathcal{R},\mathcal{W}}=\Lambda$.

Fix then $(\mathcal{Q},\mathcal{R},\sigma)\in M|\zeta[g]$ such that there is a $(\mathcal{P},\Sigma)$-supported $(\mathcal{Q},\mathcal{R},\sigma)$-b-condensation diagram on $\mathcal{P}$ as witnessed by $(\mathcal{T},\mathcal{Q}^*)\in M|\zeta[g]$ and let $(\mathcal{W},\mathcal{R})\in B^{ope}(\mathcal{P},\Sigma)\cup I^{ope}(\mathcal{P},\Sigma)$ be such that $(\mathcal{W},\mathcal{R})\in M|\zeta[g]$ and letting Λ be the σ-pullback of $\Sigma_{\mathcal{Q},\mathcal{T}}$,

(1) there is no low level disagreement between $\Sigma_{\mathcal{R},\mathcal{W}}$ and Λ but $\Sigma_{\mathcal{R},\mathcal{W}}\neq\Lambda$.

It follows from Lemma 4.7.6 that

(2) either $\mathcal{R}$ is of successor type or $\mathcal{R}$ is of lsa type and $\mathcal{J}_\omega[\mathcal{R}]\models$ "$\delta^{\mathcal{R}}$ is a Woodin cardinal".

Case 1: $\pi^{\mathcal{T}}$ is defined, and for some τ, $(\pi^{\mathcal{T}},\tau)$ supports a total $(\mathcal{Q},\mathcal{R},\sigma)$-b-condensation diagram on $\mathcal{P}$.

We thus have that $\tau:\mathcal{P}\to\mathcal{R}$ and $\pi^{\mathcal{T}}=\sigma\circ\tau$. Let then

$$\Sigma'=\begin{cases}\Sigma_{\mathcal{P}^-} & \text{if } \mathcal{P} \text{ is of successor type,}\\ \Sigma^{stc} & \text{otherwise,}\end{cases}$$

$$\mathcal{P}_0'=\begin{cases}\mathcal{P} & \text{if } \mathcal{P} \text{ is of successor type,}\\ (\mathcal{P}|\delta^{\mathcal{P}})^{\#} & \text{otherwise,}\end{cases}$$

$$\mathcal{P}_1'=\begin{cases}\mathcal{P}^- & \text{if } \mathcal{P} \text{ is of successor type,}\\ (\mathcal{P}|\delta^{\mathcal{P}})^{\#} & \text{otherwise,}\end{cases}$$

$$\Lambda'=\begin{cases}\Lambda_{\mathcal{R}^-} & \text{if } \mathcal{P} \text{ is of successor type,}\\ \Lambda^{stc} & \text{otherwise.}\end{cases}$$

Let $\mathcal{P}^+$ be the last model of

$$(\mathsf{Le}((\mathcal{P}_1',\Sigma'),\mathcal{J}_\omega[\mathcal{P}_0'])_{>\zeta})^{(M[g],\delta,\vec{G})}.$$[100]

Define $\mathcal{Q}_1'$ and $\mathcal{R}_1'$ the same way $\mathcal{P}_1'$ is defined. We then let $\Sigma'_{\mathcal{Q}_1'}$ and $\Sigma'_{\mathcal{R}_1'}$ be defined the same way Σ' is defined but relative to $\Sigma_{\mathcal{Q},\mathcal{T}}$ and $\Sigma_{\mathcal{R},\mathcal{W}}$. It follows from (2) that $\Lambda'=\Sigma'_{\mathcal{R}_1'}$.

Let

1. E be the $(\delta^{\mathcal{P}},\delta^{\mathcal{R}})$-extender derived from τ,
2. F be the $(\delta^{\mathcal{P}},\delta^{\mathcal{Q}})$-extender derived from $\pi^{\mathcal{T}}$ and

[100] See Definition 4.5.1. Here we are assuming that if $\mathcal{P}$ is of lsa type then the above construction does not break down because of the anomaly stated in clause 3(b) of Definition 4.2.1. In the sequel, we will prove that such constructions indeed converge. See Theorem 4.12.1.

3. H be the $(\delta^{\mathcal{P}}, \delta^{\mathcal{R}})$-extender derived from $\pi^{\mathcal{W}}$.[101]

We let

$$\mathcal{R}^+ = \text{Ult}(\mathcal{P}^+, E),\ \mathcal{Q}^+ = \text{Ult}(\mathcal{P}^+, F),\ \text{and} \mathcal{S}^+ = \text{Ult}(\mathcal{P}^+, H).$$

We also have $\sigma^+ : \mathcal{R}^+ \to \mathcal{Q}^+$ such that

$$\pi_F^{\mathcal{P}^+} = \sigma^+ \circ \pi_E^{\mathcal{P}^+} \quad \text{and} \quad \sigma^+ \restriction \mathcal{R} = \sigma.$$

More precisely, $\sigma^+(x) = \pi_F^{\mathcal{P}^+}(f)(\sigma(a))$ where $f \in \mathcal{P}^+$, $a \in (\mathcal{R})^{<\omega}$ and $x = \pi_E^{\mathcal{P}^+}(f)(a)$.

Notice now that both $\mathcal{Q}^+$ and $\mathcal{S}^+$ have strategies induced by Σ^* via the resurrection procedure of [23, Chapter 12] that we have outlined in Lemma 4.3.9. Let Ψ^* and Φ be these strategies. We then have that $\Psi^*_{\mathcal{Q}} = \Sigma_{\mathcal{Q},\mathcal{T}}$ and $\Phi_{\mathcal{R}} = \Sigma_{\mathcal{R},\mathcal{W}}$. Let now Ψ be the σ^+-pullback of Ψ^*. Applying the "furthermore" clause of Theorem 4.5.6 to $(\mathcal{R}^+, \Psi)$ and $(\mathcal{S}^+, \Phi)$, we conclude that $\Psi_{\mathcal{R}} = \Phi_{\mathcal{R}}$.

Case 2: There is $\tau : \mathcal{P} \to \mathcal{R}'$ such that $(\tau, \mathcal{R}') \in M|\zeta[g]$ and letting

1. F be the $\mathcal{Q}^b$-un-dropping extender of $\mathcal{T}$ if $\mathcal{Q}$ is of limit type and
2. F be the $\mathcal{Q}$-un-dropping extender otherwise,

there is a $\sigma' : \mathcal{R}' \to \mathcal{Q}' =_{\text{def}} \pi_F(\mathcal{P}^b)$ such that $((\pi_F \restriction \mathcal{P}^b, \mathcal{Q}'), (\tau, \mathcal{R}'), \sigma')$ supports a $(\mathcal{Q}, \mathcal{R}, \sigma)$-b-condensation diagram on $\mathcal{P}$.

This case is very similar. Notice that (2) implies that $\Sigma^{stc}_{\mathcal{R},\mathcal{W}} = \Lambda^{stc}$ assuming the hypothesis of clause 2 of Definition 4.9.2 holds (as there are no low level disagreements between $\Sigma_{\mathcal{R},\mathcal{W}}$ and Λ). Thus, we assume that the hypothesis of clause 2 is not applicable. Notice now that the sts conditions[102] and the fact that $\mathcal{Q}$ has to be a complete layer of $\mathcal{Q}^*$ imply that $\mathcal{Q}$ and $\mathcal{R}$ are not of lsa type. (2) then implies that $\mathcal{Q}$ and $\mathcal{R}$ must be of successor type[103]. It then follows that $\mathcal{Q} \lhd_{hod} \pi_F(\mathcal{P}^b)$.

Let now $\mathcal{P}'$ be the last model of

$$(\mathsf{Le}((\mathcal{P}^b, \Sigma_{\mathcal{P}^b}), \mathcal{J}_\omega[\mathcal{P}^b])_{>\zeta})^{(M[g], \delta, \vec{G})}.$$

Let

1. E be the $(\delta^{\mathcal{P}^b}, \delta^{\mathcal{R}'})$-extender derived from τ, and
2. H be the $\mathcal{R}$-un-dropping extender of $\mathcal{W}$;
3. $\mathcal{R}_1 = \text{Ult}(\mathcal{P}', E)$ and $\pi_0 = \pi_E^{\mathcal{P}'}$,
4. $\mathcal{Q}_1 = \text{Ult}(\mathcal{P}', F)$ and $\sigma_1 : \mathcal{R}_1 \to \mathcal{Q}_1$ is the canonical factor map, and
5. $\mathcal{S}_1 = \text{Ult}(\mathcal{P}', H)$ and $k = \pi_H^{\mathcal{P}'}$.

We then define $\mathcal{R}^+$ and $\mathcal{S}^+$ as follows. Let $\mathcal{R}^+$ be the last model of

$$(\mathsf{Le}((\mathcal{R}^-, \Lambda_{\mathcal{R}^-}), \mathcal{J}_\omega[\mathcal{R}])_{>\zeta})^{(L_{\text{ord}(M)}[\mathcal{R}_1], \delta, \vec{H}')}$$

---

[101] In the case that $(\mathcal{W}, \mathcal{R}) \in B^{ope}$, we let H be the $\mathcal{R}$-un-dropping extender of W and continue as below.

[102] See Definition 4.9.1.

[103] See Lemma 4.7.2.

where $\vec{H}' = \{K \in \vec{E}^{\mathcal{R}_1} : \nu(K) \text{ is inaccessible in } \mathcal{R}_1\}$.

Let $\mathcal{S}^+$ be the last model of

$$(\mathsf{Le}((\mathcal{R}^-, \Sigma_{\mathcal{R}^-, \mathcal{W}}), \mathcal{J}_\omega[\mathcal{R}])_{>\zeta})^{(\mathcal{L}_{\text{ord}(M)}[\mathcal{S}_1], \delta, \vec{H}'')}$$

where $\vec{H}'' = \{K \in \vec{E}^{\mathcal{S}_1} : \nu(K) \text{ is inaccessible in } \mathcal{S}_1\}$.

Notice that because $\Sigma_{\mathcal{R}^-, \mathcal{W}} = \Lambda_{\mathcal{R}^-}$ we have that both $\mathcal{R}^+$ and $\mathcal{S}^+$ are $\Sigma_{\mathcal{R}^-, \mathcal{W}}$-mice over $\mathcal{R}$. Once again, in $M[g]$, both $\mathcal{R}^+$ and $\mathcal{S}^+$ have $(\delta, \delta+1)$-iteration strategies Φ and Ψ such that $\Lambda_{\mathcal{R}} = \Phi_{\mathcal{R}}$ and $\Sigma_{\mathcal{R}, \mathcal{W}} = \Psi_{\mathcal{R}}$. It then again follows from the "furthermore" clause of Theorem 4.5.6 that $\Phi_{\mathcal{R}} = \Psi_{\mathcal{R}}$. ⊣

4.10. Positional and commuting

In this section, our goal is to show that strong branch condensation implies commuting. Recall [29, Definition 2.35]: if M is a transitive model of a fragment of ZFC and Σ is an iteration strategy for M then we say Σ is *positional* if whenever Q is a Σ-iterate of M via $\mathcal{W}$ and $(\mathcal{T}, R), (\mathcal{U}, R) \in I(Q, \Sigma_{Q, \mathcal{W}})$, $\Sigma_{R, \mathcal{W}^\frown \mathcal{T}} = \Sigma_{R, \mathcal{W}^\frown \mathcal{U}}$. Recall that commuting means that in the above scenario, $\pi^{\mathcal{T}} = \pi^{\mathcal{W}}$. If the above only holds for $Q = M$, then we say that Σ is *weakly positional* (and *weakly commuting* respectively). Using the usual proof of the Dodd-Jensen lemma, we get that (weakly) positional implies (weakly) commuting.

REMARK 4.10.1. In the previous section, we only studied branch condensation for non-gentle hod premice. This is because if $(\mathcal{P}, \Sigma)$ is a hod pair and $\mathcal{P}$ is gentle then Σ is essentially $\oplus_{\mathcal{Q} \lhd_{hod} \mathcal{P}} \Sigma_{\mathcal{Q}}$. Thus, we can say that Σ has strong branch condensation if for every complete layer $\mathcal{Q} \lhd_{hod} \mathcal{P}$, $\Sigma_{\mathcal{Q}}$ has strong branch condensation. We will state our theorems for hod pairs or sts hod pairs, but the proofs will be given for pairs $(\mathcal{P}, \Sigma)$ such that $\mathcal{P}$ is non-gentle.

PROPOSITION 4.10.2. *Suppose $(\mathcal{P}, \Sigma)$ is an allowable pair, Γ is a projectively closed pointclass and Σ has strong branch condensation and is strongly Γ-fullness preserving. Then Σ is positional. Moreover, if Σ is an iteration strategy then it is also commuting.*

PROOF. We just prove weak positionality and hence weak commuting. The proof of the general case is only notationally more complicated.

Suppose $(\mathcal{T}, \mathcal{Q}), (\mathcal{U}, \mathcal{Q}) \in I(\mathcal{P}, \Sigma)$. We want to see that $\Sigma_{\mathcal{Q}, \mathcal{T}} = \Sigma_{\mathcal{Q}, \mathcal{U}}$. Towards a contradiction, suppose not. Suppose first that $\mathcal{P}$ is of limit type and if it is of the lsa type then $\Sigma^{stc}_{\mathcal{Q}, \mathcal{T}} \neq \Sigma^{stc}_{\mathcal{Q}, \mathcal{U}}$. Let then $((\mathcal{T}_1, \mathcal{R}_1), (\mathcal{T}_2, \mathcal{R}_2), \mathcal{R}_3)$ be a minimal lower level disagreement[104] between $\Sigma_{\mathcal{Q}, \mathcal{T}}$ and $\Sigma_{\mathcal{Q}, \mathcal{U}}$. We can then apply strong branch condensation to $(\mathcal{R}_3, \mathcal{R}_3, id)$. Notice that $(\mathcal{T}^\frown \mathcal{T}_1, \mathcal{Q})$ supports a $(\mathcal{R}_3, \mathcal{R}_3, id)$-b-condensation diagram on $\mathcal{P}$ as witnessed by $((\pi, \mathcal{R}), (\pi, \mathcal{R}), id)$ where letting E be the $\mathcal{R}_3$-un-dropping extender of $\mathcal{T}^\frown \mathcal{T}_1$, $\mathcal{R} = \pi_E(\mathcal{P}^b)$ and $\pi = \pi_E \restriction \mathcal{P}^b$.

[104] See Lemma 4.7.6.

Next suppose that $\mathcal{P}$ is of successor type or of lsa type but $\Sigma^{stc}_{\mathcal{Q},\mathcal{T}} = \Sigma^{stc}_{\mathcal{Q},\mathcal{U}}$. It then follows that $(\pi^{\mathcal{T}}, \pi^{\mathcal{T}})$ supports a total $(\mathcal{Q}, \mathcal{Q}, id)$-b-condensation-diagram on $\mathcal{P}$. It then again follows that $\Sigma_{\mathcal{Q},\mathcal{T}} = \Sigma_{\mathcal{Q},\mathcal{U}}$. ⊣

The proof actually gives more.

PROPOSITION 4.10.3. *Suppose $(\mathcal{P},\Sigma)$ is an allowable pair, Γ is a projectively closed pointclass and Σ has strong branch condensation and is strongly Γ-fullness preserving. Suppose that*

$$(\mathcal{T},\mathcal{Q}) \in B^{ope}(\mathcal{P},\Sigma) \cup I^{ope}(\mathcal{P},\Sigma) \quad \textit{and} \quad (\mathcal{U},\mathcal{Q}) \in B^{ope}(\mathcal{P},\Sigma) \cup I^{ope}(\mathcal{P},\Sigma).$$

Then $\Sigma_{\mathcal{Q},\mathcal{T}} = \Sigma_{\mathcal{Q},\mathcal{U}}$.

DEFINITION 4.10.4. Suppose $(\mathcal{P},\Sigma)$ is an allowable pair and Γ is a projectively closed pointclass. Suppose further that Σ has strong branch condensation and is strongly Γ-fullness preserving. Given $\mathcal{Q} \in p[I^{ope}(\mathcal{P},\Sigma)] \cup p[B^{ope}(\mathcal{P},\Sigma)]$[105], we let $\Sigma_{\mathcal{Q}} = \Sigma_{\mathcal{Q},\mathcal{T}}$ where $\mathcal{T}$ is such that $(\mathcal{T},\mathcal{Q}) \in I^{ope}(\mathcal{P},\Sigma) \cup B^{ope}(\mathcal{P},\Sigma)$.

We need commuting not only for iteration strategies but also for short tree strategies.

DEFINITION 4.10.5. Suppose $(\mathcal{P},\Sigma)$ is an sts hod pair. We say Σ is *weakly commuting* if whenever $(\mathcal{T},\mathcal{Q}) \in I^b(\mathcal{P},\Sigma)$ and $(\mathcal{U},\mathcal{R}) \in I^b(\mathcal{P},\Sigma)$ are such that $\mathcal{R}^b \trianglelefteq_{hod} \mathcal{Q}^b$ and $\mathcal{R}^b = cHull^{\mathcal{Q}^b}(\pi^{\mathcal{T},b}[\mathcal{P}^b] \cup \delta^{\mathcal{R}^b})$, then letting

1. $k' : Hull^{\mathcal{Q}^b}(\pi^{\mathcal{T},b}[\mathcal{P}^b] \cup \delta^{\mathcal{R}^b}) \to \mathcal{R}^b$ be the transitive collapse and
2. $k =_{\mathrm{def}} k' \circ \pi^{\mathcal{T},b} : \mathcal{P}^b \to \mathcal{R}^b$,

$k = \pi^{\mathcal{U},b}$.

In the above situation we say that k is the *collapse* of $\pi^{\mathcal{T},b}[\mathcal{P}^b]$. We say $(\mathcal{P},\Sigma)$ is *commuting* if whenever $(\mathcal{T},\mathcal{Q}) \in I^{ope}(\mathcal{P},\Sigma)$, $\Sigma_{\mathcal{Q}}$ is weakly commuting.

It is not known to us if strong branch condensation and Γ-fullness preservation for sts pairs implies commuting. Nevertheless, hod pair constructions produce sts pairs that are commuting (also see Proposition 4.15.1).

THEOREM 4.10.6. *Assume* $\mathsf{AD}^+ + \mathsf{NsesS}$. *Suppose*

1. *for some α_0 such that $\vartheta_{\alpha_0} < \Theta$, $\Gamma = \{A \subseteq \mathbb{R} : w(A) < \vartheta_{\alpha_0}\}$,*
2. *$\mathsf{C} = (\mathbb{M}, (P,\Psi), \Gamma^*, A)$ Suslin, co-Suslin captures Γ,*
3. *$\mathbb{M} = (M, \delta, \vec{G}, \Sigma^*)$,*
4. *$\mathsf{hpc} = (\mathcal{M}_\gamma, \mathcal{N}_\gamma, Y_\gamma, \Phi_\gamma, F^+_\gamma, F_\gamma, b_\gamma : \gamma \leq \delta)$ is the output of the $\Gamma - \mathsf{hpc}$ of $\mathbb{M}$,*
5. *$\xi < \delta$ is such that $(\mathcal{M}_\xi, \Phi^+_\xi)$ is a hod pair, $\mathcal{M}_\xi$ is not gentle and $M \vDash (\mathcal{M}_\xi, \Phi_\xi) \in \mathsf{Hp}^\Gamma$.*

Suppose $\xi < \delta$ is such that $(\mathcal{M}_\xi, (\Phi^+_\xi)^{stc})$ is an sts pair and $M \vDash (\mathcal{M}_\xi, \Phi^{stc}_\xi) \in \mathsf{Hp}^\Gamma$. Then $(\Phi^+_\xi)^{stc}$ is commuting.

105 $p[A]$ is the projection of A. In general, the coordinate onto which we are projecting will be clear from the context.

PROOF. The proof is very similar to the proofs we have already given. Set $(\mathcal{P},\Sigma) = (\mathcal{M}_\xi, (\Phi_\xi^+)^{stc})$. We prove that Σ is weakly commuting as the general case is only notationally more complicated. Towards a contradiction assume Σ is not weakly commuting. There is then some ζ_0 such that for some $g \subseteq Coll(\omega,\zeta_0)$, $M[g]$ has a witness to the fact that Σ is not weakly commuting. Let $\zeta_1 = \sup\{\mathrm{lh}(F_\iota) : \iota < \xi\}$. Let $\zeta = \max((\zeta_0^+)^M, (\zeta_1^+)^M)$ and $\vec{G}' = \{H \in \vec{G} : \mathrm{crit}(H) > \zeta\}$.

Fix then $(\mathcal{T},\mathcal{Q}) \in I^b(\mathcal{P},\Sigma)$ and $(\mathcal{U},\mathcal{R}) \in I^b(\mathcal{P},\Sigma)$ such that

1. $\mathcal{R}^b \trianglelefteq_{hod} \mathcal{Q}^b$,
2. $\mathcal{R}^b = cHull^{\mathcal{Q}^b}(\pi^{\mathcal{T},b}[\mathcal{P}^b] \cup \delta^{\mathcal{R}^b})$ and
3. $(\mathcal{T},\mathcal{Q},\mathcal{U},\mathcal{R}) \in M[g]$.

Let $k : \mathcal{P}^b \to \mathcal{R}^b$ be the collapse of $\pi^{\mathcal{T},b}[\mathcal{P}^b]$. We want to see that $k = \pi^{\mathcal{U},b}$.

Let $\mathcal{P}'$ be the last model of $(\mathsf{Le}((\mathcal{P}^b,\Sigma_{\mathcal{P}^b}),\mathcal{J}_\omega)_{>\zeta})^{(M[g],\delta,\vec{G}')}$ and

$$\mathcal{P}^+ = \mathsf{stack}(\mathcal{P}',\Sigma_{\mathcal{P}^b}).$$

Next let E_0 be the $(\delta^{\mathcal{P}^b},\delta^{\mathcal{R}^b})$-extender derived from k and E_1 be the $(\delta^{\mathcal{P}^b},\delta^{\mathcal{R}^b})$-extender derived from $\pi^{\mathcal{U},b}$. We need to show that

(a) $\pi_{E_0} \restriction \mathcal{P}^b = \pi_{E_1} \restriction \mathcal{P}^b$.

Notice that we have that

(1) $\pi_{E_0} \restriction \delta^{\mathcal{P}^b} = \pi_{E_1} \restriction \delta^{\mathcal{P}^b}$.

This is because $k \restriction \delta^{\mathcal{P}^b} = \pi^{\mathcal{T},b} \restriction \delta^{\mathcal{P}^b}$, so if $\pi_{E_0} \restriction \delta^{\mathcal{P}^b} \neq \pi_{E_1} \restriction \delta^{\mathcal{P}^b}$, we have that for some $\mathcal{S} \triangleleft_{hod} \mathcal{P}$, $\mathcal{S}$ is a complete layer of $\mathcal{P}$ and $\Sigma_{\mathcal{S}}$ is not commuting. But since $\Sigma_{\mathcal{S}}$ has strong branch condensation and is fullness preserving[106], Proposition 4.10.2 implies that $\Sigma_{\mathcal{S}}$ is commuting.

Let then $\mathcal{N}_0 = \mathrm{Ult}(\mathcal{P}^+, E_0)$ and $\mathcal{N}_1 = \mathrm{Ult}(\mathcal{P}^+, E_1)$. Notice that it follows from Theorem 4.9.5 that

(2) both $\mathcal{N}_0$ and $\mathcal{N}_1$ are $\Sigma_{\mathcal{R}^b}$-mice over $\mathcal{R}^b$,
(3) for $i \in 2$, $\mathcal{N}_i = \mathsf{stack}(\mathcal{N}_i|\delta,\Sigma_{\mathcal{R}^b})$,
(4) both $\mathcal{N}_0$ and $\mathcal{N}_1$ have $(\delta,\delta+1)$-strategies as $\Sigma_{\mathcal{R}^b}$-mice[107].

Let then $\mathcal{M}$ be a common iterate of $\mathcal{N}_0$ and $\mathcal{N}_1$ (via these strategies). Let $j_0 : \mathcal{N}_0 \to \mathcal{M}$ and $j_1 : \mathcal{N}_1 \to \mathcal{M}$. We then have some κ such that

1. $j_0(\pi_{E_0}(\kappa)) = j_1(\pi_{E_1}(\kappa))$,
2. for $i \in 2$, $\mathrm{crit}(j_i) > \delta^{\mathcal{R}^b}$, and
3. $\mathrm{rge}(j_0 \circ \pi_{E_0}) \cap \mathrm{rge}(j_1 \circ \pi_{E_1})$ contains a κ-club C[108].

Let $D_0 = (j_0 \circ \pi_{E_0})^{-1}[C]$ and $D_1 = (j_1 \circ \pi_{E_1})^{-1}[C]$ and set $D = D_0 \cap D_1$. It follows from universality of $\mathcal{P}^+$, $\mathcal{N}_0$ and $\mathcal{N}_1$ that

[106]See Theorem 4.9.5 and Theorem 4.6.3.
[107]These strategies act on iterations below δ.
[108]See Theorem 4.5.6.

(5) $\mathcal{P}^b \subseteq Hull^{\mathcal{P}^+}(D \cup \delta^{\mathcal{P}^b})$, and for $i \in 2$, $\mathcal{R}^b \subseteq Hull^{\mathcal{N}_i}(\pi_{E_i}[D])$.

Suppose now that $A \in \mathcal{P}^b \cap \wp(\delta^{\mathcal{P}^b})$ and fix $s \in D^{<\omega}$, $t \in [\delta^{\mathcal{P}^b}]^{<\omega}$ and a term φ such that $A = \varphi^{\mathcal{P}^+}[s,t]$. We then have that $\pi_{E_0}(A) = \varphi^{\mathcal{N}_0}(\pi_{E_0}(s), \pi_{E_0}(t))$. It follows that $j_0(\pi_{E_0}(A)) \in \text{rge}(j_1 \circ \pi_{E_1})$. Let $B \in \mathcal{P}^b \cap \wp(\delta^{\mathcal{P}^b})$ be such that $j_0(\pi_{E_0}(A)) = j_1(\pi_{E_1}(B))$. Because $\text{crit}(j_i) > \delta^{\mathcal{R}^b}$, we have that $\pi_{E_0}(A) = \pi_{E_1}(B)$. But (1) now implies that $A = B$. Therefore, $\pi_{E_0}(A) = \pi_{E_1}(A)$. ⊣

It follows from Proposition 4.10.2 that iterates of $(\mathcal{P},\Sigma)$ can be successfully compared with one another. To prove it we simply compare $(\mathcal{Q},\Sigma_{\mathcal{Q}})$ with $(\mathcal{R},\Sigma_{\mathcal{R}})$ by using least-extender-disagreement comparison.

COROLLARY 4.10.7. *Suppose $(\mathcal{P},\Sigma)$ is an allowable pair, Γ is a projectively closed pointclass and Σ has strong branch condensation and is strongly Γ-fullness preserving. Suppose $(\mathcal{T},\mathcal{Q}) \in I^{ope}(\mathcal{P},\Sigma)$ and $(\mathcal{U},\mathcal{R}) \in I^{ope}(\mathcal{P},\Sigma)$. Then there is $(\mathcal{T}_1,\mathcal{Q}_1) \in I^{ope}(\mathcal{Q},\Sigma_{\mathcal{Q}})$ and $(\mathcal{U}_1,\mathcal{R}_1) \in I^{ope}(\mathcal{R},\Sigma_{\mathcal{R}})$ such that $\mathcal{T}_1$ and $\mathcal{U}_1$ are normal stacks and the following holds:*

1. *Suppose $(\mathcal{P},\Sigma)$ is a hod pair or a simple hod pair. Then one of the following holds:*
   (a) $\mathcal{Q}_1 \trianglelefteq_{hod} \mathcal{R}_1$, $\pi^{\mathcal{T}_1}$ exists and $(\Sigma_{\mathcal{R}_1})_{\mathcal{Q}_1} = \Sigma_{\mathcal{Q}_1}$.
   (b) $\mathcal{R}_1 \trianglelefteq_{hod} \mathcal{Q}_1$, $\pi^{\mathcal{U}_1}$ exists and $(\Sigma_{\mathcal{Q}_1})_{\mathcal{R}_1} = \Sigma_{\mathcal{R}_1}$.
   Moreover, if in addition $(\mathcal{T},\mathcal{Q}) \in I(\mathcal{P},\Sigma)$ and $(\mathcal{U},\mathcal{R}) \in I(\mathcal{P},\Sigma)$, then $\mathcal{Q}_1 = \mathcal{R}_1$ and both $\pi^{\mathcal{T}_1}$ and $\pi^{\mathcal{U}_1}$ are defined.
2. *Suppose $(\mathcal{P},\Sigma)$ is an sts hod pair or a simple sts hod pair. Then one of the following holds:*
   (a) $\mathcal{Q}_1 \trianglelefteq_{hod} \mathcal{R}_1$ and $(\Sigma_{\mathcal{R}_1})_{\mathcal{Q}_1} = \Sigma_{\mathcal{Q}_1}$.
   (b) $\mathcal{R}_1 \trianglelefteq_{hod} \mathcal{Q}_1$ and $(\Sigma_{\mathcal{Q}_1})_{\mathcal{R}_1} = \Sigma_{\mathcal{R}_1}$.
   Moreover, if in addition $(\mathcal{T},\mathcal{Q}) \in I(\mathcal{P},\Sigma)$ and $(\mathcal{U},\mathcal{R}) \in I(\mathcal{P},\Sigma)$, then $\mathcal{Q}_1 = \mathcal{R}_1$. Consequently, if Σ is commuting then $\pi^{\mathcal{T}^\frown\mathcal{T}_1,b} = \pi^{\mathcal{U}^\frown\mathcal{U}_1,b}$.

In clause 2 of Corollary 4.10.7, the conclusion $\mathcal{Q}_1 = \mathcal{R}_1$ is a consequence of Γ-fullness preservation and our minimality assumption. If, for example, $\mathcal{Q}_1 \triangleleft_{hod} \mathcal{R}_1$ then there is a $\Sigma_{\mathcal{Q}_1}$-sts $\mathcal{W} \triangleleft \mathcal{R}_1$ such that

1. $\mathcal{W} \models$ "$\delta^{\mathcal{Q}_1}$ is a Woodin cardinal",
2. $\mathcal{J}_\omega[\mathcal{W}] \models$ "$\delta^{\mathcal{Q}_1}$ is not a Woodin cardinal", and
3. $\mathcal{W}$ has a strategy in Γ.

It then follows that $\mathcal{W} \trianglelefteq \mathcal{Q}_1$, contradiction.

The following is a corollary of Corollary 4.13.4 and Theorem 4.10.6.

PROPOSITION 4.10.8. *Suppose $(\mathcal{P},\Sigma)$ is a hod pair or an sts hod pair, Γ is a projectively closed pointclass and Σ has strong branch condensation and is strongly Γ-fullness preserving. Then for some $(\mathcal{T},\mathcal{Q}) \in I(\mathcal{P},\Sigma)$, $\Sigma_{\mathcal{Q},\mathcal{T}}$ is commuting.*

The next lemma will be used in the proof of Theorem 6.1.4.

LEMMA 4.10.9. *Suppose*

1. $(\mathcal{P},\Sigma)$ *is an allowable pair and* $\mathcal{P}$ *is non-meek,*
2. Γ *is a projectively closed pointclass,*
3. Σ *has strong branch condensation and is strongly* Γ*-fullness preserving,*
4. *if* $(\mathcal{P},\Sigma)$ *is an sts hod pair or a simple sts hod pair then* Σ *is commuting,*
5. $(\mathcal{T},\mathcal{Q}) \in I^{ope}(\mathcal{P},\Sigma)$, $(\mathcal{U},\mathcal{R}) \in I^{ope}(\mathcal{P},\Sigma)$ *and* $(\mathcal{W},\mathcal{S}) \in I(\mathcal{R},\Sigma_{\mathcal{R}})$ *are such that* $\mathcal{W}$ *is based on* $\mathcal{R}^b$ *and* $\mathcal{S} \trianglelefteq_{hod} \mathcal{Q}$.

Then $\mathcal{S} = cHull^{\mathcal{Q}^b}(\pi^{\mathcal{T},b}[\mathcal{P}^b] \cup \delta^{\mathcal{S}})$.

PROOF. The proof is an easy corollary of commuting. Let $\mathcal{W}^+ =\uparrow(\mathcal{W},\mathcal{R})$ and let $\mathcal{S}^+$ be the last model of $\mathcal{W}^+$. Notice that $(\mathcal{S}^+)^b = \mathcal{S}$. Let $\mathcal{M}$ be a common iterate of $(\mathcal{Q},\Sigma_{\mathcal{Q}})$ and $(\mathcal{S}^+,\Sigma_{\mathcal{S}^+})$ via respectively $\mathcal{X}$ and $\mathcal{Y}$, which we can find because of Corollary 4.10.7. We have that

(1) $\mathcal{S} = cHull^{\mathcal{M}^b}(\pi^{\mathcal{Y},b}[\mathcal{S}])$ and $\mathcal{Q}^b = cHull^{\mathcal{M}^b}(\pi^{\mathcal{X},b}[\mathcal{Q}^b])$.

It follows from Proposition 4.10.3 and commuting that

(2) $\pi^{\mathcal{T}^\frown\mathcal{X},b} = \pi^{\mathcal{U}^\frown(\mathcal{W}^+)^\frown\mathcal{Y},b}$ and $\pi^{\mathcal{X},b} \restriction \delta^{\mathcal{S}} = \pi^{\mathcal{Y},b} \restriction \delta^{\mathcal{S}}$.[109]

It follows from (1), (2) and the fact that Σ is strongly Γ-fullness preserving that

(3) $\mathcal{S} = cHull^{\mathcal{M}^b}(\pi^{\mathcal{T}^\frown\mathcal{X},b}[\mathcal{P}^b] \cup \pi^{\mathcal{X},b}[\delta^{\mathcal{S}}])$.

Therefore, $\mathcal{S} = cHull^{\mathcal{Q}^b}(\pi^{\mathcal{T},b}[\mathcal{P}^b] \cup \delta^{\mathcal{S}})$. ⊣

4.11. Solidity and condensation

The main contributions of this section are Theorem 4.11.7 and Theorem 4.11.8 that can be used to show that fully backgrounded hod pair constructions are successful, which amounts to showing that clause 4 of Definition 4.3.3 never occurs. We start with the following version of Lemma 4.11.5 for phalanxes that is used in the proof of solidity and universality. We omit the actual proofs of Theorem 4.11.7 and Theorem 4.11.8 as, in the light of Lemma 4.11.6, the proofs of solidity and universality are trivial generalizations of the usual proofs of these facts (see [59, Chapter 5]).

REMARK 4.11.1. This section is devoted to showing that hod pair constructions of a background $(M,\delta,\vec{G})$ converge. We thus think of the hod pairs that appear in the statement of lemmas and propositions of this section as hod pairs constructed by hod pair constructions, and since we would like to show that hod pair constructions are successful, which amounts to showing that clause 4 of Definition 4.3.3 never occurs, the pairs we consider here are models appearing in the intermediate stages of hod pair constructions. This, in particular, means that first of all, we must deal with hod pairs (as opposed to sts hod pairs) and also the hod premouse is non-meek

[109]This equality holds as we have $(\Sigma_{\mathcal{Q}})_{\mathcal{S}} = (\Sigma_{\mathcal{S}^+})_{\mathcal{S}}$.

and not of lsa type, which is a consequence of our minimality assumption. The reason that it is enough to consider non-meek hod premice is that clause 4 of Definition 4.3.3 for meek or gentle type hod premice is not new, and the usual proofs of solidity and condensation can be used.

DEFINITION 4.11.2 (Certified phalanxes). Suppose $(\mathcal{P},\Sigma)$ is a hod pair such that $\mathcal{P}$ is non-meek and $\mathcal{R}$ is a hod premouse. Let π,ζ be such that $\pi:\mathcal{R}\to\mathcal{P}$ is a Σ_1-embedding, and $\zeta\leq\text{crit}(\pi)$. We say $(\mathcal{P},\mathcal{R},\zeta)$ is a $(\pi,\mathcal{P},\Sigma)$-certified phalanx if $\zeta>o(\mathcal{P}^b)$. We also say $(\mathcal{P},\mathcal{R},\zeta)$ is a $(\mathcal{P},\Sigma)$-certified phalanx witnessed by π.

Continuing with the set up of Definition 4.11.2, we let $\pi^+:(\mathcal{P},\mathcal{R},\zeta)\to(\mathcal{P},\mathcal{P},\zeta)$ be given by (id,π), and also, we let Σ^{π^+} be the π^+-pullback of Σ.

LEMMA 4.11.3 (No strategy disagreement). *Suppose $(\mathcal{P},\Sigma)$ is a hod pair such that $\mathcal{P}$ is non-meek, Σ has strong branch condensation and Σ is strongly Γ-fullness preserving for some pointclass Γ that is projectively closed. Suppose $(\mathcal{P},\mathcal{R},\zeta)$ is a $(\mathcal{P},\Sigma)$ certified phalanx as witnessed by $\pi:\mathcal{R}\to\mathcal{P}$. Let $\Lambda=\Sigma^{\pi^+}$. Then no strategy disagreement appears in the comparison of $\mathcal{P}$ and $(\mathcal{P},\mathcal{R},\zeta)$ where Σ is used on the $\mathcal{P}$ side and Λ is used on the $(\mathcal{P},\mathcal{R},\zeta)$ side.*

PROOF. Towards a contradiction suppose not. It follows from the proof of Lemma 4.7.2 that we can find a minimal low level disagreement $((\mathcal{T},\mathcal{Q}),(\mathcal{U},\mathcal{S}),\mathcal{W})$ between Σ and Λ. Let then $E=E^{\mathcal{U}}_{\mathcal{W}}$, be the $\mathcal{W}$-un-dropping extender of $\mathcal{U}$. We have that $\mathcal{W}\lhd_{hod}\text{Ult}(\mathcal{P},E)$. Let now $\mathcal{X}=\pi^+\mathcal{U}$, $\mathcal{P}_1$ be the last model of $\mathcal{X}$, $\sigma:\mathcal{S}\to\mathcal{P}_1$ be the copy map and F be the $\sigma(\mathcal{W})$-un-dropping extender of $\mathcal{X}$. Let $\sigma':\text{Ult}(\mathcal{P},E)\to\text{Ult}(\mathcal{P},F)$ be given by $\sigma'([a,f]_E)=[\sigma(a),f]_E$.

We now have that $(\mathcal{X},\mathcal{P}_1)$ and $((\pi_F\restriction\mathcal{P}^b,\text{Ult}(\mathcal{P},F)^b),(\pi_E\restriction\mathcal{P}^b,\text{Ult}(\mathcal{P},E)^b),\sigma')$ support a $(\sigma(\mathcal{W}),\mathcal{W},\sigma\restriction\mathcal{W})$-b-condensation diagram on $\mathcal{P}$. Because $\sigma\restriction\mathcal{W}$ pullback of $\Sigma_{\sigma(\mathcal{W}),\mathcal{X}}$ is $\Lambda_{\mathcal{W},\mathcal{U}}$, it follows from strong branch condensation that $\Sigma_{\mathcal{W},\mathcal{T}}=\Lambda_{\mathcal{W},\mathcal{U}}$. ⊣

DEFINITION 4.11.4 (Certified pairs). Suppose $(\mathcal{P},\Sigma)$ is a hod pair and $\mathcal{R}$ is a hod premouse such that both $\mathcal{P}$ and $\mathcal{R}$ are of limit type. Suppose that there is π such that $\pi:\mathcal{P}^b\to\mathcal{R}^b$ is elementary. We say the pair $(\pi,\mathcal{R})$ is $(\mathcal{P},\Sigma)$-certified by $(\sigma,\mathcal{T},\mathcal{Q},\mathcal{Q}')$ if

1. $(\mathcal{T},\mathcal{Q})\in I(\mathcal{P},\Sigma)$, $\mathcal{Q}'\trianglelefteq_{hod}\mathcal{Q}$ and $\sigma:\mathcal{R}\to\mathcal{Q}'$ is Σ_1-elementary,
2. $(\mathcal{Q}')^b=Hull^{\mathcal{Q}}(\pi^{\mathcal{T}}[\mathcal{P}^b]\cup\delta^{(\mathcal{Q}')^b})$, and
3. letting $k:\mathcal{P}^b\to(\mathcal{Q}')^b$ be the collapse of $\pi^{\mathcal{T}}[\mathcal{P}^b]$, $k=(\sigma\restriction\mathcal{R}^b)\circ\pi$.

We say $(\mathcal{R},\Lambda)$ is a $(\mathcal{P},\Sigma)$-certified hod pair if for every $(\mathcal{U},\mathcal{S})\in I(\mathcal{R},\Lambda)$, there is π and $(\sigma,\mathcal{T},\mathcal{Q},\mathcal{Q}')$ such that $(\pi,\mathcal{S})$ is $(\mathcal{P},\Sigma)$-certified by $(\sigma,\mathcal{T},\mathcal{Q},\mathcal{Q}')$ and

$$\Lambda_{\mathcal{S}^b,\mathcal{U}}=(\sigma\text{-pullback of }\Sigma_{\mathcal{Q}^b,\mathcal{T}}).$$

LEMMA 4.11.5. *Suppose $(\mathcal{P},\Sigma)$ is a hod pair such that $\mathcal{P}$ is non-lsa type non-meek hod premouse, Γ is a projectively closed pointclass and Σ has strong branch* *condensation and is strongly Γ-fullness preserving. Suppose $(\mathcal{T},\mathcal{R}) \in I^b(\mathcal{P},\Sigma)$*[110] *is such that for some Λ, $(\mathcal{R},\Lambda)$ is $(\mathcal{P},\Sigma)$-certified and there is a Σ_0-elementary embedding $\pi : \mathcal{P} \to \mathcal{R}$ such that $(\pi \restriction \mathcal{P}^b, \mathcal{R})$ is $(\mathcal{P},\Sigma)$-certified by $(\sigma,\mathcal{U},\mathcal{Q},\mathcal{Q}')$. Then $\pi^{\mathcal{T}}$ exists and $\pi^{\mathcal{T}} \leq \pi$.*

PROOF. To implement the usual proof of the Dodd-Jensen property (see [59, Chapter 4.2]), we need to know that

(a) Σ is the π-pullback of Λ.

Because $(\mathcal{R},\Lambda)$ is $(\mathcal{P},\Sigma)$-certified, (a) easily follows from strong branch condensation of Σ. ⊣

LEMMA 4.11.6 (Dodd-Jensen for certified phalanxes). *Suppose Γ is a projectively closed pointclass and $(\mathcal{P},\Sigma)$ is a hod pair such that Σ has strong branch condensation and is strongly Γ-fullness preserving. Suppose that $(\mathcal{P},\mathcal{R},\zeta)$ is a $(\mathcal{P},\Sigma)$-certified phalanx as witnessed by $\pi : \mathcal{R} \to \mathcal{P}$. Suppose that*

1. *$\mathcal{T}$ is a stack on $(\mathcal{P},\mathcal{R},\zeta)$ according to Σ^{π^+} with last model $\mathcal{Q}$,*
2. *$\mathcal{U}$ is a stack on $\mathcal{P}$ according to Σ with last model $\mathcal{S}$, and*
3. *the last branch of $\mathcal{T}$ is on $\mathcal{P}$ and either*
   (a) *$\mathcal{Q} \trianglelefteq_{hod} \mathcal{S}$ and $\pi^{\mathcal{T}}$ exists or*
   (b) *$\mathcal{S} \trianglelefteq_{hod} \mathcal{Q}$ and $\pi^{\mathcal{U}}$ exists.*

Then $\mathcal{Q} = \mathcal{S}$ and $\pi^{\mathcal{T}} = \pi^{\mathcal{U}}$.

PROOF. Let $\mathcal{T}^* = \pi^+\mathcal{T}$. Let $\mathcal{Q}^*$ be the last model of $\mathcal{T}^*$ and let $\sigma : \mathcal{Q} \to \mathcal{Q}^*$ come from the copying construction. Suppose first that $\mathcal{Q} \trianglelefteq_{hod} \mathcal{S}$ and $\pi^{\mathcal{T}}$ exists. Notice next that $\pi^{\mathcal{T}}$-pullback of $(\Sigma_{\mathcal{Q}^*})^{\sigma}$ is Σ. Hence, applying the ordinary proof of the Dodd-Jensen property we get that $\mathcal{S} = \mathcal{Q}$, $\pi^{\mathcal{U}}$ exists and $\pi^{\mathcal{U}} \leq \pi^{\mathcal{T}}$.

Suppose now $\mathcal{S} \trianglelefteq_{hod} \mathcal{Q}$ and $\pi^{\mathcal{U}}$ exists. Notice that $\pi^{\mathcal{U}}$ induces an embedding $\pi^* : (\mathcal{P},\mathcal{R},\zeta) \to (\mathcal{S},\mathcal{S},\pi^{\mathcal{U}}(\zeta))$ such that $\pi^* \restriction \mathcal{P} = \pi^{\mathcal{U}}$ and $\pi^* \restriction \mathcal{R} = \pi^{\mathcal{U}} \circ \pi$. Notice that

(1) $\Sigma^{\pi^+} = (\pi^*\text{-pullback of } \Sigma_{\mathcal{S}})$.

Applying Lemma 4.11.5 to the embedding $(\sigma \restriction \mathcal{S}) \circ \pi^{\mathcal{U}}$ and $(\mathcal{T}^*, \mathcal{Q}^*)$, we get

(2) $\sigma(\mathcal{S}) = \mathcal{Q}^*$, $\pi^{\mathcal{T}^*}$ exists and $\pi^{\mathcal{T}^*} \leq \sigma \circ \pi^{\mathcal{U}}$.

(2) now implies that $\mathcal{S} = \mathcal{Q}$. Since $\pi^{\mathcal{T}^*} = \sigma \circ \pi^{\mathcal{T}}$, we have that $\pi^{\mathcal{T}}$ exists and $\pi^{\mathcal{T}} \leq \pi^{\mathcal{U}}$. Putting the two arguments together we see that $\pi^{\mathcal{U}} = \pi^{\mathcal{T}}$. ⊣

It is clear that it follows from Lemma 4.11.6 and from Lemma 4.11.3 that the usual proofs of condensation, universality and solidity go through for hod mice. We state the results without proofs (see [59, Chapter 5] for the usual proofs of these results.)

THEOREM 4.11.7 (Solidity and universality). *Suppose Γ is a projectively closed pointclass, $k < \omega$ and $(\mathcal{P},\Sigma)$ is a hod pair such that*

[110] Thus, $\pi^{\mathcal{T},b}$ exists, see Definition 2.7.21.

1. $\mathcal{P}$ *is k-sound non-meek hod premouse,*
2. $\mathcal{P}$ *is not of lsa type and* $\rho(\mathcal{P}) > o(\mathcal{P}^b)$*, and*
3. Σ *is strongly* Γ*-fullness preserving and has strong branch condensation.*

Let r be the $k+1$*st standard parameter of* $(\mathcal{P}, u_k(\mathcal{P}))$*; then r is* $k+1$*-solid and* $k+1$*-universal over* $(\mathcal{P}, u_k(\mathcal{P}))$.

THEOREM 4.11.8 (Condensation). *Suppose* Γ *is a projectively closed pointclass and* $(\mathcal{P}, \Sigma)$ *is a hod pair such that*

1. $\mathcal{P}$ *is non-meek hod premouse,*
2. $\mathcal{P}$ *is not of lsa type and* $\rho(\mathcal{P}) > o(\mathcal{P}^b)$*, and*
3. Σ *is strongly* Γ*-fullness preserving and has strong branch condensation.*

Suppose $(\mathcal{P}, \mathcal{R}, \zeta)$ *is a* $(\mathcal{P}, \Sigma)$ *certified phalanx as witnessed by* $\pi : \mathcal{R} \to \mathcal{P}$ *such that* $\zeta = \mathrm{crit}(\pi) = \rho_\omega^{\mathcal{R}}$*. Then either*

1. $\mathcal{R} \trianglelefteq_{hod} \mathcal{P}$ *or*
2. *there is an extender E on the sequence of* $\mathcal{P}$ *such that* $\mathrm{lh}(E) = \rho_\omega^{\mathcal{R}}$ *and* $\mathcal{R} \trianglelefteq_{hod} \mathrm{Ult}(\mathcal{P}, E)$.

4.12. Backgrounded constructions relative to st-strategies

In this section, we show that if $(\mathcal{P}, \Sigma)$ is an sts pair constructed by a hod pair construction then if the fully backgrounded construction relative to Σ breaks down then it does so because it reaches an sts mouse that destroys the Woodiness of $\delta^{\mathcal{P}}$. The reader may find it helpful to review Definition 3.7.3, Definition 3.7.4, Definition 3.7.5, clause 5 of Definition 3.8.9 and Definition 3.8.16.

THEOREM 4.12.1. *Assume* AD^+*. Suppose*

1. *for some* α *such that* $\vartheta_\alpha < \Theta$, $\Gamma = \{A \subseteq \mathbb{R} : w(A) < \vartheta_\alpha\}$*,*
2. $\mathsf{C} = (\mathbb{M}, (P, \Psi), \Gamma^*, A)$ *Suslin, co-Suslin captures* Γ *and*
3. $\mathbb{M} = (M, \delta, \vec{G}, \Sigma^*)$*,*
4. $\mathsf{hpc} = (\mathcal{M}_\gamma, \mathcal{N}_\gamma, Y_\gamma, \Phi_\gamma, F_\gamma^+, F_\gamma, b_\gamma : \gamma \leq \delta)$ *is the output of the* $\Gamma-\mathsf{hpc}$ *of* M*,*
5. $\xi < \delta$ *is such that* $(\mathcal{M}_\xi, \Phi_\xi)$ *is an sts pair and* $M \vDash (\mathcal{M}_\xi, \Phi_\xi) \in \mathsf{Hp}^\Gamma$.

Set $(\mathcal{M}_\xi, \Phi_\xi^+) = (\mathcal{P}, \Sigma)$ *and let* $\zeta \geq \sup\{\mathrm{lh}(F_\nu^+) : \nu < \xi\}$*. Let*

$$(\mathsf{Le}((\mathcal{P}, \Sigma^{stc}), \mathcal{J}_\omega[\mathcal{P}])_{>\zeta})^M = (\mathcal{P}_\gamma, \mathcal{P}'_\gamma, X_\gamma^+, X_\gamma, b_\gamma : \gamma < \eta).$$

Suppose there is ξ_0 *such that the anomaly stated in clause 3(b) of Definition 4.2.1 occurs at* ξ_0*. Then* $\mathcal{P}_{\xi_0} \vDash$ *"*$\delta^{\mathcal{P}}$ *is not a Woodin cardinal".*

Consequently, if $\mathsf{Lp}^{\Gamma, \Sigma^{stc}}(\mathcal{P}) \vDash$ *"*$\delta^{\mathcal{P}}$ *is a Woodin cardinal" then the anomaly stated in clause 3(b) of Definition 4.2.1 does not occur.*

PROOF. Suppose that $\mathcal{P}_{\xi_0} \models$ "$\delta^{\mathcal{P}}$ is a Woodin cardinal". Set $\mathcal{S} = (\mathcal{P}_{\xi_0})_{\text{ex}}$[111]. We assume that ξ is the least such that our claim fails for $(\mathcal{M}_\xi, \Phi_\xi^+)$. Suppose $\mathcal{S} = (\mathcal{J}^{\vec{E},f}_{\omega\alpha_0}, \in, \vec{E}, f)$, $\alpha_0 = \beta_0 + \gamma_0$ and $t = (\mathcal{P}, \mathcal{T}) \in \lfloor \mathcal{S}|\omega\beta \rfloor \cap \text{dom}(\Sigma)$ is such that setting $w = (\mathcal{J}_\omega(t), t, \in)$, w is (f, sts)-minimal as witnessed by β_0[112], $\mathcal{T}$ is $\mathcal{P}|\omega\beta_0$-terminal[113] and $\gamma_0 = \text{lh}(t)$. Set $b = \Sigma(\mathcal{T})$, $p = \uparrow(\mathcal{T}, \mathcal{S})$, $\mathcal{P}_1 = \text{m}^+(\mathcal{T})$ and

$$\mathcal{S}_1 = \begin{cases} \mathcal{M}_b^p & \text{if } \mathcal{Q}(b,p) \text{ does not exist,} \\ \mathcal{Q}(b,p) & \text{otherwise.} \end{cases}$$

Notice that if $\mathcal{J}_\omega[\mathcal{S}] \models$ "$\delta^{\mathcal{P}}$ is not a Woodin cardinal" then $\mathcal{Q}(b,p)$ is defined.

We then have that

$$\mathcal{P}'_{\xi_0+1} = (\mathcal{J}^{\vec{E},f^+}_{\omega\beta_0+\omega\gamma_0}, \in, \vec{E}, f, \tilde{b})$$

where $\tilde{b} \subseteq \omega\beta_0 + \omega\gamma_0$ is defined by $\omega\beta_0 + \omega\nu \in \tilde{b} \iff \nu \in b$. Since we are assuming that the anomaly occurs, we have that there is $e \in \mathcal{S}|\omega\beta_0$ such that $\mathcal{S}|\omega\beta_0 \models \text{sts}_0(t,e)$[114] and $e \neq b$. Let Φ be the strategy of $\mathcal{S}$ induced by Σ^*[115]. Notice that because $\delta^{\mathcal{P}}$ is a cutpoint in $\mathcal{S}$, Φ extends Σ.

SUBLEMMA 4.12.2. *Whenever $\mathcal{S}'$ is a Φ-iterate of $\mathcal{S}$ via a stack that is above $\delta^{\mathcal{P}}$, $\mathcal{S}'$ is a Σ^{stc}-sts premouse over $\mathcal{P}$. Thus, any two Φ-iterates of $\mathcal{S}$ can be compared to each other.*

Similarly if $\mathcal{U}$ is a generalized stack on $\mathcal{S}$ according to Φ such that $\mathcal{U}$ is based on $\mathcal{P}$ and has a last normal component[116] *$\mathcal{U}'$, $d = \Sigma(\mathcal{U})$, $\mathcal{S}' = \mathcal{M}_d^{\mathcal{U}}$ and $\text{m}^+(\mathcal{U}') \in Y^{\mathcal{S}'}$ is #-lsa like then whenever $\mathcal{S}''$ is a $\Phi_{\mathcal{S}',\mathcal{U}^\frown\{d\}}$-iterate of $\mathcal{S}'$ via a stack that is above $\delta(\mathcal{U}')$, $\mathcal{S}''$ is a $\Sigma^{stc}_{\text{m}^+(\mathcal{U}'),\mathcal{U}}$-sts premouse over $\text{m}^+(\mathcal{U}')$.*

PROOF. This follows from hull condensation of Σ. We do the proof for stacks, and the more general proof is only notationally more complicated. Suppose $\mathcal{U}$ is a stack on $\mathcal{S}$ according to Φ with last model $\mathcal{S}'$. Let $\mathcal{U}^*$ be the resurrection of $\mathcal{U}$ onto M and let M^* be the last model of $\mathcal{U}^*$[117]. Set

$$\pi^{\mathcal{U}^*}((\text{Le}((\mathcal{P}, \Sigma^{stc}), \mathcal{J}_\omega[\mathcal{P}])_{>\zeta})^M) = (\mathcal{R}_\tau, \mathcal{R}'_\tau, Z_\tau^+, Z_\tau, c_\tau : \tau < \eta').$$

We then have some $\iota \leq \pi^{\mathcal{U}^*}(\xi_0)$ and $\sigma : \mathcal{S}' \to \mathcal{R}_\iota$. Let $s = (\mathcal{P}, \mathcal{X}_0, \mathcal{P}_1, \mathcal{X}_1)$ be an indexable stack such that $s \in \text{dom}(\Sigma^{\mathcal{S}'}) \cap \text{dom}(\Sigma^{stc})$. We want to see that

$$\Sigma^{stc}(s) = \Sigma^{\mathcal{S}'}(s).$$

---

[111] See Definition 2.7.3.

[112] See Definition 2.3.3. In particular, this means that we have to index the branch of t at $\omega\alpha_0$.

[113] See Definition 3.8.8.

[114] See Definition 3.8.16. This means that e is the branch of t we must choose.

[115] Notice that $\mathcal{P}$ is constructed in $M|\zeta$ and $\mathcal{S}$ above $\delta^{\mathcal{P}}$ is constructed using extenders with critical points $> \zeta$. It follows that Σ^* indeed induces a strategy Φ for $\mathcal{S}$ via the ordinary resurrection procedure of [23, Chapter 12]. See also Section 4.3.1.

[116] See Notation 2.4.4.

[117] See [23, Chapter 12] and also Section 4.3.1.

Let $d = \Sigma^{\mathcal{S}'}(s)$. Notice next that $s^\frown\{d\}$ is hull of $\sigma(s)^\frown\{\sigma(d)\}$ and that

$$\sigma(s)^\frown\{\sigma(d)\}$$

is according to Σ[118]. But Σ has hull condensation (see [29, Lemma 2.9]), implying that $\Sigma(s) = d$.

The second part of the claim is very similar. This time, letting $\mathcal{U}^*$ be the resurrection of $\mathcal{U}$, we have $\sigma : \mathcal{S}' \to \mathcal{Q}_\iota$ where

$$\pi^{\mathcal{U}^*}(\mathsf{hpc}) = (\mathcal{Q}_\iota, \mathcal{Q}'_\iota, Z_\iota, \Omega_\iota, H^+_\iota, H_\iota, e_\iota : \iota \leq \pi^{\mathcal{U}^*}(\delta))$$

is the output of the $\Gamma-\mathsf{hpc}$ of the last model of $\mathcal{U}^*$ and $\iota \leq \pi^{\mathcal{U}^*}(\xi)$. But now we repeat the same argument as before noting that $\Sigma^{stc}_{\mathrm{m}^+(\mathcal{U}'),\mathcal{U}}$ is the short-tree component of the σ-pullback of Ω^+_ι. $\dashv$

We now have that $\mathcal{Q}(e, \mathcal{T})$ exists and it is an sts premouse over $\mathrm{m}^+(\mathcal{T})$[119]. Set $\mathcal{Q} = \mathcal{Q}(e, \mathcal{T})$. Notice that because w is (f, sts)-minimal, we must have that for some $\tau_0 < \alpha_0$, $\rho(\mathcal{S}||\tau_0) \leq \delta(\mathcal{T})$ and $\mathcal{S}||\tau_0 \vDash \mathsf{sts}_0(t, e)$. Let τ_1 witnesses that $\mathcal{S}|\tau_0 \vDash \mathsf{sts}_0(t, e)$. Set $\mathcal{S}' = \mathcal{S}||\tau_0$.

Using Lemma 4.1.12, we can find a self-capturing background $(M_0, \delta_0, \vec{G}_0, \Sigma^*_0)$ which Suslin, co-Suslin captures $\mathsf{Code}(\Sigma^*)$ and

$$(\mathsf{HC}^{M_0}, \mathsf{Code}(\Sigma^*), \in) \prec^{\mathbb{R}^{M_0}} (\mathsf{HC}, \mathsf{Code}(\Sigma^*), \in)$$

where $\prec^Z$ means elementarity with respect to parameters in Z.

Let λ be the supremum of the first ω-Woodin cardinals of $\mathcal{S}'|\omega\tau_1$. Let $h \subseteq Coll(\omega, \mathbb{R}^{M_0})$ be generic and let $\mathcal{S}''$ be an $\mathbb{R}^{M_0}$-genericity iterate of $\mathcal{S}'$ via $\Phi_{\mathcal{S}'}$. Thus, we have a stack $\mathcal{U} \in M_0[h]$ on $\mathcal{S}'$ such that the following holds in $M_0[h]$:

1. $\mathrm{lh}(\mathcal{U}) = \omega_1^{M_0} + 1$,
2. $\mathcal{S}''$ is the last model of $\mathcal{U}$,
3. for every $\alpha < \omega_1^{M_0}$, $\mathcal{U}_{\leq\alpha} \in M_0$,
4. $D^{\mathcal{U}} = \emptyset$,
5. $\pi^{\mathcal{U}}(\lambda) = \omega_1^{M_0}$,
6. for some $\mathcal{S}''$-generic $k \subseteq Coll(\omega, <\omega_1^{M_0})$, $\mathbb{R}^{M_0}$ is the set of symmetric reals of $\mathcal{S}''[k]$.

We let $N = D(\mathcal{S}'', \omega_1^{M_0}, k)$[120]. We now have a strategy $\Lambda \in N$ for $\mathcal{Q}$ and some v with the property that

1. $N \vDash$ "Λ is an ω_1-iteration strategy" and

---

[118] Set $s' = \sigma(s)^\frown\{\sigma(d)\}$. We have that $\mathcal{P}|\delta^{\mathcal{P}}$ is constructed inside $M|\zeta$ while the construction producing $\mathcal{S}$ and $\mathcal{S}''$ uses extender with critical points $> \zeta$. It follows that $\mathcal{U}^*$ is above $\zeta + 1$ while Σ is determined by the pair $(M|\zeta, \Sigma^*_{M|\zeta}) = (M^*|\zeta, (\Sigma^*_{M^*})_{M|\zeta})$. Thus, $(\sigma^*(\Phi_\xi))^+ = \Sigma$. Notice also that it follows from the elementarity of $\pi^{\mathcal{U}^*}$ that $\mathcal{R}_\iota$ is indeed a Σ^{stc}-sts as the first time this breaks down is at $\pi^{\mathcal{U}^*}(\xi_0)$. Thus, any indexable stack that has been indexed in $\mathcal{R}_\iota$ is according to Σ^{stc}.

[119] See Definition 3.1.4.

[120] This is the derived model of $\mathcal{S}''$ as computed by k. See Section 3.8. We need to work inside M_0 to guarantee that $\mathcal{S}'' \in V$.

2. whenever $\mathcal{R} \in N$ is a Λ-iterate of $\mathcal{Q}$ above $\delta(\mathcal{T})$ and $s \in \mathcal{R}$ is an indexable stack on $\mathcal{P}_1 = \mathrm{m}^+(\mathcal{T})$ according to $\Sigma^{\mathcal{R}}$,

$$\mathcal{S}''[k] \models \text{“}s \text{ is } (\mathcal{P}, \Sigma^{\mathcal{S}''|\nu})\text{-authenticated”}.$$[121]

Notice that $N \in M_0$[122] implying that $\Lambda \in M_0$. Moreover, because Code(Λ) is projective in Code(Σ^*), we have that Code(Λ) is δ-universally Baire in M_0. Thus, Λ also acts on length $\omega_1^{M_0}$ iterations.

SUBLEMMA 4.12.3. *There is a Λ-iterate $\mathcal{Q}^*$ of $\mathcal{Q}$ and a $\Phi_{\mathcal{S}_1,p}$-iterate $\mathcal{S}^*$ of $\mathcal{S}_1$ such that $\mathcal{S}^* = \mathcal{Q}^*|\mathrm{ord}(\mathcal{S}^*)$ and $\mathcal{Q}^*||\mathrm{ord}(\mathcal{S}^*)$ is not a $\Sigma^{stc}_{\mathrm{m}^+(\mathcal{T}),\mathcal{T}}$-sts premouse over $\mathrm{m}^+(\mathcal{T})$ (implying that $\mathcal{Q}^*|\mathrm{ord}(\mathcal{S}^*) \neq \mathcal{Q}^*||\mathrm{ord}(\mathcal{S}^*)$).*

PROOF. Our goal now is to compare $\mathcal{S}_1$ with $\mathcal{Q}$. We use $\Phi_{\mathcal{S}_1,p}$ for $\mathcal{S}_1$ and Λ for $\mathcal{Q}$. Assume for a moment that the comparison is successful. If $\mathcal{S}_1 = \mathcal{Q}(b,p)$ then we in fact have that $\mathcal{Q}(b,p) = \mathcal{Q}$ and since $\mathcal{Q} = \mathcal{Q}(e,\mathcal{T})$, we get that $b = e$, which is a contradiction. Hence, $\mathcal{S}_1 = \mathcal{M}_b^p$ and $\mathcal{Q}(b,p)$ does not exits (and hence π_b^p is defined). In this case, we must have that $\mathcal{S}_1$ loses the comparison and if $(\mathcal{S}^*,\mathcal{Q}^*)$ are the last models of the comparison then $\mathcal{S}^* \lhd \mathcal{Q}^*$[123]. Because the $\mathcal{S}_1$-side loses we have that the iteration embedding $j_0 : \mathcal{S}_1 \to \mathcal{S}^*$ is defined. Let then $j = j_0 \circ \pi_b^p : \mathcal{S} \to \mathcal{S}^*$.

We now argue that $\mathcal{Q}^*$ is as desired. To show this we only need to show that $\mathcal{Q}^*||\mathrm{ord}(\mathcal{S}^*)$ is not a $\Sigma^{stc}_{\mathrm{m}^+(\mathcal{T}),\mathcal{T}}$-sts premouse over $\mathrm{m}^+(\mathcal{T})$. Notice that our sts-indexing scheme is so that if $\mathcal{W}$ is an sts premouse properly extending $\mathcal{S}^*$ and $\alpha^* = \mathrm{ord}(\mathcal{S}^*)$ then $(j(\mathcal{T}), j(e))$ has to be indexed in $\mathcal{W}$ at α^*. Thus, $(j(\mathcal{T}), j(e))$ must be indexed at α^* in $\mathcal{Q}^*$. Assume then that $(j(\mathcal{T}), j(e))$ is according to $\Sigma^{stc}_{\mathrm{m}^+(\mathcal{T}),\mathcal{T}}$. Because $\mathcal{T}^\frown\{e\}$ is a hull of $\mathcal{T}^\frown\{b\}^\frown j(\mathcal{T})^\frown\{j(e)\}$ as witnessed by j, it follows from hull condensation[124] of Σ that $e = \Sigma(\mathcal{T})$, contradiction.

Thus, we must have that the comparison between $\mathcal{S}_1$ and $\mathcal{Q}$ does not terminate. But then Sublemma 4.12.2 implies that this can only happen because the comparison of $\mathcal{S}_1$ and $\mathcal{Q}$ produces an iterate $\mathcal{Q}^*$ of $\mathcal{Q}$ which is not a $\Sigma^{stc}_{\mathcal{P}_1,\mathcal{T}}$-sts over $\mathcal{P}_1$. Let then $\mathcal{S}^*$ be the corresponding iterate of $\mathcal{S}_1$. We thus have some $\iota < \mathrm{ord}(\mathcal{S}^*)$ such that

(1) $\iota \notin \mathrm{dom}(\vec{E}^{\mathcal{S}^*}) \cap \mathrm{dom}(\vec{E}^{\mathcal{Q}^*})$ and $\mathcal{S}^*|\iota = \mathcal{Q}^*|\iota$ but $\mathcal{S}^*||\iota \neq \mathcal{Q}^*||\iota$.

Let $t_1 = (\mathcal{P}_1, \mathcal{T}_1, \mathcal{P}_1', \mathcal{T}_1') \in \mathrm{dom}(\Sigma^{\mathcal{Q}^*}) \cap \mathrm{dom}(\Sigma^{\mathcal{S}^*})$ be such that $\Sigma^{\mathcal{S}^*}(t_1) \neq \Sigma^{\mathcal{Q}^*}(t_1)$. Assume first that $\mathcal{T}_1'$ is defined. In this case we have that $\pi^{\mathcal{T}_1,b}$ is defined and $\mathcal{T}_1'$ is based on $(\mathcal{P}_1')^b$. We then have that $((\mathcal{P}_1')^b, \mathcal{T}_1')$ is $(\mathcal{P}, \Sigma^{\mathcal{S}''|\nu})$-authenticated

[121] See Definition 3.8.9.

[122] This follows from the fact that $\rho(\mathcal{S}') \leq \delta(\mathcal{T})$ and from Sublemma 4.12.2. See [57, Proposition 3.0.1].

[123] Equality is not possible because $\mathcal{S}^*$ is not a $\mathcal{Q}$-structure for $\mathcal{T}$.

[124] See [29, Lemma 2.9].

iteration[125]. Suppose $\mathcal{Y}$ authenticates $((\mathcal{P}_1')^b, \mathcal{T}_1')$. Set $\mathcal{R} = (\mathcal{P}_1')^b$ and let $\mathcal{W}$ be the last model of $\mathcal{Y}$. Notice that $\mathcal{T}^\frown\mathcal{T}_1$ is according to Σ^{stc} and moreover, $\mathcal{R} = \pi^{\mathcal{T}^\frown\mathcal{T}_1, b}(\mathcal{P}^b)$.

It follows that there is $\pi : \mathcal{P}^b \to \mathcal{R}$ and $\sigma : \mathcal{R} \to \mathcal{W}^b$ such that $\pi^{\mathcal{Y},b} = \sigma \circ \pi$ and $\sigma\mathcal{T}_1'$ is according to $(\Sigma^{\mathcal{S}''|\nu})_{\mathcal{W}^b,\mathcal{Y}}$[126]. It follows from Sublemma 4.12.2 that $(\Sigma^{\mathcal{S}''|\nu})_{\mathcal{W}^b,\mathcal{Y}} \subseteq \Sigma_{\mathcal{W}^b,\mathcal{Y}}$. Applying Theorem 4.9.5 to $(\mathcal{Y}, \mathcal{W}^b, \mathcal{R}, \mathcal{R}, \sigma)$ and $\mathcal{T}^\frown\mathcal{T}_1$, we get that $\mathcal{T}_1'$ is according to $\Sigma_{\mathcal{R},\mathcal{T}^\frown\mathcal{T}_1}$. Hence, we must have that $\mathcal{T}_1' = \emptyset$.

We thus have that $t_1 = (\mathcal{P}_1, \mathcal{T}_1)$. If $\mathcal{T}_1$ is uvs[127] then by arguing as above we once again prove that $\mathcal{T}_1$ is according to $\Sigma_{\mathcal{P}_1,\mathcal{T}}$. Assume then that $\mathcal{T}_1$ is nuvs. It follows that $\mathrm{m}^+(\mathcal{T}_1)$ is a #-lsa type hod premouse. In this case, our sts scheme guarantees that there are branches $c_1 \in \mathcal{S}^*$ and $c_2 \in \mathcal{Q}^*$ such that $(\mathcal{T}_1, c_1)$ is indexed at ι in $\mathcal{S}^*$ and $(\mathcal{T}_1, c_2)$ is indexed at ι in $\mathcal{Q}^*$. But because $\mathcal{S}^*|\iota = \mathcal{Q}^*|\iota$, we must have that $c_1 = c_2$ as what branch is indexed at ι in either of the models depends solely on $\mathcal{S}^*|\iota = \mathcal{Q}^*|\iota$ and not on any external factors. We thus have that $\mathcal{S}^*||\iota = \mathcal{Q}^*||\iota$ contradicting (1). This contradiction implies that in fact the comparison between $\mathcal{S}_1$ and $\mathcal{Q}$ is successful. $\dashv$

Let then $\mathcal{Q}^*$ and $\mathcal{S}^*$ be as in the sublemma above. Because $\mathcal{Q}^*$ wins the coiteration we have that the iteration embedding $j_0 : \mathcal{S}_1 \to \mathcal{S}^*$ exists. j_0 is according to $\Phi_{\mathcal{S}_1,p}$. As we have argued in the proof of the sublemma, we must also have that π_b^p exists. Set then $j = j_0 \circ \pi_b^p$. As in the proof of the sublemma, the pair $(j(\mathcal{T}), j(e))$ must be indexed in $\mathcal{Q}^*$ at $\mathrm{ord}(\mathcal{S}^*)$.

It then follows that setting $\mathcal{T}_1 = j(\mathcal{T})$, $e_1 = j(e)$ and $\mathcal{Q}_1 =_{\text{def}} (j(\mathcal{Q}(e,\mathcal{T})))_{\text{ex}} = (\mathcal{Q}(e_1,\mathcal{T}_1))_{\text{ex}}$[128], $\mathcal{Q}_1$ is $(\mathcal{P}, \Sigma^{\mathcal{S}''|\nu})$-authenticated. This means that we can find a normal stack $\mathcal{Y}$ on $\mathcal{P}$ with last model $\mathcal{W}$ and an ordinal ι such that for some normal stack $\mathcal{X}$ on $\mathcal{Q}_1$,

(2) $\mathcal{X}$ is based on $\mathcal{P}_2 =_{\text{def}} \mathrm{m}^+(j(\mathcal{T}))$,
(3) $\mathcal{W}||\iota$ is the last model of $\mathcal{X}$ and $\pi^{\mathcal{X}}$ is defined[129],
(4) $\mathcal{W}||\iota$ is a $\Sigma^{stc}_{\mathcal{W}_0,\mathcal{Y}}$-sts over $\mathcal{W}_0$ where $\mathcal{W}_0 =_{\text{def}} (\mathcal{W}||\iota)_{\#}$.

(4) follows from the fact that $\mathcal{S}''$ is a Σ^{stc}-sts premouse over $\mathcal{P}$ (see Sublemma 4.12.2). We claim that

(5) $\mathcal{X}$ is according to $\Sigma^{stc}_{\mathcal{P}_2,\mathcal{T}^\frown\mathcal{T}_1}$.

(5) follows from Sublemma 4.12.4. Assuming (5) we finish the argument. Let

1. $\mathcal{X}_0 = \downarrow(\mathcal{X}, \mathcal{P}_2)$
2. $p_1 = \uparrow(\mathcal{T}_1, \mathcal{S}_1)$,
3. $b_1 = \Sigma(p^\frown\{b\}^\frown p_1)$,

[125] In fact, an authentication exists in $\mathcal{S}''[k]$. See Definition 3.7.3.
[126] Recall that $\Sigma^{\mathcal{X}}$ is the strategy predicate of $\mathcal{X}$.
[127] See Definition 3.3.2.
[128] See Definition 2.7.3.
[129] See Definition 3.7.3. Clause 1 applies to our current situation.

4. $\mathcal{S}_2 = \mathcal{M}_{b_1}^{p^\frown\{b\}^\frown p_1}$,
5. $\mathcal{X}' = \uparrow(\mathcal{X}_0, \mathcal{S}_2)$.

Arguing just like for $(b, \mathcal{S}_1)$ we have that $\mathcal{Q}(b_1, p^\frown\{b\}^\frown p_1)$ does not exist and $\pi_{b_1}^{p^\frown\{b\}^\frown p_1}$ is defined. It follows from (5) that $\mathcal{X}'$ is according to $\Phi_{\mathcal{S}_2,q}$ where $q = p^\frown\{b\}^\frown p_1^\frown\{b_1\}$. Let $\mathcal{S}_3$ be the last model of $\mathcal{X}'$. Because $\pi^{\mathcal{X}'}$ is defined and because $\mathcal{J}_\omega[\mathcal{W}||\iota] \vDash$ "$\delta(\mathcal{X}_0)$ is not a Woodin cardinal", we have that there is a normal $\Phi_{\mathcal{S}_3,q^\frown\mathcal{X}'}$-iterate $\mathcal{S}_4$ of $\mathcal{S}_3$ and a normal $\Sigma_{\mathcal{W}||\iota,\mathcal{Y}}$-iterate $\mathcal{W}'$ of $\mathcal{W}||\iota$ such that $\mathcal{S}_4 \lhd \mathcal{W}'$ and the iteration embedding $k : \mathcal{S} \to \mathcal{S}_4$ exists. Because $\mathcal{W}'$ is a Σ-iterate of $\mathcal{P}$, we have that $(k(\mathcal{T}), k(e))$, which according to our sts indexing scheme must be indexed in $\mathcal{W}'$, is according to Σ. It then follows that $\mathcal{T}^\frown\{e\}$ is a hull of $\mathcal{T}^\frown\{b\}^\frown\mathcal{T}_1^\frown\{b_1\}^\frown\mathcal{X}_0^\frown k(\mathcal{T})^\frown\{k(e)\}$ implying that $b = e$[130].

This completes the proof of the theorem assuming (5). Sublemma 4.12.4 implies (5). To simplify the matter, the symbols used in the statement of Sublemma 4.12.4, with the exception of $(\mathcal{P}, \Sigma)$, do not have the same meaning as the same symbols in the proof given above.

SUBLEMMA 4.12.4. *Suppose*

1. *$\mathcal{X}$ is a generalized stack on $\mathcal{P}$ according to Σ^{stc},*
2. *$\mathcal{X}$ has a last normal component $\mathcal{X}'$ with the property that $\mathcal{R} = \mathrm{m}^+(\mathcal{X}')$ is a #-like lsa type hod premouse,*
3. *$\mathcal{T}$ is a stack on $\mathcal{P}$ according to Σ^{stc} that authenticates $\mathcal{R}$.*

Let $\mathcal{S}$ be the last model of $\mathcal{T}$ and let $\mathcal{U}$ be a normal stack on $\mathcal{R}$ witnessing that $\mathcal{T}$ authenticates $\mathcal{R}$[131]*. Then $\mathcal{U}$ is according to $\Sigma^{stc}_{\mathcal{R},\mathcal{X}}$.*

PROOF. The proof uses ideas from Lemma 2.10.15, Theorem 4.6.5 and Theorem 4.9.5. We will use Lemma 2.10.15 and Theorem 4.6.5 to conclude that $\mathcal{U}$ picks the branches according to $\Sigma^{stc}_{\mathcal{R},\mathcal{X}}$ in successor windows. Because the proofs are very similar to the proofs already given in the above mentioned theorems, we will sketch the arguments.

Set $\eta + 1 = \mathrm{lh}(\mathcal{U})$. Suppose $\alpha \leq \eta$ is a limit ordinal and $\mathcal{U}_{<\alpha}$ is according to $\Sigma^{stc}_{\mathcal{R},\mathcal{X}}$. We want to see that

(a) $[0,\alpha)_\mathcal{U} = \Sigma^{stc}_{\mathcal{R},\mathcal{X}}(\mathcal{U}_{<\alpha})$.

Let $\mathcal{S}' \trianglelefteq_{hod} \mathcal{S}$ be the longest such that $\mathcal{S}' \trianglelefteq \mathrm{m}(\mathcal{U}_{<\alpha})$ and either $\mathcal{S}'$ is a layer of $\mathcal{S}$ or a limit of layers of $\mathcal{S}$. There are two essential cases.

Case 1: $\mathcal{S}' \lhd_{hod} \mathcal{S}^b$ is a complete layer[132] of $\mathcal{S}^b$.

Let $\mathcal{S}''$ be the least layer of $\mathcal{S}^b$ such that $\mathcal{S}' \lhd \mathcal{S}''$ and $\delta^{\mathcal{S}''}$ is a Woodin cardinal of $\mathcal{S}^b$. If now $\delta(\mathcal{U}_{<\alpha}) < \delta^{\mathcal{S}''}$ then (a) follows from fullness preservation of Σ[133].

---

[130]Notice that $\mathcal{W}||\iota$-to-$\mathcal{W}'$ iteration is above $\delta(\mathcal{X}_0)$.

[131]See Definition 3.7.3.

[132]See Definition 2.7.14.

[133]Notice that Theorem 4.9.5 implies that $\Sigma_{\mathcal{S}',\mathcal{T}} = \Sigma_{\mathcal{S}',\mathcal{X}^\frown\mathcal{U}}$. This is because we can apply Theorem 4.9.5 to $\mathcal{Q} =_{\mathrm{def}} \mathcal{S}'$, $E =$ (the $(\delta^{\mathcal{P}^b}, \delta^{\mathcal{Q}})$-extender derived from $\pi^{\mathcal{T},b}$), $\mathcal{R} =_{\mathrm{def}} \mathcal{S}'$ and $\sigma =_{\mathrm{def}} id$.

Suppose then that $\delta(\mathcal{U}_{<\alpha}) = \delta^{\mathcal{S}''}$ then letting w be the window of $\mathcal{S}$ such that $\delta^w = \delta^{\mathcal{S}''}$, we need to see that

(b) if $b = \Sigma(\mathcal{U}_{<\alpha})$ then $s(\mathcal{T}, w) \subseteq \mathrm{rge}(\pi_b^{\mathcal{U}_{<\alpha}})$.

(b) is a consequence of the second clause of strong branch condensation[134]. Let β be the least such that $\mathcal{S}' \trianglelefteq \mathcal{M}_\beta^{\mathcal{U}}$. Set $\mathcal{N} = \mathcal{M}_\beta^{\mathcal{U}}$. Notice that both β and α are on the main branch of $\mathcal{U}$ implying that both $\pi_{\beta,\eta}^{\mathcal{U}}$ and $\pi_{\alpha,\eta}^{\mathcal{U}}$ are defined. Let $\pi = \pi_{\beta,\eta}^{\mathcal{U}} \restriction \mathcal{N}^b$ and $\sigma = \pi_{\alpha,\eta}^{\mathcal{U}} \restriction (\mathcal{M}_\alpha^{\mathcal{U}})^b$. Let $c = [0,\alpha)_{\mathcal{U}}$ and set $\mathcal{Y} = \downarrow(\mathcal{U}_{[\beta,\alpha)}, \mathcal{N}^b)$. We can now apply clause 2 of strong branch condensation to $(\pi, \sigma, \mathcal{U}_{\leq\beta}, \mathcal{N}, \mathcal{T}, \mathcal{S}, \mathcal{Y}, c)$.

Case 2: $\mathcal{S}' \triangleleft_{hod} \mathcal{S}^b$ is not a complete layer[135] of $\mathcal{S}^b$.

Let $c = [0,\alpha)_{\mathcal{U}}$. In this case, we have that $\mathcal{Q}(c, \mathcal{U}_{<\alpha})$ exists and $\mathcal{Q}(c, \mathcal{U}_{<\alpha}) \trianglelefteq \mathcal{S}$. The dificult case is when $\mathcal{Q}(c, \mathcal{U}_{<\alpha})$ is an sts premouse over $\mathrm{m}^+(\mathcal{U}_{<\alpha})$, and so we assume it. We then have that $\mathcal{Q}(c, \mathcal{U}_{<\alpha})$ is a $\Sigma^{stc}_{\mathrm{m}^+(\mathcal{U}_{<\alpha}),\mathcal{T}}$-sts premouse over $\mathrm{m}^+(\mathcal{U}_{<\alpha})$. It then follows from Proposition 4.10.2 that in fact $\Sigma_{\mathrm{m}^+(\mathcal{U}_{<\alpha}),\mathcal{T}} = \Sigma_{\mathrm{m}^+(\mathcal{U}_{<\alpha}),\mathcal{X}^\frown\mathcal{U}_{<\alpha}}$[136].

The last remaining case is when $\mathcal{S}^b \trianglelefteq \mathcal{S}'$ and this case is very similar to Case 2 above. ⊣

This finishes the proof of Theorem 4.12.1. ⊣

We finish this section by recording some consequences of the proof given above. Suppose $(\mathcal{P}, \Sigma)$ is an sts hod pair. There is one potential problem with our definition of short tree strategy indexing scheme[137]. Suppose $\mathcal{M}$ is an unambiguous[138] Σ-sts premouse and $\mathcal{T}$ is an nuvs[139] stack on $\mathcal{P}$. Suppose (γ, ξ, b) is an $\mathcal{M}$-minimal shortness witness for $\mathcal{T}$ and let $\mathcal{Q} = \mathcal{Q}(b, \mathcal{T})$. It is not clear that $\mathcal{Q}$ is a $\Sigma_{\mathrm{m}^+(\mathcal{T}),\mathcal{T}}$-sts premouse. More precisely, it is not clear that $\Sigma^{\mathcal{Q}} \subseteq \Sigma_{\mathrm{m}^+(\mathcal{T}),\mathcal{T}} \restriction \mathcal{Q}$. However, the proof of Theorem 4.12.1 shows that in many situations it is indeed the case that

(A) $\mathcal{Q}$ a $\Sigma_{\mathrm{m}^+(\mathcal{T}),\mathcal{T}}$-sts premouse, and
(B) $\Sigma(\mathcal{T}) = b$.

As the proofs are very similar to the proofs already given in the proof of Theorem 4.12.1, we will simply state our results.

PROPOSITION 4.12.5. *Suppose $(\mathcal{P}, \Sigma)$ is an sts hod pair and Γ is a projectively closed pointclass. Suppose that Σ has strong branch condensation and is strongly almost Γ-fullness preserving. Then the following holds:*

---

[134] See Definition 4.9.2.

[135] See Definition 2.7.14.

[136] Notice that we used Theorem 4.12.1 in the proof of Theorem 4.9.5, namely in the proof of Case 1. However, we use Proposition 4.10.2 for low level strategies or for the short-tree-component of our strategy, while Case 1 of Theorem 4.9.5 deals with the full lsa type hod premice.

[137] See Definition 3.6.4.

[138] See Definition 3.6.2.

[139] See Definition 3.3.2.

1. *Suppose $t = (\mathcal{P}, \mathcal{T}, \mathcal{P}_1, \mathcal{T}_1)$ is $(\mathcal{P}, \Sigma)$-authenticated indexable stack*[140]. *Then t is according to Σ.*
2. *Suppose $\mathcal{M}$ is a Σ-sts premouse over some set X and based on $\mathcal{P}$, $\mathcal{T} \in \mathcal{M}$ is an* nuvs[141] *stack on $\mathcal{P}$, (γ, ξ, b) is an $\mathcal{M}$-shortness witness for $\mathcal{T}$*[142] *and $\mathcal{Q} = \mathcal{Q}(b, \mathcal{T})$. Then $\mathcal{Q}$ is a $\Sigma_{\mathrm{m}^+(\mathcal{T}), \mathcal{T}}$-sts premouse over $\mathrm{m}^+(\mathcal{T})$.*
3. *Suppose $\mathcal{M}$ is an* hp-*indexed germane* lses *such that* $\mathsf{hl}(\mathcal{M}) = \mathcal{P}$[143] *and $\mathcal{J}_\omega[\mathcal{M}] \vDash$ "$\delta^{\mathcal{P}}$ is a Woodin cardinal". Suppose further that Λ is an $\omega_1 + 1$-iteration strategy for $\mathcal{M}$ such that $\Lambda_{\mathcal{P}} = \Sigma$ and suppose $(\mathcal{T}, b) \in \mathcal{M}$ is such that*
   (a) *$\mathcal{T}$ is an* nuvs,
   (b) *for some β and γ such that $\omega\beta + \omega\gamma \leq \mathrm{ord}(\mathcal{M})$, setting $t = (\mathcal{P}, \mathcal{T})$ and $w = (\mathcal{J}_\omega(t), t, \in)$, w is (f, sts)-minimal as witnessed by β*[144],
   (c) *$\gamma = \mathrm{lh}(\mathcal{T})$,*
   (d) *$b \in \mathcal{M}|\omega\beta$ and $\mathcal{M}|\omega\beta \vDash \mathsf{sts}_0(\mathcal{T}, b)$*[145].

   Then $\Sigma(\mathcal{T}) = b$.
4. *Suppose Σ is strongly Γ-fullness preserving and $\mathcal{M}$ is an* hp-*indexed germane* lses *such that* $\mathsf{hl}(\mathcal{M}) = \mathcal{P}$[146] *and $\mathcal{J}_\omega[\mathcal{M}] \vDash$ "$\delta^{\mathcal{P}}$ is a Woodin cardinal". Suppose further that Λ is an ω_1-iteration strategy for $\mathcal{M}$ that acts on iterations above $\delta^{\mathcal{P}}$ and suppose $(\mathcal{T}, b) \in \mathcal{M}$ is such that*
   (a) *if Λ^* is the ω_1 fragment of Λ then* $\mathsf{Code}(\Lambda^*) \in \Gamma$,
   (b) *$\mathcal{T}$ is an* nuvs,
   (c) *for some β and γ such that $\omega\beta + \omega\gamma \leq \mathrm{ord}(\mathcal{M})$, setting $t = (\mathcal{P}, \mathcal{T})$ and $w = (\mathcal{J}_\omega(t), t, \in)$, w is (f, sts)-minimal as witnessed by β*[147],
   (d) *$\gamma = \mathrm{lh}(\mathcal{T})$,*
   (e) *$b \in \mathcal{M}|\omega\beta$ and $\mathcal{M}|\omega\beta \vDash \mathsf{sts}_0(\mathcal{T}, b)$*[148].

   Then $\Sigma(\mathcal{T}) = b$.

The proof of clause 1 of Proposition 4.12.5 is contained in the proof of Sublemma 4.12.4. Clause 2 easily follows from Clause 1 and the relevant definitions. The hypothesis of Clause 3 is exactly what we have at the begining of the proof of Theorem 4.12.1 (e.g., see Sublemma 4.12.2). Clause 4 follows from the fact that Σ is strongly Γ-fullness preserving. As in the proof of Theorem 4.12.1 we have that Λ^* induces a strategy for $\mathcal{Q}(b, \mathcal{T})$. Thus, if Φ is this strategy then $\mathsf{Code}(\Phi) \in \Gamma$. Therefore, by strong Γ-fullness preservation, $\Sigma(\mathcal{T}) = \beta$.

[140] See Definition 3.7.5.
[141] See Definition 3.3.2.
[142] See Definition 3.8.9.
[143] In particular, $\mathcal{M}$ can be viewed as a Σ-sts premouse over $\mathcal{P}$.
[144] See Definition 2.3.3. In particular, this means that we have to index the branch of t at $\omega(\beta + \gamma)$.
[145] See Definition 3.8.16. This means that e is the branch of t we must choose.
[146] In particular, $\mathcal{M}$ can be viewed as a Σ-sts premouse over $\mathcal{P}$.
[147] See Definition 2.3.3. In particular, this means that we have to index the branch of t at $\omega(\beta + \gamma)$.
[148] See Definition 3.8.16. This means that e is the branch of t we must choose.

REMARK 4.12.6 (On hod pair constructions). Suppose $(\mathcal{P},\Sigma)$ is an sts hod pair and Γ is a projectively closed pointclass. Suppose that Σ has strong branch condensation and is strongly almost Γ-fullness preserving. Recall Definition 4.2.1, which introduces fully backgrounded constructions relative to Σ. In particular, recall the Important Anomaly in clause 3(b) of Definition 4.2.1. It follows from the clause 4 of Proposition 4.12.5 that, in the terminology of clause 3(b) of Definition 4.2.1, as long as $\mathcal{M}_\xi$ has an ω_1-iteration strategy (as a Σ-sts premouse over $\mathcal{P}$) the Important Anomaly cannot occur.

4.13. The normal-tree comparison theory

As in Theorem 2.2.2 of [29], under AD^+ and in several other contexts, we can prove a comparison theorem where comparison is achieved via normal trees. In this section we state a comparison theorem for hod pairs that can be applied inside models of AD^+ and also, inside models satisfying sufficiently rich extensions of ZFC, like hod mice themselves. Such comparison arguments, among other things, are useful in core model induction arguments and in the analysis of HOD of models of AD^+.

We start with some general definitions and facts. One warning is that our exposition differs from the one in [29] mainly because we would like to set up our arguments here in a more general setting than the ones stated in [29]. The notation $\trianglelefteq_{hod}$ was introduced in Definition 2.7.8.

DEFINITION 4.13.1 (Comparison). Suppose $(\mathcal{P},\Sigma)$ and $(\mathcal{Q},\Lambda)$ are two hod pairs. Then we say that *comparison holds* for $(\mathcal{P},\Sigma)$ and $(\mathcal{Q},\Lambda)$ if there are $(\mathcal{T},\mathcal{R})$ and $(\mathcal{U},\mathcal{S})$ such that

1. $\mathcal{T}$ is a stack on $\mathcal{P}$ according to Σ with last model $\mathcal{R}$,
2. $\mathcal{U}$ is a stack on $\mathcal{Q}$ according to Λ with last model $\mathcal{S}$,
   and one of the following holds:
3. $(\mathcal{Q},\Lambda)$ wins: More precisely the following clauses hold:
   (a) $\mathcal{R} \trianglelefteq_{hod} \mathcal{S}$,
   (b) $\Lambda_{\mathcal{R},\mathcal{U}} = \Sigma_{\mathcal{R},\mathcal{T}}$,
   (c) $\pi^{\mathcal{T}}$ is defined,
   (d) If $\mathcal{P}$ is meek or gentle then $\pi^{\mathcal{U}}$ is defined,
   (e) If $\mathcal{P}$ is non meek then letting $\alpha < \mathrm{lh}(\mathcal{U})$ be the least such that $\mathcal{P}^b \trianglelefteq \mathcal{M}^{\mathcal{U}}_\alpha$, $\pi^{\mathcal{U}}_{0,\alpha}$ is defined.
4. $(\mathcal{P},\Sigma)$ wins: More precisely the following clauses hold:
   (a) $\mathcal{S} \trianglelefteq_{hod} \mathcal{R}$,
   (b) $\Lambda_{\mathcal{S},\mathcal{U}} = \Sigma_{\mathcal{S},\mathcal{T}}$,
   (c) $\pi^{\mathcal{U}}$ is defined,
   (d) If $\mathcal{Q}$ is meek or gentle then $\pi^{\mathcal{T}}$ is defined,

(e) If $\mathcal{Q}$ is non meek then letting $\alpha < \mathrm{lh}(\mathcal{T})$ be the least such that $\mathcal{Q}^b \trianglelefteq \mathcal{M}_\alpha^{\mathcal{T}}$, $\pi_{0,\alpha}^{\mathcal{T}}$ is defined.

If clause 1 holds then we say that $(\mathcal{Q},\Lambda)$ *wins the comparison*, and otherwise we say that $(\mathcal{P},\Sigma)$ wins. We say normal comparison for $(\mathcal{P},\Sigma)$ and $(\mathcal{Q},\Lambda)$ holds if we can take $\mathcal{T}$ and $\mathcal{U}$ to be normal.

Similarly we define the meaning of "*comparison holds* for $(\mathcal{P},\Sigma)$ and $(\mathcal{Q},\Lambda)$" in the case $(\mathcal{P},\Sigma)$ or $(\mathcal{Q},\Lambda)$ are allowable pairs. For example, if $(\mathcal{P},\Sigma)$ is a hod pair and $(\mathcal{Q},\Lambda)$ is an sts hod pair then we say that *comparison holds* for $(\mathcal{P},\Sigma)$ and $(\mathcal{Q},\Lambda)$ if there are $(\mathcal{T},\mathcal{R})$ and $(\mathcal{U},\mathcal{S})$ such that in the case $(\mathcal{P},\Sigma)$ wins, $\Sigma_{\mathcal{S},\mathcal{T}}^{stc} = \Lambda_{\mathcal{S},\mathcal{U}}$.

As in [29], we can prove comparison for pairs whose corresponding strategies are fullness preserving. Here we show that the fully backgrounded constructions are universal in the sense that they win the comparison with hod pairs that they capture. To establish this fact, we will use *the strategy absorption* argument. The strategy absorption argument was first presented in [29] (see the proof of Theorem 2.28 of [29]) and it builds on unpublished ideas of Steel. Because we will use the strategy absorption argument several times in this paper and in the next proof, it is important to understand how it works. The general form of the argument is as follows. We have a hod pair $(\mathcal{P},\Lambda)$ captured by some *background* $(M,\delta,\vec{G},\Sigma)$. There is also an iteration tree $\mathcal{T}$ on $\mathcal{P}$ according to Λ with last model $\mathcal{Q}$ and $\mathcal{R} \trianglelefteq_{hod} \mathcal{Q}$ such that $\mathcal{R}$ is constructed via some hod pair construction of M. It is additionally required that the background extenders used to build $\mathcal{R}$ cohere Λ[149]. The goal of the argument is to show that the strategy $\mathcal{R}$ inherits from the background universe is the same as $\Lambda_{\mathcal{R},\mathcal{T}}$. In many cases, this can be done by appealing to branch condensation and the existence of minimal disagreements. Here is how a typical argument works.

Let Φ be the iteration strategy of $\mathcal{R}$ induced by the background strategy. Fix $\mathcal{U}$ on $\mathcal{R}$ that is according to both $\Lambda_{\mathcal{R},\mathcal{T}}$ and Φ but $\Lambda_{\mathcal{R},\mathcal{T}}(\mathcal{U}) \neq \Phi(\mathcal{U})$. Let $\mathcal{U}^*$ be the stack on M obtained by resurrection process. Thus, $\mathcal{U}^* = r\mathcal{U}$ (see Section 4.3.1). Let $b = \Phi(\mathcal{U}^*)$. We then have that $\pi_b^{\mathcal{U}^*}(\mathcal{T})$ is according to Λ (this is where we use coherence). Then branch condensation is applied to the equality

$$\pi^{\pi_b^{\mathcal{U}^*}(\mathcal{T})} = \sigma \circ \pi_b^{\mathcal{U}} \circ \pi^{\mathcal{T}}$$

where $\sigma : \mathcal{M}_b^{\mathcal{U}} \to \pi_b^{\mathcal{U}^*}(\mathcal{R})$ is the canonical factor map that the resurrection process gives (in particular, $\pi_b^{\mathcal{U}^*} \restriction \mathcal{R} = \sigma \circ \pi_b^{\mathcal{U}}$). The reader may wish to review Section 4.3.1. Recall that strong branch condensation and Γ-fullness preservation implies positional (see Proposition 4.10.2).

THEOREM 4.13.2 (Universality of backgrounded construction). *Assume* AD^+. *Suppose that*

[149]However, the fact that Code(Λ) is Suslin, co-Suslin captured by $(M,\delta,\vec{G},\Sigma)$ implies that all extenders in $\vec{G}$ cohere Λ.

1. *Γ is a pointclass,*
2. *$(\mathcal{P},\Lambda)$ is an allowable pair,*
3. *$k(\mathcal{P}) = \mathsf{ep}(\mathcal{P})$*[150]*,*
4. *Λ is Γ-fullness preserving and has strong branch condensation*[151]*,*
5. *$\mathsf{C} = (\mathbb{M},(P,\Psi),\Gamma^*,A)$ Suslin, co-Suslin captures both Γ and $\mathsf{Code}(\Lambda)$, and*
6. *$\mathbb{M} = (M,\delta,\vec{G},\Sigma)$.*
7. *NsesS.*

Let

$$\mathsf{hpc}^+_{\mathsf{C},\Gamma} = (\mathcal{M}_\gamma,\mathcal{N}_\gamma,Y_\gamma,\Phi^+_\gamma,F^+_\gamma,F_\gamma,b_\gamma : \gamma \leq \delta)$$

be the output of $\Gamma - \mathsf{hpc}$ of $\mathbb{M}$ with the property that each F^+_γ coheres $\Lambda \restriction M$[152]*. Then there is $\gamma \leq \delta$ such that the following holds.*

1. *If $(\mathcal{P},\Lambda)$ is not an sts hod pair then $\gamma < \delta$ and there is a normal stack $\mathcal{X}$ such that $(\mathcal{X},\mathcal{M}_\gamma) \in I(\mathcal{P},\Lambda)$ and $\Phi^+_\gamma = \Lambda_{\mathcal{M}_\gamma}$.*
2. *If $(\mathcal{P},\Lambda)$ is an sts hod pair then there is a normal stack $\mathcal{X}$ such that letting*
$$\mathcal{N} = \begin{cases} \mathcal{M}_\gamma & \text{if } \gamma < \delta, \\ \mathcal{M}^\#_\gamma & \text{if } \gamma = \delta, \end{cases}$$
$(\mathcal{X},\mathcal{N}) \in I(\mathcal{P},\Lambda)$ and $\Phi^+_\gamma = \Lambda_{\mathcal{N}}$.
3. *For every $\beta \leq \gamma$, there is a Λ-iterate $\mathcal{R}$ of $\mathcal{P}$ via a normal stack $\mathcal{T}$ such that $\mathcal{M}_\beta \trianglelefteq \mathcal{R}$ and if $\mathcal{S} \in Y_\beta$ then $(\Phi^+_\beta)_\mathcal{S} = \Lambda_\mathcal{S}$.*
4. *For every $\beta \leq \gamma$, there is a Λ-iterate $\mathcal{R}$ of $\mathcal{P}$ via a normal stack $\mathcal{T}$ such that $\mathcal{N}_\xi \trianglelefteq \mathcal{R}$.*

PROOF. As in the proof of Lemma 2.10 of [29], in the comparison of $\mathcal{P}$ with the models of $\mathsf{hpc}_{\mathsf{C},\Gamma}$ no extender disagreement appears on $\mathsf{hpc}_{\mathsf{C},\Gamma}$ side. Many of the details of the argument have appeared in [50, Lemma 3.21], and because of this we only concentrate on the new aspects of the proof. We then assume that $\mathcal{P}$ is non-meek.

We first show that clause 3 holds and then show that clause 1 and 2 hold. Clause 4 is similar to clause 3. To prove clause 3, we only verify that

(1) for every $\beta < \delta$, if $(\mathcal{T},\mathcal{R},\mathcal{S})$ are such that $\mathcal{T}$ is a normal stack on $\mathcal{P}$ according to Σ, $\mathcal{R}$ is the last model of $\mathcal{T}$, $\mathcal{M}_\beta \trianglelefteq \mathcal{R}$ and $\mathcal{S} \in Y_\beta$ then $(\Phi^+_\beta)_\mathcal{S} = \Lambda_\mathcal{S}$.

The proof of (1) is the portion of the proof that goes beyond [29, Theorem 2.28] and [50, Lemma 3.21], and so we prove (1).

---

[150] Recall that $\mathcal{P}$ is f.s $\mathcal{J}$-structure. To define $\mathsf{ep}(\mathcal{P})$ we ignore its fine-structural component $k(\mathcal{P})$ and treat $\mathcal{P}$ as just a $\mathcal{J}$-structure. See Definition 2.2.3.

[151] We could instead assume just the first two clauses of strong branch condensation and also that Λ is self-cohering. However, our proof will use self-cohering in an indirect way. Strategies with strong branch condensation are positional and therefore, self-cohering. The reader may wish to review Definition 2.10.11, Definition 4.9.2, Theorem 4.9.5 and Proposition 4.10.2.

[152] This actually follows from the fact that $\mathsf{Code}(\Lambda)$ is Suslin, co-Suslin captured.

Because Λ is self-cohering (see Definition 2.10.11) we can in fact assume that

(2) for every $\alpha+1<\mathrm{lh}(\mathcal{T})$, $\mathcal{S} \ntrianglelefteq \mathcal{M}_\alpha^{\mathcal{T}}$.

For simplicity, we prove $(\Phi_\beta^+)_{\mathcal{S}} = \Lambda_{\mathcal{S}}$ for ordinary stacks as opposed to generalized stacks. The more general proof is only notationally more complex. The reader may wish to review Section 4.3.1 and Lemma 4.3.9

Towards a contradiction, we assume that (1) fails. Let $(\beta,\mathcal{S},\mathcal{T},\mathcal{R})$ witness the failure of (1) such that β is the least possible and (2) holds. We assume that $\mathcal{S}\in Y^{\mathcal{R}}\cap Y_\beta$ is the least layer for which (1) fails. Let $\Phi=\Phi_\beta^+$ and $\mathcal{Q}=\mathcal{M}_\beta$.

Case 1: $\mathcal{S}$ is of successor type.

Then we get a contradiction using branch condensation of Λ. Let $\mathcal{U}$ be a stack on $\mathcal{S}$ such that it is according to both $\Phi_{\mathcal{S}}$ and $\Lambda_{\mathcal{S}}$ but $\Phi(\mathcal{U})\neq\Lambda_{\mathcal{S}}(\mathcal{U})$. Let $b=\Phi(\mathcal{U})$ and $c=\Lambda_{\mathcal{S}}(\mathcal{U})$. Let $\mathcal{U}^*=r\mathcal{U}$[153]. Then because extenders used to construct $\mathcal{Q}$ cohere Λ, we have that $\pi^{\mathcal{U}^*}(\mathcal{T})$ is according to Λ. Let N be the last model of $\mathcal{U}^*$.

CLAIM 4.13.3. *$\pi_b^{\mathcal{U}}$ exists.*

PROOF. The claim is a consequence of Γ-fullness preservation and the fact that $\Phi_{\mathcal{S}^-}=\Lambda_{\mathcal{S}^-}$. Towards a contradiction assume that $\pi_b^{\mathcal{U}}$ is undefined. Because $\Phi_{\mathcal{S}^-}=\Lambda_{\mathcal{S}^-}$ and because $\Phi(\mathcal{U})\neq\Lambda_{\mathcal{S}}(\mathcal{U})$, we must have some $\iota\in R^{\mathcal{U}}$ such that $\pi_{0,\iota}^{\mathcal{U}}$ is defined and $\mathcal{U}_{\geq\iota}$ is above $\mathcal{M}_\iota^-$. Moreover, because $\Phi(\mathcal{U})\neq\Lambda_{\mathcal{S}}(\mathcal{U})$, $R^{\mathcal{U}}$ must have a maximal element. Let then $\iota=\max(R^{\mathcal{U}})$ and set $\mathcal{X}=\mathcal{U}_{\geq\iota}$.

Suppose now that $\pi_c^{\mathcal{U}}$ is not defined. Γ-fullness preservation implies that $\mathcal{X}$ does not have fatal drop, and so $\delta(\mathcal{X})$ is a strong cutpoint in both $\mathcal{Q}(b,\mathcal{U})$ and $\mathcal{Q}(c,\mathcal{U})$. Hence Γ-fullness preservation implies that $b=c$. Contradiction.

Thus, $\pi_c^{\mathcal{U}}$ is defined. It then follows from Γ-fullness preservation that $\mathcal{Q}(b,\mathcal{X})\trianglelefteq\mathcal{M}_c^{\mathcal{U}}$. Therefore, $b=c$, which is a contradiction. ⊣

Let then $\mathcal{U}^+=\uparrow(\mathcal{R},\mathcal{U})$[154] be the unique stack on $\mathcal{R}$ whose tree structure and extenders are exactly those of $\mathcal{U}$. Let $\mathcal{R}^*$ be the last model of $\mathcal{U}^+$. We then have $\sigma:\mathcal{R}^*\to\pi_b^{\mathcal{U}^*}(\mathcal{R})$ such that, assuming $\pi^{\mathcal{T}}$ is defined,

$$\pi^{\pi_b^{\mathcal{U}^*}(\mathcal{T})}=\sigma\circ\pi_b^{\mathcal{U}^+}\circ\pi^{\mathcal{T}}.$$

Notice next that it follows from (2) and the fact that $\mathcal{S}$ is of successor type that $\pi^{\mathcal{T}}$ is defined. Branch condensation of Λ and the displayed equality implies that $b=c$

Case 2. $\mathcal{S}$ is of limit type.

Then by appealing to Lemma 4.7.6, we can fix some $((\mathcal{U}_1,\mathcal{S}_1),(\mathcal{U}_2,\mathcal{S}_2),\mathcal{S}_3)$ that constitutes a minimal low level disagreement between Φ and $\Lambda_{\mathcal{S}}$. Let $\mathcal{U}_1^*=r\mathcal{U}_1$, N be the last model of $\mathcal{U}_1^*$ and $\sigma:\mathcal{S}_1\to\mathcal{Q}$ be elementary such that $\mathcal{Q}$ is an $\mathcal{M}$-model appearing in the $\Gamma-\mathsf{hpc}$ of N. We then have that letting $\mathcal{T}^*=\pi^{\mathcal{U}_1^*}(\mathcal{T})$, for some ξ, $\mathcal{Q}\trianglelefteq\mathcal{M}_\xi^{\mathcal{T}^*}$. Let $\mathcal{Q}'=\sigma(\mathcal{S}_3)$. Notice next that

[153]This is the resurrection of $\mathcal{U}$. See Section 4.3.1.

[154]See Definition 2.4.10.

(3) $(\mathcal{T}^*, \mathcal{Q})$ supports a bottom type $(\mathcal{Q}', \mathcal{S}_3, \sigma \restriction \mathcal{S}_3)$-b-condensation diagram on $\mathcal{P}$.

Let Φ^* be the strategy of $\mathcal{Q}'$ induced by Σ_N. We then have that

(4) $\Phi_{\mathcal{S}_3,\mathcal{U}_1} = (\sigma\text{-pullback of } \Phi^*)$ and $\Lambda_{\mathcal{Q}'} = \Phi^*$.

$\Lambda_{\mathcal{Q}'} = \Phi^*$ follows from the fact that $\mathcal{Q}' \lhd \mathcal{Q}$ and also from the fact that β is the least satisfying (1). Thus, the strong branch condensation of Λ implies that $\Lambda_{\mathcal{S}_3} = \Phi_{\mathcal{S}_3,\mathcal{U}_1}$.

Next, we need to verify that clause 1 and 2 hold for some $\gamma < \delta$. Set $\mathcal{N} = \mathcal{M}_\delta$. Assume first that $(\mathcal{P}, \Lambda)$ is not an sts hod pair. This means that Λ is an iteration strategy. Assume then clause 1 fails. It follows that we have a normal stack $\mathcal{X}$ on $\mathcal{P}$ such that $\mathrm{lh}(\mathcal{X}) = \delta$ and $\mathrm{m}(\mathcal{X}) = \mathcal{N}$. Let $b = \Lambda(\mathcal{X})$. Let $\alpha < \mathrm{lh}(\mathcal{X})$ be least such that $\delta \in \mathrm{rge}(\pi^{\mathcal{X}}_{\alpha,b})$. Because the entire construction takes place in M and because δ is regular, we have that letting η be such that $\delta = \pi^{\mathcal{X}}_{\alpha,b}(\eta)$, η must be a measurable cardinal of $\mathcal{M}^{\mathcal{X}}_\alpha$.

Notice that $\mathcal{M}^{\mathcal{X}}_\alpha$ is germane[155] and because $\mathcal{X}$ may drop in model, $\mathcal{M}^{\mathcal{X}}_\alpha$ may not be hod-like. Let $\mathcal{R} \trianglelefteq \mathcal{N}$ be the longest such that

1. $\mathcal{R} \in Y^{\mathcal{N}}$,
2. $\mathcal{R}$ is meek or gentle,
3. $\mathcal{R} \trianglelefteq \mathcal{M}^{\mathcal{X}}_\alpha$ and
4. $\delta^{\mathcal{R}} \leq \eta$.

We now break into cases. Let $\alpha' \leq \alpha$ be the least such that $\mathcal{R} \trianglelefteq \mathcal{M}^{\mathcal{X}}_{\alpha'}$.

Suppose first that $\mathcal{R}$ is of successor type. We must then have $\mathcal{R}' \in Y^{\mathcal{M}^{\mathcal{X}}_{\alpha'}}$ such that $(\mathcal{R}')^- = \mathcal{R}$. But now $\mathcal{X}_{\geq \alpha'}$ is based on $\mathcal{R}'$ and is above $\delta^{\mathcal{R}}$. Because in this case $\mathcal{R}'$ out-iterates $\mathcal{N}$, this contradicts our assumption that NsesS[156].

Suppose then $\mathcal{R}$ is gentle. In this case, we must have $\mathcal{R}'$ such that $\mathcal{R}'$ is meek of limit type, $\delta^{\mathcal{R}'} = \delta^{\mathcal{R}}$ and $\mathcal{R}' \in Y^{\mathcal{M}^{\mathcal{X}}_{\alpha'}}$ or $\mathcal{R}' = \mathcal{M}^{\mathcal{X}}_{\alpha'}$. If $\delta^{\mathcal{R}} < \eta$ then we get that $\mathcal{X}_{\geq \alpha'}$ is based on $\mathcal{R}'$ and is above $\delta^{\mathcal{R}} + 1$, and once again this leads to a contradiction.

Suppose now that $\delta^{\mathcal{R}} = \eta$. Let now $\kappa > \eta$ be such that it reflects

1. $\mathcal{X}$, and
2. $\mathsf{hpc}_{\mathsf{C},\Gamma} = (\mathcal{M}_\gamma, \mathcal{N}_\gamma, Y_\gamma, \Phi_\gamma, F^+_\gamma, F_\gamma, b_\gamma : \gamma \leq \delta)$.

Let $\xi = o^{\mathcal{N}}(\kappa)$[157] and $\zeta + 1 \in b$ be such that $\mathcal{X}(\zeta + 1) = \kappa$. Let $E \in \vec{G}$ be an extender such that $\mathrm{crit}(E) = \kappa$, $\mathrm{lh}(E) > \xi$ and it reflects the above sets. It follows that

[155] See Definition 2.7.15.

[156] This is a consequence of the ordinary universality of the background constructions. If a mouse outiterates a fully backgrounded construction then it generates a mouse with a superstrong. See [29, Lemma 2.13].

[157] Notice that $\xi < \delta$ as otherwise δ would be a Woodin cardinal of $\mathcal{M}^{\mathcal{X}}_b$ and since it is also a measurable cardinal, we would get a contradiction to our minimality assumption on hod mice.

(5) $\kappa \in b$ and $\mathrm{crit}(\pi^{\mathcal{X}}_{\kappa,b}) = \kappa$,
(6) $E^{\mathcal{X}}_{\zeta}$ agrees with E,
(7) $\pi^{\mathcal{X}}_{\alpha',\kappa}$ is defined and $\pi^{\mathcal{X}_{\alpha',\kappa}}(\delta^{\mathcal{R}}) = \kappa$,
(8) $\mathcal{M}^{\mathcal{X}}_{\kappa}|\xi = \mathcal{N}|\xi$,
(9) $\mathcal{M}^{\mathcal{X}}_{\zeta+1}|\mathrm{ind}^{\mathcal{X}}_{\zeta} = \mathcal{N}|\mathrm{ind}^{\mathcal{X}}_{\zeta}$.

It follows from (6), (8) and (9) that $E^{\mathcal{X}}_{\zeta} \in \vec{E}^{\mathcal{N}}$ as E can serve as a background certificate for it. Clearly this is a contradiction.

Finally suppose $\mathcal{R}$ is of limit type. In this case we have that $\delta^{\mathcal{R}} < \eta$. We also have $\mathcal{R}' \in Y^{\mathcal{M}^{\mathcal{X}}_{\alpha'}}$ such that either $\mathcal{R}'$ is of successor type and $(\mathcal{R}')^{-} = \mathcal{R}$ or $\mathcal{R}'$ is of limit type and $(\mathcal{R}')^{b} = \delta^{\mathcal{R}}$. The first case leads to a contradiction via a similar argument as the one given above. Let then $\mathcal{R}'$ be a complete layer of $\mathcal{M}^{\mathcal{X}}_{\alpha'}$ (see Notation 2.7.14) such that $(\mathcal{R}')^{b} = \mathcal{R}$. It follows that $\mathcal{X}_{\geq\alpha'}$ is based on $\mathcal{R}'$, $\mathcal{X}_{\geq\alpha'}$ is above $\delta^{\mathcal{R}}$ and also, that $\mathcal{N}^{b} = \mathcal{R}^{b}$. This case once again leads to a contradiction because assuming NsesS universality implies that $\mathcal{R}'$ cannot out-iterate $\mathcal{N}$.

The case that $(\mathcal{P},\Lambda)$ is an sts hod pair is very similar. In this case, we note that $\mathcal{X}$ must be Λ-maximal as otherwise $\Lambda(\mathcal{X})$ is a branch and all of the above arguments can be repeated. If $\mathcal{X}$ is Λ-maximal then $\Lambda(\mathcal{X}) = \mathcal{N}^{\#}$, which is one of the possibilities in clause 2. ⊣

As a corollary to Theorem 4.13.2 we get that comparison holds.

COROLLARY 4.13.4. *Assume* AD^{+} *and suppose*

1. Γ *is a pointclass,*
2. $(\mathcal{P},\Sigma)$ *and* $(\mathcal{Q},\Lambda)$ *are two allowable pairs such that both* Σ *and* Λ *are* Γ-*fullness preserving and have strong branch condensation,*
3. $k(\mathcal{P}) = \mathsf{ep}(\mathcal{P})$ *and* $k(\mathcal{P}) = \mathsf{ep}(\mathcal{Q})$,
4. *there is a good pointclass* Γ' *such that* $\Gamma \cup \{\mathsf{Code}(\Lambda), \mathsf{Code}(\Sigma)\} \subseteq \Delta_{\Gamma'}$.

Then the normal comparison holds for $(\mathcal{P},\Sigma)$ *and* $(\mathcal{Q},\Lambda)$.

PROOF. Using Theorem 4.1.12 we can find $\mathsf{C} = (\mathbb{M}, (P,\Psi), \Gamma^{*}, A)$ which Suslin, co-Suslin captures Γ, $\mathsf{Code}(\Lambda)$ and $\mathsf{Code}(\Sigma)$. Let

$$\mathsf{hpc}^{+}_{\mathsf{C},\Gamma} = (\mathcal{M}_{\gamma}, \mathcal{N}_{\gamma}, Y_{\gamma}, \Phi^{+}_{\gamma}, F^{+}_{\gamma}, F_{\gamma}, b_{\gamma} : \gamma \leq \delta)$$

be the output of $\Gamma - \mathsf{hpc}$ of $\mathbb{M}$ with the property that each F^{+}_{γ} coheres both $\Sigma \restriction M$ and $\Lambda \restriction M$.

It follows from Theorem 4.13.2 that there are $\beta, \gamma \leq \delta$ and normal stacks $\mathcal{T}$ and $\mathcal{U}$ such that

1. $(\mathcal{T}, \mathcal{M}_{\beta}) \in I(\mathcal{P},\Sigma)$ and $\Phi^{+}_{\beta} = \Sigma_{\mathcal{M}_{\beta}}$ and
2. $(\mathcal{U}, \mathcal{M}_{\gamma}) \in I(\mathcal{Q},\Lambda)$, $\Phi^{+}_{\gamma} = \Lambda_{\mathcal{M}_{\gamma}}$.

If $\beta = \gamma$ then clearly the normal comparison for $(\mathcal{P},\Sigma)$ and $(\mathcal{Q},\Lambda)$ holds. Suppose then $\beta < \gamma$. Then there is $(\mathcal{U}', \mathcal{Q}')$ such that

1. $\mathcal{U}'$ is a normal stack on $\mathcal{Q}$ according to Λ,

2. $\mathcal{Q}'$ is the last model of $\mathcal{U}'$, and
3. $\mathcal{M}_\beta \trianglelefteq_{hod} \mathcal{Q}'$ and $\Phi_\beta = \Lambda_{\mathcal{M}_\beta}$.

Let $\alpha < \mathrm{lh}(\mathcal{U}')$ be the least such that $\mathcal{M}_\beta \trianglelefteq \mathcal{M}^{\mathcal{U}'}_\alpha$. Set $\mathcal{X} = \mathcal{U}'_{\leq\alpha}$ and $\mathcal{S} = \mathcal{M}^{\mathcal{X}}_\alpha$ and $\mathcal{R} = \mathcal{M}^{\mathcal{T}}_\beta$. In order to show that $(\mathcal{T}, \mathcal{R})$ and $(\mathcal{X}, \mathcal{S})$ witnesses that comparison holds for $(\mathcal{P}, \Sigma)$ and $(\mathcal{Q}, \Lambda)$, we need to show that $[0, \alpha)_{\mathcal{X}} \cap D^{\mathcal{X}} = \emptyset$. However, to obtain this condition we may need to change $\mathcal{X}$.

First observe that if $\mathcal{P}$ is meek or gentle then indeed $[0, \alpha)_{\mathcal{X}} \cap D^{\mathcal{X}} = \emptyset$. We give the argument in the case $\mathcal{P}$ is of successor type and as the rest is similar, we leave the rest to the reader. Since $\mathcal{P}$ is of successor type, we have that $\delta^{\mathcal{R}}$ is a cardinal of M. Notice that α is the least α' such that $\mathcal{M}^{\mathcal{X}}_{\alpha'}|\delta^{\mathcal{R}} = \mathcal{R}|\delta^{\mathcal{R}}$. This follows from Γ-fullness preservation, which implies that if $\mathcal{M}^{\mathcal{X}}_{\alpha'}|\delta^{\mathcal{R}} = \mathcal{S}|\delta^{\mathcal{R}}$ then $\mathcal{R} \trianglelefteq \mathcal{M}^{\mathcal{X}}_{\alpha'}$. Thus, α must be a limit ordinal. Suppose then $[0, \alpha)_{\mathcal{X}} \cap D^{\mathcal{X}} \neq \emptyset$. It follows that $\rho(\mathcal{S}) < \delta^{\mathcal{R}}$. But hod premice do not project across layers of successor type (or rather meek or gentle type)[158].

Suppose then that $\mathcal{P}$ is non-meek. Let ι be the least such that $\mathcal{M}^{\mathcal{X}}_\iota|\mathrm{ord}(\mathcal{R}^b) = \mathcal{R}^b$. It follows from the argument above that $[0, \iota)_{\mathcal{X}} \cap D^{\mathcal{X}} = \emptyset$. Moreover, $\mathcal{X}_{\geq\iota}$ is above $\mathrm{ord}(\mathcal{R}^b)$. Set $\kappa = \delta^{\mathcal{R}^b}$.

Suppose now that there is $E \in \vec{E}^{\mathcal{S}}$ such that $\mathrm{crit}(E) = \kappa$ and $\mathrm{ind}^{\mathcal{S}}(E) > \mathrm{ord}(\mathcal{R})$. Let $\mathcal{X}'$ be the continuation of $\mathcal{X}$ obtained by using E at stage α. Notice that E must be applied to $\mathcal{M}^{\mathcal{X}}_\iota$. As $\mathcal{X}_{\leq\iota}$ does not have drop on its main branch, we have that $\mathcal{X}'$ also does not have a drop on its main branch and moreover, $(\mathcal{T}, \mathcal{R})$ and $(\mathcal{X}', \mathrm{Ult}(\mathcal{M}^{\mathcal{X}}_\beta, E))$ witness that comparison holds for $(\mathcal{P}, \Sigma)$ and $(\mathcal{Q}, \Lambda)$. ⊣

Using reflection, we can eliminate the extra assumptions on Γ and the two strategies.

COROLLARY 4.13.5 (Comparison). *Assume* AD^+ *and suppose* Γ *is a pointclass. Suppose further that* $(\mathcal{P}, \Sigma)$ *and* $(\mathcal{Q}, \Lambda)$ *are two hod pairs such that*

1. *both* Σ *and* Λ *are* Γ*-fullness preserving and have branch condensation,*
2. $k(\mathcal{P}) = \mathsf{ep}(\mathcal{P})$ *and* $k(\mathcal{P}) = \mathsf{ep}(\mathcal{Q})$,

Then the normal comparison hold for $(\mathcal{P}, \Sigma)$ *and* $(\mathcal{Q}, \Lambda)$.

PROOF. Suppose not. Applying Σ^2_1-reflection, we can find Γ^* and two hod pairs $(\mathcal{P}_1, \Sigma_1)$ and $(\mathcal{Q}_1, \Lambda_1)$ such that $\Gamma^* \cup \{\mathsf{Code}(\Sigma_1), \mathsf{Code}(\Lambda_1)\} \subseteq \Delta^2_1$ and the claim of the corollary fails for $(\Gamma^*, (\mathcal{P}_1, \Sigma_1), (\mathcal{Q}_1, \Lambda_1))$. We then apply Corollary 4.13.4. We use Theorem 4.1.12 to get a $\mathsf{C} = (\mathbb{M}, (P, \Psi), \Gamma^*, A)$ that Suslin, co-Suslin captures Γ, $\mathsf{Code}(\Lambda)$ and $\mathsf{Code}(\Sigma)$. ⊣

REMARK 4.13.6. In most situations, our allowable pairs $(\mathcal{P}, \Sigma)$ will have the property that $k(\mathcal{P}) = \mathsf{ep}(\mathcal{P})$. Thus, we make a convention that unless otherwise specified, all allowable pairs have the property that $k(\mathcal{P}) = \mathsf{ep}(\mathcal{P})$. When it is necessary we will remind the reader of this.

---

[158] See Definition 2.7.1.

However, Theorem 4.13.2 can also be proven in the case that $k(\mathcal{P}) < \mathsf{ep}(\mathcal{P})$. In this case, what we get is that letting $k = k(\mathcal{P})$, $(\mathcal{X}, (\mathsf{core}_k(\mathcal{N}_\gamma), k)) \in I(\mathcal{P}, \Sigma)$. Similar results can also be proven for germane lses.

4.14. Diamond comparison

Our goal here is to provide another comparison argument, *diamond comparison*, that does not rely on branch condensation as heavily as our other argument (see Corollary 4.13.5). The new comparison argument follows the same line of thought as the proof of a similar comparison argument from [29] (see Theorem 2.47 of [29]).

As in [29], the diamond comparison argument can be used to show that $\mathsf{AD}^+ + \mathsf{LSA}$ is consistent relative to a Woodin cardinal that is a limit of Woodin cardinals. This will appear as Theorem 10.3.1. In [29], a similar argument gave the consistency of $\mathsf{AD}_{\mathbb{R}}$ + "Θ is a regular cardinal" relative to a Woodin cardinal that is a limit of Woodin cardinals.

Following the proof of Theorem 2.47 of [29], we first define a *bad block* and a *bad sequence* and show that there cannot be such a bad sequence of length ω_1. We then show that the failure of comparison produces such bad sequences of length ω_1.

4.14.1. Bad sequences. For the purposes of this subsection, we make a definition of a bad block and a bad sequence. In later subsections, we will redefine these names for different objects. Below and elsewhere, if $\mathcal{T}$ is a stack of successor length then we let $\mathcal{T}^-$ be $\mathcal{T}_{<\alpha}$ where $\alpha + 1 = \mathrm{lh}(\mathcal{T})$.

DEFINITION 4.14.1 (Bad block). Suppose $(\mathcal{P}, \Sigma)$ and $(\mathcal{Q}, \Lambda)$ are two hod pairs such that both $\mathcal{P}$ and $\mathcal{Q}$ are of limit type and are not gentle. Then

$$B = (((\mathcal{P}_i, \mathcal{Q}_i, \Sigma_i, \Lambda_i) : i < 4), (\mathcal{T}_i, \mathcal{U}_i : i < 3), (c, d))$$

is a bad block on $((\mathcal{P}, \Sigma), (\mathcal{Q}, \Lambda))$ if the following holds:

1. $(\mathcal{P}_0, \Sigma_0) = (\mathcal{P}, \Sigma)$ and $(\mathcal{Q}_0, \Lambda_0) = (\mathcal{Q}, \Lambda)$.
2. $\mathcal{T}_0$ is a stack according to Σ_0 on $\mathcal{P}$.
3. $\mathcal{U}_0$ is a stack according to Λ_0 on $\mathcal{Q}$.
4. Let $\mathcal{T}_0 = (\mathcal{M}_\beta, \mathcal{T}^*_\beta, E_\beta : \beta \leq \nu)$ and $\mathcal{U}_0 = (\mathcal{N}_\beta, \mathcal{U}^*_\beta, F_\beta : \beta \leq \nu)$. Then $\mathcal{T}^*_\nu$ and $\mathcal{U}^*_\nu$ are undefined, $\mathcal{P}_1 = \mathcal{M}_\nu$ and $\mathcal{Q}_1 = \mathcal{N}_\nu$.
5. There is $\mathcal{K}$ such that $\mathcal{K} \lhd_{hod} \mathcal{P}_1$, $\mathcal{K} \lhd_{hod} \mathcal{Q}_1$, $\mathcal{K}$ is of successor type, $\Sigma_{\mathcal{K}, \mathcal{T}_0} \neq \Lambda_{\mathcal{K}, \mathcal{U}_0}$ and $\Sigma_{\mathcal{K}, \mathcal{T}_0} = \Sigma_{\mathcal{K}, \mathcal{U}_0}$.
6. $\mathcal{T}_1$ and $\mathcal{U}_1$ are stacks on $\mathcal{P}_1$ and $\mathcal{Q}_1$ respectively with last models $\mathcal{P}_2$ and $\mathcal{Q}_2$ such that $\pi^{\mathcal{T}_1}$ and $\pi^{\mathcal{U}_1}$ are defined, $\pi^{\mathcal{T}_1}(\mathcal{K}) = \pi^{\mathcal{U}_1}(\mathcal{K})$ and setting $\mathcal{K}' = \pi^{\mathcal{T}_1}(\mathcal{K})$, $\Sigma_{\mathcal{K}', \mathcal{T}_0^\frown \mathcal{T}_1} = \Lambda_{\mathcal{K}', \mathcal{U}_0^\frown \mathcal{U}_1}$[159].

[159]Because of Theorem 4.13.5 we can take $\mathcal{T}_1$ and $\mathcal{U}_1$ to be normal trees. We will always use the diamond comparison argument in situations where Theorem 4.13.5 applies to low level strategies.

7. $\mathcal{T}_1$ and $\mathcal{U}_1$ have a last normal component of successor length whose predecessor is a limit ordinal[160] and $\mathcal{T}_1^- = \mathcal{U}_1^-$.
8. $c = \Sigma_{\mathcal{P}_1,\mathcal{T}_0}(\mathcal{T}_1^-)$, $d = \Lambda_{\mathcal{Q}_1,\mathcal{U}_0}(\mathcal{U}_1^-)$[161].
9. $\Sigma_1 = \Sigma_{\mathcal{P}_1,\mathcal{T}_0}$, $\Sigma_2 = \Sigma_{\mathcal{P}_2,\mathcal{T}_0^\frown \mathcal{T}_1}$, $\Lambda_1 = \Sigma_{\mathcal{Q}_1,\mathcal{U}_0}$, and $\Lambda_2 = \Sigma_{\mathcal{Q}_2,\mathcal{U}_0^\frown \mathcal{U}_1}$,
10. $(\mathcal{T}_2,\mathcal{P}_3) \in I(\mathcal{P}_2,\Sigma_2) \cap I^{ope}(\mathcal{P}_2,\Sigma_2)$ and $(\mathcal{U}_2,\mathcal{Q}_3) \in I(\mathcal{Q}_2,\Lambda_2) \cap I^{ope}(\mathcal{Q}_2,\Lambda_2)$,
11. $\Sigma_3 = (\Sigma_2)_{\mathcal{P}_3,\mathcal{T}_2}$ and $\Lambda_3 = (\Lambda_2)_{\mathcal{Q}_3,\mathcal{U}_2}$.
12. $\mathcal{P}_3^b = \mathcal{Q}_3^b$ and $(\Sigma_3)_{\mathcal{P}_3^b} = (\Lambda_3)_{\mathcal{Q}_3^b}$.

We set $\mathcal{T}^B = \mathcal{T}_0^\frown \mathcal{T}_1^\frown \mathcal{T}_2$ and $\mathcal{U}^B = \mathcal{U}_0^\frown \mathcal{U}_1^\frown \mathcal{U}_2$. We say $\mathcal{T}^B$ is the stack on the top of B and $\mathcal{U}^B$ is the stack in the bottom of B.

Next we show that there cannot be a bad sequence of length ω_1.

LEMMA 4.14.2 (No bad sequences, $\mathsf{ZF}+\mathsf{DC}$). *Suppose $(\mathcal{P},\Sigma)$ and $(\mathcal{Q},\Lambda)$ are two hod pairs of limit type such that $\mathcal{P}$ and $\mathcal{Q}$ are countable, and both Σ and Λ are $(\omega_1,\omega_1,\omega_1)$-strategies. There is then no bad sequence, i.e., a sequence $(B_\beta : \beta < \omega_1)$ satisfying the following conditions:*

1. *For all $\beta < \omega_1$, $B_\beta = (((\mathcal{P}_{\beta,i},\mathcal{Q}_{\beta,i},\Sigma_{\beta,i},\Lambda_{\beta,i}) : i < 4),(\mathcal{T}_{\beta,i},\mathcal{U}_{\beta,i} : i < 3),(c_\beta,d_\beta))$.*
2. *For all $\beta < \omega_1$, B_β is a bad block on $((\mathcal{P}_{\beta,0},\Sigma_{\beta,0}),(\mathcal{Q}_{\beta,0},\Lambda_{\beta,0}))$.*
3. *For all $\beta < \omega_1$, $\mathcal{P}_{\beta+1,0} = \mathcal{P}_{\beta,3}$ and $\mathcal{Q}_{\beta+1,0} = \mathcal{Q}_{\beta,3}$.*
4. *For $\beta < \alpha < \omega_1$, let $\pi_{\beta,\alpha} : \mathcal{P}_{\beta,0} \to \mathcal{P}_{\alpha,0}$ be the composition of the embeddings on the "top" and $\sigma_{\beta,\alpha} : \mathcal{Q}_{\beta,0} \to \mathcal{Q}_{\alpha,0}$ be the composition of the embeddings on the "bottom". Then for all limit $\lambda < \omega_1$, $\mathcal{P}_{\lambda,0}$ is the direct limit of $(\mathcal{P}_\alpha,\pi_{\alpha,\beta} : \alpha < \beta < \lambda)$. Similarly, for all limit $\lambda < \omega_1$, $\mathcal{Q}_{\lambda,0}$ is the direct limit of $(\mathcal{Q}_\alpha,\sigma_{\alpha,\beta} : \alpha < \beta < \lambda)$ under the maps $\sigma_{\beta,\alpha}$.*
5. *For all limit ordinals $\lambda < \omega_1$, $\mathcal{P}_{\lambda,0}^b = \mathcal{Q}_{\lambda,0}^b$.*
6. *For all $\beta < \omega_1$, $\Sigma_{\beta,0} = \Sigma_{\mathcal{P}_{\beta,0},\oplus_{\gamma<\beta}\mathcal{T}^{B_\gamma}}$ and $\Lambda_{\beta,0} = \Sigma_{\mathcal{Q}_{\beta,0},\oplus_{\gamma<\beta}\mathcal{U}^{B_\gamma}}$.*

PROOF. Towards a contradiction, suppose $\vec{B} = (B_\beta : \beta < \omega_1)$ is a bad sequence. Let $\mathcal{P}_{\omega_1}$ be the direct limit of $(\mathcal{P}_{\alpha,0},\pi_{\alpha,\beta} : \alpha < \beta < \omega_1)$ and $\mathcal{Q}_{\omega_1}$ be the direct limit of $(\mathcal{Q}_{\alpha,0},\sigma_{\alpha,\beta} : \alpha < \beta < \omega_1)$. Let $N = L((\mathcal{P},\Sigma),(\mathcal{Q},\Lambda),\vec{B},\mathbb{R})$, $\zeta = \Theta^N$ and X be a countable submodel of $N|(\zeta^+)^N$ such that letting $\tau : M \to N|(\zeta^+)^N$ be the uncollapse map, $\vec{B} \in \mathrm{rge}(\tau)$. Let $\kappa = \omega_1^M$ and notice that for every $\beta < \kappa$,

$$B_\beta^- =_{\text{def}} (((\mathcal{P}_{\beta,i},\mathcal{Q}_{\beta,i}) : i < 4),(\mathcal{T}_{\beta,i},\mathcal{U}_{\beta,i} : i < 3),(c_\beta,d_\beta)) \in M$$

and B_β^- is countable in M. It then follows that $\tau^{-1}(\mathcal{P}_{\omega_1}) = \mathcal{P}_{\kappa,0}$ and $\tau^{-1}(\mathcal{Q}_{\omega_1}) = \mathcal{Q}_{\kappa,0}$. Let

$$\pi_\beta : \mathcal{P}_{\beta,0} \to \mathcal{P}_{\omega_1} \quad \text{and} \quad \sigma_\beta : \mathcal{Q}_{\beta,0} \to \mathcal{Q}_{\omega_1}$$

be the direct limit embeddings.

[160] Recall that in Definition 4.7.8, we required that comparison stacks have a last model.

[161] Thus, $\mathcal{P}_2 = \mathcal{M}_c^{\mathcal{T}_1^-}$ and $\mathcal{Q}_2 = \mathcal{M}_d^{\mathcal{T}_1^-}$.

Standard arguments show that for all $x \in \mathcal{P}_{\kappa,0} \cap \mathcal{Q}_{\kappa,0}$,

$$\pi_\kappa(x) = \tau(x) = \sigma_\kappa(x).$$

Notice that $\mathcal{P}^b_{\kappa,0} = \mathcal{Q}^b_{\kappa,0}$ (see clause 5 of our hypothesis). Set $\delta = \delta^{\mathcal{P}^b_{\kappa,0}}$ and let $\varphi = \pi^{\mathcal{T}_{\kappa,0}}$ and $\psi = \pi^{\mathcal{U}_{\kappa,0}}$. We now have that

(1) $\mathcal{P}^b_{\kappa,0} = \mathcal{Q}^b_{\kappa,0}$ and $\pi_\kappa \restriction \mathcal{P}^b_{\kappa,0} = \sigma_\kappa \restriction \mathcal{Q}^b_{\kappa,0}$.

Let

1. $\mathcal{K}$ witness clause 5 of Definition 4.14.1 for B_κ,
2. $p = \pi^{\mathcal{T}_{\kappa,1}}_{c_\kappa}$ and $q = \pi^{\mathcal{T}_{\kappa,1}}_{d_\kappa}$,
3. $i : \mathcal{P}_{\kappa,2} \to \mathcal{P}_{\omega_1}$ and $j : \mathcal{Q}_{\kappa,2} \to \mathcal{Q}_{\omega_1}$ be the iteration embeddings along the top and bottom of $\vec{B}$.

Notice that because

$$(\Sigma_{\kappa,1})_{\mathcal{K}^-} = (\Lambda_{\kappa,1})_{\mathcal{K}^-},$$

we have that

(2) $i \circ p \restriction \mathcal{K}^- = j \circ q \restriction \mathcal{K}^-$.

Next it follows from Lemma 2.10.15 that

(3) $\delta^{\mathcal{K}} = \sup\{\varphi(f)(a) : f \in \mathcal{P}_{\kappa,0} \wedge f : \delta \to \delta \wedge a \in (\mathcal{K}^-)^{<\omega}\}$,
(4) $\delta^{\mathcal{K}} = \sup\{\psi(f)(a) : f \in \mathcal{Q}_{\kappa,0} \wedge f : \delta \to \delta \wedge a \in (\mathcal{K}^-)^{<\omega}\}$.

Because

$$\mathcal{K}' =_{\text{def}} p(\mathcal{K}) = q(\mathcal{K}) \quad \text{and} \quad (\Sigma_{\kappa,2})_{\mathcal{K}'} = (\Lambda_{\kappa,2})_{\mathcal{K}'},$$

we have that

(5) $i \restriction \mathcal{K}' = j \restriction \mathcal{K}'$.

Let then

$$s = \{\varphi(f)(a) : f \in \mathcal{P}_{\kappa,0} \wedge f : \delta \to \delta \wedge a \in (\mathcal{K}^-)^{<\omega}\},$$
$$t = \{\psi(f)(a) : f \in \mathcal{Q}_{\kappa,0} \wedge f : \delta \to \delta \wedge a \in (\mathcal{K}^-)^{<\omega}\}.$$

(1) and (2) then imply that

(6) $i \circ p[s] = j \circ q[t]$.

(5) and (6) now imply that

(7) $p[s] = q[t]$.

It follows from (3), (4) and (7) that

(8) $p[s] \cap q[t]$ is cofinal in $\delta^{\mathcal{K}'}$.

It then follows that $c_\kappa = d_\kappa$, contradiction. $\dashv$

4.14.2. The comparison argument. In this subsection we prove the following comparison theorem under the hypothesis that the *lower level comparison* holds. Suppose $(\mathcal{P},\Sigma)$ and $(\mathcal{Q},\Lambda)$ are two hod pairs of limit type such that $\Gamma(\mathcal{P},\Sigma) = \Gamma(\mathcal{Q},\Lambda) =_{\text{def}} \Gamma$, both Σ and Λ are Γ-fullness preserving.

DEFINITION 4.14.3 (Lower Level Comparison). We say low level comparison holds for hod pairs or sts hod pairs $(\mathcal{P},\Sigma)$ and $(\mathcal{Q},\Lambda)$ if

1. for every $(\mathcal{T},\mathcal{P}_1) \in B(\mathcal{P},\Sigma)$ and $(\mathcal{U},\mathcal{Q}_1) \in B(\mathcal{Q},\Lambda)$, comparison holds for $(\mathcal{P}_1,\Sigma_{\mathcal{P}_1,\mathcal{T}})$ and $(\mathcal{Q}_1,\Lambda_{\mathcal{Q}_1,\mathcal{U}})$, and
2. whenever $(\mathcal{T},\mathcal{P}_1) \in I(\mathcal{P},\Sigma)$, $(\mathcal{U},\mathcal{Q}_1) \in I(\mathcal{Q},\Lambda)$ and $\mathcal{K}$ are such that
   (a) $\mathcal{K} \trianglelefteq_{hod} \mathcal{P}_1$ and $\mathcal{K} \trianglelefteq_{hod} \mathcal{Q}_1$,
   (b) $\mathcal{K}$ is of successor type and,
   (c) $\Sigma_{\mathcal{K}^-,\mathcal{T}} = \Lambda_{\mathcal{K}^-,\mathcal{U}}$,
   there is a normal stack $\mathcal{S}$ of limit length according to both $\Sigma_{\mathcal{P}_1,\mathcal{T}}$ and $\Lambda_{\mathcal{Q}_1,\mathcal{U}}$ that is based on $\mathcal{K}$ and is such that letting $b = \Sigma_{\mathcal{P}_1,\mathcal{T}}(\mathcal{S})$ and $c = \Lambda_{\mathcal{Q}_1,\mathcal{U}}(\mathcal{S})$,
   (a) $\pi_b^{\mathcal{S}}$ and $\pi_c^{\mathcal{S}}$ exist,
   (b) $\pi_b^{\mathcal{S}}(\mathcal{K}) = \pi_c^{\mathcal{S}}(\mathcal{K})$, and
   (c) letting $\mathcal{K}' = \pi_b^{\mathcal{S}}(\mathcal{K})$, $\Sigma_{\mathcal{K}',\mathcal{T}^\frown\mathcal{S}^\frown\{b\}} = \Lambda_{\mathcal{K}',\mathcal{U}^\frown\mathcal{S}^\frown\{c\}}$.

The following is the comparison theorem we will prove in this subsection. The theorem uses concepts defined in Definition 3.3.9 and Definition 3.10.4.

THEOREM 4.14.4 (Diamond comparison). *Suppose $(\mathcal{P},\Sigma)$ and $(\mathcal{Q},\Lambda)$ are two hod pairs such that $\Gamma(\mathcal{P},\Sigma) = \Gamma(\mathcal{Q},\Lambda) =_{def} \Gamma$, both Σ and Λ are Γ-fullness preserving $(\omega_1,\omega_1,\omega_1)$-strategies, $\mathcal{P}$ and $\mathcal{Q}$ are countable and are of limit type, and lower level comparison holds between $(\mathcal{P},\Sigma)$ and $(\mathcal{Q},\Lambda)$. Then there are $(\mathcal{T},\mathcal{R}) \in I(\mathcal{P},\Sigma)$ and $(\mathcal{U},\mathcal{R}) \in I(\mathcal{Q},\Lambda)$ such that either*

1. $\mathcal{P}$ and $\mathcal{Q}$ are of lsa type and $\Sigma^{stc}_{\mathcal{R},\mathcal{T}} = \Lambda^{stc}_{\mathcal{R},\mathcal{U}}$ or
2. $\mathcal{P}$ and $\mathcal{Q}$ are not of lsa type and $\Sigma_{\mathcal{R},\mathcal{T}} = \Lambda_{\mathcal{R},\mathcal{U}}$.

There are several other variations of the above theorem that works for sts hod pairs and also for a hod pair and an sts hod pair. We will state these theorems after the proof of Theorem 4.14.4. We prove Theorem 4.14.4 by showing that the failure of its conclusion produces a bad sequence of length ω_1. Towards showing this, we prove two useful lemmas.

We say that *weak comparison* holds between $(\mathcal{P},\Sigma)$ and $(\mathcal{Q},\Lambda)$ if there is $(\mathcal{T},\mathcal{U},\mathcal{R},\mathcal{S})$ such that

1. $(\mathcal{T},\mathcal{R}) \in I(\mathcal{P},\Sigma)$,
2. $(\mathcal{U},\mathcal{S}) \in I(\mathcal{Q},\Lambda)$,
3. $\mathcal{R}^b = \mathcal{S}^b$ and $\Sigma_{\mathcal{R}^b,\mathcal{T}} = \Lambda_{\mathcal{S}^b,\mathcal{U}}$.

Our first lemma says that lower level comparison implies that weak comparison holds.

LEMMA 4.14.5. *Suppose* $(\mathcal{P},\Sigma)$ *and* $(\mathcal{Q},\Lambda)$ *are two hod pairs such that*

$$\Gamma(\mathcal{P},\Sigma)=\Gamma(\mathcal{Q},\Lambda)=_{def}\Gamma,^{162}$$

both Σ *and* Λ *are* Γ*-fullness preserving,* $\mathcal{P}$ *and* $\mathcal{Q}$ *are of limit type, and that lower level comparison holds between* $(\mathcal{P},\Sigma)$ *and* $(\mathcal{Q},\Lambda)$*. Then weak comparison holds between* $(\mathcal{P},\Sigma)$ *and* $(\mathcal{Q},\Lambda)$*.*

PROOF. We inductively construct $(\mathcal{P}_i,\mathcal{T}_i : i<\omega)$ and $(\mathcal{Q}_i,\mathcal{U}_i : i<\omega)$ such that the following conditions hold.

1. $\mathcal{P}_0=\mathcal{P}$ and $\mathcal{Q}_0=\mathcal{Q}$.
2. Suppose $i=2n$. Then the following holds.
   (a) $\mathcal{T}_i$ is a stack on $\mathcal{P}_i$ based on $\mathcal{P}_i^b$ and according to $\Sigma_{\mathcal{P}_i,\oplus_{k<i}\mathcal{T}_k}$ with last model $\mathcal{P}_{i+1}$.
   (b) $\mathcal{U}_i$ is a stack on $\mathcal{Q}_i$ according to $\Lambda_{\mathcal{Q}_i,\oplus_{k<i}\mathcal{U}_i}$ with last model $\mathcal{Q}_{i+1}$.
   (c) Letting $\mathcal{P}_i'$ be the least non gentle layer of $\mathcal{P}_{i+1}$ such that $\pi^{\mathcal{T}_i}[\mathcal{P}_i^b]\subseteq\mathcal{P}_i'$, $\mathcal{P}_i'\trianglelefteq_{hod}\mathcal{Q}_{i+1}^b$ and $\Lambda_{\mathcal{P}_i',\oplus_{k\leq i}\mathcal{U}_k}=\Sigma_{\mathcal{P}_i',\oplus_{k\leq i}\mathcal{T}_k}$.
3. Suppose $i=2n+1$. Then the following holds.
   (a) $\mathcal{T}_i$ is a stack on $\mathcal{P}_i$ according to $\Sigma_{\mathcal{P}_i,\oplus_{k<i}\mathcal{T}_k}$ with last model $\mathcal{P}_{i+1}$.
   (b) $\mathcal{U}_i$ is a stack on $\mathcal{Q}_i$ based on $\mathcal{Q}_i^b$ and according to $\Lambda_{\mathcal{Q}_i,\oplus_{k<i}\mathcal{U}_i}$ with last model $\mathcal{Q}_{i+1}$.
   (c) Letting $\mathcal{Q}_i'$ be the least non gentle layer of $\mathcal{Q}_{i+1}$ such that $\pi^{\mathcal{U}_i}[\mathcal{Q}_i^b]\subseteq\mathcal{Q}_i'$, $\mathcal{Q}_i'\trianglelefteq_{hod}\mathcal{P}_{i+1}^b$ and $\Lambda_{\mathcal{Q}',\oplus_{k\leq i}\mathcal{U}_k}=\Sigma_{\mathcal{Q}',\oplus_{k\leq i}\mathcal{T}_k}$.

We show how to carry out the inductive step. Suppose we have constructed $(\mathcal{P}_i,\mathcal{Q}_i : i\leq 2n)$ and $(\mathcal{T}_i,\mathcal{U}_i : i<2n)$. We now consider two cases.

Case 1. $\mathrm{cf}^{\mathcal{P}_{2n}}(\delta^{\mathcal{P}_{2n}^b})$ is not a measurable cardinal in $\mathcal{P}_{2n}$.

Notice that in this case, we have that $\mathcal{P}_1=\mathcal{Q}_1$ and $\Sigma_{\mathcal{P}_1,\mathcal{T}_0}=\Lambda_{\mathcal{Q}_1,\mathcal{U}_0}$. Thus, weak comparison holds for $(\mathcal{P},\Sigma)$ and $(\mathcal{Q},\Lambda)$ provided we can take care of $n=0$ case. Notice also that in this case $\mathcal{P}_0=\mathcal{P}_0^b$.

Let $(\mathcal{N}_i : i<\omega)$ be a sequence of layers of $\mathcal{P}(=\mathcal{P}_0)$ such that

1. for all $i<\omega$, $\delta^{\mathcal{N}_i}$ is a Woodin cardinal of $\mathcal{P}$,
2. for all $i<\omega$, $\mathcal{N}_i\triangleleft_{hod}\mathcal{N}_{i+1}$ and
3. $\mathcal{P}|\delta^{\mathcal{P}}=\cup_{i<\omega}\mathcal{N}_i$.

By induction we construct a sequence $(\mathcal{T}_k^*,\mathcal{W}_k,\mathcal{S}_k,\mathcal{R}_k,\mathcal{S}_k^*,\mathcal{R}_k^*,\mathcal{R}_k^{**} : k<\omega)$ such that the following hold.

1. $(\mathcal{S}_0^*,\mathcal{R}_0^*)\in I(\mathcal{Q},\Lambda_{\mathcal{Q}})$, $\mathcal{R}_0^{**}\trianglelefteq_{hod}\mathcal{R}_0^*$ and
$$\Gamma(\mathcal{N}_0,\Sigma_{\mathcal{N}_0})=\Gamma(\mathcal{R}_0^{**},\Lambda_{\mathcal{R}_0^{**},\mathcal{S}_0^*}).$$
   Also, $(\mathcal{T}_0^*,\mathcal{W}_0)\in I(\mathcal{P},\Sigma)$, $(\mathcal{S}_0,\mathcal{R}_0)\in I(\mathcal{R}_0^*,\Lambda_{\mathcal{R}_0^*,\mathcal{S}_0^*})$ and the following conditions hold:
   (a) $\mathcal{T}_0^*$ is based on $\mathcal{N}_0$ and $\mathcal{S}_0$ is based on $\mathcal{R}_0^{**}$.

---

[162] See Definition 3.10.4.

(b) $\pi^{\mathcal{T}_0^*}(\mathcal{N}_0) = \pi^{\mathcal{S}_0}(\mathcal{R}_0^{**})$ and
(c) letting $\mathcal{K} = \pi^{\mathcal{T}_0^*}(\mathcal{N}_0)$,

$$\Sigma_{\mathcal{K},\mathcal{T}_0^*} = \Lambda_{\mathcal{K},\mathcal{S}_0^{*\frown}\mathcal{S}_0}.$$

2. For $k+1 < \omega$, $(\mathcal{S}_{k+1}^*, \mathcal{R}_{k+1}^*) \in I(\mathcal{R}_k, \Lambda_{\mathcal{R}_k, \oplus_{m \leq k}(\mathcal{S}_m^{*\frown}\mathcal{S}_m)})$, $\mathcal{R}_{k+1}^{**} \lhd_{hod} \mathcal{R}_{k+1}^*$ and

$$\Gamma(\mathcal{N}_{k+1}^*, \Sigma_{\mathcal{N}_{k+1}^*, \oplus_{m\leq k}\mathcal{T}_m^*}) = \Gamma(\mathcal{R}_{k+1}^{**}, \Lambda_{\mathcal{R}_{k+1}^{**}, \oplus_{m\leq k}(\mathcal{S}_m^{*\frown}\mathcal{S}_m)}).$$

where $\mathcal{N}_{k+1}^* = \pi^{\oplus_{m\leq k}\mathcal{T}_m^*}(\mathcal{N}_{k+1})$. Also,

$$(\mathcal{T}_{k+1}^*, \mathcal{W}_{k+1}) \in I(\mathcal{W}_k, \Sigma_{\mathcal{W}_k, \oplus_{m\leq k}\mathcal{T}_m^*}),$$
$$(\mathcal{S}_{k+1}, \mathcal{R}_{k+1}) \in I(\mathcal{R}_{k+1}^*, \Lambda_{\mathcal{R}_{k+1}^*, \oplus_{m\leq k}(\mathcal{S}_m^{*\frown}\mathcal{S}_m))^\frown \mathcal{S}_{k+1}^*})$$

and the following conditions hold:
(a) $\mathcal{T}_{k+1}^*$ is based on $\mathcal{N}_{k+1}^*$ and $\mathcal{S}_{k+1}$ is based on $\mathcal{R}_{k+1}^{**}$.
(b) $\pi^{\mathcal{T}_{k+1}^*}(\mathcal{N}_{k+1}^*) = \pi^{\mathcal{S}_{k+1}}(\mathcal{R}_{k+1}^{**})$ and
(c) letting $\mathcal{K} = \pi^{\mathcal{T}_{k+1}^*}(\mathcal{N}_{k+1}^*)$,

$$\Sigma_{\mathcal{K}, \oplus_{m\leq k+1}\mathcal{T}_m^*} = \Lambda_{\mathcal{K}, \oplus_{m\leq k+1}(\mathcal{S}_m^{*\frown}\mathcal{S}_m)}.$$

We then let $\mathcal{T}_0 = \oplus_{k<\omega}\mathcal{T}_k^*$ and $\mathcal{U}_0 = \oplus_{m<\omega}\mathcal{S}_k^{*\frown}\mathcal{S}$. Also, we let $\mathcal{P}_1$ be the last model of $\mathcal{T}_0$ and $\mathcal{Q}_1$ be the last model of $\mathcal{U}_0$.

Case 2. $\mathrm{cf}^{\mathcal{P}_{2n}}(\delta^{\mathcal{P}_{2n}^b})$ is a measurable cardinal in $\mathcal{P}$.

The difference between this case and the previous case is that here we cannot start by fixing $(\mathcal{N}_i : i < \omega)$ as above. Here is the outline of the construction of $(\mathcal{T}_{2n}, \mathcal{U}_{2n}, \mathcal{P}_{2n+1}, \mathcal{Q}_{2n+1})$.

Because $\Gamma(\mathcal{P}_{2n}, \Sigma_{\mathcal{P}_{2n}, \oplus_{i<2n}\mathcal{T}_i}) = \Gamma(\mathcal{Q}_{2n}, \Lambda_{\mathcal{Q}_{2n}, \oplus_{i<2n}\mathcal{U}_i})$, we can find

$$(\mathcal{S}_0, \mathcal{R}_0) \in I(\mathcal{Q}_{2n}, \Lambda_{\mathcal{Q}_{2n}, \oplus_{i<2n}\mathcal{U}_i})$$

and $\mathcal{R}_0^* \lhd_{hod} \mathcal{R}_0$ such that letting $E \in \vec{E}^{\mathcal{P}_{2n}}$ be the extender of Mitchel order 0 with $\mathrm{crit}(E) = \mathrm{cf}^{\mathcal{P}_{2n}}(\delta^{\mathcal{P}_{2n}})$,

$$\Gamma(\mathcal{P}_{2n}^b, \Sigma_{\mathcal{P}_{2n}, (\oplus_{i<2n}\mathcal{T}_i)^\frown\{\mathrm{Ult}(\mathcal{P}_{2n}, E)\}}) = \Gamma(\mathcal{R}_0^*, \Lambda_{\mathcal{R}_0^*, (\oplus_{i<2n}\mathcal{U}_i)^\frown\{\mathcal{S}_0\}}).$$

Appealing to low level comparison, we can find

$$(\mathcal{T}_{2n}^*, \mathcal{P}_{2n+1}) \in I(\mathrm{Ult}(\mathcal{P}_{2n}, E), \Sigma_{\mathcal{P}_{2n}, (\oplus_{i<2n}\mathcal{T}_i)^\frown\{\mathrm{Ult}(\mathcal{P}_{2n}, E)\}})$$

and

$$(\mathcal{S}_1, \mathcal{R}_1) \in I(\mathcal{R}_0, \Lambda_{\mathcal{R}_0, (\oplus_{i<2n}\mathcal{U}_i)^\frown\mathcal{S}_0})$$

such that

1. $\mathcal{T}_{2n}^*$ is based on $\mathcal{P}_{2n}^b$,
2. $\mathcal{S}_1$ is based on $\mathcal{R}_0^*$,
3. $\pi^{\mathcal{T}_{2n}^*}(\mathcal{P}_{2n}^b) = \pi^{\mathcal{S}_1}(\mathcal{R}_0^*) =_{\mathrm{def}} \mathcal{K}$, and
4. $\Sigma_{\mathcal{K}, (\oplus_{i<2n}\mathcal{T}_i)^\frown\{E\}^\frown\mathcal{T}_{2n}^*} = \Lambda_{\mathcal{K}, (\oplus_{i<2n}\mathcal{U}_i)^\frown\mathcal{S}_0^\frown\mathcal{S}_1}$.

Let then $\mathcal{T}_{2n} = \{E\}^\frown \mathcal{T}^*_{2n}$, $\mathcal{U}_{2n} = \mathcal{S}_0^\frown \mathcal{S}_1$ and $\mathcal{Q}_{2n+1} = \mathcal{R}_1$.

The two cases above finish the construction of $(\mathcal{T}_{2n}, \mathcal{U}_{2n}, \mathcal{P}_{2n+1}, \mathcal{Q}_{2n+1})$. The construction of $(\mathcal{T}_{2n+1}, \mathcal{U}_{2n+1}, \mathcal{P}_{2n+2}, \mathcal{Q}_{2n+2})$ is very similar and we leave it to the reader.

Notice now that if $\mathcal{T} = \oplus_{i<\omega}\mathcal{T}_i$, $\mathcal{U} = \oplus_{i<\omega}\mathcal{U}_i$, $\mathcal{R}$ is the last model of $\mathcal{T}$ and $\mathcal{S}$ is the last model of $\mathcal{U}$ then $(\mathcal{T}, \mathcal{R})$ and $(\mathcal{U}, \mathcal{S})$ witness that weak comparison holds for $(\mathcal{P}, \Sigma)$ and $(\mathcal{Q}, \Lambda)$. ⊣

LEMMA 4.14.6. *Suppose $(\mathcal{P}, \Sigma)$ and $(\mathcal{Q}, \Lambda)$ are two hod pairs such that*

$$\Gamma(\mathcal{P}, \Sigma) = \Gamma(\mathcal{Q}, \Lambda) =_{def} \Gamma,$$

both Σ and Λ have strong branch condensation and are strongly Γ-fullness preserving, both $\mathcal{P}$ and $\mathcal{Q}$ are of limit type and low level comparison holds for $(\mathcal{P}, \Sigma)$ and $(\mathcal{Q}, \Lambda)$. Suppose further that $\mathcal{P}^b = \mathcal{Q}^b$ and $\Sigma_{\mathcal{P}^b} = \Lambda_{\mathcal{Q}^b}$. Let $(\mathcal{T}, \mathcal{R}, \mathcal{U}, \mathcal{S})$ be the trees of the extender comparison of $\mathcal{P}$ and $\mathcal{Q}$[163]*. Suppose that either*

1. *$\mathcal{R} \neq \mathcal{S}$ or*
2. *$\mathcal{R} = \mathcal{S}$ and $\Sigma_{\mathcal{R},\mathcal{T}} \neq \Lambda_{\mathcal{S},\mathcal{U}}$.*

Then there is a bad block on $((\mathcal{P}, \Sigma), (\mathcal{Q}, \Lambda))$.

PROOF. It follows from Lemma 4.7.2 that we can find minimal low level disagreement $((\mathcal{T}^*, \mathcal{P}^*), (\mathcal{U}^*, \mathcal{Q}^*), \mathcal{K})$ between $(\mathcal{R}, \Sigma_{\mathcal{R},\mathcal{T}})$ and $(\mathcal{S}, \Lambda_{\mathcal{S},\mathcal{U}})$. Let E be the $\mathcal{W}$-un-dropping extender of $\mathcal{T}^\frown\mathcal{T}^*$ and F be the $\mathcal{W}$-un-dropping extender of $\mathcal{U}^\frown\mathcal{U}^*$, and let $\mathcal{T}_0$ be the extension of $\mathcal{T}^\frown\mathcal{T}^*$ obtained by applying E and $\mathcal{U}_0$ be the extension of $\mathcal{U}^\frown\mathcal{U}^*$ obtained by applying F. We then let $\mathcal{P}_1$ and $\mathcal{Q}_1$ be the last models of $\mathcal{T}_0$ and $\mathcal{U}_0$.

Let $\mathcal{X}$ be a normal stack as in clause 2 of Definition 4.14.3. Let $b = \Sigma(\mathcal{T}_0^\frown\mathcal{S})$, $c = \Lambda(\mathcal{U}_0^\frown\mathcal{X})$, $\mathcal{P}_2 = \mathcal{M}_b^{\mathcal{T}_1}$ and $\mathcal{Q}_2 = \mathcal{M}_c^{\mathcal{X}}$. Set $\mathcal{T}_1 = \mathcal{X}^\frown\{b\}$ and $\mathcal{U}_1 = \mathcal{X}^\frown\{c\}$. We thus have that $\pi^{\mathcal{T}_1}$ and $\pi^{\mathcal{U}_1}$ exist, $\pi^{\mathcal{T}_1}(\mathcal{K}) = \pi^{\mathcal{U}_1}(\mathcal{K})$ and

$$\Sigma_{\pi^{\mathcal{T}_1}(\mathcal{K}),\mathcal{T}_0^\frown\mathcal{T}_1} = \Lambda_{\pi^{\mathcal{U}_1}(\mathcal{K}),\mathcal{U}_0^\frown\mathcal{U}_1}.$$

Next (appealing to Lemma 4.14.5) let $(\mathcal{T}_2, \mathcal{P}_3)$ and $(\mathcal{U}_2, \mathcal{Q}_3)$ witness that the weak comparison holds for

$$(\mathcal{P}_2, \Sigma_{\mathcal{P}_2,\mathcal{T}_0^\frown\mathcal{T}_1}) \quad \text{and} \quad (\mathcal{Q}_2, \Lambda_{\mathcal{Q}_2,\mathcal{U}_0^\frown\mathcal{U}_1}).$$

Next let $\mathcal{P}_0 = \mathcal{P}$, $\mathcal{Q}_0 = \mathcal{Q}$, $\Sigma_0 = \Sigma$, $\Lambda_0 = \Lambda$, and for $i \in \{1,2,3\}$ let $\Sigma_i = \Sigma_{\mathcal{P}_i,\oplus_{k<i}\mathcal{T}_k}$ and $\Lambda_i = \Lambda_{\mathcal{Q}_i,\oplus_{k<i}\mathcal{U}_k}$. It is then easy to see that

$$(((\mathcal{P}_i, \mathcal{Q}_i, \Sigma_i, \Lambda_i) : i < 4), (\mathcal{T}_i, \mathcal{U}_i : i < 3), (b, c))$$

is a bad block on $((\mathcal{P}, \Sigma), (\mathcal{Q}, \Lambda))$. ⊣

PROOF OF THEOREM 4.14.4. Suppose that the conclusion of Theorem 4.14.4 fails. This means that

(1) whenever $(\mathcal{T}, \mathcal{R}) \in I(\mathcal{P}, \Sigma)$ and $(\mathcal{U}, \mathcal{R}) \in I(\mathcal{Q}, \Lambda)$,

[163] Thus, $\mathcal{T}$ is on $\mathcal{P}$ with last model $\mathcal{R}$ and $\mathcal{U}$ is on $\mathcal{Q}$ with last model $\mathcal{S}$. See Definition 4.7.11.

1. if $\mathcal{P}$ and $\mathcal{Q}$ are of lsa type then $\Sigma^{stc}_{\mathcal{R},\mathcal{T}} \neq \Lambda^{stc}_{\mathcal{R},\mathcal{U}}$ or
2. if $\mathcal{P}$ and $\mathcal{Q}$ are not of lsa type then $\Sigma_{\mathcal{R},\mathcal{T}} \neq \Lambda_{\mathcal{R},\mathcal{U}}$.

It follows from Lemma 4.14.5 that, without loss of generality, we can assume that $\mathcal{P}^b = \mathcal{Q}^b$ and $\Sigma_{\mathcal{P}^b} = \Lambda_{\mathcal{Q}^b}$. We now by induction construct a bad sequence $(B_\alpha : \alpha < \omega_1)$ on $((\mathcal{P},\Sigma),(\mathcal{Q},\Lambda))$.

It follows from Lemma 4.14.6 that there is a bad block on $((\mathcal{P},\Sigma),(\mathcal{Q},\Lambda))$. Let B_0 be any bad block on $((\mathcal{P},\Sigma),(\mathcal{Q},\Lambda))$. Suppose next that we have constructed $(B_\beta : \beta < \lambda)$ for λ a limit. Let $\mathcal{P}_\lambda$ and $\mathcal{Q}_\lambda$ be the direct limit of respectively $(\mathcal{P}_\beta : \beta < \lambda)$ and $(\mathcal{Q}_\beta : \beta < \lambda)$ under the corresponding iteration embeddings. Then letting $\Sigma_{\lambda,0}$ and $\Lambda_{\lambda,0}$ be the corresponding tails of Σ and Λ, we have that $(\mathcal{P}_\lambda,\Sigma_\lambda)$ and $(\mathcal{Q}_\lambda,\Lambda_\lambda)$ satisfy the hypothesis of Lemma 4.14.6. Let then B_λ be a bad block on $((\mathcal{P}_\lambda,\Sigma_\lambda),(\mathcal{Q}_\lambda,\Lambda_\lambda))$.

Next suppose that we have constructed $(B_\beta : \beta < \lambda+1)$. Let $\mathcal{P}_{\lambda+1} = \mathcal{P}_{\lambda,3}$, $\mathcal{Q}_{\lambda+1} = \mathcal{Q}_{\lambda,3}$ and let $\mathcal{T}$ and $\mathcal{U}$ be the stacks respectively on the top of $(B_\beta : \beta < \lambda+1)$ and in the bottom of $(B_\beta : \beta < \lambda+1)$. We then again can find, using Lemma 4.14.6, a bad block $B_{\lambda+1}$ on $((\mathcal{P}_{\lambda+1},\Sigma_{\mathcal{P}_{\lambda+1},\mathcal{T}}),(\mathcal{Q}_{\lambda+1},\Lambda_{\mathcal{Q}_{\lambda+1},\mathcal{U}}))$. It then follows that the resulting sequence $(B_\beta : \beta < \omega_1)$ is a bad sequence on $((\mathcal{P},\Sigma),(\mathcal{Q},\Lambda))$. This is a contradiction to Lemma 4.14.2, and this contradiction completes the proof of Theorem 4.14.4. ⊣

4.15. Some concluding remarks

The proof of Theorem 4.14.4 can be used to show that fullness preserving strategies that have strong branch condensation become commuting on a tail. We end this section by an outline of this useful fact.

PROPOSITION 4.15.1. *Suppose $(\mathcal{P},\Sigma)$ is an sts hod pair and Γ is a projectively closed pointclass. Suppose that Σ has strong branch condensation and is Γ-fullness preserving. Then for some $(\mathcal{T},\mathcal{Q}) \in I^{ope}(\mathcal{P},\Sigma)$, $\Sigma_\mathcal{Q}$ is commuting*[164].

PROOF. Towards a contradiction assume not. Then we can find a sequence

$$c = (\mathcal{P}_\alpha,\mathcal{T}_\alpha,\mathcal{Q}_\alpha,\mathcal{U}_\alpha,\mathcal{R}_\alpha,k'_\alpha,k_\alpha : \alpha \leq \omega_1)$$

such that the following conditions hold:

1. For each $\alpha < \omega_1$, $\mathcal{P}_{\alpha+1}$ is the result of comparing the pairs $(\mathcal{Q}_\alpha,\Sigma_{\mathcal{Q}_\alpha})$ and $(\mathcal{R}_\alpha,\Sigma_{\mathcal{R}_\alpha})$[165]
2. For each $\alpha < \omega_1$, $(\mathcal{P}_\alpha,(\mathcal{T}_\alpha,\mathcal{Q}_\alpha),(\mathcal{U}_\alpha,\mathcal{R}_\alpha),k'_\alpha,k_\alpha)$ witnesses that $\Sigma_{\mathcal{P}_\alpha}$ is not commuting.

[164] See Definition 4.10.5.

[165] The comparison is possible because of Corollary 4.10.7.

3. For each limit ordinal $\alpha \leq \omega_1$[166], $\mathcal{P}_\alpha$ is the direct limit of $(\mathcal{P}_\beta, \pi^t_{\beta,\gamma} : \beta < \gamma < \alpha)$[167] where $\pi^t_{\beta,\gamma} : \mathcal{P}_\beta \to \mathcal{P}_\gamma$ is the embedding given by concatenating the $\mathcal{P}_\beta$-to-$\mathcal{Q}_\beta$-to-$\mathcal{P}_{\beta+1}$ stacks.

It follows from Proposition 4.10.3 that in clause 3 above we could define $\mathcal{P}_\alpha$ as the direct limit of $(\mathcal{P}_\beta, \pi^b_{\beta,\gamma} : \beta < \gamma < \alpha)$ where $\pi^b_{\beta,\gamma} : \mathcal{P}_\beta \to \mathcal{P}_\gamma$ is the embedding given by concatenating the $\mathcal{P}_\beta$-to-$\mathcal{R}_\beta$-to-$\mathcal{P}_{\beta+1}$ stacks.

Suppose now that $\sigma : H \to H_{\omega_2}$ is such that H is countable and transitive, and $c \in \text{rge}(\sigma)$. Let $\kappa = \omega_1^H$. It follows that

$$\pi^t_{\kappa,\omega_1} = \sigma \restriction \mathcal{P}_\kappa = \pi^b_{\kappa,\omega_1}.$$

Let now $j : \mathcal{Q}_\kappa \to \mathcal{P}_{\omega_1}$ and $i : \mathcal{R}_\kappa \to \mathcal{P}_{\omega_1}$ be the two iteration embeddings. It follows from strong branch condensation and in particular from Proposition 4.10.3 that letting $\delta = \delta^{\mathcal{R}^b_\kappa}$,

(1) $i \restriction \mathcal{R}_\kappa|\delta = j \restriction \mathcal{Q}_\kappa|\delta.$[168]

Hence, we have that $\pi^{\mathcal{U}_\kappa} \restriction \mathcal{P}_\kappa|\delta^{\mathcal{P}^b_\kappa} = \pi^{\mathcal{T}_\kappa} \restriction \mathcal{P}_\kappa|\delta^{\mathcal{P}^b_\kappa}$. It remains to show that for $A \in \wp(\delta^{\mathcal{P}^b_\kappa}) \cap \mathcal{P}_\kappa$, $\pi^{\mathcal{U}_\kappa}(A) = k_\kappa(A)$. But we have that

(2) $i(\pi^{\mathcal{U}_k}(A)) = j(\pi^{\mathcal{T}_\kappa}(A))$ and $k_\kappa(A) = \pi^{\mathcal{T}_\kappa}(A) \cap \delta$.

It then follows from (1) and (2) that $\pi^{\mathcal{U}_\kappa}(A) = k_\kappa(A)$. ⊣

The following proposition implies that in many situations we can construct authenticating iterations as described in Definition 3.7.3. We will use it in the proof of Theorem 6.1.4.

PROPOSITION 4.15.2. *Suppose $(\mathcal{P}, \Sigma)$ is an sts pair and Γ is a projectively closed pointclass. Suppose Σ is*

1. strongly Γ-fullness preserving,
2. has strong branch condensation and
3. is commuting[169].

Suppose $(\mathcal{T}, \mathcal{Q}) \in I^{ope}(\mathcal{P}, \Sigma)$, $(\mathcal{U}, \mathcal{R}) \in I^{ope}(\mathcal{P}, \Sigma)$ and $(\mathcal{W}, \mathcal{S}) \in I^{ope}(\mathcal{R}, \Sigma_\mathcal{R})$ are such that for some $\delta < \delta^{\mathcal{Q}^b}$ the following conditions hold:

1. $\mathcal{Q} \models$ "δ is a Woodin cardinal",
2. $\mathcal{W}$ is a normal stack, and
3. $\mathcal{S}|\delta = \mathcal{Q}|\delta$.

Let

1. $\mathcal{K} \trianglelefteq_{hod} \mathcal{Q}$ be such that $\delta^\mathcal{K} = \delta$,

[166] The rest of the objects are undefined for $\alpha = \omega_1$.

[167] Notice that in Definition 4.10.5 we can assume that $\pi^\mathcal{T}$ is defined, possibly by applying undropping extenders. This is because commuting for sts hod pairs is a principle about the bottom parts not the entire model.

[168] Notice that $k_\kappa \restriction \delta = id$.

[169] See Definition 4.10.5.

2. $\alpha < \mathrm{lh}(\mathcal{W})$ *be the least such that* $\mathcal{K}^- \trianglelefteq \mathcal{M}_\alpha^{\mathcal{W}}$,
3. $\beta < \mathrm{lh}(\mathcal{W})$ *be such that* $\mathrm{m}(\mathcal{W}_{<\beta}) = \mathcal{K}|\delta$,
4. w *is the window of* $\mathcal{Q}$ *such that* $\delta^w = \delta$[170]*, and*
5. $b = [0, \beta)_{\mathcal{W}}$.

Then $s(\mathcal{T}, w) \subseteq \pi_b^{\mathcal{W}[\alpha,\beta)}$[171]

The Proposition 4.15.2 can be proven by simply comparing $(\mathcal{S}, \Sigma_{\mathcal{S}})$ and $(\mathcal{Q}, \Sigma_{\mathcal{Q}})$ and then using commuting and Corollary 4.10.7.

[170] See Notation 2.7.14.
[171] See Definition 2.9.1.

Chapter 5

HOD MICE REVISITED

In this section we generalize the result of [29, Chapter 3] to our current context. As in [29], these results lead towards showing that given a hod pair $(\mathcal{P},\Sigma)$, $\Gamma(\mathcal{P},\Sigma)$ is an *OD*-full pointclass (see Definition 3.16 of [29]).

Recall the effect of Proposition 4.10.2; if $(\mathcal{P},\Sigma)$ is a hod pair such that Σ has strong branch condensation and if $\mathcal{Q} \in pI(\mathcal{P},\Sigma)$, then the strategy of $\mathcal{Q}$ induced by Σ is independent of the particular iteration producing $\mathcal{Q}$. In Section 4.10, this strategy was denoted by $\Sigma_{\mathcal{Q}}$. In this chapter, whenever the strategy of a hod mouse has a strong branch condensation, we will make use of the aforementioned notation without giving any further explanation.

5.1. The uniqueness of the internal strategy

The first theorem, Theorem 5.1.2, is just a direct generalization of [29, Theorem 3.3]. It says that the internal strategies are unique. First we prove a useful lemma.

LEMMA 5.1.1. *Suppose $\mathcal{P}$ is a hod premouse, $\mathcal{Q} \trianglelefteq_{hod} \mathcal{P}$, $\mathcal{U} \in \mathcal{P}$ is a stack on $\mathcal{Q}$ with last model $\mathcal{R}$ such that $\mathcal{U}$ has a one point extension*[1]*, and $\mathcal{R}' \trianglelefteq_{hod} \mathcal{R}$ is such that $\mathcal{R} \models$ "$\delta^{\mathcal{R}'}$ is a Woodin cardinal". Suppose further that if $\pi^{\mathcal{U}}$ is undefined then letting E be the $\mathcal{R}'$-un-dropping extender of $\mathcal{U}$,* $\mathrm{Ult}(\mathcal{P},E)$ *is well-founded. Then* $\mathrm{cf}^{\mathcal{P}}(\delta^{\mathcal{R}'}) > \omega$.

PROOF. Towards a contradiction, assume not. We give the proof assuming that $\pi^{\mathcal{U}}$ is defined. If not, then one could instead work with $\mathrm{Ult}(\mathcal{P},E)$ instead of $\mathcal{P}$ and π_E instead of $\pi^{\mathcal{U}}$, where E is the $\mathcal{R}'$-un-dropping extender of $\mathcal{U}$.

Notice that it cannot be the case that $\mathcal{R}' \in \mathrm{rge}(\pi^{\mathcal{U}})$ as $\pi^{\mathcal{U}}$ is continuous at the Woodin cardinals of $\mathcal{P}$. Therefore, by minimizing $\mathcal{Q}$, we can assume that $\mathcal{Q}$ is of limit type. We now apply Lemma 2.9.5 to $(\mathcal{U},w)$ where w is the window of $\mathcal{R}$ such that $\delta^w = \delta^{\mathcal{R}'}$. Let $\eta = \eta^w$. We thus have that there is a sequence $(h_i : i < \omega) \in \mathcal{Q}^b$ and a sequence $(a_i : i < \omega) \in (\eta^{<\omega})^\omega$ such that

$$\delta^{\mathcal{R}'} = \sup\{\pi^{\mathcal{U}}(h_i)(a_i) : i < \omega\}.$$

[1] See Definition 2.10.2.

Notice now that $(\pi^{\mathcal{U}}(h_i) : i < \omega) \in \mathcal{R}$. Therefore,

$$\delta^{\mathcal{R}'} = \sup\{\pi^{\mathcal{U}}(h_i)(a) : a \in [\eta]^{<\omega} \wedge \pi^{\mathcal{U}}(h_i)(a) < \delta^{\mathcal{R}'}\}.$$

It then follows that $\mathcal{R} \models \mathrm{cf}(\delta^{\mathcal{R}'}) \leq \eta$, which is a contradiction as $\delta^{\mathcal{R}'}$ is a Woodin cardinal of $\mathcal{R}$. ⊣

THEOREM 5.1.2 (Uniqueness of internal strategies). *Suppose $\mathcal{P}$ is a hod premouse such that $\mathcal{P} \models \mathsf{ZFC} - \mathsf{Powerset}$, $\delta^{\mathcal{P}}$ is a regular cardinal of $\mathcal{P}$ and $\mathcal{W} \lhd_{hod} \mathcal{P}$ is such that $\mathcal{P} \models$ "$\Sigma^{\mathcal{P}}_{\mathcal{W}}$ is a $((\delta^{\mathcal{P}})^+, (\delta^{\mathcal{P}})^+)$-strategy"[2]. Then $\mathcal{P} \models$ "$\mathcal{W}$ has a unique iteration strategy ".*

PROOF. Working in $\mathcal{P}$, suppose $\Lambda \neq \Sigma^{\mathcal{P}}_{\mathcal{W}}$ is another iteration strategy for $\mathcal{W}$. Let $\Sigma = \Sigma^{\mathcal{P}}_{\mathcal{W}}$. Notice that Lemma 5.1.1 implies that if

1. $\mathcal{U}$ is a stack on $\mathcal{W}$ according to both Λ and Σ,
2. $\mathrm{lh}(\mathcal{U})$ is a limit ordinal, and
3. $b = \Sigma(\mathcal{U})$ and $c = \Lambda(\mathcal{U})$

then

$(*)$ either
  (A) both $\mathcal{Q}(b,\mathcal{U})$ and $\mathcal{Q}(c,\mathcal{U})$ exist, or
  (B) $b = c$.

This is because if $b \neq c$ then $\mathrm{cf}^{\mathcal{P}}(\delta(\mathcal{U})) = \omega$ and hence, we have that

1. either $\pi^{\mathcal{U}}_b$ is undefined or $\delta(\mathcal{U})$ is not a Woodin cardinal of $\mathcal{M}^{\mathcal{U}}_b$, and
2. either $\pi^{\mathcal{U}}_c$ is undefined or $\delta(\mathcal{U})$ is not a Woodin cardinal of $\mathcal{M}^{\mathcal{U}}_c$.

The above clauses imply that $\delta(\mathcal{U})$ is not a Woodin cardinal neither in $\mathcal{M}^{\mathcal{U}}_b$ nor in $\mathcal{M}^{\mathcal{U}}_c$. Therefore, both $\mathcal{Q}(b,\mathcal{U})$ and $\mathcal{Q}(c,\mathcal{U})$ exist.

It then follows from the proof of Lemma 4.7.2[3] that we can find a minimal low-level disagreement $(\mathcal{T}_1, \mathcal{W}_1, \mathcal{T}_2, \mathcal{W}_2, \mathcal{Q})$ between $(\mathcal{W}, \Sigma)$ and $(\mathcal{W}, \Lambda)$. Moreover, we can assume that $\mathrm{lh}(\mathcal{T}_1) < \delta^{\mathcal{P}}$ and $\mathrm{lh}(\mathcal{T}_2) < \delta^{\mathcal{P}}$[4]. Let $\mathcal{S} \in \mathcal{P}$ be a stack on $\mathcal{Q}$ according to both $\Sigma_{\mathcal{Q},\mathcal{T}_1}$ and $\Lambda_{\mathcal{Q},\mathcal{T}_2}$ and such that $\Sigma_{\mathcal{Q},\mathcal{T}_1}(\mathcal{S}) \neq \Lambda_{\mathcal{Q},\mathcal{T}_2}(\mathcal{S})$. It then follows from $(*)$ that letting $b = \Sigma_{\mathcal{Q},\mathcal{T}_1}(\mathcal{S})$ and $c = \Sigma_{\mathcal{Q},\mathcal{T}_2}(\mathcal{S})$, both $\mathcal{Q}(b,\mathcal{S})$ and $\mathcal{Q}(c,\mathcal{S})$ exist. However, since $\Sigma_{\mathcal{Q}^-,\mathcal{T}_1} = \Lambda_{\mathcal{Q}^-,\mathcal{T}_2}$ and also both $\mathcal{Q}(b,\mathcal{S})$ and $\mathcal{Q}(c,\mathcal{S})$ are $\delta^{\mathcal{P}}+1$-iterable in $\mathcal{P}$, we have that $\mathcal{Q}(b,\mathcal{S}) = \mathcal{Q}(c,\mathcal{S})$ ⊣

5.2. Generic interpretability

We now move to generic interpretability. We start by recalling and generalizing the definition of a pre-hod pair (see [29, Definition 3.7]).

[2] If $(\delta^{\mathcal{P}})$ is the largest cardinal then we assume that $((\delta^{\mathcal{P}})^+)^{\mathcal{P}} = \mathrm{ord}(\mathcal{P})$.

[3] The use of Γ-fullness preservation can be substituted by $(*)$.

[4] If not, then we can reflect below $\delta^{\mathcal{P}}$. Recall that $\mathcal{W} \lhd_{hod} \mathcal{P}$, so the desired Skolem hull of $\mathcal{P}$ can be required to contain $\mathcal{W}$.

DEFINITION 5.2.1 (Prehod pair). $(\mathcal{P},\Sigma)$ is a prehod pair if

1. $\mathcal{P}$ is a countable hod premouse of successor type,
2. If $\mathcal{P}^-$ is not of limit type then Σ is an (ω_1,ω_1)-strategy for $\mathcal{P}$ acting on stacks based on $\mathcal{P}^-$.
3. If $\mathcal{P}^-$ is of limit type then Σ is an $(\omega_1,\omega_1,\omega_1)$-strategy for $\mathcal{P}$ acting on stacks based on $\mathcal{P}^-$.
4. If $i:\mathcal{P}\to\mathcal{Q}$ comes from an iteration according to Σ, $\Sigma^{\mathcal{Q}}_{\mathcal{Q}^-}=\Sigma_{\mathcal{Q}^-}\restriction\mathcal{Q}$[5],
5. For any $\mathcal{P}$-cardinal $\eta\in(\delta^{\mathcal{P}^-},\delta^{\mathcal{P}})$, considering $\mathcal{P}|\eta$ as a Σ-mouse over $\mathcal{P}^-$, there is an ω_1-strategy Λ for $\mathcal{P}|\eta$[6].

Notice that there must be a unique strategy Λ as in clause 5 of Definition 5.2.1.[7] Also, recall the definition of Generic Interpretability, [29, Definition 3.8]. In our current context it takes the following form.

DEFINITION 5.2.2 (Generic Interpretability). Suppose $(\mathcal{P},\Sigma)$ is a pre-hod pair, a meek hod pair of limit type or an sts hod pair. We say *generic interpretability* holds for $(\mathcal{P},\Sigma)$ if there is a function F such that

1. F is definable over $\mathcal{P}$ with no parameters,
2. dom(F) consists of pairs $(\mathcal{Q},\kappa)$ such that $\mathcal{Q}\in Y^{\mathcal{P}}$, $\mathcal{Q}\trianglelefteq\mathcal{P}|\delta^{\mathcal{P}}$ and $\kappa\in(\delta^{\mathcal{Q}},\delta^{\mathcal{P}})$ is a $\mathcal{P}$-cardinal,
3. for $(\mathcal{Q},\kappa)\in\text{dom}(F)$, $F(\mathcal{Q},\kappa)=(\dot{T},\dot{S})$ such that ,
   (a) $\dot{T},\dot{S}\in\mathcal{P}^{Coll(\omega,\text{ord}\,\mathcal{Q}))}$,
   (b) $\mathcal{P}\vDash$ "$\Vdash_{Coll(\omega,\text{ord}\,\mathcal{Q}))}\dot{T}$ and $\dot{S}$ are κ-complementing",
   (c) for any $\nu\in(\text{ord}\,\mathcal{Q}),\kappa)$ and any $\mathcal{P}$-generic $g\subseteq Coll(\omega,\nu)$,

$\mathcal{P}[g]\vDash$ "$p[\dot{T}_g]$ is an $(\omega_1,\omega_1,\omega_1)$-iteration strategy for $\mathcal{Q}$which extends $\Sigma^{\mathcal{P}}_{\mathcal{Q}}$"

and

$$(p[\dot{T}_g])^{\mathcal{P}[g]}=\Sigma_{\mathcal{Q}}\restriction HC^{\mathcal{P}[g]}.$$

The proof that the generic interpretability holds is just like the proof of [29, Theorem 3.10] using Theorem 4.13.2 and Theorem 5.1.2 instead of [29, Lemma 2.15] and [29, Theorem 3.3]. First the proof of [29, Lemma 3.9] can be used with no changes to establish the following useful lemma.

LEMMA 5.2.3. *Suppose $(\mathcal{P},\Sigma)$ is a prehod pair. Let $\kappa\in(\delta^{\mathcal{P}^-},\delta^{\mathcal{P}})$ be a $\mathcal{P}$-cardinal and Λ^* be the iteration strategy of $\mathcal{P}|\kappa$ as in 5 of Definition 5.2.1. Let Λ be the fragment of Λ^* that acts on non-dropping stacks. Let $g\subseteq Coll(\omega,\kappa)$ be $\mathcal{P}$-generic. Then $\mathcal{P}[g]$ locally Suslin, co-Suslin captures* Code(Λ) *and its complement at any cardinal of $\mathcal{P}$ greater than κ*[8].

[5]Thus, $\mathcal{P}$ is a Σ-mouse over $\mathcal{P}^-$.

[6]Thus, Λ acts on stacks above $\delta^{\mathcal{P}^-}$.

[7]Λ is the $\mathcal{Q}$-structure guided strategy.

[8]Recall that this means that for every $\mathcal{P}$-cardinal $\nu>\kappa$, there are ν-complementing trees $U,V\in\mathcal{P}[g]$ such that for any $<\nu$-generic h, Code$(\Lambda)\cap P[g][h]=(p[U])^{\mathcal{P}[g][h]}=(\mathbb{R}^{\mathcal{P}[g][h]}-p[V])^{\mathcal{P}[g][h]}$.

Fix now a prehod pair $(\mathcal{P},\Sigma)$ and let $\mathcal{Q}\in Y^{\mathcal{P}}$. Let $\kappa<\delta^{\mathcal{P}}$ be a $\mathcal{P}$-cardinal such that

1. $\kappa>\mathrm{ord}(\mathcal{Q})$ and
2. in the case $\mathcal{Q}$ is of limit type, $\mathcal{P}$ has no extenders with critical point $\delta^{\mathcal{Q}^b}$ and index greater than κ.

Let $\vec{G}=\{E\in\vec{E}^{\mathcal{P}|\delta^{\mathcal{P}}}:\nu(E)$ is an inaccessible cardinal of $\mathcal{P}$ and $\mathrm{crit}(E)>\kappa\}$. Notice that $(\mathcal{P},\delta^{\mathcal{P}},\Sigma,\vec{G})$ is a self-capturing background triple. Let

$$\mathsf{hpc}^{+}=(\mathcal{M}_\gamma,\mathcal{N}_\gamma,Y_\gamma,\Phi_\gamma^{+},F_\gamma^{+},F_\gamma,b_\gamma:\gamma\leq\delta)$$

be the output of hpc of $(\mathcal{P},\delta^{\mathcal{P}},\Sigma,\vec{G})$[9].

Here we abuse the notation and write Φ_β both for the strategy of $\mathcal{M}_\beta$ that is internal to $\mathcal{P}$ and also for the external strategy. It follows from Theorem 4.13.2 , Lemma 5.1.2 and Lemma 5.2.3 that for some β, $(\mathcal{M}_\beta,\Phi_\beta^{+})$ is a tail of $(\mathcal{Q},\Sigma_{\mathcal{Q}})$. We then set

$$\mathcal{N}_{\kappa,\mathcal{Q}}^{\mathcal{P}}=\mathcal{M}_\beta\quad\text{and}\quad\Lambda_{\kappa,\mathcal{Q}}=\Phi_\beta^{+}.$$

In what follows, we will omit superscript $\mathcal{P}$, but ask the reader to keep in mind that certain notions depend on $\mathcal{P}$. Also let $\pi_{\kappa,\mathcal{Q}}:\mathcal{Q}\to\mathcal{N}_{\kappa,\mathcal{Q}}$ be the iteration embedding according to $\Sigma_{\mathcal{Q}}$ and let $\mathcal{T}_{\kappa,\mathcal{Q}}$ be the normal stack on $\mathcal{Q}$ with last model $\mathcal{N}_{\kappa,\mathcal{Q}}$. The following is a consequence of Lemma 5.2.3, hull condensation of Σ and the proof of Theorem 4.13.2.

COROLLARY 5.2.4. *Suppose $(\mathcal{P},\Sigma)$ is a pre-hod pair such that for some projectively closed pointclass Γ, Σ has branch condensation and is Γ-fullness preserving. Suppose $\mathcal{Q}\in Y^{\mathcal{P}}$ and $\kappa>\mathrm{ord}(\mathcal{Q})$ are such that*

1. *$\kappa>\mathrm{ord}(\mathcal{Q})$ and*
2. *in the case $\mathcal{Q}$ is of limit type, $\mathcal{P}$ has no extenders with critical point $\delta^{\mathcal{Q}^b}$ and index greater than κ.*

Let $\eta\in(\mathrm{ord}(\mathcal{N}_{\kappa,\mathcal{Q}}),\delta^{\mathcal{P}})$ and $n<\omega$. Then there are names $(\dot{T},\dot{S})\in\mathcal{P}^{Coll(\omega,\eta)}$ such that

1. *$\mathcal{P}\models$ "$\Vdash_{Coll(\omega,\eta)}\dot{T}$ and $\dot{S}$ are $(\delta^{\mathcal{P}})^{+n}$-complementing",*
2. *for any $\lambda\in(\eta,((\delta^{\mathcal{P}})^{+n})^{\mathcal{P}})$ and any $\mathcal{P}$-generic $g\subseteq Coll(\omega,\lambda)$,*

$$\mathcal{P}[g]\models\text{“}p[\dot{T}_g]\text{ is an }(\omega_1,\omega_1)\text{-iteration strategy for }\mathcal{N}_{\kappa,\mathcal{Q}}\text{”}$$

and letting Φ be the $\pi_{\kappa,\mathcal{Q}}^{\mathcal{P}}$-pullback of the strategy given by $(p[\dot{T}_g])^{\mathcal{P}[g]}$ then

$$\Phi=\Sigma_{\mathcal{Q}}\restriction HC^{\mathcal{P}[g]}.$$

Our generic interpretability result can now be proved using the tree production lemma ([19, Theorem 3.3.15]) and Corollary 5.2.4. We leave the details to the reader.

[9] See Definition 4.3.3. The aforementioned definition requires a pointclass Γ but one can simply ignore all the clauses of Definition 4.3.3 that mention Γ.

THEOREM 5.2.5 (The generic interpretability). *Suppose* $(\mathcal{P},\Sigma)$ *is a prehod pair or is a non-gentle hod pair of limit type or is an sts hod pair. Also, suppose that for some projectively closed pointclass* Γ, Σ *is* Γ*-fullness preserving. Assume that for every* $\mathcal{Q}\in Y^{\mathcal{P}}$, $\Sigma_{\mathcal{Q}}$ *has strong branch condensation. Then generic interpretability holds for* $(\mathcal{P},\Sigma)$.

Next, we present our result on internal fullness preservation. The proof follows the same line of thought as the proof of [29, Theorem 3.12]. Below $\mathcal{S}^*(\mathcal{R})$ is the $*$-transform of $\mathcal{S}$ into a hybrid mouse over $\mathcal{R}$ and it is defined when $\operatorname{ord}(\mathcal{R})$ is a cutpoint of $\mathcal{S}$ (see [57, Remark 12.7] and [39]).

DEFINITION 5.2.6. Suppose $\mathcal{P}$ is a hod premouse and $\mathcal{Q}\in Y^{\mathcal{P}}$. We say $\Lambda=\Sigma^{\mathcal{P}}_{\mathcal{Q}}$ is *internally fullness preserving* if the following holds for $(\mathcal{T},\mathcal{R})\in I(\mathcal{Q},\Lambda)$[10] such that $\mathcal{P}\vDash$ "$|\mathcal{T}|^+$ exists".

1. For all limit type $\mathcal{S}\in Y^{\mathcal{R}}$, if $\mathcal{M}\in\mathcal{P}$ is a sound $\max(\delta^{\mathcal{P}}+1,(|\mathcal{T}|^+)^{\mathcal{P}})$-iterable $\Lambda_{\mathcal{S}|\delta^{\mathcal{S}^b},\mathcal{T}}$-mouse over $\mathcal{S}|\delta^{\mathcal{S}^b}$ then $\mathcal{M}\trianglelefteq\mathcal{S}$.
2. Suppose $\mathcal{W}\lhd_{hod}\mathcal{S}$ is of lsa type and $\mathcal{W}=((\mathcal{W}|\delta^{\mathcal{W}})^{\#})^{\mathcal{S}}$. Suppose $\mathcal{M}\in\mathcal{P}$ is a sound $\max(\delta^{\mathcal{P}}+1,(|\mathcal{T}|^+)^{\mathcal{P}})$-iterable $\Lambda_{\mathcal{W},\mathcal{T}}$-sts mouse over $\mathcal{W}$. Then $\mathcal{M}\trianglelefteq\mathcal{S}$.
3. Suppose $\mathcal{R}_1\lhd_{hod}\mathcal{R}$ is of successor type and $\eta\in(\operatorname{ord}(\mathcal{R}_1^-),\delta^{\mathcal{R}_1}]$ is a cutpoint cardinal of $\mathcal{R}$. Suppose $\mathcal{M}\in\mathcal{P}$ is a sound $\max(\delta^{\mathcal{P}}+1,(|\mathcal{T}|^+)^{\mathcal{P}})$-iterable $\Lambda_{\mathcal{R}_1^-,\mathcal{T}}$-mouse over $\mathcal{R}|\eta$. Then $\mathcal{M}\trianglelefteq(\mathcal{R}|(\eta^+)^{\mathcal{R}})^*(\mathcal{R}|\eta)$.

THEOREM 5.2.7 (Internal fullness preservation). *Suppose* $\mathcal{P}$ *is a hod premouse and* $\mathcal{Q}\in Y^{\mathcal{P}}$ *is such that* $(\operatorname{ord}(\mathcal{Q})^+)^{\mathcal{P}}$ *exists. Then* $\Sigma^{\mathcal{P}}_{\mathcal{Q}}$ *is internally fullness preserving.*

5.3. The derived models of hod mice

In this section, we state, without a proof, a version of [29, Theorem 3.19]. Suppose $(\mathcal{P},\Sigma)$ is an allowable pair[11] such that Σ has strong branch condensation and is fullness preserving[12]. Suppose $\mathcal{Q}\trianglelefteq_{hod}\mathcal{P}$ is such that $\mathcal{Q}$ is meek and is of limit type. Thus, $\delta^{\mathcal{Q}}$ is a limit of Woodin cardinals of $\mathcal{P}$. Suppose further that $\operatorname{cf}^{\mathcal{P}}(\delta^{\mathcal{Q}})$ is not a measurable cardinal in $\mathcal{P}$. We then let $D^*(\mathcal{P},\Sigma,\mathcal{Q})$ be the set of all $A\subseteq\mathbb{R}$ such that for some strong cutpoint $\tau<\delta^{\mathcal{Q}}$ of $\mathcal{Q}$ and $g\subseteq Coll(\omega,\tau)$-generic over $\mathcal{P}$ there are trees $T,U\in\mathcal{P}[g]$ such that

1. $\mathcal{P}[g]\vDash$ "(T,U) is $\delta^{\mathcal{Q}}$-complementing" and
2. $x\in A$ if and only if there is $(\mathcal{S},\mathcal{R})\in I(\mathcal{Q},\Sigma_{\mathcal{Q}})$ and a Woodin cardinal δ of $\mathcal{R}$ such that

[10]Thus, $(\mathcal{T},\mathcal{R})\in\mathcal{P}$.

[11]See Definition 3.10.7.

[12]See Definition 4.6.2.

(a) $\pi^{\mathcal{S}}$ is above τ,
(b) x is generic for the extender algebra of $\mathcal{R}[g]$ at δ and
(c) $\mathcal{R}[g,x] \models x \in p[\pi^{\mathcal{S}}(T)]$.

It follows from Corollary 4.13.5 and Theorem 4.10.2 that for $x \in \mathbb{R}$, the right hand side of the above equivalence is independent of the choice of $(\mathcal{S},\mathcal{R})$.

We let $D(\mathcal{P},\Sigma,\mathcal{Q})$ be the derived model of $\mathcal{Q}$ as computed by $\Sigma_{\mathcal{Q}}$, i.e., for $A \subseteq \mathbb{R}$, $A \in D(\mathcal{P},\Sigma,\mathcal{Q})$ if there is $(\mathcal{S},\mathcal{R}) \in I(\mathcal{P},\Sigma)$ such that $\mathcal{S}$ is based on $\mathcal{Q}$ and $A \in D^*(\mathcal{R},\Sigma_{\mathcal{R}},\pi^{\mathcal{S}}(\mathcal{Q}))$.

Next recall [29, Definition 3.18]. Essentially a pointclass is completely mouse-full if the next model of determinacy has the same mice relative to common iteration strategies. We introduce this notion more carefully.

Given a set of reals $A \subseteq \mathbb{R}$, we let $W_A = \{B \subseteq \mathbb{R} : w(B) < w(A)\}$. Next following Definition 3.13 of [29], we say $A \subseteq \mathbb{R}$ is a new set if

1. $L(A,\mathbb{R}) \models \mathsf{AD}^+$,
2. $\wp(\mathbb{R}) \cap L(W_A,\mathbb{R}) = W_A$,
3. $\Theta^{L(W_A,\mathbb{R})}$ is a Suslin cardinal of $L(A,\mathbb{R})$.

The following is [29, Definition 3.17].

DEFINITION 5.3.1. Given a pointclass Γ, we say Γ is *completely mouse full* if either $\Gamma = \wp(\mathbb{R})$ or there is a new set A such that

1. $\Gamma = W_A$,
2. if $(\mathcal{P},\Sigma)$ is allowable such that $\mathsf{Code}(\Sigma) \in \Gamma$ and $L(A,\mathbb{R}) \models$ "Σ has strong branch condensation and is Γ-fullness preserving" then for every $a \in HC$,
$$Lp^{\Gamma,\Sigma}(a) = (Lp^{\Sigma}(a))^{L(A,\mathbb{R})}.$$

Given two pointclasses Γ_1 and Γ_2, we write $\Gamma_1 \trianglelefteq_{mouse} \Gamma_2$ if $\Gamma_1 \subseteq \Gamma_2$ and Γ_2 has the same mice as Γ_1 relative to common iteration strategies. More precisely, if $(\mathcal{P},\Sigma) \in \Gamma_1$ is an allowable pair such that $L(\Gamma_2,\mathbb{R}) \models$ "Σ has strong branch condensation and is Γ_1-fullness preserving" then for any $a \in HC$,
$$Lp^{\Gamma_1,\Sigma}(a) = Lp^{\Gamma_2,\Sigma}(a).$$

Finally, following [29, Definition 3.18],

DEFINITION 5.3.2. Γ is *mouse full* if either it is completely mouse full or is a union of completely mouse full pointclasses $(\Gamma_\alpha : \alpha < \Omega^\Gamma)$ such that for all α, $\Gamma_\alpha \trianglelefteq_{mouse} \Gamma_{\alpha+1}$ and for all limit α, $\Gamma_\alpha = \bigcup_{\beta<\alpha}\Gamma_\beta$.

We can now state our generalization of [29, Theorem 3.19].

THEOREM 5.3.3. *Suppose $(\mathcal{P},\Sigma)$ is an allowable pair and Γ is a pointclass closed under continuous preimages.*[13] *Suppose further that $\mathcal{P}$ is non-gentle and of*

[13]We define the Solovay sequence $(\vartheta^\Gamma_\alpha : \alpha \leq \Omega)$ relative to Γ as the Solovay sequence defined in the model $L(\Gamma,\mathbb{R})$ if Γ is constructibly closed (i.e., $\wp(\mathbb{R}) \cap L(\Gamma,\mathbb{R}) = \Gamma$). We can aslo make sense of the Solovay sequence relative to Γ in the case Γ is a limit of constructibly closed pointclasses; here

limit type, and that Σ has strong branch condensation and is Γ-fullness preserving. Then

1. $\Gamma(\mathcal{P},\Sigma)=\bigcup_{\mathcal{Q}\in pI(\mathcal{P},\Sigma),\mathcal{Q}'\unlhd\mathcal{Q}} D(\mathcal{Q},\Sigma_{\mathcal{Q}},\mathcal{Q}')$.
2. *For any $\mathcal{Q}\in pI(\mathcal{P},\Sigma)$, if $\mathcal{Q}'\lhd_{hod}\mathcal{Q}$ is non-gentle and is of limit type then $D(\mathcal{Q},\Sigma_{\mathcal{Q}},\mathcal{Q}')$ is completely mouse full.*
3. *For any $\mathcal{Q}\in pI(\mathcal{P},\Sigma)$, if $\mathcal{Q}'\lhd_{hod}\mathcal{Q}''\unlhd_{hod}\mathcal{Q}$ are such that*
   (a) *$\mathcal{Q}'$ and $\mathcal{Q}''$ are non-gentle and are of limit type,*
   (b) *$\mathcal{Q}''\unlhd_{hod}\mathcal{Q}$ is the least non-gentle layer of $\mathcal{Q}$ that has ω more Woodin cardinals than $\mathcal{Q}'$,*

   then letting $\Gamma'=D(\mathcal{Q},\Sigma_{\mathcal{Q}},\mathcal{Q}'')$, if ξ is such that $\vartheta^{\Gamma'}_{\mathsf{Code}\Sigma_{\mathcal{Q}'})}=\vartheta^{\Gamma'}_{\xi}$ then for every n, letting $\mathcal{Q}'_n\unlhd_{hod}\mathcal{Q}''$ be the layer of $\mathcal{Q}''$ that has exactly n Woodin cardinals above $\mathrm{ord}(\mathcal{Q}')$,

$$\vartheta^{\Gamma'}_{\mathsf{Code}\Sigma_{\mathcal{Q}'_n})}=\vartheta^{\Gamma'}_{\xi+n}\quad \textit{and}\quad \Omega^{\Gamma'}=\xi+\omega.$$

4. *$\Gamma(\mathcal{P},\Sigma)$ is a mouse full pointclass.*

We finish with a theorem generalizing [29, Theorem 3.20]. It shows that $\Gamma(\mathcal{P},\Sigma)$ satisfies mouse capturing for any $\Sigma_{\mathcal{Q}}$ where $\mathcal{Q}\in pI(\mathcal{P},\Sigma)$. Recall from [29] (the first page of the introduction of [29]) that MC stands for mouse capturing, i.e., for the statement that for $x,y\in\mathbb{R}$, $x\in OD_y$ if and only if there is an ω_1-iterable y-mouse $\mathcal{M}$ such that $x\in\mathcal{M}$. Given an allowable pair $(\mathcal{P},\Sigma)$ such that Σ has strong branch condensation and is Γ^*-fullness preserving for some projectively closed pointclass Γ^*, we say MC holds for Σ[14] if for $x,y\in\mathbb{R}$, $x\in OD_{y,\Sigma}$ if and only if there is an ω_1-iterable Σ-mouse $\mathcal{M}$ over y such that $x\in\mathcal{M}$. Given a mouse full pointclass Γ and an allowable pair $(\mathcal{P},\Sigma)\in\Gamma$ such that Σ is Γ-fullness preserving and has strong branch condensation, we write

$$\Gamma\vDash\text{“MC for }\Sigma\text{”}$$

if one of the following holds:

1. Γ is completely mouse full and whenever A is a new set such that $\Gamma=W_A$ then $L(A,\mathbb{R})\vDash$ “MC for Σ”.
2. Γ is not completely mouse full and if $(\Gamma_\alpha:\alpha<\Omega)$ are the completely mouse full pointclasses witnessing that Γ is mouse full then for some $\alpha<\Omega$, $L(\Gamma_\alpha,\mathbb{R})\vDash$ “MC for Σ”.

---

for $A\in\Gamma$, we say a set B is $OD^{\Gamma}(A)$ if B is $OD(A)^{L(\Lambda,\mathbb{R})}$ for some constructibly closed $\Lambda\lhd\Gamma$. From here on, when we talk about the Solovay sequence relative to a pointclass Γ, Γ is assumed to have one of the two properties above. Notice that if Γ is a constructibly closed pointclass which is a union of constructibly closed pointclasses strictly contained in it, then the two ways of computing the Solovay sequence relative to Γ are equivalent.

[14]The statement “MC holds for Σ” can be made precise for an arbitrary strategy with hull condensation. Our definition also includes st-strategies.

THEOREM 5.3.4. *Suppose $(\mathcal{P},\Sigma)$ is an allowable pair of limit type and Σ has strong branch condensation and is Γ^*-fullness preserving for some projectively closed pointclass Γ^*. Suppose further that there is a good pointclass Γ such that* $\mathsf{Code}(\Sigma) \in \underset{\sim}{\Delta}_{\Gamma}$. *Then for every $\mathcal{Q} \in pB(\mathcal{P},\Sigma)$,*

$$\Gamma(\mathcal{P},\Sigma) \models \text{“}\mathsf{MC}\textit{ for } \Sigma_{\mathcal{Q}}\text{”}.$$

5.4. Anomalous hod premice

In this paper, we use anomalous hod premice the same way we used them in [29], to generate pointclasses that are mouse full but not completely mouse full. The reader may wish to review Definition 2.7.15 and Definition 3.9.2.

DEFINITION 5.4.1 (Anomalous hod premouse of type I). $\mathcal{P}$ is an *anomalous hod premouse of type I* if $\mathcal{P}$ is a germane $\mathsf{hp}-\mathsf{lses}$ such that letting $\mathcal{Q} = \mathsf{hl}(\mathcal{P})$, $\mathcal{Q}$ is of successor type, $\mathcal{P} \models$ “$\delta^{\mathcal{Q}}$ is Woodin” and either $\rho(\mathcal{P}) < \delta^{\mathcal{Q}}$ or $\mathcal{J}_\omega[\mathcal{P}] \models$ “$\delta^{\mathcal{Q}}$ is not a Woodin cardinal”.

DEFINITION 5.4.2 (Anomalous hod premouse of type II). $\mathcal{P}$ is an *anomalous hod premouse of type II* if $\mathcal{P}$ is a germane $\mathsf{hp}-\mathsf{lses}$ such that letting $\mathcal{Q} = \mathsf{hl}(\mathcal{P})$, $\mathcal{Q}$ is a gentle hod premouse, $\rho(\mathcal{P}) < \delta^{\mathcal{Q}}$ but for every $\xi \in (\delta, \mathrm{ord}\,\mathcal{P}))$, $\rho(\mathcal{P}||\omega\xi) \geq \delta^{\mathcal{Q}}$.

DEFINITION 5.4.3 (Anomalous hod premouse of type III). $\mathcal{P}$ is *an anomalous hod premouse of type III* if $\mathcal{P}$ is a germane $\mathsf{hp}-\mathsf{lses}$ such that letting $\mathcal{Q} = \mathsf{hl}(\mathcal{P})$, $\mathcal{Q}$ is non-gentle limit type hod premouse, $\rho(\mathcal{P}) < \delta^{\mathcal{Q}^b}$ but for every $\omega\xi < \mathrm{ord}(\mathcal{P})$, $\rho(\mathcal{P}||\omega\xi) > \delta^{\mathcal{Q}^b}$[15].

Thus, in the language of Definition 2.7.19, if $\mathcal{P}$ is an anomalous hod premouse then $\mathcal{P}$ is not projecting well but all of its initial segments do project well. We say $\mathcal{P}$ is an anomalous hod premouse if it is an anomalous hod premouse of type i where $i \in \{\mathrm{I}, \mathrm{II}, \mathrm{III}\}$.

DEFINITION 5.4.4 (Anomalous hod pair). $(\mathcal{P},\Sigma)$ is *an anomalous hod pair* if one of the following conditions holds:

1. $\mathcal{P}$ is an anomalous hod premouse of type I or II, Σ is an (ω_1,ω_1)-iteration strategy with hull condensation and whenever $\mathcal{Q}$ is a Σ iterate of $\mathcal{P}$, $\Sigma^{\mathcal{Q}} \subseteq \Sigma \restriction \mathcal{Q}$[16].
2. $\mathcal{P}$ is an anomalous hod premouse of type III, Σ is a $(\omega_1,\omega_1,\omega_1)$-iteration strategy[17] with hull condensation and whenever $\mathcal{Q}$ is a Σ iterate of $\mathcal{P}$, $\Sigma^{\mathcal{Q}} \subseteq \Sigma \restriction \mathcal{Q}$.

[15] It follows from the arguments on page 142 of [29] that $\rho(\mathcal{P}||\omega\xi) = \delta^{\mathcal{Q}^b}$ is not possible in situations that will arise in this book.

[16] Recall that $\Sigma^{\mathcal{Q}}$ is the internal strategy of $\mathcal{Q}$.

[17] See Definition 2.10.6.

We then say that $(\mathcal{P},\Sigma)$ is a *simple anomalous hod pair* if either

1. it is an anomalous hod pair and $\mathcal{P}$ is of type I or II, or
2. $\mathcal{P}$ is an anomalous hod premouse of type III, Σ is a (ω_1,ω_1)-iteration strategy with hull condensation and whenever $\mathcal{Q}$ is a Σ iterate of $\mathcal{P}$, $\Sigma^{\mathcal{Q}} \subseteq \Sigma \restriction \mathcal{Q}$.

The following lemma is due to Mitchell and Steel. It appears as Claim 5 in the proof of Theorem 6.2 of [23]. In the current work, the lemma is used to show that certain hod pair constructions converge, which leads to showing that generation of pointclasses holds (see Theorem 10.1.2). It was used in [29] in a similar fashion (see [29, Lemma 3.25]).

LEMMA 5.4.5. *Suppose $(\mathcal{P},\Sigma)$ is an anomalous hod pair or a simple hod pair such that for $n < k(\mathcal{P})$, $(\mathcal{P},n)$ is not anomalous. Let $k = k(\mathcal{P})$, $\mathcal{P}' = (\mathcal{P},k-1)$ and $\Sigma' = \Sigma_{\mathcal{P}'}$, and suppose $(\mathcal{T},\mathcal{Q}) \in I(\mathcal{P}',\Sigma')$. Then*

1. *if $\mathcal{P}$ is of type I or II then $\rho_k(\mathcal{Q}) < \delta^{\mathcal{Q}}$ and*
2. *if $\mathcal{P}$ is of type III then $\rho_k(\mathcal{Q}) < \delta^{\mathcal{Q}^b}$.*

The next theorem is the adaptation of [29, Theorem 3.27] to our current setting. It generalizes our results from previous sections to anomalous hod pairs.

THEOREM 5.4.6. *Suppose $(\mathcal{P},\Sigma)$ is an anomalous hod pair of type II or III. Suppose that there is a projectively closed pointclass Γ such that for any $(\mathcal{T},\mathcal{Q}) \in B(\mathcal{P},\Sigma)$ there is a hod pair $(\mathcal{R},\Lambda)$ such that Λ has (strong) branch condensation and is low-level Γ-fullness preserving*[18]*, and there is $\pi : \mathcal{Q} \to \mathcal{R}$ such that $\Lambda^\pi = \Sigma_{\mathcal{Q},\mathcal{T}}$. Then*

1. *For every $(\mathcal{T},\mathcal{Q}) \in B(\mathcal{P},\Sigma)$, $\Sigma_{\mathcal{Q},\mathcal{T}}$ has (strong) branch condensation, is positional and is commuting.*
2. *Σ is strongly low-level $\Gamma(\mathcal{P},\Sigma)$-fullness preserving and $\Gamma(\mathcal{P},\Sigma)$ is a mouse full pointclass.*

We omit the proof of Theorem 5.4.6 as it is only notationally more complicated than the proof of [29, Theorem 3.10]. We remind the reader that the proof of [29, Theorem 3.27] depended on the generic interpretability result, which appeared as [29, Theorem 3.10]. In our current context we need to use Theorem 5.2.5. The general idea is that we can translate the properties of Σ into the derived model of $\mathcal{P}$ as computed via Σ. This fact then just gets preserved under pull-back embeddings.

It is also possible to prove a version of Theorem 5.4.6 for sts hod pairs. To prove it, we again need to use Theorem 5.2.5. We state it without a proof.

THEOREM 5.4.7. *Suppose $(\mathcal{P},\Sigma)$ is an sts hod pair and Γ is a projectively closed pointclass. Suppose that for any $(\mathcal{T},\mathcal{Q}) \in B(\mathcal{P},\Sigma)$ there is a hod pair $(\mathcal{R},\Lambda)$ such that Λ has strong branch condensation and is (strongly) Γ-fullness preserving, and there is $\pi : \mathcal{Q} \to \mathcal{R}$ such that $\Lambda^\pi = \Sigma_{\mathcal{Q},\mathcal{T}}$. Then*

[18] See Definition 4.6.2.

1. *For every* $(\mathcal{T},\mathcal{Q}) \in B(\mathcal{P},\Sigma)$, $\Sigma_{\mathcal{Q},\mathcal{T}}$ *has (strong) branch condensation, is positional and is commuting.*
2. Σ *is strongly* $\Gamma^b(\mathcal{P},\Sigma)$*-fullness preserving and* $\Gamma^b(\mathcal{P},\Sigma)$ *is a mouse full pointclass.*

The following is an easy corollary of Theorem 5.4.6.

COROLLARY 5.4.8 (Branch condensation pulls back). *Suppose* $(\mathcal{P},\Sigma)$ *is a hod pair of limit type and* Σ *has (strong) branch condensation. Suppose* $\pi : \mathcal{Q} \to \mathcal{P}$ *is elementary. Then for every* $\mathcal{R} \lhd_{hod} \mathcal{Q}$ *such that* $\delta^{\mathcal{R}}$ *is a cutpoint of* $\mathcal{Q}$, $(\Sigma^\pi)_{\mathcal{R}}$ *has (strong) branch condensation.*

5.5. Branch condensation on a tail

The main theorem of this section, Theorem 5.5.3, will be used in several places (e.g., the proof of Theorem 10.1.5) in this book as well as in core model induction applications. First we need to introduce a new concept, which fortunately for us, Farmer Schlutzenberg has developed independently and much more generally.

DEFINITION 5.5.1. Suppose $(\mathcal{P},\Sigma)$ is an anomalous pair of type III. We say $(\mathcal{P},\Sigma)$ has a *supporting bicephalous* if there is a bicephalous $B = (\rho,\mathcal{M},\mathcal{P})$ in the sense of [46, Definition] such that

1. $\rho = \delta^{\mathcal{P}^b}$,
2. $\mathcal{M}$ is germane[19] and such that $\rho(\mathcal{M}) < \rho$, $\mathsf{hl}(\mathcal{M}) = \mathcal{P}|\rho$ and
$$\mathcal{M} \lhd \mathsf{Lp}^{\Sigma_{\mathcal{P}|\rho}}(\mathcal{P}|\rho),$$
3. for every $n < \omega$, $\rho_n(\mathcal{M}) \neq \rho$,
4. $k(\mathcal{M})$ is the least n such that $\rho_{n+1}(\mathcal{M}) < \rho$,
5. for every $\gamma \in [\mathsf{ord}(\mathcal{P}^b), \mathsf{ord}(\mathcal{M}))$, $\rho(\mathcal{M}||\gamma) > \rho$,
6. B has an ω_1-iteration strategy Σ^+ which extends Σ.

REMARK 5.5.2. The reader unfamiliar with [46] may treat B in Definition 5.5.1 as a pair constructed by some Γ-hod pair construction. After reaching $\mathcal{P}|\rho$ the hod pair construction aims to reach the next $\Gamma-\mathsf{cbl}^{20}$. The construction proceeds as a fully backgrounded construction relative to $\Sigma_{\mathcal{P}|\rho}$. Once $\mathcal{P}^b$ is reached it is declared to be a layer and a new strategy appears, the strategy of $\mathcal{P}^b$. To reach $\mathcal{M}$ we just simply need to continue the construction relative to $\Sigma_{\mathcal{P}|\rho}$. This will all be relevant in the proof of Theorem 10.1.5. Also notice that clause 4 implies that we iterate $\mathcal{M}$ using one fine structural level lower than one would normally do.

[19] See Definition 2.7.15.

[20] See Definition 4.3.1.

THEOREM 5.5.3 (Branch condensation on a tail). *Suppose $(\mathcal{P},\Sigma)$ is an anomalous hod pair of type II or III. Suppose that for every $(\mathcal{T},\mathcal{Q}) \in B(\mathcal{P},\Sigma)$, $\Sigma_{\mathcal{Q},\mathcal{T}}$ has strong branch condensation. Moreover, assume either*

(1) ZFC holds and $\mathcal{P}$ is of type II, or
(2) AD^+ holds and if $\mathcal{P}$ is of type III then it has a supporting bicephalous.

Then if $\mathcal{P}$ is of type II then there is $(\mathcal{T},\mathcal{Q}) \in I(\mathcal{P},\Sigma)$ such that $\Sigma_{\mathcal{Q},\mathcal{T}}$ has strong branch condensation, and if $\mathcal{P}$ is of type III then Σ has strong branch condensation.

PROOF. The case when $\mathcal{P}$ is of type II is very similar to the proof of [29, Theorem 3.28]. The case when $\mathcal{P}$ is of type III is similar to the proof of Theorem 4.9.5. In order not to repeat the entire Section 4.9, we outline the proof of branch condensation and leave the rest to reader. Let $B = (\rho,\mathcal{M},\mathcal{P})$ and Σ^+ witness that $(\mathcal{P},\Sigma)$ has a supporting bicephalous.

Suppose $(\mathcal{T},\mathcal{Q},\mathcal{U},c,\sigma)$ is such that

1. $(\mathcal{T},\mathcal{Q}) \in I(\mathcal{P},\Sigma)$,
2. $\mathcal{U}$ is a stack according to Σ and $\mathrm{lh}(\mathcal{U})$ is a limit ordinal,
3. c is a cofinal well-founded branch of $\mathcal{U}$,
4. $\sigma : \mathcal{M}_b^{\mathcal{U}} \to \mathcal{Q}$ is elementary[21] and such that $\pi^{\mathcal{T}} = \sigma \circ \pi_b^{\mathcal{U}}$.

We would like to show that $c = \Sigma(\mathcal{U})$. Let $d = \Sigma(\mathcal{U})$. The most difficult case, which also represents the difficulties involved in other cases that are left to the reader, is the case when $\pi^{\mathcal{U},b}$ exists, $R^{\mathcal{U}}$ has a maximal element, and if $\alpha = \max(R^{\mathcal{U}})$ then $\mathcal{U}_{\geq\alpha}$ is above $(\mathcal{M}_\alpha^{\mathcal{U}})^b$ and $\mathcal{Q}(c,\mathcal{U}_{\geq\alpha})$-exists. Set then $\mathcal{W} = \mathrm{m}^+(\mathcal{U})$ and let Φ_0' be the σ-pullback of $\Sigma_{\sigma(\mathcal{W}),\mathcal{T}}$ and $\Phi_1' = \Sigma_{\mathcal{W},\mathcal{U}^\frown\{d\}}$. Finally, set $\Phi_0 = (\Phi_0')^{stc}$ and $\Phi_1 = (\Phi_1')^{stc}$. We then have that if $\Phi_0 = \Phi_1$ then in fact, as $\mathcal{Q}(d,\mathcal{U})$ exists, $\mathcal{Q}(d,\mathcal{U}) = \mathcal{Q}(c,\mathcal{U})$ and therefore, $c = d$. Assume then that $\Phi_0 \neq \Phi_1$.

Let then $(\mathcal{X}_0,\mathcal{W}_0,\mathcal{X}_1,\mathcal{W}_1,\mathcal{R})$ be a minimal low level disagreement[22] between Φ_0 and Φ_1. Recall the notation $\mathcal{X}^{\mathsf{ue}}$ introduced in clause 4 and 5 of Notation 2.10.9. Let $\mathcal{Y}_0 = \mathcal{U}^\frown\{c\}^\frown(\mathcal{X}_0)^{\mathsf{ue}}_{\mathcal{R}}$ and $\mathcal{Y}_1 = \mathcal{U}^\frown\{c\}^\frown(\mathcal{X}_1)^{\mathsf{ue}}_{\mathcal{R}}$. Let $\mathcal{Y}_0^+$ and $\mathcal{Y}_1^+$ be the stacks on B obtained by applying $\mathcal{Y}_0$ and $\mathcal{Y}_1$ to B. $\mathcal{Y}_1^+$ is according to Σ^+ while $\mathcal{Y}_0^+$ is not according to a strategy but it has well-founded models because of σ. Let $B_0 = (\nu_0,\mathcal{M}_0,\mathcal{P}_0)$ and $B_1 = (\nu_1,\mathcal{M}_1,\mathcal{P}_1)$ be the last models of $\mathcal{Y}_0^+$ and $\mathcal{Y}_1^+$. Let Λ_0' be the strategy of B_0[23] and let $\Lambda_1' = \Sigma^+_{B_1,\mathcal{Y}_1^+}$.

We now have that $\pi^{\mathcal{Y}_0,b} = \pi^{\mathcal{Y}_1,b}$, which implies that $\nu_0 = \nu_1$, $\mathcal{M}_0 = \mathcal{M}_1$ and $\mathcal{P}_0^b = \mathcal{P}_1^b$. In fact, letting F be the $(\rho,\pi^{\mathcal{Y}_0,b}(\rho))$-extender[24] derived from $\pi^{\mathcal{Y}_0,b}$ then for $i \in 2$, $\mathcal{M}_i = \mathrm{Ult}(\mathcal{M},F)$ and $\mathcal{P}_i = \mathrm{Ult}(\mathcal{P}^b,F)$. Let then $\nu = \nu_0$, $\mathcal{N} = \mathcal{M}_0$, $\mathcal{S} = \mathcal{P}_0^b$, Λ^0 be the strategy of $\mathcal{N}$ induced by Λ_0' and Λ^1 be the strategy of $\mathcal{N}$ induced by Λ_1'. We have that $\Lambda^0_{\mathcal{R}} \neq \Lambda^1_{\mathcal{R}}$.

---

[21] σ may not be fully elementary, just at the right fine structural level.

[22] See Definition 4.7.1.

[23] This strategy comes from the copying procedure; B_0 embeds into a Σ^+-iterate of B that starts out by applying $\mathcal{T}$ to B and then copies $(\mathcal{X}_0)^{\mathsf{ue}}_{\mathcal{R}}$.

[24] $\rho = \delta^{\mathcal{P}^b}$.

Notice next that because $\Sigma_{\mathcal{P}|\rho}$ has branch condensation, $\rho(\mathcal{N}) < \delta^{\mathcal{R}}$[25] and moreover, letting $n = k(\mathcal{N})$[26], and $\mathcal{N}'$ be the nth reduct of $\mathcal{N}$ then

(1) $\sup(Hull_1^{\mathcal{N}'}(p_1(\mathcal{N}') \cup \mathcal{R}^-) \cap \delta^{\mathcal{R}}) = \delta^{\mathcal{R}}$.

Let then $\mathcal{K}' = cHull_1^{\mathcal{N}'}(p_1(\mathcal{N}') \cup \mathcal{R})$ and $i' : \mathcal{K}' \to \mathcal{N}'$ be the uncollapsing embedding. Let $\mathcal{K}$ be decoding of $\mathcal{K}'$ and $i : \mathcal{K} \to \mathcal{N}$ be the canonical uncoring embedding. (1) then implies that

(2) $\mathcal{K} \vDash$ "$\delta^{\mathcal{R}}$ is a Woodin cardinal" and $\mathcal{J}_\omega[\mathcal{K}] \vDash$ "$\delta^{\mathcal{R}}$ is not a Woodin cardinal".

Let now $\Lambda^2 =$ (i-pullback of Λ^0) and $\Lambda^3 =$ (i-pullback of Λ^0). Let $\mathcal{Z}$ be a normal stack[27] on $\mathcal{K}$ based on $\mathcal{R}$ such that $\mathcal{Z}^d =_{\text{def}} \downarrow(\mathcal{Z}, \mathcal{R})$ is according to both Λ^2 and Λ^3 and such that setting $e = \Lambda^2(\mathcal{Z})$ and $f = \Lambda^3(\mathcal{Z})$ then

(3) $e \neq f$.

It follows from Theorem 5.4.6 that $\mathcal{Q}(e, \mathcal{Z}^d)$ exists if and only if $\mathcal{Q}(f, \mathcal{Z}^d)$ exists, and therefore, neither exists. Let $\mathcal{K}_e = \mathcal{M}_e^{\mathcal{Z}}$ and $\mathcal{K}_f = \mathcal{M}_f^{\mathcal{Z}}$.

Let $\Lambda^e = \Lambda^2_{\mathcal{K}_e, \mathcal{Z}^\frown\{e\}}$ and $\Lambda^f = \Lambda^3_{\mathcal{K}_f, \mathcal{Z}^\frown\{f\}}$. Letting $\tau = \delta(\mathcal{Z})$, we now have that

(4) $\mathcal{K}_e, \mathcal{K}_f \vDash$ "$\delta^{\mathcal{R}}$ is a Woodin cardinal" and $\mathcal{J}_\omega[\mathcal{K}_e], \mathcal{J}_\omega[\mathcal{K}_f] \vDash$ "$\delta^{\mathcal{R}}$ is not a Woodin cardinal",

(5) $\mathcal{K}_e$ and $\mathcal{K}_f$ are τ-sound,

(6) $\pi_e^{\mathcal{Z}}(\mathcal{R}) = \pi_f^{\mathcal{Z}}(\mathcal{R}) =_{\text{def}} \mathcal{R}_1$ and $\Lambda^e_{\mathcal{R}_1} = \Lambda^f_{\mathcal{R}_1} =_{\text{def}} \Lambda_1$[28].

Thus, if we argue that $\mathcal{K}_e = \mathcal{K}_f$ then we would be done as it would show that $e = f$, contradicting (3). Set $\Gamma_e = \Gamma(\mathcal{K}_e, \Lambda^e)$ and $\Gamma_f = \Gamma(\mathcal{K}_f, \Lambda^f)$. Suppose first that $\Gamma_e = \Gamma_f$. Then because Λ^e and Λ^f both are Γ_e-fullness preserving and therefore, $(\mathcal{K}_e, \Lambda^e)$ and $(\mathcal{K}_f, \Lambda^f)$ can be compared as in Theorem 4.14.4, (5) implies that $\mathcal{K}_e = \mathcal{K}_f$.

We now assume that $\Gamma_e \neq \Gamma_f$. Without losing generality, lets suppose that $\Gamma_e \subset \Gamma_f$. It follows from the above argument that $\mathcal{K}_e$ is ordinal definable in Γ_f from Λ_1. Indeed, let $A \in \Gamma_f$ be such that every set in Γ_e has Wadge rank $< w(A)$. Then in $L(A, \mathbb{R})$, $\mathcal{K}_e$ is the unique anomalous hod premouse $\mathcal{V}$ that has an ω_1-iteration strategy Π such that

1. $\Gamma(\mathcal{V}, \Pi) = \{C \subseteq \mathbb{R} : w(C) < w(\Gamma_e)\}$,
2. $\mathcal{R}_1 \trianglelefteq^c_{hod} \mathcal{V}$[29],
3. $\Pi_{\mathcal{R}_1} = \Lambda_1$,
4. $\mathcal{V}$ is τ-sound.

It then follows from Theorem 5.4.6 that $\mathcal{K}_e \in \mathcal{K}_f$, which contradicts (4). $\dashv$

[25]See Lemma 5.4.5. This follows from the proof of Claim 5 in the proof of Theorem 6.2 of [23].

[26]We abuse our notation and think of $\mathcal{N}$ as both fine structural and non-fine structural.

[27]We can choose $\mathcal{Z}$ to be normal because of Theorem 4.13.2.

[28]This is a consequence of Theorem 4.13.2. $\mathcal{Z}$ is produced by iterating $\mathcal{R}$ into a universal model.

[29]See Definition 9.1.2.

Chapter 6

THE INTERNAL THEORY OF LSA HOD MICE

A major shortcoming of our treatment of short-tree-strategy mice is that we did not add branches to all trees. Suppose $(\mathcal{P},\Sigma)$ is an sts hod pair, X is a self-well-ordered set such that $\mathcal{P} \in X$ and $\mathcal{M}$ is a Σ-sts premouse over X based on $\mathcal{P}$. Recall the short tree strategy indexing scheme Definition 3.8.9. Recall that our strategy for indexing branches was to consider two kinds of iterations, uvs and nuvs[1]. We outright index the branches of uvs iterations. However, we only consider a subclass of nuvs iterations. If for some $\beta < o(\mathcal{M})$, $\mathcal{T} \in \mathrm{dom}(\Sigma^{\mathcal{M}|\beta})$ is an $\mathcal{M}|\beta$-ambiguous tree then (i) $\mathcal{T}$ is a result of comparing $\mathcal{P}$ with a certain background construction of $\mathcal{M}|\beta$ and (ii) we index the branch of $\mathcal{T}$ after we find a certain certificate of shortness (recall Definition 3.8.9). It is then not clear from our definition that $\Sigma \restriction \mathcal{M} \subseteq \mathcal{M}$. The main goal of this chapter is to show that, provided $\mathcal{M}$ is sufficiently closed, $\Sigma \restriction \mathcal{M}$ is a definable class of $\mathcal{M}$. Below we make our goal more precise.

> *Motivational Question.* Suppose $(\mathcal{P},\Sigma)$ is a hod pair or an sts hod pair, X is a self-well-ordered set such that $\mathcal{P} \in X$ and $\mathcal{M}$ is a Σ or Σ-sts mouse over X (see Definition 3.8.21). Is $\Sigma \restriction \mathcal{N}$ definable over $\mathcal{N}$? Is $\Sigma \restriction \mathcal{N}[g]$ definable over $\mathcal{N}[g]$ where g is $\mathcal{N}$-generic?

In Section 5.2 we gave an answer to Motivational Question in the case $\mathcal{M}$ is $\mathcal{P}$ itself (see Theorem 5.2.5). Another answer was given by [29, Lemma 3.35], where it was shown that $\Sigma \restriction \mathcal{N}[g]$ is definable over $\mathcal{N}[g]$ provided $\mathcal{P}$ does not have non-meek layers. Here, we are mainly concerned with proving a version of [29, Lemma 3.35] in the case of a non-meek hod premice. Because of this we will state many of our definitions and theorems for hod pairs or sts hod pairs $(\mathcal{P},\Sigma)$ such that $\mathcal{P}$ is non-meek[2]. To simplify our terminology, we will say $(\mathcal{P},\Sigma)$ is a non-meek hod pair if $\mathcal{P}$ is a non-meek hod premouse and Σ is either an iteration strategy or an sts-strategy (this is only allowed in the case $\mathcal{P}$ is of lsa type).

While a positive answer to the Motivational Questions is desirable, it is naive to hope that one exists for all such $\mathcal{N}$. A positive answer depends on how closed $\mathcal{N}$ is. If for instance the branch of $\mathcal{T}$ is given via a $\mathcal{Q}$-structure that is beyond the

[1] See Definition 3.3.2.
[2] See Definition 2.7.1.

#-operator while our $\mathcal{N}$ is only closed under the #-operator then, in most cases, identifying the correct branch of $\mathcal{T}$ inside $\mathcal{N}$ via a procedure that is uniform in $\mathcal{T}$ will be impossible. In this chapter, we give a positive answer to the Motivational Question provided our $\mathcal{N}$ is sufficiently closed. We make this notion more precise.

Suppose $(\mathcal{P},\Sigma)$ is a non-meek hod pair and $\mathcal{N}$ is a Σ-mouse such that $\mathcal{N} \vDash \mathsf{ZFC}-\mathsf{Replacement}$. We say $\mathcal{N}$ is Σ-closed if $\Sigma \restriction \mathcal{N} \subseteq \mathcal{N}$. We say $\mathcal{N}$ is generically Σ-closed if $\mathcal{N}$ is Σ-closed and whenever g is $\mathcal{N}$-generic, $\Sigma \restriction \mathcal{N}[g]$ is definable over $(\mathcal{N}[g],\in)$ (in the language of Σ-premice). It is worth remarking that the structure $(\mathcal{N}[g],\in)$ is a structure in the language of Σ-premice and in particular, there are names for $\vec{E}^{\mathcal{N}}$ and $\Sigma^{\mathcal{N}}$.

DEFINITION 6.0.4. We say $\mathcal{N}$ is uniformly generically Σ-closed if $\mathcal{N}$ is generically Σ-closed and there are formulas φ and ψ (in the language of Σ-premice) such that for any $\mathcal{N}$-generic g, any stack $\mathcal{T} \in \mathcal{N}[g]$ on $\mathcal{P}$ and any $b \in \mathcal{N}[g]$,

$$\begin{aligned} \mathcal{T} \in \mathrm{dom}(\Sigma) &\iff (\mathcal{N}[g],\in) \vDash \varphi[\mathcal{T}], \\ \Sigma(\mathcal{T}) = b &\iff (\mathcal{N}[g],\in) \vDash \psi[\mathcal{T},b]. \end{aligned}$$

The main theorem of this chapter is Theorem 6.1.4. It gives a positive answer to our Motivational Question in the case $\mathcal{N}$ is Σ-closed and has fullness preserving iteration strategy (see Definition 6.1.1 and Definition 6.1.3). The main idea behind the proof of Theorem 6.1.4 is that the branch of an iteration tree $\mathcal{T}$ on $\mathcal{P}$ can be identified by the authentication process introduced in Definition 3.7.4.

Given a transitive set X, we let $X^{\#}$ be the least sound active mouse over X. Also recall that if X is any set and $A \subseteq X^2$ then $p[A]$ is the projection of A onto one of the coordinates of A. The specific coordinate onto which we project will always be clear from the context.

6.1. Internally Σ-closed mice

In this section we introduce a kind of closure property of hybrid mice for which we can give a positive answer to our motivational question. The first such closure property is internal closure, which postulates that our mouse has enough of the strategy.

DEFINITION 6.1.1 (Internally Σ-closed mouse). Suppose

1. $(\mathcal{P},\Sigma)$ is an allowable pair[3],
2. $\mathcal{P}$ is a non-meek hod premouse,
3. if Σ is an iteration strategy then $\mathcal{N}$ is a Σ-premouse over some X based on $\mathcal{P}$, and
4. if Σ is an sts premouse then $\mathcal{N}$ is a Σ-sts premouse over some X based on $\mathcal{P}$.

[3] See Definition 3.10.7.

1. We say $\mathcal{N}$ is *internally Σ-closed premouse* if for every $\kappa < \text{ord}(\mathcal{N})$ there is $\mathcal{M} \trianglelefteq \mathcal{N}$ such that
   (a) $\mathcal{M} \models \mathsf{ZFC}$,
   (b) $\mathcal{N}|\kappa \trianglelefteq \mathcal{M}$,
   (c) for every $(\mathcal{T},\mathcal{S}) \in B(\mathcal{P},\Sigma^{\mathcal{M}})$[4], $\Sigma^{\mathcal{M}}_{\mathcal{S}}$ is total in $\mathcal{M}$[5],
   (d) $\mathcal{M}$ has at least three Woodin cardinals that are greater than κ,
   (e) letting $\delta_0 < \delta_1 < \delta_2$ be the first three Woodin cardinals of $\mathcal{M}$ that are greater than κ, for every $i \in 3$ and $\eta \in [\kappa, \delta_i)$, letting $((\mathcal{M}_\gamma, \mathcal{N}_\gamma : \gamma \leq \nu), (F_\gamma : \gamma < \nu), (\mathcal{T}_\gamma : \gamma \leq \nu))$ be the output of the $(\mathcal{P},\Sigma)$-coherent fully backgrounded construction of $\mathcal{M}|\delta_i$[6] in which extenders used have critical points $> \eta$, the following conditions hold:
      (i) If Σ is an iteration strategy then $\pi^{\mathcal{T}_\nu}$-exists and $\mathcal{M}_\nu$ is the last model of $\mathcal{T}_\nu$.
      (ii) If Σ is an st-strategy then $\pi^{\mathcal{T}_\nu,b}$ exists and $\mathcal{T}_\nu$ is $\mathcal{M}$-terminal[7].
      (iii) If $(\mathcal{T},\mathcal{S}) \in B(\mathcal{P},\Sigma^{\mathcal{N}|\eta})$[8] then for some β, $\mathcal{M}_\nu|\beta$ is a $\Sigma^{\mathcal{N}}_{\mathcal{S}}$-iterate of $\mathcal{S}$ via a normal stack.
2. If $\mathcal{M}, \mathcal{N}$ and κ are as above then we say $\mathcal{M}$ *witnesses* the internal Σ-closure of $\mathcal{N}$ at κ.
3. We say $\mathcal{N}$ is an *internally Σ-closed mouse* (sts mouse) if it is an internally Σ-closed premouse and has an ω_1-iteration strategy Λ witnessing that $\mathcal{N}$ is a Σ-mouse (sts mouse).

Two remarks are in order. First notice that internal Σ-closure is a first order property of $\mathcal{N}$, and in clause 3 above we do not need to require that Λ-iterates of $\mathcal{N}$ are internally Σ-closed as this is just a consequence of elementarity.

Secondly, we cannot in general hope to prove that generic interpretability holds for internally Σ-closed mice. The reason is that there might be $\mathcal{Q} \in B(\mathcal{P},\Sigma)$ such that $\Sigma_{\mathcal{Q}}$ is beyond the iteration strategy of $\mathcal{N}$ (in the sense that $\Lambda <_w \Sigma_{\mathcal{Q}}$), and if such a $\mathcal{Q}$ is generic over $\mathcal{N}$ then it is not wise to hope that $\Sigma_{\mathcal{Q}} \restriction \mathcal{N}$ would be definable over $\mathcal{N}[\mathcal{Q}]$. In order to prove generic interpretability result for internally Σ-closed premice we need to find a *fullness condition* that would let us take care of examples as above. In particular, we seem to need to require that any $\Sigma_{\mathcal{Q}}$ as above is strictly below the strategy of $\mathcal{N}$. The next couple of paragraphs make this intuitive notion more precise.

Suppose $\mathcal{N}$ is an internally Σ-closed mouse, κ is an $\mathcal{N}$-cardinal and $\mathcal{M}$ is as in Definition 6.1.1. Let $\delta_0 < \delta_1 < \delta_2$ be the first three Woodin cardinals of $\mathcal{M}$ that are greater than κ, and let $\eta \in [\kappa, \delta_2)$ and $i \in 3$ be the least such that $\eta < \delta_i$. We then let $\mathcal{S}^{\mathcal{M}}_\eta$ be the $\Sigma^{\mathcal{M}}$-iterate of $\mathcal{P}$ constructed via the $(\mathcal{P},\Sigma^{\mathcal{M}})$-coherent fully

[4] Thus, $(\mathcal{T},\mathcal{S}) \in \mathcal{M}$.
[5] Thus, $\Sigma^{\mathcal{M}}_{\mathcal{S}} = \Sigma_{\mathcal{S},\mathcal{T}} \restriction \mathcal{M}$.
[6] See Definition 3.5.1.
[7] See Definition 3.8.8.
[8] See Definition 3.3.9.

backgrounded construction of $\mathcal{M}|\delta_i$ where critical points of extenders used are $> \eta$. We let $\mathcal{U}_\eta^{\mathcal{M}}$ be the normal tree on $\mathcal{P}$ with last model $\mathcal{S}_\eta^{\mathcal{M}}$ and

$$\pi_\eta^{\mathcal{M}} = \begin{cases} \pi^{\mathcal{U}_\eta^{\mathcal{M}},b} & \text{if } \mathcal{P} \text{ is of lsa type,} \\ \pi^{\mathcal{U}_\eta^{\mathcal{M}}} & \text{otherwise.} \end{cases}$$

Notice that $\pi_\eta^{\mathcal{M}} \in \mathcal{N}$.

Keeping the notation and terminology of Definition 6.1.1, suppose Λ is an iteration strategy for $\mathcal{N}$ (witnessing that $\mathcal{N}$ is an internally Σ-closed mouse). Suppose $\xi < \text{ord}(\mathcal{N})$ and Λ^ξ is the fragment of Λ that acts on stacks above ξ. We then let $\Gamma(\mathcal{N}, \Lambda^\xi)$ be the collection of all sets $A \subseteq \mathbb{R}$ such that for some $(\mathcal{T}, \mathcal{R}) \in I(\mathcal{N}, \Lambda)$, $\kappa < \text{ord}(\mathcal{R})$ and $\mathcal{M} \trianglelefteq \mathcal{R}$ witnessing that $\mathcal{R}$ is internally Σ-closed at κ the following holds: letting $\delta_0 < \delta_1 < \delta_2$ be the first three Woodin cardinals of $\mathcal{M}$ that are greater than κ, whenever $\eta \in [\kappa, \delta_2)$, there is $\mathcal{Q} \lhd_{hod} (\mathcal{S}_\eta^{\mathcal{M}})^b$ such that

1. $\mathcal{S}_\eta^{\mathcal{M}} \models$ "$\delta^{\mathcal{Q}}$ is a Woodin cardinal" and
2. $w(A) \leq w(\mathsf{Code}(\Sigma_{\mathcal{Q}, \mathcal{U}_\eta^{\mathcal{M}}}))$.

REMARK 6.1.2. For convenience, we will use the notation $\Gamma(\mathcal{P}, \Sigma)$ for both sts pairs and hod pairs. In the case of sts hod pairs, it is just $\Gamma^b(\mathcal{P}, \Sigma)$.

DEFINITION 6.1.3. Suppose $\mathcal{N}$ is as in Definition 6.1.1 and Λ is an ω_1-iteration strategy for $\mathcal{N}$ (witnessing that $\mathcal{N}$ is an internally Σ-closed mouse). We then say that Λ is a *fullness preserving* iteration strategy for $\mathcal{N}$ if for every $\xi < \text{ord}(\mathcal{N})$, letting Λ^ξ be the fragment of Λ that acts on stacks above ξ, $\Gamma(\mathcal{N}, \Lambda^\xi) = \Gamma(\mathcal{P}, \Sigma)$.

The following is our generic interpretability result for internally Σ-closed mice $\mathcal{N}$ that have a fullness preserving iteration strategy.

THEOREM 6.1.4. *Assume* NsesN[9] *Suppose $(\mathcal{P}, \Sigma)$ is an allowable pair, Γ is a projectively closed pointclass and $\mathcal{N}$ is an internally Σ-closed premouse (possibly over some set X). Suppose Σ is*

1. *strongly Γ-fullness preserving,*
2. *has strong branch condensation and*
3. *is commuting*[10].

Then the following conditions hold.

1. *If $(\mathcal{P}, \Sigma)$ is a hod pair and $\mathcal{N}$ is a Σ-mouse then for any $\mathcal{N}$-generic g, $\mathcal{N}[g]$ is Σ-closed and $\Sigma \restriction \mathcal{N}[g]$ is uniformly in g definable over $\mathcal{N}[g]$.*
2. *If $(\mathcal{P}, \Sigma)$ is an sts hod pair and $\mathcal{N}$ is a Σ-sts mouse with a fullness preserving iteration strategy then for any $\mathcal{N}$-generic g, $\mathcal{N}[g]$ is Σ-closed and $\Sigma \restriction \mathcal{N}[g]$ is uniformly in g definable over $\mathcal{N}[g]$.*

[9] See Definition 4.0.8.

[10] See Definition 4.10.5.

In the next few sections, we will develop the terminology we need to prove Theorem 6.1.4. We will not give the proof of clause 1 of Theorem 6.1.4. It is much easier than the proof of clause 2 of Theorem 6.1.4 and it is very much like the proof of [29, Theorem 3.10]. Thus, we only concentrate on sts hod pairs.

6.2. Authentication procedure revisited

Suppose $(\mathcal{P},\Sigma)$ is an sts hod pair, $\mathcal{N}$ is an internally Σ-closed premouse, g is $\mathcal{N}$-generic and $\mathcal{T} \in \mathrm{dom}(\Sigma) \cap \mathcal{N}[g]$ is a normal stack on $\mathcal{P}$ above $\mathcal{P}^b$ such that $\mathcal{T}$ does not have fatal drops. Suppose first that $\mathcal{T} \in b(\Sigma)$. In this case, we would like to identify $\mathcal{Q}(b,\mathcal{T})$ in $\mathcal{N}[g]$ via a procedure that is uniform in $\mathcal{T}$. Here $b = \Sigma(\mathcal{T})$. Clearly if $\mathcal{Q}(b,\mathcal{T}) \trianglelefteq \mathrm{m}^+(\mathcal{T})$ then we can easily identify $\mathcal{Q}(b,\mathcal{T})$. Suppose then $\mathrm{m}^+(\mathcal{T}) \triangleleft \mathcal{Q}(b,\mathcal{T})$. We now face two problems.

The first problem is showing that $\mathcal{Q}(b,\mathcal{T}) \in \mathcal{N}[g]$ and the second is showing that $\mathcal{Q}(b,\mathcal{T})$ can be identified in $\mathcal{N}$ in a uniform manner. Both of these require more of $\mathcal{N}$ than just internal Σ-closure. To prove both of these facts, we will need that $\mathcal{N}$ has a fullness preserving iteration strategy. Our strategy for finding $\mathcal{Q}(b,\mathcal{T})$ in $\mathcal{N}$ is that if $\mathcal{N}$ is sufficiently rich then some backgrounded construction will reach $\mathcal{Q}(b,\mathcal{T})$. To execute this plan, we first need to describe the sort of backgrounded constructions that we will consider. In what follows, we borrow ideas from Section 3.7. In particular, it will be helpful to recall Definition 3.7.5 and other definitions from that section.

DEFINITION 6.2.1 ($(\mathcal{N},X)$-authenticated iteration strategy). Suppose $(\mathcal{P},\Sigma)$ is an sts hod pair, $X \subseteq \mathcal{P}^b$ and $\mathcal{N}$ is a Σ-sts premouse such that $X \in \mathcal{N}$. Suppose that $g \subseteq \mathbb{P}$ is $\mathcal{N}$-generic for some poset $\mathbb{P} \in \mathcal{N}$ and $\mathcal{R} \in \mathcal{N}[g]$ is an lsa type hod premouse. We define a partial short tree strategy $\Phi_{\mathcal{R}}^{\mathcal{N},X,g}$ without a model component for $\mathcal{R}$ as follows. $\Phi_{\mathcal{R}}^{\mathcal{N},X,g}$ acts on indexable stacks[11].

1. $t = (\mathcal{R},\mathcal{T},\mathcal{R}_1,\mathcal{T}_1) \in \mathrm{dom}(\Phi_{\mathcal{R}}^{\mathcal{N},X,g}) \cap \mathcal{N}[g]$ if and only if t is $(\mathcal{P},\Sigma^{\mathcal{N}},X)$-authenticated[12].
2. Given $t = (\mathcal{R},\mathcal{T}_0,\mathcal{R}_1,\mathcal{T}_1) \in \mathrm{dom}(\Phi_{\mathcal{R}}^{\mathcal{N},X,g}) \cap \mathcal{N}[g]$ with $\mathcal{T}_1 \neq \emptyset$, $\Phi_{\mathcal{R}}^{\mathcal{N},X,g}(t) = b$ if and only if $t^\frown\{b\}$ is $(\mathcal{P},\Sigma^{\mathcal{N}},X)$-authenticated.

When $X = \mathcal{P}^b$ we simply omit it from our terminology.

Continuing with the $\mathcal{R},\mathcal{N}$ of Definition 6.2.1, we next define an $\mathcal{N}$-authenticated backgrounded construction over $\mathcal{R}$. This is essentially a fully backgrounded construction relative to $\Phi_{\mathcal{R}}^{\mathcal{N},g}$ (see Definition 4.2.1).

DEFINITION 6.2.2. Suppose $(\mathcal{P},\Sigma)$ is an sts hod pair, $X \subseteq \mathcal{P}^b \cap \mathcal{N}$ and $\mathcal{N}$ is a Σ-sts premouse such that $X \in \mathcal{N}$. Suppose that $g \subseteq \mathbb{P}$ is $\mathcal{N}$-generic for some

[11] See Definition 3.3.3.
[12] See Definition 3.7.3.

poset $\mathbb{P} \in \mathcal{N}$ and $Y, \mathcal{R} \in \mathcal{N}[g]$ are such that Y is a self-well-ordered set and $\mathcal{R} \in Y$ is an lsa type hod premouse. Suppose further that κ is an $\mathcal{N}$-cardinal such that $\{\mathbb{P}, \mathcal{R}, Y\} \in \mathcal{N}|\kappa[g]$.

We then say that $((\mathcal{M}_\gamma, \mathcal{N}_\gamma : \gamma \leq \nu), (F_\gamma : \gamma < \nu), (\mathcal{T}_\gamma : \gamma \leq \nu))$ is the output of the $(\mathcal{N}, \kappa, X)$-authenticated fully backgrounded construction over Y based on $\mathcal{R}$ in which extenders used have critical points $> \kappa$ if $((\mathcal{M}_\gamma, \mathcal{N}_\gamma : \gamma \leq \nu), (F_\gamma : \gamma < \nu), (\mathcal{T}_\gamma : \gamma \leq \nu))$ is the output of $(\mathcal{R}, \Phi_{\mathcal{R}}^{\mathcal{N},X,g})$-coherent fully backgrounded construction[13] of $\mathcal{N}$ done over Y using extenders with critical points $> \kappa$[14].

Finally, we say $\mathcal{Q}$ is an $(\mathcal{N}, X)$-authenticated sts mouse over Y based on $\mathcal{R}$ if $\mathcal{Q} \in \mathcal{N}$ and for some ν,

1. $\{\mathbb{P}, \mathcal{R}, Y, \mathcal{Q}\} \in \mathcal{N}|\nu[g]$ and
2. $\mathcal{Q}$ appears as a model in the $(\mathcal{N}, \nu, X)$-authenticated fully backgrounded construction over Y based on $\mathcal{R}$.

When $X = \mathcal{P}^b$ we simply omit it from our terminology.

Suppose now that $(\mathcal{P}, \Sigma)$ is an sts hod pair, $X \subseteq \mathcal{P}^b$ and $\mathcal{N}$ is an internally Σ-closed mouse with a fullness preserving iteration strategy Λ such that $X \in \mathcal{N}$. Suppose $\mathbb{P} \in \mathcal{N}$ is a poset and $g \subseteq \mathbb{P}$ is $\mathcal{N}$-generic. Suppose further that $Y \in \mathcal{N}[g]$. We then let

$$\begin{aligned}\mathsf{Lp}^{\mathcal{N},g,X,sts,+}(Y, \mathcal{R}) = \bigcup\{&\mathcal{K} \in \mathcal{N}[g] : \text{there is an } \mathcal{N}\text{-cardinal } \kappa \text{ such that}\\ &\{\mathbb{P}, \mathcal{R}, Y, \mathcal{K}\} \in \mathcal{N}|\kappa[g] \text{ such that } \mathcal{K} \text{ is an } (\mathcal{N}, \kappa, X)\text{-authenticated sound}\\ &\text{sts mouse over } Y \text{ based on } \mathcal{R} \text{ such that } \rho(\mathcal{K}) = \operatorname{ord}(Y)\}.\end{aligned}$$

Again, if $X = \mathcal{P}^b$, then we omit it from the notation.

Notice that we do not know that $\mathsf{Lp}^{\mathcal{N},g,X,sts,+}(Y, \mathcal{R})$ is a meaningful object, since we do not know that if $\mathcal{Q}_0$ and $\mathcal{Q}_1$ are authenticated by $\mathcal{M}_0$ and $\mathcal{M}_1$ respectively then they are compatible. This, however, is true when $\mathcal{R}$ is an iterate of $\mathcal{P}$ and Σ has strong branch condensation and is strongly Γ-fullness preserving for some Γ. This fact will also be verified in the next section.

We can then define $(\mathsf{Lp}_\alpha^{\mathcal{N},g,X,sts,+}(Y, \mathcal{R}) : \alpha < \operatorname{ord}(\mathcal{N}))$ by induction as usual. More precisely, the sequence is defined via the following recursion.

1. $\mathsf{Lp}_0^{\mathcal{N},g,X,sts,+}(Y, \mathcal{R}) = trc(Y, \mathcal{R})$.
2. $\mathsf{Lp}_1^{\mathcal{N},g,X,sts,+}(Y, \mathcal{R}) = \mathsf{Lp}^{\mathcal{N},g,X,sts,+}(Y, \mathcal{R})$.
3. $\mathsf{Lp}_{\alpha+1}^{\mathcal{N},g,X,sts,+}(Y, \mathcal{R}) = \mathsf{Lp}^{\mathcal{N},g,X,sts,+}(\mathsf{Lp}_\alpha^{\mathcal{N},g,X,sts,+}(Y, \mathcal{R}))$.
4. $\mathsf{Lp}_\lambda^{\mathcal{N},g,X,sts,+}(Y, \mathcal{R}) = \bigcup_{\alpha<\lambda} \mathsf{Lp}_\alpha^{\mathcal{N},g,X,sts,+}(Y, \mathcal{R})$.

When $Y = \mathcal{J}_\omega[\mathcal{R}]$ or $X = \mathcal{P}^b$, we omit them from the above notation.

The "+" version of the Lp operator defined above may stack more sts mice than we need. To get the proper operator we need to only consider $\mathcal{K} \in \mathcal{N}[g]$ which

[13] See Definition 3.5.1.

[14] If the construction reaches a stack $\mathcal{T}$ such that $\Phi_{\mathcal{R}}^{\mathcal{N},X,g}(\mathcal{T})$ is undefined we stop the construction.

have an iteration strategy in $\Gamma(\mathcal{P},\Sigma)$. This fact can be expressed in a first order manner over $\mathcal{N}[g]$.

DEFINITION 6.2.3. Suppose $(\mathcal{P},\Sigma)$, $\mathcal{N}$ and $(\mathbb{P},g,X,Y,\mathcal{R})$ are as above. Let $\mathcal{K}\trianglelefteq \mathsf{Lp}^{\mathcal{N},g,X,sts,+}(Y,\mathcal{R})$. We say $\mathcal{K}$ is *simple* (in $\mathcal{N}$) if there is κ and $\mathcal{M}\trianglelefteq\mathcal{N}$ such that

1. $(\mathbb{P},X,Y,\mathcal{R},\mathcal{K})\in\mathcal{N}|\kappa[g]$,
2. $\mathcal{M}$ witnesses the internal Σ-closure of $\mathcal{N}$ at κ,
3. letting $\delta_0<\delta_1<\delta_2$ be the first three Woodin cardinals of $\mathcal{M}$ that are above κ, there is some $\mathcal{Q}\lhd_{hod}(\mathcal{S}_\kappa^{\mathcal{M}})^b$ such that if $\eta\in(\mathrm{ord}(\mathcal{Q}),\delta_0)$ is the least such that $\mathsf{Lp}^{\Gamma(\mathcal{P},\Sigma),\Sigma_{\mathcal{Q}}}(\mathcal{M}|\eta)\vDash$ "η is a Woodin cardinal"[15] then $\mathcal{K}$ is an $(\mathcal{M}|\eta,\mathrm{ord}(\mathcal{Q}),X)$-authenticated sound sts mouse over Y based on $\mathcal{R}$ such that $\rho(\mathcal{K})=\mathrm{ord}(\mathcal{R})$.

We then let

$$\mathsf{Lp}^{\mathcal{N},g,X,sts}(Y,\mathcal{R})=\bigcup\{\mathcal{K}\in\mathcal{N}[g]:\mathcal{K}\text{ is simple and }\mathcal{K}\lhd\mathsf{Lp}^{\mathcal{N},g,X,sts,+}(Y,\mathcal{R})\}.$$

We will omit g and X when they are clear from the context. The effect of clause 3 of Definition 6.2.3 is that since $\mathcal{K}$ is built by a fully backgrounded construction of $\mathcal{M}|\eta$ using extenders with critical points $>\mathrm{ord}(\mathcal{Q})$, the strategy $\mathcal{K}$ acquired from the strategy of $\mathcal{M}|\eta$ via the resurrection procedure of [23, Chapter 12] is in $\Gamma(\mathcal{P},\Sigma)$. This is because the strategy of $\mathcal{M}|\eta$ that acts on stacks above $\mathrm{ord}(\mathcal{Q})$ is in $\Gamma(\mathcal{P},\Sigma)$. Thus, the following claim is true.

PROPOSITION 6.2.4. *Suppose $(\mathcal{P},\Sigma)$, $\mathcal{N}$ and $(\mathbb{P},g,X,Y,\mathcal{R})$ are as in Definition 6.2.3. Suppose $\mathcal{R}=\mathrm{m}^+(\mathcal{T})$ where $\mathcal{T}$ is a normal stack according to Σ. Suppose further that Σ has strong branch condensation and is Γ-fullness preserving for some projectively closed pointclass Γ. Then $\mathsf{Lp}^{\mathcal{N},g,X,sts}(\mathcal{R})\trianglelefteq\mathsf{Lp}^{\Gamma(\mathcal{P},\Sigma),\Sigma_{\mathcal{R}}}(\mathcal{R})$.*

PROOF. We have already explained that every $\mathcal{K}\trianglelefteq\mathsf{Lp}^{\mathcal{N},g,X,sts}(\mathcal{R})$ has an iteration strategy in $\Gamma(\mathcal{P},\Sigma)$. Moreover, because Σ has strong branch condensation and is Γ-fullness preserving, the strategy $\mathcal{K}$ acquired from the strategy of $\mathcal{M}|\eta$ witnesses that $\mathcal{K}$ is a $\Sigma_{\mathcal{R}}$-sts[16]. ⊣

We can now describe the $\mathcal{N}$-authenticated iterations of $\mathcal{P}$.

DEFINITION 6.2.5 ($\mathcal{N}$-authenticated iteration). Suppose $(\mathcal{P},\Sigma)$ is an sts pair, Γ is a projectively closed pointclass and $\mathcal{N}$ is an internally Σ-closed mouse with a fullness preserving iteration strategy Λ. Suppose further that Σ has strong branch condensation and is strongly Γ-fullness preserving. Also suppose that $g\subseteq\mathbb{P}$ is

[15]Let $\mathcal{Q}^+$ be the least hod initial segment of $\mathcal{S}_\kappa^{\mathcal{M}}$ such that $\mathcal{Q}\lhd\mathcal{Q}^+$ and $\delta^{\mathcal{Q}^+}$ is a Woodin cardinal of $\mathcal{S}_\kappa^{\mathcal{M}}$. Since $\mathcal{M}$ is closed under $\Sigma_{\mathcal{Q}^+}$, the condition $\mathsf{Lp}^{\Gamma(\mathcal{P},\Sigma),\Sigma_{\mathcal{Q}}}(\mathcal{M}|\eta)\vDash$ "η is a Woodin cardinal" is first order over $\mathcal{N}[g]$.

[16]While this is non-trivial, most of the proof is contained in the proofs of Theorem 4.12.1 and Proposition 4.12.5.

$\mathcal{N}$-generic for some poset $\mathbb{P} \in \mathcal{N}$ and $\mathcal{T} \in \mathcal{N}[g]$ is a stack on $\mathcal{P}$. We say $\mathcal{T}$ is $\mathcal{N}$-*authenticated* if the following conditions holds.

1. For every $\alpha \in \mathsf{max}^{\mathcal{T}}$[17],

$$\mathsf{Lp}^{\mathcal{N},sts}(\mathcal{M}_\alpha^{\mathcal{T}}) \models \text{“}\delta^{\mathcal{M}_\alpha^{\mathcal{T}}} \text{ is a Woodin cardinal”}.$$

2. For every $\alpha \in \mathsf{max}^{\mathcal{T}}$, $\pi^{\mathcal{T}_{\leq\alpha},b}$ exists.
3. For all $\alpha \in R^{\mathcal{T}}$ such that $\pi^{\mathcal{T}_{\leq\alpha},b}$ exists, letting $\mathcal{W} = \mathsf{nc}_\alpha^{\mathcal{T}}$[18], if $\mathcal{W}$ is above $\mathrm{ord}((\mathcal{M}_\alpha^{\mathcal{T}})^b)$ then for all limit ordinals $\gamma < lh(\mathcal{W})$ such that $\mathcal{W} \restriction \gamma$ is nuvs,
   (a) $\mathsf{Lp}^{\mathcal{N},sts}(\mathrm{m}^+(\mathcal{W} \restriction \gamma)) \models$ “$\delta(\mathcal{W} \restriction \gamma)$ is not a Woodin cardinal”, and
   (b) letting $b = [0,\gamma)_{\mathcal{T}}$, $\mathcal{Q}(b, \mathcal{W} \restriction \gamma)$ exists and
$$\mathcal{Q}(b, \mathcal{W} \restriction \gamma) \trianglelefteq \mathsf{Lp}^{\mathcal{N},sts}(\mathrm{m}^+(\mathcal{W} \restriction \gamma)).$$
4. For every $\alpha \in R^{\mathcal{T}}$ such that $\pi^{\mathcal{T}_{\leq\alpha},b}$ exists, if $\mathsf{nc}_\alpha^{\mathcal{T}}$ is based on $\mathcal{S} =_{\mathrm{def}} (\mathcal{M}_\alpha^{\mathcal{T}})^b$ then $(\mathcal{S}, \mathsf{nc}_\alpha^{\mathcal{T}})$ is a $(\mathcal{P}, \Sigma^{\mathcal{N}})$-authenticated iteration[19].
5. For every $\alpha \in R^{\mathcal{T}}$ such that $\pi^{\mathcal{T}_{\leq\alpha},b}$ exists, letting $\mathcal{U} = \mathsf{nc}_\alpha^{\mathcal{T}}$ and $\mathcal{S} =_{\mathrm{def}} \mathcal{M}_\alpha^{\mathcal{T}}$, if $\mathcal{U}$ is above $\delta^{\mathcal{S}^b}$ and is such that for some $\eta \in (\delta^{\mathcal{S}^b}, \delta^{\mathcal{S}})$, $\mathcal{U}$ is based on $\mathcal{O}^{\mathcal{S}}_{\eta,\mathcal{S}|\eta,\eta}$[20] and is above η, then $(\mathcal{O}^{\mathcal{S}}_{\eta,\mathcal{S}|\eta,\eta}, \mathcal{U})$ is a $(\mathcal{P}, \Sigma^{\mathcal{N}})$-authenticated iteration.
6. For every $\alpha \in R^{\mathcal{T}}$ such that $\pi^{\mathcal{T}_{\leq\alpha},b}$ exists, letting $\mathcal{U} = \mathsf{nc}_\alpha^{\mathcal{T}}$ and $\mathcal{S} =_{\mathrm{def}} \mathcal{M}_\alpha^{\mathcal{T}}$, if $\mathcal{U}$ is a normal tree on $\mathcal{S}^b$ above $\delta^{\mathcal{S}^b}$, then $(\mathcal{S}^b, \mathcal{T}_{\geq\mathcal{S}})$ is a $(\mathcal{P}, \Sigma^{\mathcal{N}})$-authenticated iteration[21].

6.3. Generic interpretability in internally Σ-closed premice

In this section, we prove our main theorem, Theorem 6.1.4. As we said before, we will only prove clause 2. We start by fixing an sts hod pair $(\mathcal{P}, \Sigma)$ such that Σ has strong branch condensation, a projectively closed pointclass Γ such that Σ is strongly Γ-fullness preserving and an internally Σ-closed premouse $\mathcal{N}$ such that $\mathcal{N}$ has a fullness preserving iteration strategy Λ[22]. We want to show that $\mathcal{N}$ is uniformly generically Σ-closed.

Fix a poset $\mathbb{P} \in \mathcal{N}$ and an $\mathcal{N}$-generic $g \subseteq \mathbb{P}$. We start by defining a short tree strategy Φ for $\mathcal{P}$. Φ is defined over $\mathcal{N}[g]$ in a uniform manner. Its domain consists of $\mathcal{N}$-authenticated iterations (see Definition 6.2.5). Given an $\mathcal{N}$-authenticated iteration $\mathcal{T}$ of limit length, we set $\Phi(\mathcal{T}) = x$ if and only if one of the following conditions holds.

[17] See Definition 3.1.6.
[18] See Notation 2.4.4.
[19] See Definition 3.7.4.
[20] See Definition 2.6.11.
[21] This clause follows from the one above it.
[22] See Definition 6.1.3.

1. $\mathcal{T}$ is nuvs and letting $\alpha = \max(R^{\mathcal{T}})$, $\mathsf{Lp}^{\mathcal{N},sts}(\mathrm{m}^+(\mathcal{T}_{\geq\alpha})) \models$ "$\delta(\mathcal{T}_{\geq\alpha})$ is a Woodin cardinal" and $x = \mathrm{m}^+(\mathcal{T})$.
2. Clause 1 above fails, $x \in \mathcal{N}[g]$ is a branch of $\mathcal{T}$ such that $\mathcal{N}[g] \models$ "x is a cofinal well-founded branch of $\mathcal{T}$" and $\mathcal{T}^\frown\{x\}$ is $\mathcal{N}$-authenticated.

To complete the proof of Theorem 6.1.4 we need to show that

(a) whenever $\mathcal{T} \in \mathrm{dom}(\Phi) \cap \mathrm{dom}(\Sigma)$, $\Phi(\mathcal{T})$ is defined and is equal to $\Sigma(\mathcal{T})$.

Fix then $\mathcal{T}' \in \mathcal{N}[g]$ such that $\mathcal{T}' \in \mathrm{dom}(\Phi) \cap \mathrm{dom}(\Sigma)$. Suppose first that

($*$1) $\mathcal{T}'$ is nuvs.

Let $\alpha = \max(R^{\mathcal{T}'})$ and set $\mathcal{T} = \mathcal{T}'_{\geq\alpha}$ and $\mathcal{S} =_{\mathrm{def}} \mathcal{M}^{\mathcal{T}'}_{\alpha}$. We thus have that

$$\mathcal{J}_\omega[\mathrm{m}^+(\mathcal{T})] \models \text{“}\delta(\mathcal{T}) \text{ is a Woodin cardinal”}.$$

We want to conclude that $\Phi(\mathcal{T})$ is defined and $\Phi(\mathcal{T}) = \Sigma(\mathcal{T})$.

Suppose first that $\Sigma(\mathcal{T}) = \mathrm{m}^+(\mathcal{T})$. Then because Σ is Γ-fullness preserving, there is no $\Sigma_{\mathrm{m}^+(\mathcal{T})}$-sts $\mathcal{Q}$ over $\mathrm{m}^+(\mathcal{T})$ such that

1. $\mathcal{Q}$ is sound,
2. $\mathcal{Q}$ has a strategy $\Lambda \in \Gamma(\mathcal{P}, \Sigma)$ witnessing that $\mathcal{Q}$ is a $\Sigma_{\mathrm{m}^+(\mathcal{T})}$-sts over $\mathrm{m}^+(\mathcal{T})$,
3. $\mathcal{Q} \models$ "$\delta(\mathcal{T})$ is a Woodin cardinal" but $\mathcal{J}_\omega[\mathcal{Q}] \models$ "$\delta(\mathcal{T})$ is not a Woodin cardinal".

It then follows from Proposition 6.2.4 that $\Phi(\mathcal{T}) = \mathrm{m}^+(\mathcal{T})$.

Suppose next that $\Sigma(\mathcal{T}) = b$ where b is a cofinal well-founded branch of $\mathcal{T}$. We thus have that $\mathcal{Q}(b, \mathcal{T}) =_{\mathrm{def}} \mathcal{W}$ exists and want to conclude that $\Phi(\mathcal{T}) = b$. Notice that if

$$\mathsf{Lp}^{\mathcal{N},sts}(\mathrm{m}^+(\mathcal{T}'_{\geq\alpha})) \models \text{“}\delta(\mathcal{T}'_{\geq\alpha}) \text{ is not a Woodin cardinal”}$$

then $\mathcal{W} \trianglelefteq \mathsf{Lp}^{\mathcal{N},sts}(\mathrm{m}^+(\mathcal{T}'_{\geq\alpha}))$ and, therefore, $\Phi(\mathcal{T}) = b$. Assume then that

(1) $\mathsf{Lp}^{\mathcal{N},sts}(\mathrm{m}^+(\mathcal{T}'_{\geq\alpha})) \models$ "$\delta(\mathcal{T}'_{\geq\alpha})$ is a Woodin cardinal".

Set $\mathcal{T} = \mathcal{T}'_{\geq\mathcal{S}}$. The following claim finishes the proof of (a) assuming ($*$1).

CLAIM 6.3.1. $\mathcal{W} \trianglelefteq \mathsf{Lp}^{\mathcal{N},sts}(\mathrm{m}^+(\mathcal{T}))$.

PROOF. Recall from Definition 3.10.6 that $\mathcal{W}$ has a strategy in $\Psi \in \Gamma^b(\mathcal{P}, \Sigma)$ witnessing that $\mathcal{W}$ is a $\Sigma_{\mathrm{m}^+(\mathcal{T})}$-sts mouse over $\mathrm{m}^+(\mathcal{T})$. Let κ be an $\mathcal{N}$-cardinal such that $\{\mathbb{P}, \mathcal{T}\} \in \mathcal{N}|\kappa[g]$. Using fullness preservation of Λ, fix an iteration tree $\mathcal{U}_0$ on $\mathcal{N}$ above κ and according to Λ with last model $\mathcal{N}_1$ such that $\pi^{\mathcal{U}_0}$ exists and there is an $\mathcal{M} \trianglelefteq \mathcal{N}_1$ such that

1. $\mathcal{M}$ witnesses the internal Σ-closure of $\mathcal{N}_1$ at κ and
2. for some $\mathcal{Q} \lhd (\mathcal{S}^{\mathcal{M}}_{\kappa})^b$, $w(\mathsf{Code}(\Psi)) \leq w(\mathsf{Code}(\Sigma_{\mathcal{Q}}))$.

Fix a real x witnessing $w(\mathsf{Code}(\Psi)) \leq w(\mathsf{Code}(\Sigma_{\mathcal{Q}}))$.

Let $\delta_0 < \delta_1 < \delta_2$ be the first three Woodin cardinals of $\mathcal{M}$ that are greater than κ. Let $\mathcal{U}_1$ be an iteration tree on $\mathcal{M}$ based on $\mathcal{M}|\delta_1$ according to $\Lambda_{\mathcal{M},\mathcal{U}_0}$ and above

δ_0 that is constructed according to the rules of x-genericity iteration. Let $\pi = \pi^{\mathcal{U}_1}$ and let $\mathcal{M}_2$ be the last model of $\mathcal{U}_1$. We then have that x is generic for the extender algebra of $\mathcal{M}_2$ at $\pi(\delta_1)$. It follows that

(2) $\Psi \restriction (\mathcal{M}_2|\pi(\delta_2))[g][x] \in \mathcal{M}_2[g][x]$.[23]

Let $((\mathcal{M}_\gamma, \mathcal{N}_\gamma : \gamma \leq \nu), (F_\gamma : \gamma < \nu), (\mathcal{T}_\gamma : \gamma \leq \nu))$ be the output of the $(\mathcal{M}_2|\pi(\delta_2), \pi(\delta_1))$-authenticated fully backgrounded construction over $\mathcal{J}_\omega[\mathrm{m}^+(\mathcal{T})]$ based on $\mathrm{m}^+(\mathcal{T})$.[24] Next we have that

(3) For some $\gamma \leq \nu$, $\mathcal{W} = \mathcal{M}_\gamma$.

(3) is a consequence of the following facts:

(3A) Extenders used in the construction of $\mathcal{S}$ have critical points $> \pi(\delta_1)$ (so the construction side does not move).
(3B) For each γ, $\mathcal{M}_\gamma$ is a $\Sigma_{\mathrm{m}^+(\mathcal{T})}$-sts mouse over $\mathrm{m}^+(\mathcal{T})$.[25]
(3C) $\mathcal{W}$ side loses (because (2) implies $\mathcal{W}$ is $\pi(\delta_2)$-iterable in $\mathcal{M}_2|[g][x]$ and the construction is universal[26]).

Clearly (3) finishes the proof of the claim. $\dashv$

We now assume the following:

$(*2)$ $\mathcal{T}'$ is uvs.

Again, our goal is to show that $\Sigma(\mathcal{T}') = \Phi(\mathcal{T}')$. As many components of the proof are similar to the proofs of Theorem 4.12.1 and Proposition 4.12.4, we will not give the full proof. It is in fact enough to show the following:

(b) Suppose $\alpha \in R^{\mathcal{T}'}$ is such that $\pi^{\mathcal{T}',b}$ is defined and for all $\beta \in R^{\mathcal{T}'} \cap \alpha$,

$$(\mathcal{M}_\beta^{\mathcal{T}'})^b \neq (\mathcal{M}_\alpha^{\mathcal{T}'})^b.$$

Let $\mathcal{T} = \mathcal{T}'_{\leq\alpha}$ and $\mathcal{S} = \mathcal{M}_\alpha^{\mathcal{T}'}$. There is then $\kappa < \mathrm{ord}(\mathcal{N})$ and $\mathcal{M} \trianglelefteq \mathcal{N}$ witnessing that $\mathcal{N}$ is internally Σ-closed at κ and such that
1. $\{\mathbb{P}, \mathcal{T}\} \in \mathcal{N}|\kappa[g]$,
2. there is a normal stack $\mathcal{U}$ on $\mathcal{S}$ according to $\Sigma_{\mathcal{S}}$ such that either
   (a) $\mathcal{U}$ has a last model $\mathcal{K} \triangleleft \mathcal{S}_\kappa^{\mathcal{M}}$ or
   (b) $\mathcal{U}$ is of limit length and $\Sigma(\mathcal{U}) = \mathrm{m}^+(\mathcal{U}) = \mathcal{S}_\kappa^{\mathcal{M}}$.

The tree $\mathcal{U}$ is built using the authentication procedure described in Definition 3.7.3. Proposition 4.15.2 guarantees that the prescription for finding the branches of $\mathcal{U}$ as in clause 2 of Definition 3.7.3 produces a branch which is according to $\Sigma_{\mathcal{S}}$.

[23]To make this conclusion, we use the fact that $\mathcal{M}_2[g][x]$ is closed under $\Sigma_{\mathcal{Q}}$. This can be established using the generic interpretability results of [29].

[24]See Definition 6.2.2.

[25]See Theorem 4.12.1 and Proposition 4.12.5.

[26]See the universality clause of Theorem 4.5.6.

The fact that $\mathcal{N}$ has a fullness preserving iteration strategy implies that $\mathcal{S}$ cannot outiterate $\mathcal{S}_\kappa^{\mathcal{M}}$.

6.4. S-constructions

Our definition of sts mice makes heavy use of the fact that the set X is a self-well-ordered set. In particular, our definition cannot be used to define sts mice over $\mathbb{R}$. Another shortcoming of our definition is that it does not explain how to do S-constructions. In this short section, motivated by Section 3.38 of [29], we indicate how to use Theorem 6.1.4 to redefine hod mice in a way that one can define sts mice over $\mathbb{R}$ and perform S-constructions.

Recall the difficulty with defining hybrid mice over $\mathbb{R}$. In our definition, we always choose the least stack of some sort for which the branch has not been added and index a branch. Since $\mathbb{R}$ may not be self-well-ordered, we do not have the luxury of choosing the least such stack.

The problem with S-constructions is very similar. Suppose $(\mathcal{P},\Sigma)$ is a hod pair or an sts hod pair and N and M are two transitive models of some fragment of set theory such that $\mathcal{J}_\omega(M)\subseteq\mathcal{J}_\omega(N)$ and for some poset $\mathbb{P}\in\mathcal{J}_\omega(M)$ and some M-generic $G\subseteq\mathbb{P}$, $\mathcal{J}_\omega(N)=\mathcal{J}_\omega(M)[G]$. Suppose further that both M and N are Σ-closed and $\mathcal{P}\in N\cap M$. For us, S-constructions are constructions that translate Σ-mice over N to Σ-mice over M. For more details consult Section 3.38 of [29].[27]

The difficulty in performing S-constructions is the following. Suppose $\mathcal{N}$ is a Σ-mouse over N, and we want to translate it onto a Σ-mouse over M. Suppose our translation has produced a Σ-mouse $\mathcal{M}$ over M, and our indexing scheme demands that a branch of some stack $\mathcal{T}\in\mathcal{N}$ be indexed in the very next step in the translation procedure. The problem is that $\mathcal{T}$ may not be a stack in $\mathcal{M}$ nor may it be the stack whose branch is indexed in $\mathcal{M}$.

To solve this problem, we change the definition of hybrid premouse in a way that the iterations whose branches are indexed do not depend on generic extensions. In particular, instead of indexing iterations according to Σ, we considered generic genericity iterations on $\mathcal{M}_1^{\#,\Sigma}$. Such iterations make levels of the model generically generic and do not depend on generic extensions. This move solves both problems. In the first case what is important is that the indexed iterations do not depend on the well-ordering of the model, and in the second case what is important is that the indexed iterations do not depend on generic extensions. For more details consult Definition 3.37 of [29] or [50] for a similar construction.

Here our solution is similar. Suppose $(\mathcal{P},\Sigma)$ is an sts hod pair such that Σ has strong branch condensation and is Γ-fullness preserving for some projectively closed pointclass Γ and $\mathcal{M}$ is an Σ-sts mouse over some set X such that $\mathcal{P}\in X$. Then the iterations of $\mathcal{P}$ that are indexed in $\mathcal{M}$ are of the form $t=(\mathcal{P},\mathcal{T},\mathcal{Q},\mathcal{U})$,

[27] In [39], this process is called P-constructions.

where t is an indexable stack[28]. $\mathcal{T}$ is always the result of comparing $\mathcal{P}$ with a certain backgrounded construction. Notice that this neither depends on the well-ordering of $\mathcal{M}$ nor on small generic extensions. $\mathcal{U}$ is a stack on $\mathcal{Q}^b$ and, in Definition 3.8.9, we chose the least such stack. Thus the choice of $\mathcal{U}$ depends on both the well-ordering of $\mathcal{M}$ and small generic extensions[29]. To solve the issue, we will start considering stacks $s = (\mathcal{P}, \mathcal{T}, \mathcal{Q}, \mathcal{U})$ where $\mathcal{T}$ is as before but now $\mathcal{U}$ is a generic genericity iteration on $\mathcal{M}_2^{\#,\Sigma_{\mathcal{Q}^b}}$ to make a level of the model generically generic. We only consider such generic genericity iterations of $\mathcal{M}_2^{\#,\Sigma_{\mathcal{Q}^b}}$ that are based on the first Woodin of $\mathcal{M}_2^{\#,\Sigma_{\mathcal{Q}^b}}$.

The reason we choose $\mathcal{M}_2^{\#,\Sigma_{\mathcal{Q}^b}}$ is that we want to use clause 1 of Theorem 6.1.4. It is not hard to see that if $\delta_0 < \delta_1$ are the first two Woodin cardinals of $\mathcal{M}_2^{\#,\Sigma_{\mathcal{Q}^b}}$ and $g \subseteq Coll(\omega, \delta_0)$ is $\mathcal{M}_2^{\#,\Sigma_{\mathcal{Q}^b}}$-generic then $\mathcal{M}_2^{\#,\Sigma_{\mathcal{Q}^b}}|\delta_1[g]$ is internally $\Sigma_{\mathcal{Q}^b}$-closed. Clause 1 of Theorem 6.1.4 is a weaker result than [29, Lemma 3.35], which is what is used to reorganize hod mice in [29]. We could prove an equivalent of [29, Lemma 3.35], but doing this is much harder than proving clause 1 of Theorem 6.1.4.

To show that the resulting structure $\mathcal{M}$ is closed under Σ, we will first show that we can find branches of indexable stacks. Given such a stack $t = (\mathcal{P}, \mathcal{T}, \mathcal{Q}, \mathcal{U})$ let $\mathcal{W}$ be an iteration of $\mathcal{M}_2^{\#,\Sigma_{\mathcal{Q}^b}}$ such that $(\mathcal{P}, \mathcal{T}, \mathcal{Q}, \mathcal{W})$ is indexed in $\mathcal{M}$ and if $\mathcal{S}$ is the last model of $\mathcal{W}$ then $\mathcal{U}$ is generic over $\mathcal{S}$ for $\mathbb{B}_\delta^{\mathcal{S}}$ where δ is the least Woodin of $\mathcal{S}$ and $\mathbb{B}_\delta^{\mathcal{S}}$ is the extender algebra of $\mathcal{S}$ at δ. It then follows from Theorem 6.1.4 that $\Sigma_{\mathcal{Q}^b} \restriction \mathcal{S}|\eta[\mathcal{U}] \in \mathcal{S}[\mathcal{U}]$ where η is the second Woodin cardinal of $\mathcal{S}$. The rest of the proof is just repeating the proof of Theorem 6.1.4.

Instead of re-developing the entire theory of sts mice, we will simply give the definition of indexable stack. The rest of the definitions, those developed in Section 3.6, Section 3.7, Section 3.8, Section 3.9 and Section 3.10, stay more or less the same. It is important to note that the theory of sts mice as well as hod mice does not in general depend on particular indexing schemes.

DEFINITION 6.4.1 (Revised Indexable stack). Suppose $\mathcal{P}$ is a hod-like #-lsa type lses[30]. We say that an st-stack[31]

$$\mathcal{T} = ((\mathcal{M}_\alpha)_{\alpha<\eta}, (E_\alpha)_{\alpha<\eta-1}, D, R, (\beta_\alpha, m_\alpha)_{\alpha\in R}, \mathsf{short}, \mathsf{max}, T)$$

is a *revised indexable stack* on $\mathcal{P}$ if one of the following clauses hold:

[28]See Definition 3.3.3.

[29]Small in the sense that the generic is smaller than the critical point of the first background extender used in the construction.

[30]See Definition 2.7.3.

[31]See Definition 3.2.1.

1. $\mathsf{max} = \emptyset$ and there is $\alpha \in R^{\mathcal{T}}$ such that $\pi^{\mathcal{T}_{\leq\alpha},b}$ is defined and letting[32] $\mathcal{M} = (\mathcal{M}_2^{\#,\Sigma_{(\mathcal{M}_\alpha^{\mathcal{T}})^b}})^{\mathcal{M}_\alpha^{\mathcal{T}}}$, $\mathcal{T}_{\geq\alpha}$ is a normal stack on $\mathcal{M}$ that is above $\mathrm{ord}((\mathcal{M}_\alpha^{\mathcal{T}})^b)$ and is based on $\mathcal{M}|\delta$ where δ is the least Woodin cardinal of $\mathcal{M}$.
2. $|\mathsf{max}| = 1$, $\mathcal{T}$ is a normal stack[33] and if α is the unique element of max then $\pi_{0,\alpha}^{\mathcal{T}}$ is defined and $\mathsf{next}^{\mathcal{T}}(\alpha) = \mathrm{lh}(\mathcal{T})$[34].

We say $\mathcal{P}$ is a revised hod premouse if it is indexed according to our revised indexing scheme, which will only index *authentic revised indexable stacks*[35]. We say $(\mathcal{P},\Sigma)$ is revised hod pair if $\mathcal{P}$ is revised hod premouse and Σ is an iteration strategy for $\mathcal{P}$.

THEOREM 6.4.2. *Suppose $(\mathcal{P},\Sigma)$ is a revised hod premouse such that Σ is strongly Γ-fullness preserving for some projectively closed pointclass Γ and Σ has strong branch condensation. Then for any $\mathcal{Q} \in Y^{\mathcal{P}}$ and $\mathcal{P}$-generic g,*

1. *if $\mathcal{Q}$ is not of lsa type then $\Sigma_{\mathcal{Q}} \restriction \mathcal{P}[g]$ is uniformly in $\mathcal{Q}$ definable over $\mathcal{P}[g]$, and*
2. *if $\mathcal{Q}$ is of lsa type then the fragment of $\Sigma_{\mathcal{Q}}^{stc} \restriction \mathcal{P}[g]$ that acts on revised indexable stacks is uniformly in $\mathcal{Q}$ definable over $\mathcal{P}[g]$.*

We now just import our lemmas on S-construction from Section 3.8 of [29] to our current context. Let $(\mathcal{P},\Sigma)$ be a hod pair or an sts pair such that Σ has the strong branch condensation and is strongly Γ-fullness preserving for some pointclass Γ. Suppose $\mathcal{M}$ is a sound Σ-mouse and δ is a cutpoint cardinal of $\mathcal{M}$. Suppose further that $\mathcal{N} \in \mathcal{M}|\delta+1$ is such that $\delta \subseteq \mathcal{N} \subseteq H_\delta^{\mathcal{M}}$, $\mathcal{N}$ models a sufficiently strong fragment of $\mathsf{ZF} - \mathsf{Replacement}$, $\mathcal{N}$ is a Σ-mouse or a Σ-sts mouse and there is a partial ordering $\mathbb{P} \in L_\omega[\mathcal{N}]$ such that $\mathcal{M}|\delta$ is $\mathbb{P}$-generic over $L_\omega[\mathcal{N}]$. We would like to define S-construction of $\mathcal{M}$ over $\mathcal{N}$ relative to Σ.

DEFINITION 6.4.3. An S-construction of $\mathcal{M}$ over $\mathcal{N}$ relative to Σ is a sequence $(\mathcal{S}_\alpha, \bar{\mathcal{S}}_\alpha : \alpha \leq \eta)$ of Σ-mice over $\mathcal{N}$ such that

1. $\mathcal{S}_0 = L_\omega[\mathcal{N}]$,
2. if $\mathcal{M}|\delta$ is generic over $\bar{\mathcal{S}}_\alpha$ for a forcing in $L_\omega[\mathcal{N}]$ then
   (a) if $\mathcal{M}||(\omega\cdot\alpha)$ is active and has a last branch b then $\mathcal{S}_\alpha$ is the expansion of $\bar{\mathcal{S}}_\alpha$ by b and $\bar{\mathcal{S}}_{\alpha+1} = rud(\mathcal{S}_\alpha)$.
   (b) if $\mathcal{M}||(\omega\cdot\alpha)$ is active and has a last extender E then $\mathcal{S}_\alpha$ is the expansion of $\bar{\mathcal{S}}_\alpha$ by E and $\bar{\mathcal{S}}_{\alpha+1} = rud(\mathcal{S}_\alpha)$,
   (c) if $\mathcal{M}||(\omega\times\alpha)$ is passive then $\mathcal{S}_\alpha = \bar{\mathcal{S}}_\alpha$ and $\bar{\mathcal{S}}_{\alpha+1} = rud(\mathcal{S}_\alpha)$,
3. if λ is limit then $\bar{\mathcal{S}}_\lambda = \bigcup_{\alpha<\lambda}\mathcal{S}_\alpha$.

The following is the restatement of Lemma 3.42 of [29].

[32] $(\mathcal{M}_2^{\#,\Sigma_{(\mathcal{M}_\alpha^{\mathcal{T}})^b}})^{\mathcal{M}_\alpha^{\mathcal{T}}}$ is $(\mathcal{M}_2^{\#,\Sigma_{(\mathcal{M}_\alpha^{\mathcal{T}})^b}})$ in the sense of $\mathcal{M}_\alpha^{\mathcal{T}}$.

[33] See Definition 3.3.1.

[34] It follows that $\mathcal{T}_{\geq\alpha}$ is above $\pi_{0,\alpha}^{\mathcal{T}}(\delta^{\mathcal{P}^b})$. See also Notation 2.4.4.

[35] See Section 3.7.

LEMMA 6.4.4. *Suppose* $(\mathcal{P},\Sigma)$, $\mathcal{M},\mathcal{N}$ *are as above and* δ *is a strong cutpoint cardinal of* $\mathcal{M}$. *Suppose further that* $\mathcal{N}\in\mathcal{M}|\delta+1$ *is such that* $\delta\subseteq\mathcal{N}\subseteq H_\delta^{\mathcal{M}}$ *and there is a partial ordering* $\mathbb{P}\in L_\omega[\mathcal{N}]$ *such that whenever* $\mathcal{Q}$ *is a* Σ*-mouse over* $\mathcal{N}$ *such that* $H_\delta^{\mathcal{Q}}=\mathcal{N}$ *then* $\mathcal{M}|\delta$ *is* $\mathbb{P}$*-generic over* $\mathcal{Q}$. *Then there is a* Σ*-mouse* $\mathcal{S}$ *over* $\mathcal{N}$ *such that* $\mathcal{M}|\delta$ *is generic over* $\mathcal{S}$ *and* $\mathcal{S}[\mathcal{M}|\delta]=\mathcal{M}$.

The following is just the restatement of Lemma 3.43 of [29].

LEMMA 6.4.5. *Suppose* $(\mathcal{P},\Sigma)$, $\mathcal{M}$ *and* $\mathcal{N}$ *are as above. Suppose further that* $\mathcal{M}\vDash\mathsf{ZFC}-\mathsf{Replacement}$ *is a* Σ*-mouse and* η *is a strong cutpoint non-Woodin cardinal of* $\mathcal{M}$. *Suppose* $\gamma>\eta$ *is a cardinal of* $\mathcal{M}$ *and* $\mathcal{N}=(\mathcal{J}^{\vec{E},\Sigma})^{\mathcal{M}|\gamma}$. *Suppose* $\mathcal{J}_\omega(\mathcal{N}|\eta)\vDash$ *"*η *is Woodin". Let* $(\mathcal{S}_\alpha,\bar{\mathcal{S}}_\alpha:\alpha<\nu)$ *be the* $\mathcal{S}$*-construction of* $\mathcal{M}|(\eta^+)^{\mathcal{M}}$ *over* $\mathcal{N}|\eta$ *relative to* Σ. *Then for some* $\alpha<\nu$, $\mathcal{S}_\alpha\vDash$ *"*η *is not Woodin".*

Chapter 7

ANALYSIS OF HOD

In this chapter we analyze V_Θ^{HOD} of the minimal model of the Largest Suslin Axiom. The analysis is very much like the analysis of V_Θ^{HOD} in the minimal model of $\mathsf{AD}^+ + \vartheta_1 = \Theta$, which appeared in [29, Chapter 4]. Just like in [29, Chapter 4], we need to introduce the notion of suitable pair, B-iterable pair and etc. The proof of Theorem 7.2.2 is very much like the proof of [29, Theorem 4.24].

7.1. B-iterability

In this section, we import B-iterability technology to our current context. Most of what we will need was laid out in [29, Section 4.1 and Section 4.2]. Here we will only sketch the necessary arguments.

DEFINITION 7.1.1 (Suitable pair). $(\mathcal{P},\Sigma)$ is a *suitable pair* if the following clauses hold:

1. Either $\mathcal{P}$ is a hod premouse of successor type or $\mathcal{P}$ is a #-like lsa type hod premouse[1].
2. If $\mathcal{P}$ is not of lsa type then
   (a) $(\mathcal{P},\Sigma)$ is a pre-hod pair[2],
   (b) Σ has strong branch condensation and is strongly fullness preserving[3],
   (c) For any $\mathcal{P}$-cardinal $\eta > \delta^{\mathcal{P}}_{\lambda-1}$, if η is a strong cutpoint then $\mathcal{P}|(\eta^+)^{\mathcal{P}} = \mathsf{Lp}^\Sigma(\mathcal{P}|\eta)$.
3. If $\mathcal{P}$ is of lsa type then $(\mathcal{P},\Sigma)$ is an sts hod pair such that Σ has strong branch condensation and is strongly fullness preserving[4].

For convenience, we extend the notation $\mathcal{P}^-$ to lsa type (see Notation 2.7.14).

[1]See Definition 2.7.3. Thus, in both cases $\mathcal{P} \models$ "$\delta^{\mathcal{P}}$ is a Woodin cardinal". Also, if $\mathcal{P}$ is not of lsa type then $\mathcal{P}$ is a $\Sigma_{\mathcal{P}^-}$-mouse above $\mathcal{P}^-$.

[2]See Definition 5.2.1.

[3]Thus, $\wp(\mathbb{R})$-fullness preserving.

[4]In this fullness preservation implies that $\mathsf{Lp}^\Sigma(\mathcal{P}) \models$ "$\delta^{\mathcal{P}}$ is a Woodin cardinal".

NOTATION 7.1.2. Suppose $\mathcal{P}$[5] is either of lsa type or of successor type. We then let

$$\mathcal{P}^- = \begin{cases} \mathcal{P} & \text{if } \mathcal{P} \text{ is of lsa type,} \\ \bigcup_{\mathcal{Q} \triangleleft_{hod} \mathcal{P}} \mathcal{Q} & \text{otherwise.} \end{cases}$$

Also, if $(\mathcal{P},\Sigma)$ is a suitable sts pair then we let $\mathsf{lp}(\mathcal{P},\Sigma) = \mathsf{Lp}^{\Sigma}_{\omega}(\mathcal{P})$.

Suppose $(\mathcal{P},\Sigma)$ and $(\mathcal{Q},\Lambda)$ are hod pairs or sts hod pairs such that Σ and Λ have strong branch condensation and are strongly fullness preserving. We then let

$$(\mathcal{P},\Sigma) \leq_{DJ} (\mathcal{Q},\Lambda)$$

if and only if $(\mathcal{P},\Sigma)$ loses the coiteration with $(\mathcal{Q},\Lambda)$. Notice that $\leq_{DJ}$ is a well-founded relation. We then let $\alpha(\mathcal{P},\Sigma) = |(\mathcal{P},\Sigma)|_{\leq_{DJ}}$, and we let $[\mathcal{P},\Sigma]$ be the $=_{DJ}$ equivalence class of $(\mathcal{P},\Sigma)$, i.e.,

$$(\mathcal{Q},\Lambda) \in [\mathcal{P},\Sigma] \text{ iff } (\mathcal{Q},\Lambda) \text{ is a hod pair such that } \Lambda \text{ has branch condensation and is strongly fullness preserving and } \alpha(\mathcal{Q},\Lambda) = \alpha(\mathcal{P},\Sigma).$$

Notice that $[\mathcal{P},\Sigma]$ is independent of $(\mathcal{P},\Sigma)$. We let

$$\mathbb{B}(\mathcal{P},\Sigma) = \{B \subseteq [\mathcal{P},\Sigma] \times \mathbb{R} : B \text{ is } OD\}.$$

Note that $\mathbb{B}(\mathcal{P},\Sigma)$ is defined for hod pairs or sts hod pairs, but not for suitable pairs that are not sts hod pairs[6].

The following standard lemma features prominently in our computations of HOD. The proof is very much like the proof of Lemma 4.16 of [29]. Below SMC stands for *Strong Mouse Capturing*. More precisely, SMC states that for any hod pair or sts hod pair $(\mathcal{P},\Sigma)$ such that Σ is strongly fullness preserving and has strong branch condensation then for any $x,y \in \mathbb{R}$, $x \in OD_{y,\Sigma}$ if and only if $x \in \mathsf{Lp}^{\Sigma}(y)$.

LEMMA 7.1.3. *Assume* SMC *and suppose $(\mathcal{P},\Sigma)$ is a suitable pair and $B \in \mathbb{B}(\mathcal{P}^-,\Sigma)$. Set*

$$\mathcal{P}_+ = \begin{cases} \mathcal{P} & \text{if } \mathcal{P} \text{ is of successor type,} \\ \mathsf{lp}(\mathcal{P},\Sigma) & \text{otherwise.} \end{cases}$$

Suppose κ is a $\mathcal{P}_+$-cardinal such that if $\mathcal{P}$ is of lsa type then for some $n > 0$, $\kappa = ((\delta^{\mathcal{P}})^{+n})^{\mathcal{P}_+}$ and otherwise $\kappa > \delta^{\mathcal{P}^-}$. Then there is $\tau \in \mathcal{P}_+^{Coll(\omega,\kappa)}$ such that $(\mathcal{P}_+,\tau)$ locally term captures $B_{(\mathcal{P},\Sigma)}$ at κ for a comeager set of $\mathcal{P}_+$-genetics $g \subseteq Coll(\omega,\kappa)$.

---

[5]As was decided many pages before, we develop the theory of hod mice over $\emptyset$. Thus, $X^{\mathcal{P}} = \emptyset$.

[6]This is because if $(\mathcal{P},\Sigma)$ is a non-sts suitable pair then Σ does not act on $\mathcal{P}$ but only on $\mathcal{P}^-$. For such pairs, comparison is somewhat meaningless.

If B is locally term captured for comeager many set generics over a suitable pair $(\mathcal{P},\Sigma)$ then we let $\tau^{\mathcal{P},\Sigma}_{B,\kappa}$ be the invariant term in $\mathcal{P}_+$ locally term capturing B at κ for comeager many set generics. One way to get term capturing for all generics is to show that a suitable pair can be extended to a structure that has one more Woodin.

DEFINITION 7.1.4 (n-Suitable pair). $(\mathcal{P},\Sigma)$ is an *n-suitable pair* if there is δ such that

$$\mathcal{P} \vDash \text{“}\delta \text{ is a Woodin cardinal”}$$

and the following clauses hold:

1. Either $(\mathcal{P}|(\delta^{+\omega})^{\mathcal{P}},\Sigma)$ is a suitable pair or letting $\alpha = \min(\mathrm{dom}(\vec{E}^{\mathcal{P}}) - \delta)$, $(\mathcal{P}|\alpha,\Sigma)$ is suitable[7]. Set
$$\mathcal{P}_0 = \begin{cases} \mathcal{P}|(\delta^{+\omega})^{\mathcal{P}} & \text{if } (\mathcal{P}|(\delta^{+\omega})^{\mathcal{P}},\Sigma) \text{ is a suitable pair,} \\ \mathcal{P}|\alpha & \text{otherwise.} \end{cases}$$
2. If $\mathcal{P}_0$ is of lsa type then $\mathcal{P}$ is a Σ-sts premouse over $\mathcal{P}_0$ and otherwise $\mathcal{P}$ is a Σ-premouse over $\mathcal{P}_0$,
3. $\mathcal{P} \vDash \mathsf{ZFC} - \mathsf{Replacement} +$ “there are exactly n Woodin cardinals, $\eta_0 < \eta_1 < \cdots < \eta_{n-1}$ that are strictly greater than δ”,
4. $\mathrm{ord}(\mathcal{P}) = \sup_{i<\omega}(\eta_{n-1}^{+i})^{\mathcal{P}}$ (here we set $\eta_{-1} = \delta$),
5. For any $\mathcal{P}$-cardinal $\eta \geq \delta$, if η is a strong cutpoint then
$$\mathcal{P}|(\eta^+)^{\mathcal{P}} = \mathsf{Lp}^{\Sigma}(\mathcal{P}|\eta')$$
where $\eta' = \min(\mathrm{dom}(\vec{E}^{\mathcal{P}}) - \eta)$.

We say $\mathcal{P}$ is of lsa type if $\mathcal{P}_0$ is of lsa type. Otherwise we say that $\mathcal{P}$ is of successor type.

If $(\mathcal{P},\Sigma)$ is n-suitable then we let $\delta^{\mathcal{P}}$ be the δ of Definition 7.1.4 and $\mathcal{P}_0$ be as in Definition 7.1.4. Clearly 0-suitable pair is just a suitable pair. The following are easy consequences of Lemma 7.1.3.

LEMMA 7.1.5. *Assume* SMC. *Suppose $(\mathcal{P},\Sigma)$ is an n-suitable pair and $B \in \mathbb{B}(\mathcal{P}^-,\Sigma)$. Suppose κ is a $\mathcal{P}$-cardinal such that*

1. *if $\mathcal{P}$ is of lsa type then $\kappa > ((\delta^{\mathcal{P}})^+)^{\mathcal{P}}$ and*
2. *otherwise $\kappa > \delta^{\mathcal{P}_0^-}$.*

Then there is $\tau \in \mathcal{P}^{Coll(\omega,\kappa)}$ such that $(\mathcal{P},\tau)$ locally term captures $B_{(\mathcal{P},\Sigma)}$ at κ for comeager set of $\mathcal{P}$-generics $g \subseteq Coll(\omega,\kappa)$.

COROLLARY 7.1.6. *Assume* SMC. *Suppose $(\mathcal{P},\Sigma)$ is an n-suitable pair and $B \in \mathbb{B}(\mathcal{P}^-,\Sigma)$. Let $\nu = ((\delta^{\mathcal{P}})^{+\omega})^{\mathcal{P}}$. Suppose κ is a $\mathcal{P}$-cardinal such that*

---

[7]Because $\mathcal{P}|(\delta^{+\omega})^{\mathcal{P}}$ has infinitely many cardinals above δ, in the first case $\mathcal{P}|(\delta^{+\omega})^{\mathcal{P}}$ is of successor type and in the second, $\mathcal{P}|\alpha$ is of an lsa type.

1. *if $\mathcal{P}$ is of lsa type then $\kappa \in (((\delta^{\mathcal{P}})^+)^{\mathcal{P}}, \nu)$ and*
2. *otherwise $\kappa \in (\delta^{\mathcal{P}_0^-}, \nu)$.*

Then $(\mathcal{P}|\nu, \tau_{B,\kappa}^{\mathcal{P},\Sigma})$ locally term captures $B_{(\mathcal{P},\Sigma)}$ at κ for all $\mathcal{P}$-generic

$$g \subseteq Coll(\omega, \kappa).$$

Corollary 7.1.6 is our main method of showing that various B are term captured over the hod mice that we will construct. Suppose now that $(\mathcal{P}, \Sigma)$ is a hod pair. It is now a trivial matter to import the terminology of [29, Section 4.1] to our current context. We will have that $S(\Sigma)$ consists of those $\mathcal{Q}$ such that $\mathcal{Q}_0 \in pI(\mathcal{P}, \Sigma)$ and $(\mathcal{Q}, \Sigma_{\mathcal{Q}_0})$ is a suitable pair. Given $\mathcal{Q} \in S(\Sigma)$, we let $f_B(\mathcal{Q}) = \oplus_{\kappa<\text{ord}(\mathcal{Q})} \tau_{B,\kappa}^{\mathcal{Q},\Sigma_{\mathcal{Q}_0}}$[8]. Then the rest of the notions are defined for $F = \{f_B : B \in \mathbb{B}(\mathcal{P}, \Sigma)\}$. Therefore, in the sequel, we will freely use the terminology of [29, Section 4.1].

7.2. The computation of HOD

Throughout this section we assume $\mathsf{AD}^+ + \mathsf{SMC}$ and let $\langle \vartheta_\alpha : \alpha \leq \Omega \rangle$ be the Solovay sequence. Our goal is to compute $V_{\vartheta_\beta}^{\text{HOD}}$ for $\beta \leq \Omega$. We will do it under some additional hypothesis described below. In the next few chapters, we will prove that our additional hypothesis follows from $\mathsf{AD}^+ +$ "No initial segment of the Solovay sequence satisfies LSA".

Suppose $(\mathcal{P}, \Sigma)$ is a hod pair or an sts pair such that Σ has strong branch condensation and is strongly fullness preserving. We will continue using the notation $\alpha(\mathcal{P}, \Sigma)$, $\mathcal{P}_0$ and $\mathcal{P}^-$ from the previous section[9].

Suppose first that $\beta + 1 = \Omega$. We then let $\mathcal{I} = \{(\mathcal{Q}, \Lambda, \vec{B})$:

1. $(\mathcal{Q}, \Lambda)$ is suitable, Λ is strongly fullness preserving and has strong branch condensation, and $\alpha(\mathcal{Q}^-, \Lambda) = \beta$,
2. for some integer n, $\vec{B} = (B_0, \ldots, B_n)$ and for every $i < n$, $B_i \in \mathbb{B}(\mathcal{Q}^-, \Lambda)$, and
3. $(\mathcal{Q}, \Lambda)$ is strongly $\vec{B}$-iterable $\}$.

Theorem 8.1.15 and the results of Section 10.1 show that $\mathcal{I} \neq \emptyset$. Define $\preceq$ on $\mathcal{I}$ by

$$(\mathcal{P}, \Sigma, \vec{B}) \preceq (\mathcal{Q}, \Lambda, \vec{C}) \iff \vec{B} \subseteq \vec{C} \quad \text{and} \quad (\mathcal{Q}, \Lambda, \vec{B}) \text{ is a } \vec{B}\text{-tail of } (\mathcal{P}, \Sigma, \vec{B}).$$

When $(\mathcal{R}, \Psi, \vec{B}) \preceq (\mathcal{Q}, \Lambda, \vec{C})$, there is a canonical map

$$\pi : H_{\vec{B}}^{\mathcal{R},\Psi} \to H_{\vec{B}}^{\mathcal{Q},\Lambda},$$

which is independent of $\vec{B}$-iterable branches. We let $\pi_{(\mathcal{R},\Psi,\vec{B}),(\mathcal{Q},\Lambda,\vec{B})}$ be this map. We then have that $(\mathcal{I}, \preceq)$ is directed. Let

$$\mathcal{F} = \{H_{\vec{B}}^{\mathcal{Q},\Lambda} : (\mathcal{Q}, \Lambda, \vec{B}) \in \mathcal{I}\}.$$

[8] Here $\mathcal{Q}_0$ is defined in Definition 7.1.4.

[9] See Definition 7.1.2.

and also let $\mathcal{M}_\infty$ be the direct limit of $\mathcal{F}$ under the iteration maps $\pi_{(\mathcal{R},\Psi,\vec{B}),(\mathcal{Q},\Lambda,\vec{B})}$. Let $\delta_\infty = \delta^{\mathcal{M}_\infty}$. For $(\mathcal{Q},\Lambda,B) \in \mathcal{I}$, we let $\pi_{(\mathcal{Q},\Lambda,B),\infty} : H_B^{\mathcal{Q},\Lambda} \to \mathcal{M}_\infty$. Standard arguments show that $\mathcal{M}_\infty$ is well-founded.

Following [29, Section 4.4], we let φ be the following sentence: for every $\beta+1<\Omega$ there is a hod pair $(\mathcal{P},\Sigma)$ such that

1. $\mathcal{P}$ is of successor type,
2. $\alpha(\mathcal{P}^-,\Sigma_{\mathcal{P}^-}) = \beta$,
3. Σ is strongly fullness preserving and has strong branch condensation,
4. for any $\mathcal{Q} \in pI(\mathcal{P},\Sigma) \cup pB(\mathcal{P},\Sigma)$, if $\mathcal{Q}$ is of successor type then
   (a) there is a sequence $\langle B_i : i<\omega\rangle \in \mathbb{B}(\mathcal{Q}^-,\Sigma_{\mathcal{Q}^-})^\omega$ which guides $\Sigma_\mathcal{Q}$ and
   (b) for any $B \in \mathbb{B}(\mathcal{Q}^-,\Sigma_{\mathcal{Q}^-})$ there is $\mathcal{R} \in pI(\mathcal{Q},\Sigma_\mathcal{Q})$ such that $\Sigma_\mathcal{R}$ respects B.

Our additional hypothesis, ψ, is the conjunction of φ with the following statement: If $\Omega = \beta+1$ then there is a suitable $\emptyset$-iterable $(\mathcal{P},\Sigma)$ such that

1. $\alpha(\mathcal{P}^-,\Sigma_{\mathcal{P}^-}) = \beta$ and $\Sigma_{\mathcal{P}^-}$ is strongly fullness preserving and has strong branch condensation,
2. for any $B \in \mathbb{B}(\mathcal{P}^-,\Sigma_{\mathcal{P}^-})$ there is an $\emptyset$-iterate $(\mathcal{Q},\Phi)$ of $(\mathcal{P},\Sigma)$ such that $(\mathcal{Q},\Phi)$ is strongly B-iterable.
3. $\mathcal{M}_\infty$ is well-founded and $\delta_\infty = \Theta = \vartheta_{\beta+1}$.

We will use the following lemma to establish ψ. It can be proved exactly the same way as [29, Lemma 4.23].

LEMMA 7.2.1. *Suppose $\Gamma \subseteq \wp(\mathbb{R})$ is such that*

$$L(\Gamma,\mathbb{R}) \vDash \mathsf{AD}^+ + \mathsf{SMC} + \Omega = \beta+1 \quad \textit{and} \quad \Gamma = \wp(\mathbb{R}) \cap L(\Gamma,\mathbb{R}).$$

Suppose $\Gamma^* \subseteq \wp(\mathbb{R})$ is such that $\Gamma \subseteq \Gamma^*$, $L(\Gamma^*,\mathbb{R}) \vDash \mathsf{AD}^+$ and there is a hod a pair $(\mathcal{P},\Sigma) \in \Gamma^*$ such that the following hold.

1. *Σ has strong branch condensation and is strongly Γ-fullness preserving.*
2. *Either $\mathcal{P}$ is of successor type or of lsa type.*
3. *$\mathsf{Code}(\Sigma_{\mathcal{P}^-}) \in \Gamma$ and*
   (a) if $\mathcal{P}$ is of successor type then $L(\Gamma,\mathbb{R}) \vDash$ "$(\mathcal{P},\Sigma_{\mathcal{P}^-})$ is a suitable pair such that $\alpha(\mathcal{P}^-,\Sigma_{\mathcal{P}^-}) = \beta$" and
   (b) if $\mathcal{P}$ is of lsa type then $L(\Gamma,\mathbb{R}) \vDash$ "$(\mathcal{P},\Sigma^{stc}_{\mathcal{P}})$ is a suitable pair such that $\alpha(\mathcal{P}^-,\Sigma_{\mathcal{P}^-}) = \beta$".

   Letting

$$\Lambda = \begin{cases} \Sigma_{\mathcal{P}^-} & \textit{if } \mathcal{P} \textit{ is of successor type,} \\ \Sigma^{stc} & \textit{otherwise} \end{cases}$$

   the following clauses hold:
4. *There is a sequence $\langle B_i : i<\omega\rangle \in [(\mathbb{B}(\mathcal{P}^-,\Lambda))^{L(\Gamma,\mathbb{R})}]^\omega$ guiding Σ.*
5. *For any $B \in (\mathbb{B}(\mathcal{P}^-,\Lambda))^{L(\Gamma,\mathbb{R})}$ there is $\mathcal{R} \in pI(\mathcal{P},\Sigma)$ such that $\Sigma_\mathcal{R}$ respects B.*

Then $L(\Gamma,\mathcal{R}) \models \psi$ *and* $\mathcal{M}_\infty^{L(\Gamma,\mathbb{R})} = \mathcal{M}_\infty^+(\mathcal{P},\Sigma)$.[10]

The next theorem is the adaptation of [29, Theorem 2.24] to our current context. It can be proved via exactly the same proof. Because of this, we omit the proof.

THEOREM 7.2.2 (Computation of HOD). *Assume* AD^+. *Suppose* $\Gamma \subseteq \wp(\mathbb{R})$ *is such that* $\Gamma = \wp(\mathbb{R}) \cap L(\Gamma,\mathbb{R})$. *Set* $W = L(\Gamma,\mathbb{R})$ *and let* $(\vartheta_\beta : \beta \leq \Omega)$ *be the Solovay sequence of* W. *Then the following holds:*

1. *Suppose* $W \models \varphi$ *and* $\beta + 1 < \Omega$. *Let* $(\mathcal{P},\Sigma)$ *witness* φ *for* β. *Then letting* $\mathcal{M} = \mathcal{M}_\infty^+(\mathcal{P},\Sigma)$, $\vec{E} = \vec{E}^{\mathcal{M}}$ *and* $\Lambda = \Sigma^{\mathcal{M}}$, *for every* $\alpha \leq \beta$ *there* $\mathcal{Q} \lhd_{hod} \mathcal{M}$ *such that*
   (a) $\delta^{\mathcal{Q}} = \vartheta_\alpha$,
   (b) $\delta^{\mathcal{Q}}$ *is either a Woodin cardinal of* $\mathcal{M}$ *or a limit of Woodin cardinals of* $\mathcal{M}$, *and*
   (c) $\mathcal{M}|\vartheta_\alpha = (V_{\vartheta_\alpha}^{\mathrm{HOD}^W}, \vec{E} \restriction \vartheta_\alpha, \Lambda \restriction V_{\vartheta_\alpha}^{\mathrm{HOD}^W}, \in)$.
2. *If* $W \models \psi$ *then letting* $\mathcal{M} = \mathcal{M}_\infty^W$ $\vec{E} = \vec{E}^{\mathcal{M}}$ *and* $\Lambda = \Sigma^{\mathcal{M}}$, *for every* $\alpha \leq \Omega$ *there is* $\mathcal{Q} \trianglelefteq_{hod} \mathcal{M}$ *such that*
   (a) $\delta_{\mathcal{Q}} = \vartheta_\alpha$,
   (b) $\delta^{\mathcal{Q}}$ *is either a Woodin cardinal of* $\mathcal{M}$ *or a limit of Woodin cardinals of* $\mathcal{M}$, *and*
   (c) $\mathcal{M}|\vartheta_\alpha = (V_{\vartheta_\alpha}^{\mathrm{HOD}^W}, \vec{E} \restriction \vartheta_\alpha, \Lambda \restriction V_{\vartheta_\alpha}^{\mathrm{HOD}^W}, \in)$.

Thus, working in a model of AD^+, if $\alpha < \Omega$ then to compute $\mathrm{HOD}|\vartheta_\alpha$ we only need to produce a hod pair $(\mathcal{P},\Sigma)$ satisfying Lemma 7.2.1. In the next chapter, in particular in Theorem 8.1.15 and Section 10.1, we will show that this is indeed true in the minimal model of the Largest Suslin Axiom.

[10] Recall that $\mathcal{M}_\infty^+(\mathcal{P},\Sigma)$ is the direct limit of all Σ-iterates of $\mathcal{P}$.

Chapter 8

MODELS OF LSA AS DERIVED MODELS

In this chapter, we show that certain derived models satisfy the LSA. We also prove results that are important elsewhere. The results of Section 10.1 and Theorem 8.1.15 are needed to carry out the computation of HOD (see Theorem 7.2.2). We start with introducing the pointclass $\Gamma(\mathcal{P},\Sigma)$ where $(\mathcal{P},\Sigma)$ is an sts hod pair.

8.1. $\Gamma(\mathcal{P},\Sigma)$ revisited

In this section, we translate the results of [29, Section 5.6] to our current context. Suppose $(\mathcal{P},\Sigma)$ is a hod pair such that either $\mathcal{P}$ is of successor type or of #-lsa type[1] and Σ is strongly fullness preserving and has strong branch condensation. Recall the notation $\mathcal{P}^-$.

Suppose first that $\mathcal{P}$ is of successor type. We now generalize the results of [29, Section 5.6]. Recall the notation Mice_Σ (see Notation 4.1.14). Because $\mathcal{P}$ is not of lsa type, it follows that $\mathsf{Code}(\Sigma)$ is Suslin, co-Suslin (this can be proved using the proof of [29, Lemma 5.9]). It follows that there is a scaled pointclass closed under continuous images and pre-images and under $\exists^{\mathbb{R}}$, and also contains $\mathsf{Mice}_{\Sigma_{\mathcal{P}^-}}$. We then let Γ^*_Σ be the least such pointclass. Also, let

$$\Gamma_\Sigma = (\Sigma^2_1(\mathsf{Code}(\Sigma_{\mathcal{P}^-})))^{L(\mathsf{Mice}_{\Sigma_{\mathcal{P}^-}},\mathbb{R})}.$$

Notice that Γ_Σ is a lightface good pointclass, and so we set

$$\underset{\sim}{\Gamma}_\Sigma = \cup_{x\in\mathbb{R}}(\Sigma^2_1(\mathsf{Code}(\Sigma_{\mathcal{P}^-}),x))^{L(\mathsf{Mice}_{\Sigma_{\mathcal{P}^-}},\mathbb{R})}.$$

Also $\mathsf{Mice}_{\Sigma_{\mathcal{P}^-}}$ belongs to Γ_Σ and is a universal Γ_Σ set. We let

$$\Gamma(\mathcal{P},\Sigma) = \{A : \text{for cone of } x\in\mathbb{R},\ A\cap C_{\Gamma_\Sigma}(x)\in C_{\Gamma_\Sigma}(C_{\Gamma_\Sigma}(x))\} = Env(\Gamma_\Sigma).^2$$

Notice that if $(\mathcal{Q},\Lambda)$ is a tail of $(\mathcal{P},\Sigma)$ then $\Gamma(\mathcal{Q},\Lambda) = \Gamma(\mathcal{P},\Sigma)$. The next theorem is essentialy the conjunction of [29, Lemma 5.13-5.16].

[1] See the discussion after Definition 2.7.3.

[2] Here, $C_\Gamma(x)$ is the largest countable $\Gamma(x)$ set. It is defined to be the set of $y\in\mathbb{R}$ such that for some set $A\in\Gamma\cap\wp(\mathbb{R}^3)$ and some ordinal $\alpha<\omega_1$, for every $z\in\mathbb{R}$ coding α, y is the unique real such that $(y,x,z)\in A$.

THEOREM 8.1.1. *Assume* AD^+ *and suppose* $(\mathcal{P},\Sigma)$ *is a hod pair of successor type and* Σ *is strongly fullness preserving and has strong branch condensation. Then the following holds.*

1. *There is a tail* $(\mathcal{Q},\Lambda)$ *of* $(\mathcal{P},\Sigma)$ *such that* $\Gamma^*_\Lambda = \underset{\sim}{\Gamma}_\Lambda$.
2. *Suppose* $\Gamma^*_\Sigma = \underset{\sim}{\Gamma}_\Sigma$. *Then for any real x coding* $\mathcal{P}^-$,

$$C_{\Gamma_\Sigma}(x) = \mathsf{Lp}^{\Gamma,\Sigma_{\mathcal{P}^-}}(x).$$

3. *Suppose* $\Gamma^*_\Sigma = \underset{\sim}{\Gamma}_\Sigma$. *Then* $\mathsf{Code}(\Sigma) \notin \Gamma(\mathcal{P},\Sigma)$.
4. *Suppose* $\Gamma^*_\Sigma = \underset{\sim}{\Gamma}_\Sigma$. *Then there is a tail* $(\mathcal{Q},\Lambda)$ *of* $(\mathcal{P},\Sigma)$ *such that*

$$\Gamma(\mathcal{Q},\Lambda) = \wp(\mathbb{R}) \cap L(\Gamma(\mathcal{Q},\Lambda),\mathbb{R}).$$

Because $\Gamma(\mathcal{Q},\Lambda) = \Gamma(\mathcal{P},\Sigma)$, *it follows that* $\Gamma(\mathcal{P},\Sigma) = \wp(\mathbb{R}) \cap L(\Gamma(\mathcal{P},\Sigma),\mathbb{R})$.

We spend the rest of this section defining $\Gamma(\mathcal{P},\Sigma)$ in the case $\mathcal{P}$ is of #-lsa type. The reader may wish to review Definition 3.4.2, Definition 3.10.5 and Definition 3.10.6. The difficulty with representing the LSA pointclasses as $\Gamma(\mathcal{P},\Sigma)$ is the following. Suppose Γ is an LSA pointclass, i.e., $\Gamma = \wp(\mathbb{R}) \cap L(\Gamma,\mathbb{R})$ and $L(\Gamma,\mathbb{R}) \vDash \mathsf{AD}^+ + \mathsf{LSA}$. Let α be such that $\alpha + 1 = \Omega^\Gamma$ and set $\Gamma^b = \{A \subseteq \mathbb{R} : w(A) < \vartheta_\alpha\}$[3]. LSA pointclasses are peculiar because the pair that generates Γ^b is essentially the same as the pair that generates Γ. More precisely, if $(\mathcal{P},\Sigma)$ generates Γ then $(\mathcal{P},\Sigma^{stc})$ generates Γ^b.

DEFINITION 8.1.2. Suppose $(\mathcal{P},\Sigma)$ is a hod pair such that $\mathcal{P}$ is of #-lsa type and Σ has strong branch condensation and is strongly fullness preserving[4]. We then let

$$\Gamma(\mathcal{P},\Sigma) = \{A : \text{for cone of } x \in \mathbb{R},\ A \cap \mathsf{Lp}^{\Sigma^{stc}}(x) \in \mathsf{Lp}_2^{\Sigma^{stc}}(x)\}.$$

Notice that the definition of $\Gamma(\mathcal{P},\Sigma)$ depends on Σ^{stc} and hence, can also be defined for sts pairs. It is not immediately clear that $L(\Gamma(\mathcal{P},\Sigma)) \cap \wp(\mathbb{R}) = \Gamma(\mathcal{P},\Sigma)$. Theorem 8.1.14 shows that it is indeed true. Before we prove it, we prove some useful lemmas. The first lemma shows that various Σ-sts mice are internally Σ-closed.

LEMMA 8.1.3. *Assume* $\mathsf{AD}^+ + \mathsf{NsesS}$[5]. *Suppose* $(\mathcal{R},\Phi)$ *is an sts hod pair such that* Φ *has strong branch condensation and is strongly fullness preserving*[6] *and* $\mathcal{M}$ *is a* Φ*-sts mouse over* $\mathcal{R}$. *Suppose* η *is a Woodin cardinal of* $\mathcal{M}$ *and* $(\eta^+)^{\mathcal{M}}$ *exists. Suppose further that whenever* $\mathcal{Q} \in B(\mathcal{R},\Sigma^{\mathcal{M}|\eta})$ *and* $\mathcal{Q}$ *is of successor type, then* $\Sigma^{\mathcal{M}}_{\mathcal{Q}} = \Phi_{\mathcal{Q}} \restriction \mathcal{M}$. *Given* $\nu < \eta$, *let* $\mathcal{S}^{\mathcal{M}}_\nu$ *be the last model of* $(\mathcal{R},\Sigma^{\mathcal{M}})$*-coherent fully backgrounded construction of* $\mathcal{M}|\eta$ *that uses extenders with critical points*

[3]The superscript "b" stands for "bottom".

[4]See Definition 4.9.2 and Definition 4.6.4.

[5]See Definition 4.0.8.

[6]See Proposition 4.10.2.

$> \nu$[7] and let $\mathcal{T}_\nu$ on $\mathcal{R}$ be the normal tree leading to $\mathcal{S}_\nu^{\mathcal{M}}$. Then for all $\nu < \eta$, $\pi^{\mathcal{T}_\nu,b}$ exists and $\pi^{\mathcal{T}_\nu,b}(\delta^{\mathcal{R}^b}) = \delta^{(\mathcal{S}_\nu^{\mathcal{M}})^b}$.

PROOF. Towards a contradiction assume that for some ν our claim fails. Suppose first that $\pi^{\mathcal{T}_\nu,b}$ is undefined. We omit ν and $\mathcal{M}$ from subscripts and superscripts. Let B be the set of layers $\mathcal{P}$ of $\mathcal{S}^{\mathcal{M}}$ such that $\pi^{\mathcal{T}}_{\leq\delta^{\mathcal{P}},b}$ exists. We then have that $\cup_{\mathcal{P}\in B}\mathcal{P} \neq \mathcal{S}$, and so letting $\alpha = \sup\{\delta^{\mathcal{P}} : \mathcal{P} \in B\}$, $\alpha < \eta$ and $\pi^{\mathcal{T}_{\leq\alpha},b}$ is defined.

Suppose first that $\delta^{(\mathcal{M}_\alpha^{\mathcal{T}})^b} > \alpha$. Let $\mathcal{Q} \lhd_{hod} \mathcal{M}_\alpha^{\mathcal{T}}$ be the least complete layer[8] of $\mathcal{M}_\alpha^{\mathcal{T}}$ such that $\cup B \lhd \mathcal{Q}$. It follows that $\mathcal{T}_{\geq\alpha}$ is a normal tree based on $\mathcal{Q}$. But since $\Sigma_{\mathcal{Q}}^{M} = \Phi \restriction \mathcal{M}$, it follows from universality[9] that $\mathrm{lh}(\mathcal{T}_{\geq\alpha}) < \eta$ and $\pi^{\mathcal{T}_{\geq\alpha}}$ is defined. This is a contradiction, as it implies that there is $\mathcal{Q}' \in B$ such that $\delta^{\mathcal{Q}'} > \alpha$.

Assume now that $\delta^{(\mathcal{M}_\alpha^{\mathcal{T}})^b} = \alpha$. Since $\pi^{\mathcal{T},b}$ does not exist, $\mathcal{T}_{\geq\alpha}$ must be based on $(\mathcal{M}_\alpha^{\mathcal{T}})^b$ and be above $\delta^{(\mathcal{M}_\alpha^{\mathcal{T}})^b}$. However, it follows from our assumption that $\Sigma_{\mathcal{Q}}^{\mathcal{M}} = \Phi_{\mathcal{Q}} \restriction \mathcal{M}$, and once again we get a counterexample to the universality of $\mathcal{S}$.

The proof that $\pi^{\mathcal{T}_\nu,b}(\delta^{\mathcal{R}^b}) = \delta^{(\mathcal{S}_\nu^{\mathcal{M}})^b}$ is very similar and we leave it to the reader. $\dashv$

The following set up will be used in Lemma 8.1.5, Corollary 8.1.6, Corollary 8.1.7, Lemma 8.1.8, Corollary 8.1.9, Lemma 8.1.10, Corollary 8.1.12 and Lemma 8.1.13.

Assume $\mathsf{AD}^+ + NsesS$. Suppose $(\mathcal{P},\Sigma)$ is a hod pair such that $\mathcal{P}$ is of lsa type, $\mathcal{P} = (\mathcal{P}|\delta^{\mathcal{P}})^\#$ and Σ has strong branch condensation and is strongly fullness preserving. Suppose $\mathsf{Code}(\Sigma)$ is Suslin, co-Suslin. Let Γ be any good pointclasses such that $\mathsf{Code}(\Sigma) \in \underset{\sim}{\Delta}_\Gamma$. Let $\mathbb{M} = (M,\delta,\vec{G},\Omega)$ and let $\mathsf{C} = (\mathbb{M},(P_0,\Psi_0),\Gamma^*,A)$ Suslin, co-Suslin capture both Γ[10] and $\mathsf{Code}(\Sigma)$. We then have that the fully backgrounded hod pair construction of $\mathbb{M}$ reaches a tail of $(\mathcal{P},\Sigma)$ (see Theorem 4.13.5). Let $(\mathcal{Q},\Lambda)$ be this tail. Let $\mathcal{N}$ be the last model of[11]

$$(\mathsf{Le}((\mathcal{Q},\Lambda^{stc}),\mathcal{J}_\omega[\mathcal{Q}]))^{(M,\delta,\vec{G})}.$$

Because Σ is fullness preserving we have that $\mathcal{N} \models$ "$\delta^{\mathcal{Q}}$ is a Woodin cardinal". Let Φ be the strategy of $\mathcal{N}$ induced by Ψ. Notice that Φ is fullness preserving in the sense of Lp operator, i.e., whenever $\mathcal{M}$ is a Φ-iterate of $\mathcal{N}$ and η is a strong cutpoint of $\mathcal{M}$ then $\mathcal{M}|(\eta^+)^{\mathcal{M}} = \mathsf{Lp}^{\Lambda^{stc}}(\mathcal{M}|\eta)$. This can be shown using the proof of Theorem 4.6.3. We now prove several lemmas about $(\mathcal{N},\Phi)$ leading up to showing that $\Gamma(\mathcal{Q},\Lambda^{stc})$ can be realized as a derived model of $\mathcal{N}$. Let κ be the least strong cardinal of $\mathcal{N}$. The first lemma is quite standard.

[7] See Definition 3.5.1. See also Section 4.12.

[8] See Notation 2.7.14.

[9] See Theorem 4.5.6 and Theorem 4.13.2.

[10] See Definition 4.1.5, Definition 4.1.8 and Lemma 4.1.12.

[11] See Definition 4.2.1.

LEMMA 8.1.4. *$\mathcal{N} \models$ "κ is a limit of Woodin cardinals".*

PROOF. It is enough to show that δ is a limit of M-cardinals η such that $\mathsf{Lp}^{\Lambda^{stc}}(M|\eta) \models$ "η is a Woodin cardinal". Fix $\nu < \delta$. Because Code$(\Sigma) \in \underset{\sim}{\Delta}_\Gamma$, we have that for cone of z, $\mathsf{Lp}^{\Sigma^{stc}}(z) \in C_\Gamma(z)$. We can assume, using absoluteness[12], that the base of this cone is in M. Let $T, S \in M$ be δ-complementing trees witnessing that A is Suslin, co-Suslin captured by $(M, \delta, \vec{G}, \Omega)$. Let $\pi : R \to H_{(\delta^+)^M}$ be a Skolem hull such that crit$(\pi) > \nu$ is an M-cardinal and $\{T, S\} \in$ rge(π). Let $\eta =$ crit(π). Then it follows that $C_\Gamma(M|\eta) \in M$ and hence, $C_\Gamma(M|\eta) \models$ "η is a Woodin cardinal". It follows that $\mathsf{Lp}^{\Lambda^{stc}}(M|\eta) \models$ "η is a Woodin cardinal". ⊣

The next lemma uses language introduced in Definition 6.1.3.

LEMMA 8.1.5. *Φ is fullness preserving, i.e., Φ witnesses that $\Gamma(\mathcal{N}|\kappa, \Phi) = \Gamma^b(\mathcal{Q}, \Lambda^{stc})$.*

PROOF. Clearly, because Φ witnesses that $\mathcal{N}$ is a Λ^{stc}-sts mouse, $\Gamma(\mathcal{N}|\kappa, \Phi) \subseteq \Gamma^b(\mathcal{Q}, \Lambda^{stc})$. Fix then $(\mathcal{T}, \mathcal{R}) \in B(\mathcal{Q}, \Lambda^{stc})$. We want to see that

(1) there is a Φ-iterate $\mathcal{N}_1$ of $\mathcal{N}|\kappa$ such that for some $t = (\mathcal{Q}, \mathcal{T}, \mathcal{S}, \mathcal{U}) \in \mathcal{N}_1$, t is according to $\Sigma^{\mathcal{N}_1}$ and $\Lambda_\mathcal{R} \leq_w \Lambda_{\mathcal{S}^b}$.

Suppose (1) fails. We can then assume, without loss of generality, that for some $\nu < \delta$ and some $g \subseteq Coll(\omega, \nu)$, $(\mathcal{T}, \mathcal{R}) \in M[g]$. Again without losing generality we can assume that $\mathcal{R}$ is of successor type. Let now $\mathcal{S}$ be the output of the $(\mathcal{Q}, \Sigma^\mathcal{N})$-coherent fully backgrounded construction of $\mathcal{N}$ that uses extenders with critical points $> \nu$. Let $\mathcal{U}$ be a normal tree on $\mathcal{Q}$ with last model $\mathcal{S}$. We claim that

(2) $\pi^{\mathcal{U},b}$ exists, $\pi^{\mathcal{U},b}(\delta^{\mathcal{Q}^b}) = \delta^{\mathcal{S}^b}$ and $\mathcal{S}' \lhd_{hod} \mathcal{S}^b$ is a $\Lambda_\mathcal{R}$-iterate of $\mathcal{R}$.

The first two clauses of (2) are consequences of Lemma 8.1.3. The third is a straightforward consequence of the fact that Λ is both positional and fullness preserving and of the fact that $\mathcal{S}$ side never moves in the comparison with $\mathcal{R}$[13]. This finishes the proof of Lemma 8.1.5. ⊣

Before we proceed, we record some lemmas that can now be proved. Since these lemmas are standard, we will state these results without proofs and instead will give references. The next lemma can be proved following the proof of clause 2 of Theorem 6.1.4 and also standard arguments like the proofs of Corollary 1.2 and Proposition 1.4, 1.5 of [33] and [29, Chapter 3.1].

LEMMA 8.1.6. *Suppose $\pi : \mathcal{N}|(\kappa^+)^\mathcal{N} \to \mathcal{M}$ is an iteration via $\Phi_{\mathcal{N}|\kappa}$ and g is $\mathcal{M}$-generic. Then letting F be the function $F(X) = \mathsf{Lp}^{\Lambda^{stc}}(X)$, $F \restriction \mathcal{M}[g]$ is definable over $\mathcal{M}[g]$ uniformly in $(\mathcal{M}, g)$[14].*

---

[12]See Lemma 4.1.11.

[13]See Proposition 4.10.2 and [29, Lemma 2.6].

[14]That is, the definition works for any such $\mathcal{M}$ and g.

Below HC stands for the set of all hereditarily countable sets.

COROLLARY 8.1.7. *Suppose* $\pi : \mathcal{N}|(\kappa^+)^{\mathcal{N}} \to \mathcal{M}$ *is an iteration via* $\Phi_{\mathcal{N}|(\kappa^+)^{\mathcal{N}}}$ *and* F *is as in Lemma 8.1.6. Then if* $h \subseteq Coll(\omega, <\pi(\kappa))$ *is* $\mathcal{M}$*-generic then* $F \restriction \mathsf{HC}^{\mathcal{M}[h]} \in \mathcal{M}[\mathbb{R}^{\mathcal{M}[h]}]$[15].

Lemma 8.1.6 can be used to prove the following lemma. See also the proof of clause 2 of Theorem 6.1.4, Proposition 6.2.4, Definition 6.2.5 and [33, Proposition 1.5].

LEMMA 8.1.8. *Suppose* $\pi : \mathcal{N}|(\kappa^+)^{\mathcal{N}} \to \mathcal{M}$ *is an iteration via* $\Phi_{\mathcal{N}|(\kappa^+)^{\mathcal{N}}}$ *and* δ *is a cutpoint Woodin cardinal of* $\mathcal{M}$*. Let* ξ *be a cutpoint cardinal of* $\mathcal{M}$ *such that* $\mathcal{M}$ *has no Woodin cardinals in the interval* (ξ, δ)*. Let* $\eta \in (\xi, \delta)$ *be an* $\mathcal{M}$*-cardinal and let* Ψ *be the fragment of* Φ *that acts on normal non-dropping trees based on* $\mathcal{M}|(\eta^+)^{\mathcal{M}}$ *that are above* ξ*. Then letting* $h \subseteq Coll(\omega, (\eta^+)^{\mathcal{M}})$ *be* $\mathcal{M}$*-generic,* $\Phi \restriction \mathcal{M}|\pi(\kappa)[h] \in \mathcal{M}$ *and is* $\pi(\kappa)$*-universally Baire in* $\mathcal{M}[h]$.

COROLLARY 8.1.9. *Suppose* $\pi : \mathcal{N}|(\kappa^+)^{\mathcal{N}} \to \mathcal{M}$ *is an iteration via* $\Phi_{\mathcal{N}|(\kappa^+)^{\mathcal{N}}}$*. Suppose* g *is* $\mathcal{M}|\pi(\kappa)$*-generic,* $X \in (\mathcal{M}|\pi(\kappa))[g]$ *and* $\mathcal{R} \in \mathsf{Lp}^{\Lambda^{stc}}(X)$ *is such that* $\rho(\mathcal{R}) = \mathsf{ord}(X)$*. Let* $h \subseteq Coll(\omega, |X|)$ *be* $(\mathcal{M}|\pi(\kappa))[g]$*-generic. Then* $\mathcal{R} \in \mathcal{M}[g][h]$ *and* $\mathcal{M}[g][h] \models$ *"*$\mathcal{R}$ *has a* $\pi(\kappa)$*-universally Baire iteration strategy* Ψ *witnessing that* $\mathcal{R}$ *is a* Λ^{stc}*-sts mouse over* X *based on* $\mathcal{Q}$*".*

Moreover, if $\mathcal{R} \in (\mathcal{M}|\pi(\kappa))[g]$ *is a sound* Λ^{stc}*-sts premouse over* X *such that* $\rho(\mathcal{R}) = \mathsf{ord}(X)$ *and for some* $(\mathcal{M}|\pi(\kappa))[g]$*-generic* $h \subseteq Coll(\omega, |X|)$*,* $\mathcal{M}[g][h] \models$ *"*$\mathcal{R}$ *has a* $\pi(\kappa)$*-iteration strategy" then* $\mathcal{R} \trianglelefteq \mathsf{Lp}^{\Lambda^{stc}}(X)$[16].

The next lemma shows that $\Gamma(\mathcal{Q}, \Lambda^{stc})$ can be realized as the set of reals of *a derived model* of a Φ-iterate of $\mathcal{N}$. We introduced the notation $D(\mathcal{M}, \lambda, h)$ in Section 3.8. The derived model theorem says that $D(\mathcal{M}, \lambda, h) \models \mathsf{AD}^+$, but we need a stronger version of this theorem.

Suppose $\mathcal{V}$ is a transitive inner model of $\mathsf{ZFC} - \mathsf{Powerset}$, λ is a limit of Woodin cardinals of $\mathcal{V}$, $(\lambda^{++})^{\mathcal{V}}$ exists and $h \subseteq Coll(\omega, <\lambda)$ is $\mathcal{V}$-generic. Let

$$\mathbb{R}^* = \cup_{\alpha<\lambda} \mathbb{R}^{\mathcal{V}[g \cap Coll(\omega, <\alpha)]}$$

and $\Gamma = \{A \in \mathcal{V}(\mathbb{R}^*) : \mathcal{V}(\mathbb{R}^*) \models \text{"}L(A, \mathbb{R}^*) \models \mathsf{AD}^+\text{"}\}$. Set

$$D^+(\mathcal{V}, \lambda, h) =_{\text{def}} (L(\Gamma, \mathbb{R}^*))^{\mathcal{V}(\mathbb{R}^*)}.$$

Then the stronger version of Woodin's derived model theorem says that

$$D^+(\mathcal{V}, \lambda, h) \models \mathsf{AD}^+.$$

Sometimes $D^+(\mathcal{V}, \lambda, h)$ is called the *new* derived model.

Suppose now that in addition to the above, $\mathcal{V}$ is countable and Φ is an $\omega_1 + 1$-iteration strategy for $\mathcal{V}$. Let $g : \omega \to \mathbb{R}$ be generic for $Coll(\omega, \mathbb{R})$ and let $(x_i : i < \omega)$

[15] Because κ is a regular cardinal in $\mathcal{N}$, we have that $\mathbb{R}^{M[h]} = (\mathbb{R}^*)^{M[h]}$.

[16] Notice that $\mathcal{R}$ has a unique $\pi(\kappa)$-iteration strategy in $\mathcal{M}[g][h]$.

be the enumeration of $\mathbb{R}$ given by $x_i = g(i)$. We can now perform an $\mathbb{R}$-genericity iteration of $\mathcal{V}$ via Φ much like it is done in [59, Chapter 7.2 and Corollary 7.17]. Let $\mathcal{V}'$ be this iterate of $\mathcal{V}$ and let $h \subseteq Coll(\omega, < \omega_1^{\mathcal{V}})$ be $\mathcal{V}'$-generic such that $(\mathbb{R}^*)^{\mathcal{V}'[g]} = \mathbb{R}^V$. We then let $D^+(\mathcal{V}, \Phi, \lambda, g) = D^+(\mathcal{V}', \omega_1^{\mathcal{V}}, h)$.

LEMMA 8.1.10. *The new derived model of $\mathcal{N}|(\kappa^+)^{\mathcal{N}}$ as computed via $\Psi =_{def} \Phi_{\mathcal{N}|(\kappa^+)^{\mathcal{N}}}$ is $L(\Gamma(\mathcal{Q}, \Lambda^{stc}))$. More precisely, for any $g \subseteq Coll(\omega, \mathbb{R})$,*

$$D^+(\mathcal{N}|(\kappa^+)^{\mathcal{N}}, \Psi, \kappa, g) = L(\Gamma(\mathcal{Q}, \Lambda^{stc}))$$

and $\wp(\mathbb{R}) \cap D^+(\mathcal{V}, \Psi, \lambda, g) = \Gamma(\mathcal{Q}, \Lambda^{stc})$. In particular, $\Gamma(\mathcal{Q}, \Lambda^{stc}) = \wp(\mathbb{R}) \cap L(\Gamma(\mathcal{Q}, \Lambda^{stc}))$.

PROOF. We will use clause 2 of Theorem 6.1.4. First we verify that clause 2 of Theorem 6.1.4 applies. For this we need to verify that

(1) $\mathcal{N}$ is internally Λ^{stc}-closed, and
(2) Φ is a fullness preserving strategy for $\mathcal{N}$.

Notice that (1) is a consequence of Lemma 8.1.3 and (2) is just Lemma 8.1.5. We thus have that clause 2 of Theorem 6.1.4 applies.

To prove Lemma 8.1.10 we need to show that given an $\mathbb{R}$-genericity iteration $\pi : \mathcal{N}|(\kappa^+)^{\mathcal{N}} \to \mathcal{W}$ according to $\Phi_{\mathcal{N}|(\kappa^+)^{\mathcal{N}}}$,

(3) if $A \in \Gamma(\mathcal{Q}, \Lambda^{stc})$ then $A \in \mathcal{W}(\mathbb{R})$, and
(4) if $A \in \mathcal{W}(\mathbb{R})$ is such that $L(A, \mathbb{R}) \vDash \mathsf{AD}^+$ then $A \in \Gamma(\mathcal{Q}, \Lambda^{stc})$.

We start with (3). Towards a contradiction, assume not and let $A \in \Gamma(\mathcal{Q}, \Lambda^{stc})$ witness this. We have that for cone of $z \in \mathbb{R}$, $A \cap \mathsf{Lp}^{\Lambda^{stc}}(z) \in \mathsf{Lp}_2^{\Lambda^{stc}}(z)$. Let z be some base of the aforementioned cone. Let $\xi > \Theta$ be such that $L_\xi(\wp(\mathbb{R})) \vDash \mathsf{ZF} - \mathsf{Replacement}$ and $\sigma : M \to L_\xi(\wp(\mathbb{R}))$ be a countable hull such that $\mathcal{N}, z \in \mathsf{HC}^M$ and $\{\Phi, A\} \in \mathrm{rge}(\sigma)$. Let $A^M = \sigma^{-1}(A)$.

Let $g \in L(\wp(\mathbb{R}))$ be M-generic for $Coll(\omega, \mathbb{R}^M)$. Let $(y_i : i < \omega)$ be the generic sequence enumerating $\mathbb{R}^M$ and let $(\delta_i : i < \omega)$ be a sequence of cutpoint Woodin cardinals of $\mathcal{N}|(\kappa^+)$ with sup κ. Let $(\mathcal{N}_i, \mathcal{T}_i : i < \omega)$ be the $\mathbb{R}^M$-genericity iteration. Thus, $\mathcal{N}_0 = \mathcal{N}|(\kappa^+)^{\mathcal{N}}$, $\mathcal{T}_i$ is a tree on $\mathcal{N}_i$ that is based on $\mathcal{N}_i|\pi^{\oplus_{j<i}\mathcal{T}_j}(\delta_i)$ and is above $\pi^{\oplus_{j<i}\mathcal{T}_j}(\delta_{i-1})$[17] and $\mathcal{T}_i$ is built according to the rules of y_i-genericity iteration. Let $\pi_{i,k} : \mathcal{N}_i \to \mathcal{N}_k$ be the composition of the iteration embeddings. Let $\mathcal{N}_\omega$ be the direct limit of $\mathcal{N}_i$ under $\pi_{i,k}$.

Because $z \in \mathbb{R}^M$, we have that $A \cap (\mathcal{N}_\omega|\omega_1^M)(\mathbb{R}^M) \subset \mathsf{Lp}^{\Lambda^{stc}}((\mathcal{N}_\omega|\omega_1^M)(\mathbb{R}^M)))$. Notice that it follows from Lemma 8.1.6 that if $\mathcal{N}_\omega^+$ is the last model of $\uparrow(\oplus_{i<\omega}\mathcal{T}_i, \mathcal{N})$[18] then

$$\mathsf{Lp}^{\Lambda^{stc}}((\mathcal{N}_\omega|\omega_1^M)(\mathbb{R}^M)) \in \mathcal{N}_\omega^+(\mathbb{R}^M).$$

[17]Let $\delta_{-1} = 0$.

[18]See Definition 2.4.10.

It follows that $A^M \in D(\mathcal{N}_\omega, \omega_1^M, h)$ where $h \subseteq Coll(\omega, <\omega_1^M)$ is an $\mathcal{N}_\omega$-generic such that $\mathbb{R}^{\mathcal{N}_\omega[h]} = \mathbb{R}^M$. This finishes the proof of (3).

We keep the notation used to prove (3) and start proving (4). To prove (4), we need to show that if A is as in (4) and M, σ etc were defined as before relative to A then

(5) $\sigma^{-1}(A) \in (\Gamma(\mathcal{Q}, \Lambda^{stc}))^M$.

Suppose that (5) fails. We then have that there is $B \in \mathcal{N}_\omega(\mathbb{R}^M)$ such that $L(B, \mathbb{R}^M) \models \mathsf{AD}^+$ and $B \notin (\Gamma(\mathcal{Q}, \Lambda^{stc}))^M$. We first claim that

CLAIM 8.1.11. *In $L(B, \mathbb{R}^M)$, for cone of y, $B \cap \mathsf{Lp}^{\Lambda^{stc}}(y) \in \mathsf{Lp}_2^{\Lambda^{stc}}(y)$.*

PROOF. Suppose not. Working in $L(B, \mathbb{R}^M)$, fix $y \in \mathbb{R}^M$ such that for any $y^* \in \mathbb{R}^M$ Turing above y, $B \cap \mathsf{Lp}^{\Lambda^{stc}}(y) \notin \mathsf{Lp}_2^{\Lambda^{stc}}(y)$. Fix $i < \omega$ such that $y \in \mathcal{N}_\omega[h \cap Coll(\omega, \delta_i)]$. Notice that

(6) for every $y \in \mathbb{R}^M$, $(\mathsf{Lp}^{\Lambda^{stc}}(y))^{L(B,\mathbb{R}^M)} = \mathsf{Lp}^{\Lambda^{stc}}(y)$.

(6) is a consequence of Corollary 8.1.9. This is because if $\mathcal{R} \trianglelefteq (\mathsf{Lp}^{\Lambda^{stc}}(y))^{L(B,\mathbb{R}^M)}$ is such that $\rho(\mathcal{R}) = \omega$ then $\mathcal{R}$ has an iteration strategy in $\mathcal{N}_\omega[y]$ as the iteration strategy of $\mathcal{R}$ is ordinal definable from Λ^{stc}, y in the derived model of $\mathcal{N}_\omega$.

Let $k < \omega$ be such that there is a symmetric name τ for B in $\mathcal{N}_\omega[h \cap Coll(\omega, \delta_k)]$. Let $j = \max(i,k) + 1$. We then have that

(7) in $L(B, \mathbb{R}^M)$, $B \cap (\mathcal{N}_\omega|\delta_j)[h \cap Coll(\omega, \delta_j)] \notin \mathsf{Lp}^{\Lambda^{stc}}((\mathcal{N}_\omega|\delta_j)[h \cap Coll(\omega, \delta_j)])$.

However, it follows from Lemma 6.4.4 that

(8) $\mathsf{Lp}^{\Lambda^{stc}}((\mathcal{N}_\omega|\delta_j)[h \cap Coll(\omega, \delta_j)]) = \mathcal{N}_\omega|(\delta_j^+)^{\mathcal{N}_\omega}[h \cap Coll(\omega, \delta_j)]$.

(8) and (7) contradict (6) (as $\tau_{h \cap Coll(\omega, \delta_j)} = B \cap (\mathcal{N}_\omega|\delta_j)[h \cap Coll(\omega, \delta_j)]$). ⊣

We will now make use of [34, Theorem 0.1]. It follows from the proof of the aforementioned theorem (applied to all sets of reals in $L(B, \mathbb{R}^M)$) that

(9) $\wp(\mathbb{R})^{L(B,\mathbb{R}^M)} \subseteq (\mathsf{Lp}^{\Lambda^{stc}}(\mathbb{R}^M))^{L(B,\mathbb{R}^M)}$ and

(10) if $\mathcal{K} \trianglelefteq (\mathsf{Lp}^{\Lambda^{stc}}(\mathbb{R}^M))^{L(B,\mathbb{R}^M)}$ is such that $\rho(\mathcal{K}) = \mathbb{R}$ and $k : \mathcal{K}' \to \mathcal{K}$ is such that $\mathcal{K}'$ is countable in $L(B, \mathbb{R}^M)$ then $\mathcal{K}' \trianglelefteq \mathsf{Lp}^{\Lambda^{stc}}(k^{-1}(\mathbb{R}^M))$.

A Skolem hull argument done inside $\mathcal{N}_\omega^+$ shows that (10) implies that,

(11) $(\mathsf{Lp}^{\Lambda^{stc}}(\mathbb{R}^M))^{L(B,\mathbb{R}^M)} \trianglelefteq \mathsf{Lp}^{\Lambda^{stc}}(\mathbb{R}^M)$.

Suppose now that

(a) $\mathsf{Lp}^{\Lambda^{stc}}(\mathbb{R}^M) \in M$.

Then clearly (11) implies that

(12) $B \in M$.

(12) and the Claim imply (5). Thus, it is enough to prove that (a) holds. (a) easily follows from the fact that $\mathsf{Lp}^{\Lambda^{stc}}(\mathbb{R}^M) \in \mathcal{N}^+_\omega(\mathbb{R})$ implying that $\mathsf{Lp}^{\Lambda^{stc}}(\mathbb{R}^M) \in M[g]$. But since g is generic for a homogenous poset, it follows that $\mathsf{Lp}^{\Lambda^{stc}}(\mathbb{R}^M) \in M$. ⊣

The following is a simple corollary of the proof of Lemma 8.1.10.

COROLLARY 8.1.12. *Suppose $(\eta_i : i < \omega)$ is a sequence of consecutive Woodin cardinals of $\mathcal{N}|\kappa$ and $\lambda = \sup_{i<\omega} \eta_i$. The derived model of $\mathcal{R} =_{def} \mathcal{N}|(\lambda^+)^{\mathcal{N}}$ as computed via $\Phi_{\mathcal{R}}$ is $L(\Gamma(\mathcal{Q}, \Lambda^{stc}))$. In particular,*

$$\Gamma(\mathcal{Q}, \Lambda^{stc}) = \wp(\mathbb{R}) \cap L(\Gamma(\mathcal{Q}, \Lambda^{stc})).$$

Let $\mathcal{Q}_\infty$ be the direct limit of all Λ-iterates of $\mathcal{Q}$ and let $\pi : \mathcal{Q} \to \mathcal{Q}_\infty$ be the iteration embedding. Let Ω be the (ω_1, ω_1) fragment of $\Lambda_{\mathcal{Q}^b_\infty}$[19]. Notice that $\pi \restriction \mathcal{Q}^b$ depends only on Λ^{stc}[20] and hence (by the coding lemma), it is in $L(\Gamma(\mathcal{Q}, \Lambda^{stc}))$. Also, because Λ^{stc} is fullness preserving, it follows that $\pi[\mathcal{Q}^b]$ can be coded as a subset of $w(\Gamma^b(\mathcal{Q}, \Lambda))$. This is because $\mathcal{Q}^b_\infty|\delta^{\mathcal{Q}^b_\infty} = \bigcup\{\mathcal{M}_\infty(\mathcal{R}, \Lambda_{\mathcal{R}}) : \mathcal{R} \in pB(\mathcal{Q}, \Lambda)\}$ and $\delta^{\mathcal{Q}^b} = w(\Gamma^b(\mathcal{Q}, \Lambda))$.

LEMMA 8.1.13. $\Lambda^{stc} \in \mathcal{J}_\omega(\pi[\mathcal{Q}^b], \mathcal{Q}^b_\infty, \Gamma^b(\mathcal{Q}, \Lambda))$.

PROOF. Set $\Psi = \Lambda^{stc}$. Notice that if $(\mathcal{T}, \mathcal{S}) \in I(\mathcal{Q}, \Psi)$ and $\mathcal{W}$ is a tree on $\mathcal{S}$ of limit length according to $\Psi_{\mathcal{S}}$ such that $\mathcal{W}$ is above $\delta^{\mathcal{S}^b}$ and $\mathcal{W} \in b(\Psi_{\mathcal{S}})$ then letting $b = \Psi_{\mathcal{S}}(\mathcal{W})$[21], $\mathcal{Q}(b, \mathcal{W})$ exists and has an iteration strategy in $\Gamma^b(\mathcal{Q}, \Lambda)$. This is simply because there is an extender $E \in \vec{E}^{\mathcal{M}^{\mathcal{W}}_b}$ with critical point $\delta^{\mathcal{S}^b}$ such that $\mathcal{Q}(b, \mathcal{W}) \triangleleft (\mathrm{Ult}(\mathcal{M}^{\mathcal{W}}_b, E))^b$. We can then define Ψ in $\mathcal{J}_\omega(\pi[\mathcal{Q}^b], \mathcal{Q}^b_\infty, \Gamma^b(\mathcal{Q}, \Lambda))$ with the following procedure. We work in $\mathcal{J}_\omega(\pi[\mathcal{Q}^b], \mathcal{Q}^b_\infty, \Gamma^b(\mathcal{Q}, \Lambda))$.

Suppose first X is a transitive set and $\mathcal{R} \in X$ is an lsa type hod mouse. Suppose that there is an embedding $\tau : \mathcal{Q}^b \to \mathcal{R}^b$. Suppose further that $\mathcal{M}$ is an sts mouse over X based on $\mathcal{R}$. We say $\mathcal{M}$ is *good* if it has an iteration strategy $\Delta \in \Gamma^b(\mathcal{Q}, \Lambda)$ such that if $\mathcal{S}$ is a Δ-iterate of $\mathcal{M}$, $t = (\mathcal{R}, \mathcal{T}, \mathcal{R}^*_1, \mathcal{U}) \in \mathcal{S}$ is according to $\Sigma^{\mathcal{S}}$, and $\mathcal{R}_1 = \pi^{\mathcal{T},b}(\mathcal{R}^b)$ then letting $\Delta_1 = \Delta_{\mathcal{R}_1}$,

1. $(\mathcal{R}_1, \Delta_1)$ is a hod pair such that Δ_1 has strong branch condensation and is strongly fullness preserving,
2. $\mathcal{R}_1 = Hull^{\mathcal{R}_1}(\pi^{\mathcal{T},b} \circ \tau[\mathcal{Q}^b] \cup \delta^{\mathcal{R}_1})$,
3. letting $\sigma : \mathcal{R}_1 \to \mathcal{Q}^b_\infty$ be given by
$$\sigma(x) = \pi(f)(\pi^{\Delta_1}_{\mathcal{R}_1,\infty}(a)),$$
where $f \in \mathcal{Q}^b$ and $a \in (\delta^{\mathcal{R}^b_1})^{<\omega}$ are such that $x = \pi^{\mathcal{T},b} \circ \tau(f)(a)$,
$$\pi \restriction \mathcal{Q}^b = \sigma \circ \pi^{\mathcal{T},b} \circ \tau.$$
4. $\mathcal{U}$ is according to Δ_1.

[19]See [32] where it is shown that Λ is $< \Theta$-uB.
[20]If $\mathcal{T}$ is the $\mathcal{Q}$-to-$\mathcal{Q}_\infty$ stack then $\pi \restriction \mathcal{Q} = \pi^{\mathcal{T},b}$.
[21]Thus, b is a branch.

We can now define $\mathsf{Lp}^{good,sts,\tau}(X)$ which is the stack of good sts mice over X that are based on $\mathcal{R}$. Then we can define $\mathsf{Lp}_{\omega}^{good,sts,\tau}(X)$.

Suppose next that $\mathcal{R}$ is an lsa type hod premouse and $\tau : \mathcal{Q}^b \to \mathcal{R}^b$ is an embedding. Suppose $\mathcal{U}$ is a stack on $\mathcal{R}^b|\delta^{\mathcal{R}^b}$. We say $(\mathcal{R}^b,\mathcal{U})$ is a τ-*good* iteration if there is $k : \mathcal{R}^b \to \mathcal{Q}^b_\infty$ such that $\pi \restriction \mathcal{Q}^b = k \circ \tau$ and for some $(\mathcal{S},\Delta) \in \Gamma^b(\mathcal{Q},\Lambda)$ such that Δ has strong branch condensation and is strongly fullness preserving, $k \restriction (\mathcal{R}^b|\delta^{\mathcal{R}^b}) \subseteq \pi^{\Delta}_{\mathcal{S},\infty}[\mathcal{S}]$ and if $\sigma : \mathcal{R}|\delta^{\mathcal{R}^b} \to \mathcal{S}$ is given by

$$\sigma(x) = (\pi^{\Delta}_{\mathcal{S},\infty})^{-1}(k(x))$$

then $\mathcal{U}$ is according to σ-pullback of Δ.

We can similarly define τ-*good* iterations when $\mathcal{U}$ is above $\delta^{\mathcal{R}^b}$. In this case, we simply demand that $\mathcal{U}$ be according to the unique strategy of Ψ' of $\mathcal{R}^b$ which acts on stacks that are above $\delta^{\mathcal{R}^b}$ and letting $\Delta' = (\sigma\text{-pullback of }\Delta)$[22], Ψ' witnesses that $\mathcal{R}^b$ is a Δ'-premouse above $\delta^{\mathcal{R}^b}$.

Suppose now that

$$\mathcal{T} = ((\mathcal{M}_\alpha)_{\alpha<\eta}, (E_\alpha)_{\alpha<\eta-1}, D, R, (\beta_\alpha, m_\alpha)_{\alpha\in R}, \mathsf{short}, \mathsf{max}, T)$$

is an st-stack[23] on $\mathcal{Q}$ of countable length. Recall Remark 3.2.3 and Definition 2.7.24. These conventions stipulate that R consists of cutpoints of $\mathcal{T}$. Also recall Notation 2.4.4. We say $\mathcal{T}$ is π-*b-realizable* if there is a sequence $(\sigma_\alpha : \alpha \in R)$ such that the following clauses hold[24]:

1. $\mathcal{T}$ does not have a fatal drop[25],
2. $\sigma_\alpha : (\mathcal{M}_\alpha)^b \to \mathcal{Q}^b_\infty$ is an elementary embedding.
3. For all $\alpha, \alpha' \in R$ with $\alpha < \alpha'$, $\sigma_\alpha = \sigma_{\alpha'} \circ \pi^{\mathcal{T},b}_{\alpha,\alpha'}$.
4. For all $\alpha \in R$, letting $\Lambda_\alpha = (\sigma_\alpha \restriction \mathcal{M}_\alpha|\delta^{\mathcal{M}^b_\alpha}\text{-pullback of }\Omega)$, for each complete layer $\mathcal{R} \lhd \mathcal{M}^b_\alpha$, $\sigma_\alpha \restriction \mathcal{R} = \pi^{\Lambda_\alpha}_{\mathcal{R},\infty}$ where $\pi^{\Lambda_\alpha}_{\mathcal{R},\infty} : \mathcal{R} \to \mathcal{M}_\infty(\mathcal{R},(\Lambda_\alpha)_{\mathcal{R}})$ is the iteration map according to $(\Lambda_\alpha)_{\mathcal{R}}$.
5. For all $\alpha \in R$ such that $\alpha \neq \max(R)$, letting $\alpha' = \min(R - (\alpha+1))$, if $\mathcal{T}_{\alpha,\alpha'}$ is based on $\mathcal{M}^b_\alpha|\delta^{\mathcal{M}^b_\alpha}$ then $\mathcal{T}_{\alpha,\alpha'}$ is according to Λ_α.
6. For all $\alpha \in R$ such that $\alpha \neq \max(R)$, letting $\alpha' = \min(R - (\alpha+1))$, if $\mathcal{T}_{\alpha,\alpha'}$ is based on $\mathcal{M}^b_\alpha$ and is above $\delta^{\mathcal{M}^b_\alpha}$ then $\mathcal{T}_{\alpha,\alpha'}$ is according to the unique strategy of $\mathcal{M}^b_\alpha$ that acts on stacks above $\delta^{\mathcal{M}^b_\alpha}$ and witnesses that $\mathcal{M}^b_\alpha$ is a $(\Lambda_\alpha)_{\mathcal{M}_\alpha|\delta^{\mathcal{M}^b_\alpha}}$-mouse over $\mathcal{M}_\alpha|\delta^{\mathcal{M}^b_\alpha}$.

---

[22] Here, σ is as above.

[23] See Definition 3.2.1.

[24] For the definition of $\pi^{\mathcal{T},b}_{\alpha,\alpha'}$, see Section 2.8.

[25] See Definition 2.6.8.

7. For every $\alpha \in R$ such that $\alpha+1 < \mathrm{lh}(\mathcal{T})$, letting $\mathcal{W} = \mathsf{nc}_\alpha^{\mathcal{T}}$[26], for all limit ordinal $\gamma < lh(\mathcal{W})$ such that $\mathcal{W} \restriction \gamma$ is nuvs, letting $\tau = \pi_{0,\alpha}^{\mathcal{T},b}$,
   (a) if $\mathsf{Lp}^{good,sts,\tau}(\mathrm{m}^+(\mathcal{W} \restriction \gamma)) \vDash$ "$\delta(\mathcal{W} \restriction \gamma)$ is a Woodin cardinal" then $\mathrm{lh}(\mathcal{W}) = \gamma+1$ and $\gamma \in R$ and
   (b) if $\mathsf{Lp}^{good,sts,\tau}(\mathrm{m}^+(\mathcal{W} \restriction \gamma)) \vDash$ "$\delta(\mathcal{W} \restriction \gamma)$ is not a Woodin cardinal" then setting $b = [0,\gamma)_{\mathcal{W}}$, b is a cofinal branch for $\mathcal{W} \restriction \gamma$ such that $\mathcal{Q}(b, \mathcal{W} \restriction \gamma)$ exists and $\mathcal{Q}(b, \mathcal{W} \restriction \gamma) \trianglelefteq \mathsf{Lp}^{good,sts,\tau}(\mathrm{m}^+(\mathcal{W} \restriction \gamma))$.

Let then Δ be an iteration strategy for $\mathcal{Q}$ such that its domain consists of st-stacks $\mathcal{T}$ which are π-b-realizable and for $\mathcal{T} \in \mathrm{dom}(\Delta)$, $\Delta(\mathcal{T}) = b$ if and only if $\mathcal{T}^\frown\{b\}$ is a π-b-realizable st-stack. It can now be shown that Δ is the fragment of Ψ that acts on st-stacks that do not have a fatal drop. The proof is very much like the proof of clause 2 of Theorem 6.1.4 and it also uses Definition 3.10.6[27]. We leave it to the reader.

To compute Λ^{stc}, notice that Λ^{stc} is the unique short-tree strategy Λ' of $\mathcal{Q}$ such that Λ' is fullness preserving and Δ as defined above is the fragment of Λ' that acts on st-stacks without fatal drops. This easily follows from Lemma 4.7.2. ⊣

We are now in a position to state the main theorem of this section.

THEOREM 8.1.14. *Assume* $\mathsf{AD}^+ + NsesS$. *Suppose* $(\mathcal{P},\Sigma)$ *is a hod pair such that* $\mathcal{P}$ *is of #-lsa type*[28] *and* Σ *has strong branch condensation and is strongly fullness preserving. Suppose* $\mathsf{Code}(\Sigma)$ *is Suslin, co-Suslin. Then for some* $\mathcal{Q} \in pI(\mathcal{P},\Sigma)$,

1. $L(\Gamma(\mathcal{Q},\Sigma_\mathcal{Q})) \cap \wp(\mathbb{R}) = \Gamma(\mathcal{Q},\Sigma_\mathcal{Q})$,
2. *the set* $\{(x,y) : x \in \mathbb{R}$ *and* $y \notin \mathsf{Lp}^{\Sigma_\mathcal{Q}^{stc}}(x)\}$ *cannot be uniformized in* $L(\Gamma(\mathcal{Q}, \Sigma_\mathcal{Q}))$, *and*
3. $L(\Gamma(\mathcal{Q},\Sigma_\mathcal{Q})) \vDash \mathsf{LSA}$.

PROOF. Assume that one of 1-3 above is false. Let $\Gamma_0 = \{A \subseteq \mathbb{R} : A$ is ordinal definable from Σ and a real$\}$. Then one of 1-3 is false inside $L(\Gamma_0,\mathbb{R})$, which means we can assume that $V = L(\Gamma_0,\mathbb{R})$. Let (α_0,β_0) be lexicographically least such that letting $\Gamma_0^* = \{A \subseteq \mathbb{R} : w(A) < \alpha_0\}$ the following holds:

1. $W =_{\mathrm{def}} L_{\beta_0}(\Gamma_0^*,\mathbb{R}) \vDash \mathsf{ZF} - \mathsf{Powerset} +$ "Θ exists",
2. $\Sigma \in \Gamma_0^*$ and $\alpha_0 = \Theta^{L_{\beta_0}(\Gamma_0^*,\mathbb{R})}$, and
3. one of clauses 1-3 fails in $L_{\beta_0}(\Gamma_0^*,\mathbb{R})$.

Let $\Gamma_0 = (\Sigma_1^2(\mathsf{Code}(\Sigma)))^W$ and let Γ be any good pointclass such that $\Gamma_0 \subseteq \Delta_\Gamma$. Using Theorem 4.1.12 we can find $\mathbb{M} = (M,\delta,\vec{G},\Omega)$ and $\mathsf{C} = (\mathbb{M},(P_0,\Psi_0),\Gamma^*,A)$ such that C Suslin, co-Suslin capture both Γ[29], $\mathsf{Code}(\Sigma)$ and the set D consisting

[26] See Notation 2.4.4.
[27] In particular, see the conclusion of Definition 3.10.6.
[28] See Definition 2.7.3.
[29] See Definition 4.1.5, Definition 4.1.8 and Lemma 4.1.12.

of triples $(u,v,w) \in \mathbb{R}^3$ where u codes $\mathcal{Q} \in pI(\mathcal{P},\Sigma)$, v codes a self-well-ordered $X \in \mathsf{HC}$ with $\mathcal{Q} \in X$ and w codes $\mathsf{Lp}^{\Sigma^{stc}}(X)$.

We then have that the fully backgrounded hod pair construction of $\mathbb{M}$ reaches a tail of $(\mathcal{P},\Sigma)$ (see Theorem 4.13.5). Let $(\mathcal{Q},\Lambda)$ be this tail (so $\Lambda = \Sigma_{\mathcal{Q}}$). Let $\mathcal{N}$ be the last model of

$$(\mathsf{Le}((\mathcal{Q},\Lambda^{stc}),\mathcal{J}_\omega[\mathcal{Q}]))^{(M,\delta,\vec{G})}.$$

Because Σ is fullness preserving we have that $\mathcal{N} \models$ "$\delta^{\mathcal{Q}}$ is a Woodin cardinal". Let Φ be the strategy of $\mathcal{N}$ induced by Ω. We now start proving that $(\mathcal{Q},\Lambda)$ is as desired.

Clause 1 is just Lemma 8.1.10. We prove clause 2 of Theorem 8.1.14, which amounts to showing that the set $B = \{(x,y) : x \in \mathbb{R} \wedge y \notin \mathsf{Lp}^{\Lambda^{stc}}(x)\}$ as computed in W cannot be uniformized in $L(\Gamma(\mathcal{Q},\Lambda^{stc}))$. Towards a contradiction assume that B can be uniformized in $L(\Gamma(\mathcal{Q},\Lambda^{stc}))$. It follows that we can find a set of reals $A \in \Gamma(\mathcal{Q},\Lambda^{stc})$ such that A codes a sjs $(A_i : i < \omega)$ with the property that $A_0 = B$.

Let $\pi : \mathcal{N}|(\kappa^+) \to \mathcal{M}$ be an $\mathbb{R}$-genericity iteration. We then have that A is in the (new) derived model of $\mathcal{M}$. Fix then a $< \pi(\kappa)$-generic g over $\mathcal{M}$ such that there is a term relation $\tau \in \mathcal{M}[g]$ realizing A. Let $\delta < \pi(\kappa)$ be a cutpoint Woodin cardinal of $\mathcal{M}$ which is not a limit of Woodin cardinals of $\mathcal{M}$ and such that g is a $< \delta$-generic. Let $\xi < \delta$ be a cutpoint $\mathcal{M}$-cardinal such that $\mathcal{M}$ has no Woodin cardinals in the interval (ξ,δ). Let $\mathcal{M}^* \lhd \mathcal{M}$ be such that $\tau \in \mathcal{M}^*$, $\mathcal{M}^* \models \mathsf{ZFC} - \mathsf{Powerset}$ and $\mathcal{M}|\pi(\kappa) \trianglelefteq \mathcal{M}^*$. Let now $\sigma : \mathcal{S} \to \mathcal{M}^*$ be such that $\mathrm{crit}(\sigma) \in (\xi,\delta)$, $\sigma(\mathrm{crit}(\sigma)) = \delta$, $\mathrm{crit}(\sigma)$ is an $\mathcal{M}$-cardinal and $\tau \in \mathrm{rge}(\sigma)$. It follows that $\mathsf{Lp}^{\Lambda^{stc}}(\mathcal{M}|\mathrm{crit}(\sigma)) \in \mathcal{S}$ and $\mathsf{Lp}^{\Lambda^{stc}}(\mathcal{M}|\mathrm{crit}(\sigma)) \models$ "$\mathrm{crit}(\sigma)$ is a Woodin cardinal", contradiction! This finishes the proof of clause 2 of Theorem 8.1.14.

To finish the proof of Theorem 8.1.14 we need to show that $L(\Gamma(\mathcal{Q},\Lambda)) \models \mathsf{LSA}$. Suppose first that

(a) for every transitive $X \in \mathsf{HC}$ such that $\mathcal{Q} \in X$ and for every $\mathcal{R} \trianglelefteq \mathsf{Lp}^{\Lambda^{stc}}(X)$ such that $\rho(\mathcal{R}) = \mathsf{ord}(X)$, $\mathcal{R}$ has an iteration strategy $\Phi^* \in \Gamma^b(\mathcal{Q},\Lambda)$ such that Φ^* witnesses that $\mathcal{R}$ is a Λ^{stc}-sts mouse over X based on $\mathcal{Q}$.

We claim that (a) implies $L(\Gamma(\mathcal{Q},\Lambda)) \models \mathsf{LSA}$. Towards a contradiction assume not and set $B = \{(x,y) : x \in \mathbb{R} \wedge y \notin \mathsf{Lp}^{\Lambda^{stc}}(x)\}$. We claim that

(1) B is Suslin, co-Suslin in $L(\Gamma(\mathcal{Q},\Lambda))$.

Clearly (2) contradicts clause 2 of Theorem 8.1.14. Set $\Psi = \Lambda^{stc}$. It follows from (a) that

(2) B is projective in Ψ.

Let $\mathcal{Q}_\infty$ be the direct limit of all Λ-iterates of $\mathcal{Q}$ and let $\pi : \mathcal{Q} \to \mathcal{Q}_\infty$ be the iteration embedding. Notice that $\pi \restriction \mathcal{Q}^b$ depends only on Ψ and hence, because of Lemma 8.1.13, it is in $L(\Gamma(\mathcal{Q},\Lambda))$. Also, because Ψ is fullness preserving, it follows that $\pi[\mathcal{Q}^b]$ can be coded as a subset of $w(\Gamma^b(\mathcal{Q},\Lambda))$. This is because $\mathcal{Q}^b_\infty|\delta^{\mathcal{Q}^b_\infty} = \bigcup\{\mathcal{M}_\infty(\mathcal{R},\Lambda_\mathcal{R}) : \mathcal{R} \in pB(\mathcal{Q},\Lambda)\}$ and $\delta^{\mathcal{Q}^b} = w(\Gamma^b(\mathcal{Q},\Lambda))$.

It follows from (2) and Lemma 8.1.13 that $B \in \mathcal{J}_\omega(\pi[\mathcal{Q}^b],\mathcal{Q}^b,\Gamma^b(\mathcal{Q},\Lambda))$. Since we are assuming $L(\Gamma(\mathcal{Q},\Lambda)) \vDash \neg\mathsf{LSA}$ and since, in $L(\Gamma(\mathcal{Q},\Lambda))$, $\delta^{\mathcal{Q}^b_\infty}$ is both $< \Theta$ and is a limit of Suslin cardinals, B must be Suslin, co-Suslin in $L(\Gamma(\mathcal{Q},\Lambda))$, implying (1). Thus, it is enough to prove (a).

Suppose (a) fails. We can then assume that the witness is in some $Coll(\omega,\mathcal{Q})$-generic extension of M. Let $g \subseteq Coll(\omega,\mathcal{Q})$ be M-generic and let $X \in \mathsf{HC}^{M[g]}$ be a counterexample to (a). We then have that X is $< \kappa$-generic over $\mathcal{N}$. In fact, if $\eta \in (\mathsf{ord}(\mathcal{Q}),\kappa)$ is any Woodin cardinal of $\mathcal{N}$, then X can be added to $\mathcal{N}$ by the extender algebra of $\mathcal{N}$ at η. Let then $\mathcal{R} \trianglelefteq \mathsf{Lp}^{\Lambda^{stc}}(X)$ be the least such that $\rho(\mathcal{R}) = o(X)$ yet if Δ is the strategy of $\mathcal{R}$ witnessing that $\mathcal{R}$ is a Λ^{stc}-sts mouse over X based on $\mathcal{Q}$ then $\Delta \notin \Gamma^b(\mathcal{Q},\Lambda)$. Notice that we have that

(3) $\mathsf{Code}(\Delta)$ is Suslin, co-Suslin in $L(\Gamma(\mathcal{Q},\Lambda))$ (this follows from Corollary 8.1.9).

It is then enough to show that the Suslin, co-Suslin sets of $L(\Gamma(\mathcal{Q},\Lambda))$ are exactly those of $\Gamma^b(\mathcal{Q},\Lambda)$. Assume otherwise. Let $\mathcal{Q}_\infty = \mathcal{M}_\infty(\mathcal{Q},\Lambda)$. Because every set in $\Gamma^b(\mathcal{Q},\Lambda)$ is $\delta^{\mathcal{Q}^b_\infty}$-Suslin, co-Suslin we have that there is some $\eta < \kappa$ such that if $h \subseteq Coll(\omega,\eta)$ is $\mathcal{N}|\kappa$-generic then there is

$$(\mathcal{T},\mathcal{S}) \in I(\mathcal{Q},\Lambda^{stc}) \cap \mathsf{HC}^{\mathcal{N}|\kappa[h]}$$

such that $\Lambda_\mathcal{S} \in L(\Gamma(\mathcal{Q},\Lambda))$[30]. It then follows that

$$\Lambda_\mathcal{S} \restriction \mathcal{N}|\kappa[h] \in \mathcal{N}[h].^{31}$$

Let now $\nu > o(\mathcal{S})$ be a cutpoint Woodin cardinal of $\mathcal{N}|\kappa$. Let $\mathcal{S}_1$ be an iterate of $\mathcal{S}$ above $\delta^\mathcal{S}$ that is built according to the rules of $\mathcal{N}|\nu$-genericity iteration[32]. For this genericity iteration we use the extender algebra at $\delta^\mathcal{S}$ that uses extenders with critical points $> \delta^{\mathcal{S}^b}$. Thus, the $\mathcal{S}$-to-$\mathcal{S}_1$ iteration is above $\delta^{\mathcal{S}^b}$. We have that $\mathcal{S}_1 \in \mathcal{N}[h]|(\nu^+)^\mathcal{N}$. Let $\mathcal{N}_1$ be the output of $(\mathsf{Le}((\mathcal{Q},\Lambda^{stc}),\mathcal{J}_\omega[\mathcal{Q}]))^{(L[\mathcal{N}],\delta,\vec{G}')}$ where $\vec{G}'$ consists of those extenders of $\mathcal{N}$ that have an inaccessible length (in $\mathcal{N}$) and a critical point $> \nu^+$. It follows from fullness preservation that $\mathcal{N}_1 \vDash$ "$\delta^\mathcal{S}$ is a Woodin cardinal".

Let $\mathcal{N}_2$ be the $(\mathcal{N}_1,\nu,\pi^{\mathcal{T},b}[\mathcal{Q}^b])$-authenticated backgrounded construction over $\mathcal{N}|\nu$ based on $\mathcal{Q}$[33]. Then it follows from universality of $\mathcal{N}_2$ that $\mathcal{N}|(\nu^+)^\mathcal{N} \subseteq \mathcal{N}_2 \subseteq \mathcal{N}_1[\mathcal{N}|\nu]$. However, $\delta^{\mathcal{S}_1}$ is not a cardinal of $\mathcal{N}$ yet it is a cardinal of $\mathcal{N}_1[\mathcal{N}|\nu]$, contradiction! This finishes the proof of (a) and hence, the proof of Theorem 8.1.14. ⊣

---

[30]This can be shown using Theorem 4.13.2 and the fact that Λ^{stc} is Suslin, co-Suslin in $L(\Gamma(\mathcal{Q},\Lambda))$, which follows from our assumption and Lemma 8.1.13.

[31]$\Lambda_\mathcal{S} \restriction \mathcal{N}|\kappa[h] \in \mathcal{N}[h]$ because of Lemma 8.1.6.

[32]This iteration starts by iterating the least measurable of $\mathcal{S}$ that is $> \delta^{\mathcal{S}^b}$ $\nu+1$ times.

[33]This makes sense as $\mathcal{N}|\nu$ is generic over $\mathcal{S}_1$ and $\pi^{\mathcal{T},b} \in \mathcal{N}|\delta$, see Definition 6.2.2.

The next theorem can now be proved using Corollary 8.1.12 and the proof of Theorem 5.20 of [29].

THEOREM 8.1.15. *Assume* $\mathsf{AD}^+ + NsesS$. *Suppose* $(\mathcal{P},\Sigma)$ *is a hod pair such that* $\mathcal{P}$ *is either of successor type or of #-lsa type and* Σ *has branch condensation and is fullness preserving. Suppose* $B \in \mathbb{B}(\mathcal{P}^-,\Sigma_{\mathcal{P}^-})$. *There is then* $\mathcal{Q} \in pI(\mathcal{P},\Sigma)$ *and* $\vec{B} = \langle B_i : i < \omega\rangle \subseteq \mathbb{B}(\mathcal{P},\Sigma_{\mathcal{P}^-})$ *such that* $\vec{B}$ *strongly guides* $\Sigma_{\mathcal{Q}}$ *and* $B_0 = B$.

8.2. A hybrid upper bound for LSA

The main theorem of this section, Theorem 8.2.6, is a corollary to the proofs given in the previous section. It can be used in core model induction applications to show that certain hypotheses imply that there is a model of LSA. We give a fairly detailed proof of Theorem 8.2.6.

DEFINITION 8.2.1. Suppose $(\mathcal{P},\Sigma)$ is an sts hod pair[34]. We let $\mathcal{N}^{\#}_{\omega.2,lsa}(\mathcal{P},\Sigma)$ be the minimal active Σ-sts mouse $\mathcal{M}$ over $\mathcal{P}$ such that $\mathcal{M}$ has $\omega.2$ many Woodin cardinals greater than $\delta^{\mathcal{P}}$.

Recall Definition 2.3.15 and Definition 2.5.2. Suppose now that

$$\mathcal{M} = \mathcal{N}^{\#}_{\omega.2,lsa}(\mathcal{P},\Sigma).$$

Let λ be the supremum of the Woodin cardinals of $\mathcal{M}$. Because the only total extender of $\mathcal{M}$ whose critical point is $> \lambda$ is the last extender of $\mathcal{M}$, the strategy predicate above λ is empty. Thus, $\mathcal{M} = (\mathcal{M}|\lambda)^{\#}$. We use $\omega.2$ many Woodin cardinals because we need to produce proper initial segments of $\mathcal{M}$ that are unambiguous and satisfy the properties listed in clause 5 of Definition 3.8.9. Notice that the way we stated clause 5 of Definition 3.8.9 implies that the strategy predicate of $\mathcal{M}|\gamma$ cannot be empty above γ. We remark that we strongly believe that one could re-organize the book in a way that we could prove all the lemmas in this section for $\mathcal{N}^{\#}_{\omega,lsa}(\mathcal{P},\Sigma)$ which is the minimal active Σ-sts premouse over $\mathcal{P}$ that has ω Woodin cardinals above $\delta^{\mathcal{P}}$.

DEFINITION 8.2.2. We say $\mathcal{N}$ is *an active* $\omega.2$ *Woodin lsa mouse* if it has an iteration strategy Σ such that

1. $\mathcal{N}$ has a Woodin cardinal δ such that letting $\mathcal{P} = ((\mathcal{N}|\delta)^{\#})^{\mathcal{N}}$, $(\mathcal{P},\Sigma^{stc}_{\mathcal{P}})$ is an sts hod pair such that $\Sigma^{stc}_{\mathcal{P}}$ has strong branch condensation and is strongly $\Gamma^b(\mathcal{P},\Sigma^{stc}_{\mathcal{P}})$-fullness preserving,
2. $\mathcal{N} = \mathcal{N}^{\#}_{\omega.2,lsa}(\mathcal{P},\Sigma^{stc}_{\mathcal{P}})$,
3. for every $\mathcal{P}' \lhd_{hod} \mathcal{P}$ such that $\mathcal{P}'$ is of #-lsa type[35] layer of $\mathcal{P}$,
$$\mathcal{N}^{\#}_{\omega.2,lsa}(\mathcal{P}',\Sigma^{stc}_{\mathcal{P}'}) \trianglelefteq \mathcal{P}$$

[34] See Definition 3.10.6.

[35] See Definition 2.7.3. This means that $\mathcal{P}' = ((\mathcal{P}'|\delta^{\mathcal{P}'})^{\#})^{\mathcal{P}}$.

and

$$\mathcal{N}^{\#}_{\omega.2,lsa}(\mathcal{P}', \Sigma^{stc}_{\mathcal{P}'}) \models \text{“}\xi \text{ is not a Woodin cardinal”}.$$

We say $\mathcal{P}$ is the lsa part of $\mathcal{N}$. We say $(\mathcal{N}, \Sigma)$ is *an active $\omega.2$ Woodin lsa pair.*

Notice that if $(\mathcal{N}, \Sigma)$ is an active $\omega.2$ Woodin lsa pair then $\rho(\mathcal{N}) \leq (\kappa^+)^{\mathcal{N}}$ where, letting $\mathcal{P}$ be the lsa part of $\mathcal{N}$, κ is the least $<\delta^{\mathcal{P}}$-strong cardinal of $\mathcal{P}$[36].

In what follows, we let the statement *there is an active $\omega.2$ Woodin lsa pair* be shortening for the statement that there is a pair $(\mathcal{N}, \Sigma)$ such that $\mathcal{N}$ is an active ω Woodin lsa mouse and Σ witnesses the clauses of Definition 8.2.2.

Notice that it follows from Theorem 4.14.4 that if $(\mathcal{N}, \Sigma)$ and $(\mathcal{M}, \Lambda)$ are two active ω Woodin lsa pairs with common lsa part $\mathcal{P}$ such that $\Sigma^{stc} = \Lambda^{stc}$ then $\mathcal{N} = \mathcal{M}$ and $\Sigma = \Lambda$. Let $I = \omega.2 - \{\omega\}$.

LEMMA 8.2.3. *Suppose $(\bar{\mathcal{N}}, \Sigma)$ is an active $\omega.2$ Woodin lsa pair and $\mathcal{P}$ is the lsa part of $\bar{\mathcal{N}}$. Let $\mathcal{N}$ be the result of iterating the last extender of $\bar{\mathcal{N}}$ through the ordinals. Let $(\delta_i : i \in I)$ be the Woodin cardinals of $\mathcal{N}$ above $\delta^{\mathcal{P}}$ and let λ be their supremum. Let $\pi : \mathcal{N} \to \mathcal{M}$ be an iteration via Σ that is above $\delta^{\mathcal{P}}$. Suppose g is $< \pi(\lambda)$-generic over $\mathcal{M}$ and $\mathcal{W} \in (\mathcal{M}|\lambda[g]) \cap pB(\mathcal{P}, \Sigma^{stc})$[37]. Let $k \in \omega$ be such that g is generic for a poset in $\mathcal{M}|\pi(\delta_k)$ and let $\mathcal{S}^{\mathcal{M}}_k$ be the last model of the $(\mathcal{P}, \Sigma^{\mathcal{M}})$-coherent fully backgrounded construction of $\mathcal{M}|\pi(\delta_{k+1})$ using critical points $> \delta_k$[38]. Then the following holds:*

1. *Suppose $\mathcal{T} =_{def} \mathcal{T}^{\mathcal{M}}_k$ is the normal $\mathcal{P}$-to-$\mathcal{S}^{\mathcal{M}}_k$ stack. Then*
   (a) *$\mathrm{lh}(\mathcal{T})$ is a limit ordinal,*
   (b) *$\mathcal{T}$ is nuvs*[39]*,*
   (c) *$\pi^{\mathcal{T},b}$ exists,*
   (d) *$\pi^{\mathcal{T}b}(\mathcal{P}^b) = (\mathcal{S}^{\mathcal{M}}_k)^b$, and*
   (e) *$\mathcal{N}^{\#}_{\omega.2,lsa}(\mathrm{m}^+(\mathcal{T}), \Sigma^{stc}_{\mathrm{m}^+(\mathcal{T})}) \models$ “$\delta(\mathcal{T})$ is a Woodin cardinal”.*
2. *There is $\mathcal{U} \in \mathcal{M}[h]$ such that $\mathcal{U}$ is according to $\Sigma_{\mathcal{W}}$ and the last model of $\mathcal{U}$ is a layer of $(\mathcal{S}^{\mathcal{M}}_k)^b$.*

PROOF. To make the proof notationally more pleasant, we ignore π and assume $\mathcal{N} = \mathcal{M}$. The general case is very similar.

Clause 2 above follows from clause 1 and from the fact that Σ is positional[40] and that $(\mathcal{S}^{\mathcal{N}}_k)^b$-side does not move in the comparison of $\mathcal{W}$ and $(\mathcal{S}^{\mathcal{N}}_k)^b$. As proofs like this have appeared in the book many times before we omit most of it. The

[36] The fact that $\rho(\mathcal{N}) \leq (\kappa^+)^{\mathcal{N}}$ can be proved as follows. Suppose that $\rho(\mathcal{N}) > (\kappa^+)^{\mathcal{N}}$. Let $\mathcal{M} = Hull^{\mathcal{N}}((\kappa^+)^{\mathcal{N}})$. Clearly $\mathcal{M}$ is also an active ω Woodin lsa mouse. We would be done if we had $\mathcal{M} \trianglelefteq \mathcal{N}$. To show this, we use the proof of Theorem 4.11.8, and compare $(\mathcal{N}, \mathcal{M}, (\kappa^+)^{\mathcal{N}})$ with $\mathcal{N}$. We need to verify that a version of Lemma 4.11.6 holds for $(\mathcal{N}, \mathcal{M}, (\kappa^+)^{\mathcal{N}})$. However, this can be done via exactly the same proof. We leave the details to the reader.

[37] See Definition 3.3.9. Recall that pT is the projection of T.

[38] See Definition 3.5.1.

[39] See Definition 3.3.2.

[40] See Section 4.10.

exact procedure used to recover $\mathcal{U} \in \mathcal{N}[h]$ is the authentication process used to define sts mice[41].

Clause 1(a) and clause 1(b) follows from standard arguments. Clause 1(b) is a consequence of the fact that assuming $\mathcal{T}$ is a uvs, $(\mathcal{P},\mathcal{T})$ is an indexable stack and since $\mathcal{N}$ has more than $\delta_{\kappa+1}$ many inaccessible cardinals, $\mathcal{T} \in \mathrm{dom}(\Sigma^{\mathcal{N}})$ and hence, the construction producing $\mathcal{S}_k^{\mathcal{N}}$ can go further[42]. Clause 1(c) and 1(d) are straightforward consequences of clause 1(b)[43]. We verify clause 1(e).

Let $\mathcal{P}_1 = \mathrm{m}^+(\mathcal{T})$. Notice that $\mathcal{P}_1|\delta^{\mathcal{P}_1} = \mathcal{S}_k^{\mathcal{N}}$ and also $\mathcal{P}_1$ is a #-lsa type. We want to see that $\mathcal{N}^{\#}_{\omega.2,lsa}(\mathcal{P}_1,\Sigma^{stc}_{\mathcal{P}_1}) \models$ "$\delta^{\mathcal{P}_1}$ is a Woodin cardinal". Towards a contradiction suppose

(∗) $\mathcal{N}^{\#}_{\omega.2,lsa}(\mathcal{P}_1,\Sigma^{stc}_{\mathcal{P}_1}) \models$ "$\delta^{\mathcal{P}_1}$ is not a Woodin cardinal".

Let $b = \Sigma(\mathcal{T})$. (∗) then implies that $\mathcal{Q}(b,\mathcal{T})$ exists and is a $\Sigma^{stc}_{\mathcal{P}_1}$-sts mouse over $\mathcal{P}_1$. We now work towards showing that $\mathcal{N}$ has a branch indexed for $\mathcal{T}$, which is a contradiction as then the construction of $\mathcal{S}_k^{\mathcal{N}}$ can go further.

Working in $\mathcal{N}$, let Σ_1 be the $\mathcal{N}$-authenticated st-iteration strategy[44] of $\mathcal{P}_1$ and let $\mathcal{K}'$ be the output of the fully backgrounded construction of $\mathcal{N}|\lambda$ relative to Σ_1 done over $\mathcal{J}_\omega[\mathcal{P}_1]$ using extenders with critical point $> \delta_k$[45] and let $\mathcal{K} = \mathcal{J}[\mathcal{K}']$. Notice that $\Sigma_1 = \Sigma_{\mathcal{P}_1} \restriction \mathcal{N}|\lambda$[46].

CLAIM 1. $\mathcal{K}$ has $\omega.2$ Woodin cardinals. In fact, for every $k' > k$, $\delta_{k'}$ is a Woodin cardinal of $\mathcal{K}$.

PROOF. Suppose not. This means that the construction producing $\mathcal{K}$ does not reach λ. As iterability cannot be an issue (recall that $\mathcal{N}$ is iterable), the construction fails to reach λ because the construction reaches a model $\mathcal{K}^*$ such that there is an indexable stack $t = (\mathrm{m}^+(\mathcal{T}),\mathcal{T}_1,\mathcal{P}_2,\mathcal{U}) \in \mathcal{K}^*$ whose branch must be indexed but $t \notin \mathrm{dom}(\Sigma_1)$. Notice now that t cannot be nuvs as branches of such iterations are determined internally in $\mathcal{K}^*$[47]. Thus, t must be uvs. Notice, however, that because $\mathcal{P}_2^b \in pB(\mathcal{P}_1,\Sigma_{\mathcal{P}_1})$, we have that $\mathcal{P}_2$ is $(\mathcal{P},\Sigma^{\mathcal{N}})$-authenticated and so, we must have that $(\mathcal{P}_2^b,\mathcal{U})$ is an $(\mathcal{P},\Sigma^{\mathcal{N}})$-authenticated iteration. ⊣

Our goal now is to compare the construction producing $\mathcal{K}$ and $\mathcal{Q}(b,\mathcal{T})$. Let Ψ be the strategy of $\mathcal{Q}(b,\mathcal{T})$ witnessing that $\mathcal{Q}(b,\mathcal{T})$ is a $\Sigma^{stc}_{\mathcal{P}_1}$-sts mouse. Notice that we do not know that $\mathcal{Q}(b,\mathcal{T}) \in \mathcal{N}[h]$. The comparison that we use is the one used in [49].

---

[41] See Section 3.7 and also the proof of Sublemma 4.12.4.

[42] See Definition 3.3.3.

[43] See Lemma 2.7.25.

[44] See Definition 6.2.1.

[45] See Definition 4.2.1.

[46] See Theorem 6.1.4.

[47] Recall that there can be an issue here. It can be the case that the branch determined by $\mathcal{K}^*$ does not agree with the branch determined by Σ_1. To show this, we use an argument like the one used in the proof of Theorem 4.12.1.

CLAIM 2. The comparison of the construction producing $\mathcal{K}$ and $\mathcal{Q}(b,\mathcal{T})$ is successful.

PROOF. Towards a contradiction assume not. We can then find a normal tree $\mathcal{T}_1$ on $\mathcal{Q}(b,\mathcal{T})$ with last model $\mathcal{Q}_1$ and a normal tree $\mathcal{U}_1$ on $\mathcal{N}$ with last model $\mathcal{N}_1$ such that

1. $\mathcal{T}_1$ is according to Ψ,
2. $\mathcal{U}_1$ is according to Σ and has no drops,
3. for some $\beta \notin \text{dom}(\vec{E}^{\mathcal{Q}_1})$, letting $\mathcal{K}_1 = \pi^{\mathcal{U}_1}(\mathcal{K})$,
   (a) $\mathcal{Q}_1|\beta = \mathcal{K}_1|\beta$,
   (b) $\beta \notin \text{dom}(\vec{E}^{\mathcal{K}_1})$ and
   (c) $\mathcal{Q}_1||\beta \neq \mathcal{K}_1||\beta$.

Let then $t = (\mathcal{P}_1, \mathcal{W}, \mathcal{R}, \mathcal{W}_1) \in \mathcal{Q}_1|\beta$ be an indexable stack whose branch is indexed at β (either in $\mathcal{Q}_1$ or $\mathcal{K}_1$). As our indexing schema is local, it follows that a branch of t must be indexed at β in both $\mathcal{K}_1$ and $\mathcal{Q}_1$. Since both $\mathcal{Q}(b,\mathcal{T})$ and $\mathcal{K}_1$ are $\Sigma^{stc}_{\mathcal{P}_1}$-sts mice over $\mathcal{P}_1$, we have that $\mathcal{K}_1$ and $\mathcal{Q}_1$ cannot disagree on the branch of t. ⊣

Because $\mathcal{K}$ has $\omega.2$ Woodin cardinals and is a proper class model, it follows from Claim 2 and clause 3 of Definition 8.2.2 that $\mathcal{Q}(b,\mathcal{T}) \trianglelefteq \mathcal{K}$[48]. We thus have that $\mathcal{Q}(b,\mathcal{T}) \in \mathcal{N}$. It follows that to show that $\mathcal{N}$ has a branch indexed for $\mathcal{T}$, it is enough to show that clause 5 of Definition 3.8.9 holds for $\mathcal{W} =_{\text{def}} \mathcal{Q}(b,\mathcal{T})$. To do this, we need to show that

(a) there is $\mathcal{M} \trianglelefteq \mathcal{N}$ and a pair (β,γ) such that,
   1. $\beta < o(\mathcal{M})$ and $b \in \mathcal{M}|\beta$,
   2. $\mathcal{M}|\beta$ is unambiguous (see Definition 3.6.2) and $\mathcal{M}|\beta \models \mathsf{ZFC} +$ "there are infinitely many Woodin cardinals $> \delta(\mathcal{T})$",
   3. letting $(\eta_i : i < \omega)$ be the first ω Woodin cardinals $> \delta(\mathcal{T})$ of $\mathcal{M}|\beta$, $\mathcal{M}|\beta \models$ "$\mathcal{W}$ is $< Ord$-iterable above $\delta(\mathcal{T})$ via a strategy Φ such that letting $\nu = \sup_{i<\omega} \eta_i$, for every generic $g \subseteq Coll(\omega, <\nu)$, Φ has an extension $\Phi^+ \in D(\mathcal{M}|\beta,\nu,g)$ such that $D(\mathcal{M},\nu,g) \models$ "Φ^+ is an ω_1-iteration strategy" and whenever $\mathcal{R} \in D(\mathcal{M}|\beta,\nu,g)$ is a Φ^+-iterate of $\mathcal{W}$ and $t \in \mathcal{R}$ is an indexable stack on $\mathcal{P}_1$ then t is $(\mathcal{P},\Sigma^{\mathcal{M}})$-authenticated.

To show the existence of such an $\mathcal{M}$, it is enough to show that $\mathcal{N}|\delta_{\omega+1}$ satisfies clauses 1-3 and first two clauses are straightforward. Let $(\eta_i : i \in \omega)$ enumerate $(\delta_i : i \in (k+2,\omega))$ in increasing order. We show that $(\eta_i : i \subset \omega)$ witnesses clause 3 holds.

CLAIM 3. If $\mathcal{K}_1$ is the $(\mathcal{N},\delta_{k+1})$-authenticated[49] construction of $\mathcal{N}|\delta_{k+2}$ done over $\mathcal{J}_\omega[\mathcal{P}_1]$ based on $\mathcal{P}_1$ then $\text{ord}(\mathcal{K}_1) = \delta_{k+2}$ and $\mathcal{K}_1 \trianglelefteq \mathcal{K}$.

[48]This is because $\mathcal{Q}(b,\mathcal{T})$ is $\omega.2$ small.
[49]See Definition 6.2.2.

PROOF. Suppose not. It follows from the proof of Claim 2 that $\mathcal{K}_1$ has height δ_{k+2}. If $\mathcal{K}_1 \not\trianglelefteq \mathcal{K}$ then there is some model $\mathcal{Q}$ appearing in the construction producing $\mathcal{K}$ such that $\rho(\mathcal{Q}) < \delta_{k+2}$. Let p be the standard parameter of $\mathcal{Q}$. Let $X \prec \mathcal{Q}$ be such that $\rho(\mathcal{Q}) < X \cap \delta_{k+2} \in \delta_{k+2}$ and $X \cap \delta_{k+2}$ is a cardinal in $\mathcal{N}$[50] and $\bar{\mathcal{Q}}$ be the transitive collapse of X. By condensation (using the fact that X contains solidity witnesses for p), $\bar{\mathcal{Q}} \lhd \mathcal{Q}$. Since $\bar{\mathcal{Q}}$ is sound and $\rho(\bar{\mathcal{Q}}) = \rho(\mathcal{Q}) < X \cap \delta$, $X \cap \delta$ is not a cardinal in $\mathcal{N}$. Contradiction. ⊣

It follows from Claim 3 that $\mathcal{W} \trianglelefteq \mathcal{K}_1$. To complete the proof of Clause 3 of (a), it is now enough to show the following claim.

CLAIM 4. Suppose $\eta \in (\delta_{k+1}, \delta_{k+2})$ is an $\mathcal{N}$-cardinal and $g \subseteq Coll(\omega, (\eta^+)^{\mathcal{N}})$. Let Φ be the fragment of Σ that acts on non-dropping trees that are based on $\mathcal{N}|(\eta^+)^{\mathcal{N}}$ and are above δ_{k+1}. Then $\Phi \restriction \mathcal{N}|\lambda[g] \in \mathcal{N}|\lambda[g]$ and if $\Lambda = \Phi \restriction \mathsf{HC}^{\mathcal{N}|\lambda[g]}$ then in $\mathcal{N}[g]$, Λ is a $< \lambda$-universally Baire iteration strategy such that for any poset $\mathbb{P} \in \mathcal{N}|\lambda[g]$, if $k \subseteq \mathbb{P}$ is $\mathcal{N}[g]$-generic and Λ^k is the canonical extension of Λ to $\mathsf{HC}^{\mathcal{N}|\lambda[g*k]}$ then $\Lambda^k = \Phi \restriction \mathsf{HC}^{\mathcal{N}|\lambda[g*k]}$.

PROOF. We only prove that $\Phi \restriction \mathcal{N}|\lambda[g] \in \mathcal{N}|\lambda[g]$ and leave the rest to the reader. Let $\mathcal{Q} = \mathcal{N}|(\eta^+)^{\mathcal{N}}$ and let $\mathcal{W}_1 \in \mathcal{N}[g]$ be a tree on $\mathcal{Q}$ of limit length and according to Φ. Let $e = \Phi(\mathcal{W}_1)$. We want to show that $e \in \mathcal{N}[g]$ and $\mathcal{N}[g]$ has uniform way of identifying e. Notice that $\mathcal{Q}(e, \mathcal{W}_1)$ exists. Let $\mathcal{K}_2$ be the $\mathcal{N}$-authenticated background construction over $\mathcal{M}(\mathcal{W}_1)$. The proof of Claim 1 and Claim 2 show that $\mathcal{Q}(e, \mathcal{W}_1) \trianglelefteq \mathcal{K}_2$. It is now easy to find the uniform definition of e. The reader may wish to consult the proof of [33, Proposition 1.4]. ⊣

Claim 4 finishes the proof of Lemma 8.2.3. ⊣

COROLLARY 8.2.4. *Suppose $(\bar{\mathcal{N}}, \Sigma)$ is an active $\omega.2$ Woodin lsa pair and $\mathcal{P}$ is the lsa part of $\bar{\mathcal{N}}$. Let $\mathcal{N}$ be the result of iterating the last extender of $\bar{\mathcal{N}}$ through the ordinals. Let Φ be the fragment of Σ that acts on stacks above $\delta^{\mathcal{P}}$. Then Φ is $\Gamma^b(\mathcal{P}, \Sigma^{stc}_{\mathcal{P}})$-fullness preserving*[51].

PROOF. Given $\mathcal{S} \in pB(\mathcal{P}, \Sigma^{stc}_{\mathcal{P}})$, let $\pi : \mathcal{N} \to \mathcal{M}$ be a Σ-iterate of $\mathcal{N}$ above $\delta^{\mathcal{P}}$ such that $\mathcal{S}$ is generic over $\mathcal{M}$ for the extender algebra at the first Woodin of $\mathcal{M}$ that is larger than $\delta^{\mathcal{P}}$. It follows from Lemma 8.2.3 that $\mathcal{S}$ is $\mathcal{M}$-authenticated[52]. ⊣

LEMMA 8.2.5. *Suppose $(\bar{\mathcal{N}}, \Sigma)$ is an active $\omega.2$ Woodin lsa pair and $\mathcal{P}$ is the lsa part of $\bar{\mathcal{N}}$. Let $\mathcal{N}$ be the result of iterating the last extender of $\bar{\mathcal{N}}$ through the ordinals. Let $\delta < \eta$ be two consecutive Woodin cardinals of $\mathcal{N}$ such that $\delta > \delta^{\mathcal{P}}$. Let $\mathcal{N}^*$ be the output of $(\mathcal{N}, \delta)$-authenticated*[53] *background construction of $\mathcal{N}|\eta$ done over $\mathcal{J}_\omega[\mathcal{P}]$ based on $\mathcal{P}$. Then*

1. *$\mathcal{N}^*$ has height η and*

---

[50]This is possible because δ_{k+2} is strongly inaccessible in $\mathcal{N}$.

[51]See Definition 6.1.3.

[52]See Definition 6.2.1.

[53]See Definition 6.2.2.

2. *if $\mathcal{N}_1$ is the result of translating $\mathcal{N}$ onto a structure over $\mathcal{N}^*$ via S-constructions*[54] *then $\mathcal{N}_1$ is a normal iterate of $\mathcal{N}$ via a tree that is based on $\mathcal{N}|\delta_0$ where δ_0 is the least Woodin cardinal of $\mathcal{N}$ above $\delta^{\mathcal{P}}$.*

PROOF. We start by verifying clause 1. Suppose $\mathcal{N}^*$ fails to reach height η. This can only happen if at some stage of the construction we reach a model $\mathcal{M}$ such that there is some indexable stack $t = (\mathcal{P},\mathcal{T},\mathcal{P}_1,\vec{\mathcal{U}}) \in \mathcal{M}$ that is according to $\Sigma^{\mathcal{M}}$, it is required by the rules of the sts indexing scheme that we add a branch of t to $\mathcal{M}$ but t does not have an $\mathcal{N}$-authenticated branch. Notice that this can only happen when t is uvs but Lemma 8.2.3 implies that any uvs has an $\mathcal{N}$-authenticated branch.

We now verify clause 2. Notice that $\mathcal{N}_1[\mathcal{N}|\eta] = \mathcal{N}$. Thus $\mathcal{N}_1$ is η-sound $\omega.2$ Woodin mouse. It is then enough to show that there is a tree $\mathcal{U} \in \mathcal{N}$ on $\mathcal{N}|\delta_0$ such that $\mathrm{m}(\mathcal{U}) = \mathcal{N}^*$.

Suppose not. Let $\mathcal{U} \in \mathcal{N}$ be the normal stack on $\mathcal{N}|\delta_0$ that is built by comparing $\mathcal{N}|\delta_0$ with the construction producing $\mathcal{N}^*$. Since the aforementioned comparison fails, we must have that $\Sigma(\mathcal{U}) \notin \mathcal{N}$. Let $b = \Sigma(\mathcal{U})$. It follows from Lemma 6.4.4 that $\mathcal{Q}(b,\mathcal{U}) \in \mathcal{N}$[55]. Hence, $b \in \mathcal{N}$, contradiction.

We must have that $\mathcal{Q}(b,\mathcal{U})$ exists and $\mathcal{Q}(b,\mathcal{U}) \ntrianglelefteq \mathcal{N}^*$. It follows that $\mathcal{N}^* \models$ "$\delta(\mathcal{U})$ is a Woodin cardinal". Thus, in the further comparison of $\mathcal{Q}(b,\mathcal{U})$ and the construction producing $\mathcal{N}^*$, $\mathcal{N}^*$ side does not move. ⊣

THEOREM 8.2.6. *Suppose $(\bar{\mathcal{N}},\Sigma)$ is an active $\omega.2$ Woodin lsa pair and $\mathcal{P}$ is the lsa part of $\bar{\mathcal{N}}$. Let $\mathcal{N}$ be the result of iterating the last extender of $\bar{\mathcal{N}}$ through the ordinals and let $\Sigma^{>}$ be the strategy of $\mathcal{N}$ that acts on iterations above $\delta^{\mathcal{P}}$. Let λ be the supremum of the Woodin cardinals of $\mathcal{N}$ and let λ' be the supremum of the first ω Woodin cardinals of $\mathcal{N}$. Then whenever $g \subseteq Coll(\omega,\mathbb{R})$ is generic, $D^+(\mathcal{N},\Sigma^{>},\lambda',g) \models \mathsf{LSA}$*[56].

PROOF. Let $(\delta_i : i < \omega.2)$ be the Woodin cardinals of $\mathcal{N}$ and their limits that are greater than $\delta^{\mathcal{P}}$. It follows from Lemma 8.1.3 that $\mathcal{N}|\lambda$ is internally $\Sigma^{stc}_{\mathcal{P}}$-closed. It follows from Corollary 8.2.4 that $\Sigma^{>}$ is $\Gamma^b(\mathcal{P},\Sigma^{stc}_{\mathcal{P}})$-fullness preserving.

Suppose X is a transitive countable set such that $\mathcal{P} \in X$. Let for $i \in 2$, $\pi_i : \mathcal{N} \to \mathcal{M}_i$ be an iteration according to Σ such that $\mathrm{crit}(\pi_i) > \delta^{\mathcal{P}}$ and X is $< \pi(\lambda')$-generic over $\mathcal{M}_i$.

CLAIM 1. $\mathsf{Lp}^{\mathcal{M}_0,sts}(X,\mathcal{P}) = \mathsf{Lp}^{\mathcal{M}_1,sts}(X,\mathcal{P})$[57].

PROOF. Let $\mathcal{K}_0$ be the $\mathcal{M}_0$-authenticated background construction over X based on $\mathcal{P}$ and $\mathcal{K}_1$ be the $\mathcal{M}_1$-authenticated background construction over X based on $\mathcal{P}$. We compare the construction producing $\mathcal{K}_0$ with the one producing $\mathcal{K}_1$. Notice that

[54] See Definition 6.4.3.

[55] $\mathcal{Q}(b,\mathcal{U})$ can be obtained via an S-construction, translating $\mathcal{N}$ to an sts mouse over $\mathrm{m}(\mathcal{U})$.

[56] See Definition 8.1.10.

[57] See Definition 6.2.3 and the discussion following it.

it follows from the proof of Claim 1 of Lemma 8.2.3 that both constructions reach proper class models. It then follows from the proof of Claim 2 of Lemma 8.2.3 that the aforementioned comparison produces $\sigma_0 : \mathcal{M}_0 \to \mathcal{M}_2$ and $\sigma_1 : \mathcal{M}_1 \to \mathcal{M}_3$ such that $\mathsf{crit}(\sigma_i) > \mathsf{ord}(X)$ and $\sigma_0(\mathcal{K}_0)$ and $\sigma_1(\mathcal{K}_1)$ are lined up (i.e., one is an initial segment of the other). Because they both have exactly $\omega.2$ Woodin cardinals it follows from our minimality assumption on $\mathcal{N}$ that $\sigma_0(\mathcal{K}_0) = \sigma_1(\mathcal{K}_1)$. The claim now follows. ⊣

Given a transitive $X \in \mathsf{HC}$, we let $\mathcal{W}(X) = \mathsf{Lp}^{\mathcal{M},sts}(X,\mathcal{P})$ where $\mathcal{M}$ is such that there is an iteration $\pi : \mathcal{N} \to \mathcal{M}$ according to Σ such that $\mathsf{crit}(\pi) > \delta^{\mathcal{P}}$ and X is $< \pi(\lambda')$-generic over $\mathcal{M}$. Suppose now that $\mathcal{S}' \in pI(\mathcal{P},\Sigma)$ and $\mathcal{S}$ is a #-lsa type[58] proper layer of $\mathcal{S}'$. Let $\eta = \delta^{\mathcal{S}}$. We then claim that

CLAIM 2. $\mathcal{W}(\mathcal{S}) \vDash$ "η is not a Woodin cardinal".

PROOF. Suppose otherwise. Notice that $\mathcal{S}' \vDash$ "η is not a Woodin cardinal". Let $\mathcal{Q} \trianglelefteq \mathcal{S}'$ be the longest initial segment $\mathcal{Q}^*$ of $\mathcal{S}'$ such that $\mathcal{Q}^* \vDash$ "η is a Woodin cardinal". Then $\mathcal{Q}$ is a $\Sigma^{stc}_{\mathcal{S}}$-sts mouse. Let now $\pi : \mathcal{N} \to \mathcal{M}$ be an iteration according to $\Sigma^{>}$ such that $\mathcal{S}'$ is $< \pi(\lambda')$-generic over $\mathcal{M}$. Let $\mathcal{K}$ be the $\mathcal{M}$-authenticated background construction done over $\mathcal{J}_\omega[\mathcal{S}]$ based on $\mathcal{S}$. Because we are assuming that the claim fails, we must have that $\mathcal{K} \vDash$ "η is a Woodin cardinal".

We now compare $\mathcal{Q}$ with the construction of $\mathcal{M}$ producing $\mathcal{K}$. Notice that this comparison halts (this follows from the proof of Claim 2 that appears in the proof of Lemma 8.2.3). Now, $\mathcal{Q}$ has to win this comparison. Since $\mathcal{K}$ is proper class and has $\omega.2$ Woodin cardinals, the fact that $\mathcal{Q}$ wins contradicts the minimality assumption on $\mathcal{N}$ (more precisely, contradicts clause 3 of Definition 8.2.2). ⊣

Suppose next that $\mathcal{S} \in pI(\mathcal{P},\Sigma)$ and $\eta = \delta^{\mathcal{S}}$. We then have that

CLAIM 3. $\mathcal{W}(\mathcal{S}) \vDash$ "η is a Woodin cardinal".

PROOF. Let $\sigma : \mathcal{N} \to \mathcal{S}^+$ be the result of applying the iteration producing $\mathcal{S}$ to the entire model $\mathcal{N}$. Thus $\mathcal{S}$ is the lsa part of $\mathcal{S}^+$. Let now $\pi : \mathcal{N} \to \mathcal{M}$ be an iteration according to Σ above $\delta^{\mathcal{P}}$ such that $\mathcal{S}$ is $< \pi(\lambda)$-generic over $\mathcal{M}$. Let $\mathcal{K}$ be the $\mathcal{M}$-authenticated background construction done over $\mathcal{J}_\omega[\mathcal{S}]$ based on $\mathcal{S}$. We now compare the construction producing $\mathcal{K}$ with $\mathcal{S}^+$. As before this construction has to halt. It then follows from our minimality condition on $\mathcal{N}$ that $\mathcal{W}(\mathcal{S}) \vDash$ "η is a Woodin cardinal". ⊣

The next claim computes the powerset of the Woodin cardinals of $\mathcal{N}$. The proof is very similar to the proof of Claim 3 and we omit it.

CLAIM 4. Let $\pi : \mathcal{N} \to \mathcal{M}$ be an iteration according to Σ above $\delta^{\mathcal{P}}$. Then for any $k < \omega$, $\mathcal{M}|(\delta_k^+)^{\mathcal{M}} = \mathcal{W}(\mathcal{M}|\delta_k)$.

The next claim can be proved using the proof of Claim 3 and the proof of Lemma 8.1.9. Also see the proof of Claim 4 of Lemma 8.2.3.

---

[58] See Definition 2.7.3.

CLAIM 5. Suppose $X \in HC$ is a transitive set and $\mathcal{R} \trianglelefteq \mathcal{W}(X)$ is such that $\rho(\mathcal{R}) = o(X)$. Let $\pi : \mathcal{N} \to \mathcal{M}$ be an iteration according to Σ above $\delta^{\mathcal{P}}$ such that X is $< \pi(\lambda')$-generic over $\mathcal{M}$. Let $k < \omega$ be such that for some $g \subseteq Coll(\omega, < \pi(\delta_k))$, $X \in \mathsf{HC}^{\mathcal{M}|\pi(\delta_k)[g]}$. Then $\mathcal{R}$ has a $< \pi(\lambda')$-universally Baire iteration strategy in $\mathcal{M}[g]$.

Suppose $g \subseteq Coll(\omega, \mathbb{R})$ generic. Let $(x_i : i < \omega)$ be an enumeration of $\mathbb{R}$ in $V[g]$. Let $\pi : \mathcal{N} \to \mathcal{M}$ be $\mathbb{R}$-genericity iteration according to Σ that is below λ' and is guided by $(x_i : i < \omega)$. The next claim is a corollary to Claim 5 and clause 2 of Theorem 6.1.4.

CLAIM 6. Set $B = \{(x,y) \in \mathbb{R}^2 : y \notin \mathcal{W}(x)\}$. Then $B \in \mathcal{M}(\mathbb{R})$ and $\Sigma^{stc}_{\mathcal{P}} \in \mathcal{M}(\mathbb{R})$.

Let $\mathcal{P}_\infty$ be the direct limit of all $\Psi =_{\text{def}} \Sigma_{\mathcal{P}}$-iterates of $\mathcal{P}$ and let $\pi : \mathcal{P} \to \mathcal{P}_\infty$ be the iteration embedding. Notice that $\pi \restriction \mathcal{P}^b$ depends only on Ψ. Also, because Ψ is strongly $\Gamma^b(\mathcal{P}, \Psi)$-fullness preserving, it follows that $\pi[\mathcal{P}^b]$ can be coded as a subset of $w(\Gamma^b(\mathcal{P}, \Sigma^{stc}))$[59]. This is because $\mathcal{P}^b_\infty | \delta^{\mathcal{P}^b_\infty} = \bigcup\{\mathcal{M}_\infty(\mathcal{R}, \Lambda_{\mathcal{R}}) : \mathcal{R} \in pB(\mathcal{P}, \Sigma^{stc})\}$ and $\delta^{\mathcal{P}^b} = w(\Gamma^b(\mathcal{P}, \Sigma^{stc}))$. It follows from Lemma 8.1.13 that

CLAIM 7. $\Psi \in \mathcal{J}_\omega(\mathcal{P}^b_\infty, \pi[\mathcal{P}^b], \Gamma^b(\mathcal{P}, \Psi))$.

Next we establish a crucial claim.

CLAIM 8. $\mathcal{J}(\mathcal{P}^b_\infty, \pi[\mathcal{P}^b], \Gamma^b(\mathcal{P}, \Psi)) \vDash \mathsf{AD}^+$.

PROOF. Suppose not. Let $A \in \mathcal{J}(\mathcal{P}^b_\infty, \pi[\mathcal{P}^b], \Gamma^b(\mathcal{P}, \Psi))$ be a set of reals that is not determined. Let $X = \pi[\mathcal{P}^b]$. Fix $x \in \mathbb{R}$ and $\mathcal{Q} \in pB(\mathcal{P}, \Psi)$ such that A is definable from $(X, x, (\mathcal{Q}, \Psi_{\mathcal{Q}}), \mathcal{P}^b_\infty)$ and a finite sequence of ordinals over $\mathcal{J}(\mathcal{P}^b_\infty, \pi[\mathcal{P}^b], \Gamma^b(\mathcal{P}, \Psi))$. By minimizing the sequence of ordinals we can suppose that A is definable without ordinal parameters.

Let $(\mathcal{M}_i, \mathcal{T}_i : i < \omega)$ be the $\mathbb{R}$-genericity iteration of $\mathcal{N}$ relative to a generic enumeration $(x_i : i < \omega)$ of $\mathbb{R}$ (this iteration is according to $\Sigma^>$, is below λ' (the sup of the first ω-Woodins of $\mathcal{N}$ and is above $\delta^{\mathcal{P}}$). For $i < \omega$ let $\pi_i = \pi^{\oplus_{j \leq i} \mathcal{T}_j}$ and for $i < j \leq \omega$ let $\pi_{i,j} : \mathcal{M}_i \to \mathcal{M}_j$ be the composition of iteration embeddings. We then have that $A \in \mathcal{M}(\mathbb{R})$, where $\mathcal{M}$ is the direct limit of $\mathcal{M}_i$'s under the embeddings $\pi_{i,j}$.

Let i be large enough so that $x, \mathcal{Q} \in \mathsf{HC}^{\mathcal{M}_i[(x_j : j \leq i)]}$ and $\Sigma_{\mathcal{Q}} \restriction \mathsf{HC}^{\mathcal{M}_i[(x_j : j \leq i)]}$ is $< \pi_i(\lambda')$-universally Baire. Let $\tau \in \mathcal{M}_i[(x_j : j \leq i)]$ be a name such that $\pi_{i,\omega}(\tau)$ is a term relation for A. Let $\eta = \pi_i(\delta_{i+1})$. We claim that if

$$\mathcal{R} = (\mathcal{M}_i | (\eta^+)^{\mathcal{M}}))[(x_j : j \leq i)]$$

then letting Φ be the fragment of $\Sigma_{\mathcal{M}_i}$ that acts on trees based on $\mathcal{R}$ that are above $\pi_i(\delta_i)$, $(\mathcal{R}, \Phi, \tau_{\mathcal{R}})$ term captures A where $(p, u) \in \tau_{\mathcal{R}}$ if and only if $p \in Coll(\omega, \eta)$,

---

[59] $\delta^{\mathcal{P}^b_\infty}$ is the largest cardinal of $\mathcal{P}^b_\infty$ and $\pi[\mathcal{P}^b]$ is cofinal in $\mathcal{P}^b_\infty$. Thus, we have $A \subseteq \delta^{\mathcal{P}^b_\infty}$ which codes the pair $(\mathcal{P}^b, \pi[\mathcal{P}^b])$.

$u \in \mathcal{R}^{Coll(\omega,\eta)}$ and $p \Vdash$ "$\Vdash_{Coll(\omega,<\pi_i(\lambda'))} u \in \tau$". It then follows from a result of Neeman that A is determined (see [25]).

Let then $\mathcal{T}$ be an iteration tree on $\mathcal{M}_i$ based on $\mathcal{R}$ according to Φ. Let $\mathcal{S}$ be the last model of $\mathcal{T}$. We want to see that

(a) if $h \subseteq Coll(\omega, \pi^{\mathcal{T}}(\eta))$ is $\mathcal{S}$-generic then $(\pi^{\mathcal{T}}(\tau_{\mathcal{R}}))_h = A \cap \mathcal{S}[h]$.

Let $k > i$ be large enough that $\mathcal{S} \in \mathcal{M}_k[(x_j : j \leq k)]$. Let $\mathcal{S}^*$ be the output of $\mathcal{M}_k|\pi_k(\delta_{k+1})$-authenticated backgrounded construction done over $\mathcal{S}|\pi^{\mathcal{T}}(\eta)$ based on $\mathcal{P}$. We then have that $\mathcal{S}^*$ is an iterate of $\mathcal{S}|\pi^{\mathcal{T}}(\pi_i(\delta_{i+2}))$[60]. Let $\mathcal{S}^{**} = \pi_{k,k+1}(\mathcal{S}^*)$. Finally, let $\mathcal{S}_1$ be the result of translating $\mathcal{M}_{k+1}$ over $\mathcal{S}^{**}$ via S-constructions. We then have that

(1) $\mathcal{S}_1[\mathcal{M}_{k+1}|\pi_{k+1}(\delta_{k+1})] = \mathcal{M}_{k+1}$
(2) $\mathcal{S}_1$ is an iterate of $\mathcal{S}$ such that if $\nu : \mathcal{S} \to \mathcal{S}_1$ is the iteration embedding then $\text{crit}(\nu) > \pi^{\mathcal{T}}(\eta)$.

(2) is a consequence of the fact that $\mathcal{S}^{**} = \text{m}(\mathcal{U}')$ where $\mathcal{U}' = \pi_{k,k+1}(\mathcal{U})$ and $\mathcal{U}$ is as in the footnote above. It then follows that if b is the branch of $\mathcal{U}'$ given by $\Sigma_{\mathcal{M}_i}$ then $\mathcal{M}_b^{\mathcal{U}}$ is $\pi_k(\delta_{k+1})$-sound.

It follows that we can think of $p = (\mathcal{T}_j : j \in (k+1, \omega))$ as an $\mathbb{R}$-genericity iteration on $\mathcal{S}_1$ guided by $(x_j : j \in (k+1, \omega))$. Let then $\mathcal{S}_2$ be the last model of this genericity iteration and let $m : \mathcal{N} \to \mathcal{S}_2$ be the iteration embedding. More precisely, $m = \pi^p \circ \nu \circ \pi^{\mathcal{T}} \circ \pi_i$.

Because $\mathcal{M}|\pi_{k+1}(\delta_{k+1}) = \mathcal{M}_{k+1}|\pi_{k+1}(\delta_{k+1})$, we have that

$$\mathcal{S}_2[\mathcal{M}|\pi_{k+1}(\delta_{k+1})] = \mathcal{M}.$$

Let $\sigma : \mathcal{M}_i \to \mathcal{S}_2$ be the iteration embedding. It then follows that in $\mathcal{S}_2[(x_j : j \leq i)]$, $\sigma(\tau)$ is the term relation that is forced by $Coll(\omega, < m(\lambda'))$ to be the least set in $\mathcal{J}(\mathcal{P}_\infty^b, \pi[\mathcal{P}^b], \Gamma^b(\mathcal{P}, \Psi))$ which is not determined and is definable from $(X, x, (\mathcal{Q}, \Psi_{\mathcal{Q}}), \mathcal{P}_\infty^b)$. It then follows that

(3) $\sigma(\tau)$ is realized as A.

(a) now follows from (2) and the fact that $\text{crit}(\pi^p) > \pi^{\mathcal{T}}(\eta)$. $\dashv$

The proof of the next claim is exactly like the proof of (a) that appeared in the proof of Theorem 8.1.14 and Lemma 8.2.3. We leave it to the reader.

CLAIM 9. For any transitive $X \in \mathsf{HC}$ such that $\mathcal{P} \in X$ and for any $\mathcal{R} \trianglelefteq \mathcal{W}(X)$ such that $\rho(\mathcal{R}) = o(X)$, $\mathcal{R}$ has an iteration strategy in $\Gamma^b(\mathcal{P}, \Psi)$.

It follows from Claim 9 that the set $B = \{(x,y) \in \mathbb{R}^2 : y \notin \mathcal{W}(x)\}$ is projective in Ψ and hence, $B \in \mathcal{J}(\mathcal{P}_\infty^b, \pi[\mathcal{P}^b], \Gamma^b(\mathcal{P}, \Sigma))$. It follows from Claim 9 that $\mathcal{J}(B) \vDash \mathsf{AD}^+$. We now have the following:

CLAIM 10. In $\mathcal{M}(\mathbb{R})$, let $\Gamma = \{A \subseteq \mathbb{R} : L(A, \mathbb{R}) \vDash \mathsf{AD}^+\}$. Then $\Psi, B \in L(\Gamma, \mathbb{R})$.

---

[60] See Lemma 8.2.5. More precisely, there is a normal stack $\mathcal{U}$ on $\mathcal{S}|\pi^{\mathcal{T}}(\pi_i(\delta_{i+2}))$ that is above $\pi^{\mathcal{T}}(\eta)$, $\text{lh}(\mathcal{U})$ is a limit ordinal and $\text{m}(\mathcal{U}) = \mathcal{S}^*$.

It follows from the proof of clause 2 of Theorem 8.1.14 that B cannot be uniformized in $L(\Gamma, \mathbb{R})$. Hence, $L(\Gamma, \mathbb{R}) \models \mathsf{LSA}$. $\dashv$

Chapter 9

CONDENSING SETS

The goal of this chapter is to introduce the theory of condensing sets. Such sets were first considered in [31, Section 10, 11.1], where they were presented in the form of a condensation property for elementary embeddings (see [31, Definition 11.14]). The current presentation dates back to an unpublished note by the first author.

Prior to this work, condensing sets have been used in the context of the core model induction. As a convenience to the reader, we recap some of the basic machinery used in the core model induction. We model our presentation on [31] but we will also use the set up of [65]. A typical situation is as follows. We have an embedding $j : M \to N$ with critical point κ and such that $H^M_{\kappa^+} = H^N_{\kappa^+}$. In M, we consider the maximal model of determinacy that has been built via core model induction. While the exact definition of the maximal model is somewhat case specific, it can be essentially described as follows.

Let $g \subseteq Coll(\omega, < j(\kappa))$ be N-generic. For $\nu < j(\kappa)$ let $g_\nu = g \cap Coll(\omega, < \nu)$. We then can extend j to act on $M[g_\kappa]$. We denote this extension by j again and we have that $j : M[g_\kappa] \to N[g]$.

Consider the set of hod pairs $(\mathcal{Q}, \Lambda)$ such that

1. $\mathcal{Q} \in \mathsf{HC}^{M[g_\kappa]}$,
2. for some $\nu < \kappa$ such that $\mathcal{Q} \in M[g_\nu]$, letting $\Psi = \Lambda \restriction \mathsf{HC}^{M[g_\nu]}$, $\Psi \in M[g_\nu]$ and $M[g_\nu] \models$ "$\mathsf{Code}(\Psi)$ is κ-uB" and
3. if $T, S \in M[g_\nu]$ witness that $\mathsf{Code}(\Psi)$ is κ-uB then $\mathsf{Code}(\Lambda) = p[T]^{M[g_\kappa]}$.

Let Γ be the set of such pairs $(\mathcal{Q}, \Lambda)$. An additional requirement is that Λ is fullness preserving and has branch condensation. While the branch condensation is the same as before, fullness preservation is not the same as the definition given in earlier chapters. We refer the interested reader to [31] for more details on how to define Γ. It is in fact somewhat more involved.

The goal of a core model induction is to show that Γ is rich. This is done as follows. First a target theory is fixed. The theory used in [31] is "$\mathsf{AD}_{\mathbb{R}}$ + "Θ is regular". In Chapter 12, our target is LSA. Suppose then there is no lsa type hod pair $(\mathcal{Q}, \Lambda) \in \Gamma$. Preliminary arguments, such as those used in [35, Theorem 4.1], show that Γ is of limit type, i.e., for any $(\mathcal{Q}, \Lambda) \in \Gamma$ there is $(\mathcal{R}, \Psi) \in \Gamma$ such that $\Gamma(\mathcal{Q}, \Lambda) \subset \Gamma(\mathcal{R}, \Psi)$.

Next we let $\mathcal{P}^- = \bigcup_{(\mathcal{Q},\Lambda)\in\Gamma} \mathcal{M}_\infty(\mathcal{Q},\Lambda)$. Fixing a complete layer[1] $\mathcal{R}$ of $\mathcal{P}^-$ and $(\mathcal{Q},\Lambda)\in\Gamma$ such that $\mathcal{R} = \mathcal{M}_\infty(\mathcal{Q},\Lambda)$, we let $\Sigma_\mathcal{R} = \Lambda_\mathcal{R}$. It follows from comparison that Σ_R is independent of $(\mathcal{Q},\Lambda)$. Let $\Sigma = \oplus_\mathcal{R}\Sigma_\mathcal{R}$ where the joint ranges over the complete layers of $\mathcal{P}^-$.

We now define $\mathcal{P}$ as follows. Suppose next that there is $\mathcal{M} \trianglelefteq \mathsf{Lp}^\Sigma(\mathcal{P}^-)$ such that $\rho(\mathcal{M}) < \mathsf{ord}(\mathcal{P}^-)$. We then let $\mathcal{P}$ be the least such $\mathcal{M}$. Otherwise we let $\mathcal{P} = \mathsf{Lp}^\Sigma(\mathcal{P}^-)$.

The next major step is to build an iteration strategy for $\mathcal{P}$ that extends Σ. We let Σ^+ be this new strategy. Σ^+ is constructed as follows.

DEFINITION 9.0.7 (The construction of the strategy). Suppose $\mathcal{T} \in \mathsf{HC}^{N[g]}$ is a stack on $\mathcal{P}$ where

$$\mathcal{T} = ((\mathcal{M}_\alpha)_{\alpha<\eta}, (E_\alpha)_{\alpha<\eta-1}, D, R, (\beta_\alpha, m_\alpha)_{\alpha\in R}, T).$$

Recall Notation 2.4.4. Suppose $j \restriction \mathcal{P} \in N[g]$. Working in $N[g]$, we say $\mathcal{T}$ is *j-realizable* if there is a sequence $(\sigma_\alpha : \alpha \in R)$ such that the following clauses hold[2]:

1. $\mathcal{T}$ does not have a fatal drop[3],
2. $\sigma_\alpha : \mathcal{M}_\alpha \to j(\mathcal{P})$ is an elementary embedding.
3. For all $\alpha, \alpha' \in R$ with $\alpha < \alpha'$, $\sigma_\alpha = \sigma_{\alpha'} \circ \pi^{\mathcal{T}}_{\alpha,\alpha'}$.
4. For all $\alpha \in R$, letting $\Lambda_\alpha = (\sigma_\alpha \restriction \mathcal{M}_\alpha|\delta^{\mathcal{M}_\alpha}$-pullback of $j(\Sigma))$, for each complete layer $\mathcal{R} \triangleleft \mathcal{M}_\alpha$, $\sigma_\alpha \restriction \mathcal{R} = \pi^{\Lambda_\alpha}_{\mathcal{R},\infty}$ where $\pi^{\Lambda_\alpha}_{\mathcal{R},\infty} : \mathcal{R} \to \mathcal{M}_\infty(\mathcal{R}, (\Lambda_\alpha)_\mathcal{R})$ is the iteration map according to $(\Lambda_\alpha)_\mathcal{R}$[4].
5. For all $\alpha \in R$ such that $\alpha \neq \max(R)$, letting $\alpha' = \min(R - (\alpha+1))$, $\mathcal{T}_{\alpha,\alpha'}$ is according to Λ_α[5].

Given a stack $\mathcal{T} \in \mathsf{HC}^{N[g]}$, we set $\mathcal{T} \in \mathrm{dom}(\Sigma^+)$ if $\mathcal{T}$ is j-realizable. For $\mathcal{T} \in \mathrm{dom}(\Sigma)$, we set $\Sigma^+(\mathcal{T}) = b$ if $\mathcal{T}^\frown\{\mathcal{M}^\mathcal{T}_b\}$ is j-realizable.

It is not hard to extend Σ^+ to act on all stacks, not just those without fatal drops. Σ^+ may not be a total strategy simply because we may not be able to satisfy clauses 4 and 5 of Definition 9.0.7. Moreover, it may also depend on the realization maps. However, the proof of [31, Lemma 11.6] gives the following.

THEOREM 9.0.8. *Suppose $|\mathcal{P}| < (\kappa^+)^M$. Then $j \restriction \mathcal{P} \in N[g]$ and Σ^+ is a total (ω_1, ω_1)-strategy in $N[g]$.*

Then there are two arguments that we run as part of the proof of Theorem 9.0.8. First we show that $\mathcal{P} = \mathsf{Lp}^\Sigma_\omega(\mathcal{P}^-)$. The reader can see, for example [65, Lemma 3.78], for an argument. Roughly, if not, suppose n is such that $\rho_{n+1}(\mathcal{P}) < \delta^\mathcal{P} \leq$

[1]See Definition 2.7.14.

[2]For the definition of $\pi^{\mathcal{T},b}_{\alpha,\alpha'}$, see Section 2.8.

[3]See Definition 2.6.8.

[4]This condition assumes that Λ_α is fullness preserving.

[5]Notice that because we are assuming $\mathcal{T}$ does not have fatal drops, $\mathcal{T}_{\alpha,\alpha'}$ is based on $\mathcal{M}_\alpha|\delta^{\mathcal{M}_\alpha}$.

$\rho_n(\mathcal{P})$, then in $j(\Gamma)$, we can find a complete layer $\mathcal{R}$ of $\mathcal{P}$ and an $OD^{j(\Gamma)}_{\Sigma_{\mathcal{R}}}$ set $A \subseteq \delta^{\mathcal{R}}$ such that $A \notin \mathcal{P}$. By fullness of $\mathcal{P}$ and SMC in $j(\Gamma)$, $A \in \mathcal{P}$. Contradiction.

The next argument attempts to show that $\mathcal{P} \models$ "$\delta^{\mathcal{P}}$ is regular". Showing this finishes the proof of the main theorem of [31]. In this book we present an argument for obtaining a model of LSA from PFA (see Theorem 12.0.21). To prove Theorem 12.0.21, we need to do more in order to finish the argument. It is in this step that the theory of condensing sets is used. A reader interested in more details may consult [31, Section 10, 11.1] and [65, Lemma 3.81].

9.1. Condensing sets

We introduce the notion of condensing set in the most general setting. Suppose φ is a formula in the language of set theory and A is a set. We let $\mathcal{F}_{\varphi,A}$ be a collection of hod pairs $(\mathcal{Q},\Lambda)$ such that $\mathcal{Q}$ is countable, Λ is an $(\omega_2,\omega_2,\omega_2)$-iteration strategy having strong branch condensation and such that $\varphi[A,(\mathcal{Q},\Lambda)]$ holds.

TERMINOLOGY 9.1.1.
1. We say (φ,A) is *bottom part closed* if whenever $(\mathcal{Q},\Lambda) \in \mathcal{F}_{\varphi,A}$ and $\mathcal{R} \in pB(\mathcal{Q},\Lambda)$ then $(\mathcal{R},\Lambda_{\mathcal{R}}) \in \mathcal{F}_{\varphi,A}$.
2. We say (φ,A) is of *limit type* if for every $(\mathcal{Q},\Lambda) \in \mathcal{F}_{\varphi,A}$, there is $(\mathcal{R},\Psi) \in \mathcal{F}_{\varphi,A}$ such that $\mathcal{R}$ is of limit type and $\mathsf{Code}(\Lambda) \in \Gamma^b(\mathcal{R},\Psi)$.
3. Let $\Gamma_{\varphi,A} = \bigcup\{\Gamma(\mathcal{R},\Psi) : (\mathcal{R},\Psi) \in \mathcal{F}_{\varphi,A} \wedge \mathcal{R}$ is of limit type$\}$. We say (φ,A) is *stable* if whenever $(\mathcal{R},\Psi) \in \mathcal{F}_{\varphi,A}$, Ψ is strongly $\Gamma_{\varphi,A}$-fullness preserving.
4. We say (φ,A) is *directed* if whenever $(\mathcal{Q},\Lambda),(\mathcal{P},\Sigma) \in \mathcal{F}_{\varphi,A}$, there are $\mathcal{R} \in pI(\mathcal{Q},\Lambda)$ and $\mathcal{S} \in pI(\mathcal{P},\Sigma)$ such that either
   (a) $\mathcal{R} \trianglelefteq_{hod} \mathcal{S}$ and $\Sigma_{\mathcal{R}} = \Lambda_{\mathcal{R}}$ or
   (b) $\mathcal{S} \trianglelefteq_{hod} \mathcal{R}$ and $\Lambda_{\mathcal{S}} = \Sigma_{\mathcal{S}}$.

NOTATION 9.1.2. Given a hod premouse $\mathcal{P}$, we write $\mathcal{R} \trianglelefteq^c_{hod} \mathcal{P}$ if $\mathcal{R}$ is a complete[6] layer of $\mathcal{P}$.

NOTATION 9.1.3. Suppose (φ,A) is bottom part closed, is of limit type, is stable and is directed.
1. Let $\mathcal{P}^-_{\varphi,A} = \bigcup_{(\mathcal{Q},\Lambda)\in\mathcal{F}_{\varphi,A}} \mathcal{M}_\infty(\mathcal{Q},\Lambda)$.
2. Fix $\mathcal{R} \triangleleft^c_{hod} \mathcal{P}^-_{\varphi,A}$ and $(\mathcal{Q},\Lambda) \in \mathcal{F}_{\varphi,A}$ such that $\mathcal{R} = \mathcal{M}_\infty(\mathcal{Q},\Lambda)$. Let $\Sigma_{\mathcal{R},\varphi,A} = \Lambda_{\mathcal{R}}$ and let $\Sigma_{\varphi,A} = \bigoplus_{\mathcal{R}\triangleleft^c_{hod}\mathcal{P}^-_{\varphi,A}} \Sigma_{\mathcal{R},\varphi,A}$.
3. Suppose there is $\mathcal{M} \trianglelefteq \mathsf{Lp}^{\Gamma_{\varphi,A},\Sigma_{\varphi,A}}(\mathcal{P}^-_{\varphi,A})$ such that $\rho(\mathcal{M}) < \mathsf{ord}(\mathcal{P}^-_{\varphi,A})$. Then let $\mathcal{P}_{\varphi,A}$ be the least such $\mathcal{M}$. Otherwise let $\mathcal{P}_{\varphi,A} = \mathsf{Lp}^{\Gamma_{\varphi,A},\Sigma_{\varphi,A}}(\mathcal{P}^-_{\varphi,A})$.

In clause 3 above $\mathcal{M} \trianglelefteq \mathsf{Lp}^{\Gamma_{\varphi,A},\Sigma_{\varphi,A}}(\mathcal{P}^-_{\varphi,A})$ if and only if whenever $\pi : \mathcal{M}' \to \mathcal{M}$ is an elementary embedding and $\mathcal{M}'$ is countable, $\mathcal{M}' \trianglelefteq \mathsf{Lp}^{\Gamma_{\varphi,A},\Sigma^\pi_{\varphi,A}}(\pi^{-1}(\mathcal{P}^-_{\varphi,A}))$.

[6]See Notation 2.7.14.

DEFINITION 9.1.4. Suppose (φ,A) is bottom-part closed, is of limit type, is stable and is directed. We say (φ,A) is *full* if $\mathcal{P}_{\varphi,A} = \mathsf{Lp}^{\Gamma_{\varphi,A},\Sigma_{\varphi,A}}(\mathcal{P}^-_{\varphi,A})$.

DEFINITION 9.1.5. We say *lower part* (φ,A)*-covering* holds if (φ,A) is full and $cf(\mathrm{ord}(\mathcal{P}_{\varphi,A})) \geq \omega_1$.

NOTATION 9.1.6. Suppose now that lower part (φ,A)-covering fails. Given $X \subset \wp_{\omega_1}(\mathcal{P})$, we let

1. $\mathcal{P}_X$ be the transitive collapse of $Hull^{\mathcal{P}_{\varphi,A}}(X)$,
2. $\tau_X : \mathcal{P}_X \to \mathcal{P}_{\varphi,A}$ be the inverse of the transitive collapse,
3. Σ_X be the τ_X-pullback of $\Sigma_{\varphi,A}$,
4. $\delta_X = \delta^{\mathcal{P}_X}$.

REMARK 9.1.7. Thus, Σ_X is a strategy that acts on stacks that are based on $\mathcal{P}_X|\delta_X$. It follows that if $\mathcal{P}_X \models$ "δ_X is a regular cardinal" then Σ_X is (essentially[7]) a strategy for $\mathcal{P}_X$.

DEFINITION 9.1.8 (Weakly condensing set). Suppose (φ,A)-covering fails and set $\Gamma = \Gamma_{\varphi,A}$, $\mathcal{P} = \mathcal{P}_{\varphi,A}$ and $\Sigma = \Sigma_{\varphi,A}$. We say that $X \in \wp_{\omega_1}(\mathcal{P})$ is a (φ,A)*-weakly condensing set* if $\mathcal{P} = Hull^{\mathcal{P}}(X \cup \delta^{\mathcal{P}})$ and whenever $X \subseteq Y \in \wp_{\omega_1}(\mathcal{P})$, Σ_Y is a strongly Γ-fullness preserving iteration strategy with strong branch condensation.

NOTATION 9.1.9. Suppose (φ,A)-covering fails and set $\Gamma = \Gamma_{\varphi,A}$, $\mathcal{P} = \mathcal{P}_{\varphi,A}$ and $\Sigma = \Sigma_{\varphi,A}$. Suppose $X \subseteq Y \in \wp_{\omega_1}(\mathcal{P})$. Let $\tau_{X,Y} : \mathcal{P}_X \to \mathcal{P}_Y$ be $\tau_Y^{-1} \circ \tau_X$. Let

1. $\sigma_Y^{X,-} = \bigcup_{\mathcal{R} \lhd^c_{hod} \mathcal{P}_Y} \pi^{\Sigma_Y}_{\mathcal{R},\infty}$,
2. $\sigma_Y^X : \mathcal{P}_Y \to \mathcal{P}$ be given by: for any $f \in \mathcal{P}_X$ and any $a \in (\mathcal{P}_Y|\delta_Y)^{<\omega}$, and $x = \tau_{X,Y}(f)(a)$,
$$\sigma_Y^X(x) = \tau_X(f)(\sigma_Y^{X,-}(a)).$$

DEFINITION 9.1.10. Suppose (φ,A)-covering fails and set $\Gamma = \Gamma_{\varphi,A}$, $\mathcal{P} = \mathcal{P}_{\varphi,A}$ and $\Sigma = \Sigma_{\varphi,A}$. Let $X \subseteq Y \in \wp_{\omega_1}(\mathcal{P})$. We say that Y *extends* X or Y is an *extension of* X if

1. $\tau_{X,Y} \restriction (\mathcal{P}_X|\delta^{\mathcal{P}_X})$ is the iteration map via Σ_X,
2. letting $\nu = \sup \tau_{X,Y}[\delta^{\mathcal{P}_X}]$, $\tau_Y \restriction \mathcal{P}_Y|\nu$ is the iteration embedding according to $(\Sigma_X)_{\mathcal{P}_Y|\nu}$, and
3. $\mathcal{P}_Y = Hull_1^{\mathcal{P}_Y}(\delta^{\mathcal{P}_Y} \cup \tau_{X,Y}[\mathcal{P}_X])$.

DEFINITION 9.1.11. Suppose (φ,A)-covering fails and set $\Gamma = \Gamma_{\varphi,A}$, $\mathcal{P} = \mathcal{P}_{\varphi,A}$ and $\Sigma = \Sigma_{\varphi,A}$. Suppose Y is an extension of a weakly condensing set X. Let $\delta_Y = \delta^{\mathcal{P}_Y}$. We say that Y is *an honest extension* of X if

(a) $\mathcal{P}_Y|\sup(\tau_{X,Y}[\delta_X])$ is a Σ_X-iterate of $\mathcal{P}_X|\delta_X$,
(b) $\tau_{X,Y} \restriction (\mathcal{P}_X|\delta_X) = \pi^{\Sigma_X}_{\mathcal{P}_X|\delta_X, \mathcal{P}_Y|\sup(\tau_{X,Y}[\delta_X])}$ and

[7]As defined, Σ_X still does not act on iterations that are above δ_X.

(c) σ_Y^X is an elementary embedding[8].

REMARK 9.1.12. X is obviously an honest extension of itself, but there are other (non-trivial) honest extensions of X. For example, if $X = X' \cap \mathcal{P}$ where $X' \prec H_\lambda^V$ for some regular λ (this will be the case for our intended X) and $Y = Y' \cap \mathcal{P}$ for some $X' \prec Y'$, then Y is an honest extension of X.

DEFINITION 9.1.13 (Condensing set). Suppose (φ, A)-covering fails and set $\Gamma = \Gamma_{\varphi,A}$, $\mathcal{P} = \mathcal{P}_{\varphi,A}$ and $\Sigma = \Sigma_{\varphi,A}$. Suppose $X \in \wp_{\omega_1}(\mathcal{P})$ is a (φ, A)-weakly condensing set. We say that X is a *(φ, A)-condensing set* if whenever Y extends X, Y is an honest extension of X.

We say that X is a *strongly (φ, A)-condensing set* if whenever Y extends X, Y is a (φ, A)-condensing set.

We expect that under many hypothesis such as PFA lower part (φ, A)-covering fails. We also expect that under many hypothesis, failure of lower part (φ, A)-covering implies the existence of (φ, A)-condensing sets. In the next few chapters, we explore some specific situations where we know how to prove the existence of (φ, A)-condensing sets.

We finish by remarking that (φ, A) depends on the specific situation we are in. For instance, in [31], φ isolates those hod pairs that have certain extendability and self-determining properties (see [31, Definition 3.1, 3.5, 3.8]).

We finish here by showing that below LSA, pullback strategies are unique.

LEMMA 9.1.14 (Uniqueness of strategies). *Suppose (φ, A, X) is such that φ is a formula in the language of set theory, (φ, A) is full, lower part (φ, A)-covering fails and X is a (φ, A)-condensing set. Suppose further that whenever $(\mathcal{Q}, \Lambda) \in \Gamma_{\varphi,A}$, $\mathcal{Q}$ is not of lsa type. Then whenever Y_1 and Y_2 are two honest extensions of X such that $\mathcal{P}_{Y_1} = \mathcal{P}_{Y_2}$, then $\Sigma_{Y_1} = \Sigma_{Y_2}$.*

PROOF. Suppose that $\Sigma_{Y_1} \neq \Sigma_{Y_2}$. Let $\mathcal{P}_1 = \mathcal{P}_{Y_1}$, $\mathcal{P}_2 = \mathcal{P}_{Y_2}$, $\Phi_1 = \Sigma_{Y_1}$ and $\Phi_2 = \Sigma_{Y_2}$. Because we can trace disagreement of strategies to minimal disagreements (using our smallness assumption on hod mice)[9], we can find a minimal low level disagreement[10] $(\mathcal{T}_1, \mathcal{Q}_1', \mathcal{T}_2, \mathcal{Q}_2', \mathcal{R})$ between Φ_1 and Φ_2[11]. Let E be the $\mathcal{R}$-undropping extender of $\mathcal{T}_1$ and $\mathcal{T}_2$[12], and set for $i = 1, 2$, $\mathcal{W}_i = \mathrm{Ult}(\mathcal{P}_i, E)$. We thus have that

(1) $\mathcal{W}_1 = \mathcal{W}_2$, $\mathcal{R}$ is of successor type and $(\Phi_1)_{\mathcal{R}^-} = (\Phi_2)_{\mathcal{R}^-}$.

Because both Y_1 and Y_2 are extensions of X, we have that both $\tau_{X,Y} \restriction (\mathcal{P}_X|\delta_X)$ and $\tau_{X,Z} \restriction (\mathcal{P}_X|\delta_X)$ are the iteration embedding according to Σ_X. Because Σ_X has strong branch condensation and is strongly $\Gamma_{\varphi,A}$-fullness preserving, we have

[8] We clearly have that $\tau_X = \sigma_Y^X \circ \tau_{X,Y}$.

[9] See Lemma 4.7.2.

[10] See Definition 4.7.1.

[11] See Remark 9.1.7. It follows that $\mathcal{T}_1$ and $\mathcal{T}_2$ are based on a proper initial segment of $\mathcal{P}_1 = \mathcal{P}_2$.

[12] See clause 5d of Definition 4.7.1.

that $\tau_{X,Y} \restriction (\mathcal{P}_X|\delta_X) = \tau_{X,Z} \restriction (\mathcal{P}_X|\delta_X)$[13]. Let then $\tau =_{\text{def}} \tau_{X,Y} \restriction (\mathcal{P}_X|\delta_X) = \tau_{X,Z} \restriction (\mathcal{P}_X|\delta_X)$.

Next, because of the smallness assumption on hod pairs in $\Gamma_{\varphi,A}$, it follows from (φ,A)-condensation of X that

(2) for $i \in \{1,2\}$, $\sup(Hull^{\mathcal{W}_i}(\pi_E \circ \tau[\mathcal{P}_X]) \cap \delta^{\mathcal{R}}) = \delta^{\mathcal{R}}$[14].

Set for $i = 1,2$, $\mathcal{X}_i = \mathcal{T}_i^\frown\{E\}$. We can now find, using Theorem 4.13.5, a normal stack $\mathcal{U}_1$ on $\mathcal{W}_1$ according to $(\Phi_1)_{\mathcal{W}_1,\mathcal{X}_1}$ and a normal stack $\mathcal{U}_2$ on $\mathcal{W}_2$ according to $(\Phi_2)_{\mathcal{W}_2,\mathcal{X}_2}$ such that setting $b_1 = (\Phi_1)_{\mathcal{W}_1,\mathcal{X}_1}(\mathcal{U}_1)$, $b_2 = (\Phi_2)_{\mathcal{W}_2,\mathcal{X}_2}(\mathcal{U}_2)$, $\mathcal{R}_1 = \mathcal{M}_{b_1}^{\mathcal{U}_1}$ and $\mathcal{R}_2 = \mathcal{M}_{b_2}^{\mathcal{U}_2}$ then setting

$$\Psi_1 = (\Phi_1)_{\mathcal{R}_1,\mathcal{X}_1^\frown\mathcal{U}_1^\frown\{b\}} \quad \text{and} \quad \Psi_2 = (\Phi_2)_{\mathcal{R}_2,\mathcal{X}_2^\frown\mathcal{U}_2^\frown\{b_2\}},$$

(3) for $i \in \{1,2\}$, $\mathcal{U}_i$ is based on $\mathcal{R}$, $\downarrow(\mathcal{U}_1,\mathcal{R}) = \downarrow(\mathcal{U}_2,\mathcal{R})$, $b_1 \neq b_2$ and $\pi_{b_1}^{\mathcal{U}_1}(\mathcal{R}) = \pi_{b_2}^{\mathcal{U}_2}(\mathcal{R})$,

(4) letting $\mathcal{S} = \pi_{b_1}^{\mathcal{U}_1}(\mathcal{R})$, $(\Psi_1)_{\mathcal{S}} = (\Psi_2)_{\mathcal{S}}$.

Notice now that we can find $k_1 : \mathcal{R}_1 \to \mathcal{P}$ and $k_2 : \mathcal{R}_2 \to \mathcal{P}$ such that letting for $i = 1,2$, $\tau_{Y_i} = \tau_i$,

(5) for $i = 1,2$, $\tau_i = k_i \circ (\pi_{b_i}^{\mathcal{U}_i} \circ \pi_E)$,

(6) for $i = 1,2$, $k_i \restriction \mathcal{S} = \pi_{\mathcal{S},\infty}^{(\Psi_i)_{\mathcal{S}}}$[15], and

(7) for $i = 1,2$, $\pi_{b_i}^{\mathcal{U}_i} \circ \pi_E \circ \tau$ is according to Σ_X.

It follows that letting for $i = 1,2$, $Z_i = Y_i \cup \operatorname{rge}(\pi_{\mathcal{S},\infty}^{\Psi_i})$, Z_i extends X, and moreover, because X is a condensing set, for $i = 1,2$, $k_i = \sigma_{Z_i}^X$[16].

Notice that it follows from (4) that $k_1 \restriction \mathcal{S} = k_2 \restriction \mathcal{S}$. Also, notice that

(8) $k_1 \restriction (Hull^{\mathcal{R}_1}(\mathcal{S}^- \cup \pi_{b_1}^{\mathcal{U}_1} \circ \pi_E \circ \tau[\mathcal{P}_X])) = k_2 \restriction (Hull^{\mathcal{R}_2}(\mathcal{S}^- \cup \pi_{b_2}^{\mathcal{U}_2} \circ \pi_E \circ \tau[\mathcal{P}_X]))$.

Combining (2) and (8) we get that (using (3))

(9) $\operatorname{rge}(\pi_{b_1}^{\mathcal{U}_1}) \cap \operatorname{rge}(\pi_{b_2}^{\mathcal{U}_2})$ is cofinal in $\delta^{\mathcal{S}}$.

Clearly (9) and parts of (3)[17] imply that $b_1 = b_2$, while other parts of (3) state that $b_1 \neq b_2$. ⊣

The following is a useful corollary of the definition of a condensing set. We will apply this corollary in many applications later.

---

[13]See Proposition 4.10.2.

[14]It is easier to first establish that $\sup(Hull^{\mathcal{W}_i}(\pi_E \circ \tau[\mathcal{P}_X] \cup \delta^{\mathcal{R}^-}) \cap \delta^{\mathcal{R}}) = \delta^{\mathcal{R}}$. See Lemma 2.9.5.

[15](5) and (6) easily follows from the fact that $\mathcal{T}_1$ and $\mathcal{T}_2$ are based on proper initial segment of $\mathcal{P}_1|\delta^{\mathcal{P}_1}$.

[16]Again, this easily traces back to the fact that $\mathcal{T}_i$ is based on a proper initial segment of $\mathcal{P}_i|\delta^{\mathcal{P}_i}$.

[17]That for $i \in \{1,2\}$, $\mathcal{U}_i$ is based on $\mathcal{R}$, $\downarrow(\mathcal{U}_1,\mathcal{R}) = \downarrow(\mathcal{U}_2,\mathcal{R})$.

COROLLARY 9.1.15. *Suppose $Y \prec Z$ are extensions of a (φ,A)-condensing set X and Z is an extension of Y. Suppose $B \in \wp(\delta^{\mathcal{P}}) \cap \mathcal{P}$ and $B \in Y$. Let $a \in (\delta^{\mathcal{Q}_Y})^{<\omega}$. Then $\pi^{\Sigma_Y}_{\mathcal{Q}_Y,\infty}(a) \in B$ if and only if $\pi^{\Sigma_Z}_{\mathcal{Q}_Z,\infty}(\tau_{Y,Z}(a)) \in B$.*

9.2. Condensing sets from elementary embeddings

The following two theorems can be proved using the proof of [31, Lemma 11.15]. First we introduce some terminology.

TERMINOLOGY 9.2.1. Suppose κ is an inaccessible cardinal and $G \subseteq Col(\omega, <\kappa)$ is V-generic. Suppose (φ,A) is such that $V[G] \models$ "(φ,A) is full and lower part (φ,A)-covering fails". We say (φ,A) is *homogenous* if $\mathcal{P}_{\varphi,A} \in V$, $\Sigma_{\varphi,A} \restriction V \in V$ and for any $(\mathcal{Q},\Lambda) \in \mathcal{F}_{\varphi,A}$, there is $(\mathcal{R},\Psi) \in \mathcal{F}_{\varphi,A}$ such that $\mathcal{R} \in V$, $\Psi \restriction H^V_\kappa \in V$ and $V[G] \models \Gamma(\mathcal{Q},\Lambda) \subseteq \Gamma(\mathcal{R},\Psi)$.

THEOREM 9.2.2. *Suppose $N \subseteq M$ are transitive models of set theory and $j : M \to N$ is an elementary embedding with critical point κ such that j is amenable to M, i.e., for every $X \in M$, $j(X) \in M$. Suppose $g \subseteq Coll(\omega, < j(\kappa))$ is N-generic. Let $j^+ : M[g_\kappa] \to N[g]$ be the extension of j where for $\alpha < j(\kappa)$, $g_\alpha = g \cap Coll(\omega, <\alpha)$. Suppose φ is a formula in the language of set theory and $A \in M[g]$. Suppose further that $M[g_\kappa] \models$ "(φ,A) is full, (φ,A) is homogenous and lower part (φ,A)-covering fails". Then $j[\mathcal{P}_{\varphi,A}]$ is a strongly $(\varphi, j(A))$-condensing set in $N[g]$. Hence, $M[g] \models$ "there is a strongly (φ,A)-condensing set".*

TERMINOLOGY 9.2.3. We say (φ,A) is *maximal* if there is no hod pair or an sts hod pair $(\mathcal{Q},\Lambda)$ such that $\mathcal{Q}$ is of limit type, Λ has strong branch condensation and is strongly $\Gamma_{\varphi,A}$-fullness preserving and $\Gamma(\mathcal{Q},\Lambda) = \Gamma_{\varphi,A}$.

THEOREM 9.2.4. *Assume* ZF + DC *and suppose (φ,A) is maximal and full, lower part (φ,A)-covering fails and X is a (φ,A)-condensing set. Then $\mathcal{P}_{\varphi,A} \models$ "$\delta^{\mathcal{P}_{\varphi,A}}$ is regular".*

We will not prove Theorem 9.2.4 but will give a fairly complete proof of Theorem 9.2.2.

The proof of Theorem 9.2.2. We fix $(M,N,j,\kappa,g,\varphi,A)$ as in the statement of Theorem 9.2.2. The proof follows the proof of [31, Theorem 10.3]. Throughout this section we will use the following notation:

NOTATION 9.2.5. Working in $M[g_\kappa]$, let

1. $\mathcal{P}^- = \mathcal{P}^-_{\varphi,A}$,
2. $\mathcal{P} = \mathcal{P}_{\varphi,A}$,
3. $\Sigma = \Sigma_{\varphi,A}$,
4. $\mathcal{F} = \mathcal{F}_{\varphi,A}$,
5. $\Gamma = \Gamma_{\varphi,A}$.

THEOREM 9.2.6. $N[g] \models$ "$j[\mathcal{P}]$ *is a weakly condensing set*".

PROOF. Notice that $j[\mathsf{ord}(\mathcal{P})]$ is cofinal in $\mathsf{ord}(j(\mathcal{P}))$. Below, we often confuse strategies with their interpretations in relevant generic extensions or in relevant inner models. However, in some cases, the distinction between the two strategies is important, and in those situations we will either separate the two strategies or point out that the distinction is important. Also, below if $Y \in \wp_{\omega_1}(j(\mathcal{P}))$ then we let $\mathcal{P}_Y = j(\mathcal{P})_Y$ and $\Sigma_Y = j(\Sigma)_Y$.

We want to show that

(a) if $Y \in (\wp_{\omega_1}(j(\mathcal{P})))^{N[g]}$ is such that $j[\mathcal{P}] \subseteq Y$ then $L(j^+(\Gamma)) \models$ "$\mathcal{P}_Y$ is Σ_Y-full"[18].

Towards a contradiction assume that (a) is false. Notice that if Y witnesses that (a) is false then $\mathcal{P}_Y$ may not be in $M[g_\kappa]$. Fix one such Y that is a counterexample to (a), and let $\mathcal{M}$ be a sound Σ_Y-mouse over $\mathcal{P}_Y|\delta_Y$ that has an iteration strategy in $j^+(\Gamma)$ but such that $\mathcal{M} \ntrianglelefteq \mathcal{P}_Y$ and $\rho(\mathcal{M}) = \delta_Y$. Let $\dot{Y} \in N^{Coll(\omega,<(\kappa,\lambda))}$ be a name for Y and $\dot{\mathcal{M}}$ be a name for $\mathcal{M}$. We can then find some $\Sigma_{j[\mathcal{P}]}$-hod pair $(\mathcal{P}^+,\Pi) \in N$ and a hod pair $(\mathcal{S},\Phi) \in N$ such that

1. $\mathcal{P}^+ \in H^N_{j(\kappa)}$,
2. Π has strong branch condensation,
3. $\mathcal{P}^+$ is meek and of limit type,
4. $\mathrm{cf}^{\mathcal{P}^+}(\delta^{\mathcal{P}^+}) = \omega$,
5. $(Y \cap j(\mathcal{P}|\delta^{\mathcal{P}})) \subseteq \mathrm{rge}(\pi^{\Phi}_{\mathcal{S},\infty})$ and no proper complete layer of $\mathcal{S}$ has this property[19],
6. $\Pi \in N$ is a $(j(\kappa), j(\kappa))$-strategy for $\mathcal{P}^+$ that can be uniquely extended to a strategy $\Pi^g \in j^+(\Gamma)$[20], and moreover, Π witnesses that $\mathcal{P}^+$ is a $\Sigma_{j[\mathcal{P}]}$-hod mouse,
7. $N[g_\kappa] \models$ it is forced by $Coll(\omega, <(\kappa, j(\kappa))$ that
   (a) $\dot{\mathcal{M}}$ is a sound $\Sigma_{\dot{Y}}$-mouse over $\mathcal{P}_{\dot{Y}}|\delta_{\dot{Y}}$ that projects to $\delta_{\dot{Y}}$.
   (b) $\dot{\mathcal{M}}$ has an iteration strategy in the derived model of $(\mathcal{P}^+,\Pi)$[21] as computed by any $\mathbb{R}$-genericity iteration,
   (c) Φ is in the derived model of $(\mathcal{P}^+,\Pi)$ as computed by any $\mathbb{R}$-genericity iteration,
   (d) $\dot{\mathcal{M}}$ is not an initial segment of $\mathcal{P}_{\dot{Y}}$.

Because $\mathcal{P}^+$ might have cardinality $> \kappa$, when we form $\mathcal{P}_Y^+ =_{\text{def}} Ult(\mathcal{P}^+, E)$, where E is the $(\mathrm{crit}(\tau_{j[\mathcal{P}],Y}), \delta_Y)$-extender derived from $\tau_{j[\mathcal{P}],Y}$, we cannot conclude

[18]That is, $\mathcal{P}_Y = \mathsf{Lp}^{j^+(\Gamma),\Sigma_Y}(\mathcal{P}_Y|\delta_Y)$.

[19]That is, if $\mathcal{S}' \triangleleft^c_{hod} \mathcal{S}$ then $(Y \cap j(\mathcal{P}|\delta^{\mathcal{P}})) \nsubseteq \mathrm{rge}(\pi^{\Phi}_{\mathcal{S}',\infty})$. See Notation 9.1.2.

[20]Π can be obtained by computing the direct limit of all those hod pairs $(\mathcal{Q},\Lambda) \in j^+(\mathcal{F}_{\varphi,A})$ with the property that $\mathcal{Q} \in N[g_\nu]$ where $\nu \in (\kappa, j(\kappa))$ is chosen in a way that Σ_Y and the strategy of $\mathcal{M}$ appear in $N[g_\nu]$. We might then have to take some initial segment of this direct limit to satisfy clauses 3-5 above.

[21]We confuse the extension of Π to this extension with Π-itself.

that $\mathcal{P}_Y^+$ is iterable in $N[g]$. This is because we do not know that $j \restriction \mathcal{P}^+ \in N$. To resolve this issue we take a hull of size κ. Let $\kappa_1 = (\kappa^+)^M$.

We work in $N[g_\kappa]$. We can now find $\pi : W[g_\kappa] \to (H_{j(\kappa_1)})^{N[g_\kappa]}$ (in $N[g_\kappa]$) such that

1. $W \in M$ is transitive and $\kappa + 1 \subseteq W$,
2. $(j(\mathcal{P}), j \restriction \mathcal{P}, \dot{Y}, (\mathcal{P}^+, \Pi), (\mathcal{S}, \Phi)) \in \mathrm{rge}(\pi)$.

Let $\dot{Z} = \pi^{-1}(\dot{Y})$, $\dot{\mathcal{N}} = \pi^{-1}(\dot{\mathcal{M}})$, $\mathcal{R} = \pi^{-1}(j(\mathcal{P}))$ and $k : \mathcal{P} \to \mathcal{R}$ be $\pi^{-1}(j \restriction \mathcal{P})$. Working in $N[g_{\kappa_1}]$, let $h \subseteq Coll(\omega, < (\kappa, k(\kappa))$ be W-generic, and set

1. $Z = \dot{Z}_h$, $\dot{\mathcal{N}}_h = \mathcal{N}$, $\mathcal{Q} = (\mathcal{P}_Z)^{W[h]}$,
2. $\sigma = (\tau_{k[\mathcal{P}],Z})^{W[h]}$ and $\tau = (\tau_Z)^{W[h]}$,
3. $\overline{\mathcal{P}^+} = \pi^{-1}(\mathcal{P}^+)$ and $\overline{\Pi} = \pi^{-1}(\Pi)$,
4. $(\overline{\mathcal{S}}, \overline{\Phi}) = \pi^{-1}(\mathcal{S}, \Phi)$.

Thus, we have that

(A) $k = \tau \circ \sigma$, $\sigma : \mathcal{P} \to \mathcal{Q}$ and $\tau : \mathcal{Q} \to \mathcal{R}$,

(B) in $W[g_\kappa * h]$,

1. $\mathcal{N}$ is a sound Σ_Z-mouse over $\mathcal{Q}|\delta^{\mathcal{Q}}$ that projects to $\delta^{\mathcal{Q}}$.
2. in any derived model of $(\overline{\mathcal{P}^+}, \overline{\Pi})$ as computed by an $\mathbb{R}$-genericity iteration, $\mathcal{N}$ has an ω_1-iteration strategy witnessing that it is a Σ_Z-mouse,
3. $\mathcal{N}$ is not an initial segment of $\mathcal{Q}$.
4. $\overline{\Phi}$ is in the derived model of $(\overline{\mathcal{P}^+}, \overline{\Pi})$ as computed by any $\mathbb{R}$-genericity iteration,
5. letting $\xi : \mathcal{Q}|\delta^{\mathcal{Q}} \to \overline{\mathcal{S}}|\delta^{\overline{\mathcal{S}}}$ be such that $\xi = (\pi^{\overline{\Phi}}_{\overline{\mathcal{S}},\infty})^{-1} \circ \tau$, $\Sigma_Z = (\xi$-pullback of $\overline{\Phi}_{\overline{\mathcal{S}}|\delta^{\overline{\mathcal{S}}}}$.

Let now F be the $(\mathrm{crit}(\sigma), \delta^{\mathcal{Q}})$-extender derived from σ, and set $\mathcal{Q}^+ = \mathrm{Ult}(\overline{\mathcal{P}^+}, F)$. Let $\sigma^+ = \pi_F^{\overline{\mathcal{P}^+}}$. Notice that because $\pi \circ k = j \restriction \mathcal{P}$, we have $\varphi^+ : \mathcal{Q}^+ \to j(\overline{\mathcal{P}^+})$ such that

(C) $j \restriction \overline{\mathcal{P}^+} = \varphi^+ \circ \sigma^+$.

Let $\overline{\Pi}^+$ be the $\pi \restriction \overline{\mathcal{P}^+}$-pullback of Π[22] and let $\overline{\Phi}^+$ be the π-pullback of Φ. Notice that

(D1) $\overline{\Pi}^+ \restriction \mathsf{HC}^{W[g_\kappa * h]} = \overline{\Pi}$[23],

(D2) $\overline{\Pi}^+$ witnesses that $\overline{\mathcal{P}^+}$ is a Σ-hod mouse[24],

(D3) $\overline{\Phi}^+ \restriction \mathsf{HC}^{W[g_\kappa * h]} = \overline{\Phi}$.

Notice now that we have

---

[22]We confuse Π with its extension to $N[g]$. Similarly, we think of $\overline{\Pi}^+$ as a strategy in N as well as in $N[g]$. Same comment applies below to $\overline{\Pi}$ and $\overline{\Phi}$.

[23]See proof of Claim 2 in the proof of [31, Lemma 10.4]. The same equation for $W[g]$ follows easily from hull condensation of $\overline{\Pi}^+$, but this equation for $W[g_\kappa * h]$ needs more work.

[24]This follows from the fact that Π witnesses that $\mathcal{P}^+$ is a Σ-hod mouse and $\pi \restriction \mathcal{P} = id$.

(F) in $N[g]$, $j^+(\overline{\Pi}^+ \restriction (H^{M[g_\kappa]}_{\kappa_1}))$ is a $(j(\kappa), j(\kappa))$-iteration strategy witnessing that $j(\overline{\mathcal{P}^+})$ is a $j(\Sigma)$-hod mouse, and moreover, $j \restriction \overline{\mathcal{P}^+} \in N[g]$[25].

We let $\Psi = (\Sigma_Z)^{W[g_\kappa * h]}$. Notice that in $W[g_\kappa * h]$, Ψ is the τ-pullback of $\pi^{-1}(j(\Sigma))$. Let Ψ^+ be the $\varphi^+ \restriction (\mathcal{Q}|\delta^{\mathcal{Q}}) = \pi \circ \tau \restriction (\mathcal{Q}|\delta^{\mathcal{Q}})$-pullback of $j(\Sigma)$. It follows that

(G) Ψ^+ is the $\pi \circ \xi$-pullback of Φ, and it is also ξ pullback of $\overline{\Phi}^+$.

We now claim that

(b) in $N[g]$, in any derived model of $(\overline{\mathcal{P}^+}, \overline{\Pi}^+)$ as computed by an $\mathbb{R}$-genericity iteration, $\mathcal{N}$ has an ω_1-iteration strategy witnessesing that $\mathcal{N}$ is a Ψ^+-mouse.

The proof of (b) is like the proof of Claim 1 of [31, Lemma 10.4]. We outline it below. Working in $W[g_\kappa]$, let $\mathcal{W} = \mathcal{M}^{\#,\overline{\Pi},\overline{\Phi}}_\omega$ and let Λ be the unique iteration strategy of $\mathcal{W}$. Because $\pi(\mathcal{W}) = \mathcal{M}^{\#,\Pi,\Phi}_\omega$, we have that letting Λ^+ be the π-pullback of $\pi(\Lambda)$,

(H) $\mathcal{W} = \mathcal{M}^{\#,\overline{\Pi}^+,\overline{\Phi}^+}_\omega$, and Λ^+ witnesses that $\mathcal{W}$ is a $\overline{\Pi}^+ \oplus \overline{\Phi}^+$-mouse.

Working in $W[g_\kappa * h]$ and using (B2), we can find a Λ-iterate $\mathcal{W}_1$ of $\mathcal{W}$, a Woodin cardinal η of $\mathcal{W}_1$ and $\mathcal{W}_1$-generic $m \subseteq Coll(\omega, \eta)$ such that letting λ be the sup of the Woodin cardinals of $\mathcal{W}_1$,

(I1) $\mathcal{N}, \xi \in \mathcal{W}_1[m]$,
(I2) $\mathcal{W}_1[m] \models$ "the derived model at λ satisfies that any derived model of $(\overline{\mathcal{P}^+}, \overline{\Pi})$ as computed by an $\mathbb{R}$-genericity iteration has an ω_1-iteration strategy for $\mathcal{N}$ witnessing that $\mathcal{N}$ is a mouse relative to the ξ-pullback of $\overline{\Phi}$".

Let then $\mathcal{W}_\infty$ be a Λ^+-iterate of $\mathcal{W}_1$ which is obtained via some $\mathbb{R}^{\mathcal{N}[g]}$-genericity iteration in such a way that letting $i : \mathcal{W}_1 \to \mathcal{W}_\infty$ be the iteration embedding, $\mathrm{crit}(i) > \eta$. It then follows from (I1), (I2) and (H) that

(J) $\mathcal{W}_\infty[\mathcal{N}, \xi] \models$ "the derived model at λ satisfies that any derived model of $(\overline{\mathcal{P}^+}, \overline{\Pi}^+)$ as computed by an $\mathbb{R}$-genericity iteration has an ω_1-iteration strategy for $\mathcal{N}$ witnessing that $\mathcal{N}$ is a mouse relative to the ξ-pullback of $\overline{\Phi}^+$".

(b) now easily follows from (J), (H) and (G).

To finish the proof of Theorem 9.2.6, it remains to implement the last portion of the proof of [31, Theorem 10.3]. Let Δ_0 be φ^+-pullback of $j^+(\overline{\Pi}^+ \restriction (H^{M[g_\kappa]}_{\kappa_1}))$. Notice that it follows from (F) that Δ_0 witnesses that $\mathcal{Q}^+$ is a Ψ^+-hod mouse. It then follows from (b) that

(K) in $N[g]$, in any derived model of $(\mathcal{Q}^+, \Delta_0)$ as computed by an $\mathbb{R}$-genericity iteration, $\mathcal{N}$ has an ω_1-iteration strategy Δ witnessing that $\mathcal{N}$ is a Ψ^+-mouse.

(K) gives contradiction, as it implies that

[25] Because $|\overline{\mathcal{P}^+}|^M = \kappa$.

(L) $\mathcal{Q}^+ \models$ "$\mathsf{ord}(\mathcal{Q})$ is not a cardinal"[26],

while clearly $\overline{\mathcal{P}^+} \models$ "$\mathsf{ord}(\mathcal{P})$ is a cardinal", contradicting the elementarity of φ^+ ⊣

The main theorem of this chapter is.

THEOREM 9.2.7. *In $N[g]$, $j[\mathcal{P}]$ is a strongly condensing set.*

PROOF. We will show that $j[\mathcal{P}]$ is a condensing set. A very similar proof, which is only notationally more complicated, shows that $j[\mathcal{P}]$ is strongly condensing. To prove the theorem, we need the following definition, due to the first author (cf. [31] or [65]). The proof is based on [31, Lemma 11.15]. For completeness, we give a fairly detailed argument here. The reader may wish to recall Notation 9.1.2. Below if $Y \in \wp_{\omega_1}(j(\mathcal{P}))$ then we let $\mathcal{P}_Y = j(\mathcal{P})_Y$ and $\Sigma_Y = j(\Sigma)_Y$.

We work in $N[g]$. Suppose $X \in \wp_{\omega_1}(j(\mathcal{P}))$ is a weakly condensing set and $B \in X \cap \wp(\delta^{j(\mathcal{P})})$. We say that X has *B-condensation* if whenever $Y \in \wp_{\omega_1}(j(\mathcal{P}))$ is such that $X \prec Y$, $\tau_{X,Y}(T_{X,B}) = T_{Y,B}$, where for $Z \in \wp_{\omega_1}(j(\mathcal{P}))$,

$$T_{Z,B} = \{(\varphi, s) \mid s \in [\delta^{\mathcal{R}}]^{<\omega} \text{ for some } \mathcal{R} \triangleleft^c_{hod} \mathcal{P}_Z \wedge j(\mathcal{P}) \models \varphi[\pi^{\Sigma_Z}_{\mathcal{R},\infty}(s), B]\}.$$

We say X has *term condensation* if it has B-condensation for every $B \in X \cap \wp(\delta^{j(\mathcal{P})})$.

To prove that a weakly condensing set X is condensing, it is enough to prove that τ_X has term condensation. It is not hard to show that if for every $A \in j(\mathcal{P})$ there is X with A-condensation then $j[\mathcal{P}]$ has term condensation[27]. We say, working in $N[g]$, $X \in \wp_{\omega_1}(j(\mathcal{P}))$ is *good* if for every $\mathcal{R} \triangleleft^c_{hod} \mathcal{P}_X$, $\tau_X \restriction \mathcal{R} = \pi^{\Sigma_X}_{\mathcal{R},\infty}$. It follows from [31, Lemma 11.9] that the set of good X is a club. Notice that $j[\mathcal{P}]$ is good.

Towards a contradiction, assume that (in $N[g]$) there is a set $A \in \mathcal{P}$ such that no $X \in \wp_{\omega_1}(j(\mathcal{P}))$ with the property $j[\mathcal{P}] \subseteq X$, has A-condensation. We now fix such a set A. We say (in $N[g]$) that a tuple $\{\langle \mathcal{P}_i, \mathcal{Q}_i, X_i, Y_i, \xi_i, \pi_i, \varphi_i \mid i < \omega\rangle, B, \mathcal{M}\}$ is a *bad tuple* (relative to A) if

1. $X_0 = Y_0 = j[\mathcal{P}]$,
2. for all $i < \omega$, $X_i \in \wp_{\omega_1}(j(\mathcal{P}))$ is good,
3. for all $i < \omega$, $\mathcal{P}_i = \mathcal{P}_{X_i}$ and $\mathcal{Q}_i = \mathcal{P}_{Y_i}$;
4. for all $i < j < \omega$, $X_i \prec Y_i \prec X_j$;
5. for all $i < \omega$, $\xi_i = \tau_{X_i,Y_i}$, $\pi_i = \tau_{Y_i,X_{i+1}}$ and $\varphi_i = \tau_{X_i,X_{i+1}}$[28];
6. $\mathcal{M} \in j(\mathcal{P}|\delta^{\mathcal{P}})$ and letting $\eta = \sup_{i<\omega}(X_i \cap \delta^{j(\mathcal{P})})$, $\mathcal{M}|\eta = j(\mathcal{P})|\eta$;
7. letting η be as above, for every formula φ and for every $s \in \eta^{<\omega}$, $\mathcal{M} \models \varphi[B,s]$ if and only if $j(\mathcal{P}) \models \varphi[A,s]$;
8. for all $i \in [1,\omega)$, $\xi_i(T_{X_i,A}) \neq T_{Y_i,A}$.

CLAIM 9.2.8. *There is a bad tuple.*

---

[26]This is because (K) implies that $\mathcal{N}$ is ordinal definable in $\mathcal{Q}^+$ and therefore, $\mathcal{N} \in \mathcal{Q}$.

[27]See the proof of [31, Lemma 11.15]. This essentially follows from the elementarity of j.

[28]Thus, $\xi_i : \mathcal{P}_i \to \mathcal{Q}_i$, $\pi_i : \mathcal{Q}_i \to \mathcal{P}_{i+1}$ and $\varphi_i = \pi_i \circ \xi_i$.

PROOF. It is easy to construct a bad tuple $\{\langle\mathcal{P}_i,\mathcal{Q}_i,X_i,Y_i,\xi_i,\pi_i,\varphi_i \mid i<\omega\rangle,A, j(\mathcal{P})\}$ with $j(\mathcal{P})$ playing the role of $\mathcal{M}$ and A playing the role of B. Once this is done, letting $\eta=\sup(X_i\cap\delta^{j(\mathcal{P})})$, we set $\mathcal{M}=cHull^{j(\mathcal{P})}(\{A\},\eta)$. B then is the transitive collapse of A. ⊣

Fix a bad tuple $\mathcal{A}^*=\{\langle\mathcal{P}_i,\mathcal{Q}_i,X_i,Y_i,\xi_i,\pi_i,\varphi_i \mid i<\omega\rangle,B,\mathcal{M}\}$. Let

1. $\eta=\sup(X_i\cap\delta^{j(\mathcal{P})})$,
2. $Z_i=X_i\cap\eta$, $W_i=Y_i\cap\eta$,
3. $\Phi_i=\Sigma_{X_i}$ and $\Psi_i=\Sigma_{Y_i}$,
4. $T_i=\{(\varphi,s): s\in[\delta^{\mathcal{P}_i}]^{<\omega}\wedge\mathcal{M}\vDash\varphi[B,\pi^{\Phi_i}_{\mathcal{P}_i|\delta^{\mathcal{P}_i},\infty}(s)]\}$,
5. $S_i=\{(\varphi,s): s\in[\delta^{\mathcal{Q}_i}]^{<\omega}\wedge\mathcal{M}\vDash\varphi[B,\pi^{\Psi_i}_{\mathcal{Q}_i|\delta^{\mathcal{Q}_i},\infty}(s)]\}$.

and set

$$\mathcal{A}=\{\langle\mathcal{P}_i,\mathcal{Q}_i,\Phi_i,\xi_i,\pi_i,\varphi_i,T_i,S_i \mid i<\omega\rangle,B,\mathcal{M}\}.$$

Notice that it follows that for all $i<\omega$, $T_i=T_{X_i,A}$ and $S_i=T_{Y_i,A}$. Let $C\in j^+(\Gamma)$ be such that $\eta<\Theta^{L(C,\mathbb{R})}$ and $\mathcal{M}\subseteq\mathrm{HOD}^{L(C,\mathbb{R})}$. Then because Φ_i and Ψ_i can be recovered from $j(\Sigma)_{\mathcal{M}|\eta}$ and respectively Z_i and W_i, $\mathcal{A}\in L(C,\mathbb{R})$ and $L(C,\mathbb{R})\vDash *(\mathcal{A})$ where $*(\mathcal{A})$ is the conjunction of the following clauses:

1. for all $i<\omega$, $\xi_i:\mathcal{P}_i\to\mathcal{Q}_i$, $\pi_i:\mathcal{Q}_i\to\mathcal{P}_{i+1}$ and $\varphi_i=\pi_i\circ\xi_i$;
2. for all $i<\omega$, $T_i\in\mathcal{P}_i$ and $\xi_i(T_i)\neq S_i$;
3. for all $i<\omega$, letting Ψ_i be the π_i-pullback of Φ_{i+1}, $S_i=\{(\varphi,s): s\in[\delta^{\mathcal{Q}_i}]^{<\omega}\wedge\mathcal{M}\vDash\varphi[B,\pi^{\Psi_i}_{\mathcal{Q}_i|\delta^{\mathcal{Q}_i},\infty}(s)]\}$;
4. for all $i<\omega$, $T_i=\{(\varphi,s): s\in[\delta^{\mathcal{P}_i}]^{<\omega}$ and $\mathcal{M}\vDash\varphi[B,\pi^{\Phi_i}_{\mathcal{P}_i|\delta^{\mathcal{P}_i},\infty}(s)]\}$;
5. $\eta<\Theta$ and $\mathcal{M}\subseteq\mathrm{HOD}^{L(C,\mathbb{R})}$.

Notice that $\mathcal{P}_0=\mathcal{P}$ and $\Phi_0=\Sigma$. Let now $(\mathcal{P}_0^+,\Pi_0)$ be a Φ_0-hod pair such that

$$L(\Gamma(\mathcal{P}_0^+,\Pi_0),\mathbb{R})\vDash *(\mathcal{A}).$$

We may also assume $(\mathcal{P}_0^+,\Pi_0\restriction N)\in N$, $\mathcal{P}_0^+$ is of limit type and $\mathrm{cf}^{\mathcal{P}_0^+}(\delta^{\mathcal{P}_0^+})$ is not a measurable cardinal of $\mathcal{P}_0^+$. This type of reflection is possible because we replaced $j(\mathcal{P})$ by $\mathcal{M}$. Let $u=(\varphi_i,T_i:i<\omega)$ and set

$$\mathcal{W}=\mathcal{M}_\omega^{\sharp,\Pi_0,\oplus_{i<\omega}\Phi_i}(u).$$

Let Λ be the unique strategy of $\mathcal{W}$ witnessing that $\mathcal{W}$ is a $\Pi_0\oplus(\oplus_{i<\omega}\Phi_i)$-mouse over u. We now have that

(A) in $N[g]$, whenever D is obtained as a derived model of $(\mathcal{W},\Lambda)$ via some $\mathbb{R}$-genericity iteration, $D\vDash *(\mathcal{A})$.

We remark that the following objects are in N:

1. $\mathcal{M}$, $\mathcal{W}$ and $\Lambda\restriction N$.
2. $\langle\mathcal{P}_i,\Phi_i\restriction N,\varphi_i,T_i \mid i<\omega\rangle$.

However, $(\mathcal{Q}_i, \xi_i, \pi_i, S_i)$ are not in V. Notice also that the objects listed above are in $D(\mathcal{Z}, \omega_1^{N[g]}, h)$. We set $\mathcal{B} = \{\langle \mathcal{P}_i, \Phi_i, \varphi_i, T_i \mid i < \omega \rangle, B, \mathcal{M}\}$ and given $t = (\mathcal{N}_i, \psi_i, \sigma_i, U_i)$ we write $\mathcal{B}^t$ for the set $\{\langle \mathcal{P}_i, \mathcal{N}_i, \Phi_i, \psi_i, \sigma_i, \varphi_i, T_i, U_i \mid i < \omega \rangle, B, \mathcal{M}\}$. Thus, we have the following:

(B) in $N[g]$, whenever D is obtained as a derived model of $(\mathcal{W}, \Lambda)$ via some $\mathbb{R}$-genericity iteration,

$$D \models \text{“there is } t = (\mathcal{N}_i, \psi_i, \sigma_i, U_i) \text{ such that } * (\mathcal{B}^t)\text{”}.$$

Let now $\pi : W[g_\kappa] \to (H_{j(\kappa_1)}^{N[g_\kappa]})$ be such that all the relevant objects are in the range of π, $W \in N$, $|W|^N = \kappa$, $j \restriction \mathcal{P} \in \mathrm{rge}(\pi)$ and $\mathrm{crit}(\pi) > \kappa$. By “all relevant objects” we mean those objects that are in N, and in particular those listed above. For $a \in \mathrm{rge}(\pi)$, let $\overline{a} = \pi^{-1}(a)$. For $i < \omega$, let $\overline{\Phi_i}^+$ be the π-pullback of Φ_i, and also Let $\overline{\Lambda}^+$ be the π-pullback of Λ.

We thus have that

(C) $\overline{\mathcal{W}} = \mathcal{M}_\omega^{\sharp, \overline{\Pi_0}^+, \oplus_{i<\omega} \overline{\Phi_i}^+}$, and $\overline{\Lambda}^+$ witnesses that $\overline{\mathcal{W}}$ is a $\overline{\Pi_0}^+ \oplus (\oplus_{i<\omega} \overline{\Phi_i}^+)$-mouse,

(D) in $N[g]$, whenever D is obtained as a derived model of $(\overline{\mathcal{W}}, \overline{\Lambda}^+)$ via some $\mathbb{R}$-genericity iteration, there is $\mathcal{S} \in \mathrm{HOD}^D$ and $F \in \mathcal{S}$ such that letting $\mathcal{B}_0 = \{\langle \overline{\mathcal{P}_i}, \overline{\Phi_i}^+, \overline{\varphi_i}, \overline{T_i} \mid i < \omega \rangle, F, \mathcal{S}\}$

$$D \models \text{“there is } t = (\mathcal{N}_i, \psi_i, \sigma_i, U_i) \text{ such that } * (\mathcal{B}_0^t)\text{”},$$

(E) in $N[g]$, whenever D is obtained as a derived model of $(\overline{\mathcal{P}_0^+}, \overline{\Pi_0}^+)$ via some $\mathbb{R}$-genericity iteration, there is $\mathcal{S} \in \mathrm{HOD}^D$ and $F \in \mathcal{S}$ such that letting $\mathcal{B}_0 = \{\langle \overline{\mathcal{P}_i}, \overline{\Phi_i}^+, \overline{\varphi_i}, \overline{T_i} \mid i < \omega \rangle, F, \mathcal{S}\}$

$$D \models \text{“there is } t = (\mathcal{N}_i, \psi_i, \sigma_i, U_i) \text{ such that } * (\mathcal{B}_0^t)\text{”}.$$

The proof of (E) is like the proof of (b) in the proof of Theorem 9.2.6. Notice that in $N[g]$, the derived models of $(\overline{\mathcal{P}_0^+}, \overline{\Pi_0}^+)$ obtained via $\mathbb{R}$-genericity iteration have the form $D =_{\mathrm{def}} L(\Gamma(\overline{\mathcal{P}_0^+}, \overline{\Pi_0}^+))$. Fix then some $(F, \mathcal{S}) \in D$ and $t = (\mathcal{N}_i, \psi_i, \sigma_i, U_i) \in D$ such that letting $\mathcal{B}_0 = \{\langle \overline{\mathcal{P}_i}, \overline{\Phi_i}^+, \overline{\varphi_i}, \overline{T_i} \mid i < \omega \rangle, F, \mathcal{S}\}$, $D \models *(\mathcal{B}_0^t)$.

We thus have that the following clauses hold:

1. for all $i < \omega$, $\psi_i : \overline{\mathcal{P}_i} \to \mathcal{N}_i$, $\sigma_i : \mathcal{N}_i \to \overline{\mathcal{P}_{i+1}}$ and $\overline{\varphi_i} = \sigma_i \circ \psi_i$;
2. for all $i < \omega$, $\overline{T_i} \in \overline{\mathcal{P}_i}$ and for all $i \in [1, \omega)$, $\psi_i(\overline{T_i}) \neq U_i$;
3. for all $i < \omega$, $U_i = \{(\varphi, s) : s \in [\delta^{\mathcal{N}_i}]^{<\omega} \wedge \mathcal{S} \models \varphi[F, \pi^{\Sigma_i}_{\mathcal{N}_i | \delta^{\mathcal{N}_i}, \infty}(s)]\}$ where Σ_i is the σ_i-pullback of $\overline{\Phi_i}^+$;
4. for all $i < \omega$, $\overline{T_i} = \{\varphi, s) \in : s \in [\delta^{\overline{\mathcal{P}_i}}]^{<\omega}$ and $\mathcal{S} \models \varphi[F, \pi^{\overline{\Phi_i}^+}_{\overline{\mathcal{P}_i} | \delta^{\overline{\mathcal{P}_i}}, \infty}(s)]\}$;
5. $\mathcal{S} \in \mathrm{HOD}^{L(\Gamma(\overline{\mathcal{P}_0^+}, \overline{\Pi_0}^+))}$.

Now we define by induction $\psi_i^+ : \overline{\mathcal{P}_i^+} \to \mathcal{N}_i^+$, $\sigma_i^+ : \mathcal{N}_i^+ \to \overline{\mathcal{P}_{i+1}^+}$, $\overline{\varphi_i^+} : \overline{\mathcal{P}_i^+} \to \overline{\mathcal{P}_{i+1}^+}$ as follows. $\overline{\varphi_0^+} : \overline{\mathcal{P}_0^+} \to \overline{\mathcal{P}_1^+}$ is the ultrapower map by the $(\text{crit}(\overline{\varphi_0}), \delta^{\mathcal{P}_1})$-extender derived from $\overline{\varphi_0}$. Note that $\overline{\varphi_0^+}$ extends $\overline{\varphi_0}$. Let $\psi_0^+ : \overline{\mathcal{P}_0^+} \to \mathcal{N}_0^+$ be the ultrapower map by the $(\text{crit}(\psi_0), \delta^{\mathcal{N}_0})$-extender derived from ψ_0. Again ψ_0^+ extends ψ_0. Finally let $\sigma_0^+ = (\overline{\varphi_0^+})^{-1} \circ \psi_0^+$. The maps $\psi_i^+, \sigma_i^+, \overline{\varphi_i^+}$ are defined similarly. Let $\overline{\mathcal{P}_\omega^+}$ be the direct limit of the linear system $(\overline{\mathcal{P}_i^+}, \overline{\varphi_{i,k}^+} : i < k < \omega)$ where $\overline{\varphi_{i,k}^+}$ is the composition of $(\overline{\varphi_m^+} : m \in [i,k))$. Let $\overline{\varphi_{i,\omega}^+} : \overline{\mathcal{P}_i^+} \to \overline{\mathcal{P}_\omega^+}$ and $\sigma_{i,\omega}^+ : \mathcal{N}_i^+ \to \overline{\mathcal{P}_\omega^+}$ be the direct limit embeddings.

Let now Hypo be the following statement:

Hypo: There is an (ω_1, ω_1+1)-iteration strategy for $\overline{\Pi_\omega}$ for $\overline{\mathcal{P}_\omega^+}$ such that the following clauses hold:

Hypo1: $\overline{\Pi_\omega}$ acts on stacks that are above $\delta^{\overline{\mathcal{P}_\omega}}$ where $\overline{\mathcal{P}_\omega} = \overline{\varphi_{0,\omega}}(\overline{\mathcal{P}_0})$.

Hypo2: For every $i < \omega$[29], letting $\overline{\Pi_i}$ be the $\overline{\varphi_{i,\omega}^+}$-pullback of $\overline{\Pi_\omega}$, $\overline{\Pi_i}$ witnesses that $\overline{\mathcal{P}_i^+}$ is a $\overline{\Phi_i}^+$-hod mouse over $\overline{\mathcal{P}_i}$.

We have that $N[g] \models$ Hypo. Indeed, let for $i < \omega$, $m_i = \tau_{X_i} \circ (\pi \restriction \overline{\mathcal{P}_i})$ and let $m_\omega : \overline{\mathcal{P}_\omega} \to j(\mathcal{P})$ be the canonical embedding built via the direct limit construction. We thus have that for each i, $m_i = m_\omega \circ \overline{\varphi_{i,\omega}}$. Just like $\overline{\varphi_i}$, we can extend m_i (for $i \le \omega$) to $m_i^+ : \overline{\mathcal{P}_i^+} \to j(\overline{\mathcal{P}_0^+})$. The desired strategy $\overline{\Pi_\omega}$ is m_ω^+-pullback of $j(\overline{\Pi_0}^+ \restriction (H_{\kappa^+}^M))$. Because m_i^+ extends $\tau_{X_i} \circ (\pi \restriction \overline{\mathcal{P}_i})$, we have that Hypo holds.

We now show how to finish the proof assuming Hypo. By a similar argument as in [64, Theorem 3.1.25] or as in [31, Page 663, just before (8) in the proof of Lemma 11.15], we can use the strategies $\overline{\Pi_i}^+$'s to simultaneously execute a $\mathbb{R}^{V[G]}$-genericity iterations. The process yields a sequence of models $\langle \overline{\mathcal{P}_{\omega,i}^+}, \mathcal{N}_{\omega,i}^+ \mid i \le \omega \rangle$ and maps $\psi_{\omega,i}^+ : \overline{\mathcal{P}_{\omega,i}^+} \to \mathcal{N}_{\omega,i}^+$, $\sigma_{\omega,i}^+ : \mathcal{N}_{\omega,i}^+ \to \overline{\mathcal{P}_{\omega,i+1}^+}$, and $\overline{\varphi_{\omega,i}^+} = \psi_{\omega,i}^+ \circ \sigma_{\omega,i}^+$[30]. The iteration described above uses σ_i^+-pullback of $\overline{\Pi_i}$ to iterate $\mathcal{N}_i^+$. We denote this strategy by Σ_i^+.

Because the genericity iterations are above $\text{ord}(\overline{\mathcal{P}_i})$ and $\text{ord}(\mathcal{N}_i)$ for all $i \le \omega$ and by [29, Theorem 3.26], the interpretation of the strategy of $\overline{\mathcal{P}_i}$ ($\mathcal{N}_i$ respectively) in the derived model of $\overline{\mathcal{P}_{\omega,i}^+}$ ($\mathcal{N}_{\omega,i}^+$, respectively) is $\overline{\Phi_i}^+$ (Σ_i, respectively). Let C_i be the derived model of $\overline{\mathcal{P}_{\omega,i}^+}$ and D_i be the derived model of $\mathcal{N}_{\omega,i}^+$ (at the sup of the Woodin cardinals of each model). Then $\mathbb{R}^{V[G]} = \mathbb{R}^{C_i} = \mathbb{R}^{D_i}$. Furthermore, $C_i \cap \wp(\mathbb{R}) \subseteq D_i \cap \wp(\mathbb{R}) \subseteq C_{i+1} \cap \wp(\mathbb{R})$ for all i.

Notice that we in fact have that $C_i = L(\Gamma(\overline{\mathcal{P}_i^+}, \overline{\Pi_i}))$ and $D_i = L(\Gamma(\mathcal{N}_i^+, \Sigma_i^+))$. Therefore, it follows from our choice of Π_0, $(F, \mathcal{S}) \in \cap_{i \le \omega}(C_i \cap D_i)$, and since $\mathcal{S}$ is

[29] Including $i = 0$.

[30] This embedding should not be confused with $\overline{\varphi_{i,\omega}^+}$.

ordinal definable in each of C_i and D_i, $(F,\mathcal{S}) \in \bigcap_{i\leq\omega}(\overline{\mathcal{P}^+_{\omega,i}} \cap \mathcal{N}^+_{\omega,i})$. It follows that for each $i<\omega$, $\overline{T_i}$ and U_i are definable respectively in $\overline{\mathcal{P}^+_{\omega,i}}$ and $\mathcal{N}^+_{\omega,i}$ from $(F,\mathcal{S})$. Indeed, we have that

(F) for every $i<\omega$, $(\varphi,s) \in \overline{T_i}$ if and only if $s \in [\delta^{\overline{\mathcal{P}_i}}]^{<\omega}$ and in the derived model of $\overline{\mathcal{P}^+_{\omega,i}}$ at $\delta^{\overline{\mathcal{P}^+_{\omega,i}}}(=\omega_1^{N[g]})$, $\mathcal{S} \models \varphi[F, \pi^{\overline{\Phi_i}^+}_{\overline{\mathcal{P}_i}|\delta^{\overline{\mathcal{P}_i}},\infty}(s)]\}$[31],

(G) for every $i<\omega$, $(\varphi,s) \in U_i$ if and only if $s \in [\delta^{\mathcal{N}_i}]^{<\omega}$ and in the derived model of $\mathcal{N}^+_{\omega,i}$ at $\delta^{\mathcal{N}^+_{\omega,i}}(=\omega_1^{N[g]})$, $\mathcal{S} \models \varphi[F, \pi^{\Sigma_i}_{\mathcal{N}_i|\delta^{\mathcal{N}_i},\infty}(s)]\}$[32].

Notice next that

(H) for every $i<\omega$, $\psi^+_{\omega,i}(\overline{\mathcal{P}_i}) = \mathcal{N}_i$ and $\psi^+_{\omega,i}(\overline{\Phi_i}) = \Sigma_i$[33],

(I) there is $i_0<\omega$ such that for every $i \geq i_0$, $\varphi^+_{\omega,i}(\mathcal{S},F) = (\mathcal{S},F)$ and $\psi^+_{\omega,i}(\mathcal{S},F) = (\mathcal{S},F)$[34].

Thus, if i_0 is as in (I) then $\psi_{i_0}(\overline{T_{i_0}}) = U_{i_0}$, contradiction! ⊣

It is now easy to derive Theorem 9.2.2 from Theorem 9.2.7 and Theorem 9.2.6.

Theorem 9.2.2 is typically applied in core model induction applications where there exists a mild large cardinal (e.g., a measurable cardinal) that gives rise to the embedding j as in the hypothesis of the theorem. Below, we outline the proof of the following theorem, which gives the existence of condensing sets in some situations where large cardinals may not exist (e.g., under PFA). One applies Theorem 9.2.2 in applications where the core model induction is carried out in $V^{Coll(\omega,<\kappa)}$ and applies Theorem 9.2.9 applications where the core model induction is carried out in $V^{Coll(\omega,\kappa)}$.

In the following, we use the notations as in 9.2.5 and in the previous section (in particular, $\mathcal{P} = \mathcal{P}_{(\varphi,A)}, \Sigma = \Sigma_{(\varphi,A)}$ etc. In the case $A \in V$, we define $\mathcal{P}^+ = \mathcal{P}^+_{\varphi,A} = \mathsf{Lp}^{\Sigma,\Gamma,c}(\mathcal{P}^-)$ to be the union of sound Σ-premice $\mathcal{M}$ such that $\rho_\omega(\mathcal{M}) \leq \mathsf{ord}(\mathcal{P}^-)$ such that whenever $\pi : \mathcal{M}^* \to \mathcal{M}$ is elementary and $\mathcal{M}^* \in V$ is countable, then there is a unique iteration strategy $\Lambda \in \Gamma$ witnessing $\mathcal{M}^*$ is a Σ^π-mouse. We note that $\mathcal{P}, \mathcal{P}^+ \in V$ and $\mathcal{P} \trianglelefteq \mathcal{P}^+$ though in general, equality may not hold.

THEOREM 9.2.9. *Suppose κ is a cardinal such that $\kappa^\omega = \kappa$, and $\kappa \geq 2^{2^{\aleph_0}}$. Let $g \subseteq Coll(\omega,\kappa)$ be V-generic. Suppose φ is a formula in the language of set theory and $A \in V$ such that $V[g] \models$ "(φ,A) is full, homogeneous and lower part (φ,A)-covering fails". Furthermore, suppose* $\mathrm{cof}(\mathsf{ord}(\mathcal{P}^+)) \leq \kappa$. *Then $V[g] \models$ "there is a strongly (φ,A)-condensing set".*

---

[31] Here we abuse notation and use $\overline{\Phi_i}^+$ for the extension of $\overline{\Phi_i}$ to the derived model of $\overline{\mathcal{P}^+_{\omega,i}}$.

[32] Similar comments like above apply here as well.

[33] By this equation we mean that the internal strategy of $\overline{\mathcal{P}^+_i}$ is mapped to the internal strategy of $\mathcal{N}_i$.

[34] Because $\mathcal{S}$ and F are ordinal definable in the respective derived models.

REMARK 9.2.10. The assumption "cof(ord($\mathcal{P}^+$)) $\leq \kappa$" in the above theorem holds in many situations, e.g., PFA. If $\mathcal{P} = \mathcal{P}^+$ then this clause is superfluous as it is implied by the failure of lower part (φ, A)-covering.

The rest of the section is dedicated to outlining the proof of the theorem. We assume the hypothesis of the theorem from now to the end of this section. Let $\lambda >> \kappa$ and $X \prec (H_\lambda, \varepsilon)$. We say that X is *good* if $|X| = \kappa$, $\kappa \subset X$, $X^\omega \subseteq X$, $\{\mathcal{P}, \mathcal{P}^+, \Sigma \restriction V\} \in X$, and $X \cap \mathcal{P}^+$ is cofinal in $\mathcal{P}^+$.

Let X be good. Let $\pi_X : M_X \to (H_\lambda, \varepsilon)$ be the uncollapse map. π_X extends uniquely to a map $\pi_X^+ : M_X[g] \to H_\lambda[g]$. We let γ_X be the critical point of π_X and $\pi_X^+(\mathcal{P}_X^-, \mathcal{P}_X, \mathcal{P}_X^+, \Sigma_X, \Gamma_X) = (\mathcal{P}^-, \mathcal{P}, \mathcal{P}^+, \Sigma, \Gamma)$; in general, if $a \in H_\lambda[g]$ is in the range of π_X^+, then we let $a_X = \pi_X^{+,-1}(a)$. We say that a good X is *c*Γ*-full* if $\mathcal{P}_X^+ = \mathsf{Lp}^{\Sigma^{\pi_X}, \Gamma, c}(\mathcal{P}_X^-)$ and is Γ*-full* if $\mathcal{P}_X = \mathsf{Lp}^{\Sigma^{\pi_X}, \Gamma}(\mathcal{P}_X^-)$. It is clear that

$$\mathsf{Lp}^{\Sigma^{\pi_X}, \Gamma}(\mathcal{P}_X^-) \trianglelefteq \mathsf{Lp}^{\Sigma^{\pi_X}, \Gamma, c}(\mathcal{P}_X^-).$$

LEMMA 9.2.11. *The set S of cΓ-full X is stationary. Furthermore, there is a stationary $T \subseteq S$ such that for each $X \in T$, X is Γ-full.*

PROOF. We first show the first clause implies the second. Suppose the set S of $c\Gamma$-full X is stationary and for contradiction, suppose that there is a club C such that for all $X \in C \cap S$, X is not Γ-full. Let $(X_\alpha, \mathcal{M}_\alpha : \alpha < \kappa^+)$ be such that

1. $(X_\alpha : \alpha < \kappa^+)$ is an increasing and continuous sequence in C such that for all successor α, $X_\alpha \in S$.
2. For each α, $\mathcal{M}_\alpha \in \mathsf{Lp}^{\Sigma^{\pi_{X_\alpha}}, \Gamma}(\mathcal{P}_{X_\alpha}^-) \backslash \mathcal{P}_{X_\alpha}$.

Letting $\mathcal{P}_\alpha = \mathcal{P}_{X_\alpha}, \mathcal{P}_\alpha^+ = \mathcal{P}_{X_\alpha}^+, \pi_{X_\alpha} = \pi_\alpha$ etc., we note that for any successor α,

$$\mathcal{M}_\alpha \triangleleft \mathcal{P}_\alpha^+.$$

This easily implies that letting $\pi_{\alpha,\beta} = \pi_\beta^{-1} \circ \pi_\alpha$ for $\alpha < \beta$, then for all successor ordinals $\alpha < \beta$,

$$\pi_{\alpha,\beta}(\mathcal{M}_\alpha) = \mathcal{M}_\beta.$$

Now let $\mathcal{M}$ be the direct limit of $(\pi_{\alpha,\beta}, \mathcal{M}_\alpha : \alpha < \beta \wedge X_\alpha, X_\beta \in S)$, then

$$\mathcal{P} \trianglelefteq \mathcal{M} \trianglelefteq \mathcal{P}^+.$$

In particular, $\mathcal{M}$ is not an initial segment of $\mathcal{P}$. Now we show $\mathcal{M} \triangleleft \mathcal{P}$, which is a contradiction. Let $\pi : \mathcal{M}^* \to \mathcal{M}$ be elementary with $\mathcal{M}^* \in V[g]$ countable, transitive. Then there is $X_\alpha \in S$ and $\tau : \mathcal{M}^* \to \mathcal{M}_\alpha$ such that $\pi_\alpha \circ \tau = \pi$. This implies:

1. $\Sigma^\pi = \Sigma_\alpha^\tau$, and
2. by the definition of $\mathcal{M}_\alpha$, $\mathcal{M}^*$ is a Σ^π-mouse with unique strategy in Γ.

This shows $\mathcal{M} \triangleleft \mathcal{P}$. Contradiction.

Now we prove the first clause; the idea of the proof is basically that of [12, Theorem 3.4]. Suppose the set W of good X such that X is not $c\Gamma$-full contains a κ-club. Let $\eta = cof^V(\mathsf{ord}(\mathcal{P}^+))$ and $(\mathcal{M}_i : i < \eta)$ be an enumeration of a cofinal sequence of sound $\mathcal{M}$ such that $\rho_1(\mathcal{M}) \leq \mathrm{ord}(\mathcal{P}^-)$ and $\mathcal{P}^- \lhd \mathcal{M} \lhd \mathcal{P}^+$. Let $(X_\alpha : \alpha < \kappa^+)$ enumerate an increasing, continuous sequence such that for successor ordinal α or limit α of cofinality $\geq \omega_1$, $X_\alpha \in W$. We use the notation as above, writing for example $\mathcal{P}_\alpha = \mathcal{P}_{X_\alpha}$. We also write Θ for $\mathrm{ord}(\mathcal{P}^-)$, ϑ_α for $\pi_\alpha^{-1}(\Theta)$, and γ_α for γ_{X_α}. We assume $(\mathcal{M}_i : i < \eta) \in X_0$ and let $(\mathcal{M}_i^\alpha : i < \eta) = \pi_\alpha^{-1}((\mathcal{M}_i : i < \eta)$.

For each α, let $\mathcal{N}_\alpha$ be the least sound $\mathcal{N}$ such that $\mathcal{P}_\alpha^+ \lhd \mathcal{N} \lhd \mathrm{Lp}^{\Sigma^{\pi_X},\Gamma,c}(\mathcal{P}_X^-)$ and $\rho_\omega(\mathcal{N}_\alpha) \leq \mathrm{ord}(\mathcal{P}_\alpha^-)$. Let n_α be the least n such that $\rho_{n+1}(\mathcal{N}_\alpha) \leq \mathrm{ord}(\mathcal{P}_\alpha^-)$. Let $\pi_\alpha^* : \mathcal{N}_\alpha \to \mathcal{Q}_\alpha$ be the corresponding $r\Sigma_{n_\alpha}$ ultrapower map given by the extender of length Θ derived from π_α; similarly, we define $\pi_{\alpha,\beta}^* : \mathcal{N}_\alpha \to \mathcal{Q}_\alpha^\beta$ from $\pi_{\alpha,\beta}$. Note that the objects $\mathcal{Q}_\alpha, \mathcal{Q}_\alpha^\beta$ are all well-founded and hence we identify them with their transitive isomorph. By the assumption that $\eta < \kappa$ and X_α is good, the map π_α is cofinal in $\mathrm{ord}(\mathcal{P}^+)$ and therefore, $\neg(\mathcal{Q}_\alpha \lhd \mathcal{P}^+)$.

So there is an n such that

$$C = \{\alpha < \kappa^+ : \gamma_\alpha = \alpha \wedge n_\alpha = n \wedge \mathrm{cof}(\alpha) \geq \omega_1\}$$

contains a ω_1-club. Fix an $\alpha \in C$ for now and let $(Y_\beta : \beta < \kappa^+)$ be a continuous, increasing sequence of $Y \prec H_\lambda$ such that $Y \cap \kappa^+ \in \kappa^+$, $Y^\omega \subset Y$ and $(\mathcal{N}_\alpha, \mathcal{Q}_\alpha, \pi_\alpha^*) \in Y$. For each β, let $\sigma_\beta : H_\beta \to Y_\beta$ be the uncollapse map and $\kappa_\beta = \mathrm{crt}(\sigma_\beta)$. As in the proof of [12, Theorem 3.4], we get a club

$$C_\alpha = \{\beta < \kappa^+ : \kappa_\beta = \beta \wedge \mathrm{rng}(\pi_\beta) \cap \mathcal{P}^+ = Y_\beta \cap \mathcal{P}^+\}$$

and furthermore, using the agreement between σ_β with π_β on points in $\mathcal{P}^+$, we also get for $\beta \in C_\alpha$,

$$\mathcal{Q}_\alpha^\beta = \sigma_\beta^{-1}(\mathcal{Q}_\alpha).^{35}$$

Fix $\beta \in \triangle_{\alpha<\kappa^+} C_\alpha \cap C$ such that $\mathrm{cof}(\beta) \neq \eta$. This is possible because $\eta \leq \kappa$ and $\kappa \geq \omega_2$. For simplicity, let us assume $\rho_1(\mathcal{N}_\beta) = \Theta_\beta$[36]. So we have $\rho_0(\mathcal{N}_\beta) = \mathrm{ord}(\mathcal{N}_\beta)$ and

$$\eta = \mathrm{cof}(\mathrm{ord}(\mathcal{P}_\beta^+)) = \mathrm{cof}(\mathrm{ord}(\mathcal{N}_\beta)) \leq \kappa.^{37}$$

Now we let $(\delta_i : i < \eta)$ be cofinal in $\mathrm{ord}(\mathcal{N}_\beta)$ and for each $i < \eta$, let $\overline{\sigma_i} : \overline{\mathcal{N}_i} \to Hull_1^{\mathcal{N}_\beta|\delta_i}(\Theta_\beta \cup p_1(\mathcal{N}_\beta))$ be the uncollapse map. By condensation, $\overline{\mathcal{N}_i} \lhd \mathcal{P}_\beta^+$ for each i.

[35] As in the proof of [12, Theorem 3.4], the map $\varphi : \mathcal{Q}_\alpha^\beta \to \sigma_\beta^{-1}(\mathcal{Q}_\alpha)$ defined as: $\varphi(\pi_{\alpha,\beta}^*(f)(a)) = \sigma_\beta^{-1} \circ \pi_\alpha^*(f)(a)$ for $a \in [\Theta_\beta]^{<\omega}$ and letting δ be such that $\pi_{\alpha,\beta}(\delta) > \max(a)$, $f : [\delta]^{|a|} \to \mathcal{N}_\alpha$ come from the level n Skolem term over $\mathcal{N}_\alpha$, is a well-defined elementary map and surjective. Therefore, it must be the identity.

[36] The general case where $n > 1$ is the least such that $\rho_n(\mathcal{N}_\beta) = \Theta_\beta$ is handled just as in [12, Theorem 3.4] by working with the n-reduct.

[37] The second equality follows from [12, Lemma 1.2].

Since $\mathrm{cof}(\beta) \neq \eta$ and $\Theta_\beta = \bigcup_{\alpha<\beta} \pi''_{\alpha,\beta}\Theta_\alpha$, so $\mathcal{P}^+_\beta$ is the direct limit of the $\mathcal{P}^+_\alpha$ under the maps $\pi_{\alpha,\beta}$, there is an $\alpha < \beta$ and there are cofinal sets $T, T' \subset \eta$ such that

$$i \in T \implies \mathcal{M}^\beta_i, p(\mathcal{M}^\beta_i) \in Hull_1^{\mathcal{N}_\beta}(\pi''_{\alpha,\beta}\Theta_\alpha \cup p_1(\mathcal{N}_\beta))$$

and

$$i \in T' \longrightarrow \overline{\mathcal{N}_i}, \overline{\sigma_i}^{-1}(p_1(\mathcal{N}_\beta)) \in \mathrm{rng}(\pi_{\alpha,\beta}).$$

Now we claim that for the α above,

$$Hull_1^{\mathcal{N}_\beta}(\pi''_{\alpha,\beta}\Theta_\alpha \cup p_1(\mathcal{N}_\beta)) \cap \mathsf{ord}(\mathcal{P}^+_\beta) = \mathrm{rng}(\pi_{\alpha,\beta}) \cap \mathsf{ord}(\mathcal{P}^+_\beta). \qquad (1)$$

Suppose $\xi \in \mathrm{rng}(\pi_{\alpha,\beta}) \cap \mathsf{ord}(\mathcal{P}^+_\beta)$. Let $\pi_{\alpha,\beta}(\overline{\xi}) = \xi$ for some $\overline{\xi} < \mathsf{ord}(\mathcal{P}^+_\alpha)$. There is some $i \in T$ such that

$$\overline{\xi} \in Hull_1^{\mathcal{M}^\alpha_i}(\Theta_\alpha \cup p_1(\mathcal{M}^1_\alpha)).$$

This implies

$$\xi \in Hull_1^{\mathcal{M}^\beta_i}(\pi''_{\alpha,\beta}\Theta_\alpha \cup p_1(\mathcal{M}^i_\beta)) \subset Hull_1^{\mathcal{N}^\beta}(\pi''_{\alpha,\beta}\Theta_\alpha \cup p_1(\mathcal{N}_\beta)).$$

Conversely, let $\xi \in Hull_1^{\mathcal{N}_\beta}(\pi''_{\alpha,\beta}\Theta_\alpha \cup p_1(\mathcal{N}_\beta)) \cap \mathsf{ord}(\mathcal{P}^+_\beta)$. So there is $i \in T'$, a Skolem term τ, and parameter $\vec{\varepsilon} \in [\pi''_{\alpha,\beta}\Theta_\alpha]^{<\omega}$ such that $\xi = \tau^{\mathcal{N}_\beta|\delta_i}[\vec{\varepsilon}, p_1(\mathcal{N}_\beta)]$. We may also assume $\Theta_\beta \in Hull_1^{\mathcal{N}^\beta|\delta_i}(\pi''_{\alpha,\beta}\Theta_\alpha \cup p_1(\mathcal{N}_\beta))$; this is possible by the "$\supseteq$" direction of (1), which we just proved. We easily get that $\xi \in Hull_1^{\mathcal{N}_\beta|\delta_i}(\Theta_\beta \cup p_1(\mathcal{N}_\beta))$, hence $\xi \in Hull_1^{\overline{\mathcal{N}_i}}(\Theta_\beta \cup \overline{\sigma_i}^{-1}(p_1(\mathcal{N}_\beta)))$. So in fact,

$$\xi = \tau^{\overline{\mathcal{N}_i}}[\vec{\varepsilon}, \overline{\sigma_i}^{-1}(p_1(\mathcal{N}_\beta))].$$

This implies

$$\xi \in Hull_1^{\overline{\mathcal{N}_i}}(\pi''_{\alpha,\beta}\Theta_\alpha \cup \overline{\sigma_i}^{-1}(p_1(\mathcal{N}_\beta))) \subset \mathrm{rng}(\pi_{\alpha,\beta})$$

as desired.

Now we finish the proof of the theorem. Let $\overline{\sigma} : \overline{\mathcal{N}} \to \mathcal{N}_\beta$ be the uncollapse map. By (1), $\overline{\sigma}$ and $\pi_{\alpha,\beta}$ agree on $\mathrm{rng}(\pi_{\alpha,\beta}) \cap \mathsf{ord}(\mathcal{P}^+_\beta)$. Therefore, $\overline{\mathcal{N}}|\mathsf{ord}(\mathcal{P}^+_\alpha) = \mathcal{N}_\alpha|\mathsf{ord}(\mathcal{P}^+_\alpha)$ and $\mathcal{Q}^\beta_\alpha|\mathsf{ord}(\mathcal{P}^+_\beta) = \mathcal{N}_\beta|\mathsf{ord}(\mathcal{P}^+_\beta)$. Now let $\pi : M \to H_\lambda$ be elementary with M countable transitive and $\mathrm{rng}(\pi)$ containing all relevant objects. We let $\pi(\overline{\mathcal{M}}) = \overline{\mathcal{N}}$ and $\pi(\overline{\mathcal{M}}^*) = \mathcal{N}_\alpha$. Then note that $\overline{\mathcal{M}}$ is a $\Sigma_\beta^{\pi\circ\pi_{\alpha,\beta}}$-mouse and $\overline{\mathcal{M}}^*$ is a Σ^π_α-mouse. But $\Sigma_\alpha = \Sigma_\beta^{\pi_{\alpha,\beta}}$ so in fact, $\overline{\mathcal{M}}$ is a Σ^π_α-mouse. This easily implies $\overline{\mathcal{M}} = \overline{\mathcal{M}}^*$. By elementarity, $\overline{\mathcal{N}} = \mathcal{N}_\alpha$. Finally, using $\overline{\mathcal{N}} = \mathcal{N}_\alpha$ and the agreement between $\overline{\sigma}$ and $\pi_{\alpha,\beta}$, we have

$$\mathcal{Q}^\beta_\alpha = \mathcal{N}_\beta.$$

By pressing down, there is an α and a stationary set Y of β such that for all $\beta \in Y$, $\mathcal{Q}_\alpha^\beta = \mathcal{N}_\beta$ is the ultrapower of $\mathcal{N}_\alpha$ using the extender of length Θ_β derived from $\pi_{\alpha,\beta}$. Let $\mathcal{N}$ be the direct limit of such the $\mathcal{N}_\beta$ under these ultrapower maps. Then we easily get $\mathcal{N} = \mathcal{Q}_\alpha$ and hence

$$\mathcal{Q}_\alpha \lhd \mathcal{P}^+.$$

Contradiction. ⊣

REMARK 9.2.12. The proof of the above lemma just requires a bit less of κ than the hypothesis of Theorem 9.2.9, namely we just need $\kappa \geq \omega_2$ and $\kappa^\omega = \kappa$.

The following theorems are the corresponding versions of Theorem 9.2.6 and Theorem 9.2.7 and immediately imply Theorem 9.2.9. Before proving Theorem 9.2.14, we note that the set of good X contains an ω_1-club and if X is good, then X contains $\wp(\mathbb{R})$ because $\kappa \geq 2^{2^\omega}$. The following lemma will be used in the proof of both theorems.

LEMMA 9.2.13. *Suppose X is good and Γ-full. Suppose $(\mathcal{P}_X^*, \Pi)$ is a Σ_X-hod pair in Γ such that $(\mathcal{P}_X^*, \Pi \restriction V) \in V$. Let $\mathcal{P}^* = \mathrm{Ult}(\mathcal{P}_X^*, E)$ be the ultrapower of $\mathcal{P}_X^*$ via the extender of length Θ derived from π_X and π_X^* be the ultrapower map. Let $k : \mathcal{R}^* \to \mathcal{P}^*$ be elementary and $\mathcal{R}^*$ is countable, transitive in V. Then there is a Σ_0-elementary map $\overline{\pi} : \overline{\mathcal{R}^*} \to \mathcal{P}_X^+$ such that letting $\mathcal{R} = k^{-1}(\mathcal{P})$, then $\Sigma^{k \restriction \mathcal{R}} = \Sigma_X^{\overline{\pi} \restriction \mathcal{R}}$.*[38]

PROOF. First we note that E is a total extender over $\mathcal{P}_X^*$ because Lemma 9.2.11 implies that $\wp^{\mathcal{P}_X^*}(\Theta_X) \subset M_X$. So the definition of $\mathcal{P}^*$ makes sense.

The proof of this lemma is essentially that of [41, Lemma 8.12] but with an additional detail. We will use the notations as introduced in [41, Section 8] regarding extenders. First let $W = \{(\mathcal{P}_\alpha, \Sigma_\alpha) : \alpha < 2^{2^\omega}\}$ enumerate all countable hod pairs in V such that $\Sigma_\alpha \in \Gamma$[39]. Since X is good, $W \subseteq X$; this is where we use $\kappa \geq 2^{2^\omega}$ in an essential way. Let α be such that $\Sigma_\alpha = \Sigma^{\pi^* \restriction \mathcal{R}}$ and $\mathcal{R} = \mathcal{Q}_\alpha$.

Let $U = \mathrm{rng}(\overline{\pi})$ and $((a_i, A_i) : i < \omega)$ enumerate all pairs (c, A) such that there is a Σ_0-formula ψ and $[a^1, f_1]_E, \ldots, [a^k, f_k]_E \in U$ such that

$$A = \{u \in \mathrm{ord}(\mathcal{P}_X)^{|c|} : \mathcal{P}_X^* \models \psi[f_1^{a^1,c}(u), \ldots, f_k^{a_k,c}(u)]\} \in E_c.$$

Let $a \subset \omega$ be the set of n such that $[a_n, f_n]_E$ represents some element of $\mathcal{P}$. Let $\{\tau_n : n < \omega\}$ enumerate all the Skolem functions of $\mathcal{P}_X^*$ and $b = \{i : \exists n \in a\ f_n = \tau_{n_i}\}$. So $\{\pi_X^*(f_n)(a_n) : n \in a\}$ is an elementary substructure of $\mathcal{P}$. In H_λ, the following first order statement with parameters $(\mathcal{Q}_\alpha, \Sigma_\alpha), (\mathcal{P}, \Sigma)$ holds: "there is a sequence $(a_n : n < \omega)$ of finite sets of ordinals such that for each n, $a_n \in \pi_X(A_n)$ and $\mathcal{Q}_\alpha = \{\pi_X(\tau_{n_i})(a_n) : i \in b \wedge n \in a\} \prec_{\Sigma_0} \mathcal{P}$ and Σ_α is the pullback of Σ under

[38] D. Adolf has observed that this lemma holds and can be used to prove Lemma 9.2.11. However, Lemma 9.2.13 uses essentially that $\kappa \geq 2^{2^\omega}$, while Lemma 9.2.11 holds with less required of κ.

[39] We confuse Σ_α with its canonical extension in $V[g]$.

the uncollapse map". So by elementarity, the corresponding statement holds in M_X: "there is a sequence $(a_n : n < \omega)$ of finite sets of ordinals such that for each n, $a_n \in A_n$ and $\mathcal{Q}_\alpha = \{\tau_{n_i}(a_n) : i \in b \wedge n \in a\} \prec_{\Sigma_0} \mathcal{P}_X$ and Σ_α is the pullback of Σ_X under the uncollapse map". Let $(\bar{a}_n : n < \omega)$ witness the above statement.

The embedding $\overline{\pi}$ is defined by: $\overline{\pi}([a_n, f_n]_E) = f_n(\bar{a}_n)$ is the desired embedding with the property that

$$\Sigma_X^{\overline{\pi}\restriction\mathcal{R}} = \Sigma^{k\restriction\mathcal{R}}. \qquad \dashv$$

THEOREM 9.2.14. *There is a stationary set $S' \subset S$ such that whenever $X \in S'$, $X \cap \mathcal{P}$ is a weakly (φ, A)-condensing set.*

PROOF. Suppose not. Fix a good X such that X is Γ-full but $X \cap \mathcal{P}$ is not a weakly condensing set. Note that $\pi_X \restriction \mathcal{P}_X$ is cofinal in $\mathcal{P}$. Let Y be an extension of $X \cap \mathcal{P}$ such that $(\mathcal{Q}_Y, \Sigma_Y)$ has the following properties:

(i) letting $k = \tau_Y$, $\Sigma_Y = \Sigma^k$;
(ii) $\mathcal{Q}_Y$ is not Γ-full, so letting $\mathcal{R} = \mathcal{Q}_Y$, there is a sound Σ_Y-mouse $\mathcal{M}$ such that $\neg(\mathcal{M} \trianglelefteq \mathcal{R})$ and $\rho_\omega(\mathcal{M}) = \delta^{\mathcal{R}}$.

By definition, $\tau_X = \tau_Y \circ \tau_{X,Y}$ ($\tau_X = \pi_X \restriction \mathcal{P}_X$ here). Let $(\mathcal{P}_X^*, \Lambda_X) \in V$ be a Σ_X-hod pair such that

1. $\Gamma(\mathcal{P}_X^*, \Lambda_X) \vDash \mathcal{R}$ is not full as witnessed by $\mathcal{M}$.[40]
2. $\Lambda_X \in \Gamma$ is Γ-fullness preserving and has strong branch condensation.
3. $\mathcal{P}_X^*$ is meek, is of limit type, and $\mathrm{cof}^{\mathcal{P}_X^*}(\delta^{\mathcal{P}_X^*}) = \omega$.

Such a pair $(\mathcal{P}_X^*, \Lambda_X)$ exists by boolean comparisons. In particular, $\mathcal{P}_X^*$ is a Σ_X-hod premouse over $\mathcal{P}_X$.

By arguments similar to before or that used in [65, Lemma 3.78], no $\mathcal{M} \triangleleft \mathcal{P}_X^*$ is such that $\rho_\omega(\mathcal{M}) < \mathrm{ord}(\mathcal{P}_X^-)$ and in fact, $\mathrm{ord}(\mathcal{P}_X)$ is a cardinal of $\mathcal{P}_X^*$.

By the above argument, $\mathcal{P}_X^*$ thinks $\mathcal{P}_X$ is full. Let

$$\pi_X^* : \mathcal{P}_X^* \to \mathcal{P}^*$$

be the ultrapower map by the extender E of length Θ induced by π_X. Note that π_X^* extends $\pi_X \restriction \mathcal{P}_X$ (since π_X is cofinal in $\mathcal{P}$) and $\mathcal{P}^*$ is wellfounded since X is closed under ω-sequences. Let

$$i^* : \mathcal{P}_X^* \to \mathcal{R}^+$$

be the ultrapower map by the extender of length $\delta^{\mathcal{R}}$ induced by $i =_{\mathrm{def}} \tau_{X,Y}$. Note that $\mathcal{R} \triangleleft \mathcal{R}^+$ and $\mathcal{R}^+$ is wellfounded since there is a natural map

$$k^* : \mathcal{R}^+ \to \mathcal{P}^*$$

[40]For brevity, we suppress mentioning the pair $(\mathcal{S}, \Phi)$ as in the proof of Theorem 9.2.6 and instead focus on the main points of the proof.

extending k such that $\tau_X^* = k^* \circ i^*$. Without loss of generality, we may assume $\mathcal{M}$'s unique strategy $\Sigma_{\mathcal{M}} \leq_w \Lambda_X$. Also, let $(\dot{\mathcal{R}}, \dot{\mathcal{M}})$ be the canonical $Col(\omega, \kappa)$-names for $(\mathcal{R}, \mathcal{M})$. Let K be the transitive closure of $H_\kappa^V \cup (\dot{\mathcal{R}}, \dot{\mathcal{M}})$.

Let $\mathcal{W} = \mathcal{M}_\omega^{\Lambda_X,\sharp}$ and Λ be the unique strategy of $\mathcal{W}$. Let $\mathcal{W}^*$ be a Λ-iterate of $\mathcal{W}$ below its first Woodin cardinal that makes K-generically generic. Then in $\mathcal{W}^*[K]$, the derived model $D(\mathcal{W}^*[K])$ satisfies

$$L(\Gamma(\mathcal{P}_X^*, \Lambda_X), \mathbb{R}) \vDash \dot{\mathcal{R}} \text{ is not full as witnessed by } \mathcal{M}.^{41}$$

So the above fact is forced over $\mathcal{W}^*[K]$ for $\dot{\mathcal{R}}$.

Let $H \prec H_\lambda$ be countable (in V) such that all relevant objects are in H. Let $\pi : M \to H$ invert the transitive collapse and for all $a \in H$, let $\bar{a} = \pi^{-1}(a)$. By Lemma 9.2.13, there is a map $\overline{\pi} : \overline{\mathcal{R}^+} \to \mathcal{P}_X^*$[42] such that letting Λ_0 be the π-pullback of Λ_X and Λ_1 be the $\overline{\pi}$-pullback of Λ_X, then

$$\Lambda_1 \restriction \overline{\mathcal{R}} = \Sigma^{\overline{\pi} \restriction \overline{\mathcal{P}} \circ \bar{k}},^{43}$$

and furthermore since $\pi \restriction \overline{\mathcal{P}_X^*} = \overline{\pi} \circ \bar{i}^*$,[44]

$$\Lambda_0 = \Lambda_1^{\bar{i}}.$$

In particular, $\Lambda_0 \leq_w \Lambda_1$ and letting $\Sigma_{\overline{\mathcal{R}}} = \Lambda_1 \restriction \overline{\mathcal{R}} = \Sigma^{\overline{\pi} \restriction \overline{\mathcal{P}} \circ \bar{k}}$, $(\overline{\mathcal{R}^+}, \Lambda_1)$ is a $\Sigma_{\overline{\mathcal{R}}}$-hod pair and that $\mathrm{ord}(\overline{\mathcal{R}})$ is a cardinal in $\overline{\mathcal{R}^+}$.

We also confuse $\overline{\Lambda}$ with the π-pullback of Λ. Hence $\Gamma(\overline{\mathcal{P}_X^*}, \Lambda_0)$ witnesses that $\overline{\mathcal{R}}$ is not full and this fact is forced over $\bar{\mathcal{W}}^*[\bar{K}]$ for the name $\bar{\dot{\mathcal{R}}}$. This means if we further iterate $\overline{\mathcal{W}^*}$ via $\overline{\Lambda}$ to $\mathcal{Y}$ such that $\mathbb{R}^{V[G]}$ can be realized as the symmetric reals over $\mathcal{Y}$ then in the derived model $D(\mathcal{Y})$,

$$L(\Gamma(\overline{\mathcal{P}_X^*}, \Lambda_0)) \vDash \overline{\mathcal{R}} \text{is not full.} \tag{2}$$

In the above, we have used the fact that the interpretation of the UB-code of the strategy for $\overline{\mathcal{P}_X^*}$ in $\mathcal{Y}$ to its derived model is $\Lambda_0 \restriction \mathbb{R}^{V[G]}$; this key fact is proved in [29, Theorem 3.26] and Chapter 6.

Now we iterate $\overline{\mathcal{R}^*}$ to $\mathcal{S}$ via Λ_1 to realize $\mathbb{R}^{V[G]}$ as the symmetric reals for the collapse $Col(\omega, < \delta^{\mathcal{S}})$, where $\delta^{\mathcal{S}}$ is the sup of $\mathcal{S}$'s Woodin cardinals. By the fact

41 This is because we can continue iterating $\mathcal{W}^*$ above the first Woodin cardinal to $\mathcal{W}^{**}$ such that letting λ be the sup of the Woodin cardinals of $\mathcal{W}^{**}$, then there is a $Col(\omega, < \lambda)$-generic h such that $\mathbb{R}^{V[G]}$ is the symmetric reals for $\mathcal{W}^{**}[h]$. And in $\mathcal{W}^{**}(\mathbb{R}^{V[G]})$, the derived model satisfies that $L(\Gamma(\mathcal{P}_X^+, \Lambda_X)) \vDash \mathcal{R}$ is not full.

42 We abuse notation a bit here. Technically, $\mathcal{R}$ is not in V. $\overline{\mathcal{R}}$ is the interpretation the name $\bar{\dot{\mathcal{R}}}$ over a $M[h]$ where $h \in V$ is $Coll(\omega, \bar{\kappa})$ generic over M. A similar comment applies to the maps $\overline{i^*}, \overline{k^*}$.

43 This fact was missing from the proof of [65, Lemma 3.80]. We need this to know that $(\overline{\mathcal{R}^+}, \Lambda_1)$ is a $\Sigma^{\overline{\pi} \restriction \overline{\mathcal{P}} \circ \bar{k}}$-hod pair.

44 This follows from the definition of $\overline{\pi}$ and the fact that $\pi_X^* \circ \pi \restriction \overline{\mathcal{P}_X^*} = \pi \restriction \bar{\mathcal{P}}^+ \circ \overline{k^*} \circ \overline{i^*}$.

that $\Lambda_0 \leq_w \Lambda_1$ and $(\overline{\mathcal{R}^*}, \Lambda_1)$ is a $\Sigma_{\overline{\mathcal{R}}}$-hod pair, we get that in the derived model $D(\mathcal{S})$,

$$\overline{\mathcal{R}} \text{ is not full as witnessed by } \bar{\mathcal{M}}.$$

So $\Sigma_{\bar{\mathcal{M}}}$ is $\mathrm{OD}_{\Sigma_{\overline{\mathcal{R}}}}$ in $D(\mathcal{S})$ and hence $\overline{\mathcal{M}} \in \overline{\mathcal{R}^*}$.[45] This contradicts internal fullness of $\overline{\mathcal{R}}$ in $\overline{\mathcal{R}^*}$ since $\overline{\mathcal{M}}$ collapses $\mathsf{ord}(\overline{\mathcal{R}})$ in $\overline{\mathcal{R}^*}$ but $\mathsf{ord}(\overline{\mathcal{R}})$ is a cardinal in $\overline{\mathcal{R}^*}$. $\dashv$

For a good X, using the embedding π_X we can define a π_X-realizable strategy Σ_X^+ for $\mathcal{P}_X$ using the construction of Definition 9.0.7. We have that Σ_X^+ is such that

1. Σ_X^+ extends Σ_X;
2. for any Σ_X^+ iterate $\mathcal{Q}$ of $\mathcal{P}_X$ via stack $\vec{\mathcal{T}}$ such that the iteration embedding $\pi^{\vec{\mathcal{T}}}$ exists, there is an embedding $\sigma : \mathcal{Q} \to \mathcal{P}$ such that $\pi_X = \sigma \circ \pi^{\vec{\mathcal{T}}}$. Furthermore, letting $\Psi = (\Sigma_X^+)_{\vec{\mathcal{T}},\mathcal{Q}}$, for all $\mathcal{S} \lhd_{hod}^c \mathcal{Q}$, $\Psi_{\mathcal{S}}$ has branch condensation.
3. Σ_X^+ is $\Gamma(\mathcal{P}_X, \Sigma_X^+)$-fullness preserving.

Theorem 9.2.14 then implies that Σ_X^+ is Γ-fullness preserving.

THEOREM 9.2.15. *There is a stationary set $S' \subset S$ such that whenever $X \in S'$, $X \cap \mathcal{P}$ is a strongly (φ, A)-condensing set.*

PROOF. The proof of this theorem is an adaptation of the proof of Theorem 9.2.14 in a similar way one adapts the proof of Theorem 9.2.6 to prove Theorem 9.2.7. For completeness, we give a fairly detailed argument here. We will omit (φ, A) from our notations.

Suppose X is a weakly condensing set and $B \in \mathcal{P}_X \cap \wp(\Theta_X)$.[46] We say that τ_X has *B-condensation* if whenever $\mathcal{Q} = \mathcal{Q}_Y$ (where Y is an extension of X) is such that there are elementary embeddings $\upsilon : \mathcal{P}_X \to \mathcal{Q}$, $\tau : \mathcal{Q} \to \mathcal{P}$ such that $\mathcal{Q}$ is countable in $V[g]$ and $\tau_X = \tau \circ \upsilon$, then $\upsilon(T_{\mathcal{P}_X,B}) = T_{\mathcal{Q},\tau,B}$, where

$$T_{\mathcal{P}_X,B} = \{(\psi, s) \mid s \in [\Theta_X]^{<\omega} \wedge \mathcal{P}_X \vDash \psi[s, B]\},$$

and

$$T_{\mathcal{Q},\tau,B} = \{(\psi, s) \mid s \in [\delta_\alpha^{\mathcal{Q}}]^{<\omega} \text{ for some } \alpha < \lambda_{\mathcal{Q}} \wedge \mathcal{P} \vDash \varphi[\pi^{\Sigma_{\mathcal{Q}}^{\tau}}_{\mathcal{Q}(\alpha),\infty}(s), \tau_X(B)]\},$$

where $\Sigma_{\mathcal{Q}}^{\tau}$ is the τ-pullback strategy of Σ. We say τ_X has *condensation* if it has B-condensation for every $B \in \mathcal{P}_X \cap \wp(\delta_X)$.

As before, we just prove the condensing part. To prove that a weakly condensing set X is condensing, it is enough to prove that τ_X has condensation. Suppose for contradiction that the set T of $X' \in S$ such that $X = X' \cap \mathcal{P}$ is cofinal in $\mathcal{P}$ and is not a condensing set is stationary. For each $X' \in T$, let $X = X' \cap \mathcal{P}$ (we will use this type of notations throughout this proof without mentioning again) and A_X be the $\lesssim_X$-least such that τ_X fails to have A_X-condensation, where $\lesssim_X$ is the canonical

[45]We note that it is crucial here that both $\overline{\mathcal{M}}$ and $\overline{\mathcal{R}^*}$ are $\Sigma_{\overline{\mathcal{R}}}$-mice.

[46]For the rest of this proof, whenever X is weakly condensing, we automatically assume that $X = X' \cap \mathcal{P}$ for some good X'.

well-ordering of $\mathcal{P}_X$. We say that a tuple $\{\langle \mathcal{P}_i, \mathcal{Q}_i, \tau_i, \xi_i, \pi_i, \sigma_i \mid i < \omega\rangle, \mathcal{M}_{\infty,Y}\}$ is a *bad tuple* if

1. $Y \in T$;
2. $\mathcal{P}_i = \mathcal{P}_{X_i}$ for all i, where $X_i' \in T$ and $\mathcal{Q}_i = \mathcal{Q}_{Y_i}$ for Y_i an extension of X_i;
3. for all $i < j$, $X_i \prec Y_i \prec X_j \prec Y$;
4. $\mathcal{M}_{\infty,Y}$ be the direct limit of iterates $(\mathcal{Q}, \Lambda)$ of $(\mathcal{P}_Y, \Sigma_Y^+)$ such that Λ has branch condensation;
5. for all i, $\xi_i : \mathcal{P}_i \to \mathcal{Q}_i$, $\sigma_i : \mathcal{Q}_i \to \mathcal{M}_{\infty,Y}$, $\tau_i : \mathcal{P}_{i+1} \to \mathcal{M}_{\infty,Y}$, and $\pi_i : \mathcal{Q}_i \to \mathcal{P}_{i+1}$;
6. for all i, $\tau_i = \sigma_i \circ \xi_i$, $\sigma_i = \tau_{i+1} \circ \pi_i$, and $\tau_{X_i,X_{i+1}} \restriction \mathcal{P}_i =_{\text{def}} \varphi_{i,i+1} = \pi_i \circ \xi_i$;
7. $\varphi_{i,i+1}(A_{X_i}) = A_{X_{i+1}}$;
8. for all i, $\xi_i(T_{\mathcal{P}_i,A_{X_i}}) \neq T_{\mathcal{Q}_i,\sigma_i,A_{X_i}}$.

In (8), $T_{\mathcal{Q}_i,\sigma_i,A_{X_i}}$ is computed relative to $\mathcal{M}_{\infty,Y}$, that is

$$T_{\mathcal{Q}_i,\sigma_i,A_{X_i}} = \{(\varphi, s) \mid s \in [\delta_\alpha^{\mathcal{Q}_i}]^{<\omega} \text{ for some } \alpha < \lambda^{\mathcal{Q}_i} \wedge \\ \mathcal{M}_{\infty,Y} \models \varphi[\pi^{\Sigma^{\sigma_i}_{\mathcal{Q}_i}}_{\mathcal{Q}_i(\alpha),\infty}(s), \tau_i(A_{X_i})]\}$$

CLAIM 9.2.16. *There is a bad tuple.*

PROOF. For brevity, we first construct a bad tuple $\{\langle \mathcal{P}_i, \mathcal{Q}_i, \tau_i, \xi_i, \pi_i, \sigma_i \mid i < \omega\rangle, \mathcal{P}\}$ with $\mathcal{P}$ playing the role of $\mathcal{M}_{\infty,Y}$. We then simply choose a sufficiently large, good Y and let $i_Y : \mathcal{P}_Y \to \mathcal{M}_{\infty,Y}$ be the direct limit map, $m_Y : \mathcal{M}_{\infty,Y} \to \mathcal{P}$ be the natural factor map, i.e., $m_Y \circ i_Y = \pi_Y$. It is easy to see that for all sufficiently large Y, the tuple $\{\langle \mathcal{P}_i, \mathcal{Q}_i, m_Y^{-1} \circ \tau_i, m_Y^{-1} \circ \xi_i, m_Y^{-1} \circ \pi_i, m_Y^{-1} \circ \sigma_i \mid i < \omega\rangle, \mathcal{M}_{\infty,Y}\}$ is a bad tuple.

The key point is (6). Let $A_X^* = \tau_X(A_X)$ for all $X \in T$. By Fodor's lemma, there is an A^* such that $\exists^* X \in T\ A_X^* = A^*$.[47] So there is an increasing and cofinal sequence $\{X_\alpha \mid \alpha < \kappa^+\} \subseteq T$ such that for $\alpha < \beta$, $\tau_{X_\alpha,X_\beta}(A_{X_\alpha}) = A_{X_\beta} = \tau_{X_\beta}^{-1}(A)$. This easily implies the existence of such a tuple $\{\langle \mathcal{P}_i, \mathcal{Q}_i, \tau_i, \xi_i, \pi_i, \sigma_i \mid i < \omega\rangle, \mathcal{P}\}$. ⊣

Fix a bad tuple $\mathcal{A} = \{\langle \mathcal{P}_i, \mathcal{Q}_i, \tau_i, \xi_i, \pi_i, \sigma_i \mid i < \omega\rangle, \mathcal{M}_{\infty,Y}\}$. Let $(\mathcal{P}_0^+, \Pi)$ be a (g-organized) $\Sigma_{\mathcal{P}_0}$-hod pair (cf. [50]) such that

$$\Gamma(\mathcal{P}_0^+, \Pi) \models \mathcal{A} \text{ is a bad tuple.}$$

We may also assume $(\mathcal{P}_0^+, \Pi \restriction V) \in V$, $\delta^{\mathcal{P}_0^+}$ is limit of Woodin cardinals and is of nonmeasurable cofinality in $\mathcal{P}_0^+$ and there is some $\alpha < \lambda^{\mathcal{P}_0^+}$ such that $\Sigma_Y \leq_w \Pi_{\mathcal{P}_0^+(\alpha)}$. This type of reflection is possible because we replace $\mathcal{P}$ by $\mathcal{M}_{\infty,Y}$. Let $\mathcal{W} = \mathcal{M}_\omega^{\sharp,\Sigma_Y,\Pi,\oplus_{n<\omega}\Sigma_{X_n}}$ and Λ be the unique strategy of $\mathcal{W}$. If $\mathcal{Z}$ is the result of iterating $\mathcal{W}$ via Λ to make $\mathbb{R}^{V[G]}$ generic, then letting h be $\mathcal{Z}$-generic for the Levy

[47] "$\exists^* X \in T$" means "stationarily many $X \in T$".

collapse of the sup of $\mathcal{Z}$'s Woodin cardinals to ω such that $\mathbb{R}^{V[G]}$ is the symmetric reals of $\mathcal{Z}[h]$, then in $\mathcal{Z}(\mathbb{R}^{V[G]})$,

$$\Gamma(\mathcal{P}_0^+,\Pi) \models \mathcal{A} \text{ is a bad tuple.}$$

Now we define by induction $\xi_i^+ : \mathcal{P}_i^+ \to \mathcal{Q}_i^+$, $\pi_i^+ : \mathcal{Q}_i^+ \to \mathcal{P}_{i+1}^+$, $\varphi_{i,i+1}^+ : \mathcal{P}_i^+ \to \mathcal{P}_{i+1}^+$ as follows. $\varphi_{0,1}^+ : \mathcal{P}_0^+ \to \mathcal{P}_1^+$ is the ultrapower map by the extender derived from π_{X_0,X_1} of length Θ_{X_1}. Note that $\varphi_{0,1}^+$ extends $\varphi_{0,1}$. Let $\xi_0^+ : \mathcal{P}_0^+ \to \mathcal{Q}_0^+$ extend ξ_0 be the ultrapower map by the extender derived from ξ_0 of length $\delta^{\mathcal{Q}_0}$. Finally let $\pi_0^+ = (\varphi_{0,1}^+)^{-1} \circ \xi_0^+$. The maps $\xi_i^+, \pi_i^+, \varphi_{i,i+1}^+$ are defined similarly. Let also $\mathcal{M}_Y = \text{Ult}(\mathcal{P}_0^+, E)$, where E is the extender derived from $\pi_{X,Y}$ of length Θ_Y. There are maps $\varepsilon_{2i} : \mathcal{P}_i^+ \to \mathcal{M}_Y$, $\varepsilon_{2i+1} : \mathcal{Q}_i^+ \to \mathcal{M}_Y$ for all i such that $\varepsilon_{2i} = \varepsilon_{2i+1} \circ \xi_i^+$ and $\varepsilon_{2i+1} = \varepsilon_{2i+2} \circ \pi_i^+$. When $i = 0$, ε_0 is simply i_E. Letting $\Sigma_i = \Sigma_{\mathcal{P}_i}$ and $\Psi = \Sigma_{\mathcal{Q}_i}$, $A_i = A_{X_i}$, there is a finite sequence of ordinals t and a formula $\vartheta(u,v)$ such that in $\Gamma(\mathcal{P}_0^+,\Pi)$

9. for every $i < \omega$, $(\varphi, s) \in T_{\mathcal{P}_i, A_i} \iff \vartheta[\pi^{\Sigma_i}_{\mathcal{P}_i(\alpha),\infty}, t]$, where α is least such that $s \in [\delta_\alpha^{\mathcal{P}_i}]^{<\omega}$;
10. for every i, there is $(\varphi_i, s_i) \in T_{\mathcal{Q}_i, \xi_i(A_i)}$ such that $\neg\vartheta[\pi^{\Psi_i}_{\mathcal{Q}_i(\alpha)}(s_i), t]$ where α is least such that $s_i \in [\delta_\alpha^{\mathcal{Q}_i}]^{<\omega}$.

The pair (ϑ, t) essentially defines a Wadge-initial segment of $\Gamma(\mathcal{P}_0^+,\Pi)$ that can define the pair $(\mathcal{M}_{\infty,Y}, A^*)$, where $\tau_i(A_i) = A^*$ for some (any) i.

Now let $X \prec H_\lambda$ be countable that contains all relevant objects and $\pi : M \to X$ invert the transitive collapse. For $a \in X$, let $\overline{a} = \pi^{-1}(a)$. By countable completeness of the extender E and Lemma 9.2.13, there is a map $\pi^* : \overline{\mathcal{M}_Y} \to \mathcal{P}_0^+$ with the property specified in Lemma 9.2.13. Let $\overline{\Pi_i}$ be the $\pi^* \circ \overline{\varepsilon_i}$-pullback of Π, so in $V[g]$,

$$(\overline{\mathcal{M}_Y}, \Pi^{\pi^*}) \text{ is a } \Sigma^\pi_{\mathcal{M}_{\infty,Y}}\text{-hod pair,}^{48}$$

$$\forall i < \omega,\ (\overline{\mathcal{P}_i^+}, \overline{\Pi_i}) \text{ is a } \Sigma^\pi_{\mathcal{P}_i}\text{-hod pair,}$$

and

$$\overline{\Sigma_Y} \leq_w \overline{\Pi_0} \leq_w \overline{\Pi_1} \cdots \leq_w \Pi^{\pi^*}.$$

Let $\dot{\mathcal{A}} \in (H_{\bar{\kappa}})^M$ be the canonical name for $\bar{\mathcal{A}}$. It is easy to see (using the assumption on $\mathcal{W}$) that if $\mathcal{W}^*$ is a result of iterating $\bar{\mathcal{W}}$ via $\bar{\Lambda}$ (we confuse $\bar{\Lambda}$ with the π-pullback of Λ; they coincide on M) in M below the first Woodin of $\bar{\mathcal{W}}$ to make H-generically generic, where H is the transitive closure of $H^M_{\omega_2} \cup \dot{\mathcal{A}}$, then in $\mathcal{W}^*[H]$, the derived model of $\mathcal{W}^*[H]$ at the sup of $\mathcal{W}^*$'s Woodin cardinals satisfies:

$$L(\bar{\mathcal{P}}_0, \mathbb{R}) \models \dot{\mathcal{A}} \text{ is a bad tuple.}$$

---

[48] [65, Lemma 3.82] concludes this by claiming $\pi \restriction \overline{\mathcal{M}_Y} = \varepsilon_0 \circ \pi^*$, which is not true. One needs Lemma 9.2.13 to conclude this.

Now we stretch this fact out to $V[G]$ by iterating $\mathcal{W}^*$ to $\mathcal{W}^{**}$ to make $\mathbb{R}^{V[G]}$-generic. In $\mathcal{W}^{**}(\mathbb{R}^{V[G]})$, letting $i : \mathcal{W}^* \to \mathcal{W}^{**}$ be the iteration map then

$$\Gamma(\bar{\mathcal{P}_0}^+, \bar{\Pi}) \vDash i(\bar{\mathcal{A}}) \text{ is a bad tuple.}^{49}$$

By a similar argument as in [64, Theorem 3.1.25], we can use the strategies $\overline{\Pi_i^+}$'s to simultanously execute a $\mathbb{R}^{V[G]}$-genericity iterations. The last branch of the iteration tree is wellfounded. The process yields a sequence of models $\langle \overline{\mathcal{P}_{i,\omega}^+}, \overline{\mathcal{Q}_{i,\omega}^+} \mid i < \omega \rangle$ and maps $\overline{\xi_{i,\omega}^+} : \overline{\mathcal{P}_{i,\omega}^+} \to \overline{\mathcal{Q}_{i,\omega}^+}$, $\overline{\pi_{i,\omega}^+} : \overline{\mathcal{Q}_{i,\omega}^+} \to \overline{\mathcal{P}_{i+1,\omega}^+}$, and $\overline{\varphi_{i,i+1,\omega}^+} = \overline{\pi_{i,\omega}^+} \circ \overline{\pi_{i,\omega}^+}$. Furthermore, each $\overline{\mathcal{P}_{i,\omega}^+}, \overline{\mathcal{Q}_{i,\omega}^+}$ embeds into a Π^{π^*}-iterate of $\overline{\mathcal{M}_Y}$ and hence the direct limit $\mathcal{P}_\infty$ of $(\overline{\mathcal{P}_{i,\omega}^+}, \overline{\mathcal{Q}_{j,\omega}^+} \mid i, j < \omega)$ under maps $\overline{\pi_{i,\omega}^+}$'s and $\overline{\xi_{i,\omega}^+}$'s is wellfounded. As mentioned above, $\overline{\mathcal{P}_{i,\omega}^+}$ is a (g-organized) Σ_i^π-premouse and $\overline{\mathcal{Q}_{i,\omega}^+}$ is a ${}^g\Psi_i^\pi$-premouse. Let C_i be the derived model of $\overline{\mathcal{P}_{i,\omega}^+}$, D_i be the derived model of $\overline{\mathcal{Q}_{i,\omega}^+}$ (at the sup of the Woodin cardinals of each model), then $\mathbb{R}^{V[G]} = \mathbb{R}^{C_i} = \mathbb{R}^{D_i}$. Furthermore, $C_i \cap \wp(\mathbb{R}) \subseteq D_i \cap \wp(\mathbb{R}) \subseteq C_{i+1} \cap \wp(\mathbb{R})$ for all i.

(9), (10) and the construction above give us that there is a $t \in [\mathrm{OR}]^{<\omega}$, a formula $\vartheta(u,v)$ such that

11. for each i, in C_i, for every (φ, s) such that $s \in \delta^{\overline{\mathcal{P}_i}}$, $(\varphi, s) \in T_{\overline{\mathcal{P}_i}, \overline{A_i}} \iff \vartheta[\pi^{\overline{\Sigma_i}}_{\overline{\mathcal{P}_i}(\alpha),\infty}(s), t]$ where α is least such that $s \in [\delta_\alpha^{\overline{\mathcal{P}_i}}]^{<\omega}$.

Let n be such that for all $i \geq n$, $\overline{\xi_{i,\omega}^+}(t) = t$. Such an n exists because the direct limit $\mathcal{P}_\infty$ is wellfounded as we can arrange that $\mathcal{P}_\infty$ is embeddable into a Π^{π^*}-iterate of $\overline{\mathcal{M}_Y}$. By elementarity of $\overline{\xi_{i,\omega}^+}$ and the fact that $\overline{\xi_{i,\omega}^+} \restriction \overline{\mathcal{P}_i} = \overline{\xi_i}$,

12. for all $i \geq n$, in D_i, for every (φ, s) such that $s \in \delta^{\overline{\mathcal{Q}_i}}$, $(\varphi, s) \in T_{\overline{\mathcal{Q}_i}, \overline{\xi_i}(\overline{A_i})} \iff \vartheta[\pi^{\overline{\Psi_i}}_{\overline{\mathcal{Q}_i}(\alpha),\infty}(s), t]$ where α is least such that $s \in [\delta_\alpha^{\overline{\mathcal{Q}_i}}]^{<\omega}$.

However, using (10), we get

13. for every i, in D_i, there is a formula φ_i and some $s_i \in [\delta^{\overline{\mathcal{Q}_i}}]^{<\omega}$ such that $(\varphi_i, s_i) \in T^{\overline{\mathcal{Q}_i}, \overline{\xi_i}(\overline{A_i})}$ but $\neg\varphi[\pi^{\overline{\Psi_i}}_{\overline{\mathcal{Q}_i}(\alpha),\infty}(s_i), t]$ where α is least such that $s \in [\delta_\alpha^{\overline{\mathcal{Q}_i}}]^{<\omega}$.

Clearly (12) and (13) give us a contradiction. This completes the proof of the lemma. ⊣

---

49 We abuse the notation slightly here. Technically, $\bar{\mathcal{A}}$ is not in $\mathcal{W}^*$ but $\mathcal{W}^*$ has a canonical name $\dot{\mathcal{A}}$ for $\bar{\mathcal{A}}$. Hence by $i(\bar{\mathcal{A}})$, we mean the interpretation of $i(\dot{\mathcal{A}})$.

9.3. Condensing sets in models of AD^+

Thus far we have built condensing sets while working in models of ZFC. In this section, we prove their existence in models of AD^+. The material presented in this section will be used in the proof of generation of pointclasses (see Theorem 10.1.2). Throughout this section we assume $\mathsf{AD}^+ + V = L(\wp(\mathbb{R}))$. Recall the notation $\Gamma_1 \trianglelefteq_{mouse} \Gamma_2$ (see [29, Page 82] or Section 5.3).

Suppose Γ is a mouse full pointclass (Definition 5.3.2) such that:

$(*)_\Gamma$ there is a good pointclass Γ^* containing Γ and there is a sequence $(\Gamma_\alpha : \alpha < \Omega)$ with the property that

1. Ω is a limit ordinal,
2. $\Gamma_\alpha \triangleleft_{mouse} \Gamma$,
3. for $\alpha < \beta < \Omega$, $\Gamma_\alpha \triangleleft_{mouse} \Gamma_\beta$,
4. $\forall -1 \leq \alpha < \Omega$, $\Gamma_{\alpha+1}$ is completely mousefull[50],
5. there is no completely mouse-full pointclass $\Psi \triangleleft_{mouse} \Gamma$ such that for some α, $\Gamma_\alpha \triangleleft_{mouse} \Psi \triangleleft_{mouse} \Gamma_{\alpha+1}$,
6. if $\alpha < \Omega$ is a limit ordinal then $\Gamma_\alpha = \bigcup_{\beta<\alpha} \Gamma_\beta$,
7. $\Gamma = \bigcup_{\alpha<\Omega} \Gamma_\alpha$.

Recall the definitions of HP^Γ and Mice^Γ (see Notation 4.1.14). Let $\mathcal{F} = \{(\mathcal{P},\Sigma) \in \mathsf{HP}^\Gamma : \Sigma$ is strongly Γ-fullness preserving and has strong branch condensation$\}$. We then let $\mathcal{M}^- = \bigcup_{(\mathcal{P},\Sigma)\in\mathcal{F}} \mathcal{M}_\infty(\mathcal{P},\Sigma)$. It follows from AD^+ theory that if $(\mathcal{P},\Sigma) \in \mathcal{F}$ then Σ can be extended to a (Θ,Θ,Θ)-iteration strategy[51]. In what follows, we assume that if $(\mathcal{P},\Sigma) \in \mathcal{F}$ then Σ is a (Θ,Θ,Θ)-iteration strategy.

Recall Notation 9.1.2. Given $\mathcal{R} \triangleleft^c_{hod} \mathcal{M}^-$, we let $\Sigma_\mathcal{R}$ be the strategy of $\mathcal{R}$ such that whenever $(\mathcal{P},\Lambda) \in \mathcal{F}$ is such that $\mathcal{M}_\infty(\mathcal{P},\Sigma) = \mathcal{R}$ then $\Lambda_\mathcal{R} = \Sigma_\mathcal{R}$. Next we let $\mathsf{Lp}^{\Gamma,\oplus_{\mathcal{R}\triangleleft^c_{hod}\mathcal{M}^-}\Sigma_\mathcal{R}}(\mathcal{M}^-)$ be the stack of all sound $\oplus_{\mathcal{R}\triangleleft^c_{hod}\mathcal{M}^-}\Sigma_\mathcal{R}$-premice $\mathcal{N}$ over $\mathcal{M}^-$ such that $\rho(\mathcal{N}) \leq \mathsf{ord}(\mathcal{M}^-)$ and whenever $\pi : \mathcal{S} \to \mathcal{N}$ is elementary and $\mathcal{S}$ is countable then $\mathcal{S}$, as a $\oplus_{\mathcal{R}\triangleleft^c_{hod}\pi^{-1}(\mathcal{M}^-)}\Sigma^\pi_\mathcal{R}$-mouse, has an ω_1-iteration strategy in Γ. Finally, if there is $\mathcal{N} \trianglelefteq \mathsf{Lp}^{\Gamma,\oplus_{\mathcal{R}\triangleleft^c_{hod}\mathcal{M}^-}\Sigma_\mathcal{R}}(\mathcal{M}^-)$ such that $\rho(\mathcal{N}) < \mathsf{ord}(\mathcal{M}^-)$ then let $\mathcal{M}$ be the least such $\mathcal{N}$ and otherwise let $\mathcal{M} = \mathsf{Lp}^{\Gamma,\oplus_{\mathcal{R}\triangleleft^c_{hod}\mathcal{M}^-}\Sigma_\mathcal{R}}(\mathcal{M}^-)$.

We let $\varphi(u,v)$ be the formula that expresses the fact that u is a mouse full pointclass such that $(*)_u$ holds and v is a hod pair $(\mathcal{Q},\Lambda)$ such that $\mathsf{Code}(\Lambda) \in u$ and Λ has strong branch condensation and is strongly u-fullness preserving.

REMARK 9.3.1. We have developed the concept of a hod mouse below LSA. In the next theorem, hod pairs are all lsa small. However, the proof is general enough and uses this hypothesis only because we have not set up a general theory of hod

[50] Set $\Gamma_{-1} = \emptyset$.

[51] For example, see [32].

mice. Because of this, we omit the extra hypothesis that we are in the minimal model of LSA.

THEOREM 9.3.2. *Assume $\mathsf{ZF}+\mathsf{AD}^{+}$*[52]. *Suppose Γ is a mouse full pointclass such that $(*)_\Gamma$ holds. Let*

1. *$\mathcal{F}=\mathcal{F}_{\varphi,\Gamma}=\{(\mathcal{P},\Sigma)\in\mathsf{HP}^\Gamma : \Sigma$ is strongly Γ-fullness preserving and has strong branch condensation$\}$,*
2. *$\mathcal{M}^- = \bigcup_{(\mathcal{P},\Sigma)\in\mathcal{F}}\mathcal{M}_\infty(\mathcal{P},\Sigma)$, and*
3. *let $\mathcal{M}$ be defined as follows: if there is $\mathcal{N}\trianglelefteq \mathsf{Lp}^{\Gamma,\oplus_{\mathcal{R}\triangleleft^c_{hod}\mathcal{M}^-}\Sigma_{\mathcal{R}}}(\mathcal{M}^-)$ such that $\rho(\mathcal{N})<\mathrm{ord}(\mathcal{M}^-)$ then let $\mathcal{M}$ be the least such $\mathcal{N}$ and otherwise let $\mathcal{M}=\mathsf{Lp}^{\Gamma,\oplus_{\mathcal{R}\triangleleft^c_{hod}\mathcal{M}^-}\Sigma_{\mathcal{R}}}(\mathcal{M}^-)$.*

Then one of the following holds[53].

1. *There is a hod pair or an anomalous hod pair $(\mathcal{P},\Sigma)$ such that Σ has strong branch condensation and is strongly Γ-fullness preserving, and $\Gamma(\mathcal{P},\Sigma)=\Gamma$ (i.e., (φ,Γ) is not maximal).*
2. *$\mathcal{M}=\mathsf{Lp}^{\Gamma,\oplus_{\mathcal{R}\triangleleft^c_{hod}\mathcal{M}^-}\Sigma_{\mathcal{R}}}(\mathcal{M}^-)$, lower part (φ,Γ)-covering fails and there is a strongly (φ,Γ)-condensing set $X\in\wp_{\omega_1}(\mathcal{M})$.*
3. *For some $(\mathcal{Q},\Lambda)\in\mathsf{HP}^\Gamma$ such that Λ has strong branch condensation and is strongly Γ-fullness preserving and for some $x\in\mathbb{R}$, $\mathsf{Lp}^\Lambda(x)\neq\mathsf{Lp}^{\Gamma,\Lambda}(x)$.*

PROOF. Towards a contradiction assume that all three clauses are false. We drop (φ,Γ) from our terminology. We will abuse our terminology and will say "Γ-hod pair construction of M". Whenever we do this we mean the Γ-hod pair construction of $\mathbb{M}$ as defined in Definition 4.3.3. Here, $\mathbb{M}$ is a background whose universe is M, and it will always be clear exactly what $\mathbb{M}$ should be.

Let $A_0\subseteq\mathbb{R}$ be such that $A_0\in lub(\Gamma)$. Let $\Gamma_0,\Gamma_0^*,(N_0,\Phi_0),A_0^*,\Gamma_1$ be such that

1. Γ_0, Γ_0^* and Γ_1 are good pointclasses,
2. $\Gamma\subseteq\underset{\sim}{\Delta}_{\Gamma_0}$,
3. $\Gamma_0\subseteq\underset{\sim}{\Delta}_{\Gamma_0^*}$,
4. $A_0^*\in lub(\Gamma_0^*)$,
5. (N_0,Φ_0) is a Γ_0^*-Woodin Suslin, co-Suslin capturing the sequence $(T_n(A_0):n\in\omega)$.[54]

Let F_0 be as in Theorem 4.1.12 for $(\Gamma_0,\Gamma_0^*,(N_0,\Phi_0),A_0^*)$, and fixing some (N_1,Φ_1), Γ_1^*, A_1^* let F_1 be as in Theorem 4.1.12 for $(\Gamma_1,\Gamma_1^*,(N_1,\Phi_1),A_1^*)$.

1. Let $x\in\mathrm{dom}(F_0)$ be such that if $F_0(x)=(\mathcal{N}',\mathcal{M}',\delta',\Psi')$ then letting $\vec{G}$ be as in clause 7 of Theorem 4.1.12 and setting $\mathbb{M}_0=(\mathcal{N},\delta,\vec{G},\Psi')$, $(\mathbb{M}_0,(N_0,\Phi_0),\Gamma_0,A_0^*)$ Suslin, co-Suslin captures Γ and A_0.

---

[52] Also, see the above remark.

[53] What follows is not intended as an "either or" conclusion.

[54] See Section 4.1.1. There $T_n(X)$ is defined for X a strategy but the same definition can be applied to any set of reals.

2. Let $y \in \mathrm{dom}(F_1)$ be such that if $F_1(y) = (\mathcal{N}_y^*, \mathcal{M}_y, \delta_y, \Psi_y)$ then letting $\vec{G}_y$ be as in clause 7 of Theorem 4.1.12 and setting $\mathbb{M}_y = (\mathcal{N}_y^*, \delta_y, \vec{G}_y, \Sigma_y)$,

$$(\mathbb{M}_y, (N_1, \Phi_1), \Gamma_1^*, A_1^*)$$

Suslin, co-Suslin captures Γ and (N_1, Φ_1) Suslin, co-Suslin captures $\mathsf{Code}(\Psi^*)$ where Ψ^* is the ω_1-strategy of $\mathcal{M}_2^{\#,\Phi_0}$.

We record the following fact, which is a consequence of the proof of Lemma 4.1.11[55].

LEMMA 9.3.3. *Suppose u is a set, $\mathcal{W} = \mathcal{M}_1^{\#,\Phi_0}(u)$ and Λ is the unique strategy of $\mathcal{W}$ witnessing that $\mathcal{W}$ is a Ψ-mouse. Let δ be the least Woodin cardinal of $\mathcal{W}$ and let $\mathcal{W}'$ be a Λ-iterate of $\mathcal{W}$ such that the iteration embedding $j : \mathcal{W} \to \mathcal{W}'$ exists. Let $h \subseteq Coll(\omega, j(\delta))$ be $\mathcal{W}'$-generic. Then for any real $\tau \in \mathcal{W}'[h]$,*

$$(\mathsf{HC}^{\mathcal{W}'[h]}, A_0 \cap \mathcal{W}'[h], \tau, \in) \prec (\mathsf{HC}, A_0, \tau, \in).$$

Let κ be the least $< \delta_y$-strong cardinal of $\mathcal{N}_y^*$. Let $g \subseteq Coll(\omega, < \kappa)$ be $\mathcal{N}_y^*$-generic. Let $\mathcal{F}_0 \in \mathcal{N}_y^*[g]$ be the set of $(\mathcal{Q}, \Lambda) \in \mathcal{N}_y^*[g]$ such that

1. $\mathcal{Q} \in \mathsf{HC}^{\mathcal{N}_y^*[g]}$,
2. $\mathcal{N}_y^*[g] \vDash (\mathcal{Q}, \Lambda) \in \mathsf{HP}^\Gamma$,
3. $\mathcal{N}_y^*[g] \vDash$ "Λ is Γ-fullness preserving and has strong branch condensation".

We use the methodology of Section 4.1.3 to obtain (D, ψ) such that

$$\mathcal{F}_0 = (\mathcal{F}_{\psi,D})^{\mathcal{N}_y^*[g]}.$$

Notice that $(\mathcal{Q}, \Lambda) \in \mathcal{F}_0$ if and only if there is a real $\sigma \in \mathcal{N}_y^*[g]$ such that $\sigma(0)$ is a Gödel number for some formula ζ and (in $\mathcal{N}_y^*[g]$) letting $A_0^{y,g} = A_0 \cap \mathcal{N}_y^*[g]$,

(A) $\mathsf{Code}(\Lambda)$ is definable over $(\mathsf{HC}, A_0^{y,g}, \sigma, \in)$ via ζ without parameters and
(B) $(\mathsf{HC}, A_0^{y,g}, \sigma, \mathsf{Code}(\Lambda), \in) \vDash$ "Λ is Γ-fullness preserving and has strong branch condensation",
(C) (A) and (B) hold in all further generic extensions of $\mathcal{N}_y^*[g]$.

We have that (ψ, D) is lower part closed and stable. The next claim shows that it is directed.

CLAIM 1. $\mathcal{N}_y^*[g] \vDash$ "(ψ, D) is directed".

PROOF. Fix $(\mathcal{Q}_0, \Lambda_0), (\mathcal{Q}_1, \Lambda_1) \in \mathcal{F}_0$. We now compare $(\mathcal{Q}_i, \Lambda_i)$ with the hod pair construction of $\mathcal{N}_y^*$. It follows from Theorem 4.13.5 that for each $i < 2$, $\mathcal{Q}_i$ iterates, via Λ_i, to some model $\mathcal{Q}_i^+$ in the aforementioned hod pair construction such that $(\Lambda_i)_{\mathcal{Q}_i^+}$ is the strategy $\mathcal{Q}_i$ inherits from the background construction.

[55]The lemma follows because letting δ_1 be the second Woodin cardinal of $\mathcal{W}'$, Ψ allows us to define a δ_1-uB representation for $T_n(\Phi_0)$ (see Lemma 4.1.11).

Let $\nu_i < \kappa$ be such that $\mathcal{Q}_i \in \mathcal{N}_y^*[g \cap Coll(\omega, \nu_i)]$, and let $g_i = g \cap Coll(\omega, \nu_i)$. To complete the proof it is enough to show that[56]

(a) for each i, $(\mathcal{Q}_i^+, (\Lambda_i)_{\mathcal{Q}_i^+})$ appears in the Γ-hod pair construction of $\mathcal{N}_y^*|\kappa[g_i]$ in which all extenders used have critical point $> \max(\nu_0, \nu_1)$.

Let $\eta \in (\kappa, \delta_y)$ be such that $(\mathcal{Q}_i^+, \Lambda_i)$ appears in the Γ-hod pair construction of $\mathcal{N}_y^*|\eta[g_i]$. Let then $E \in \vec{E}^{\mathcal{N}_y^*}$ be such that $\text{crit}(E) = \kappa$ and $\nu_E > \nu_i$. It follows that in $\text{Ult}(\mathcal{N}_y^*, E)[g_i]$, $(\mathcal{Q}_i^+, \Lambda_i)$ appears in the Γ-hod pair construction of $(\text{Ult}(\mathcal{N}_y^*, E)|\pi_E(\kappa))[g_i]$. (a) now follows from elementarity. ⊣

Working in $\mathcal{N}_y^*$, let $\mathcal{P}^- = \mathcal{P}_{\psi,D}^-$. For $\alpha < \kappa$, let $g_\alpha = g \cap Coll(\omega, <\alpha)$. Our next claim implies that (ψ, D) is of limit type.

CLAIM 2. $\mathcal{P}^-$ is of a limit type.

PROOF. Suppose not. It follows that there is $(\mathcal{Q}, \Lambda) \in \mathcal{F}_0$ such that $\mathcal{P}^- = \mathcal{M}_\infty(\mathcal{Q}, \Lambda)$. Let $\nu < \kappa$ be a cutpoint cardinal of $\mathcal{N}_y^*$ such that $\mathcal{Q} \in \mathsf{HC}^{\mathcal{N}_y^*[g_\nu]}$[57]. It follows from the proof of (a) in Claim 1 above that the Γ-hod pair construction of $\mathcal{N}_y^*|\kappa$ in which extenders used have critical point $> \nu$ reaches a pair $(\mathcal{R}, \Phi)$ such that $\mathcal{R}$ is a Λ-iterate of $\mathcal{Q}$ and $\Phi = \Lambda_\mathcal{R}$.

Because of our condition on Γ (namely that Ω is a limit ordinal) there is $\alpha + 1 < \Omega$ such that $\Gamma_\alpha = \Gamma(\mathcal{Q}, \Lambda)$. It follows that the Γ-hod pair construction of $\mathcal{N}_y^*$ using extenders with critical point $> \nu$ reaches $(\mathcal{S}, \Delta) \in \mathcal{F}$ such that $\Gamma(\mathcal{S}, \Delta) = \Gamma_{\alpha+1}$. It follows from the proof of (a) in Claim 1 above that the Γ-hod pair construction of $\mathcal{N}_y^*|\kappa$ in which extenders used have critical point $> \nu$ reaches such a pair $(\mathcal{S}, \Delta)$. It is then enough to show that $\mathcal{N}_y^*[g] \vDash (\mathcal{S}, \Delta) \in \mathsf{HP}^\Gamma$[58]. Let $\nu_1 \in (\nu, \kappa)$ be an $\mathcal{N}_y^*$-cutpoint cardinal such that $\mathcal{S} \in \mathcal{N}_y^*|\nu_1$.

Let $\eta \in (\nu_1, \kappa)$ be the least $\mathcal{N}_y^*$-cardinal such that $\mathcal{M}_1^{\#, \Phi_0}(\mathcal{N}_y^*|\eta) \vDash$ "η is a Woodin cardinal". Let $\mathcal{N}_1$ be the output of the fully backgrounded construction of $\mathcal{N}_y^*|\eta$ relative to Φ_1 using extenders with critical points $> \nu_1$[59]. We now compare $(\mathcal{S}, \Delta)$ with the Γ-hod pair construction of $\mathcal{N}_1$. Notice that all extenders of $\mathcal{N}_1$ have critical points $> \nu_1$. Let $\mathcal{S}_1$ be the output of the aforementioned Γ-hod pair construction. We claim that

(b) some proper initial segment of $\mathcal{S}_1$ is a Δ-iterate of $\mathcal{S}$.

Suppose not. Let $z \in \text{dom}(F_1)$ be such that $y <_T z$ and letting

---

[56]This is because then by a Skolem hull argument we can obtain common iterates of $(\mathcal{Q}_0, \Lambda_0)$, $(\mathcal{Q}_1, \Lambda_1)$ that are in $\mathsf{HC}^{\mathcal{N}_y^*[g]}$, and apply Lemma 4.1.11.

[57]It follows that $\Lambda \restriction \mathsf{HC}^{\mathcal{N}_y^*[g_\nu]} \in \mathcal{N}_y^*[g_\nu]$ and $\Lambda = (\Lambda \restriction \mathsf{HC}^{\mathcal{N}_y^*[g_\nu]} \in \mathcal{N}_y^*[g_\nu])^g$. We leave the details of such calculations to the reader. The methodology behind such calculations is presented in Section 4.1.3.

[58]Here we confuse Δ with its extension to $\mathcal{N}_y^*[g]$. Fullness preservation and branch condensation follow from Theorem 4.6.3 and Theorem 4.9.5. Recall that we are assuming that clause 3 of Theorem 9.3.2 is false.

[59]See Remark 4.2.3.

1. $F_1(z) = (\mathcal{N}_z^*, \mathcal{M}_z, \delta_z, \Psi_z)$,
2. $\vec{G}_z$ be as in clause 7 of Theorem 4.1.12 and
3. $\mathbb{M}_z = (\mathcal{M}_z, \delta_z, \vec{G}_z, \Psi_z)$,

then $(\mathbb{M}_z, (N_1, \Phi_1), \Gamma_1^*, A_1^*)$ Suslin, co-Suslin captures $\mathsf{Code}(\Delta)$ and $\mathcal{N}_y^* \in \mathsf{HC}^{\mathcal{N}_z^*}$.

Working in $\mathcal{N}_z^*$, let η_1 be the least $\mathcal{N}_z^*$-cardinal such that $\mathcal{M}_1^{\#,\Phi_0}(\mathcal{N}_z^*|\eta_1) \models$ "η_1 is a Woodin cardinal". Let $\mathcal{N}^*$ be the output of the fully backgrounded construction of $\mathcal{N}_z^*|\eta_1$ relative to Φ_1 done over $\mathcal{N}_y^*|\nu_1$. Comparing $\mathcal{N}_y^*$ with the construction producing $\mathcal{N}^*$ we get a normal stack $\mathcal{T}$ on $\mathcal{N}_y^*$ according to Ψ_y such that $\mathcal{T}$ is based on $\mathcal{N}_y^*|\eta$ and if $\mathcal{T}^-$ is $\mathcal{T}$ without its last branch then $\mathrm{m}(\mathcal{T}^-) = \mathcal{N}^*|\eta_1$.

We now have that $\mathcal{M}_1^{\#,\Phi_0}(\mathcal{N}^*|\eta_1) \models$ "η_1 is a Woodin cardinal" (this can be shown by considering S-constructions). Yet, by elementarity $(\mathcal{S}, \Delta)$ wins the comparison against the Γ-hod pair construction of $\mathcal{N}^*|\eta_1$, contradicting universality of the latter. This contradiction implies that some initial segment of $\mathcal{S}_1$ is a Δ-iterate of $\mathcal{S}$. Let $\mathcal{S}_2$ be this initial segment. This finishes the proof of (b).

We now want to show that there is a real $q \in \mathbb{R}^{\mathcal{N}_y^*[g]}$ such that $\mathsf{Code}(\Delta_{\mathcal{S}_2})$ is definable over $(\mathsf{HC}, A_0, q, \in)$ without parameters. Fix $r \in \mathbb{R}$ such that $\mathsf{Code}(\Delta_{\mathcal{S}_2})$ is definable over $(\mathsf{HC}, A_0, r, \in)$ without parameters, and let ζ be the formula defining $\mathsf{Code}(\Delta_{\mathcal{S}_2})$. Let ξ be a cutpoint of $\mathcal{N}_1$ such that $\mathcal{S}_2 \in \mathcal{N}_1|\xi$. Let $\mathcal{N}_1^+ = \mathcal{M}_1^{\#,\Phi_0}(\mathcal{N}_1|\eta)$ and let Ψ^+ be the strategy of $\mathcal{N}_1^+$. Let $\pi : \mathcal{N}_1^+ \to \mathcal{N}_2$ be an iteration of $\mathcal{N}_1^+$ via Ψ^+ such that r is generic over $\mathcal{N}_1^+$ for the extender algebra at $\pi(\eta)$. We now have that

(1) $\mathcal{N}_2[r] \models$ "$\mathsf{Code}(\Delta_{\mathcal{S}_2})$ is definable over $(\mathsf{HC}, A_0 \cap \mathcal{N}_2[r], r, \in)$[60] via formula ζ"[61].

It follows from elementarity of π that

(2) $\mathcal{N}_1^+ \models$ "it is forced by $Coll(\omega, \eta)$ that there is a real s such that $\mathsf{Code}(\Delta_{\mathcal{S}_2})$ is definable via ζ over $(\mathsf{HC}, A_0, s \in)$".

Because $\mathcal{N}_1^+$ is countable in $\mathcal{N}_y^*[g]$, we can fix $q \in \mathbb{R}^{\mathcal{N}_y^*[g]}$ such that

(3) q is in some $\leq \eta$-generic extension of $\mathcal{N}_1^+$ and $\mathcal{N}_1^+[q] \models$ "$\mathsf{Code}(\Delta_{\mathcal{S}_2})$ is definable via ζ over $(\mathsf{HC}, A_0, q, \in)$".

Now δ is a Woodin cardinal in $\mathcal{N}_1^+[q]$, and so using genericity iterations we can show that $\mathsf{Code}(\Delta_{\mathcal{S}_2})$ is definable over $(\mathsf{HC}, A_0, q, \in)$ via ζ. This finishes the proof of Claim 2.[62] $\dashv$

[60] Here and below, we confuse A_0 with its interpretations in relevant models.

[61] See Lemma 9.3.3.

[62] The proof is a bit more involved. Notice that $\mathcal{N}_1^+[q]$ captures Suslin, co-Suslin captures $\mathsf{Code}(\Delta_{\mathcal{S}_2})$. This is because for some $\mathcal{N}_1^+$-successor cardinal $\nu' \in [\nu_1, \eta)$, $\mathsf{Code}(\Delta_{\mathcal{S}_2})$ is determined by the fragment of $\Psi^+_{\mathcal{N}_1^+|\nu'}$ that acts on iteration that are above ν_1. It follows that $\mathcal{N}_1^+[q]$ has a way of determining $\mathsf{Code}(\Delta_{\mathcal{S}_2})$ in its generic extensions. Lemma 9.3.3 then gives what we want.

Our discussion before Claim 1, Claim 1 and Claim 2 show that (ψ, D) is lower part closed, is of limit type, is stable and is directed. We now work in $\mathcal{N}_y^*[g]$.

NOTATION 9.3.4. Let

1. $\Sigma = \Sigma_{\psi,D}$ (see clause 2 of Notation 9.1.3) and
2. $\mathcal{P} = \mathcal{P}_{\psi,D}$.

Notice that if h is $Coll(\omega, \mathbb{R}^{\mathcal{N}_y^*[g]})$-generic over $\mathcal{N}_y^*[g]$ then there is a real $z \in \mathcal{N}_y^*[g]$ such that $z(0)$ is a Gödel number for a formula ζ such that Σ is definable over $(\mathsf{HC}^{\mathcal{N}_y^*[g*h]}, A_0 \cap \mathsf{HC}^{\mathcal{N}_y^*[g*h]}, z, \in)$ via ζ without parameters. Notice that if Σ^+ is the strategy for $\mathcal{P}|\delta^{\mathcal{P}}$ definable over $(\mathsf{HC}, A_0, z, \in)$ via ζ without parameters then $\Sigma^+ \restriction (\mathcal{N}_y^*|\delta_y)[g] = \Sigma$. We will confuse Σ^+ with Σ.

CLAIM 3. $\mathsf{Code}(\Sigma) \in \Gamma$.

PROOF. Towards a contradiction assume $\mathsf{Code}(\Sigma) \notin \Gamma$. It then follows that $\Gamma(\mathcal{P}|\delta^{\mathcal{P}}, \Sigma) = \Gamma$, and hence clause 1 of Theorem 9.3.2 holds. ⊣

Since $\Gamma_1^* \neq \wp(\mathbb{R})$, there is a $C \subseteq \mathbb{R}$ such that $\Gamma_1^*, F_1 \in L(C, \mathbb{R})$. We then have that $L(C, \mathbb{R}) \vDash \mathsf{DC}$. Work in $W = L(C, \mathbb{R})$ and let $G \subseteq Coll(\omega_1, \mathbb{R})$ be W-generic. Notice that $W[G] \vDash \mathsf{ZFC}$. Recall $\mathcal{F}$ from the statement of Theorem 9.3.2. Let $((\mathcal{Q}_\alpha, \Lambda_\alpha) : \alpha < \omega_1) \in W[G]$ be an enumeration of $\mathcal{F}$ and $(z_\alpha : \alpha < \omega_1) \in W[G]$ be an enumeration of $\mathbb{R}$. In $W[G]$, choose a sequence $(y_\alpha : \alpha < \omega_1)$ of reals such that

1. $y_0 = y$ and $g \in \mathcal{N}_{y_1}^*$,
2. for all $\alpha < \omega_1$, letting $F_1(y_\alpha) = (\mathcal{N}_{y_\alpha}^*, \mathcal{M}_{y_\alpha}, \delta_{y_\alpha}, \Psi_{y_\alpha})$, $(z_\beta : \beta \leq \alpha) \in \mathcal{N}_{y_\alpha}^*$ and $\oplus_{\beta \leq \alpha} \Lambda_\alpha$ is Suslin, co-Suslin captured by $(\mathcal{N}_{y_\alpha}^*, \delta_{y_\alpha}, \Psi_{y_\alpha})$, and
3. for $\beta < \omega_1$, $(\mathcal{N}_{y_\gamma}^* : \gamma < \beta) \in \mathsf{HC}^{\mathcal{N}_{y_\beta}^*}$.

We now construct a sequence of Φ_1-mice $(\mathcal{M}_\alpha, \mathcal{N}_\alpha : \mathcal{N}_\alpha \lhd \mathcal{M}_\alpha \wedge \alpha < \omega_1)$ and a sequence of commuting embeddings $\pi_{\alpha,\beta} : \mathcal{M}_\alpha \to \mathcal{M}_\beta$ such that $\pi_{\alpha,\beta}(\mathcal{N}_\alpha) = \mathcal{N}_\beta$ and if $\kappa_\alpha = \mathrm{crit}(\pi_{\alpha,\beta})$ then $\mathcal{N}_\alpha = \mathcal{M}_\alpha|\kappa_\alpha$. For $\alpha > 0$ we will have that $\mathcal{M}_\alpha$ is the output of a fully backgrounded construction of $\mathcal{N}_{y_\alpha}^*$ relative to Φ_1 and also that $\mathcal{N}_\alpha \trianglelefteq \mathcal{M}_\alpha$, and $\mathcal{M}_\alpha$ will be a Ψ_y-iterate of $\mathcal{N}_y^*$. Below we describe the construction.

1. Set $\mathcal{M}_0 = \mathcal{N}_{y_0}^*$ and $\mathcal{N}_0 = \mathcal{M}_0|\kappa$.
2. For $\alpha < \omega_1$, let $\vec{G}_\alpha$ consist of those $E \in \vec{E}^{\mathcal{N}_{y_\alpha}^*}$ such that $\mathrm{crit}(E) > \mathrm{ord}(\mathcal{N}_\alpha)$ and $\nu(E)$ is an inaccessible cardinal of $\mathcal{N}_{y_\alpha}^*$.
3. Given $\mathcal{M}_\alpha$ and $\mathcal{N}_\alpha$, let $\mathcal{M}_{\alpha+1} = (\mathsf{Le}((N_1, \Phi_1), \mathcal{N}_\alpha))^{(\mathcal{N}_{y_\alpha}^*, \delta_\alpha, \vec{G}_\alpha)}$.
4. Let $\pi_{\alpha,\alpha+1} : \mathcal{M}_\alpha \to \mathcal{M}_{\alpha+1}$ be the iteration embedding according to $(\Psi_y)_{\mathcal{M}_\alpha}$.[63]
5. Let $\kappa_{\alpha+1}$ be the least $\delta_{y_{\alpha+1}}$-strong cardinal of $\mathcal{M}_{\alpha+1}$ and let
$$\mathcal{N}_{\alpha+1} = \mathcal{M}_{\alpha+1}|\kappa_{\alpha+1}.$$

[63] Notice that because $\mathcal{N}_{y_\alpha}^* \in \mathsf{HC}^{\mathcal{N}_{y_{\alpha+1}}^*}$, $\mathcal{M}_{\alpha+1}$ is a $(\Psi_y)_{\mathcal{M}_\alpha}$-iterate of $\mathcal{M}_\alpha$.

It follows that $\mathcal{N}_{\alpha+1} = \pi_{\alpha,\alpha+1}(\mathcal{N}_\alpha)$[64].

6. Suppose now that $\lambda < \omega_1$ is a limit ordinal and we have constructed a sequence $(\mathcal{M}_\alpha, \mathcal{N}_\alpha : \mathcal{N}_\alpha \triangleleft \mathcal{M}_\alpha \wedge \alpha < \lambda)$ and a sequence of commuting embedding $\pi_{\alpha,\beta} : \mathcal{M}_\alpha \to \mathcal{M}_\beta$ for $\alpha < \beta < \lambda$. Let $\mathcal{M}_\lambda^*$ be the direct limit of $\mathcal{M}_\alpha$ under $\pi_{\alpha,\beta}$. Let $\pi_{\alpha,\lambda}^* : \mathcal{M}_\alpha \to \mathcal{M}_\lambda^*$ be the embedding given by the direct limit construction.

   Let then $\mathcal{N}_\lambda = \pi_{0,\lambda}(\mathcal{N}_0)$ and let $\mathcal{M}_\lambda = (\mathsf{Le}((N_1, \Phi_1), \mathcal{N}_\lambda))^{(\mathcal{N}_{y_\lambda}^*, \delta_\lambda, G_\lambda)}$. Letting $k : \mathcal{M}_\lambda^* \to \mathcal{M}_\lambda$ be the iteration embedding according to $(\Psi_y)_{\mathcal{M}_\lambda^*}$, we set $\pi_{\alpha,\lambda} = k \circ \pi_{\alpha,\lambda}^*$.

Finally let $\mathcal{M}_{\omega_1}$ be the direct limit of the system $(\mathcal{M}_\alpha, \pi_{\alpha,\beta} : \alpha < \beta < \omega_1)$ and let $\pi_{\alpha,\omega_1} : \mathcal{M}_\alpha \to \mathcal{M}_{\omega_1}$ be the direct limit embedding. Let $\mathcal{P}_{\omega_1} = \pi_{0,\omega_1}(\mathcal{P})$ and $\mathcal{P}_{\omega_1}^- = \pi_{0,\omega_1}(\mathcal{P}^-)$.

CLAIM 4. Fix $\alpha < \omega_1$ and let $h \subseteq Coll(\omega, < \kappa_\alpha)$ be $\mathcal{N}_{y_\alpha}^*$-generic. Then

$$\pi_{0,\alpha}(\mathcal{P}) = (\mathcal{P}_{\psi,D})^{\mathcal{N}_{y_\alpha}^*[h]}$$

and $\pi_{0,\alpha}(\Sigma) = (\Sigma_{\psi,D})^{\mathcal{N}_{y_\alpha}^*[h]}$.

We leave the proof of Claim 4 to the reader as it is very similar to the proofs of Claim 1 and Claim 2. For $\alpha \leq \omega_1$, we let $\mathcal{P}_\alpha = \pi_{0,\alpha}(\mathcal{P}_\alpha)$, $\mathcal{P}_\alpha^- = \pi_{0,\alpha}(\mathcal{P}^-)$ and $\Sigma^\alpha = \pi_{0,\alpha}(\Sigma)$.

CLAIM 5. $\mathcal{P}_{\omega_1} = \mathcal{M}$.

PROOF. Notice that

(1) for $\alpha < \beta < \omega_1$ and for $\mathcal{R} \triangleleft_{hod}^c \mathcal{P}_\alpha$, $\pi_{\alpha,\beta} \restriction \mathcal{R}$ is the iteration embedding according to $(\Sigma^\alpha)_\mathcal{R}$, and
(2) if $\alpha < \omega_1$, $\mathcal{R} \triangleleft_{hod}^c \mathcal{P}_\alpha$ and $\mathcal{Q}$ is a $(\Sigma^\alpha)_\mathcal{R}$-iterate of $\mathcal{R}$ then there is $\beta < \omega_1$ such that some $\pi_{\alpha,\beta}(\mathcal{R})$ is a $(\Sigma^\alpha)_\mathcal{Q}$-iterate of $\mathcal{Q}$.
(3) for all $\alpha < \omega_1$ there is $\mathcal{R} \triangleleft_{hod}^c \mathcal{P}_\alpha$ such that $\mathcal{R}$ is a Λ_α-iterate of $\mathcal{Q}_\alpha$.

To see (2), let β be large enough such that $(\mathcal{Q}_\beta, \Lambda_\beta) = (\mathcal{Q}, (\Sigma^\alpha)_\mathcal{Q})$. It then follows that $\pi_{\alpha,\beta}(\mathcal{R})$ is a $(\Sigma^\alpha)_\mathcal{Q}$-iterate of $\mathcal{Q}$. It follows from (1) and (2) that $\mathcal{P}_{\omega_1} | \delta^\mathcal{M} = \mathcal{M} | \delta^\mathcal{M}$.

If $\rho(\mathcal{P}_{\omega_1}) < \mathsf{ord}(\mathcal{P}_{\omega_1}^-)$ then we must have that $\mathcal{P}_{\omega_1} = \mathcal{M}$. Suppose then

$$\rho(\mathcal{P}_{\omega_1}) > \mathsf{ord}(\mathcal{P}_{\omega_1}^-).$$

Clearly $\mathcal{P}_{\omega_1} \trianglelefteq \mathsf{Lp}^{\Gamma,\Sigma}(\mathcal{M}^-)$. Suppose then $\mathcal{P}_{\omega_1} \triangleleft \mathsf{Lp}^{\Gamma,\Sigma}(\mathcal{M}^-)$. By a standard Skolem hull argument, it follows that for some $\alpha < \omega_1$, $\mathcal{P}_\alpha \triangleleft \mathsf{Lp}^{\Gamma,\Sigma^\alpha}(\pi_{0,\alpha}(\mathcal{P}^-))$. However, because $\rho(\mathcal{P}_{\omega_1}) > \mathsf{ord}(\mathcal{P}_{\omega_1}^-)$, $\mathcal{N}_{y_\alpha}^* \models$ "$\mathcal{P}_\alpha = \mathsf{Lp}^{\Gamma,\Sigma^\alpha}(\pi_{0,\alpha}(\mathcal{P}^-))$", contradiction. $\dashv$

[64] Notice that if $E \in \vec{E}^{\mathcal{M}_\alpha}$ is the extender with the least index on the extender sequence of $\mathcal{M}_\alpha$ such that $\mathrm{crit}(E) = \kappa_\alpha$ then E is the first extender used in the $\mathcal{M}_\alpha$-to-$\mathcal{M}_{\alpha+1}$ iteration. Here we assume that all extenders with $\mathrm{crit}(E)$ are total. Otherwise we can translate them away as is done in [57, Remark 12.7].

CLAIM 6. $\rho(\mathcal{M}) > \mathrm{ord}(\mathcal{M}^-)$.

PROOF. Assume $\rho(\mathcal{M}) < \mathrm{ord}(\mathcal{M}^-)$ (it follows from the definition of $\mathcal{M}$ that equality is impossible). We now have that $\rho(\mathcal{P}) < \delta^{\mathcal{P}}$. The argument now takes place in $\mathcal{N}_y^*[g]$. Let $N = \mathcal{N}_y^*$ and let $U \in N$ be the Mitchell order 0 ultrafilter on κ. Let $j : N \to \mathrm{Ult}(N,U)$ be the ultrapower embedding and $j^+ : N[g] \to \mathrm{Ult}(N,U)[g']$ be its lift up to $N[g]$. Notice that $j^+(\Gamma)$ makes sense. As in core model induction applications Σ can be extended to a strategy Σ' for $\mathcal{P}$[65]. It follows from clause 2 of Theorem 5.5.3 that there is a tail $(\mathcal{Q},\Lambda) \in \mathrm{Ult}(N,U)[g]$ of $(\mathcal{P},\Sigma')$ such that Λ has strong branch condensation. Because we are assuming that (in $\mathrm{Ult}(N,U)[g]$) clause 1 of Theorem 9.3.2 fails and because Σ' is a j-realizable strategy, $\Gamma(\mathcal{Q},\Lambda) \subset j^+(\Gamma)$ and $\mathrm{Code}(\Lambda) \in \Gamma$. Notice next that $\rho(\mathcal{Q}) < \delta^{\mathcal{Q}}$. We can then finish by using the argument given on page 143 of [29][66]. ⊣

We thus have that $\mathcal{P} = \mathrm{Lp}^{\Gamma,\Sigma}(\mathcal{P}^-)$.

CLAIM 7. $\mathcal{N}_y^* \models |\mathcal{P}| = \kappa$.

PROOF. Recall the real z introduced before the statement of Claim 3. We have that $z \in \mathcal{N}_y^*[g][h]$ where h is $Coll(\omega, \mathbb{R}^{\mathcal{N}_y^*[g]})$-generic. It then follows that $\mathcal{P}$ is definable over $(\mathrm{HC}^{\mathcal{N}_y^*[g][h]}, A_0 \cap \mathrm{HC}^{\mathcal{N}_y^*[g][h]}, z, \in)$ and hence, $\mathcal{P} \in \mathrm{HC}^{\mathcal{N}_y^*[g][h]}$. Thus $|\mathcal{P}|^{\mathcal{N}_y^*[g]} = \kappa$. ⊣

Notice that Claim 7 implies that lower part (φ,Γ)-covering fails as it implies that $\mathrm{cf}(\mathrm{ord}(\mathcal{P}_{\omega_1})) = \omega$[67]. It follows from Theorem 9.2.7 that $X =_{\mathrm{def}} \pi_{0,1}[\mathcal{P}] \in \wp(\mathcal{P}_1) \cap \mathcal{M}_1$ is such that

(A) for any $\mathcal{M}_1$-generic $h \subseteq Coll(\omega, <\kappa_1)$, $\mathcal{M}_1[h] \models$ "X is countable and is a (ψ, D)-condensing set".

It follows from Claim 7 that

(B) for every $\alpha \in [1,\omega_1)$ and for every $\mathcal{M}_{y_\alpha}$-generic $h \subseteq Coll(\omega, <\kappa_\alpha)$, $\mathcal{M}_{y_\alpha}[h] \models$ "$\pi_{1,\alpha}[X]$ is a (ψ, D)-condensing set".

CLAIM 8. For every $\alpha \in [1,\omega_1)$ and for every $\mathcal{N}_{y_\alpha}^*$-generic $h \subseteq Coll(\omega, <\kappa_\alpha)$, $\mathcal{N}_{y_\alpha}^*[h] \models$ "$\pi_{1,\alpha}[X]$ is a weakly (ψ, D)-condensing set".

PROOF. We give the proof for $\alpha = 1$ and leave the rest to the reader. Let $h \subseteq Coll(\omega, <\kappa_1)$ be $\mathcal{N}_{y_1}^*$-generic and let $Y \in (\wp_{\omega_1}(\mathcal{P}_1))^{\mathcal{N}_{y_1}^*[h]}$ be an extension of X. In what follows we will use the notation introduced in Section 9.1 relative to $\mathcal{N}_{y_1}^*[h]$. Thus, $\Sigma_Y \in \mathcal{N}_{y_1}^*[h]$ is the $\tau_Y : \mathcal{P}_Y \to \mathcal{P}_1$-pullback of $\pi_{0,1}(\Sigma)$. However, we will also confuse Σ_Y and $\pi_{0,1}(\Sigma)$ with their canonical extensions that act on all stacks.

[65]For example, see Definition 9.0.7, Section 10.2.5, [29, Definition 6.14] and also [31] and [65]. Recall that Σ is a strategy for $\mathcal{P}^-$ which in this case is just $\mathcal{P}|\delta^{\mathcal{P}}$.

[66]This is a standard argument in core model induction. The reader can also consult [31] and [65].

[67]This is because π_{0,ω_1} is continuous at $\mathrm{ord}(\mathcal{P})$.

The proof of the claim follows the steps of Theorem 9.2.6. Recall that in that proof the key step is to find a universal model extending $\mathcal{P}$ such that $\pi_{0,1}$ acts on it. Here, we describe how to find this universal model and leave the rest, which is just like the proof of Theorem 9.2.6, to the reader. To simplify, we only show that if $\mathcal{Q}_Y^- = \tau_Y^{-1}(\mathcal{P}_1^-)$ then $\mathcal{Q}_Y = \mathsf{Lp}^{\Gamma,\Sigma_Y}(\mathcal{Q}_Y^-)$. The rest of the proof is very similar.

Suppose then that $\mathcal{Q}_Y \lhd \mathsf{Lp}^{\Gamma,\Sigma_Y}(\mathcal{Q}_Y^-)$ and let $\mathcal{S} \lhd \mathsf{Lp}^{\Gamma,\Sigma_Y}(\mathcal{Q}_Y^-)$ be the least such that $\rho(\mathcal{S}) \leq \mathsf{ord}(\mathcal{Q}_Y^-)$ and $\mathcal{S} \not\trianglelefteq \mathcal{Q}_Y$. Let $(\mathcal{R},\Lambda) \in \mathsf{HP}^{\Gamma}$ be such that

1. $\mathcal{R}$ is meek and of limit type,
2. $(\mathcal{R},\Lambda)$ be a Σ-hod pair,
3. $L(\Gamma(\mathcal{R},\Lambda),\mathbb{R}) \models$ "$\mathcal{S}$, as a Σ_Y-mouse, has an ω_1-iteration strategy."

Let $\alpha < \omega_1$ be such that $\mathsf{Code}(\Lambda)$ is Suslin, co-Suslin captured by $(\mathcal{N}^*_{y\alpha}, \delta_{y\alpha}, \Psi_{y\alpha})$. Recalling Definition 4.2.1 and Remark 4.2.3, let

1. $\mathcal{W}^*$ be the output of $(\mathsf{Le}((N_1,\Phi_1)\oplus(\mathcal{P},\Sigma),\mathcal{J}_\omega(N_1,\mathcal{P})))_{>\kappa}^{(\mathcal{N}^*_y,\delta_y,\vec{G}_y)}$ and
2. $\mathcal{W}^{**}$ be the output of $(\mathsf{Le}((N_1,\Phi_1)\oplus(\mathcal{P},\Sigma),\mathcal{J}_\omega(N_1,\mathcal{P})))_{>\kappa}^{(\mathcal{N}^*_{y\alpha},\delta_{y\alpha},\vec{G}_{y\alpha})}$.

Notice that it follows that $\mathsf{ord}(\mathcal{W}^*) = \delta_y$ and $\mathsf{ord}(\mathcal{W}^{**}) = \delta_{y\alpha}$. We now compare the construction producing $\mathcal{W}^*$ and the construction producing $\mathcal{W}^{**}$. The comparison produces a tree $\mathcal{T}$ on $\mathcal{N}^*_y$ of limit length such that

(T1) $\mathcal{T} \in \mathcal{N}^*_{y\alpha}$,
(T2) setting $b = \Psi_y(\mathcal{T})$, $\pi_b^{\mathcal{T}}(\mathcal{W}^*) = \mathcal{W}^{**}$,
(T3) $\mathcal{T}$ is above κ.

Let $\mathcal{W}$ be the Γ-hod pair construction of $(\mathcal{W}^{**},\delta_{y\alpha},\vec{G}^*,\Sigma^*)$ done over $\mathcal{P}$ and relative to Σ[68] where

1. $\vec{G}^*$ is the set of those extenders from $\vec{E}^{\mathcal{W}^{**}}$ whose critical point is $> \mathsf{ord}(\mathcal{P})$ and $\nu(E)$ is an inaccessible cardinal and
2. Σ^* is the strategy of $\mathcal{W}^{**}$ induced by $\Psi_{y\alpha}$.

It follows from Theorem 4.13.2 that there is $\mathcal{K} \lhd_{hod} \mathcal{W}$ which is a Λ-iterate of $\mathcal{R}$, and hence,

(1) $L(\Gamma(\mathcal{K},\Lambda_{\mathcal{K}})) \models$ "$\mathcal{S}$, as a Σ_Y-mouse, has an ω_1-iteration strategy".

$\mathcal{K}$ is our universal model but we cannot yet apply $\pi_{0,1}$ to it. To do this, let $\mathcal{U} = \pi_{0,1}\mathcal{T}$. The copying construction produces $\sigma : \mathcal{M}_b^{\mathcal{T}} \to \mathcal{M}_b^{\mathcal{U}}$ such that $\pi_b^{\mathcal{U}} \circ \pi_{0,1} = \sigma \circ \pi_b^{\mathcal{T}}$. Moreover, because of (T3) above, $\mathrm{crit}(\sigma) = \kappa$, $\sigma(\mathcal{P}) = \mathcal{P}_1$ and $\sigma \restriction \mathcal{P} = \pi_{0,1} \restriction \mathcal{P}$. It then follows that

(2) $\sigma(\mathcal{K})$ is a $\pi_{0,1}(\Sigma)$-hod premouse over $\mathcal{P}_1$, and
(3) $\Lambda_{\mathcal{K}}$ is the σ-pullback of the strategy of $\sigma(\mathcal{K})$ induced by $(\Psi_y)_{\mathcal{M}_b^{\mathcal{U}}}$.

The reason (3) holds is the following. First notice that $\Lambda_{\mathcal{K}}$ is the strategy of $\mathcal{K}$ induced by Σ^*. But for some $\nu < \delta_{y\alpha}$, we build $\mathcal{K}$ via Γ-hod pair construction of

[68] See Definition 4.3.3.

$\mathcal{W}^{**}|\nu$, and hence $\Lambda_{\mathcal{K}}$ is the strategy of $\mathcal{K}$ induced by $\Sigma^*_{\mathcal{W}^{**}|\nu}$. $\mathcal{W}^{**}|\nu$ has a unique ω_1-strategy as a $\Phi_1 \oplus \Sigma$-mouse, and therefore $\Sigma^*_{\mathcal{W}^{**}|\nu}$ is the strategy induced by $(\Psi_y)_{\mathcal{M}_b^{\mathcal{T}}}$, and this strategy is the σ-pullback of the strategy of $\sigma(\mathcal{W}^{**}|\nu)$ which is induced by $(\Psi_y)_{\mathcal{M}_b^{\mathcal{U}}}$.

It now follows that we can lift $\pi_{X,Y}$ to $\mathcal{K}$ and obtain $\pi^+_{X,Y} : \mathcal{K} \to \mathcal{Y}$ and $\tau^+_Y : \mathcal{Y} \to \sigma(\mathcal{K})$ such that

(4) $\sigma \restriction \mathcal{K} = \tau^+_Y \circ \pi^+_{X,Y}$,
(5) $\sigma \restriction \mathcal{P} = \pi_{0,1} \restriction \mathcal{P}$.

The rest of the proof follows very closely to the proof of Lemma 9.2.6 and uses (1). This finishes our outline of the proof of Claim 8. ⊣

Working in V, given $A \in \mathcal{P}_{\omega_1}$ and $X \in \wp_{\omega_1}(\mathcal{P}_{\omega_1})$, we let $T_{X,A}$ be the set of (φ, s) such that $s \in [\delta_X]^{<\omega}$ and $\mathcal{P}_{\omega_1} \models \varphi[A, \pi^{\Sigma_X}_{\mathcal{P}_X|\delta_X,\infty}(s)]$. We then say that X has *A-condensation* if for every $Y \in \wp_{\omega_1}(\mathcal{P}_{\omega_1})$, $\tau_{X,Y}(T_{X,A}) = T_{Y,A}$. To show that there is a strongly (φ, Γ)-condensing set, it is enough to show that for each A there is an $X \in \wp_{\omega_1}(\mathcal{P}_{\omega_1})$ with A-condensation. Assuming this, it is not hard to show that for some $\alpha < \omega_1$, $\pi_{\alpha,\omega_1}[\mathcal{P}_\alpha]$ is a condensing set.

CLAIM 9. Suppose $A \in \mathcal{P}_{\omega_1}$. There is $\alpha_0 < \omega_1$ such that $A \in \mathrm{rge}(\pi_{\alpha_0,\omega_1})$ and for every $\alpha \in (\alpha_0, \omega_1)$ and for every $\mathcal{N}^*_{y_\alpha}$-generic $h \subseteq Coll(\omega, <\kappa_\alpha)$, $\mathcal{N}^*_{y_\alpha}[h] \models$ "$\pi_{\alpha_0,\alpha}[\mathcal{P}_{\alpha_0}]$ is an A_α-condensing set" where $A_\alpha = \pi^{-1}_{\alpha,\omega_1}(A)$.

PROOF. Towards a contradiction assume otherwise. Let $(\alpha_i : i < \omega)$ be such that

1. for all $i < \omega$, $\alpha_i < \omega_1$,
2. for all $i < \omega$ and for every $\mathcal{N}^*_{y_{\alpha_{i+1}}}$-generic $h \subseteq Coll(\omega, <\kappa_{\alpha_{i+1}})$, $\mathcal{N}_{y_{\alpha_{i+1}}}[h] \models$ "$\pi_{\alpha_i,\alpha_{i+1}}[\mathcal{P}_{\alpha_i}]$ is not a $A_{\alpha_{i+1}}$-condensing set".

Let $\alpha = \sup_{i<\omega} \alpha_i$. Let $\nu_i < \kappa_{\alpha_{i+1}}$ be such that for some $\mathcal{N}^*_{y_{\alpha_{i+1}}}$-generic $h \subseteq Coll(\omega, \nu_i)$ there is $W \in (\wp_{\omega_1}(\mathcal{P}_{\alpha_{i+1}}))^{\mathcal{N}^*_{y_{\alpha_{i+1}}}[h]}$ such that $\mathcal{N}^*_{y_{\alpha_{i+1}}}[h] \models$ "it is forced by $Coll(\omega, <\kappa_{\alpha_{i+1}})$ that W witness that $\pi_{\alpha_i,\alpha_{i+1}}[\mathcal{P}_{\alpha_i}]$ is not a $A_{\alpha_{i+1}}$-condensing set". Fix then (h_i, W_i) that play the role of (h, W) and such that $h_i \in \mathcal{N}^*_{y_\alpha}$. Set $Z_i = \pi_{\alpha_i,\alpha_{i+1}}[\mathcal{P}_{\alpha_i}]$. We thus have that

(1) in $\mathcal{N}^*_{y_{\alpha_{i+1}}}[h_i]$, $\tau_{Z_i,W_i}(T^i_{Z_i,A_{\alpha_{i+1}}}) \neq T^i_{W_i,A_{\alpha_{i+1}}}$ where $T^i_{U,A_{\alpha_{i+1}}}$ is defined like $T_{U,A}$ above only inside $\mathcal{N}^*_{y_{\alpha_{i+1}}}[h_i]$[69].

Let $X'_i = \pi_{\alpha_i,\alpha}[\mathcal{P}_{\alpha_i}]$ and $Y'_i = \pi_{\alpha_{i+1},\alpha}[W_i]$. It is not hard to verify that

[69]More precisely, $(s, \varphi) \in T^i_{U,A_{\alpha_{i+1}}}$ if and only if $s \in [\delta_U]^{<\omega}$ and $\mathcal{P}_{\alpha_{i+1}} \models \varphi[A_{\alpha_{i+1}}, \pi^{\Sigma_U}_{\mathcal{P}_U|\delta_U,\infty}(s)]$. All of the relevant objects are computed in $\mathcal{N}^*_{y_{\alpha_{i+1}}}[h_i]$.

(2) in $\mathcal{N}^*_{y\alpha}$, $\tau_{X'_i,Y'_i}(T^{\alpha}_{X'_i,A\alpha}) \neq T^{\alpha}_{Y'_i,A\alpha}$ where $T^{\alpha}_{U,A\alpha}$ is defined like $T_{U,A}$ above only inside $\mathcal{N}^*_{y\alpha}$.

Like in the proof of Theorem 9.2.7, we can find some $\mathcal{Y} \in \mathcal{P}_\alpha|\delta^{\mathcal{P}_\alpha}$ and $B \in \mathcal{Y}$ with the property that letting $\sup_{i<\omega} X'_i =_{\text{def}} \eta < \delta^{\mathcal{P}_\alpha}$, $\mathcal{Y}|\eta = \mathcal{P}_\alpha|\eta$ and for every $s \in [\eta]^{<\omega}$ and every φ, $\mathcal{Y} \vDash \varphi[B,s]$ if and only if $\mathcal{P}_\alpha \vDash \varphi[A_\alpha, s]$. Let now (in $\mathcal{N}^*_{y\alpha}$) $\mathcal{P}_i = \mathcal{P}_{X'_i}$, $\mathcal{Q}_i = \mathcal{P}_{Y'_i}$, $\xi_i = \tau_{X'_i,Y'_i} : \mathcal{P}_i \to \mathcal{Q}_i$, $\pi_i = \tau_{Y'_i,X'_{i+1}} : \mathcal{Q}_i \to \mathcal{P}_{i+1}$ and $\varphi_i = \tau_{X'_i,X'_{i+1}}$. Finally, set $(C,\mathcal{X}) = \pi_{\alpha,\omega_1}(B,\mathcal{Y})$, $X_i = \pi_{\alpha,\omega_1}(X'_i)$ and $Y_i = \pi_{\alpha,\omega_1}(Y'_i)$. It is not hard to verify that

(3) in V, $\mathcal{A} = \{(\mathcal{P}_i, \mathcal{Q}_i, X_i, Y_i, \xi_i, \pi_i, \varphi_i), C, \mathcal{X}\}$ is a bad tuple relative to A in the sense that
   1. for all $i < \omega$, $X_i \in \wp_{\omega_1}(\mathcal{P}_{\omega_1})$ is such that $\tau_{X_i} \restriction \mathcal{P}_{X_i} = \pi^{\Sigma_{X_i}}_{\mathcal{P}_{X_i}|\delta_{X_i},\infty}$,
   2. for all $i < \omega$, $\mathcal{P}_i = \mathcal{P}_{X_i}$ and $\mathcal{Q}_i = \mathcal{P}_{Y_i}$;
   3. for all $i < j < \omega$, $X_i \prec Y_i \prec X_j$;
   4. for all $i < \omega$, $\xi_i = \tau_{X_i,Y_i}$, $\pi_i = \tau_{Y_i,X_{i+1}}$ and $\varphi_i = \tau_{X_i,X_{i+1}}$[70];
   5. $\mathcal{X} \in \mathcal{P}_{\omega_1}$ and letting $\nu = \sup_{i<\omega}(X_i \cap \delta^{\mathcal{P}_{\omega_1}})$, $\mathcal{X}|\nu = \mathcal{P}_{\omega_1}|\nu$;
   6. letting ν be as above, for every formula φ and for every $s \in \nu^{<\omega}$, $\mathcal{X} \vDash \varphi[C,s]$ if and only if $j(\mathcal{P}) \vDash \varphi[A,s]$;
   7. for all $i \in [1,\omega)$, $\xi_i(T_{X_i,A}) \neq T_{Y_i,A}$.

As in the proof of Claim 8 we can find some normal stack $\mathcal{T}$ on $\mathcal{N}^*_y$ with last model $\mathcal{W}$ such that

(3) $\mathcal{T}$ is above κ and $\pi^{\mathcal{T}}$ is defined,
(4) the Γ-hod pair construction of

$$(\mathcal{W}, \pi^{\mathcal{T}}(\delta_y), \pi^{\mathcal{T}}(\vec{G}_y))$$

done over $\mathcal{P}$ and relative to Σ using extenders with critical point $> \kappa$ reaches a Σ-hod pair $(\mathcal{K}, \Lambda)$ such that

$L(\Gamma(\mathcal{K},\Lambda)) \vDash$ "the sequence $\mathcal{A} = \{(\mathcal{P}_i, \mathcal{Q}_i, X_i, Y_i, \xi_i, \pi_i, \varphi_i), C, \mathcal{X}\}$ is a bad tuple".

Let $\mathcal{U} = \pi_{0,\alpha}\mathcal{T}$, $\mathcal{W}'$ be the last model of $\mathcal{U}$ and let $\sigma : \mathcal{W} \to \mathcal{W}'$ be the copy map. We have that σ-pullback of $\sigma(\Lambda)$ is Λ[71]. We now finish the proof by performing the following steps:

Step 1: lift $\mathcal{K}$ to each $\mathcal{P}_i$ via π_{0,α_i} and obtain $\mathcal{P}_i^+$,
Step 2: lift $\mathcal{K}$ to each $\mathcal{Q}_i$ via $\xi_i \circ \pi_{0,\alpha_i}$ and obtain $\mathcal{Q}_i^+$,
Step 3: extend $(\xi_i, \pi_i, \varphi_i)$ to $\xi_i^+ : \mathcal{P}_i^+ \to \mathcal{Q}_i^+$, $\pi_i^+ : \mathcal{Q}_i^+ \to \mathcal{P}_{i+1}^+$ and $\varphi_i^+ : \mathcal{P}_i^+ \to \mathcal{P}_{i+1}^+$,
Step 4: find maps $p_i : \mathcal{P}_i^+ \to \sigma(\mathcal{K})$ and $q_i : \mathcal{Q}_i^+ \to \sigma(\mathcal{K})$ such that $p_i = q_i \circ \xi_i^+ = p_{i+1} \circ \pi_i^+$,

[70] Thus, $\xi_i : \mathcal{P}_i \to \mathcal{Q}_i$, $\pi_i : \mathcal{Q}_i \to \mathcal{P}_{i+1}$ and $\varphi_i = \pi_i \circ \xi_i$.
[71] Here we confuse the local strategies with their interpretations in V.

Step 5: let $\mathcal{P}_\omega^+$ be the direct limit of $(\mathcal{P}_i^+, \varphi_{i,k} : i < k < \omega)$ where $\varphi_{i,k} : \mathcal{P}_i^+ \to \mathcal{P}_k^+$ be the composition of $(p_n : n \in [i,k))$,
Step 6: let $\varphi_{i,\omega} : \mathcal{P}_i^+ \to \mathcal{P}_\omega^+$ and $\psi_{i,\omega} : \mathcal{Q}_i^+ \to \mathcal{P}_\omega^+$ be the direct limit embeddings,
Step 7: let $p_\omega^+ : \mathcal{P}_\omega^+ \to \sigma(\mathcal{K})$ be constructed via the direct limit construction,
Step 8: set for $i \leq \omega$, $\Pi_i^p = (p_i\text{-pullback of } \sigma(\Lambda))$ and $\Pi_i^q = (q_i\text{-pullback of } \sigma(\Lambda))$,
Step 9: apply the three dimensional argument from the last portion of the proof of Theorem 9.2.7 to derive a contradiction.

The above steps finish the proof of Claim 9. ⊣

The discussion before Claim 9 implies that there is a (φ, Γ)-condensing set, which is clause 3 of Theorem 9.3.2. Since we were assuming all three clauses of Theorem 9.3.2 are false, this is clearly a contradiction and finishes the proof of Theorem 9.3.2. ⊣

Chapter 10

APPLICATIONS

10.1. The generation of the mouse full pointclasses

In this section, our goal is to show that under Strong Mouse Capturing (SMC) if Γ is a mouse full pointclass (see Definition 5.3.2) such that $\Gamma \neq \wp(\mathbb{R})$ and there is a good pointclass Γ^* with the property that $\Gamma \subset \Gamma^*$ then there is a hod pair or an sts pair $(\mathcal{P}, \Sigma)$ such that $\Gamma(\mathcal{P}, \Sigma) = \Gamma$. Recall that SMC states that for any hod pair or sts hod pair $(\mathcal{P}, \Sigma)$ such that Σ is strongly fullness preserving and has strong branch condensation then for any $x, y \in \mathbb{R}$, $x \in OD_{y,\Sigma}$ if and only if $x \in Lp^{\Sigma}(y)$. We work under the following two minimality assumptions.

DEFINITION 10.1.1. $\#_{\mathsf{lsa}}$ is the following statement: There is a pointclass $\Gamma \subset \wp(\mathbb{R})$ such that there is a Suslin cardinal bigger than $w(\Gamma)$ and $L(\Gamma, \mathbb{R}) \vDash \mathsf{LSA}$.

NWLW is the following statement: There is no iteration strategy for an active mouse with a Woodin cardinal that is a limit of Woodin cardinals.

As in [29, Section 6.1], we will construct $(\mathcal{P}, \Sigma)$ as above via a hod pair construction of some sufficiently strong background universe. Here is our theorem on generation of pointclasses.

THEOREM 10.1.2 (The generation of the mouse full pointclasses I). *Assume*

$$\mathsf{AD}^+ + \neg\#_{\mathsf{lsa}} + \mathsf{NWLW}.^1$$

Suppose $\Gamma \neq \wp(\mathbb{R})$ is a mouse full pointclass such that $\Gamma \vDash \mathsf{SMC}$. Then one of the following holds:

1. *For some $(\mathcal{Q}, \Lambda) \in \mathsf{HP}^{\Gamma}$ such that Λ has strong branch condensation and is strongly Γ-fullness preserving and for some $x \in \mathbb{R}$, $\mathsf{Lp}^{\Lambda}(x) \neq \mathsf{Lp}^{\Gamma,\Lambda}(x)$.*
2. *Γ is completely mouse full and letting $A \subseteq \mathbb{R}$ witness the fact that Γ is completely mouse full, the following holds in $L(A, \mathbb{R})$:*
   (a) $\neg\mathsf{LSA}$ and there is a hod pair $(\mathcal{P}, \Sigma)$ such that Σ has strong branch condensation and is strongly fullness preserving and $\Gamma(\mathcal{P}, \Sigma) = \Gamma$.

[1] See Theorem 10.3.1, which removes the hypothesis that NWLW holds.

(b) LSA *and there is an sts hod pair* $(\mathcal{P},\Sigma)$ *such that* Σ *has branch condensation and is fullness preserving,* $\mathcal{P}$ *is of #-lsa type*[2] *and* $\Gamma^b(\mathcal{P},\Sigma)=\Gamma$.

Additionally, assuming (i) clause 1 fails, (ii) if A is as in clause 2 then $L(A,\mathbb{R})\models$ LSA, *and (iii) there is a good pointclass* Γ^* *such that* $\Gamma\subset\underset{\sim}{\Delta}_{\Gamma^*}$, *then there is a hod pair* $(\mathcal{P},\Sigma)$ *such that* $\mathcal{P}$ *is of #-lsa type,* $(\mathcal{P},\Sigma^{stc})\in L(A,\mathbb{R})$ *and* $(\mathcal{P},\Sigma^{stc})$ *satisfies the conditions in clause 2(b).*

PROOF. Our proof has the same structure as the proof of [29, Theorem 6.1]. However, unlike that proof, we will make an important use of Theorem 9.3.2. The proof is again by induction. Suppose $\Gamma\neq\wp(\mathbb{R})$ is a mouse full pointclass such that whenever Γ^* is properly contained in Γ and is a mouse full pointclass then there is a hod pair $(\mathcal{P},\Sigma)$ as in 1 or 2. We want to show that the claim holds for Γ. Towards a contradiction assume the conclusion of Theorem 10.1.2 is false. We examine several cases.

Case 1. There is a sequence of mouse full pointclass $(\Gamma_\alpha:\alpha<\Omega)$ such that $\Gamma_\alpha\subseteq\Gamma$, $\Gamma=\bigcup_{\alpha<\Omega}\Gamma_\alpha$ and for $\alpha<\beta<\Omega$, $\Gamma_\alpha\trianglelefteq_{mouse}\Gamma_\beta$.

We will use the terminology of Section 9.3. Let $\varphi(u,v)$ be the formula that expresses the fact that u is a mouse full pointclass having the properties that Γ has and v is a hod pair $(\mathcal{Q},\Lambda)$ such that $\mathsf{Code}(\Lambda)\in u$ and Λ has strong branch condensation and is strongly u-fullness preserving.

Let $\mathcal{M}^-=\mathcal{P}^-_{\varphi,\Gamma}$ and $\mathcal{M}=\mathcal{P}_{\varphi,\Gamma}$. Because we are assuming that Γ is not generated by a hod pair, it follows from clause 2 of Theorem 9.3.2 that $\rho(\mathcal{M})>o(\mathcal{M}^-)$ and that there is a condensing set $X\in\wp_{\omega_1}(\mathcal{M})$. In what follows we will use the notation introduced in Section 9.1. In particular, recall the definition of τ_Y and σ^X_Y.

Following the proof of Theorem 9.3.2 let Γ_0, Γ_0^*, Γ_1, Γ_1^*, A_0, A_1^*, (N_0,Φ_0), (N_1,Φ_1), F_0, and F_1 be as in that proof. We introduce two more kinds of set of reals that we need to be captured.

Let $(\alpha_i:i<\omega)$ be an enumeration of X and let $x_i=(\alpha_k:k\leq i)$. Let $(\varphi_i:i<\omega)$ be an enumeration of formulas in the language of hod mice. Let $B_{i,k}$ be the set of pairs $((\mathcal{Q},\Lambda,\beta),(\mathcal{R},\Psi,\gamma))$ such that $(\mathcal{Q},\Lambda),(\mathcal{R},\Psi)\in\mathsf{HP}^\Gamma$, $\beta<\delta^{\mathcal{Q}}$, $\gamma<\delta^{\mathcal{R}}$ and $\pi^{\Psi}_{\mathcal{R},\infty}(\gamma)$ is the unique ordinal ξ such that $\mathcal{M}\models\varphi_k[x_i,\pi^{\Lambda}_{\mathcal{Q},\infty}(\beta),\xi]$. We then let $A_{i,k}$ be the set of reals σ such that $\sigma(0)$ is a Gödel number of some formula ζ such that $B_{i,k}$ is definable over $(\mathsf{HC},A_0,\sigma,\in)$ via ζ without parameters.

Next, let B' be the set of $(\mathcal{Q},\Lambda)\in\mathsf{HP}^\Gamma$ such that $X\cap\delta^{\mathcal{M}}\subseteq\pi^{\Lambda}_{\mathcal{Q},\infty}[\mathcal{Q}|\delta^{\mathcal{Q}}]$ and the transitive collapse of $Hull^{\mathcal{M}}(X\cup\pi^{\Lambda}_{\mathcal{Q},\infty}[\mathcal{Q}|\delta^{\mathcal{Q}}])$ is $\mathcal{Q}$. Given $(\mathcal{Q},\Lambda)\in B'$, let $Y_{\mathcal{Q},\Lambda}=\pi^{\Lambda}_{\mathcal{Q},\infty}[\mathcal{Q}|\delta^{\mathcal{Q}}]$. Let B'' be the set of $((\mathcal{Q},\Lambda),X_{\mathcal{Q},\Lambda})$ such that $(\mathcal{Q},\Lambda)\in B'$ and $X_{\mathcal{Q},\Lambda}=\tau^{-1}_{Y_{\mathcal{Q},\Lambda}}[X]$. Let B be the set of reals σ such that $\sigma(0)$ is a Gödel number of some formula ζ such that B is definable over $(\mathsf{HC},A_0,\sigma,\in)$ via ζ without parameters.

---

[2] See Definition 2.7.3.

We now define our final set C. Given $x \in \mathbb{R}$, let $A_x = \{u \in \mathbb{R} : \{u\} \text{ is } OD^{\Gamma}_{x,x}\}$. We let $C = \{(x,y) \in \mathbb{R}^2 : y \text{ codes } A_x\}$.[3] Let now $x \in \text{dom}(F_0)$ be such that if $F_0(x) = (\mathcal{N}, \mathcal{M}, \delta, \Psi)$ then letting $\vec{G}$ be as in clause 7 of Theorem 4.1.12, $((\mathcal{N}, \delta, \vec{G}, \Psi), (N_0, \Phi_0), \Gamma_0^*, A_0)$ Suslin, co-Suslin captures Γ, A_0, B and C. Let Ψ^* be the iteration strategy of $\mathcal{M}_3^{\#,\Phi_0}$ and let $y \in \text{dom}(F_1)$ be such that if $F_1(y) = (\mathcal{N}_y^*, \mathcal{M}_y, \delta_y, \Sigma_y)$ and letting $\mathbb{M}_y$ be as in clause 3 of Theorem 4.1.12, then $(\mathbb{M}_y, (\mathcal{N}_1, \Phi_1), \Gamma_1^*, A_1^*)$ captures $\mathsf{Code}(\Psi^*)$.

We claim that some hod pair appearing on the Γ-hod pair construction of $\mathcal{N}_y^*|\delta_y$ generates Γ. Here the proof is somewhat different than the proof of Theorem 6.1 of [29]. There the contradictory assumption that such constructions do not reach Γ led to a construction of a hod pair $(\mathcal{P}, \Sigma)$ such that $\lambda^{\mathcal{P}} = \delta^{\mathcal{P}}$ and $\mathcal{P} \models$ "$\delta^{\mathcal{P}}$ is regular". This meant that a pointclass satisfying $\mathsf{AD}_{\mathbb{R}}$ + "Θ is a regular cardinal" had been reached giving the desired contradiction. In our current situation, if the constructions never stops then we will reach an lsa type hod premouse $\mathcal{P}$ of height δ_y. We need techniques to argue that this cannot happen.

We proceed by assuming that the Γ-hod pair construction of $\mathcal{N}_y^*|\delta_y$ does not reach a pair generating Γ. Let $\mathcal{P}^*$ be the final model of the Γ-hod pair construction of $\mathcal{N}_y^*|\delta_y$. By this we mean that either (i) $\mathsf{ord}(\mathcal{P}^*) = \delta_y$ and $\mathcal{P}^*)^{\#}$ is a hod premouse or (ii) $\mathsf{ord}(\mathcal{P}^*) < \delta_y$ and the Γ-hod pair construction of $\mathcal{N}_y^*|\delta_y$ after reaching $\mathcal{P}^*$ does not produce a hod premouse $\mathcal{Q}$ such that $\delta^{\mathcal{Q}} = \delta^{\mathcal{P}^*}$. If (i) is true then set $\mathcal{P} = (\mathcal{P}^*)^{\#}$ and otherwise set $\mathcal{P} = (\mathcal{P}^*|\delta^{\mathcal{P}^*})^{\#}$. Let Σ be the strategy of $\mathcal{P}$ induced by Σ_y. Note that δ_y is not a limit of Woodin cardinals in $\mathcal{P}$ as otherwise $\mathcal{P} \models$ "δ_y is a Woodin cardinal that is limit of Woodin cardinals", contradicting our smallness assumption.

CLAIM 1. $\mathsf{ord}(\mathcal{P}^*) = \delta_y$.

PROOF. Suppose not. It follows from Theorem 4.9.5, Theorem 4.6.3 and Theorem 4.12.1 that the only way our construction could break down before reaching δ_y is if

1. $\delta^{\mathcal{P}} < \delta_y$ and $\rho(\mathcal{P}) = \mathsf{ord}(\mathcal{P}^*)$,
2. $\mathcal{P}$ is of #-lsa type, and
3. letting $\Lambda = \Sigma_{\mathcal{P}}$, Λ has strong branch condensation and is strongly Γ-fullness preserving,
4. $\mathsf{Lp}^{\Lambda,sts}(\mathcal{P}) \models$ "$\delta^{\mathcal{P}}$ is a Woodin cardinal".

Because $\Gamma^b(\mathcal{P}, \Lambda^{stc}) \subseteq \Gamma$ and $\Gamma^b(\mathcal{P}, \Lambda^{stc}) \neq \Gamma$, we can fix $(\mathcal{R}, \Phi) \in \mathsf{HP}^{\Gamma}$ such that $\mathcal{R}$ is meek and of limit type and $(\mathcal{P}, \Lambda) \in L(\Gamma(\mathcal{R}, \Phi))$. We have that in $L(\Gamma(\mathcal{R}, \Phi))$, Λ has strong branch condensation and is strongly fullness preserving. It now follows from Theorem 8.1.14 applied in $L(\Gamma(\mathcal{R}, \Phi))$ that for some $\mathcal{S} \in pI(\mathcal{P}, \Lambda)$, $L(\Gamma(\mathcal{S}, \Lambda_{\mathcal{S}})) \models \mathsf{LSA}$, contradicting our assumption that $\neg\#_{lsa}$ holds. ⊣

[3] A_x is countable. Here we just mean that y lists the members of A_x via the coding introduced in Definition 4.1.2.

We thus have that $\delta^{\mathcal{P}} = \delta_y$. Let κ be the least $< \delta^{\mathcal{P}}$-strong cardinal of $\mathcal{P}$. For $\alpha < \kappa$, let $g_\alpha = g \cap Coll(\omega, < \alpha)$.

CLAIM 2. $(\mathcal{P}^b, \Sigma_{\mathcal{P}^b}) \in B$.

PROOF. Let $g \subseteq Coll(\omega, < \kappa)$ be $\mathcal{N}_y^*$-generic. We let $(\psi(u,v), D)$ be as in the proof of Theorem 9.3.2. Following the notation used in the proof of Theorem 9.3.2, let $\mathcal{S} = \mathcal{P}_{\psi,D}$[4] and $\mathcal{S} = \mathcal{P}_{\psi,D}$. It follows from the proof of Theorem 9.3.2 that $\rho(\mathcal{S}) > o(\mathcal{S})$.

We claim that $\mathcal{S}$ is an iterate of $\mathcal{P}^b$. Let, in $\mathcal{N}_y^*[g]$, $\mathcal{M}'_\infty(\mathcal{P}^b, \Sigma_{\mathcal{P}^b})$ be the direct limit of $\Sigma_{\mathcal{P}^b}$-iterates $\mathcal{Q}$ of $\mathcal{P}^b$ such that $\mathcal{P}^b$-to-$\mathcal{Q}$ iteration has countable length. Notice that $\mathcal{M}'_\infty(\mathcal{P}^b, \Sigma_{\mathcal{P}^b}) \trianglelefteq \mathcal{S}$. This is simply because for every $\mathcal{Q} \trianglelefteq_{hod} \mathcal{P}^b$, $(\mathcal{Q}, \Sigma_{\mathcal{Q}}) \in \mathsf{HP}^\Gamma$. Suppose then that $\mathcal{M}_\infty(\mathcal{P}^b, \Sigma_{\mathcal{P}^b}) \triangleleft \mathcal{S}$. Let $(\mathcal{R}, \Pi) \in \mathsf{HP}^\Gamma \cap \mathcal{N}_y^*[g]$[5] be such that $\mathcal{M}_\infty(\mathcal{P}^b, \Sigma_{\mathcal{P}^b}) \triangleleft \mathcal{M}_\infty(\mathcal{R}, \Pi)$. Let $\eta < \kappa$ be such that $\mathcal{R} \in \mathcal{N}_y^*[g_\eta]$ and there is a $\sigma \in \mathbb{R} \cap \mathcal{N}_y^*[g_\eta]$ such that $\sigma(0)$ is a Gödel number for a formula ζ with the property that $\mathsf{Code}(\Pi)$ is definable over $(\mathsf{HC}, A_0, \sigma, \in)$ via ζ without parameters.

Let $\mathcal{Q}$ be the output of the hod pair construction of $\mathcal{P}$[6] in which extenders used have critical points $> \kappa$. It follows from Theorem 4.13.2 that for some $\mathcal{Q}' \triangleleft_{hod} \lambda^{\mathcal{Q}}$, $\mathcal{Q}'$ is a Π-iterate of $\mathcal{R}$. Let $E \in \vec{E}^{\mathcal{P}}$ be an extender with critical point κ such that ν_E is an inaccessible cardinal of $\mathcal{P}$ and $\mathcal{Q}'$ is constructed by the hod pair construction of $\mathcal{P}|\nu_E$. Let $E^* \in \vec{E}^{\mathcal{N}_y^*}$ be the resurrection of E. It follows that in $\mathrm{Ult}(\mathcal{N}_y^*, E^*)$, some hod pair appearing on the hod pair construction of $\pi(\mathcal{P}^b)$ in which extenders used are bigger than κ is a Π-iterate of $\mathcal{R}$. It then follows that some hod pair appearing on some hod pair construction of $\mathcal{P}^b$ is a Π-iterate of $\mathcal{R}$. It then follows that $\mathsf{Code}(\Pi) <_w \mathsf{Code}(\Sigma_{\mathcal{P}^b})$ implying that $\mathcal{M}'_\infty(\mathcal{R}, \Pi) \triangleleft \mathcal{M}'_\infty(\mathcal{P}^b, \Sigma_{\mathcal{P}^b})$, contradiction (here $\mathcal{M}'_\infty(\mathcal{R}, \Pi)$ is defined similarly to $\mathcal{M}'_\infty(\mathcal{P}^b, \Sigma_{\mathcal{P}^b})$. This contradiction proves the claim. $\dashv$

It is not hard to see, by a simple Skolem hull argument using the fact that $\mathcal{P} \in \mathcal{N}_y^*$, that

(1) for a club of $\eta < \delta_y$, $(\mathcal{P}|\eta)^\# \vDash$ "η is a Woodin cardinal".

Let C be the club in (1). For $\eta \in C$, let $\mathcal{R}_\eta = (\mathcal{P}|\eta)^\#$, $\Sigma_\eta = \Sigma^{stc}_{\mathcal{R}_\eta}$ and $\mathcal{Q}_\eta \trianglelefteq \mathcal{P}$ be the longest Σ_η-sts mouse such that $\mathcal{Q}_\eta \vDash$ "η is a Woodin cardinal". Using Lemma 6.4.4, we can translate $\mathcal{Q}_\eta$ into Σ_η-sts mouse $\bar{\mathcal{Q}}_\eta$ over $(\mathcal{N}_y^*|\eta)^\#$. Notice that

[4]Recall that this is defined in $\mathcal{N}_y^*[g]$.

[5]Here we are abusing the notation and use Π for both the strategy in $\mathcal{N}_y^*[g]$ as well as its extension in V.

[6]Following Section 6.2 it can be shown that for each $\nu \in (\kappa, \delta_y)$ which is a successor cardinal of $\mathcal{P}$, the fragment of Σ that acts on non-droping iterations based on $\mathcal{P}|\nu$ and are above κ is in $\mathcal{P}$. This allows us to make sense of hod pair constructions. See also the results of [33, Section 1]. The above outline uses Theorem 4.5.6.

(2) for every η, $\bar{\mathcal{Q}}_\eta$ has an iteration strategy Δ witnessing that $\bar{\mathcal{Q}}_\eta$ is a Σ_η-sts mouse over $\mathcal{J}_\omega[(\mathcal{N}_y^*|\eta)^\#]$ based on $\mathcal{R}_\eta$.

(2) is a consequence of the fact that $\mathcal{Q}_\eta$ appears on a Γ-hod pair construction of $\mathcal{N}_y^*$. Moreover,

(3) for every η and for every real x coding $\mathcal{N}_y^*|\eta$, $\bar{\mathcal{Q}}_\eta$ is $OD^\Gamma_{x,X}$.

(3) follows from proofs that have already appeared in the book. For instance, see the notion of goodness that appeared in the proof of Lemma 8.1.13. We now claim that

CLAIM 3. for a club of $\eta \in C-(\kappa+1)$, $\mathcal{Q}_\eta \in \mathcal{J}_{\nu_\eta}^{\Psi^*}(\mathcal{N}_y^*|\eta)$ where ν_η is the least ordinal such that $\mathcal{J}_{\nu_\eta}^{\Psi^*}(\mathcal{N}_y^*|\eta) \vDash \mathsf{ZFC}$.

PROOF. Let λ be least such that $\mathcal{J}_\lambda^{\Psi^*}(\mathcal{N}_y^*|\delta_y) \vDash \mathsf{ZFC}$. Let $\eta \in C$ be such that there is a map $\pi : \mathcal{J}_{\nu_\eta}^{\Psi^*}(\mathcal{N}_y^*|\eta) \to \mathcal{J}_\lambda^{\Psi^*}(\mathcal{N}_y^*|\delta_y)$. Thus

(4) $\mathcal{J}_{\nu_\eta}^{\Psi^*}(\mathcal{N}_y^*|\eta) \vDash$ "η is a Woodin cardinal".

Using genericity iterations, we can find $\mathcal{N} \in \mathcal{J}_{\nu_\eta}^{\Psi^*}(\mathcal{N}_y^*|\eta)$ such that $\mathcal{N}$ is a Ψ^*-iterate of $\mathcal{M}_3^{\#,\Phi_0}$ such that $(\mathcal{N}_y^*|\eta)^\#$ is generic over the extender algebra $\mathbb{B}_{\delta_0}^{\mathcal{N}}$ where δ_0 is the least Woodin cardinal of $\mathcal{N}$ that is $> \eta$. Let $h \subseteq Coll(\omega,\eta)$ be $\mathcal{N}_y^*$-generic. Fix a real $x \in \mathcal{N}[\mathcal{N}_y^*|\eta][h]$ coding $\mathcal{N}_y^*|\eta$. It follows that there is $y \in \mathbb{R}$ such that $(x,y) \in C \cap \mathcal{N}[\mathcal{N}_y^*|\eta][h]$. Therefore $\mathcal{Q}_\eta \in \mathcal{N}[\mathcal{N}_y^*|\eta][h]$. As x is arbitrary, we have that $\mathcal{Q}_\eta \in \mathcal{J}_{\nu_\eta}^{\Psi^*}(\mathcal{N}_y^*|\eta)$. It follows that $\mathcal{J}_{\nu_\eta}^{\Psi^*}(\mathcal{N}_y^*|\eta) \vDash$ "η is not a Woodin cardinal", contradicting (4). ⊣

The rest of the proof is easy. It follows from Claim 3 that we can find an η such that $\mathcal{Q}_\eta \in \mathcal{J}_{\nu_\eta}^{\Psi}(\mathcal{N}_y^*|\eta)$ and there is an elementary embedding $\pi : \mathcal{J}_{\nu_\eta}^{\Psi^*}(\mathcal{N}_y^*|\eta) \to \mathcal{J}_\lambda^{\Psi^*}(\mathcal{N}_y^*|\delta_y)$ where λ is the least such that $\mathcal{J}_\lambda^{\Psi^*}(\mathcal{N}_y^*|\delta_y) \vDash \mathsf{ZFC}$. Because $\mathcal{Q}_\eta \in \mathcal{J}_{\nu_\eta}^{\Psi}(\mathcal{N}_y^*|\eta)$, we have that $\mathcal{J}_{\nu_\eta}^{\Psi^*}(\mathcal{N}_y^*|\eta) \vDash$ "η is not a Woodin cardinal", and because $\pi : \mathcal{J}_{\nu_\eta}^{\Psi^*}(\mathcal{N}_y^*|\eta) \to \mathcal{J}_\lambda^{\Psi^*}(\mathcal{N}_y^*|\delta_y)$, we have that $\mathcal{J}_\lambda^{\Psi^*}(\mathcal{N}_y^*|\delta_y) \vDash$ "δ_y is not a Woodin cardinal". This is an obvious contradiction! Thus, we must have that the Γ hod pair construction of $\mathcal{N}_y^*$ reaches a generator for Γ. We now move to case 2.

Case 2. Γ is a completely mouse full pointclass such that for some α, $L(\Gamma,\mathbb{R}) \vDash \vartheta_{\alpha+1} = \Theta$.

Because we are assuming $\neg\#_{lsa}$, we must have that $L(\Gamma,\mathbb{R}) \vDash \neg\mathsf{LSA}$. The rest of the proof is very much like the proof of [29, Theorem 6.1]. To complete it, we need to use Theorem 7.2.2 instead of [29, Theorem 4.24]. We leave the details to the reader. The proof of "additionally" clause is similar to Case 1 and uses Theorem 4.13.2 and Lemma 8.1.13. ⊣

Theorem 10.1.2 has one shortcoming. It cannot be used to compute HOD of the minimal model of LSA as it only generates pointclasses whose Wadge ordinal is strictly smaller than the largest Suslin cardinal. To compute HOD of the minimal model of LSA we will need the following theorem.

THEOREM 10.1.3. *Assume* $\mathsf{AD}^+ + \mathsf{LSA}$ *and suppose* $\neg\#_{lsa} + \mathsf{NWLW}$.[7] *Let* α *be such that* $\vartheta_{\alpha+1} = \Theta$, *and suppose that there is a hod pair or an sts hod pair* $(\mathcal{P},\Sigma)$ *such that*

1. Σ *is strongly fullness preserving and has strong branch condensation and*
2. $\Gamma^b(\mathcal{P},\Sigma) = \{A \subseteq \mathbb{R} : w(A) < \vartheta_\alpha\}$.

Then $(\mathcal{P},\Sigma)$ *is an sts hod pair and for any* $B \in \mathbb{B}[\mathcal{P},\Sigma]$ *there is* $\mathcal{Q} \in pI(\mathcal{P},\Sigma)$ *such that* $(\mathcal{Q},\Sigma_{\mathcal{Q}})$ *is strongly B-iterable.*

PROOF. Towards a contradiction, assume not. We reflect the failure of our claim to $\utilde{\Delta}^2_1$. Let (β,γ) be lexicographically least such that letting $\Gamma = \{A \subseteq \mathbb{R} : w(A) < \gamma\}$,

1. $\Gamma = \wp(\mathbb{R}) \cap \mathcal{J}_\beta(\Gamma,\mathbb{R})$ and $L_\beta(\Gamma,\mathbb{R}) \vDash \mathsf{LSA} + \mathsf{ZF} - \mathsf{Powerset}$,
2. letting α be such that $L_\beta(\Gamma,\mathbb{R}) \vDash$ "$\vartheta_{\alpha+1} = \Theta$", $L_\beta(\Gamma,\mathbb{R}) \vDash$ "there is a hod pair or an sts hod pair $(\mathcal{P},\Sigma)$ such that Σ is strongly fullness preserving and has strong branch condensation and $\Gamma^b(\mathcal{P},\Sigma) = \{A \subseteq \mathbb{R} : w(A) < \vartheta_\alpha\}$ but either
   (a) $(\mathcal{P},\Sigma)$ is not an sts hod pair or
   (b) there is a $B \in \mathbb{B}[\mathcal{P},\Sigma]$ such that whenever $\mathcal{Q} \in pI(\mathcal{P},\Sigma)$, $(\mathcal{Q},\Sigma_{\mathcal{Q}})$ is not strongly B-iterable".

Because (β,γ) is minimized, we have that $\Gamma \subset \utilde{\Delta}^2_1$. Fix $(\mathcal{P},\Sigma)$ as above. First we claim that

CLAIM 10.1.4. Σ *is not an iteration strategy.*

PROOF. Towards a contradiction suppose not. Let $A_0 \in lub(\Gamma)$, Γ^* be a good pointclass beyond Γ and (N_0,Ψ_0) be a Γ^*-Woodin which Suslin, co-Suslin captures $(T_n(A_0) : n < \omega)$[8]. Let F be as in Theorem 4.1.12 for Γ^*, and let $x \in \mathrm{dom}(F)$ be such that letting $F(x) = (\mathcal{N}^*_x, \mathcal{M}_x, \delta_x, \Sigma_x)$ and $\mathbb{M}_x$ be as in clause 7 of Theorem 4.1.12, $(\mathbb{M}_x, (N_0,\Psi_0), \Gamma^*, A_0)$ Suslin, co-Suslin captures $\mathsf{Code}(\Sigma)$ and Γ. It follows that $\mathsf{Le}((\mathcal{P},\Sigma),\mathcal{J}_\omega[\mathcal{P}])^{\mathcal{N}^*_x|\delta_x}$ reaches $\mathcal{M}_2^{\#,\Sigma}$. Let Ψ be the iteration strategy of $\mathcal{M}_2^{\#,\Sigma}$. Notice that

(1) $\Psi \in L_\beta(\Gamma,\mathbb{R})$.

Because Σ is an iteration strategy, it follows from clause 1 of Theorem 6.1.4 that there are trees $(T,S) \in \mathcal{M}_2^{\#,\Sigma}$ such that letting $\delta_0 < \delta_1$ be the Woodin cardinals of $\mathcal{M}_2^{\#,\Sigma}$

1. $\mathcal{M}_2^{\#,\Sigma} \vDash$ "(T,S) are δ_1-complementing",

---

[7] As in the previous theorem, Theorem 10.3.1 removes the extra assumption that NWLW holds.

[8] Here, $T_n(A_0)$ is defined the way $T_n(\Psi)$ is defined in Section 4.1.1.

2. whenever $\pi : \mathcal{M}_2^{\#,\Sigma} \to \mathcal{N}$ is an iteration according to Ψ and $g \subseteq Coll(\omega, \pi(\delta_0))$ is $\mathcal{N}$-generic then $\mathsf{Code}(\Sigma) \cap \mathbb{R}^{\mathcal{N}|\delta_1[g]} = p[\pi(T)]$ and $(\mathsf{Code}(\Sigma))^c \cap \mathbb{R}^{\mathcal{N}|\delta_1[g]} = p[\pi(S)]$.

Let $\mathcal{M}_\infty$ be the direct limit of all Ψ-iterates of $\mathcal{M}_2^{\#,\Sigma}$ and let $\pi : \mathcal{M}_2^{\#,\Sigma} \to \mathcal{M}_\infty$ be the direct limit embedding. It then follows that $\mathsf{Code}(\Sigma) = p[\pi(T)]$ and $(\mathsf{Code}(\Sigma))^c = p[\pi[S]]$. It follows from (1) that $\pi(T), \pi(S) \in L(\Gamma, \mathbb{R})$, implying that $L(\Gamma, \mathbb{R}) \vDash$ "$\mathsf{Code}(\Sigma)$ is Suslin, co-Suslin". It follows that $\mathsf{Code}(\Sigma) \in \Gamma(\mathcal{P}, \Sigma)$, contradiction! ⊣

It follows from Claim 1 that $(\mathcal{P}, \Sigma)$ is an sts hod pair. Hence, we must have that

(2) there is $B \in \mathbb{B}[\mathcal{P}, \Sigma]$ such that whenever $\mathcal{Q} \in pI(\mathcal{P}, \Sigma)$, $(\mathcal{Q}, \Sigma_\mathcal{Q})$ is not strongly B-iterable.

We can now finish by following the proof of Theorem 8.1.15. The only issue is that in Theorem 8.1.15 we require that Σ be a strategy, but this is only needed externally. In our current context, we need a strategy Σ^* that extends Σ and is $L_\beta(\Gamma, \mathbb{R})$-fullness preserving and has branch condensation. Obtaining such a Σ^* might require passing to a Σ-iterate of $\mathcal{P}$. We can obtain such a strategy by further iterating $\mathcal{P}$ via Σ to a hod pair construction of a sufficiently strong background triple using the theory developed in Section 4.13. Notice that (2) holds even for this new pair, and so without loss of generality we may just as well assume that Σ^* exists. The rest is just like in the proof of Theorem 8.1.15. ⊣

THEOREM 10.1.5 (The generation of the mouse full pointclasses II). *Assume*

$$\mathsf{AD}^+ + \neg\#_{\mathsf{lsa}} + \mathsf{NWLW}.^9$$

Suppose that

1. *$\Gamma \neq \wp(\mathbb{R})$ is a mouse full pointclass such that $\Gamma \vDash \mathsf{SMC}$ and*
2. *for some $(\mathcal{Q}, \Lambda) \in \mathsf{HP}^\Gamma$ such that Λ has strong branch condensation and is strongly Γ-fullness preserving and for some $x \in \mathbb{R}$, $\mathsf{Lp}^\Lambda(x) \neq \mathsf{Lp}^{\Gamma,\Lambda}(x)$.*

Then there is an anomalous pair[10] *$(\mathcal{P}, \Sigma)$ such that*

1. *Σ has strong branch condensation and is strongly Γ-fullness preserving and*
2. *$\Gamma(\mathcal{P}, \Sigma) = \Gamma$.*

PROOF. The proof is similar to the proof of Theorem 10.1.2 but here we need to revise Theorem 4.6.3 and Theorem 4.9.5. There, the strong fullness preservation and strong branch condensation are proved using the method of thick hulls developed in Section 4.5. Here, we need to use the fact that

(∗) for some $(\mathcal{Q}, \Lambda) \in \mathsf{HP}^\Gamma$ such that Λ has strong branch condensation and is strongly Γ-fullness preserving and for some $x \in \mathbb{R}$, $\mathsf{Lp}^\Lambda(x) \neq \mathsf{Lp}^{\Gamma,\Lambda}(x)$.

---

[9]See Theorem 10.3.1, which removes the hypothesis that NWLW holds.

[10]See Definition 5.4.4.

The following is the main way $(*)$ affects the proof of Theorem 4.6.3: For example, we can no longer assume that if τ and $\mathcal{N}$ are as in that proof (just after (b)) then (4) of that proof holds. Below we outline our method of dealing with the aforementioned issue.

Towards contradiction assume not and suppose Γ is the least pointclass satisfying our hypothesis which is not generated as stated above. The following are the two lemmas that we need to prove.

LEMMA 10.1.6. *Suppose* $\mathsf{C} = (\mathbb{M}, (P, \Psi), \Gamma^*, A)$ *Suslin, co-Suslin captures* Γ *and* $\mathbb{M} = (M, \delta, \vec{G}, \Sigma)$. *Set*

$$\mathsf{hpc}^{+}_{\mathsf{C},\Gamma} = (\mathcal{M}_\gamma, \mathcal{N}_\gamma, Y_\gamma, \Phi^+_\gamma, F^+_\gamma, F_\gamma, b_\gamma : \gamma \leq \delta).$$

Suppose $\beta < \delta$, $\mathcal{P} \in Y_\beta$ *and* $M \models$ "$(\mathcal{P}, (\Phi_\beta)_\mathcal{P}) \in \mathsf{HP}^\Gamma$". *Then* $(\Phi^+_\beta)_\mathcal{P}$ *is almost low-level strongly* Γ*-fullness preserving.*[11].

LEMMA 10.1.7. *Suppose* $\mathsf{C} = (\mathbb{M}, (P, \Psi), \Gamma^*, A)$ *Suslin, co-Suslin captures* Γ *and* $\mathbb{M} = (M, \delta, \vec{G}, \Sigma)$. *Set*

$$\mathsf{hpc}^{+}_{\mathsf{C},\Gamma} = (\mathcal{M}_\gamma, \mathcal{N}_\gamma, Y_\gamma, \Phi^+_\gamma, F^+_\gamma, F_\gamma, b_\gamma : \gamma \leq \delta).$$

Suppose $\beta < \delta$, $\mathcal{P} \in Y_\beta$ *and* $M \models$ "$(\mathcal{P}, (\Phi_\beta)_\mathcal{P}) \in \mathsf{HP}^\Gamma$". *Then* $(\Phi^+_\beta)_\mathcal{P}$ *has branch condensation.*

In both of those cases, the hard case is when $\mathcal{P}$ is of limit type and Γ is of limit type. We assume this and set $\Lambda = (\Phi_\beta)_\mathcal{P}$. We will only outline the proof of fullness preservation. Strong fullness preservation can be established by a very similar argument. We then have that there is a witness to non fullness preservation or non branch condensation added by a collapse of some $\nu < \delta$. Let then $g \subseteq Coll(\omega, \nu)$ be this collapse. We now outline how to proceed assuming that $\mathcal{P}$ is not gentle (there is not much to prove in this case). This in particular implies that $\mathcal{P} = \mathsf{Lp}^{\Gamma,\Lambda}(\mathcal{P}|\delta^{\mathcal{P}^b})$.

In the case of fullness preservation this witness is a tuple $(\mathcal{T}, \mathcal{M})$ such that $\mathcal{T}$ is according to Λ and $\mathcal{M}$ is a mouse witnessing the failure of one of the clauses of fullness preservation. The key is that $\mathcal{M}$ has an iteration strategy coded by a set in Γ. Let Γ' be a good pointclass contained in Γ and such that the iteration strategy of $\mathcal{M}$ is in $\Delta_{\Gamma'}$. Let η be the least Γ'-Woodin cardinal of M above $\iota =_{\text{def}} \max(\nu, \zeta)$ where $\zeta = \sup\{\mathrm{lh}(F^+_\gamma) : \gamma < \beta\}$. We now repeat the proof of Theorem 4.6.3 while working inside $M' = C_{\Gamma'}(M|\eta)$. Let $\tau = \delta^{\mathcal{P}^b}$ and $\vec{G}' = \{F \in \vec{G} : \mathrm{crit}(F) > \iota\}$. The key point is that when in that proof we let $\mathcal{N}$ be the last model of

$$(\mathsf{Le}((\mathcal{P}|\tau, \Lambda_{\mathcal{P}|\tau}), \mathcal{P}^b)_{>\zeta})^{(M'[g], \eta, \vec{G}')},$$

[11]In this context, it may not be the case that $\mathcal{P}$ is Γ-full at the top. Meaning, if $\mathcal{P}$ is meek of limit type then it may not be the case that $\mathcal{P} = \mathsf{Lp}^{\Gamma,(\Phi_\beta)_{\mathcal{P}|\delta^\mathcal{P}}}(\mathcal{P}|\delta^\mathcal{P})$. For example, this may happen if $\mathcal{P} = (\mathcal{P}|\delta^\mathcal{P})^\#$ and $\Gamma = \Gamma(\mathcal{P}, (\Phi_\beta)_\mathcal{P})$. In this context, fullness preservation is meant to be for lower level strategies. See Definition 4.6.2.

we have that no level of $\mathcal{N}^*$ projects across P^b. This is because if Σ' is the fragment of $\Sigma_{M'}$ that acts on stacks that are above ι then $\mathsf{Code}(\Sigma') \in \Gamma$.

The issue with branch condensation can be resolved similarly. In the case of branch condensation, the witness is $(\mathcal{T}, \mathcal{U}, b, \sigma) \in M[g]$ such that

(i) $\mathcal{T}$ is according to Λ, $\pi^{\mathcal{T}}$ exists and $\mathcal{T}$ has a last model $\mathcal{R}$,
(ii) $\mathcal{U}$ is according to Λ and is of limit length,
(iii) b is a cofinal branch of $\mathcal{U}$ and $\Lambda(\mathcal{U}) \neq b$,
(iv) $\sigma : \mathcal{M}_b^{\mathcal{U}} \to \mathcal{R}$ is such that $\pi^{\mathcal{T}} = \sigma \circ \pi_b^{\mathcal{U}}$.

The dificult case is when $\mathcal{P}$ is non-meek, and so we assume this. We assume Λ is an iteration strategy as the other case is very similar. Let ζ and τ be as above.

The most dificult case is when $\pi^{\mathcal{U},b}$ is defined, $\mathcal{Q}(b,\mathcal{U})$ exists and is an sts premouse over $\mathrm{m}^+(\mathcal{U})$. Other cases follow the same pattern but this one is the most involved. Here we need to show that $\mathcal{Q}(b,\mathcal{U})$ is in fact $(\Lambda_{\mathrm{m}^+(\mathcal{U}),\mathcal{U}})^{stc}$-sts mouse over $\mathrm{m}^+(\mathcal{U})$. Let η be such that $\sigma \restriction \mathcal{Q}(b,\mathcal{U}) : \mathcal{M}_b^{\mathcal{U}} \to \mathcal{R}||\eta$. Because $\mathsf{Code}((\Lambda_{\mathcal{R}|\eta,\mathcal{T}})^{stc}) \in \Gamma$, we have that if Φ is the $\sigma \restriction \mathcal{Q}(b,\mathcal{U})$-pullback of $(\Lambda_{\mathcal{R}|\eta,\mathcal{T}})^{stc}$ then $\mathsf{Code}(\Phi) \in \Gamma$. Thus it is enough to show that, setting $\Psi = (\Lambda_{\mathrm{m}^+(\mathcal{U}),\mathcal{U}})^{stc}$,

(a) $\Phi = \Psi$.

Let $\mathcal{W} = \mathrm{m}^+(\mathcal{U})$. Suppose $\Phi \neq \Psi$, and let $(\mathcal{X}_0, \mathcal{W}_0, \mathcal{X}_1, \mathcal{W}_1, \mathcal{Y})$ be a minimal disagreement[12] between Φ and Ψ. We have that $\mathsf{Code}(\Phi_{\mathcal{Y},\mathcal{U}^\frown\mathcal{X}_0})$ and $\mathsf{Code}(\Psi_{\mathcal{Y},\mathcal{U}^\frown\mathcal{X}_1})$ are in Γ. Let then Γ' be a good pointclass contained in Γ such that

$$\{\mathsf{Code}(\Phi_{\mathcal{Y},\mathcal{U}^\frown\mathcal{X}_0}), \mathsf{Code}(\Psi_{\mathcal{Y},\mathcal{U}^\frown\mathcal{X}_1})\} \subseteq \Delta_{\Gamma'}.$$

Let now $\eta, M', \vec{G}', \Sigma'$ be as above defined relative to the new meaning of Γ'. Again the proof now simply follows the proof of Theorem 4.9.5.

The next major issue to deal with is when we pass from gentle stage to the next hod premouse. Just like in the proof of Theorem 10.1.2, we build our desired generator for Γ via a hod pair construction of some background triple. We find some $\mathsf{C} = (\mathbb{M}, (P,\Psi), \Gamma^*, A)$ that Suslin, co-Suslin captures Γ where $\mathbb{M} = (M, \delta, \vec{G}, \Sigma)$. Fix a hod pair $(\mathcal{Q}, \Phi) \in \mathsf{HP}^\Gamma$ such that Φ has strong branch condensation and is Γ-fullness preserving and for some $x \in \mathbb{R}$, $\mathsf{Lp}^\Phi(x) \neq \mathsf{Lp}^{\Gamma,\Phi}(x)$. We now get that

(1) whenever $i : \mathcal{Q} \to \mathcal{Q}'$ is an iteration according to Φ and whenever $y \in \mathbb{R}$ is a real Turing above x,

$$\mathsf{Lp}^{\Phi_{\mathcal{Q}'}}(y) \neq \mathsf{Lp}^{\Gamma,\Phi_{\mathcal{Q}'}}(y).$$

(1) can be established via more or less standard arguments. For example, see [29, Lemma 6.21]. The key ingredient of the proof is that if (1) fails for $\mathcal{Q}'$

[12] See Definition 4.7.1.

and y then any universal $\Phi_{\mathcal{Q}}$-mouse over which i and $\mathcal{Q}$ are set generic is also Φ-universal. To find such a universal $\Phi_{\mathcal{Q}}$-mouse, we can choose some good pointclass Γ' such that $\Phi, \Phi' \in \Delta_{\Gamma'}$ where Φ' is the iteration strategy of $\mathsf{Lp}^{\Phi}(x)$. Let then F be as in Theorem 4.1.12 for Γ' and let $z \in \mathrm{dom}(F)$ be such that $(\mathbb{M}_z, (N, \Psi), \Gamma^*, A)$ Suslin, co-Suslin captures some set of reals coding $(\mathcal{Q}, \Phi_{\mathcal{Q}'}, y, \Phi', i, x)$, and where the rest of the objects are defined as in clause 7 of Theorem 4.1.12. The desired universal $\Phi_{\mathcal{Q}'}$-mouse is

$$(\mathsf{Le}((\mathcal{Q}', \Phi_{\mathcal{Q}'}), y))^{(M,\delta,\vec{G})}.$$

Letting $\mathcal{N}$ be that model, we have that $\mathcal{N}$ has a Woodin cardinal and $(i, \mathcal{Q}, x)$ is set generic over $\mathcal{N}$. It then follows from the universality of $\mathcal{N}$ that $\mathsf{Lp}^{\Phi}(x) \in \mathcal{N}[(i, \mathcal{Q}, x)]$ and if $\mathcal{K} \trianglelefteq \mathsf{Lp}^{\Phi}(x)$ then $\mathcal{K}$ appears in some fully backgrounded construction of $\mathcal{N}[(i, \mathcal{Q}, x)]$. We leave the details to the reader.

For each $\mathcal{Q}' \in pI(\mathcal{Q}, \Phi)$ and for each $y \in \mathbb{R}$ Turing above x let $\mathcal{M}_{\mathcal{Q}',y} \trianglelefteq \mathsf{Lp}^{\Phi_{\mathcal{Q}'}}(y)$ be the least such that $\mathcal{M}_{\mathcal{Q}',y}$ does not have an iteration strategy in Γ (as a $\Phi_{\mathcal{Q}'}$-mouse). Let $\Psi_{\mathcal{Q}',y}$ be the unique iteration strategy of $\mathcal{M}_{\mathcal{Q}',y}$. In addition to the requirements mentioned above, we demand that A that appeared in C codes the set of all triple that have the form $(\mathcal{Q}', y, \Psi_{\mathcal{Q}',y'})$. In particular, $\mathcal{Q}, x \in M$ and $\Phi_{\mathcal{Q},x}$ is Suslin, co-Suslin captured by C.

Set now

$$\mathsf{hpc}^{+}_{\mathsf{C},\Gamma} = (\mathcal{M}_\gamma, \mathcal{N}_\gamma, Y_\gamma, \Phi^{+}_{\gamma}, F^{+}_{\gamma}, F_\gamma, b_\gamma : \gamma \leq \delta).$$

Because of our set up, there is $\gamma < \delta$ such that $\mathcal{M}_\gamma \in pI(\mathcal{Q}, \Phi)$. This implies that the construction cannot last δ steps. If it did, then because $\Phi^{+}_{\gamma} = \Phi_{\mathcal{M}_\gamma}$[13], letting $\mathcal{N}$ be the last model of

$$(\mathsf{Le}((\mathcal{M}_\gamma, \Phi_{\mathcal{M}_\gamma})))^{(\mathcal{N}_\delta,\delta,\vec{G}')},$$

where $\vec{G}' = \{F \in \vec{E}^{\mathcal{N}_\delta} : \forall \gamma' < \gamma(\nu(F^{+}_{\gamma'}) < \mathrm{crit}(F))$ and $\nu(F)$ is an inaccessible cardinal of $\mathcal{N}_\delta\}$, some fully backgrounded construction of some set generic extension of $\mathcal{N}$ would reach $\mathcal{M}_{\mathcal{M}_\gamma,x}$. This would imply that $\mathsf{Code}(\Phi_{\mathcal{M}_\gamma,x}) \in \Gamma$, contradiction. We thus have that the construction has to stop.

Because clause 4(a) of Definition 4.3.3 never occurs, we must have that clause 4(b) occurs. Let then ξ be such that $\mathcal{N}_\xi$ has the property described in clause 4(b) of Definition 4.3.3. We have that $\mathcal{N}_\xi$ is germane[14]. Let $\mathcal{P} = \mathcal{N}_\xi$. Let $\Psi = \Phi^{+}_{\xi}$. The following is our main claim.

CLAIM 10.1.8. $\Gamma(\mathcal{P}, \Psi) = \Gamma$.

PROOF. Because Ψ-iterates of $\mathcal{P}$, via the resurrection process, embed into hod mice whose iteration strategies are in Γ, we have that $\Gamma(\mathcal{P}, \Psi) \subseteq \Gamma$. It remains to show that $\Gamma \subseteq \Gamma(\mathcal{P}, \Psi)$. Assume then that $\Gamma(\mathcal{P}, \Psi) \subset \Gamma$.

---

[13] See Theorem 4.13.2.

[14] See Definition 2.7.15.

Using Theorem 5.5.3[15], we can find some tail $(\mathcal{P}',\Psi')$ of $(\mathcal{P},\Psi)$ such that $\mathcal{P}' \in pI(\mathcal{P},\Psi')$, Ψ' has branch condensation and $\Gamma(\mathcal{P}',\Psi') = \Gamma(\mathcal{P},\Psi)$. It then follows that $\mathsf{Code}(\Psi) \in \Gamma$, which can be shown by using the proof of Lemma 8.1.13. It now follows that $\mathsf{Code}(\Psi) \in \Gamma$.

Notice next that by induction we can assume that if $\mathcal{S} \lhd^c_{hod} \mathcal{P}$ then $\Psi_{\mathcal{S}}$ is Γ-fullness preserving and has branch condensation. This means that we can now apply the proof of 2a and 2b on page 142 of [29] to conclude that $\rho(\mathcal{P}) \geq \mathsf{ord}(\mathsf{hl}(\mathcal{P}))$, which is a contradiction[16]. ⊣

The desired hod pair generating Γ is the tail of $(\mathcal{P},\Psi)$ provided by Theorem 5.5.3. ⊣

10.2. A proof of the Mouse Set Conjecture below LSA

Throughout we will assume $\mathsf{AD}^{++} =_{\text{def}} \mathsf{AD}^+ + V = L(\wp(\mathbb{R}))$. Recall the definition of $\#_{\mathsf{lsa}}$ and NWLW defined in the previous section[17]. Recall that Strong Mouse Capturing (SMC) is the statement that for any hod pair or an sts hod pair $(\mathcal{P},\Sigma)$ such that Σ has strong branch condensation and is strongly fullness preserving, and for any reals x, y, x is ordinal definable from Σ and y if and only if x is in some Σ-mouse over y. The following is the main theorem of this section.

THEOREM 10.2.1. *Assume* $\mathsf{AD}^{++} + \neg\#_{lsa} + NWLW$. *Then the* Strong Mouse Capturing *holds.*

The rest of this section is devoted to the proof of Theorem 10.2.1. We assume familiarity with the proof of [29, Theorem 6.19] and build directly on it. We start by stating the main steps of [29, Theorem 6.19]. We will follow these steps and provide proofs only for the new cases.

Towards a contradiction assume that SMC is false. Our first step is to locate the minimal level of the Wadge hierarchy over which SMC becomes false. For simplicity we assume that the Mouse Capturing, instead of the Strong Mouse Capturing, is false. Mouse Capturing is the same as SMC when the pair $(\mathcal{P},\Sigma) = \emptyset$. The general case is only different in one aspect, it needs to be relativized to some strategy or a short tree strategy Σ.

NOTATION 10.2.2. Throughout this section, we let Γ be the least Wadge initial segment such that for some α

[15]Theorem 5.5.3 is applicable because $(\mathcal{P},\Psi)$ is an anomalous hod pair and if it is of type III then we can produce a supporting bicephalous via fully backgrounded construction done over $\mathcal{P}^b$ relative to $\Psi_{\mathcal{P}|\delta^{\mathcal{P}^b}}$. The proof of (1) above shows that this construction will reach an $\mathcal{M}$ with the property that $\rho(\mathcal{M}) \leq \delta^{\mathcal{P}^b}$. The arguments presented on page 142 of [29] then show that if $\mathcal{M}$ is the least such level of the aformentioned backgrounded construction then in fact $\rho(\mathcal{M}) < \delta^{\mathcal{P}^b}$.

[16]hl is defined in Definition 2.7.15.

[17]See Definition 10.1.1.

1. $\Gamma = \wp(\mathbb{R}) \cap L_\alpha(\Gamma, \mathbb{R})$,
2. $L_\alpha(\Gamma, \mathbb{R}) \vDash \mathsf{SMC}$,
3. there are reals x and y such that $L_{\alpha+1}(\Gamma, \mathbb{R}) \vDash$ "y is $OD(x)$" yet no x-mouse has y as a member.

For the purposes of this section we make the following definition.

DEFINITION 10.2.3. Suppose $(\mathcal{P}, \Sigma)$ is a hod pair and Γ^* is a projectively closed pointclass[18]. We say $(\mathcal{P}, \Sigma)$ is Γ^*-*perfect* if the following conditions are met.

1. Σ is Γ^*-strongly fullness preserving and has strong branch condensation.
2. For every $\mathcal{Q} \in pI(\mathcal{P}, \Sigma) \cup pB(\mathcal{P}, \Sigma)$ such that $\mathcal{Q}$ is of successor type, there is $\vec{B} = (B_i : i \leq \omega) \subseteq \mathbb{B}[\mathcal{Q}^-, \Sigma_{\mathcal{Q}^-})$ such that $\vec{B}$ strongly guides $\Sigma_{\mathcal{Q}}$.

If $\Gamma^* = \wp(\mathbb{R})$ then we omit Γ^* from our notation.

The following theorem was heavily used in [29]. It is essentially due to Steel and Woodin (see [54]).

THEOREM 10.2.4. *Assume* AD^+ *and suppose* $(\mathcal{P}, \Sigma)$ *is a hod pair or an sts hod pair such that* $L(\Sigma, \mathbb{R}) \vDash$ *"*$(\mathcal{P}, \Sigma)$ *is perfect". Then* $L(\Sigma, \mathbb{R}) \vDash \mathsf{MC}(\Sigma)$.

A key theorem used in the proof of Theorem 10.2.1 is the following capturing theorem. Its precursor is stated as [29, Theorem 6.5].

THEOREM 10.2.5. *Suppose* $(\mathcal{P}, \Sigma)$ *is a perfect hod pair and* Γ_1 *is a good pointclass such that* $\mathsf{Code}(\Sigma) \in \utilde{\Delta}_{\Gamma_1}$. *Suppose* F *is as in Theorem 4.1.12 for* Γ_1 *and* $z \in \mathrm{dom}(F)$ *is such that if* $F(z) = (\mathcal{N}_z^*, \mathcal{M}_z, \delta_z, \Sigma_z)$ *then* $(\mathcal{N}_z^*, \delta_z, \Sigma_z)$ *Suslin, co-Suslin captures* $\mathsf{Code}(\Sigma)$[19]. *Let* $\mathcal{N} = (\mathsf{Le}(\emptyset))^{\mathcal{N}_z^*|\delta_z}$[20]. *Then there is* $\mathcal{Q} \in pI(\mathcal{P}, \Sigma) \cap \mathcal{N}$ *such that* $\Sigma_{\mathcal{Q}} \restriction \mathcal{N} \in \mathcal{J}[\mathcal{N}]$.

The next key lemma that is used in the proof of Theorem 10.2.1 is the following generation lemma that can be traced to [29, Lemma 6.23]. Below Γ is as in Notation 10.2.2.

LEMMA 10.2.6. *There is a perfect pair* $(\mathcal{P}, \Sigma)$ *such that*

$$\Gamma(\mathcal{P}, \Sigma) \subseteq \Gamma \subseteq L(\Sigma, \mathbb{R}).$$

Our goal now is to give an outline of the way Theorem 10.2.5 and Lemma 10.2.6 are used to prove Theorem 10.2.1.

10.2.1. The structure of the proof of the Mouse Set Conjecture. First we outline the proof of the following general theorem.

THEOREM 10.2.7. *Suppose* $(\mathcal{P}, \Sigma)$ *is a perfect pair. Then* $L(\Sigma, \mathbb{R}) \vDash$ *"for every* $\mathcal{R} \triangleleft_{hod}^c \mathcal{P}$[21], $\mathsf{Mouse\ Capturing}$ *holds for* $\Sigma_{\mathcal{R}}$*".*

[18] See Definition 4.6.1.

[19] We abuse the terminology and omit the other object used to express this type of capturing. In the sequel, if the nature of these other objects, like the pair (N, Ψ), is not important we will omit them from the discussions.

[20] This is just the ordinary fully backgrounded construction. See Definition 4.2.1.

[21] See Notation 9.1.2.

PROOF. We only outline the proof as the full proof is presented in [29, Section 6.4]. For simplicity we outline the proof for the least complete layer of $\mathcal{P}$. Let $\mathcal{R} \triangleleft_{hod}^{c} \mathcal{P}$ be the least layer of $\mathcal{P}$. We want to show that

(1) $L(\Sigma,\mathbb{R}) \models$ "Mouse Capturing holds for $\Sigma_{\mathcal{R}}$".

The general case is only notationally more complex. Suppose $x, y \in \mathbb{R}$ are such that $L(\Sigma,\mathbb{R}) \models$ "$y \in OD_{\Sigma_{\mathcal{R}}}(x)$". It follows from Theorem 10.2.4 that there is a Σ-mouse $\mathcal{M}$ over $(\mathcal{P},x)$ containing y such that $\mathcal{M}$ has an iteration strategy in $L(\Sigma,\mathbb{R})$. In fact, it follows from Theorem 10.2.4 that

(2) for every $\mathcal{Q} \in pI(\mathcal{P},\Sigma)$ there is a $\Sigma_{\mathcal{Q}}$-mouse $\mathcal{M}$ over $(\mathcal{Q},x)$ such that $y \in \mathcal{M}$ and $\mathcal{M}$ has an iteration strategy in $L(\Sigma,\mathbb{R})$.[22]

Let $\mathcal{M}_{\mathcal{Q}}$ be the least $\Sigma_{\mathcal{Q}}$-mouse over $(\mathcal{Q},x)$ such that y is definable over $\mathcal{M}_{\mathcal{Q}}$. Let $\Lambda_{\mathcal{Q}}$ be the iteration strategy of $\mathcal{M}_{\mathcal{Q}}$ (witnessing that $\mathcal{M}_{\mathcal{Q}}$ is a $\Sigma_{\mathcal{Q}}$-mouse). Let $\Gamma^* \in L(\Sigma,\mathbb{R})$ be a good pointclass such that the set

$$A = \{(z,u) \in \mathbb{R}^2 : z \text{ codes some } \mathcal{M}_{\mathcal{Q}} \text{ and } u \text{ is an iteration according to } \Lambda_{\mathcal{Q}}\}$$

is in $\utilde{\Delta}_{\Gamma^*}$. Let F be as in Theorem 4.1.12 for Γ^* and let $z \in \mathrm{dom}(F)$ be such that if $F(z) = (\mathcal{N}_z^*, \mathcal{M}_z, \delta_z, \Sigma_z)$ then $(\mathcal{N}_z^*, \delta_z, \Sigma_z)$ Suslin, co-Suslin captures Σ and the set A. Let $\mathcal{N} = (\mathsf{Le}(\emptyset,x))^{\mathcal{N}_z^*|\delta_z}$. It follows from Theorem 10.2.5 that

(3) there is a $\mathcal{Q} \in \mathcal{N}$ such that $\Sigma_{\mathcal{Q}} \restriction \mathcal{N} \in \mathcal{J}[\mathcal{N}]$.

It follows from the universality of $\mathcal{N}$ that $\mathcal{M}_{\mathcal{Q}} \in \mathcal{N}$ (this is because $(\mathsf{Le}((\mathcal{Q},\Sigma_{\mathcal{Q}}))^{\mathcal{N}}$ is universal in $\mathcal{N}_z^*$ and the strategy $\Lambda_{\mathcal{Q}}$ of $\mathcal{M}_{\mathcal{Q}}$ is captured by $\mathcal{N}_z^*$ (via A)). It then follows that $y \in \mathcal{N}$. As $\mathcal{N}$ is an x-mouse, this completes the proof. ⊣

Suppose now that $(\mathcal{P},\Sigma)$ is a Γ-perfect pair such that $\Gamma(\mathcal{P},\Sigma) \subseteq \Gamma \subseteq L(\Sigma,\mathbb{R})$. Such a pair is given to us by Lemma 10.2.6.

We now apply Theorem 10.2.4. For each $\mathcal{Q} \in pI(\mathcal{P},\Sigma)$ there is a $\Sigma_{\mathcal{Q}}$-mouse $\mathcal{M}_{\mathcal{Q}}$ over $(\mathcal{Q},x)$ such that y is definable over $\mathcal{M}_{\mathcal{Q}}$. We then again can find an x-mouse $\mathcal{N}$ such that for some $\mathcal{Q} \in \mathcal{N} \cap pI(\mathcal{P},\Sigma)$, $\mathcal{M}_{\mathcal{Q}} \in \mathcal{N}$. It follows that $y \in \mathcal{N}$. Thus, to finish the proof of Theorem 10.2.1, it is enough to establish Theorem 10.2.5 and Lemma 10.2.6.

10.2.2. Review of basic notions. In this subsection we review basic notions introduced in [29, Theorem 6.5] for proving a version of Theorem 10.2.5.

TERMINOLOGY 10.2.8. We are in fact working towards the proof of Theorem 10.2.5, and the notation and the terminology of this subsection will be used in the later subsections. Fix $(\mathcal{P},\Sigma)$, Γ_1, F and z as in the statement of Theorem 10.2.5. Let $\mathcal{N} = (\mathsf{Le}(\emptyset))^{\mathcal{N}_z^*}$.

Goal: We are looking for $\mathcal{Q} \in pI(\mathcal{P},\Sigma) \cap \mathcal{N}$ such that $\Sigma_{\mathcal{Q}} \restriction \mathcal{N} \in \mathcal{J}[\mathcal{N}]$.

We start working in $\mathcal{N}_z^*$. Without loss of generality we can assume that

[22] This is because $L(\Sigma_{\mathcal{Q}},\mathbb{R}) = L(\Sigma,\mathbb{R})$ and $L(\Sigma_{\mathcal{Q}},\mathbb{R}) \models \mathsf{MC}(\Sigma_{\mathcal{Q}})$.

(1) whenever $\mathcal{R} \in pB(\mathcal{P},\Sigma) \cap (\mathcal{N}_z^*|\delta_z)$ there is $\mathcal{S} \in pI(\mathcal{R},\Sigma_{\mathcal{R}}) \cap \mathcal{N}$ such that $\Sigma_{\mathcal{S}} \restriction \mathcal{N} \in \mathcal{J}[\mathcal{N}]$.

As in [29], there are several cases.

1. $\mathcal{P}$ is of successor type.
2. $\mathcal{P}$ is of limit type and $\mathcal{P}$ is meek.
3. $\mathcal{P}$ is non-meek but $\mathcal{P}$ is not of #-lsa type.
4. $(\mathcal{P},\Sigma)$ is an sts hod pair.

The first two cases are just like the cases considered in [29, Theorem 6.5], we leave those to the reader. Here we analyze the remaining two cases. To start, we need to import some notions from [29, Section 6.3].

DEFINITION 10.2.9. Suppose for a moment that we are working in some model of ZFC. Suppose κ is an inaccessible cardinal. We say that $(\mathcal{Q},\Lambda)$ is a *hod pair at* κ if

1. $(\mathcal{Q},\Lambda)$ is a hod pair,
2. $\mathcal{Q} \in \mathsf{HC}$[23]
3. Λ is a (κ,κ)-iteration strategy,
4. $\mathsf{Code}(\Lambda)$ is a κ-universally Baire set of reals.

Suppose $(\mathcal{Q},\Lambda)$ is a hod pair at κ. Then we let

$$Lp^{\Lambda,\kappa}(a) = \bigcup\{\mathcal{M} : \mathcal{M} \text{ is a sound } \Lambda\text{-mouse over } a \text{ such that } \rho_\omega(\mathcal{M}) = \mathsf{ord}(a) \text{ and } \mathcal{M} \trianglelefteq (\mathsf{Le}((\mathcal{Q},\Lambda),a)^{V_\kappa}\}.$$

As is customary, we let $\mathsf{Lp}_\alpha^{\Lambda,\kappa}(a)$ be the αth iterate of $\mathsf{Lp}^{\Lambda,\kappa}(a)$. Below $\mathcal{S}^*(\mathcal{R})$ is the $*$-transform of $\mathcal{S}$ into a hybrid mouse over $\mathcal{R}$, it is defined when $\mathcal{R}$ is a cutpoint of $\mathcal{S}$ (cf. [39]).

DEFINITION 10.2.10 (Fullness preservation in models of ZFC). Suppose now that $(\mathcal{P},\Sigma)$ is a hod pair at κ. We then say Σ is *κ-fullness preserving* if the following holds for all $(\mathcal{T},\mathcal{Q}) \subset I(\mathcal{P},\Sigma) \cap V_\kappa$.

1. For all meek[24] layers $\mathcal{R}$ of $\mathcal{Q}$ such that $\mathcal{R}$ is of successor type[25], letting $\mathcal{S} = \mathcal{R}^{-}$[26], for all $\eta \in (\mathrm{ord}(\mathcal{S}),\mathrm{ord}(\mathcal{R}))$ if η is a cutpoint cardinal of $\mathcal{R}$ then
$$(\mathcal{R}|(\eta^+)^{\mathcal{R}})^* = \mathsf{Lp}^{\Sigma_{\mathcal{S},\mathcal{T}},\kappa}(\mathcal{R}|\delta).$$
2. For all meek[27] layers $\mathcal{R}$ of $\mathcal{Q}$ such that $\mathcal{R}$ is of limit type,
$$\mathcal{R} = \mathsf{Lp}^{\Sigma_{\mathcal{R}|\delta^{\mathcal{R}}},\mathcal{T}},\kappa}(\mathcal{R}|\delta^{\mathcal{R}}).$$

[23]We will later apply this definition to $\mathcal{Q}$ which are not countable. The reason we make this assumption is so that we can have clause 4 below. It follows that the current definition makes sense in a variety of situations, and in particular when clause 4 holds after collapsing $\mathcal{Q}$ to be countable.

[24]See Definition 2.7.1.

[25]See Definition 2.7.17.

[26]This is the longest proper layer of $\mathcal{R}$. See Notation 2.7.14.

[27]See Definition 2.7.1.

3. If $\mathcal{P}$ is of #-lsa type then $\mathsf{Lp}^{\Sigma^{stc}_{\mathcal{Q},\mathcal{T}},\kappa}(\mathcal{Q}) \models$ "$\delta^{\mathcal{Q}}$ is a Woodin cardinal"[28].

We continuing our work inside some model of ZFC.

DEFINITION 10.2.11 (Universal tail). Suppose $(\mathcal{Q},\Lambda)$ is a hod pair at κ such that Λ has branch condensation and is κ-fullness preserving. Suppose $\lambda < \kappa$ is an inaccessible cardinal. Then we say $(\mathcal{Q}^*,\Lambda^*)$ is a λ-*universal tail* of $(\mathcal{Q},\Lambda)$ if there is a (possibly generalized) stack

$$\mathcal{T} = (\mathcal{M}_\beta, \mathcal{T}_\beta, E_\beta : \beta < \lambda)$$

on $\mathcal{Q}$ according to Λ with last model $\mathcal{Q}^*$ such that $\mathrm{lh}(\mathcal{T}) = \lambda$ and for any $(\mathcal{S},\mathcal{R}) \in I(\mathcal{Q},\Lambda) \cap V_\lambda$ there is a stack $\mathcal{U}$ on $\mathcal{R}$ according to $\Lambda_{\mathcal{R},\mathcal{S}}$ such that for some $\alpha < \lambda$, $\mathcal{M}_\alpha$ is the last model of $\mathcal{U}$.

If $\mathcal{T}$ is as above then we say $\mathcal{T}$ is a λ-*universal stack* on $\mathcal{Q}$ according to Λ.

We now resume the proof of Theorem 10.2.5, and continue with the objects introduced in Terminology 10.2.8. and start working in $\mathcal{N}_z^*$. Observe that because of our assumption on $(\mathcal{P},\Sigma)$, whenever $\mathcal{Q},\mathcal{R} \in pI(\mathcal{P},\Sigma)$, $(\mathcal{Q},\Sigma_\mathcal{Q})$ and $(\mathcal{R},\Sigma_\mathcal{R})$ have a common tail in $\mathcal{N}_z^*|\delta_z$. In fact more is true. Suppose κ is a strong cardinal of $\mathcal{N}_z^*$. Then it follows from Corollary 4.13.5 that if $\mathcal{Q},\mathcal{R} \in pI(\mathcal{P},\Sigma) \cap \mathcal{N}_z^*|\kappa$ then $(\mathcal{Q},\Sigma_\mathcal{Q})$ and $(\mathcal{R},\Sigma_\mathcal{R})$ have a common tail in $\mathcal{N}_z^*|\kappa$. This means that whenever $\kappa < \delta_z$ is a cardinal of $\mathcal{N}_z^*$ and $\mathcal{Q} \in (pI(\mathcal{P},\Sigma) \cup pB(\mathcal{P},\Sigma)) \cap \mathcal{N}_z^*|\kappa$, we can form the direct limit of all $\Sigma_\mathcal{Q}$ iterates of $\mathcal{Q}$ that are in $\mathcal{N}_z^*|\kappa$. Let $\mathcal{R}_\kappa^{\mathcal{Q},\Sigma_\mathcal{Q}}$ be this direct limit. The next lemma shows that the universal tails are unique. It appeared as [29, Lemma 6.8].

In what follows, we will often abuse the terminology introduced in Definition 10.2.11. Usually when, working inside $\mathcal{N}_z^*$, we talk about κ-universal tail of some $(\mathcal{Q},\Lambda)$ with $\mathcal{Q} \in \mathcal{N}_z^*|\kappa$ then we mean that we are working in $\mathcal{N}_z^*[h]$ where $h \subseteq Coll(\omega,\mathcal{Q})$ is $\mathcal{N}_z^*$-generic. Then take our current κ be λ of Definition 10.2.11 and κ of Definition 10.2.11 be δ_z.

LEMMA 10.2.12 (Uniqueness of universal tails). *Suppose*

$$\mathcal{Q} \in pI(\mathcal{P},\Sigma) \cap \mathcal{N}_z^*|\delta_z.$$

Then for each $\mathcal{S} \trianglelefteq^c_{hod} \mathcal{Q}$[29] *and $\mathcal{N}$-strong $\kappa < \delta_z$ such that $\mathcal{S} \in \mathcal{N}_z^*|\kappa$, there is a unique κ-universal tail of $(\mathcal{S},\Sigma_\mathcal{S})$. In fact, letting $\mathcal{R} = \mathcal{R}_\kappa^{\mathcal{S},\Sigma_\mathcal{S}}$, $(\mathcal{R},\Sigma_\mathcal{R})$ is the unique κ-universal tail of $(\mathcal{S},\Sigma_\mathcal{S})$*

DEFINITION 10.2.13. Suppose $\mathcal{Q} \in (pI(\mathcal{P},\Sigma) \cup pB(\mathcal{P},\Sigma)) \cap \mathcal{N}_z^*|\delta_z$ and κ is an $\mathcal{N}$-strong cardinal such that $\mathcal{Q} \in \mathcal{N}_z^*|\kappa$. Then we say $\mathcal{N}$ *captures* a tail of $(\mathcal{Q},\Sigma_\mathcal{Q})$ below κ if there is a hod pair $(\mathcal{R},\Lambda) \in \mathcal{N}$ such that Λ is a (κ,κ)-iteration strategy and there is a term relation $\tau \in \mathcal{N}^{Coll(\omega,<\kappa)}$ such that whenever $g \subseteq Coll(\omega,|\mathcal{R}|^+)$ is $\mathcal{N}$-generic,

[28] Here, if Σ is a short tree strategy then $\Sigma^{stc} = \Sigma$.

[29] See Definition 9.1.2.

1. $\mathcal{N}[g] \models$ "$(\mathcal{R}, \tau_g)$ is a hod pair at κ such that τ_g is κ-fullness preserving" and $\tau_g \restriction \mathcal{N} = \Lambda$,
2. for some $\lambda < \kappa$, $\mathcal{R} = \mathcal{R}_\lambda^{\mathcal{Q},\Lambda}$ and letting $T, U \in \mathcal{N}[g]$ witness that τ_g is κ-uB, whenever $h \subseteq Coll(\omega, <\kappa)$ is $\mathcal{N}[g]$-generic, $(p[T])^{\mathcal{N}[g][h]} = \mathsf{Code}(\Sigma_{\mathcal{R}}) \cap \mathcal{N}[g][h]$.

We say $\mathcal{N}$ captures $B(\mathcal{Q}, \Sigma_{\mathcal{Q}})$ below κ if whenever $\mathcal{R} \in pB(\mathcal{Q}, \Sigma_{\mathcal{Q}}) \cap \mathcal{N}_z^*|\kappa$, $\mathcal{N}$ captures $(\mathcal{R}, \Sigma_{\mathcal{R}})$ below κ.

Towards a contradiction, we assume that $\mathcal{N}$ does not capture a tail of $(\mathcal{P}, \Sigma)$ and that either

1. $\mathcal{P}$ is non-meek but $\mathcal{P}$ is not of #-lsa type,
2. $(\mathcal{P}, \Sigma)$ is an sts hod pair.

NOTATION 10.2.14. For each $\mathcal{Q} \in pB(\mathcal{P}, \Sigma)$, we let $\lambda_{\mathcal{Q}}$ be the least $\mathcal{N}$-strong cardinal ν such that $\mathcal{N}$ captures the ν-universal tail of $(\mathcal{Q}, \Sigma_{\mathcal{Q}})$. We let $(\mathcal{R}^{\mathcal{Q},\Sigma}, \Phi^{\mathcal{Q},\Sigma})$ be the $\lambda_{\mathcal{Q}}$-universal tail of $(\mathcal{Q}, \Sigma_{\mathcal{Q}})$. For each inaccessible cardinal ν such that $\mathcal{Q} \in \mathcal{N}|\nu$, we let $(\mathcal{R}_\nu^{\mathcal{Q},\Sigma}, \Phi_\nu^{\mathcal{Q},\Sigma})$ be the ν-universal tail of $(\mathcal{Q}, \Sigma_{\mathcal{Q}})$. If $\lambda \geq \lambda_{\mathcal{Q}}$ then $\pi^{\Sigma_{\mathcal{Q}}}_{\mathcal{Q}, \mathcal{R}_\lambda^{\mathcal{P},\Sigma}}$ is the iteration map $\pi^{\Sigma_{\mathcal{Q}}}_{\mathcal{Q}, \mathcal{R}^{\mathcal{Q},\Sigma}}$.

10.2.3. The ideas behind the proof. The notation and the terminology introduced in this subsection will be used in the next few subsections. We are continuing with the set up of Terminology 10.2.8.

NOTATION 10.2.15. Suppose now that κ_0 is an $\mathcal{N}$-strong cardinal that reflects the set of $\mathcal{N}$-strong cardinals. Let

$$\begin{aligned}\mathcal{E}_0 = \{E \in \vec{E}^{\mathcal{N}} : \mathcal{N} \models \text{“}\nu(E) \text{ is inaccessible” and for all } \eta \in (\kappa_0, \nu(E)),\\ \mathcal{N} \models \text{“}\eta \text{ is a strong cardinal” if and only if}\\ \mathrm{Ult}(\mathcal{N}, E) \models \text{“}\eta \text{ is a strong cardinal”}\}.\end{aligned}$$

NOTATION 10.2.16. Working in $\mathcal{N}$, let

$$\begin{aligned}\mathcal{F} = \{(\mathcal{Q}, \Lambda) : \mathcal{Q} \in \mathcal{N}|\delta_z \wedge \mathcal{J}[\mathcal{N}] \models \text{“}(\mathcal{Q}, \Lambda) \text{ is a hod pair at } \delta_z \text{ and}\\ \Lambda \text{ has branch condensation and is } \delta_z\text{-fullness preserving”}\}.\end{aligned}$$

We have that $\mathcal{F}$ is a directed system. Let for $\lambda \leq \delta_z$,

$$\mathcal{F} \restriction \lambda = \{(\mathcal{Q}, \Lambda) \in \mathcal{F} : \mathcal{Q} \in \mathcal{N}|\lambda\}.$$

We let $\mathcal{R}^*$ be the direct limit of $\mathcal{F} \restriction \kappa_0$ under the iteration maps.

Recall that we fixed an $\mathcal{N}$-strong cardinal κ_0 that reflects the set of strong cardinals of $\mathcal{N}$. The equality below is computed inside $\mathcal{N}_z^*$.

DEFINITION 10.2.17. Let $\mathcal{R}_0 = (\mathcal{R}_{\kappa_0}^{\mathcal{P},\Sigma})^b$.

The next lemma summarizes what was proved in [29].

LEMMA 10.2.18. *The following holds.*

1. *Suppose* $\mathcal{Q} \in pB(\mathcal{P},\Sigma) \cap \mathcal{N}_z^*|\kappa_0$. *Then* $\mathcal{R}^{\mathcal{Q},\Sigma_{\mathcal{Q}}} \in \mathcal{N}|\kappa_0$.
2. *Suppose* $\mathcal{Q} \in pB(\mathcal{P},\Sigma)$, $\lambda > \kappa_0$ *is a strong cardinal of* $\mathcal{N}$ *such that* $\mathcal{Q} \in \mathcal{N}|\lambda$, *and* $E \in \mathcal{E}_0$ *is an extender with critical point* κ_0 *such that* $\nu(E) > (\lambda^+)^{\mathcal{N}_z^*}$. *Then* $\Phi^{\mathcal{Q},\Sigma} \restriction \mathrm{Ult}(\mathcal{N},E) \in \mathrm{Ult}(\mathcal{N},E)$.
3. *Let* $\mathcal{R}^*$ *be as in Notation 10.2.16. Then either* $\mathcal{R}_0 \trianglelefteq_{hod} \mathcal{R}^*$ *or* $\mathcal{R}_0|\delta^{\mathcal{R}_0} = \mathcal{R}^*$. *Moreover,* $\mathcal{R}_0 \in \mathcal{N}$ *and* $\Sigma_{\mathcal{R}_0} \restriction \mathcal{N} \in \mathcal{J}[\mathcal{N}]$.

Clause 1 is just [29, Lemma 6.11], clause 2 is [29, Lemma 6.12] and clause 3 is [29, Lemma 6.13].

Below we will develop a technology for recovering the full iterate of $\mathcal{P}$. Let $\mathcal{R}_0^+ = \mathcal{R}_{\kappa_0}^{\mathcal{P},\Sigma}$ be the iterate of $\mathcal{P}$ extending $\mathcal{R}_0$ and let $i : \mathcal{P} \to \mathcal{R}_0^+$ be the iteration embedding. We will recover an iterate of $\mathcal{R}_0^+$ inside $\mathcal{N}$ as an output of a back-grounded construction that is done over $\mathcal{R}_0$. Such constructions are called mixed hod pair constructions. The details of this construction appear in Section 10.2.8.

There are two kinds of extenders that we will use in this construction. The extenders with critical point $> \delta^{\mathcal{R}_0}$ will have traditional background certificates. We will use the total extenders on the sequence of $\mathcal{N}$ to certify such extenders. The extenders with critical point $\delta^{\mathcal{R}_0}$ will come from a different source. The following key lemma illustrates the idea.

LEMMA 10.2.19. *Let* $\delta = \delta^{\mathcal{R}_0}$. *Suppose* $\mathcal{S} \in pI(\mathcal{R}_0^+,\Sigma_{\mathcal{R}_0^+})$ *is a normal iterate of* $\mathcal{R}_0^+$ *that is obtained by iterating entirely above* $\delta^{\mathcal{R}_0}$. *Suppose that* $\alpha \in \mathrm{dom}(\vec{E}^{\mathcal{S}})$ *is such that letting* $E =_{def} \vec{E}^{\mathcal{S}}(\alpha)$, $\mathrm{crit}(E) = \delta$, $\mathcal{S}|\alpha \in \mathcal{N}$ *and* $\Sigma_{\mathcal{S}|\alpha} \restriction \mathcal{N} \in \mathcal{J}[\mathcal{N}]$. *Then* $E \in \mathcal{N}$. *Moreover,* $(a,A) \in E$ *if and only if* $a \in \nu_E^{<\omega}$, $A \in [\delta]^{|a|}$ *and whenever* $F \in \mathcal{E}_0$ *is such that* $\mathrm{crit}(F) = \kappa_0$ *and*

$$\mathcal{N} \models \text{"there is a strong cardinal } \lambda \text{ in the interval } (\kappa_0, \nu_F) \text{ such that } \mathcal{S} \in \mathcal{N}|\lambda\text{"},$$

$\pi_{\mathcal{S}|\alpha,\pi_F(\mathcal{R}_0)}^{\Sigma_{\mathcal{S}|\alpha}}(a) \in \pi_F(A)$[30].

PROOF. Set $\mathcal{M}^+ = \mathrm{Ult}(\mathcal{R}_0^+,E)$ and $\mathcal{M} = \mathrm{Ult}(\mathcal{R}_0,E)$. Let F^* be the resurrection of F and let $\sigma : \mathrm{Ult}(\mathcal{N},F) \to \pi_{F^*}(\mathcal{N})$ be the canonical factor map. We have that $\sigma \restriction \nu_F = id$. Thus, $\pi_{F^*} \restriction \mathcal{N} = \sigma \circ \pi_F$. It follows that $\pi_{F^*} \restriction \mathcal{R}^+$ is the iteration embedding implying

$$(1)\quad \pi_{F^*} \restriction \mathcal{R}_0^+ = \pi_{\mathcal{M}^+,\pi_{F^*}(\mathcal{R}_0^+)}^{\Sigma_{\mathcal{M}^+}} \circ \pi_E^{\mathcal{R}_0^+}.$$

We now have that

$$\begin{aligned} (a,A) \in E &\iff a \in \pi_E^{\mathcal{R}_0^+}(A) \\ &\iff \pi_{\mathcal{M}^+,\pi_{F^*}(\mathcal{R}_0^+)}^{\Sigma_{\mathcal{M}^+}}(a) \in \pi_{\mathcal{M}^+,\pi_{F^*}(\mathcal{R}_0^+)}^{\Sigma_{\mathcal{M}^+}}(\pi_E^{\mathcal{R}_0^+}(A)) \end{aligned}$$

---

[30] The embedding $\pi_{\mathcal{S}|\alpha,\pi_F(\mathcal{R}_0)}^{\Sigma_{\mathcal{S}|\alpha}}$ is just $\pi^{\Sigma_{\mathcal{S}|\alpha}}_{\mathcal{S}|\alpha,\mathcal{R}_{\pi_F(\kappa_0)}^{\mathcal{S}|\alpha,\Sigma_{\mathcal{S}\alpha}}}$. We will often abuse our notation this way.

$$\iff \sigma(\pi^{\Sigma_{\mathcal{M}}}_{\mathcal{M},\pi_F(\mathcal{R}_0)}(a)) \in \pi_{F^*}(A)$$
$$\iff \sigma(\pi^{\Sigma_{\mathcal{M}}}_{\mathcal{M},\pi_F(\mathcal{R}_0)}(a)) \in \sigma(\pi_F(A))$$
$$\iff \pi^{\Sigma_{\mathcal{M}}}_{\mathcal{M},\pi_F(\mathcal{R}_0)}(a) \in \pi_F(A)$$

Therefore,

$$(a,A) \in E \iff \pi^{\Sigma_{\mathcal{M}}}_{\mathcal{M},\pi_F(\mathcal{R}_0)}(a) \in \pi_F(A).$$

By our assumption, the right hand side of the equivalence can be computed in $\mathcal{N}$. Hence $E \in \mathcal{N}$. ⊣

Thus, the extenders with critical point $\delta^{\mathcal{R}_0}$ that we will use in our mixed hod pair construction have the following property. If $\mathcal{Q}$ is the current level of the construction and Λ is its strategy then let E be the set of pairs (a,A) such that $a \in (\delta^{\mathcal{R}_0})^{<\omega}$ and for every $F \in \mathcal{E}_0$ such that $\mathrm{crit}(F) = \kappa_0$ and

$\mathcal{N} \models$ "there is a strong cardinal λ in the interval (κ_0, ν_F) such that $\mathcal{Q} \in \mathcal{N}|\lambda$",

$\pi^{\Lambda}_{\mathcal{Q},\pi_F(\mathcal{R}_0)}(a) \in \pi_F(A)$.

There is one problem with this approach. We need to know the strategy Λ of $\mathcal{Q}$ before we can find the relevant extender. To resolve this problem, we will first define the strategy Λ. Essentially Λ will pick branches that, for some η, are π_E-realizable for all $E \in \mathcal{E}_0$ such that $\nu_E > \eta$. We will call such strategies $\mathcal{E}_0$-certified.

In the sequel, we will first introduce the $\mathcal{E}_0$-certified strategies. Then we will prove basic facts about them. Then we will introduce the mixed hod pair constructions and show that some model appearing on this construction is an iterate of $\mathcal{R}_0^+$ via an iteration that is entirely above $\delta^{\mathcal{R}_0}$.

10.2.4. A condensing set in $\mathcal{N}$. The following is the main lemma of this section. Our goal is still to prove Theorem 10.2.5 and our set up is as in Terminology 10.2.8 and Definition 10.2.17.

LEMMA 10.2.20. *Suppose $\eta > \kappa_0$ is such that $\mathcal{N} \models$ "η is a strong cardinal that reflects the set of strong cardinals". Set $\mathcal{S}^+ = \mathcal{R}^{\mathcal{P},\Sigma}_{\eta}$, $i^+ = \pi^{\Sigma_{\mathcal{R}_0^+}}_{\mathcal{R}_0^+,\mathcal{S}^+}$, $\mathcal{S} = (\mathcal{S}^+)^b$ and $i = i^+ \restriction \mathcal{R}_0$. Then $i \in \mathcal{N}$ and $\mathcal{N} \models |\mathcal{S}| < (\eta^+)^{\mathcal{N}}$. Moreover, the following holds:*

1. *If $E \in \vec{E}^{\mathcal{N}_z^*}$ is such that $\mathrm{crit}(E) = \eta$ and E is total over $\mathcal{N}_z^*$ then $\pi_E \restriction \mathcal{S}$ is a strongly condensing set in $\mathrm{Ult}(\mathcal{N}_z^*, E)[g]$ where $g \subseteq Coll(\omega, \pi_E(\eta))$ is any $\mathrm{Ult}(\mathcal{N}_z^*, E)$-generic[31].*
2. *If $E \in \vec{E}^{\mathcal{N}}$ is such that $\mathrm{crit}(E) = \eta$ and $\nu(E)$ is an inaccessible cardinal of $\mathcal{N}$ then $\pi_E \restriction \mathcal{S}$ is a strongly condensing set in $\mathrm{Ult}(\mathcal{N}, E)[g]$ where $g \subseteq Coll(\omega, \pi_E(\eta))$ is any $\mathrm{Ult}(\mathcal{N}_z^*, E)$-generic.*

[31] See Definition 9.1.13. This definition uses a formula φ and a parameter A. In the current case, (φ, A) is chosen in a way that the resulting directed system is $\mathcal{F} \restriction \eta$ where $\mathcal{F}$ is as in Notation 10.2.16.

PROOF. The fact that $\mathcal{S} \in \mathcal{N}$ follows from [29] and it is the same argument that shows that $\mathcal{R}_0 \in \mathcal{N}$ (see Lemma 10.2.18). Suppose now that $i \in \mathcal{N}$. It then follows that $\mathcal{N} \models |\mathcal{S}| < \eta^+$. Hence, Theorem 9.2.2 implies both clause 1 and 2 above. Thus, it is enough to show that $i \in \mathcal{N}$.

Let $F \in \vec{E}^{\mathcal{N}}$ be any extender such that $\text{crit}(F) = \kappa_0$ and $\text{Ult}(\mathcal{N}, F) \models$ "η is a strong cardinal". Let F^* be the background certificate of F and let $k : \text{Ult}(\mathcal{N}, F) \to \pi_{F^*}(\mathcal{N})$ be the canonical factor map. We now have that $\pi_{F^*} \restriction \mathcal{R}_0^+ = \pi^{\Sigma_{\mathcal{R}_0^+}}_{\mathcal{R}^+, \pi_{F^*}(\mathcal{R}_0^+)}$. We thus have that

(1) $\pi_{F^*} \restriction \mathcal{R}_0^+ = \pi^{\Sigma_{\mathcal{S}^+}}_{\mathcal{S}^+, \pi_{F^*}(\mathcal{R}_0^+)} \circ \pi^{\Sigma_{\mathcal{R}_0^+}}_{\mathcal{R}_0^+, \mathcal{S}^+}$.

Let $m = \pi^{\Sigma_{\mathcal{S}^+}}_{\mathcal{S}^+, \pi_{F^*}(\mathcal{R}_0^+)} \restriction \mathcal{S}|\delta^{\mathcal{S}}$. We have that

(2) $m = \pi^{\Sigma_{\mathcal{S}|\delta^{\mathcal{S}}}}_{\mathcal{S}|\delta^{\mathcal{S}}, \pi_{F^*}(\mathcal{R}_0)|\xi}$ where $\xi = \sup(m[\delta^{\mathcal{S}}])$.

Because $\Sigma_{\mathcal{S}|\delta^{\mathcal{S}}} \restriction \mathcal{N} \in \mathcal{N}$, we have that $k(\Sigma_{\mathcal{S}|\delta^{\mathcal{S}}} \restriction \mathcal{N}) = \Sigma_{\mathcal{S}|\delta^{\mathcal{S}}} \restriction \pi_{F^*}(\mathcal{N})$ and therefore, $m \in \pi_{F^*}(\mathcal{N})$ and $m \in \text{rge}(k)$. Let $n = k^{-1}(m)$. Thus,

(3) $n = \pi^{\Sigma_{\mathcal{S}|\delta^{\mathcal{S}}}}_{\mathcal{S}|\delta^{\mathcal{S}}, \pi_F(\mathcal{R}_0)|k^{-1}(\xi)}$.

Notice now that for $x \in \mathcal{R}_0$,

(4) $\pi_{F^*}(x) = \pi^{\Sigma_{\mathcal{S}^+}}_{\mathcal{S}^+, \pi_{F^*}(\mathcal{R}_0^+)}(i(x))$

implying that

(5) $\mathcal{S}$ is the transitive collapse of $\{\pi_{F^*}(f)(m(a)) : f \in \mathcal{R}_0 \wedge a \in (\delta^{\mathcal{S}})^{<\omega}\}$ and $\pi^{\Sigma_{\mathcal{S}^+}}_{\mathcal{S}^+, \pi_{F^*}(\mathcal{R}_0^+)} \restriction \mathcal{S}$ is the inverse of the transitive collapse.

(5) now implies that

(6) $\mathcal{S}$ is the transitive collapse of $\{\pi_F(f)(n(a)) : f \in \mathcal{R}_0 \wedge a \in (\delta^{\mathcal{S}})^{<\omega}\}$.

Since $\{\pi_F(f)(n(a)) : f \in \mathcal{R}_0 \wedge a \in (\delta^{\mathcal{S}})^{<\omega}\} \in \mathcal{N}$, we have that if $\sigma : \mathcal{S} \to \pi_F(\mathcal{R}_0)$ is the inverse of the transitive collapse then $\sigma \in \mathcal{N}$. Moreover,

(7) $\pi_{F^*} \restriction \mathcal{R}_0 = k \circ \sigma \circ i$ and $k \circ \sigma = \pi^{\Sigma_{\mathcal{S}^+}}_{\mathcal{S}^+, \pi_{F^*}(\mathcal{R}_0^+)} \restriction \mathcal{S}$.

It now follows that

(8) $i(x) = \sigma^{-1}(\pi_F(x))$.

Since both σ and π_F are in $\mathcal{N}$, we get that $i \in \mathcal{N}$. ⊣

NOTATION 10.2.21. Suppose now that κ is an $\mathcal{N}$-strong cardinal that reflects the set of $\mathcal{N}$-strong cardinals such that $\kappa > \kappa_0$. Let

$$\begin{aligned}\mathcal{E} = \{E \in \vec{E}^{\mathcal{N}} : \mathcal{N} \models \text{"}\nu(E) \text{ is inaccessible" and for all } \eta \in (\kappa, \nu(E)),\\ \mathcal{N} \models \text{"}\eta \text{ is a strong cardinal" iff } \text{Ult}(\mathcal{N}, E) \models \text{"}\eta \text{ is a strong cardinal"}\}.\end{aligned}$$

Set $\mathcal{R}^+ = \mathcal{R}_\kappa^{\mathcal{P},\Sigma}$ and let $\mathcal{R} = (\mathcal{R}^+)^b$. It follows from Lemma 10.2.18 that $\mathcal{R} \in \mathcal{N}$. Let $\Phi^+ = (\Phi_\kappa^{\mathcal{P},\Sigma})_{\mathcal{R}|\delta^\mathcal{R}}$ and $\Phi = \Phi^+_\mathcal{R}$[32].

Notice that $\Phi \restriction \mathcal{N} \in L[\mathcal{N}]$. We will abuse our notation and write Φ for both $\Phi \restriction \mathcal{N}_z^*$ and $\Phi \restriction \mathcal{N}$, but we encourage the reader to keep this subtle difference between the three versions of Φ in mind.

DEFINITION 10.2.22. Working in $\mathcal{N}$, we say $(\sigma, \mathcal{Q})$ is $\mathcal{E}$*-realizable* if

1. $\sigma : \mathcal{R} \to \mathcal{Q}$ is an elementary embedding,
2. for some $\mathcal{N}$-strong cardinal $\xi \in (\kappa, \delta_z)$, $\mathcal{Q} \in \mathcal{N}|\xi$ and for all $E \in \mathcal{E}$ such that $\xi < \nu(E)$ and for all $\mathcal{N}$-generic $g \subseteq Coll(\omega, \mathcal{Q})$, there is $j : \mathcal{Q} \to \pi_E(\mathcal{R})$ such that $j \in \mathrm{Ult}(\mathcal{N}, E)[g]$ and $\pi_E \restriction \mathcal{R} = \sigma \circ j$.[33]

We say that j is (π_E, σ)-realizable. Continuing our work in $\mathcal{N}$, let $\mathcal{F}'_\mathcal{E}$ be the set of π_E-realizable pairs $(\sigma, \mathcal{Q})$. Given $(\sigma, \mathcal{Q}) \in \mathcal{F}'_\mathcal{E}$, let $\xi(\sigma, \mathcal{Q}) < \delta_z$ witness that clause 2 above holds for $(\sigma, \mathcal{Q})$. Given $E \in \mathcal{E}$ such that $\xi(\sigma, \mathcal{Q}) < \nu(E)$, letting $j : \mathcal{Q} \to \pi_E(\mathcal{R})$ be any (π_E, σ)-realizable embedding, set $\Psi_{\sigma,\mathcal{Q},E,j} = (j$-pullback of $\pi_E(\Phi))$[34].

The following is an easy consequence of Lemma 9.1.14.

LEMMA 10.2.23. *Suppose $(\sigma, \mathcal{Q}, E)$ are as in Definition 10.2.22 and that $\sigma \restriction \mathcal{R}|\delta^\mathcal{R}$ is an iteration embedding according to $\Phi_{\mathcal{R}|\delta^\mathcal{R}}$. Let $j_0 : \mathcal{Q} \to \pi_E(\mathcal{R})$ and $j_1 : \mathcal{Q} \to \pi_E(\mathcal{R})$ be two (π_E, σ)-realizable embeddings. Then $\Psi_{\sigma,\mathcal{Q},E,j_0} = \Psi_{\sigma,\mathcal{Q},E,j_1}$.*

Given $(\sigma, \mathcal{Q}, E)$ as in Lemma 10.2.23, we let $\Psi_{\sigma,\mathcal{Q},E}$ be the common value of all $\Psi_{\sigma,\mathcal{Q},E,j}$ where j is any (π_E, σ)-realizable embedding. The next definition integrates $\Psi_{\sigma,\mathcal{Q},E}$ with respect to E.

DEFINITION 10.2.24. Working in $\mathcal{N}$, we say $(\sigma, \mathcal{Q})$ is *neatly* $\mathcal{E}$*-realizable* if $(\sigma, \mathcal{Q})$ is $\mathcal{E}$-realizable and for all $E_0, E_1 \in \mathcal{E}$ with $\nu(E_0) \leq \nu(E_1)$,

$$\Psi_{\sigma,\mathcal{Q},E_0} \restriction \mathcal{N}|\nu(E_0) = \Psi_{\sigma,\mathcal{Q},E_1} \restriction \mathcal{N}|\nu(E_0).$$

Let $\mathcal{F}_\mathcal{E}$ be the set of neatly $\mathcal{E}$-realizable pairs, and for $(\sigma, \mathcal{Q}) \in \mathcal{F}_\mathcal{E}$, let

$$\Psi_{\sigma,\mathcal{Q}} = \cup\{\Psi_{\sigma,\mathcal{Q},E} : E \in \mathcal{E} \wedge \xi(\sigma, \mathcal{Q}) < \nu(E)\}.$$[35]

The following is the key lemma of this section.

LEMMA 10.2.25. *Suppose $\mathcal{S}$ is a Φ^+-iterate of $\mathcal{R}^+$ via $\mathcal{T}$ such that $\pi^{\mathcal{T},b}$ is defined and $\mathcal{S}^b \in \mathcal{N}$. Then $\pi^{\mathcal{T},b} \in \mathcal{N}$, $(\pi^{\mathcal{T},b}, \mathcal{S}^b)$ is neatly $\mathcal{E}$-realizable and*

$$\Psi_{\pi^{\mathcal{T},b},\mathcal{S}^b} = \Phi^+_{\mathcal{S}^b} \restriction \mathcal{N}.$$

[32] See Notation 10.2.14.

[33] Notice that $\pi_E \restriction \mathcal{R} \in \mathrm{Ult}(\mathcal{N}, E)$, see Lemma 10.2.20.

[34] $\Psi_{\sigma,\mathcal{Q},E,j}$ is defined in $\mathrm{Ult}(\mathcal{N}, E)$.

[35] $\Psi_{\sigma,\mathcal{Q}}$ is defined in $\mathcal{J}[\mathcal{N}]$ and $\Psi_{\sigma,\mathcal{Q}} \restriction \mathcal{N}$ is total.

PROOF. The proof of $\pi^{\mathcal{T},b} \in \mathcal{N}$ is exactly the proof of Lemma 10.2.20. The proof of the fact that $(\pi^{\mathcal{T},b}, \mathcal{S}^b)$ is neatly $\mathcal{E}$-realizable is via a simple absoluteness argument. Let $E \in \mathcal{E}$ be such that $\mathcal{S}^b \in \mathcal{N}|\nu(E)$ and let E^* be the background certificate of E. Let $k : Ult(\mathcal{N}, E) \to \pi_{E^*}(\mathcal{N})$ be the canonical factor map. We have that $\mathrm{crit}(k) \geq \nu(E)$. Set $\sigma = \pi^{\mathcal{T},b}$. Notice that

(1) in $\pi_{E^*}(\mathcal{N})$, it is forced by $Coll(\omega, \mathcal{S}^b)$ that there is a (π_{E^*}, σ)-realizable $j : \mathcal{S}^b \to \pi_{E^*}(\mathcal{R})$, and
(2) if $g \subseteq Coll(\omega, \mathcal{S}^b)$ is $\pi_{E^*}(\mathcal{N})$-generic and $j : \mathcal{S}^b \to \pi_{E^*}(\mathcal{R})$ is any (π_{E^*}, σ)-realizable embedding, then the j-pullback of $\pi_{E^*}(\Phi)$ is $\Phi^+_{\mathcal{S}^b}$.

It follows that

(3) in $\mathcal{N}$, it is forced by $Coll(\omega, \mathcal{S}^b)$ that there is a (π_{E^*}, σ)-realizable $j : \mathcal{S}^b \to \pi_{E^*}(\mathcal{R})$, and
(4) if $g \subseteq Coll(\omega, \mathcal{S}^b)$ is $\mathcal{N}$-generic and $j : \mathcal{S}^b \to \pi_{E^*}(\mathcal{R})$ is any (π_{E^*}, σ)-realizable embedding, then the j-pullback of $\pi_E(\Phi)$ is independent of j.

Let Π in $\mathrm{Ult}(\mathcal{N}, E)$ be the strategy of $\mathcal{S}^b$ such that it is forced by $Coll(\omega, \mathcal{S}^b)$, that for some (π_E, σ)-realizable j, Π is the j-pullback of $\pi_E(\Phi)$. Let $\tau : \mathcal{S}^b \to \pi_E(\mathcal{R})$ be defined by setting $\tau(x) = \pi_E(f)(\pi^{\Pi}_{\mathcal{S}|\delta^{\mathcal{S}^b}, \pi_E(\mathcal{R})}(a))$ where $x = \sigma(f)(a)$, $f \in \mathcal{R}$ and $a \in (\delta^{\mathcal{S}^b})^{<\omega}$. It follows from clause 2 of Lemma 10.2.20 that τ is a (π_E, σ)-realization and $\tau \in \mathrm{Ult}(\mathcal{N}, E)$. It then follows from (2) that $k(\Pi) = \Phi^+_{\mathcal{S}^b}$ and therefore, $\Pi = \Phi^+_{\mathcal{S}^b} \restriction \mathrm{Ult}(\mathcal{N}, E)$. $\dashv$

DEFINITION 10.2.26. Suppose $(\sigma, \mathcal{Q}) \in \mathcal{F}_{\mathcal{E}}$ and $E \in \mathcal{E}$ is such that $\xi(\sigma, \mathcal{Q}) < \nu(E)$. We say that $\tau = \tau^E_{\sigma,\mathcal{Q}}$ is the *canonical E-realization* of $(\sigma, \mathcal{Q})$ if $\tau : \mathcal{Q} \to \pi_E(\mathcal{R})$ and $\tau(x) = \pi_E(f)(\pi^{\Psi_{\sigma,\mathcal{Q},E}}_{\mathcal{Q}|\delta^{\mathcal{Q}}, \mathcal{R}(\sigma,\mathcal{Q})}(a))$ where $\mathcal{R}(\sigma, \mathcal{Q}) \lhd_{hod} \mathcal{R}$ is the $\Psi_{\sigma,\mathcal{Q},E}$-iterate of $\mathcal{Q}|\delta^{\mathcal{Q}}$, $f \in \mathcal{R}$, $a \in (\delta^{\mathcal{Q}})^{<\omega}$ and $x = \sigma(f)(a)$.

10.2.5. $\mathcal{E}$-certified iteration strategies. Our goal is still to prove Theorem 10.2.5 and our set up is as in Terminology 10.2.8, Definition 10.2.17 and Notation 10.2.21. The following is a modification of [29, Definition 6.14].

DEFINITION 10.2.27 (π_E-realizable iterations). Suppose

1. $\mathcal{V} \in \mathcal{N}$ is a hod premouse extending $\mathcal{R}$ such that $\mathcal{R} = \mathcal{V}^b$,
2. $\mathcal{T} \in \mathcal{N}$ is either a stack on $\mathcal{V}$ or an st-stack[36] on $\mathcal{V}$[37],
3. $E \in \mathcal{E}$.

Recall Definition 2.7.24, Remark 2.7.27 and Notation 2.4.4. Suppose that

$$\mathcal{T} = ((\mathcal{M}_\alpha)_{\alpha<\eta}, (E_\alpha)_{\alpha<\eta-1}, D, R, (\beta_\alpha, m_\alpha)_{\alpha \in R}, T)$$

is a stack. Set $R^b = \{\alpha \in R : \pi^{\mathcal{T},b}_{0,\alpha} \text{ is defined}\}$. We say $\mathcal{T}$ is π_E-*realizable* if the following holds:

---

[36] See Definition 3.2.1.

[37] If $\mathcal{T}$ is an st-stack then $\mathcal{V}$ must be of #-lsa type.

1. $\mathcal{N} \models$ "λ is a strong cardinal".
2. $\mathcal{T} \in \mathcal{N}|\mathrm{lh}(E)$.
3. For all $\alpha \in R^b$, $(\pi^{\mathcal{T}_{\leq\alpha},b}, \mathcal{M}^b_\alpha) \in \mathcal{F}_{\mathcal{E}}$.[38]
4. For all $\alpha < \beta$ such that $\alpha, \beta \in R^b$, setting $\tau_\alpha = \tau^E_{\sigma,\mathcal{Q}}$, $\tau_\alpha = \tau_\beta \circ \pi^{\mathcal{T},b}_{\alpha,\beta}$.[39]
5. For all $\alpha \in R^b$, letting $\Psi_\alpha = \Psi_{\sigma_\alpha, \mathcal{M}^b_\alpha}$,
   (a) if $\alpha \neq \max(R^b)$ and $\mathsf{nc}^{\mathcal{T}}_\alpha$ is based on $\mathcal{M}^b_\alpha|\delta^{\mathcal{M}^b_\alpha}$ then $\mathsf{nc}^{\mathcal{T}}_\alpha$ is according to Ψ_α,
   (b) if $\alpha = \max(R^b)$ and $\mathcal{U} =\downarrow (\mathcal{T}_{\geq\alpha}, \mathcal{M}^b_\alpha)$[40] then
      (i) if $\mathcal{U}$ is based on $\mathcal{M}^b_\alpha$ and is above $\delta^{\mathcal{M}^b_\alpha}$ then it is according to the unique strategy Π of $\mathcal{M}^b_\alpha$ witnessing that $\mathcal{M}^b_\alpha$ is a Ψ_α-mouse over $\mathcal{M}^b_\alpha|\delta^{\mathcal{M}^b_\alpha}$, and
      (ii) if $\mathcal{U}$ is based on $\mathcal{M}^b_\alpha|\delta^{\mathcal{M}^b_\alpha}$ then $\mathcal{U}$ is according to Ψ_α.

We say that $(\sigma_\alpha : \alpha \in R^b)$ are the *π_E-realizable embeddings* of $\mathcal{T}$ and $(\Psi_\alpha : \alpha \in R^b)$ are the *π_E-realizable strategies* of $\mathcal{T}$. We say $\mathcal{T}$ is *$\mathcal{E}$-realizable* if for some η, $\mathcal{T}$ is π_E-realizable for every $E \in \mathcal{E}$ with the property that $\mathrm{lh}(E) > \eta$.

The definition of the above concepts for st-stacks is very similar. The embeddings σ_α are once again defined for $\alpha \in R^b$ which once again consists of those $\alpha < \mathrm{lh}(\mathcal{T})$ with the property that $\pi^{\mathcal{T},b}_{0,\alpha}$ is defined. We leave the details to the reader.

We now introduce a kind of backgrounded constructions reminiscent to the backgrounded construction introduced in Definition 4.2.1. We will use them to find the $\mathcal{Q}$-structures of various iterations.

DEFINITION 10.2.28 ($\mathcal{E}$-realizable backgrounded constructions).
Suppose $\mathcal{V}, \mathcal{T}$,[41] $\tau, \mathcal{Q}, \eta$ are such that

1. $\mathcal{V} \in \mathcal{N}$ is a hod premouse extending $\mathcal{R}$ such that $\mathcal{V}^b = \mathcal{R}$,
2. $\mathcal{T}$ is a $\mathcal{E}$-realizable stack or an st-stack on $\mathcal{V}$ such that $\pi^{\mathcal{T},b}$ exists,
3. $\tau \in (R^b)^{\mathcal{T}}$[42] and $\mathcal{U} =_{\mathrm{def}} \mathsf{nc}^{\mathcal{T}}_\tau$ is based on $\mathcal{M}_\tau$ and is above $\delta^{\mathcal{M}^b_\tau}$,
4. if $\mathcal{U}$ is of limit length then $\mathcal{Q} = \mathrm{m}(\mathcal{U})$ and otherwise for some $\xi < \mathrm{lh}(\mathcal{U})$, $\mathcal{Q} = \mathcal{M}^{\mathcal{U}}_\xi$,
5. $(\mathcal{Q}|\eta)^\# \models$ "η is a Woodin cardinal".

Then

$$\mathsf{Le}^{\mathcal{E},c}((\mathcal{Q}|\eta)^\#)(\mathcal{X}_\gamma, \mathcal{Y}_\gamma, F^+_\gamma, F_\gamma, b_\gamma : \gamma \leq \delta_z)$$

is the *fully backgrounded $\mathcal{E}$-realizable construction over $(\mathcal{Q}|\eta)^\#$ done in $\mathcal{N}$* if the following is true.

---

[38]See Definition 10.2.24.
[39]See Section 2.8.
[40]See Definition 2.4.9.
[41]We will omit $\mathcal{T}$ from most superscripts. Thus, $\mathcal{M}_\alpha$ is just $\mathcal{M}^{\mathcal{T}}_\alpha$.
[42]R^b was introduced in Definition 10.2.27.

1. $\mathcal{X}_0 = \mathcal{J}_\omega((\mathcal{Q}|\eta)^\#)$, and for all $\xi < \delta_z$, $\mathcal{X}_\xi$ and $\mathcal{Y}_\xi$ are sts premice such that if $\mathcal{W}$ is a stack indexed either in $\mathcal{X}_\xi$ or $\mathcal{Y}_\xi$ then $\mathcal{T}_{\leq \mathcal{Q}}{}^\frown \mathcal{W}$ is $\mathcal{E}$-realized.

2. If for some $\xi \leq \delta_z$, $\mathcal{Y}_\xi$ is defined but is not a reliable sts premouse over $(\mathcal{Q}|\eta)^\#$ then all other objects with index $\geq \xi$ are undefined.

3. Suppose for some $\xi < \delta_z$, for all $\gamma \leq \xi$, both $\mathcal{X}_\gamma$ and $\mathcal{Y}_\gamma$ are defined. Then $\mathcal{X}_{\xi+1}$, $\mathcal{Y}_{\xi+1}$, F^+_ξ, F_ξ and b_ξ are determined as follows.

(a) Suppose $\mathcal{X}_\xi = (\mathcal{J}^{\vec{E},f}_{\omega\alpha}, \in, \vec{E}, f)$ is a passive ses[43] and there is an extender $F^* \in \vec{G}$, an extender F over $\mathcal{X}_\xi$, and an ordinal $\nu < \omega\alpha$ such that
  (i) $\nu < \nu(F^*)$,
  (ii) $F = F^* \cap ([\nu]^\omega \times \lfloor \mathcal{X}_\xi \rfloor)$, and
  (iii) setting

$$\mathcal{Y}_{\xi+1} = (\mathcal{J}^{\vec{E},f}_{\omega\alpha}, \in, \vec{E}, f, \tilde{F})$$

  where $\tilde{F}$ is the amenable code of F[44], clause 2 fails for $\xi + 1$.

  Then $\mathcal{X}_{\xi+1} = \mathsf{core}(\mathcal{Y}_{\xi+1})$[45], $F^+_\xi = \vec{G}(\xi)$ where ξ is the least such that $F^* = \vec{G}(\xi)$ has the above properties, $F_\xi = F^+ \cap ([\nu]^\omega \times \lfloor \mathcal{X}_\xi \rfloor)$ where ν is chosen so that the above clauses hold and $b_\xi = \emptyset$.

(b) Suppose $\mathcal{X}_\xi = (\mathcal{J}^{\vec{E},f}_{\omega\alpha}, \in, \vec{E}, f)$ is a passive ses, $\alpha = \beta + \gamma$ and there is $t = (\mathcal{P}_0, \mathcal{T}, \mathcal{P}_1, \mathcal{U}) \in \lfloor \mathcal{X}_\xi | \omega\beta \rfloor \cap \mathrm{dom}(\Lambda)$ such that setting $w = (\mathcal{J}_\omega(t), t, \in)$, w is (f, sts)-minimal as witnessed by β[46] and $\gamma = \mathrm{lh}(t)$.
  Suppose there is a branch b of t (in $\mathcal{N}$) such that $(\mathcal{T}_{\leq \mathcal{Q}})^\frown t^\frown \{b\}$[47] is $\mathcal{E}$-certified.
  Then set

$$\mathcal{Y}_{\xi+1} = (\mathcal{J}^{\vec{E},f^+}_{\omega\beta+\omega\gamma}, \in, \vec{E}, f, \tilde{b})$$

  where $\tilde{b} \subseteq \omega\beta + \omega\gamma$ is defined by $\omega\beta + \omega\nu \in \tilde{b} \iff \nu \in b$. Assuming clause 2 fails for $\xi + 1$, $\mathcal{X}_{\xi+1} = \mathsf{core}(\mathcal{Y}_{\xi+1})$, $F^+_\xi = F_\xi = \emptyset$ and $b_\xi = \tilde{b}$.

  Important Anomaly: Suppose t is nuvs and suppose $e \in \mathcal{X}_\xi | \omega\beta$ is such that $\mathcal{X}_\xi | \omega\beta \models \mathsf{sts}_0(t, e)$[48]. If $e \neq b$ then $\mathcal{Y}_{\xi+1}$ is not an sts premouse over X based on $\mathcal{V}$, and so clause 2 holds.

(c) If $\mathcal{X}_\xi$ does not satisfy clause 3(a) or 3(b) then set $\mathcal{Y}_{\xi+1} = \mathcal{J}_\omega[\mathcal{X}_\xi]$. Assuming clause 2 fails for $\xi + 1$, $\mathcal{X}_{\xi+1} = \mathsf{core}(\mathcal{Y}_{\xi+1})$, $F^+_\xi = F_\xi = b_\xi = \emptyset$.

---

[43] That is, with no last predicate.

[44] For the definition of the "amenable code" see the last paragraph on page 14 of [59].

[45] Recall that $\mathsf{core}(\mathcal{M})$ is the core of $\mathcal{M}$.

[46] See Definition 2.3.3. In particular, this means that we have to index the branch of t at $\omega\alpha$.

[47] When re-organizing $(\mathcal{T}_{\leq \mathcal{Q}})^\frown t^\frown \{b\}$ as a stack, there maybe a drop at $\mathcal{Q}$, as we have to drop to $(\mathcal{Q}|\eta)^\#$.

[48] See Definition 3.8.16. This means that e is the branch of t we must choose.

4. Suppose $\xi \leq \delta_z$ is a limit ordinal and for all $\gamma < \xi$, both $\mathcal{X}_\gamma$ and $\mathcal{Y}_\gamma$ are defined. Then $\mathcal{X}_\xi$ and $\mathcal{Y}_\xi$ are determined as follows[49]. Set $\mathcal{Y}_\xi = lim_{\alpha\to\xi}\mathcal{X}_\alpha$. Assuming clause 2 fails for ξ, $\mathcal{X}_\xi = \mathsf{core}(\mathcal{Y}_\xi)$.

5. $\mathcal{X}_{\delta_z} = \mathcal{Y}_{\delta_z}$ and $F^+_{\delta_z} = F_{\delta_z} = b_{\delta_z} = \emptyset$.

We say that $\mathsf{Le}^{\mathcal{E},c}((\mathcal{Q}|\eta)^\#)$ is *successful* if either for all $\xi < \delta_z$ clause 2 above fails or letting ξ_0 be the least for which clause 2 holds, there is $\xi < \xi_0$ such that $\mathcal{J}_\omega[\mathcal{X}_\xi] \models$ "η is not a Woodin cardinal". The "c" in $\mathsf{Le}^{\mathcal{E},c}$ stands for "certified". Given $\kappa < \delta_z$, we can also define $\mathsf{Le}^{\mathcal{E},c}((\mathcal{Q}|\eta)^\#)_{\geq\kappa}$ by requiring that in clause 3(a), $\mathrm{crit}(F) \geq \kappa$.

We will use the following terminology. We say $\mathcal{K}$ is a $\mathcal{Y}$*-model* of

$$\mathsf{Le}^{\mathcal{E},c}((\mathcal{Q}|\eta)^\#)_{\geq\kappa}$$

if for some $\gamma \leq \delta_z$, $\mathcal{Q} = \mathcal{Y}_\gamma$. Similarly we define $\mathcal{X}$-model and other such expressions. We say $\mathcal{K}$ is *the last model* of $\mathsf{Le}^{\mathcal{E},c}((\mathcal{Q}|\eta)^\#)_{\geq\kappa}$ if $\mathcal{K} = \mathcal{Y}_{\delta_z}$.

We can now define the $\mathcal{E}$-certified iterations.

DEFINITION 10.2.29. Suppose $\mathcal{V} \in \mathcal{N}$ is a hod premouse extending $\mathcal{R}$ such that $\mathcal{R} = \mathcal{V}^b$. Suppose $\mathcal{T} \in \mathcal{N}$ is a stack or an st-stack on $\mathcal{V}$ and $E \in \mathcal{E}$. We say $\mathcal{T}$ is *E-certified* if the following conditions are satisfied.

1. $\mathcal{T}$ is π_E-realizable.
2. Suppose $\tau \in (R^b)^\mathcal{T}$ is such that letting $\mathcal{U} =_{\mathrm{def}} \mathsf{nc}^\mathcal{T}_\tau$, $\mathcal{U}$ is above $\mathcal{M}^b_\tau$. Let $\alpha < \mathrm{lh}(\mathcal{U})$ be a limit ordinal and let $c = [0,\alpha)_\mathcal{U}$. Then the following conditions hold.
   (a) If $\mathrm{m}^+(\mathcal{U}\restriction\alpha) \models$ "$\delta(\mathcal{U}\restriction\alpha)$ is not a Woodin cardinal"[50] then $\mathcal{Q}(c,\mathcal{U}\restriction\alpha)$ exists and $\mathcal{Q}(c,\mathcal{U}\restriction\alpha) \trianglelefteq \mathrm{m}^+(\mathcal{U}\restriction\alpha)$.
   (b) If $\mathrm{m}^+(\mathcal{U}\restriction\alpha) \models$ "$\delta(\mathcal{U}\restriction\alpha)$ is a Woodin cardinal" and there is $\mathcal{W}$ such that
      (i) $\mathcal{W}$ appears on the $\mathsf{Le}^{\mathcal{E},c}(\mathrm{m}^+(\mathcal{U}|\restriction\alpha))$ construction of $\mathcal{N}$ and
      (ii) $\mathcal{W} \models$ "$\delta(\mathcal{U}\restriction\alpha)$ is a Woodin cardinal" but $\mathcal{J}_\omega[\mathcal{W}] \models$ "$\delta(\mathcal{U}\restriction\alpha)$ is not a Woodin cardinal",

      then $\mathcal{Q}(c,\mathcal{U}\restriction\alpha)$ exists and $\mathcal{Q}(c,\mathcal{U}\restriction\alpha) = \mathcal{W}$.
   (c) The above two clauses fail. Then $\mathcal{T}$ is an st-stack, $\alpha+1 = \mathrm{lh}(\mathcal{U})$ and $\tau+\alpha \in R^\mathcal{T} \cap \mathsf{max}^\mathcal{T}$.

We say that $\mathcal{T}$ is $\mathcal{E}$*-certified* if for some λ, $\mathcal{T}$ is E-certified for every $E \in \mathcal{E}$ such that $\mathrm{lh}(E) > \lambda$.

And finally we define $\mathcal{E}$-certified strategies.

DEFINITION 10.2.30. Suppose $\mathcal{V} \in \mathcal{N}$ is a hod premouse extending $\mathcal{R}$ such that $\mathcal{R} = \mathcal{V}^b$. We let $\Lambda_\mathcal{V}$ be the partial strategy of $\mathcal{V}$ with the property that

---

[49] F_ξ, b_ξ will be defined at the next stage of the induction as in clause 2.

[50] See Definition 3.1.4.

1. $\mathrm{dom}(\Lambda_{\mathcal{V}})$ consists of $\mathcal{E}$-certified stacks $\mathcal{T}$ of limit length, and
2. for all $\mathcal{T} \in \mathrm{dom}(\Lambda_{\mathcal{V}})$, $\Lambda_{\mathcal{V}}(\mathcal{T}) = b$ if b is the unique x such that $\mathcal{T}^\frown\{x\}$ is $\mathcal{E}$-certified.

We say $\Lambda_{\mathcal{V}}$ is the $\mathcal{E}$-certified strategy of $\mathcal{V}$.

REMARK 10.2.31. According to our definition, the $\mathcal{E}$-certified strategy of $\mathcal{V}$ is an ordinary strategy acting on smooth iterations. However, it is straightforward to generalize our definition to obtain a strategy acting on generalized stacks[51]. Below we assume that the strategy described in Definition 10.2.30 acts on generalized stacks, and is a partial $(\delta_z, \delta_z, \delta_z)$-iteration strategy. However, there is one subtle point that we address.

$\Lambda_{\mathcal{V}}$, the $\mathcal{E}$-certified strategy of $\mathcal{V}$, must be self-cohering[52]. To achieve this, we use the following idea. Suppose $\mathcal{T}$ is a π_E-realizable generalized stack and F is the un-dropping extender of $\mathcal{T}$. Let $\mathcal{Q}$ be the last model of $\mathcal{T}$. We then have $\sigma : \mathcal{Q}^b \to \mathcal{R}' \trianglelefteq \pi_E(\mathcal{R})$ coming from the definition of π_E-realizability. To ensure that $\Lambda_{\mathcal{V}}$ will be self-cohering it is enough to use $\sigma' : Ult(\mathcal{V}, F)^b \to \pi_E(\mathcal{R})$ given by $\sigma'(x) = \pi_E(f)(\sigma(a))$ where $f \in \mathcal{V}$ and $a \in [\delta^{\mathcal{Q}^b}]^{<\omega}$ is such that $x = \pi_F(f)(a)$.

Suppose now that

$$\mathcal{T} = ((\mathcal{M}_\alpha)_{\alpha<\eta}, (E_\alpha)_{\alpha<\eta-1}, D, R, (\beta_\alpha, m_\alpha)_{\alpha\in R}, T)$$

is π_E-realizable as witnessed by $(\sigma_\alpha : \alpha \in R^b)$ and $(\Psi_\alpha : \alpha \in R^b)$. Using the language of Section 9.1 applied in $\mathrm{Ult}(\mathcal{N}, E)$ to $\pi_E(\mathcal{F} \restriction \kappa)$, it is not hard to see that for $\alpha \in (R^b)^{\mathcal{T}}$, $\mathcal{M}_\alpha^b = \mathcal{P}_{Y_\alpha}$ where $Y_\alpha = \sigma_\alpha[\mathcal{M}_\alpha^b]$ and $\Psi_\alpha = \Sigma_{Y_\alpha}$. It now follows from Lemma 9.1.14 that there are unique $\mathcal{E}$-certified strategies.

LEMMA 10.2.32. *Suppose $\mathcal{V} \in \mathcal{N}$ is a hod premouse extending $\mathcal{R}$ such that $\mathcal{R} = \mathcal{V}^b$. Suppose Λ and Ψ are two $\mathcal{E}$-certified strategies for $\mathcal{V}$. Then $\Lambda = \Psi$.*

It is also not hard to show that $\mathcal{E}$-certified iterations are according to Φ^+, which is the topic of the next subsection.

10.2.6. Correctness of certified strategies. Our goal is still to prove Theorem 10.2.5 and our set up is as in Terminology 10.2.8 and Definition 10.2.17.

LEMMA 10.2.33. *Suppose $\mathcal{S}^*$ is a Φ^+-iterate of $\mathcal{R}^+$ via an iteration that is entirely above $\delta^{\mathcal{R}}$. Suppose further that $\mathcal{S} \trianglelefteq_{hod} \mathcal{S}^*$ is such that $\mathcal{S}^b = \mathcal{R}$ and $\mathcal{S} \in \mathcal{N}$. Let $\mathcal{T} \in \mathcal{N}$ be a stack on $\mathcal{S}$[53]. Suppose $\mathcal{T}$ is $\mathcal{E}$-certified. Then $\mathcal{T}$ is according to $\Phi^+_{\mathcal{S}}$. Thus, $\Lambda_{\mathcal{S}} = \Phi^+_{\mathcal{S}} \cap \mathcal{N}^2$.*[54]

---

[51] See Section 2.10.

[52] See Definition 2.10.11.

[53] We assume that $\mathcal{T}$ is a stack, but the proof works for generalized stacks as well.

[54] This equation does not imply that $\Lambda_{\mathcal{S}} = \Phi^+_{\mathcal{S}} \restriction \mathcal{N}$, simply because it does not imply that if $x \in \mathrm{dom}(\Phi^+_{\mathcal{S}}) \cap \mathcal{N}$ then $\Phi^+_{\mathcal{S}}(x) \in \mathcal{N}$. To get the aforementioned equality, we need to show that $\Lambda_{\mathcal{S}}$ is total.

PROOF. Suppose

$$\mathcal{T} = ((\mathcal{M}_\alpha)_{\alpha<\eta}, (E_\alpha)_{\alpha<\eta-1}, D, R, (\beta_\alpha, m_\alpha)_{\alpha\in R}, T),$$

and suppose $\alpha \in R^b$ is such that $\mathcal{T}_{\leq\alpha}$ is according to $\Phi^+_{\mathcal{S}}$. We want to show that $\mathcal{U} = \mathsf{nc}^{\mathcal{T}}_\alpha$ is according to $\Phi^+_{\mathcal{M}_\alpha}$.

Suppose first that $\mathcal{U}$ is based on $\mathcal{M}^b_\alpha$[55]. Let $E \subset \mathcal{E}$ be such that $\mathcal{T}$ is π_E-realizable as witnessed by $(\sigma_\alpha : \alpha \in R^b)$ and $(\Psi_\alpha : \alpha \in R^b)$. Let E^* be the background certificate of E and let $k : \mathrm{Ult}(\mathcal{N}, E) \to \pi_{E^*}(\mathcal{N})$ be the canonical factor map. Notice that for $\alpha \in R^b$,

(1) $k \restriction \mathcal{N}|\xi = id$ where ξ is the least such that $\mathcal{T} \in \mathcal{N}|\xi$.
(2) In $\pi_{E^*}(\mathcal{N})$, $k(\sigma_\alpha) : \mathcal{M}^b_\alpha \to \pi_{E^*}(\mathcal{N})$ and $k(\Psi_\alpha)$ is the $k(\sigma_\alpha)$-pullback of $\pi_{E^*}(\Phi)$.
(3) $k(\sigma_\alpha) \restriction \mathcal{M}_\alpha|\delta^{\mathcal{M}^b_\alpha}$ is the iteration embedding according to $k(\Psi_\alpha)$.

Let F be the un-dropping extender of $\mathcal{T}_{\leq\alpha}$ and set $\mathcal{K}^+ = \mathrm{Ult}(\mathcal{R}^+, F)$ and $j = \pi^{\Phi^+_{\mathcal{K}^+}}_{\mathcal{K}^+, \pi_{E^*}(\mathcal{R}^+)} \restriction \mathcal{M}_\alpha|\delta^{\mathcal{M}^b_\alpha}$. Notice now that

(4) $\Phi_{\mathcal{M}_\alpha|\delta^{\mathcal{M}^b_\alpha}}$ is the j-pullback of $\pi_{E^*}(\Phi)$ and j is the iteration embedding according to $\Phi_{\mathcal{M}_\alpha|\delta^{\mathcal{M}^b_\alpha}}$.

As the pairs $(k(\sigma_\alpha), k(\Psi_\alpha))$ and $(j, \Phi_{\mathcal{M}_\alpha|\delta^{\mathcal{M}^b_\alpha}})$ have the same property, it follows from Lemma 9.1.14 that $k(\sigma_\alpha) = j$ and $k(\Psi_\alpha) = \Phi_{\mathcal{M}_\alpha|\delta^{\mathcal{M}^b_\alpha}} \restriction \pi_{E^*}(\mathcal{N})$. Since $k(\mathcal{U}) = \mathcal{U}$, in the case $\mathcal{U}$ is based on $\mathcal{M}_\alpha|\delta^{\mathcal{M}^b_\alpha}$, we have that $\mathcal{U}$ is according to $k(\Psi_\alpha)$ and therefore, $\mathcal{U}$ is according to $\Phi_{\mathcal{M}_\alpha|\delta^{\mathcal{M}^b_\alpha}}$, and in the case $\mathcal{U}$ is above $\delta^{\mathcal{M}^b_\alpha}$, we have that $\mathcal{U}$ is according to the unique strategy of $\mathcal{M}^b_\alpha$ that witnesses the fact that $\mathcal{M}^b_\alpha$ is a $\Phi_{\mathcal{M}_\alpha|\delta^{\mathcal{M}^b_\alpha}}$-mouse over $\mathcal{M}_\alpha|\delta^{\mathcal{M}^b_\alpha}$.

Suppose now that $\mathcal{U}$ is above $\mathsf{ord}(\mathcal{M}^b_\alpha)$. Here, we need to see that

(a) if $\beta < \mathrm{lh}(\mathcal{U})$ is a limit ordinal then letting $b = [0,\beta)_{\mathcal{U}}$, either $\mathcal{Q}(b,\mathcal{U}) \trianglelefteq \mathrm{m}^+(\mathcal{U})$ or else $\mathcal{Q}(b,\mathcal{U}) \trianglelefteq \mathsf{Lp}^{\Gamma,(\Phi^+)^{sts}_{\mathrm{m}^+(\mathcal{U})}}(\mathrm{m}^+(\mathcal{U}))$.

The following lemma establishes (a). For convenience, we will ignore the objects introduced above and treat next lemma in a general context. Thus $\mathcal{T}$ in the next lemma is not the $\mathcal{T}$ fixed above.

LEMMA 10.2.34. *Suppose $\mathcal{T}$ is an $\mathcal{E}$-certified iteration of $\mathcal{S}$, $\alpha \in R^b$ and $\mathcal{U} = \mathsf{nc}^{\mathcal{T}}_\alpha$ is above $\mathsf{ord}(\mathcal{M}^b_\alpha)$. Suppose further that $\beta < \mathrm{lh}(\mathcal{U})$ is a limit ordinal and $\mathcal{U}_{<\beta}$ is according to $\Phi^+_{\mathcal{M}_\alpha}$. Let $\mathcal{Q} = \mathcal{M}^{\mathcal{U}}_\beta$ and $\eta > \delta^{\mathcal{Q}^b}$ be such that $\mathcal{J}_\omega[(\mathcal{Q}|\eta)^\#] \vDash$ "η*

[55] There is yet another case: namely, $\alpha = \max R^b$ and $\mathcal{U} = \mathcal{T}_{\geq\alpha}$. But this case is very similar to our two cases.

is a Woodin cardinal" and let $\mathcal{W} \trianglelefteq \mathcal{Q}$ be an sts mouse over $(\mathcal{Q}|\eta)^{\#}$. Then $\mathcal{W}$ is a $(\Phi^+)^{stc}_{(\mathcal{Q}|\eta)^{\#}}$-sts mouse.

PROOF. Towards a contradiction assume that $\mathcal{W}$ is not a $(\Phi^+)^{stc}_{(\mathcal{Q}|\eta)^{\#}}$-sts mouse. It follows that $b = [0,\beta)_{\mathcal{U}}$ is not the branch chosen by $\Phi^+_{\mathcal{S}}$. For convenience, we change our notation and let $\mathcal{U}$ be $\mathcal{U} \restriction \beta$ and $\mathcal{Q} = \mathrm{m}^+(\mathcal{U})$. It follows from Definition 10.2.29 that

(1) $\mathcal{W}$ is a model appearing in the fully backgrounded $\mathcal{E}$-realizable construction over $(\mathcal{Q}|\eta)^{\#}$ done in $\mathcal{N}$.

What we need to see is that $\mathcal{W}$ is a $(\Phi^+)^{stc}_{\mathcal{Q}}$-sts mouse over $\mathcal{Q}$. To show this it is enough to show that every stack indexed in $\mathcal{W}$ is according to $(\Phi^+)^{stc}_{\mathcal{Q}}$. To show this later fact, it is enough to show that

(b) if $t = (\mathcal{Q}, \mathcal{U}_0, \mathcal{Q}_1, \mathcal{U}_1)$ is an indexable stack[56] on $\mathcal{Q}$ appearing in the fully backgrounded $\mathcal{E}$-realizable construction over $\mathcal{Q}$ (done in $\mathcal{N}$) and c is the branch of t indexed in this construction then $t^\frown\{c\}$ is according to $(\Phi^+)^{stc}_{\mathcal{Q}}$.

(b) is indeed enough. To see this, notice that if $s = (\mathcal{Q}, \mathcal{U}'_0, \mathcal{Q}'_1, \mathcal{U}'_1)$ is indexed in $\mathcal{W}$ and c' is the branch of s indexed in $\mathcal{W}$ then for some stack $t = (\mathcal{Q}, \mathcal{U}_0, \mathcal{Q}_1, \mathcal{U})$ as in (b) if e is the branch of t then $s^\frown\{c\}$ is a hull of $t^\frown\{e\}$. If t is according to $(\Phi^+)^{stc}_{\mathcal{Q}}$ then it follows from hull condensation of $(\Phi^+)^{stc}_{\mathcal{Q}}$ that s is also according to $(\Phi^+)^{stc}_{\mathcal{Q}}$. We now work towards showing that t is according to $(\Phi^+)^{stc}_{\mathcal{Q}}$.

Suppose first that $\mathcal{U}_0$ is according to $(\Phi^+)^{stc}_{\mathcal{Q}}$. We then have that $\mathcal{U}_1$ is a stack based on $\mathcal{Q}^b_1$. Because $(\mathcal{T}_{\leq\alpha})^\frown t$ is $\mathcal{E}$-certified, we can fix an extender $E \in \mathcal{E}$ such that $(\mathcal{T}_{\leq\alpha})^\frown t$ is π_E-realizable. We then have $\sigma : \mathcal{Q}^b_1 \to \pi_E(\mathcal{R})$ such that $\pi_E \restriction \mathcal{R} = \sigma \circ \pi^{\mathcal{U}_0,b} \circ \pi^{\mathcal{T}_{\leq\mathcal{Q}},b}$. We also have that $\mathcal{U}_1^\frown\{c\}$ is according to the σ-pullback of $\pi_E(\Phi_{\mathcal{R}})$. Therefore, t is according to $(\Phi^+)^{stc}_{\mathcal{Q}}$.

It remains to show that $\mathcal{U}_0$ is according to $(\Phi^+)^{stc}_{\mathcal{Q}}$. Without loss of generality, we assume that

1. $\mathrm{lh}(\mathcal{U}_0) = \gamma + 1$,
2. γ is a limit ordinal,
3. $\mathcal{U}_0 \restriction \gamma$ is according to $(\Phi^+)^{stc}_{\mathcal{Q}}$,
4. $[0,\gamma)_{\mathcal{U}_0} \neq (\Phi^+)^{stc}_{\mathcal{Q}}(\mathcal{U}_0)$,
5. there is $\zeta \in R^{\mathcal{U}_0 \restriction \gamma}$ such that $(\mathcal{U}_0)_{\geq\zeta} = \mathsf{nc}^{\mathcal{U}_0}_{\zeta}$ and $\pi^{\mathcal{U}_0,b}$ exists,
6. $\mathcal{J}_\omega[\mathrm{m}^+(\mathcal{U}_0)] \models$ "$\delta(\mathcal{U}_0)$ is a Woodin cardinal".

The last two clauses can be shown by examining the proof given for $\mathcal{U}_1$. Set $c_0 = [0,\gamma)_{\mathcal{U}_0}$, $\mathcal{Q}_0 =_{\mathrm{def}} \mathcal{Q}$, $\mathcal{Q}_2 = \mathrm{m}^+(\mathcal{U}_0)$, $\mathcal{W}_0 =_{\mathrm{def}} \mathcal{W}$ and $\mathcal{W}_2 = \mathcal{Q}(c_0, \mathcal{U}_0)$. We then have that

(2) $\mathcal{W}_2$ appears in the fully backgrounded $\mathcal{E}$-realizable construction over $\mathcal{Q}_2$ (done in $\mathcal{N}$).

[56] See Definition 3.7.5.

Clearly (2) leads to an infinite descend. ⊣

This completes the proof of Lemma 10.2.33. ⊣

REMARK 10.2.35. It is important to note that Lemma 10.2.34 does not resolve the Important Anomaly stated in clause 3(b) of Definition 10.2.28.

The following straightforward yet important lemmas are steps towards showing that the Important Anomaly stated in clause 3(b) of Definition 10.2.28 does not occur. The reader may wish to review Definition 3.7.3 and Definition 3.3.2.

LEMMA 10.2.36. *Suppose*

1. *$\mathcal{V} \in \mathcal{N}$ is a hod premouse extending $\mathcal{R}$ such that $\mathcal{R} = \mathcal{V}^b$,*
2. *$\mathcal{T} \in \mathcal{N}$ is either a stack on $\mathcal{V}$ or an st-stack*[57] *on $\mathcal{V}$*[58]*,*
3. *$\pi^{\mathcal{T},b}$ is defined, $\mathcal{T}$ has a last model and $\mathcal{E}$-realizable.*

Let $\mathcal{S}$ be the last model of $\mathcal{T}$ and suppose $\mathcal{Q}$ is authenticated[59] *by $\mathcal{T}$ and is meek and of limit type*[60]*. Then $\mathcal{W},\mathcal{U},\sigma$ be as in Definition 3.7.3 and letting $k : \mathcal{R} \to \mathcal{Q}$ be given by $k(x) = y$ if and only $\sigma^{-1}(\pi^{\mathcal{T},b}(x)) = \pi^{\mathcal{U}}(y)$, $(k,\mathcal{Q})$ is $\mathcal{E}$-realizable.*

The reader may wish to review Notation 2.4.4 and Definition 10.2.30. The lemma below implies that the Important Anomaly stated in clause 3(b) of Definition 10.2.28 does not occur.

LEMMA 10.2.37. *Suppose*

1. *$\mathcal{V} \in \mathcal{N}$ is a hod premouse extending $\mathcal{R}$ such that $\mathcal{R} = \mathcal{V}^b$,*
2. *$\mathcal{T} \in \mathcal{N}$ is either a stack on $\mathcal{V}$ or an st-stack*[61] *on $\mathcal{V}$*[62]*,*
3. *$\pi^{\mathcal{T},b}$ is defined, $\mathcal{T}$ has a last model and $\mathcal{E}$-realizable.*

Let $\mathcal{S}'$ be the last model of $\mathcal{T}$ and suppose $\eta < \mathrm{ord}(\mathcal{S}')$ is such that

$$\mathcal{J}_\omega[(\mathcal{S}'|\eta)^\#] \models \text{“}\eta \text{ is a Woodin cardinal”}.$$

Suppose $\mathcal{S} =_{def} (\mathcal{S}'|\eta)^\#$ is such that $\mathcal{S} \trianglelefteq_{hod} \mathcal{S}'$ and $\mathcal{U} \in \mathcal{N}$ is an nuvs *stack according to $(\Lambda_\mathcal{V})_\mathcal{S}$ such that $\pi^{\mathcal{U},b}$ is defined. Let $\mathcal{Q} = \mathrm{m}^+(\mathcal{U})$ and suppose $t \in \mathcal{N}$ be an indexable stack on $\mathcal{Q}$ which is $(\mathcal{S},(\Lambda_\mathcal{V})_\mathcal{S})$-authenticated*[63]*. Then $\mathcal{T}^\frown\mathcal{U}^\frown t$ is according to $\Lambda_\mathcal{V}$.*

PROOF. Suppose $t = (\mathcal{Q}_0, \mathcal{X}_0, \mathcal{Q}_1, \mathcal{X}_1)$. Assume first that $\mathcal{X}_0$ is according to $(\Lambda_\mathcal{V})_\mathcal{Q}$. Set $p = \mathcal{T}^\frown\mathcal{U}^\frown\mathcal{X}_0$ and let $\sigma = \pi^{p,b}$. It follows from Lemma 10.2.33 that p is according to $\Lambda_\mathcal{V}$, and the previous lemma implies that $(\sigma, \mathcal{Q}_1^b) \in \mathcal{F}_\mathcal{E}$. Because

[57] See Definition 3.2.1.

[58] If $\mathcal{T}$ is an st-stack then $\mathcal{M}$ must be of lsa type.

[59] See Definition 3.7.3.

[60] Thus, clause 3 of Definition 3.7.3 holds.

[61] See Definition 3.2.1.

[62] If $\mathcal{T}$ is an st-stack then $\mathcal{V}$ must be of #-lsa type.

[63] Notice that we, at this point, do not know that $\Lambda_\mathcal{V}$ is a total strategy in $\mathcal{J}[\mathcal{N}]$.

$(\mathcal{Q}_1^b, \mathcal{X}_1)$ is a $(\mathcal{S}, (\Lambda_\mathcal{V})_\mathcal{S})$-authenticated iteration, it follows from Lemma 10.2.33 that $\mathcal{X}_1$ is according to $\Psi_{\sigma,\mathcal{Q}_1^b}$, and therefore, $\mathcal{T}^\frown\mathcal{U}^\frown t$ is according to $\Lambda_\mathcal{V}$.

Thus, it is enough to show that $\mathcal{X}_0$ is according to $(\Lambda_\mathcal{V})_\mathcal{Q}$. The argument given above implies that it is enough to show that for every $\alpha \in R^{\mathcal{X}_0}$ such that $\pi_{0,\alpha}^{\mathcal{X}_0,b}$ is defined and $\mathsf{nc}_\alpha^{\mathcal{X}_0}$ is a stack on $\mathcal{M}_\alpha^{\mathcal{X}_0}$ above $\mathsf{ord}((\mathcal{M}_\alpha^{\mathcal{X}_0})^b)$ then $\mathsf{nc}_\alpha^{\mathcal{X}_0}$ is according to $(\Lambda_\mathcal{V})_{\mathcal{M}_\alpha^{\mathcal{X}_0}}$.

Assume then α is as above and $(\mathcal{X}_0)_{\leq\alpha}$ is according to $(\Lambda_\mathcal{V})_\mathcal{Q}$. Set $\mathcal{M} = \mathcal{M}_\alpha^{\mathcal{X}_0}$, $\mathcal{X} = (\mathcal{X}_0)_{\leq\alpha}$ and let $\mathcal{Y} = \mathsf{nc}_\alpha^{\mathcal{X}_0}$. We want to see that $\mathcal{Y}$ is according to $(\Lambda_\mathcal{V})_\mathcal{M}$. Let $\beta < \mathrm{lh}(\mathcal{Y})$ be a limit ordinal such that $\mathcal{Y}_{<\beta}$ is according to $(\Lambda_\mathcal{V})_\mathcal{M}$. We want to see that if $b = [0,\beta)_\mathcal{Y}$ then $b = (\Lambda_\mathcal{V})_\mathcal{M}(\mathcal{Y}_{<\beta})$. The dificult case is when $\mathcal{Q}(b, \mathcal{Y}_{<\beta})$ exists and is an sts mouse over $\mathrm{m}^+(\mathcal{Y}_{<\beta})$. In this case, we want to see that $\mathcal{Q}(b, \mathcal{Y}_{<\beta})$ is a model appearing in the fully backgrounded $\mathcal{E}$-realizable construction over $\mathrm{m}^+(\mathcal{U}_0)$ (done in $\mathcal{N}$). This would follows from the proof of the previous lemma. Our strategy for showing this is by showing (a) and (b) where these are the following statements:

(a) $\mathcal{Q}(b, \mathcal{Y}_{<\beta})$ is a $(\Phi^+_{\mathrm{m}^+(\mathcal{Y}_{<\beta})})^{stc}$-mouse over $\mathrm{m}^+(\mathcal{Y}_{<\beta})$.

(b) If $\mathcal{W}$ is a $(\Phi^+_{\mathrm{m}^+(\mathcal{Y}_{<\beta})})^{stc}$-mouse over $\mathrm{m}^+(\mathcal{Y}_{<\beta})$ then $\mathcal{W}$ appears in the fully backgrounded $\mathcal{E}$-realizable construction over $\mathrm{m}^+(\mathcal{U}_0)$ (done in $\mathcal{N}$). More precisely, letting

$$\mathsf{Le}^{\mathcal{E},c}(\mathrm{m}^+(\mathcal{Y}_{<\beta})) = (\mathcal{Z}_\gamma, \mathcal{K}_\gamma, F_\gamma^+, F_\gamma, b_\gamma : \gamma \leq \delta_z)$$

be the fully backgrounded $\mathcal{E}$-realizable construction over $\mathrm{m}^+(\mathcal{Y}_{<\beta})$ done in $\mathcal{N}$ then for some $\gamma < \delta_z$, $\mathcal{Z}_\gamma = \mathcal{W}$.

(a) is a consequence of strong branch condensation of Φ^+ and can be shown using the proof of Sublemma 4.12.4 and Lemma 10.2.33.

(b) is a consequence of the fact that $\Lambda_\mathcal{V}$ is total, and hence $\Phi_\mathcal{V}^+ \restriction \mathcal{J}[\mathcal{N}] = \Lambda_\mathcal{V}$ (see Lemma 10.2.33). Assuming that $\Lambda_\mathcal{V}$ is total, (b) can be proven by simply comparing $\mathcal{W}$ with the $\mathsf{Le}^{\mathcal{E},c}(\mathrm{m}^+(\mathcal{Y}_{<\beta}))$ construction. The stationarity of $\mathsf{Le}^{\mathcal{E},c}(\mathrm{m}^+(\mathcal{Y}_{<\beta}))$ implies that the construction side does not move, and the fact that $\Lambda_\mathcal{V}$ is total implies that the construction does not break down because in clause 3(b) of Definition 10.2.28 we are unable to find the desired branch. The Important Anomaly stated in clause 3(b) of Definition 10.2.28 does not occur (at least does not occur before reaching $\mathcal{W}$) because these type of branches are chosen internally and both the construction side and the $\mathcal{W}$-side must be choosing the same branch. But on the $\mathcal{W}$-side, the branch is according to $\Phi_\mathcal{V}^+$ and therefore, according to $\Lambda_\mathcal{V}$. In the next subsection, we will prove that $\Lambda_\mathcal{V}$ is total, and more details will be given. $\dashv$

10.2.7. $\Lambda_\mathcal{V}$ is total. The goal of this subsection is to show that $\Lambda_\mathcal{V}$, the $\mathcal{E}$-certified strategy of $\mathcal{V}$, is total.

LEMMA 10.2.38. *$\mathcal{V} \in \mathcal{N}$ is a hod premouse extending $\mathcal{R}$ such that $\mathcal{R} = \mathcal{V}^b$. Then $\Lambda_{\mathcal{V}}$ is total, and hence, $\Phi^+_{\mathcal{V}} \restriction \mathcal{N} = \Lambda_{\mathcal{V}}$.*

PROOF. The equality $\Phi^+_{\mathcal{V}} \restriction \mathcal{N} = \Lambda_{\mathcal{V}}$ follows from Lemma 10.2.33. Suppose $\mathcal{T} \in \mathcal{N}$ is a stack according to $\Lambda_{\mathcal{V}}$ and of limit length. We want to show that $\Lambda_{\mathcal{V}}(\mathcal{T})$ is defined. Let $b = \Phi^+_{\mathcal{V}}(\mathcal{T})$. It is enough to show that $b \in \mathcal{N}$ and $\mathcal{T}^\frown\{b\}$ is $\mathcal{E}$-certified. When discussing objects in $\mathcal{T}$, we omit $\mathcal{T}$ from superscripts. The two non-straightforward cases are the following:

Case 1: There is $\alpha \in R$ such that $\pi^{\mathcal{T},b}_{0,\alpha}$ is defined and $\mathcal{T}_{\geq\alpha}$ is based on $\mathcal{M}^b_\alpha$.

Case 2: There is $\alpha \in R$ such that $\alpha = \max(R)$, $\pi^{\mathcal{T},b}_{0,\alpha}$ is defined, $\mathcal{T}_{\geq\alpha}$ is above $\mathsf{ord}(\mathcal{M}^b_\alpha)$, $\mathcal{T}_{\geq\alpha}$ does not have a fatal drop and $\mathcal{J}_\omega[\mathrm{m}^+(\mathcal{T}_{\geq\alpha}))] \vDash$ "$\delta(\mathcal{T}_{\geq\alpha})$ is a Woodin cardinal".

Lemma 10.2.25 implies that under Case 1, $b \in \mathcal{N}$ and $\mathcal{T}$ is $\mathcal{E}$-certified. Assume then Case 2. Let $\mathcal{W} = \mathcal{Q}(b, \mathcal{T})$. We have that $\mathcal{W}$ is a $(\Phi^+_{\mathrm{m}^+(\mathcal{T})})^{stc}$-mouse over $\mathrm{m}^+(\mathcal{T})$. It is now enough to show that $\mathcal{W}$ appears in the $\mathsf{Le}^{\mathcal{E},c}(\mathrm{m}^+(\mathcal{T}))$ construction of $\mathcal{N}$. Let

$$\mathsf{Le}^{\mathcal{E},c}(\mathrm{m}^+(\mathcal{T})) = (\mathcal{X}_\gamma, \mathcal{Y}_\gamma, F^+_\gamma, F_\gamma, b_\gamma : \gamma \leq \delta_z)$$

be the $\mathsf{Le}^{\mathcal{E},c}(\mathrm{m}^+(\mathcal{T}))$ construction of $\mathcal{N}$. For each $\gamma \leq \delta_z$, we let $\mathcal{U}_\gamma$ be the normal stack on $\mathcal{W}$ above $\delta(\mathcal{T})$ that is constructed by comparing $\mathcal{W}$ with $\mathcal{Y}_\gamma$. What we need to show is that

(a) the construction side never moves, i.e., for any $\gamma < \delta_z$, letting $\mathcal{W}_\gamma$ be the last model of $\mathcal{U}_\gamma$, if for some τ, $\mathcal{W}_\gamma|\tau = \mathcal{Y}_\gamma|\tau$ and $\mathcal{W}_\gamma||\tau \neq \mathcal{Y}_\gamma||\tau$ then $\tau \notin \mathrm{dom}(\vec{E})^{\mathcal{Y}_\gamma}$ and τ is not an index of a branch in $\mathcal{W}_\gamma$.

The fact that $\tau \notin \mathrm{dom}(\vec{E})^{\mathcal{Y}_\gamma}$ follows from the stationarity of the fully backgrounded constructions[64]. It is then enough to show that τ is not an index of a branch in $\mathcal{W}_\gamma$. Suppose to the contrary, and let

(1) $t \in \mathcal{W}_\gamma$ be an indexable stack whose branch is indexed in $\mathcal{W}_\gamma$ at τ.

Because $\mathcal{W}_\gamma|\tau = \mathcal{Y}_\gamma|\tau$, it follows that all initial segments of t are according to $(\Lambda_{\mathcal{V}})_{\mathrm{m}^+(\mathcal{T})}$. We need to show that

(b) there is a branch indexed at τ in $\mathcal{Y}_\gamma$ and that branch is according to $(\Phi^+_{\mathrm{m}^+(\mathcal{T})})^{stc}$.

Again the two difficult cases are the cases stated under Case 1 and Case 2, and Case 1 can be analyzed as above, and so we only state Case 2. We thus have that

(2) setting $\mathcal{K} = \mathrm{m}^+(\mathcal{T})$, $t = (\mathcal{K}, \mathcal{Z})$ and $\mathcal{J}_\omega[\mathrm{m}^+(\mathcal{Z}_0)] \vDash$ "$\delta(\mathcal{Z}_0)$ is a Woodin cardinal".

[64] See Theorem 4.5.6 and the references given there.

Notice that $t \in \mathcal{W}_\gamma$ and the branch chosen for t both in $\mathcal{W}_\gamma$ and in $\mathcal{Y}_\gamma$ depends on $\mathcal{W}_\gamma|\tau = \mathcal{Y}_\gamma|\tau$. Hence, both $\mathcal{W}_\gamma$ and $\mathcal{Y}_\gamma$ must have the same branch of t indexed in their strategy predicates. ⊣

10.2.8. Mixed hod pair constructions. We devote this entire subsection to the definition of a construction producing the iterate of $\mathcal{R}^+$. In this construction, we use $\mathcal{E}$-certification method to acquire extenders with critical point $\delta^{\mathcal{R}}$, and we use the total extenders on the sequence of $\mathcal{N}$ to generate extenders with critical points $> \delta^{\mathcal{R}}$. First we define $\mathcal{E}$-certified extenders. The reader may wish to review Definition 10.2.26.

DEFINITION 10.2.39. Suppose $\mathcal{Q} \in \mathcal{N}$ is a hod premouse such that $\Lambda_{\mathcal{Q}}$ (see Definition 10.2.30) is total and $\mathcal{Q}^b = \mathcal{R}$. Suppose F is an extender such that $(\mathcal{Q}, \tilde{F})$ is a reliable lses where $\tilde{F}$ is the amenable code of F. We say F is *$\mathcal{E}$-certified* if

1. $(\pi_F \restriction \mathcal{R}, \pi_F(\mathcal{R})) \in \mathcal{F}_{\mathcal{E}}$ and
2. for some $\mathcal{N}$-strong cardinal λ, for any $E \in \mathcal{E}$ such that $\mathrm{lh}(E) > \lambda$, setting $\tau = \tau_{\pi_F \restriction \mathcal{R}, \pi_F(\mathcal{R})}$[65],

$$(a, A) \in F \iff \tau(a) \in \pi_E(A).$$

We say that τ is the *E-realizability map* of F.

The next lemma shows that $\mathcal{E}$-certified extenders are on the sequence of $\mathcal{R}^+$ and its iterates.

LEMMA 10.2.40. *Suppose $\mathcal{S}^* \in pI(\mathcal{R}^+, \Phi^+)$ and $\mathcal{S} \lhd_{hod} \mathcal{S}^*$ is such that $\mathcal{S} \in \mathcal{N}$ and $\mathcal{S}^b = \mathcal{R}$. Suppose F is such that $(\mathcal{S}, \tilde{F})$ is a reliable lses where $\tilde{F}$ is the amenable code of F and F is $\mathcal{E}$-certified. Then $F \in \vec{E}^{\mathcal{S}^*}$.*

PROOF. Let $\gamma = \mathrm{ord}(\mathcal{S})$ and suppose $F^* \in \vec{E}^{\mathcal{S}^*}(\gamma)$. Then F^* has exactly the same property as F and therefore, $F = F^*$. Thus, it is enough to show that $\gamma \in \mathrm{dom}(\vec{E}^{\mathcal{S}^*})$. Suppose first that there is $\gamma' \in \mathrm{dom}(\vec{E}^{\mathcal{S}^*})$ such that $\mathcal{S} \lhd_{hod} \mathcal{S}^*|\gamma'$ and if $G' = \vec{E}^{\mathcal{S}^*}(\gamma')$ then $\mathrm{crit}(G') = \delta^{\mathcal{R}}$. Let γ^* be the least such γ' and set $G = \vec{E}^{\mathcal{S}^*}(\gamma^*)$. As F and G both have the property described in Definition 10.2.39, F is an initial segment of G, and therefore, $\gamma = \gamma^*$ and $\gamma \in \mathrm{dom}(\vec{E}^{\mathcal{S}^*})$. Suppose then that

(1) there is no $\gamma' \in \mathrm{dom}(\vec{E}^{\mathcal{S}^*})$ such that $\mathrm{crit}(\vec{E}^{\mathcal{S}^*}(\gamma')) = \delta^{\mathcal{R}}$.

Because F is $\mathcal{E}$-certified, we have that for some $\mathcal{N}$-strong cardinal λ, whenever $E \in \mathcal{E}$ is such that $\mathrm{lh}(E) > \lambda$, some proper initial segment of $\pi_E(\mathcal{R})$ is a $\Phi^+_{\mathcal{S}}$-iterate of $\mathcal{S}$. Therefore, $(\mathcal{S}, \Phi^+_{\mathcal{S}})$ is in HP^{Γ}, and hence, $(\mathcal{S}^*, \Phi^+_{\mathcal{S}^*}) \in \mathsf{HP}^{\Gamma}$[66]. This is because (1) implies that $\mathcal{S}^* \lhd \mathsf{Lp}^{\Gamma, \Phi^+_{\mathcal{S}}}(\mathcal{S})$ or $\mathcal{S}^* \lhd \mathsf{Lp}^{\Gamma, (\Phi^+_{\mathcal{S}})^{stc}}(\mathcal{S})$. ⊣

Next we introduce the mixed hod pair constructions.

[65] See Definition 10.2.26.
[66] Γ was introduced in Notation 10.2.2.

DEFINITION 10.2.41. We say that

$$\mathsf{mhpc} = (\mathcal{M}_\gamma, \mathcal{N}_\gamma, Y_\gamma, \Phi_\gamma, F_\gamma^+, F_\gamma, b_\gamma : \gamma \leq \delta)$$

is the output of the *mixed hod pair construction* of $\mathbb{N}$ over $\mathcal{R}$ if the following conditions hold.

1. $\mathcal{M}_0 = \mathcal{J}_\omega[\mathcal{R}]$, and for all $\gamma \leq \delta$, each of $\mathcal{M}_\gamma$ and $\mathcal{N}_\gamma$ is either undefined or is an hp-indexed lses (see Definition 3.9.2).
2. For all $\gamma \leq \delta$, if $\mathcal{M}_\gamma$ is defined then $Y_\gamma = Y^{\mathcal{M}_\gamma}$ (see Definition 2.3.13).
3. For all $\gamma \leq \delta$, if $\mathcal{M}_\gamma$ is defined then $\Phi_\gamma = \Phi_{\mathcal{M}_\gamma}$ is the $\mathcal{E}$-certified strategy of $\mathcal{M}_\gamma$[67].
4. For all $\gamma \leq \delta$, if $\mathcal{N}_\gamma$ is defined and either

(a) $\mathcal{N}_\gamma$ is not a reliable hp-indexed lses[68] or

(b) $\mathcal{N}_\gamma$ is a reliable hp-indexed lses but for some $\mathcal{Q} \in Y^{\mathcal{N}_\gamma}$ such that $\mathcal{Q}$ is meek or gentle[69] and for some $n < \omega$, $\rho_n(\mathcal{N}_\gamma) \leq \delta^{\mathcal{Q}}$, or

(c) Φ_γ is not total,

then all remaining objects with index $\geq \gamma$ are undefined.

For all $\gamma \leq \eta$ for which clause 4 (the above statement) fails, $\pi_\gamma : \mathsf{core}(\mathcal{N}_\gamma) \to \mathcal{N}_\gamma$ is the uncollapse map.

5. Suppose for some $\xi < \delta$, for all $\gamma \leq \xi$, both $\mathcal{M}_\gamma, \mathcal{N}_\gamma$ are defined. Then $\mathcal{M}_{\xi+1}, \mathcal{N}_{\xi+1}, Y_{\xi+1}, \Phi_{\xi+1}, F_\xi^+, F_\xi$ and b_ξ are determined as follows.

(a) Suppose $\mathcal{M}_\xi = (\mathcal{J}_{\omega\alpha}^{\vec{E},f}, \in, \vec{E}, f, Y_\xi, \in)$ is a passive hp-indexed lses[70], there is an extender $H^* \in \mathcal{E}$ an extender H over $\mathcal{M}_\xi$, and an ordinal $\nu < \omega\alpha$ such that $\nu < \mathrm{lh}(H^*)$ and setting

$$H = H^* \cap ([\nu]^\omega \times \lfloor \mathcal{M}_\xi \rfloor) \quad \text{and} \quad \mathcal{N}_{\xi+1} = (\mathcal{J}_{\omega\alpha}^{\vec{E},f}, \in, \vec{E}, f, Y_\xi, \tilde{H}, \in)$$

where $\tilde{H}$ is the amenable code of H, clause 4(a) fails for $\xi+1$. Then letting $\iota \in \mathrm{dom}(\vec{E}^{\mathcal{N}})$ be the least such that $H^* =_{\mathrm{def}} \vec{E}^{\mathcal{N}}(\iota) \in \mathcal{E}$ has the above properties,

$$\mathcal{N}_{\xi+1} = (\mathcal{J}_{\omega\alpha}^{\vec{E},f}, \in, \vec{E}, f, Y_\xi, \tilde{H}, \in)$$

where $\tilde{H}$ is the amenable code of H[71]. Assuming clause 4 fails for $\xi+1$, the remaining objects are defined as follows.

(i) $\mathcal{M}_{\xi+1} = \mathsf{core}(\mathcal{N}_{\xi+1})$[72],

(ii) $F_\xi^+ = H^*$ and $F_\xi = H$,

(iii) $b_\xi = \emptyset$ and

[67] See Definition 10.2.30.

[68] Recall clause 2 of Definition 2.5.4. To verify that $\mathcal{N}_\gamma$ is lses, we need to verify that clause 2 of Definition 2.5.4 holds.

[69] See Definition 2.7.1.

[70] That is, with no last predicate. See Definition 3.9.2.

[71] Here H is what is determined by H^*. For the definition of the "amenable code" see the last paragraph on page 14 of [59].

[72] Recall that $\mathsf{core}(\mathcal{M})$ is the core of $\mathcal{M}$.

(iv) $Y_{\xi+1} = \pi_{\xi+1}^{-1}(Y_\xi)$.

(b) Suppose $\mathcal{M}_\xi = (\mathcal{J}_{\omega\alpha}^{\vec{E},f}, \in, \vec{E}, f, Y_\xi, \in)$ is a passive hp-indexed lses[73] and there is an extender H over $\mathcal{M}_\xi$ such that setting

$$\mathcal{N}_{\xi+1} = (\mathcal{J}_{\omega\alpha}^{\vec{E},f}, \in, \vec{E}, f, Y_\xi, \tilde{H}, \in)$$

where $\tilde{H}$ is the amenable code of H, clause 4(a) fails for $\xi+1$ and H is $\mathcal{E}$-certified as defined in Definition 10.2.39. Assuming clause 4 fails for $\xi+1$, the remaining objects are defined as follows.

(i) $\mathcal{M}_{\xi+1} = \mathsf{core}(\mathcal{N}_{\xi+1})$[74],

(ii) $F_\xi^+ = H^*$ and $F_\xi = H$,

(iii) $b_\xi = \emptyset$ and

(iv) $Y_{\xi+1} = \pi_{\xi+1}^{-1}(Y_\xi)$.

(c) Suppose $\mathcal{M}_\xi = (\mathcal{J}_{\omega\alpha}^{\vec{E},f}, \in, \vec{E}, f, Y_\xi, \in)$ is a passive hp-indexed lses, $\mathcal{M}_\xi$ is strategy-ready[75], $\alpha = \beta + \gamma$ and there is $t \in \lfloor \mathcal{M}_\xi | \omega\beta \rfloor$ such that setting $w = (\mathcal{J}_\omega(t), t, \in)$, w is (f, hp)-minimal as witnessed by β[76] and $\gamma = \mathrm{lh}(t)$. Set $b = \Phi_\xi(t)$ and

$$\mathcal{N}_{\xi+1} = (\mathcal{J}_{\omega\beta+\omega\gamma}^{\vec{E},f^+}, \in, \vec{E}, f, Y_\xi, \tilde{b}, \in)$$

where $\tilde{b} \subseteq \omega\beta + \omega\gamma$ is defined by $\omega\beta + \omega\nu \in \tilde{b} \iff \nu \in b$. Assuming clause 4 fails for $\xi+1$, the remaining objects are defined as follows.

(i) $\mathcal{M}_{\xi+1} = \mathsf{core}(\mathcal{N}_{\xi+1})$,

(ii) $F_\xi = F_\xi^+ = \emptyset$,

(iii) $b_\xi = \tilde{b}$ and

(iv) $Y_{\xi+1} = \pi_{\xi+1}^{-1}(Y_\xi)$.

Important Anomaly: Suppose $\cup Y_\xi$ is #-lsa type[77] and t is nuvs. Suppose $e \in \mathcal{M}_\xi | \omega\beta$ is such that $\mathcal{M}_\xi | \omega\beta \vDash \mathsf{sts}_0(t, e)$[78]. If $e \neq b$ then $\mathcal{N}_{\xi+1}$ is not an sts premouse over $\mathcal{J}_\omega(\cup Y_\xi)$ based on $\cup Y_\xi$, and so the construction must stop.

(d) If $\mathcal{M}_\xi$ does not satisfy clause 2(a), 2(b) or 2(c) then set $\mathcal{N}_{\xi+1} = \mathcal{J}_\omega[\mathcal{M}_\xi]$ (this presupposes that $Y^{\mathcal{N}_{\xi+1}} = Y_\xi$). Assuming clause 4 fails for $\xi+1$, the remaining objects are defined as follows.

(i) $\mathcal{M}_{\xi+1} = \mathsf{core}(\mathcal{N}_{\xi+1})$[79],

(ii) $F_\xi = F_\xi^+ = \emptyset$,

[73] That is, with no last predicate.

[74] Recall that $\mathsf{core}(\mathcal{M})$ is the core of $\mathcal{M}$.

[75] See Definition 3.9.1.

[76] See Definition 2.3.3. In particular, this means that we have to index the branch of t at $\omega\alpha$.

[77] See Definition 2.7.3.

[78] See Definition 3.8.16. This means that e is the branch of t we must choose.

[79] Recall that $\mathsf{core}(\mathcal{M})$ is the core of $\mathcal{M}$.

(iii) $b_\xi = \emptyset$,
and $Y_{\xi+1} = \pi^{-1}_{\xi+1}(Y_\xi) \cup \{\pi^{-1}_{\xi+1}(\mathcal{M}_\xi)$ in the case $\mathcal{M}_{\xi+1}$ is a hod premouse and otherwise, $Y_{\xi+1} = \pi^{-1}_{\xi+1}(Y_\xi)$.

6. Suppose $\xi \leq \delta$ is a limit ordinal and for all $\gamma < \xi$, both $\mathcal{M}_\gamma$ and $\mathcal{N}_\gamma$ are defined. Then $\mathcal{M}_\xi$ and $\mathcal{N}_\xi$ are determined as follows[80]. Set $\mathcal{N}_\xi = lim_{\alpha\rightarrow\xi}\mathcal{M}_\alpha$. Assuming clause 4 fails for $\zeta + 1$, the remaining objects are defined as follows.
(a) $\mathcal{M}_\xi = \mathsf{core}(\mathcal{N}_\xi)$ and
(b) $Y_\xi = \pi_\xi^{-1}(Y^{\mathcal{N}_\xi})$[81].

7. $\mathcal{M}_\delta = \mathcal{N}_\delta$ and $Y_\delta, \Phi_\delta, F^+_\delta, F_\delta$, and b_δ are undefined.

We say that the mhpc is *successful* if for some γ, $\mathcal{M}_\gamma$ is a Φ^+-iterate of $\mathcal{R}^+$.

The following is the main fact we need, which is a corollary to several lemmas established before.

LEMMA 10.2.42. mhpc *is successful.*

PROOF. The lemma follows easily from Lemma 10.2.40, Lemma 10.2.38 and (b) that appears in the proof of Lemma 10.2.37 (which was also established in the proof of Lemma 10.2.38).

To prove the lemma, we simply compare $\mathcal{R}^+$ with mhpc-construction of $\mathcal{N}$ and argue that mhpc side reaches an iterate of $\mathcal{R}^+$. As all extender used in mhpc with critical point $> \mathsf{ord}(\mathcal{R})$ have background certificates, the usual stationarity argument shows that such extenders cannot be part of a disagreement in the resulting comparison process. Lemma 10.2.40 shows that extenders with critical point $\delta^{\mathcal{R}}$ also cannot be part of a disagreement, while Lemma 10.2.38 shows that there cannot be a strategy disagreement. Therefore, $\mathcal{R}^+$ iterates to some model appearing on the mhpc-construction. ⊣

Lemma 10.2.42 and Lemma 10.2.38 now imply Theorem 10.2.5, and this finishes our proof of Theorem 10.2.5.

10.2.9. A proof of Lemma 10.2.6. In this subsection we outline the proof of Lemma 10.2.6. The proof is very similar to the proof of [29, Lemma 6.23]. Suppose that there is no hod pair or an sts hod pair $(\mathcal{P}, \Sigma)$ such that

1. Σ has strong branch condensation and is strongly fullness preserving,
2. $\Gamma(\mathcal{P}, \Sigma) \subseteq \Gamma \subseteq L(\Sigma, \mathbb{R})$

Just like in the proof of [29, Lemma 6.23], it follows from Theorem 10.1.2 that Γ is not a mouse full pointclass (as we are assuming that $L_\alpha(\Gamma, \mathbb{R}) \models \mathsf{SMC}$). Following the proof of [29, Lemma 6.23], we let A be the set of hod pairs or sts hod pairs $(\mathcal{P}, \Sigma)$ such that $\mathsf{Code}(\Sigma) \in \Gamma$ and Σ has strong branch condensation and is strongly fullness preserving. It follows from Claim 1 on page 158 of [29] that $A \neq \emptyset$. It

[80]The rest of the objects will be defined at the next stage of the induction as in clause 4.
[81]F_ξ and b_ξ are defined at step $\xi + 1$.

follows from Claim 2 on the same page of [29] that if

$$\Gamma_1 = \bigcup_{(\mathcal{P},\Sigma)\in A} \Gamma(\mathcal{P},\Sigma)$$

then

(1) Γ_1 is a mouse full pointclass such that for some limit ordinal α there is a sequence of mouse full pointclasses $(\Gamma_\beta : \beta < \alpha)$ such that for $\beta < \gamma < \alpha$, $\Gamma_\beta \trianglelefteq_{mouse} \Gamma_\gamma$ and $\Gamma_1 = \bigcup_{\beta<\alpha}\Gamma_\beta$.

It follows from Theorem 10.1.2 that there is a possibly anomalous hod pair $(\mathcal{P},\Sigma)$ such that either

1. $\mathcal{P}$ is of lsa type and $\Gamma^b(\mathcal{P},\Sigma) = \Gamma_1$ or
2. $\mathcal{P}$ is not of lsa type and $\Gamma(\mathcal{P},\Sigma) = \Gamma_1$.

Because $\Gamma \models \mathsf{SMC}$ and because $\Gamma_1 \triangleleft_{mouse} \wp(\mathbb{R})$, we must have that Σ is strongly fullness preserving (for instance see [29, Lemma 6.21]). Notice that even if clause 1(b) of Theorem 10.1.2 applies, we still get a hod pair as opposed to an sts pair. This is because we have good pointclasses beyond Γ.

Notice also that $\mathsf{Code}(\Sigma) \notin \Gamma$, as otherwise it follows from Claim 2 on page 158 of [29] that $(\mathcal{P},\Sigma) \in A$. Thus, it must be the case that $\mathcal{P}$ is an anomalous hod premouse. We now get a contradiction as in page 159 of [29], where it is argued that the computation of $\mathrm{HOD}^{L(\Sigma,\mathbb{R})}$ gives a contradiction.

10.3. A proof of LSA from large cardinals

In this section, we generalize [29, Theorem 6.26].

THEOREM 10.3.1. *The theory* $\mathsf{AD}^+ + \mathsf{LSA} + V = L(\wp(\mathbb{R}))$ *is consistent relative to a Woodin cardinal that is a limit of Woodin cardinals.*

PROOF. Woodin showed that it is consistent relative to a Woodin cardinal that is a limit of Woodin cardinals that there are divergent models of AD^+, i.e., there are sets of reals $A, B \subseteq \mathbb{R}$ such that $L(A,\mathbb{R}) \models \mathsf{AD}^+$, $L(B,\mathbb{R}) \models \mathsf{AD}^+$, $A \notin L(B,\mathbb{R})$ and $B \notin L(A,\mathbb{R})$. Moreover, his construction shows that we can assume that both $L(A,\mathbb{R})$ and $L(B,\mathbb{R})$ satisfy $\mathsf{MC} + \Theta = \vartheta_0 + \mathsf{NWLW}$[82]. Thus, we assume that such a pair of models exists.

Suppose towards a contradiction that there is no inner model satisfying $\mathsf{AD}^+ + \mathsf{LSA} + V = L(\wp(\mathbb{R}))$. Let $\Gamma = L(A,\mathbb{R}) \cap L(B,\mathbb{R}) \cap \wp(\mathbb{R})$. It is an unpublished theorem of Woodin that $L(\Gamma,\mathbb{R}) \models \mathsf{AD}_{\mathbb{R}}$ but see [67, Theorem 8.1]. We also have that $\Gamma = \wp(\mathbb{R}) \cap L(\Gamma,\mathbb{R})$. Applying Lemma 10.1.2 in $L(A,\mathbb{R})$ and in $L(B,\mathbb{R})$ we

[82]See Definition 10.1.1. The proof of Woodin's theorem appeared in [7] as Theorem 6.1 and it is obtained as a forcing extension of the minimal active mouse with a Woodin cardinal that is a limit of Woodin cardinal. If we also would like to have NWLW then we simply take the $\in$-minimal model in which divergent models exists.

get two hod pairs or sts hod pairs $(\mathcal{P},\Sigma)\in L(A,\mathbb{R})$ and $(\mathcal{Q},\Lambda)\in L(B,\mathbb{R})$ such that both $\mathcal{P}$ and $\mathcal{Q}$ are of limit type and $\Gamma=\Gamma(\mathcal{P},\Sigma)=\Gamma(\mathcal{Q},\Lambda)$.

Working in $L(A,\mathbb{R})$, let $\mathcal{M}^*=\bigcup_{(\mathcal{S},\Psi)\in B(\mathcal{P},\Sigma)}\mathcal{M}_\infty(\mathcal{S},\Psi)$ and for $\mathcal{S}'\lhd^c_{hod}\mathcal{M}^*$[83] let $\Psi_{\mathcal{S}'}$ be the iteration strategy of $\mathcal{S}'$ obtained from any $(\mathcal{S},\Psi)$ such that $\mathcal{S}'=\mathcal{M}_\infty(\mathcal{S},\Psi)$. Notice that $\mathcal{M}^*$ and the strategies $(\Psi_{\mathcal{S}'}:\mathcal{S}'\lhd^c_{hod}\mathcal{M}^*)$ are independent of $(\mathcal{P},\Sigma)$ i.e., working in $L(B,\mathbb{R})$ and using $(\mathcal{Q},\Lambda)$ instead of $(\mathcal{P},\Sigma)$ yields the same model $\mathcal{M}^*$ and the same strategies $(\Psi_{\mathcal{S}'}:\mathcal{S}'\lhd^c_{hod}\mathcal{M}^*)$. Let[84]

$$\mathcal{M}_A=(\mathsf{Lp}^{\Gamma,\oplus_{\mathcal{S}\lhd^c_{hod}\mathcal{M}^*}\Psi_{\mathcal{S}}}(\mathcal{M}^*))^{L(A,\mathbb{R})}$$

and

$$\mathcal{M}_B=(\mathsf{Lp}^{\Gamma,\oplus_{\mathcal{S}\lhd^c_{hod}\mathcal{M}^*}\Psi_{\mathcal{S}}}(\mathcal{M}^*))^{L(B,\mathbb{R})}.$$

We then have that either $\mathcal{M}_A\trianglelefteq\mathcal{M}_B$ or $\mathcal{M}_B\trianglelefteq\mathcal{M}_A$. Without loss of generality we assume that $\mathcal{M}_A\trianglelefteq\mathcal{M}_B$.

Let $\pi:\mathcal{P}^b\to\mathcal{M}_A$ be the iteration embedding given by Σ. It follows from the proof of Claim 7 appearing in the proof of Theorem 8.2.6 that $\Sigma\in L(\pi[\mathcal{P}],\mathcal{M}_A,\Gamma)$. By our assumption, $\mathcal{M}_A\in L(B,\mathbb{R})$. Because $\pi[\mathcal{P}]$ is a countable set we have that $\pi[\mathcal{P}]\in L(B,\mathbb{R})$. It follows that $\Sigma\in L(B,\mathbb{R})$. Therefore, $\mathsf{Code}(\Sigma)\in\Gamma$ implying that $\Gamma(\mathcal{P},\Sigma)\subset\Gamma$, contradiction! $\dashv$

We remark that just like in the proof of [29, Theorem 6.26], we could have used Theorem 4.14.4 instead of Theorem 8.2.6.

---

[83] See Definition 9.1.2.

[84] $\mathcal{W}\trianglelefteq\mathsf{Lp}^{\Gamma,\oplus_{\mathcal{S}\lhd^c_{hod}\mathcal{M}^*}\Psi_{\mathcal{S}}}(\mathcal{M}^*)$ if and only if $\mathcal{W}$ is a sound $\oplus_{\mathcal{S}\lhd^c_{hod}\mathcal{M}^*}\Psi_{\mathcal{S}}$-premouse over $\mathcal{M}^*$ such that $\rho(\mathcal{W})\leq\mathsf{ord}(\mathcal{M}^*)$ and whenever $\pi:\mathcal{W}'\to\mathcal{W}$ is elementary and $\mathcal{W}'$ is countable, $\mathcal{W}'\trianglelefteq\mathsf{Lp}^{\Gamma,\Psi'}(\pi^{-1}(\mathcal{M}^*))$ where Ψ' is the π-pullback of $\oplus_{\mathcal{S}\lhd^c_{hod}\mathcal{M}^*}\Psi_{\mathcal{S}}$.

Chapter 11

A PROOF OF SQUARE IN LSA-SMALL HOD MICE

DEFINITION 11.0.2. For a cardinal κ and a cardinal $\gamma \leq \kappa$, the principle $\square_{\kappa,\gamma}$ states that there is a sequence $\langle \mathcal{C}_\alpha : \alpha < \kappa^+ \rangle$ such that for each $\alpha < \kappa^+$

1. $\mathcal{C}_\alpha \neq \emptyset$ and for each $C \in \mathcal{C}_\alpha$, C is a closed unbounded subset of α of order type at most κ,
2. $|\mathcal{C}_\alpha| \leq \gamma$,
3. for each $C \in \mathcal{C}_\alpha$, for each $\beta \in \lim(C_\alpha)$, $C \cap \beta \in \mathcal{C}_\beta$.

If $\gamma = 1$, then the principle $\square_{\kappa,\gamma}$ is simply $\square_\kappa$.

Pure extender models are models constructed from a canonical sequence of extenders. Jensen (cf. [10]) initiated the program of understanding square principles in pure extender models by proving $L \models \forall\kappa\, \square_\kappa$. Building on works of several people, Schimmerling and Zeman (cf. [38]) give the most optimal characterization of $\square$ in (short) extender models, namely they prove that in an iterable, short extender model, $\square_\kappa$ holds if and only if κ is not subcompact. Results on squares in extender models are important in understanding structure theory of such models and have found many applications in set theory. The reader can see, for instance, [37] and [12], for some of the applications of square in extender models in computing lower-bound consistency strength of theories like PFA. Recent advances in the core model induction methods have indicated that to improve the lower-bounds of combinatorial principles like PFA, failure of square at a singular cardinal, the existence of guessing models etc., one way is to prove square holds in the hod mice that are currently being studied and constructed.

All known square proofs in extender models rely heavily on the fine-structure of such models, in particular, they make essential use of condensation properties of these models (cf. [38, Lemma 1.6]). Unfortunately, the full condensation lemma,[38, Lemma 1.6], does not hold in hod mice. However, it is possible to overcome this shortcoming. We present here a proof of $\square_{\kappa,2}$ in an *lsa-small hod mouse* $\mathcal{P}$ for all cardinals κ of $\mathcal{P}$. In this chapter by lsa-small hod mouse, we mean that $\mathcal{P}$ does not contain an active ω Woodin lsa mouse as defined in Definition 8.2.2.

We first set up some terminology. Our hod premice $\mathcal{P}$ are *lsa-small* and hence for no $\alpha < \lambda^{\mathcal{P}}$, $\mathcal{P}(\alpha)$ is an lsa hod premouse, though $\mathcal{P}$ can be of lsa type.

Throughout this paper, if $\mathcal{Q}$ is an initial segment of $\mathcal{P}$, we let $\Sigma_{\mathcal{Q}}$ denote the restriction of Σ to $\mathcal{Q}$. If $\mathcal{P}$ is of limit type and has a top window $[\delta_\alpha^{\mathcal{P}}, \delta_{\alpha+1}^{\mathcal{P}})$, then we let $\mathcal{P}^b = \mathcal{P}|(\delta_\alpha^{\mathcal{P}+})^{\mathcal{P}}$. See Section 11.1 for a more detailed discussion of hod mice along with the definitions used in statements of this section. In the definitions below, we adapt the Σ^*-language (see [38]) to hod mice in the obvious way. Let $\rho_{\mathcal{Q}}^n$ be the nth-projectum of $\mathcal{Q}$, and $p_{\mathcal{Q}}^n$ be the nth-standard parameter of $\mathcal{Q}$.[1] Semantically, suppose $\mathcal{Q}$ is an initial segment of $\mathcal{P}$, a relation $A \subset |\mathcal{Q}|$ is $\Sigma_l^{(n)}(\mathcal{Q})$ from p, or $\Sigma_l^{(n)}(\mathcal{Q})$, if it is Σ_l from p (or Σ_l) over the nth-reduct $\langle H_{\mathcal{Q}}^n, A_{\mathcal{Q}}^n\rangle$ of $\mathcal{Q}$, where $H_{\mathcal{Q}}^n = |\mathcal{Q}|\rho_{\mathcal{Q}}^n|$ and $A_{\mathcal{Q}}^n$ is the nth standard master code (with respect to $p_{\mathcal{Q}}^n$) of $\mathcal{Q}$.

DEFINITION 11.0.3. Suppose Σ is an iteration strategy for a hod premouse $\mathcal{P}$. Suppose Γ is an inductive-like pointclass. We say that Σ is *locally strongly Γ-fullness preserving* if Σ is Γ-fullness preserving and if $\mathcal{P}$ is of limit type with a top window and whenever $(\vec{\mathcal{T}}, \mathcal{S}) \in I(\mathcal{P}, \Sigma)$, and

$$\pi^{\vec{\mathcal{T}},b} : \mathcal{P}^b \to \mathcal{S}^b \text{ exists,}$$

then letting $\pi = \pi^{\vec{\mathcal{T}},b}$, whenever $\mathcal{S}^b \lhd \mathcal{W} \unlhd \mathcal{S}$ is such that for some n

$$o(\mathcal{S}^b) \leq \omega\rho_{\mathcal{W}}^{n+1} < \omega\rho_{\mathcal{W}}^n,$$

$\mathcal{W}$ is n-sound, and $\tau : \mathcal{R} \to \mathcal{W}$ is cardinal preserving and $\Sigma_0^{(n)}$ and $\omega\rho_{\mathcal{R}}^n > \mathrm{cr}(\tau) \geq \omega\rho_{\mathcal{R}}^{n+1} = \omega\rho_{\mathcal{W}}^{n+1}$, then the τ-pullback of the strategy $\Sigma_{\mathcal{W},\vec{\mathcal{T}}}$ is Γ-fullness preserving in the following sense: whenever $\vec{\mathcal{U}}$ is according to $\Sigma_{\mathcal{W},\vec{\mathcal{T}}}^\tau$, then letting $\mathcal{R}^*$ be the last model of $\vec{\mathcal{U}}$, $(\mathcal{R}^*)^b$ is Γ-full.

DEFINITION 11.0.4. Suppose Σ is an iteration strategy for a hod premouse $\mathcal{P}$. We say that Σ has *locally strong branch condensation* if Σ has branch condensation and if $\mathcal{P}$ is of limit type with a top window and $\mathcal{Q}$ is such that $\mathcal{P}^b \lhd \mathcal{Q} \unlhd \mathcal{P}$, and n is such that $\omega\rho_{\mathcal{Q}}^{n+1} < \omega\rho_{\mathcal{Q}}^n$, $\mathcal{Q}$ is n-sound, $\omega\rho_{\mathcal{Q}}^{n+1}$ is a cardinal of $\mathcal{P}$, and $\mathcal{S}$ is a $\Sigma_{\mathcal{Q}}$-iterate along a stack $\vec{\mathcal{T}}$ such that $\pi^{\vec{\mathcal{T}},b}$ exists, and $\tau : \mathcal{Q} \to \mathcal{R}$ is a cardinal preserving, $\Sigma_0^{(n)}$-embedding such that $\mathcal{R} \unlhd \mathcal{S}$ and $(\mathcal{Q}^*)^b = \mathcal{R}^b$ for some non-dropping $\Sigma_{\mathcal{Q}}$-iterate $\mathcal{Q}^*$ of $\mathcal{Q}$. Suppose also that letting $j : \mathcal{Q} \to \mathcal{Q}^*$ be the iteration map, then $j \restriction \mathcal{Q}^b = \tau \restriction \mathcal{Q}^b$. Then $\Sigma_{\mathcal{R},\vec{\mathcal{T}}}^\tau = \Sigma_{\mathcal{Q}}$.

The proof of Theorem 4.6.3 can be modified to get hod mice $(\mathcal{P}, \Sigma)$ with Σ being locally strongly Γ-fullness preserving. Similarly the proof of strong branch condensation can be modified to obtain strategies with locally strong branch condensation. We leave the easy proofs to the reader. The term "locally" refers to the fact that these forms of fullness preservation and branch condensation apply

[1] Other notations for the nth-projectum and n^{th}-standard parameter of $\mathcal{Q}$ used elsewhere in this book are $\rho_n(\mathcal{Q})$ and $p_n(\mathcal{Q})$ respectively. For this chapter, we stick to the more compact notations $\rho_{\mathcal{Q}}^n$ and $p_{\mathcal{Q}}^n$; this notation is compatible with the notation used in [38].

to initial segments of $\mathcal{P}$ (like $\mathcal{W}$ and $\mathcal{Q}$ in these definitions). In the proof of Theorem 11.0.5, initial segments like $\mathcal{Q} \lhd \mathcal{P}$ are typically collapsing structures of some $\tau \in (\kappa, (\kappa^+)^{\mathcal{P}})$, where $\kappa = \rho_{\mathcal{Q}}^{n+1}$ is a cardinal of $\mathcal{P}$. We seem to need to modify the usual notions of fullness preservation and branch condensation (as in Definitions 11.0.3 and 11.0.4) to ensure that various phalanx comparison arguments involving initial segments like $\mathcal{Q}$ (which is not $\mathcal{P}$ in most cases of interest) go through in the proof of Theorem 11.0.5. In most (but not all) applications, the map π in Definitions 11.0.3 and 11.0.4, is the identity and τ is the uncollapse map associated to a sufficiently elementary hull. The main theorem is the following.

THEOREM 11.0.5. *Suppose $(\mathcal{P}, \Sigma)$ is an lsa-small hod pair such that Σ has locally strong branch condensation and is locally strongly Γ-fullness preserving for some inductive-like pointclass Γ that satisfies "$\mathsf{AD}^+ + \mathsf{SMC}$". Then $\mathcal{P} \models \forall\kappa\, \square_{\kappa,2}$.*[2]

Many techniques in the proof of Theorem 11.0.5 come from the Schimmerling-Zeman's proof in [38]. In Section 11.1, we import some results from the theory of hod mice we need. In Section 11.2, we will import some terminology, results from [38] that we need here. We also explain in this section why a straightforward adaptation of [38] fails in the context of hod mice. In Section 11.3, we give the actual proof of Theorem 11.0.5.

Finally, we remark that hod pairs constructed in practice (those constructed in sufficiently strong AD^+ models or in the core model induction settings) do have the properties in the hypothesis of Theorem 11.0.5. The main application of Theorem 11.0.5 in this book is to improve the lower-bound consistency strength of various theories such as PFA to that of LSA (see Chapter 12).

11.1. Ingredients from hod mice theory

We summarize some definitions and results of the hod mice theory developed above that we need to prove Theorem 11.0.5. The language for hod premice's is $\mathcal{L}_1$ with symbols $\in, \dot{Y}, \dot{E}, \dot{F}, \dot{\Sigma}, \dot{B}, \dot{\gamma}$.

Suppose $(\mathcal{P}, \Sigma)$ is an lsa-small hod pair. $\mathcal{P}$ is constructible from a sequence of extenders and a sequence of strategies of its own hod initial segments. As before $\dot{Y}^{\mathcal{P}}$ codes the layers of $\mathcal{P}$. In the following, all extenders in $\dot{E}^{\mathcal{P}}$ and $\{\dot{F}^{\mathcal{P}}\}$ will be indexed on the $\mathcal{P}$-sequence according to the λ-indexing scheme described in [70]. $\dot{\gamma}^{\mathcal{P}}$, just as in [70], codes where $\dot{F}^{\mathcal{P}}$ restricted to the largest cutpoint of $\dot{F}^{\mathcal{P}}$ would be indexed. There are two ways in which an initial segment $\mathcal{Q}$ of $\mathcal{P}$ can be active: B-active and E-active. $\mathcal{Q}$ is *B-active* the top predicate $\dot{B}^{\mathcal{Q}}$ for $\mathcal{Q}$ (amenably) codes a branch for some tree on an initial segment of $\mathcal{Q}$. $\mathcal{Q}$ is *E-active* if the top predicate $\dot{F}^{\mathcal{Q}}$ of $\mathcal{Q}$ codes an extender. Otherwise, we say that $\mathcal{Q}$ is *passive*.

[2]The assumption that $\mathcal{P}$ is lsa-small implies that there are no subcompact cardinals in $\mathcal{P}$ and all extenders on the $\mathcal{P}$-sequence are short.

B-active levels and passive levels are more or less treated the same way in the proof of Theorem 11.0.5. In our situation, we note that since all initial segments of $\mathcal{P}$ and $\mathcal{P}$ itself are lsa-small, all extenders on the $\mathcal{P}$-sequence are of type A, i.e., they have no cutpoints. So $\dot{\gamma}^{\mathcal{P}} = \emptyset$. This aspect somewhat simplifies our proof, compared to [38].

A few words about how the B-predicate codes up branches for an iteration tree $\mathcal{T}$ in $\mathcal{P}$ are in order. Suppose $\lambda = \mathrm{lh}(\mathcal{T})$ is limit and $\mathcal{P}|\gamma$ is B-active such that $B^{\mathcal{P}|\gamma}$ codes a cofinal branch b of $\mathcal{T}$. The traditional way that B codes b is that letting $\gamma^* + \lambda = \gamma$, $B^{\mathcal{P}|\gamma} = \{\gamma^* + \alpha \mid \alpha \in b\}$. While this approach is sufficient for developing the basic theory of strategic premice and certainly is sufficient for the theory of hod mice we have developed so far, it seems to create significant obstructions in the proof of $\square$ in this chapter. So instead, we use the coding method developed in [50]. Using [50, Definition 2.26], we let $\mathcal{P}|\gamma = \mathfrak{B}(\mathcal{P}|\gamma^*, \mathcal{T}, b)$. The reader is advised to consult [50] for the precise definition of $\mathfrak{B}(\mathcal{P}|\gamma^*, \mathcal{T}, b)$. Roughly, for every $0 < \alpha < \lambda$, $\mathcal{P}|(\gamma^* + \omega\alpha)$ is B-active and $B^{\mathcal{P}|(\gamma^*+\omega\alpha)}$ codes the branch $[0,\alpha)_{\mathcal{T}}$ and $B^{\mathcal{P}|\gamma}$ codes b in the manner described above. We make this a bit more precisely here.

$\dot{\Sigma}^{\mathcal{P}}$ and $\dot{B}^{\mathcal{P}}$ are used to record information about an iteration strategy Ω of $\mathcal{P}$. $\dot{\Sigma}^{\mathcal{P}}$ codes the strategy information added at earlier stages; $\dot{\Sigma}^{\mathcal{P}}$ acts on stacks in $\mathcal{P}$ based on a hod layer $\mathcal{Q} \in \dot{Y}^{\mathcal{P}}$. $\dot{\Sigma}^{\mathcal{P}}(s,b)$ implies that $s = \mathcal{T}$, where $\mathcal{T}$ is a stack on $\mathcal{Q}$ in $\mathcal{P}$ of limit length and $\mathcal{T}^\frown b$ is according to the strategy. We say that s is an $\mathcal{P}$-tree, and write $s = \mathcal{T}(s)$. We write $\dot{\Sigma}^{\mathcal{P}}_{\nu,k}$ for the partial iteration strategy for $\mathcal{P}|(\nu,k)$ determined by $\dot{\Sigma}$ (here $\mathcal{Q} = \mathcal{P}|(\nu,k)$ for some ν,k). We write $\Sigma^{\mathcal{P}}(s) = b$ when $\dot{\Sigma}^{\mathcal{P}}(s,b)$, and we say that s is according to $\Sigma^{\mathcal{P}}$ if $\mathcal{T}(s)$ is according to $\dot{\Sigma}^{\mathcal{P}}_{\mathcal{Q}}$.

We say $\mathcal{P}$ is *branch-active* (or just B-active) iff

(a) there is a largest $\eta < o(\mathcal{P})$ such that $\mathcal{P}|\eta \models \mathsf{KP}$, and letting $N = \mathcal{P}|\eta$,
(b) there is a $<_N$-least N-tree s such that s is by Σ^N, $\mathcal{T}(s)$ has limit length, and $\Sigma^N(s)$ is undefined.
(c) for N and s as above, $o(\mathcal{P}) \leq o(N) + \mathrm{lh}(\mathcal{T}(s))$.

Note that being branch-active can be expressed by a Σ_2 sentence in $\mathcal{L}_1 - \{\dot{B}\}$. This contrasts with being extender-active, which is not a property of the premouse with its top extender removed. In contrast with extenders, we know when branches must be added before we do so.

DEFINITION 11.1.1. Suppose that $\mathcal{P}$ is branch-active. We set

$$\eta^P = \text{the largest } \eta \text{ such that } \mathcal{P}|\eta \models \mathsf{KP},$$
$$\nu^{\mathcal{P}} = \text{unique } \nu \text{ such that } \eta^{\mathcal{P}} + \nu = o(\mathcal{P}),$$
$$s^{\mathcal{P}} = \text{least } \mathcal{P}|\eta^M\text{-tree such that } \dot{\Sigma}^{\mathcal{P}|\eta^{\mathcal{P}}} \text{ is undefined, and}$$
$$b^{\mathcal{P}} = \{\alpha \mid \eta + \alpha \in \dot{B}^{\mathcal{P}}\}.$$

Moreover,

(1) $\mathcal{P}$ is a *potential hod B-active premouse* iff b^M is a cofinal branch of $\mathcal{T}(s) \upharpoonright \nu^{\mathcal{P}}$.
(2) $\mathcal{P}$ is *honest* iff $\nu^{\mathcal{P}} = \mathrm{lh}(\mathcal{T}(s))$, or $\nu^{\mathcal{P}} < \mathrm{lh}(\mathcal{T}(s))$ and $b^{\mathcal{P}} = [0, \nu^{\mathcal{P}})_{T(s)}$.
(3) $\mathcal{P}$ is a hod premice iff $\mathcal{P}$ is an honest potential lpm.
(4) $\mathcal{P}$ is *strategy active* iff $\nu^{\mathcal{P}} = \mathrm{lh}(\mathcal{T}(s))$.

Note that $\eta^{\mathcal{P}}$ is a $\Sigma_0^{\mathcal{P}}$ singleton, because it is the least ordinal in $\dot{B}^{\mathcal{P}}$ (because 0 is in every branch of every iteration tree), and thus s^M is also a $\Sigma_0^{\mathcal{P}}$ singleton. We have separated honesty from the other conditions because it is not expressible by a Q-sentence, whereas the rest is. Honesty is expressible by a Boolean combination of Σ_2 sentences.

The definition of B-active hod premice defined in previous chapters required that when $o(\mathcal{P}) < \eta^{\mathcal{P}} + \mathrm{lh}(\mathcal{T}(s))$, $\dot{B}^{\mathcal{P}}$ is empty, whereas here we require that it code $[0, o(\mathcal{P}))_{T(s)}$, in the same way that $\dot{B}^{\mathcal{P}}$ will have to code a new branch when $o(\mathcal{P}) = \eta^M + \mathrm{lh}(\mathcal{T}(s))$. Of course, $[0, \nu^{\mathcal{P}})_{T(s)} \in \mathcal{P}$ when $o(\mathcal{P}) < \eta^{\mathcal{P}} + \mathrm{lh}(\mathcal{T}(s))$ and $\mathcal{P}$ is honest, so the current $\dot{B}^{\mathcal{P}}$ seems equivalent to the original $\dot{B}^{\mathcal{P}} = \emptyset$. However, $\dot{B}^{\mathcal{P}} = \emptyset$ leads to $\Sigma_1^{\mathcal{P}}$ being too weak, with the consequence that a Σ_1 hull of $\mathcal{P}$ might collapse to something that is not a hod premouse.[3] Our current choice for $\dot{B}^{\mathcal{P}}$ solves that problem. Furthermore, the indexing of branches for $\mathcal{P}$-trees $\mathcal{T}$ using the $\mathfrak{B}$-operator is done for all eligible $\mathcal{T}$ regardless of whether $\mathrm{cof}(\mathrm{lh}(\mathcal{T}))$ is measurable in $\mathcal{T}$ as we required using the original definition. This makes the definition more uniform.

REMARK 11.1.2. Suppose N is an hod premouse, and $N \models \mathsf{KP}$. It is very easy to see that $\dot{\Sigma}^N$ is defined on all N-trees s that are by $\dot{\Sigma}^N$ iff there are arbitrarily large $\xi < o(N)$ such that $N|\xi \models \mathsf{KP}$. Thus if M is branch-active, then η^M is a successor admissible; moreover, we do add branch information, related to exactly one tree, at each successor admissible. Waiting until the next admissible to add branch information is just a convenient way to make sure we are done coding in the branch information for a given tree before we move on to the next one. One could go faster.

As mentioned above and discussed in more details in [50], we can prove stronger forms of condensation for hod mice where the $\mathfrak{B}$-operator is used to index branches. For instance, we have the following condensation lemma for hod pairs.

LEMMA 11.1.3. *Let (M, Ω) be a hod pair*[4] *with $k(M) = k$*[5]*, and let $\pi \colon H \to M$ be $\Sigma_0^{(k)}$ and cardinal preserving.*[6] *Suppose that one of the following holds:*

(a) M is passive or branch-active, or
(b) H is a hod premouse.

Then (H, Ω^π) is a hod pair.

[3] The hull could satisfy $o(H) = \eta^H + \mathrm{lh}(\mathcal{T}(s^H))$, even though $o(\mathcal{P}) < \eta^{\mathcal{P}} + \mathrm{lh}(\mathcal{T}(s^{\mathcal{P}}))$. But then being a hod premouse requires $\dot{B}^H \neq \emptyset$.

[4] Recall Ω has hull condensation.

[5] $k(M)$ is the largest k such that M is k-sound.

[6] This is the corresponding version of what is called "weak embeddings" in [50].

PROOF. We show first that H is a hod premouse. If (b) holds, this is rather easy (in fact, a tautology). If M is passive, then so is H, noting that Q sentences are preserved downwards under π and being a passive hod premouse can be expressed by a Q sentence. So let us assume that M is branch-active.

Note that H is a potential branch active hod premouse by the same reasoning as above: the property of being a potential branch active hod premouse is expressible by a Q sentence and this is preserved downwards by π. So we just need to see that H is honest. Let $\nu = \nu^H$, $b = b^H$, and $\mathcal{T} = \mathcal{T}(s^H)$. If $\nu = \mathrm{lh}(\mathcal{T})$, there is nothing to show, so assume $\nu < \mathrm{lh}(\mathcal{T})$. We must show that $b = [0, \nu)_{\mathcal{T}}$. We have by induction that for $N = H|\eta^H$, (N, Ω^π_N) is a hod pair. Thus $\mathcal{T}$ is by Ω^π, and so we just need to see that for $\mathcal{U} = \mathcal{T} \restriction \nu$ and $\mathcal{W} = \mathcal{U}^\frown b$, $\mathcal{W}$ is by Ω^π, or equivalently, that $\pi\mathcal{W}$ is by Ω. But it is easy to see that $\pi\mathcal{W}$ is a hull of $\pi(U)^\frown b^M$, and Ω has hull condensation, so we are done. ⊣

The use of the $\mathfrak{B}$-operator in coding branches of iteration trees in the $\square$-construction will be explained in Section 11.3 (see, for example Lemma 11.3.13, where Lemma 11.1.3 is used).

We briefly discuss indexing schemes for extenders on the $\mathcal{P}$-sequence. We recall the mixed indexing scheme for hod premice being used in the previous chapters. Suppose κ is a cardinal limit of cutpoint Woodin cardinals of $\mathcal{P}$, and if E is an extender on the $\mathcal{P}$'s sequence such that $\mathrm{cr}(E) = \kappa$, then the index of E is γ where γ is the successor cardinal of the least cutpoint above κ in $\mathrm{Ult}(\mathcal{P}|\xi, E)$ (we call this *cutpoint indexing scheme*), where $\xi \leq o(\mathcal{P})$ is the largest such that E measures all sets in $\mathcal{P}|\xi$. It turns out that such extenders are all total over $\mathcal{P}$ (see Chapters 2, 3 for more discussions). Suppose E is an extender with critical point ξ and E is indexed according to the cutpoint indexing scheme. Then according to [60], for all $\gamma < \mathrm{lh}(E)$, $E \restriction \gamma$ is not on the $\mathcal{P}$-sequence, though $E \restriction \gamma \in \mathcal{P}$ (for γ below the sup of the generators of E) and the trivial completion of $E \restriction \gamma$ is on the $\mathcal{P}$-sequence for various γ (this is similar to the initial segment condition for Jensen indexing). Also, the set of indices of extenders with a fixed critical point ξ indexed according to the cutpoint indexing scheme is nowhere continuous. For other extenders on $\mathcal{P}$'s sequence, we use *the Mitchell-Steel indexing scheme* (ms-indexing).

In the following, all extenders on a hod premouse's extender sequence will be indexed according to the *Jensen indexing scheme* (λ-indexing). By results of [8, 9], one can translate a hod pair $(\mathcal{P}, \Sigma)$ in the mixed indexing scheme described above to a hod pair $(\mathcal{P}^*, \Sigma^*)$ in the λ-indexing scheme. The hod pair $(\mathcal{P}^*, \Sigma^*)$ is obtained by applying the Λ-function to $(\mathcal{P}, \Sigma)$. See for example, [9, Lemmas 3.4, 6.3, 6.4].[7] As in [9], one can show that the universe of $\mathcal{P}$ and $\mathcal{P}^*$ are the same and much more. Σ^* is essentially the same as Σ, so it has all the properties Σ has. The change in indexing schemes is for convenience of importing terminology and results from

[7]Technically speaking, Fuchs shows the intertranslatability for ms-mice and λ-mice. But the same proof techniques can be used without virtually any change to translated mice with mixed-indexing to λ-mice.

[38]. The result we prove here for hod mice with the Jensen indexing scheme will also hold for hod mice with the Mitchell-Steel indexing scheme by [9] and the above discussion, i.e., if κ is a cardinal of $\mathcal{P}^*$ (equivalently of $\mathcal{P}$) then

$$\mathcal{P}^* \models \square_{\kappa,2} \text{ implies } \mathcal{P} \models \square_{\kappa,2}.$$

Suppose E is an extender with critical point ξ on the sequence of $\mathcal{P}$ and E is indexed by the Jensen indexing scheme, that is the index of E in $\mathcal{P}$ is the successor cardinal of $i_E(\xi)$ in $\mathrm{Ult}(\mathcal{P},E)$. For a summary of the fine structure, see [38, Section 1]. A couple of remarks regarding the adaptation of [38, Section 1] into our situation are in order. First, we still demand extenders indexed according to the Jensen indexing to satisfy the *initial segment condition* (ISC) in the sense of [38, Section 1.4]; that is for all $\gamma < \mathrm{lh}(E)$, if γ is a cutpoint of E, then $E \upharpoonright \gamma \in |\mathcal{P}|\mathrm{lh}(E)|$. Secondly, under this initial segment condition, using the assumption that our hod premice are lsa-small, it is easy to see that these extenders E are all of type A, that is the set of cutpoints is empty; this is because there are no superstrong cardinals in lsa-small hod mice. The initial segment condition is needed to prove comparisons terminate.

We recall some concepts related to layers discussed in previous chapters. If $\mathcal{P}$ is a hod premouse, we let $\lambda^{\mathcal{P}}$ denote the order type of the layer Woodin cardinals of $\mathcal{P}$ and $(\delta^{\mathcal{P}}_\alpha : \alpha < \lambda^{\mathcal{P}})$ enumerate the closure of the set of Woodin cardinals and indices of extenders whose critical point is a limit of layer Woodin cardinals of $\mathcal{P}$. Recall that δ is a layer Woodin cardinal of $\mathcal{P}$ if there is some $\mathcal{Q} \in Y^{\mathcal{P}}$ (i.e., $\mathcal{Q}$ is a layer of $\mathcal{P}$) such that $\delta = \delta^{\mathcal{Q}}$. Intuitively, $Y^{\mathcal{P}}$ is the set of initial segments $\mathcal{Q}$ of $\mathcal{P}$ such that the strategy (or sts strategy) of $\mathcal{Q}$ is activated in $\mathcal{P}$. See 2.7.8. If $\mathcal{P}$ has a largest Woodin cardinal, we denote that $\delta^{\mathcal{P}}$. Recall we use $\mathcal{P}^b$ to denote the "bottom part" of $\mathcal{P}$ in the case that $\mathcal{P}$ has a top window $[\kappa = \delta^{\mathcal{P}}_\alpha, \delta^{\mathcal{P}})$, where recall that by definition, κ is either a Woodin or limit of Woodins in $\mathcal{P}$. In this case, $\mathcal{P}^b = \mathrm{Lp}^{\Sigma^{\mathcal{P}}_\kappa, \mathcal{P}}(\mathcal{P}|\kappa)$, where $\Sigma^{\mathcal{P}}_\kappa = \oplus_{\beta<\alpha}\Sigma^{\mathcal{P}}_{\mathcal{P}(\beta)}$. In the case α is a limit ordinal, $\mathcal{P}^b = \mathcal{P}|((\kappa)^+)^{\mathcal{P}}$. In this case, if κ happens to be measurable in $\mathcal{P}$, then all extenders E on the $\mathcal{P}$ sequence with critical point κ are indexed according to the cutpoint indexing scheme. Notice that since $\mathcal{P}$ is lsa-small, κ is a cutpoint (but not a strong cutpoint) in $\mathcal{P}$, though κ is a strong cutpoint in $\mathcal{P}^b = \mathcal{P}|(\kappa^+)^{\mathcal{P}}$. Let $o(\kappa)$ be the supremum of the indices of extenders on the $\mathcal{P}$ sequence with critical point κ. If $\mathcal{P}$ is of *lsa type* (i.e., $o(\kappa) = \delta^{\mathcal{P}}$) then there may be local large cardinals in the interval $(\kappa, o(\kappa))$, e.g., there may be a $\gamma \in (\kappa, o(\kappa))$ which is Woodin in some initial segment $\mathcal{Q}$ of $\mathcal{P}$; such large cardinals are witnessed by the extender sequence and *the short tree strategy* of initial segments of $\mathcal{Q}$, but not the full strategy. This point is crucial in many arguments given below (see Lemma 11.1.4).

Suppose $(\mathcal{P},\Sigma)$ is a hod pair such that Σ is Γ-fullness preserving for some inductive-like pointclass Γ and has branch condensation. Suppose $\mathcal{R} \lhd \mathcal{P}$ is an initial segment of $\mathcal{P}$, then we let $\Sigma_{\mathcal{R}}$ denote the restriction of Σ to trees based on $\mathcal{R}$. Let $I(\mathcal{P},\Sigma)$ denote the set of $(\vec{\mathcal{T}},\mathcal{R})$ where $\vec{\mathcal{T}}$ is a stack according to Σ with last model $\mathcal{R}$. In this case, the "$\vec{\mathcal{T}}$-tail" of Σ, denoted $\Sigma_{\vec{\mathcal{T}},\mathcal{R}}$, is a strategy for $\mathcal{R}$. We let $B(\mathcal{P},\Sigma)$ denote the set of $(\vec{\mathcal{T}},\mathcal{R})$ where $\vec{\mathcal{T}}$ is according to Σ and $\mathcal{R}$ is a strict hod-initial segment of $\mathcal{N}^{\vec{\mathcal{T}}}$, the last model of $\vec{\mathcal{T}}$. We let $\Gamma(\mathcal{P},\Sigma)$ be the set of $A \subseteq \mathbb{R}$ such that $A <_w \Sigma_{\vec{\mathcal{T}},\mathcal{R}}$ for some $(\vec{\mathcal{T}},\mathcal{R}) \in B(\mathcal{P},\Sigma)$. Note that $\Gamma(\mathcal{P},\Sigma)$ is a Wadge initial segment of Γ. We say that $\mathcal{P}$ *generates* Ω if $\Gamma(\mathcal{P},\Sigma) = \Omega$.

The following fact will be used in many places throughout this chapter, and whose proof is essentially that of Lemma 4.11.3.

LEMMA 11.1.4 (No strategy disagreement). *Suppose $(\mathcal{P},\Sigma)$ is an lsa-small hod pair such that $\mathcal{P}$ has a top window $[\delta^{\mathcal{P}}_\alpha, \delta^{\mathcal{P}})$ and $\delta^{\mathcal{P}}_\alpha$ is not a strong cutpoint of $\mathcal{P}$, Σ has locally strong branch condensation and is locally strongly Γ-fullness preserving for some constructibly closed pointclass $\Gamma \vDash$ "$\mathsf{AD}^+ + \mathsf{SMC}$". Suppose $\pi : \mathcal{P}' \to \mathcal{P}^*$ for some cardinal preserving, $\Sigma_0^{(n)}$ map π such that $\mathcal{P}^b \lhd \mathcal{P}^* \unlhd \mathcal{P}$, and $\omega\rho^n_{\mathcal{P}^*} > \mathrm{cr}(\pi) =_{def} \gamma > \omega\rho^{n+1}_{\mathcal{P}'} = \omega\rho^{n+1}_{\mathcal{P}^*} \geq o(\mathcal{P}^b)$ and $\rho^{n+1}_{\mathcal{P}^*}$ is a cardinal of $\mathcal{P}$. Then letting $\Lambda = \Sigma^\pi_{\mathcal{P}^*}$, the comparison of the phalanx $(\mathcal{P}^*, \mathcal{P}', \gamma)$ (using Λ) versus $\mathcal{P}^*$ (using $\Sigma_{\mathcal{P}^*}$) does not involve disagreements of strategies.*

Lemma 11.1.4 is useful since it reduces such comparisons to ordinary extender comparisons. Such phalanx comparisons will appear in many places in the proof of Theorem 11.0.5. A corollary of this is the following version of the Condensation Lemma for hod mice (cf. [70, Lemma 9.3.2]). For notations used in the statement of the lemma, see [38, Section 1.3].

THEOREM 11.1.5. *Suppose $(\mathcal{P},\Sigma)$ is an lsa-small hod pair such that $\mathcal{P}$ has a top window $[\delta^{\mathcal{P}}_\alpha, \delta^{\mathcal{P}})$ and $\delta^{\mathcal{P}}_\alpha$ is not a strong cutpoint of $\mathcal{P}$, Σ has locally strong branch condensation and is locally strongly Γ-fullness preserving for some constructibly closed pointclass $\Gamma \vDash$ "$\mathsf{AD}^+ + \mathsf{SMC}$". Suppose $\mathcal{P}^b \lhd \mathcal{M} \unlhd \mathcal{P}$, $\tilde{\mathcal{M}}$ is a hod premouse, and $\sigma : \tilde{\mathcal{M}} \to \mathcal{M}$ is a cardinal preserving and $\Sigma_0^{(n)}$ embedding such that $\sigma \restriction \omega\rho^{n+1}_{\tilde{\mathcal{M}}} = \mathrm{id}$, where $\omega\rho^{n+1}_{\tilde{\mathcal{M}}} = \omega\rho^{n+1}_{\mathcal{M}} \geq o(\mathcal{P}^b)$ is a cardinal of $\mathcal{P}$.*[8,9] *Then $\tilde{\mathcal{M}}$ is solid and $p_{\mathcal{M}}$ is k-universal for all $k \in \omega$. Furthermore, if $\tilde{\mathcal{M}}$ is sound above $\nu = \mathrm{cr}(\sigma)$ then one of the following holds:*

(a) $\tilde{\mathcal{M}}$ is ν-core of $\mathcal{M}$ and σ is the uncollapse map.

(b) $\tilde{\mathcal{M}}$ is an initial segment of $\mathcal{M}$.

(c) $\tilde{\mathcal{M}} = \mathrm{Ult}^*(\mathcal{M}||\eta, E^{\mathcal{M}}_\alpha)$ where $\nu \leq \eta < o(\mathcal{M})$, $\alpha \leq \omega\eta$ and $\nu = (\kappa^+)^{\mathcal{M}||\eta}$ where $\kappa = \mathrm{cr}(E^{\mathcal{M}}_\alpha)$; moreover, $E^{\mathcal{M}}_\alpha$ has a single generator κ.

REMARK 11.1.6. If δ_α is a strong cutpoint of $\mathcal{P}$, then it follows simply from the definition of hod premice that for all $\kappa \in [\delta_\alpha, \delta)$, $\square_\kappa$ holds in $\mathcal{P}$; this is because $\mathcal{P}$ is a $\Sigma^{\mathcal{P}}_\alpha$-premouse ($\Sigma_\alpha$ is the strategy for $\mathcal{P}(\alpha)$) and the $\square$ proof of [38] adapts straightforwardly. On the other hand, if δ_α is not a strong cutpoint of $\mathcal{P}$, then

[8] In the Mitchell-Steel language, one requires σ to be a weak n-embedding such that $\sigma''T^{\tilde{\mathcal{M}}}_n \subseteq T^{\mathcal{M}}_n$.

[9] If $\omega\rho^{n+1}_{\tilde{\mathcal{M}}} = \omega\rho^{n+1}_{\mathcal{M}} = o(\mathcal{P}^b)$, then since $o(\mathcal{P}^b)$ is a cardinal of $\mathcal{P}$, $\mathrm{cr}(\sigma) > o(\mathcal{P}^b)$. Equality can happen in other cases.

Theorem 11.1.5 is false if for example one required that $\mathcal{M}=\mathcal{P}$, the embedding σ have critical point δ_α, δ_α is a limit of Woodin cardinals, $\tilde{\mathcal{M}}\in\mathcal{M}$ is sound, and the cardinality $\tilde{\mathcal{M}}$ in $\mathcal{M}$ is less than δ_α^+. In this case, $\rho_\omega^{\tilde{\mathcal{M}}}=\delta_\alpha=\nu$. Case (a) is immediately ruled out because $\tilde{\mathcal{M}}\in\mathcal{M}$. Case (c) cannot happen because of the fact that no extenders on the $\mathcal{M}$-sequence can be indexed at δ_α, which is a cardinal of $\mathcal{P}$. Case (b) also fails because otherwise, $\tilde{\mathcal{M}}\lhd\mathcal{M}|\delta_\alpha^+$. $\tilde{\mathcal{M}}$ and hence $\mathcal{M}$ has extenders indexed in the interval $[\delta_\alpha,o(\tilde{\mathcal{M}}))\subset[\delta_\alpha,\delta_\alpha^+)$. But by the definition of hod premice, no extenders on the $\mathcal{M}$-sequence can be indexed in the interval $[\delta_\alpha,\delta_\alpha^+)$.

Since we are below superstrong and the extenders on the model are indexed according to the Jensen indexing, the possibility that

$$\tilde{\mathcal{M}} \text{ is a proper initial segment of } \mathrm{Ult}(\mathcal{M},E^{\mathcal{M}}_{\mathrm{cr}(\sigma)})$$

cannot happen. In [70, Lemma 9.3.2], the aforementioned case can occur; in that case, $E^{\mathcal{M}}_{\mathrm{cr}(\sigma)}$ is a superstrong extender.

The proof of the theorem is essentially that of [70, Theorem 9.3.2]. The idea is one compares the phalanx $(\mathcal{M},\tilde{\mathcal{M}},\mathrm{cr}(\sigma))$ against $\mathcal{M}$. Depending on how the comparison terminates, one gets one of the four possibilities in the statement of the theorem. Using locally strong Γ-fullness preservation and the fact that $\mathrm{cr}(\sigma)>\omega\rho^{n+1}_{\mathcal{M}}$, Lemma 11.1.4 shows that the comparison is an extender comparison (no strategy disagreements are encountered). This puts us the in the situation to apply the proof of [70, Theorem 9.3.2] (the Dodd-Jensen-like property we assume as part of locally strong branch condensation is enough to carry out the proof of [70, Theorem 9.3.2]). To illustrate the main ideas, we present a proof of a special case, which often shows up in the $\square$-constructions.

PROOF. We assume $\tilde{\mathcal{M}}$ is sound. Let $\tilde{\tau}=\mathrm{cr}(\sigma)$ and let $\tau=\sigma(\tilde{\tau})$. We further assume that: letting $\kappa=\omega\rho^{n+1}_{\mathcal{M}}$, $\tau=(\kappa^+)^{\mathcal{M}}$, and hence $\tilde{\tau}=(\kappa^+)^{\tilde{\mathcal{M}}}$. In this case, we prove that $\tilde{\mathcal{M}}\lhd\mathcal{M}$. The reader can see [70, Theorem 9.3.2] for the full argument.

CLAIM 11.1.7. *Let $\Lambda=\Sigma^\sigma_{\mathcal{M}}$. Then the comparison of the phalanx $(\mathcal{M},\tilde{\mathcal{M}},\tilde{\tau})$ and $\mathcal{M}$ using Λ and $\Sigma_{\mathcal{M}}$ respectively is successful. Furthermore, the main branch on the phalanx side does not drop (in model or degree) and is above $\tilde{\mathcal{M}}$, and the $\mathcal{M}$ side does not move.*

PROOF. Using locally strong fullness preservation of Σ, Λ is fullness preserving; so the comparison can be carried out (see Chapter 4). By Lemma 11.1.4, the comparison is an extender comparison (no strategy disagreements show up in the comparison). Now we use locally strong branch condensation to prove the claim. The proof is a fairly standard argument. Let $\mathcal{T}$ and $\mathcal{U}$ be the trees on $(\mathcal{M},\tilde{\mathcal{M}},\tilde{\tau})$ and $\mathcal{M}$ respectively that are generated by the comparison (via Λ and $\Sigma_{\mathcal{M}}$ respectively). The comparison terminates successfully with $\mathcal{Q}$ being the last model of $\mathcal{T}$ and $\mathcal{S}$ being the last model of $\mathcal{U}$.

Let $\sigma\mathcal{T}$ be the copy tree and $\sigma^*: \mathcal{Q} \to \mathcal{Q}^*$ be the copy map, where $\mathcal{Q}^*$ is the last model of $\sigma\mathcal{T}$. Note then that $\sigma\mathcal{T}$ is via $\Sigma_{\mathcal{M}}$.

Suppose $\mathcal{Q}$ is above $\mathcal{M}$. We prove this case is impossible. Suppose $\mathcal{Q} \lhd \mathcal{S}$ and hence the branch embedding $\pi^{\mathcal{T}}$ exists. Note that $(\mathcal{Q}^*)^b \lhd \mathcal{Q}$ and $\mathcal{Q}^*$ is a non-dropping $\Sigma_{\mathcal{M}}$ iterate. Hence by strong branch condensation,

$$\Sigma^{\pi^{\mathcal{T}}}_{\mathcal{Q},\mathcal{T}} = \Sigma_{\mathcal{M}}.$$

The usual Dodd-Jensen argument yields a contradiction. The main point is that the tree $\pi^{\mathcal{T}}\mathcal{U}$ is via $\Sigma_{\mathcal{Q},\mathcal{T}}$.

Suppose now $\mathcal{S} \lhd \mathcal{Q}$ and hence the branch embedding $\pi^{\mathcal{U}}$ exists. Note then that $\sigma^*(\mathcal{S}) \lhd \mathcal{Q}^*$. Again, by strong branch condensation,

$$\Sigma^{\sigma^* \restriction \mathcal{S} \circ \pi^{\mathcal{U}}}_{\sigma^*(\mathcal{S}),\sigma\mathcal{T}} = \Sigma_{\mathcal{M}}.$$[10]

The usual Dodd-Jensen argument then yields a contradiction. The main point is that $(\sigma^* \restriction \mathcal{S} \circ \pi^{\mathcal{U}})\sigma\mathcal{T}$ is according to $\Sigma_{\mathcal{Q}^*,\sigma\mathcal{T}}$.

The above arguments easily give us that: $\mathcal{Q} = \mathcal{S}$ and $\pi^{\mathcal{T}}$, $\pi^{\mathcal{U}}$ both exist and they are equal. We can then find a pair of extenders (E,F) used on $\mathcal{T}$ and $\mathcal{U}$ respectively such that E and F are compatible. By a standard argument using the ISC, this is not possible.

Hence $\mathcal{Q}$ is on the main branch above $\tilde{\mathcal{M}}$. Note then that if $\pi^{\mathcal{T}}$ exists, then $\mathrm{cr}(\pi^{\mathcal{T}}) > \tilde{\tau}$. Say b is the main branch of $\mathcal{T}$. Then b cannot drop (in model or degree) as otherwise, we have $\mathcal{S} \lhd \mathcal{Q}$ and $\pi^{\mathcal{U}}$ exists. As before, $\sigma^*(\mathcal{S}) \lhd \mathcal{Q}^*$ and $\Sigma_{\mathcal{M}} = \Sigma^{\sigma^* \restriction \mathcal{S} \circ \pi^{\mathcal{U}}}_{\mathcal{Q}^*,\sigma\mathcal{T}}$. We get a contradiction as before.

So b does not drop. Since $\tilde{\mathcal{M}}$ is κ-sound, $\rho^{\omega}_{\tilde{\mathcal{M}}} = \kappa < \tilde{\tau}$ and the branch b is above $\tilde{\tau}$ and does not drop in model or degree, we get that $b = \{0\}$. And hence $\mathcal{Q} = \tilde{\mathcal{M}}$. Now it is not the case that $\mathcal{S}$ is a strict segment of $\mathcal{Q} = \tilde{\mathcal{M}}$; otherwise, $\pi^{\mathcal{U}}$ exists and

$$\sigma^* \circ \pi^{\mathcal{U}} : \mathcal{M} \to \sigma^*(\mathcal{S}) \lhd \mathcal{Q}^*.$$

We get a contradiction as before.

If $\mathcal{S} = \mathcal{Q} = \tilde{\mathcal{M}}$, then $\mathcal{U}$'s main branch does not drop. This is because $\tilde{\mathcal{M}}$ is sound. Note also that $\mathcal{U} \neq \emptyset$ since otherwise, $\mathcal{M} = \tilde{\mathcal{M}}$ which is impossible (after all, $\tau = (\kappa^+)^{\mathcal{M}} > \tilde{\tau} = (\kappa^+)^{\tilde{\mathcal{M}}}$). Now $\rho^{\omega}_{\mathcal{M}} = \rho^{\omega}_{\tilde{\mathcal{M}}} = \kappa$ and if there is an extender E used along the main branch of $\mathcal{U}$ such that $\nu(E) > \kappa$[11] then either $\mathcal{S}$ is not κ-sound or $\rho(\mathcal{S}) > \kappa$. Contradiction.

So for all E used along the main branch of $\mathcal{U}$, $\nu(E) \leq \kappa$. If for all such E, $\nu(E) < \kappa$, then since $\mathcal{M}|\kappa = \tilde{\mathcal{M}}|\kappa = \mathcal{Q}|\kappa$, $\mathcal{S}|\kappa \neq \mathcal{Q}|\kappa$. Contradiction. If there is some such E such that $\nu(E) = \kappa$, then $\mathcal{U}$ must drop since otherwise, $\rho^{\omega}_{\mathcal{S}} > \kappa$. Contradiction.

---

[10]Note that in this case, $\mathcal{S}^b = \sigma^*(\mathcal{S})^b$. The last equality follows from the fact that σ^* has critical point $> o(\mathcal{Q}^b) \geq o(\mathcal{S}^b)$.

[11]$\nu(E)$ is the sup of the generators of E.

So $\mathcal{Q} \lhd \mathcal{S}$. We claim that $\mathcal{Q} = \tilde{\mathcal{M}} \lhd \mathcal{M}$. It suffices to show $\mathcal{U} = \emptyset$. Otherwise, let $E = E_0^{\mathcal{U}}$. Then

$$\mathrm{lh}(E) \geq \tilde{\tau} \quad \text{and} \quad \mathrm{lh}(E) < o(\mathcal{Q}) = o(\tilde{\mathcal{M}}). \tag{3}$$

Note that $\mathrm{lh}(E)$ is a cardinal of $\mathcal{S}$ strictly larger than κ and $|o(\mathcal{Q})|^{\mathcal{S}} = \rho^{\omega}_{\tilde{\mathcal{M}}} = \kappa$. This contradicts (3). This completes the proof of Claim 11.1.7. ⊣

Using the claim, it is easy to see that $\tilde{\mathcal{M}} \lhd \mathcal{M}$ (that is, case (b) holds). This is because the branch embedding on the phalanx side must have critical point $> \kappa$ and $\tilde{\mathcal{M}}$ is κ-sound, so the branch is trivial with end model $\tilde{\mathcal{M}}$. ⊣

11.2. Ingredients from the Schimmerling-Zeman construction

In this section, we briefly remind the reader of the $\square$-construction in [38]. First, the reader should recall from [38] the notions of a *protomouse* and a *pluripotent level* of $L[E]$ (we give definitions of these notions in the context of hod premice in Section 11.3.1). See the beginning of [38, Section 2] for a fairly detailed discussion on how protomice appear in interpolation arguments. Basically, protomice arise in interpolation arguments where the target structure is a pluripotent level. The reader should see the definition of *divisor*, [38, Section 2.1], and *strong divisor*, [38, Section 2.4] (these notions are also defined in Section 11.3.1 for hod premice). Divisors identify protomice in interpolation arguments and (canonical) strong divisors in some sense are those (amongst many possible divisors of a given collapsing structure) that one uses in the course of the construction.

We proceed to briefly outline the proof of $\square_\kappa$ in $L[E]$ as done in [38]. To get the main ideas across in a reasonable amount of space, we will be imprecise at various places. The reader can see [38, Section 3] for a precise construction of the $\square_\kappa$-sequence $(C_\tau : \tau < \kappa^+)$. The proof starts by choosing the collapsing structure $\mathcal{N}_\tau$ for $\kappa < \tau < (\kappa^+)^{L[E]}$: $\mathcal{N}_\tau$ is the first level of $L[E]$ that satisfies "$\tau = \kappa^+$" and $\rho^{\omega}_{\mathcal{N}_\tau} = \kappa$. There is a club $\mathcal{S} \subset \kappa^+$ of such τ in $L[E]$. We further require that for each $\tau \in \mathcal{S}$, $\mathcal{J}^E_\tau \prec \mathcal{J}^E_{\kappa^+}$. For each $\tau \in \mathcal{S}$, let $\mathcal{S}_1 \subseteq \mathcal{S}$ be the set of τ for which the strong divisors of $\mathcal{N}_\tau$ exists (and let $(\mu(\mathcal{N}_\tau), q(\mathcal{N}_\tau))$ be the canonical strong divisor and $\mathcal{N}_\tau(\mu(\mathcal{N}_\tau), q(\mathcal{N}_\tau))$ be the unique associated protomouse as defined at the end of [38, Section 2]). Let $\mathcal{S}_0 = \mathcal{S} - \mathcal{S}_1$.

For $\tau \in \mathcal{S}_0$, the associated club $C_\tau \subset \tau$ can be constructed by Jensen's method of constructing $\square$-sequences in L. In this case, C_τ is obtained canonically from the set B_τ of all $\bar{\tau} \in \mathcal{S}_0 \cap \tau$ such that:

1. $\mathcal{N}_{\bar{\tau}}$ is a premouse of the same type as $\mathcal{N}_\tau$ and $n_{\bar{\tau}} = n_\tau$, where for a $\sigma \in \mathcal{S}$, n_σ is the least n such that $\omega\rho^{n+1}_{\mathcal{N}_\sigma} \leq \kappa < \omega\rho^{n}_{\mathcal{N}_\sigma}$.
2. There is a map $\sigma_{\bar{\tau},\tau} : \mathcal{N}_{\bar{\tau}} \to \mathcal{N}_\tau$ that is $\Sigma_0^{(n_\tau)}$-preserving with respect to the language of premice and such that: $\bar{\tau} = \mathrm{cr}(\sigma_{\bar{\tau},\tau})$, $\sigma_{\bar{\tau},\tau}(\bar{\tau}) = \tau$, $\sigma_{\bar{\tau},\tau}(p(\mathcal{N}_{\bar{\tau}})) = p(\mathcal{N}_\tau)$, and each $\alpha \in p(\mathcal{N}_\tau)$ has a generalized witness with respect to

$(\mathcal{N}_\tau, p(\mathcal{N}_\tau))$ in the range of $\sigma_{\bar{\tau},\tau}$. Here, and later, $p(\mathcal{N}_\tau)$ is the n_τ^{th}-standard parameter of $\mathcal{N}_\tau$.

For $\tau \in \mathcal{S}_1$, the set C_τ is obtained canonically from the set B_τ of $\bar{\tau} \in \mathcal{S}_1 \cap \tau$ that satisfies:

1. $(\mu(\mathcal{N}_{\bar{\tau}}), |q(\mathcal{N}_{\bar{\tau}})|) = (\mu(\mathcal{N}_\tau), |q(\mathcal{N}_\tau)|)$; here by definition of divisors, $q(\mathcal{N}_\tau)$ is a bottom initial segment of $d(\mathcal{N}_\tau)$, the Dodd-parameter of $\mathcal{N}_\tau$.
2. There is a map $\sigma_{\bar{\tau},\tau} : \mathcal{N}_{\bar{\tau}}(\mu(\mathcal{N}_{\bar{\tau}}), q(\mathcal{N}_{\bar{\tau}})) \to \mathcal{N}_\tau(\mu(\mathcal{N}_\tau), q(\mathcal{N}_\tau))$ that is Σ_0-preserving with respect to the language for coherent structures such that: $\bar{\tau} = \mathrm{cr}(\sigma_{\bar{\tau},\tau})$, $\sigma_{\bar{\tau},\tau}(\bar{\tau}) = \tau$, $\sigma_{\bar{\tau},\tau}(q(\mathcal{N}_{\bar{\tau}})) = q(\mathcal{N}_\tau)$, and each $\alpha \in q(\mathcal{N}_\tau)$ has a generalized witness (with respect to $(\mathcal{N}_\tau(\mu(\mathcal{N}_\tau), q(\mathcal{N}_\tau)), q(\mathcal{N}_\tau))$ in the range of $\sigma_{\bar{\tau},\tau}$.

See [38, Section 3] for how C_τ is defined from B_τ. Now we focus on the key point: the proof that

$$B_\tau \text{ is unbounded in } \tau \text{ if } \tau \in \mathcal{S}^1 \text{ and } \mathrm{cof}(\tau) > \omega \text{ in } L[E].$$

Fix such a τ and let $\kappa < \gamma < \tau$ be arbitrary. We want to find a $\gamma < \bar{\tau} < \tau$ in B_τ. Working in $L[E]$, fix some $\vartheta >> \kappa$ and let $X \prec H_\vartheta$ be countable such that all relevant objects are in X, in particular $\{\kappa, \tau, \gamma\} \in X$. Let $\sigma : \bar{\mathcal{M}} \to \mathcal{M}$ be the uncollapse map of $X \cap \mathcal{M}$, where $\mathcal{M} = \mathcal{N}(\mu(\mathcal{N}_\tau), q(\mathcal{N}_\tau))$. We write $\sigma^{-1}(x) = \bar{x}$ for each x in the range of σ. Let $\tilde{\tau} = \sup(\sigma''\bar{\tau})$. Let $\tilde{\sigma} : \bar{\mathcal{M}} \to \tilde{\mathcal{M}}$ come from the $(\mathrm{cr}(\sigma), \tilde{\tau})$-extender derived from σ. Also, let $\sigma' : \tilde{\mathcal{M}} \to \mathcal{M}$ be given by the interpolation lemma [38, Lemma 1.2]. In this case, $\tilde{\mathcal{M}} = (\mathcal{N}, \tilde{F})$ is a protomouse (even if $\mathcal{N}(\mu(\mathcal{N}_\tau), q(\mathcal{N}_\tau)) = \mathcal{N}_\tau$ since in this case, $\mathcal{N}_\tau$ is a pluripotent level of $L[E]$ and the map σ' is not cofinal). The way one shows $\tilde{\tau} \in B_\tau$ is as follows. Let $\mathcal{M}^*$ be the largest segment of $\mathcal{N}$ such that $\tilde{F}$ measures all sets in $\mathcal{M}^*$. One then shows that $\mathrm{Ult}(\mathcal{M}^*, \tilde{F})$ is $\mathcal{N}_{\tilde{\tau}}$. Say $\mathcal{M} = (\mathcal{M}^-, F)$. This is accomplished by applying the condensation lemma [38, Lemma 1.6] to $\varphi : \mathrm{Ult}(\mathcal{M}^*, \tilde{F}) \to \pi_F(\mathcal{M}^*)$ such that

$$\varphi(\pi_{\tilde{F}}(f)(a)) = \sigma'(f)(\pi_F(a))$$

where π_F is the F-ultrapower embedding applied to the largest initial segment of $\mathcal{M}^-$ that makes sense.

The key for the proof above is that we can always compare two iterable pure extender models; in this case, we compare the phalanx $(\pi_F(\mathcal{M}^*), \mathrm{Ult}(\mathcal{M}^*, \tilde{F}), \tilde{\tau})$ against $\pi_F(\mathcal{M}^*)$. If one adapted this argument to hod mice, it fails because the hod mice $\pi_F(\mathcal{M}^*)$ and $\mathrm{Ult}(\mathcal{M}^*, \tilde{F})$ generally belong to two different pointclasses, and hence cannot be directly compared. This traces back to the fact that if $\mathcal{P}$ is a hod premouse of limit type with a top window, then $(\delta)^{\mathcal{P}^b}$ is a strong cutpoint of $\mathcal{P}^b$. The fix for this, as done in the next section, is to sometimes allow for the collapsing structure of τ, $\mathcal{N}_\tau$, to *not be an initial segment of* the hod mouse and incorporate this kind of collapsing structures into the construction. It is this aspect that forces the construction to yield a weaker result, i.e., $\square_{\kappa,2}$, rather than $\square_\kappa$.

One other new situation in the hod mouse case that does not come up in the $L[E]$ case is the following. Suppose in the above, $\mathcal{M} = \mathcal{N}_\tau$ is B-active. Then the way branches are coded into the model (using the $\mathfrak{B}$-operator as discussed in the previous section) allows us to show that $\tilde{M}$ is a B-active hod premouse. If one used the traditional coding of branches, then $\tilde{M}$ may fail to be a hod premouse due to the lack of condensation one can prove with the traditional coding device; this is the reason we switch to the coding of branches via the $\mathfrak{B}$-operator. We will discuss this in more details in the next section.

11.3. The proof

We give a proof of Theorem 11.0.5, making use of the notions, notations, and proofs in [38] whenever applicable. We only focus on the details that are new in our situation and direct the reader to constructions in [38] that are obviously generalizable to our situation. To make the situation more concrete, we make the following assumption about $\mathcal{P}$:

$$\mathcal{P} \text{ is of lsa type.} \tag{4}$$

So under (4), $\mathcal{P}$ has the top window $(\mu, \delta^{\mathcal{P}})$, μ is strong to $\delta^{\mathcal{P}}$. This is the hardest case and we will focus on the $\square$-construction under this assumption. The other cases where $\mathcal{P}$ is not of lsa type are easier and the proof is a simpler version of what we are about to give.

11.3.1. Some set-up. We will use the fine-structure terminology and notations from [38, Section 1], generalized to our context in an obvious way. For example, notions in [38] that are defined using the language of premice are defined here using the language of hod premice; when we talk about a *coherent structure* in this paper, we mean a structure M of the form $(\mathcal{Q}, F)$ where $\mathcal{Q}$ is an amenable structure in the language of hod premice and F is a whole extender at (κ, λ) (in the language of [38, Section 1]) and $\mathcal{Q} = \text{Ult}(\mathcal{Q}|\bar{\alpha}, F)$, where $\bar{\alpha}$ is the largest $\tau \leq \kappa^{+\mathcal{Q}}$ such that $\text{dom}(F) = \wp(\kappa) \cap \mathcal{Q}|\tau$. We say N is *the hod premouse associated with M*. The notion of *a generalized witness* for some ordinal α with respect to a pair (M, s) where M is a coherent structure, s is a finite set of ordinals (or a generalized witness for an ordinal α with respect to a hod premouse N associated with M and some finite set of ordinals $r \cup s$) in [38] is generalized in an obvious way to our context.[12] A *protomouse* $\mathcal{P} = (\mathcal{Q}, F)$ is a coherent structure where F is an extender

[12]Let M, N, κ, λ be as above and $s \subset \lambda$ is finite. The *standard witness* $W_M^{\alpha,s}$ for α with respect to M and s to be the transitive collapse of $h_M(\alpha \cup \{s\})$, where h_M is the canonical Σ_1-Skolem function of the coherent structure M. Similarly, $W_N^{\alpha, r\cup s}$ denotes the standard witness for α with respect to N and $r \cup s$ and is the transitive collapse of $\tilde{H}_N^{n+1}(\alpha \cup \{r \cup s\})$, where $\tilde{h}_N^{n+1}$ is the canonical $\Sigma_1^{(n)}$-Skolem function of the hod premouse N. A *generalized witness* for α with respect to M and s is a pair (Q, t), where $t \subset Q$ is a finite set of ordinals and such that for any $\xi_1, \ldots, \xi_l < \alpha$, if $M \models \Phi(i, \xi_1, \ldots, \xi_l, s)$ then $Q \models \Phi(i, \xi_1, \ldots, \xi_l, t)$, where Φ is the universal Σ_1-formula. A generalized witness for α with respect

with critical κ such that F does not measure $\wp(\kappa)^{\mathcal{Q}}$. A *pluripotent level* of a hod premouse $\mathcal{P}$ is an E-active initial segment $\mathcal{Q}$ of $\mathcal{P}$ such that $\mathrm{cr}(E^{\mathcal{Q}}_{\mathrm{top}}) < \kappa$ and $\omega\rho^1_{\mathcal{Q}} = \kappa$, where κ is a cardinal of $\mathcal{P}$. The language used for treating protomice and pluripotent levels is the language of coherent structures, namely $\mathcal{L}_1 - \{\dot{\gamma}\}$.[13]

Fix $(\mathcal{P}, \Sigma)$ as in the hypothesis of Theorem 11.0.5. Fix $\kappa \geq \delta^{\mathcal{P}^b}$, a cardinal of $\mathcal{P}$. Working in $\mathcal{P}$, let $\mu = \delta^{\mathcal{P}^b}$ and $\mathcal{S} \subset \kappa^+$ be the club of $\kappa < \tau < \kappa^+$ such that for some $\tau < \tau' < \kappa^+$, $\mathcal{P}|\tau' \prec \mathcal{P}|\kappa^{++}$ and $\tau = (\kappa^+)^{\mathcal{P}|\tau'}$. We note that since we are below superstrong, the set of indices below κ^+ is non-stationary in κ^+; therefore, we can assume the club $\mathcal{S}$ consists of τ such that $\mathcal{P}|\tau$ is passive.

Let $\mathcal{N}^*_\tau \lhd \mathcal{P}$ be the collapsing level for τ, that is $\mathcal{N}^*_\tau$ the least initial segment $\mathcal{N}$ of $\mathcal{P}$ such that $\mathcal{N} \models \tau = \kappa^+$ and $\rho^\omega_{\mathcal{N}} = \kappa$. Let γ_τ be the supremum of indexes γ of extenders $E \in \dot{E}^{\mathcal{N}^*_\tau}$ such that $\mathrm{cr}(E) = \mu$. Note that $\gamma_\tau < o(\mathcal{N}^*_\tau)$ and if $\tau < \sigma$ in $\mathcal{S}$, then $\mathcal{N}^*_\tau \lhd \mathcal{N}^*_\sigma$ and therefore, $\gamma_\tau \leq \gamma_\sigma$. Without loss of generality, we may assume throughout this chapter that

$$\kappa \geq o(\mathcal{P}^b) \quad \text{and} \quad \sup_{\tau \in \mathcal{S}}(\gamma_\tau) \geq \kappa^+.$$[14]

The following statements can be easily verified using the definitions and our assumption.

PROPOSITION 11.3.1. *For a club of $\tau \in \mathcal{S}$:*

1. $o(\mathcal{N}^*_\tau) > \tau$,
2. $\gamma_\tau \geq \tau$.

PROOF. By the construction of $\mathcal{S}$, for any $\tau \in \mathcal{S}, \mathcal{P}|\tau$ is passive and is a ZFC^--model, and therefore $\rho^{\mathcal{P}|\tau}_\omega = \tau$. This means $o(\mathcal{P}|\tau) < o(\mathcal{N}^*_\tau)$. This proves (1).

For (2), suppose not. Then there is a stationary set of $\tau \in \mathcal{S}$ such that $\gamma_\tau < \tau$. By pressing down, there is a stationary set $T \subset \mathcal{S}$ and a $\gamma < \kappa^+$ such that for all $\tau \in T$,

$$\gamma_\tau = \gamma < \tau.$$

But then since the sequence $(\gamma_\tau : \tau \in \mathcal{S})$ is monotonically non-decreasing, this means that

$$\sup_{\tau \in \mathcal{S}}(\gamma_\tau) = \gamma < \kappa.$$

---

to N and $r \cup s$ is a pair (Q, t), where $t \subset Q$ is a finite set of ordinals such that given any $\xi_1, \ldots, \xi_l < \alpha$, if $N \models \Phi(i, \xi_1, \ldots, \xi_l, r \cup s)$ then $Q \models \Phi(i, \xi_1, \ldots, \xi_l, t)$, where Φ is the universal $\Sigma^{(n)}_1$-formula.

[13] We note again that for type A hod premice, which are all the hod premice that we encounter in this book, $\dot{\gamma} = \emptyset$, so there is essentially no distinction between the language of hod premice and coherent structures.

[14] If $\kappa = \delta^{\mathcal{P}}_\alpha$, where $\mu = \delta^{\mathcal{P}}_\alpha$, then since $\mathcal{P}^b = \mathcal{P}|(\kappa^+)^{\mathcal{P}} = \mathrm{Lp}^{\oplus_{\beta<\alpha}\Sigma^{\mathcal{P}}_{\mathcal{P}(\beta)}}(\mathcal{P}|\delta^{\mathcal{P}}_\alpha)$, then $\mathcal{P} \models \square_\kappa$ since κ is a strong cutpoint cardinal of $\mathcal{P}^b$. If $\kappa > \delta^{\mathcal{P}}_\alpha$ and $\sup_{\tau \in \mathcal{S}}(\gamma_\tau) < \kappa^+$, then the proof is significantly easier. One constructs the $\square_\kappa$-sequence using points $\tau \in \mathcal{S}$ above $\sup_{\tau \in \mathcal{S}}(\gamma_\tau)$ mimicking essentially the Schimmerling-Zeman construction and use Theorem 11.1.5. If $\kappa < \mu$, we will then be showing $\mathcal{P}(\alpha) \models \square_{\kappa,2}$, where $\mathcal{P}(\alpha)$ plays the role of $\mathcal{P}$.

This contradicts our assumption. ⊣

Extenders E with $\mathrm{cr}(E) = \mu$ play a special role in this construction. Note that μ is a strong cutpoint of $\mathcal{P}^b$, that is, there are no partial extenders with critical point μ on the sequence of $\mathcal{P}$. This is the main difference between our situation and the $L[E]$-situation. It is this situation that forces us to consider collapsing structures that are not initial segments of our hod mouse $\mathcal{P}$.

Some discussions regarding protomice and divisors are in order. Following [38], for a hod premouse $\mathcal{N}$ such that $\omega\rho_{\mathcal{N}}^{n+1} \leq \kappa < \omega\rho_{\mathcal{N}}^{n}$, we say that (ν, q) is a *divisor* of $\mathcal{N}$ if and only if there is an ordinal $\lambda = \lambda_{\mathcal{N}}(\nu, q)$ such that letting $p_{\mathcal{N}}$ be the $(n+1)$-st standard parameter of $\mathcal{N}$, setting $r = p_{\mathcal{N}} - q$, the following hold:

(a) $\nu \leq \kappa < \lambda < \omega\rho_{\mathcal{N}}^{n}$;
(b) $q = p_{\mathcal{N}} \cap \lambda$;
(c) $\tilde{h}_{\mathcal{N}}^{n+1}(\nu \cup \{r\}) \cap \omega\rho_{\mathcal{N}}^{n}$ is cofinal in $\omega\rho_{\mathcal{N}}^{n}$;
(d) $\lambda = \min(\mathrm{OR} \cap \tilde{h}_{\mathcal{N}}^{n+1}(\nu \cup \{r\}) - \nu)$.

As in [38], both ν and λ are (inaccessible) cardinals in $\mathcal{N}$. Let $\mathcal{N}^*(\nu, q)$ be the transitive collapse of $\tilde{h}_{\mathcal{N}}^{n+1}(\nu \cup \{r\})$.

The notion of strong divisors in [38] generalize in an obvious way to our context. We recall it now. A divisor (ν, q) of $\mathcal{N}$ is *strong* if and only if for every $\xi < \nu$ and every x of the form $\tilde{h}_{\mathcal{N}}^{n+1}(\xi, p_{\mathcal{N}})$ we have $x \cap \nu \in \mathcal{N}^*(\nu, q)$. If $\mathcal{N}$ is pluripotent, we define the notion of strong divisor in the same way, but with $h_{\mathcal{N}}^*$ (the Σ_1-Skolem function of $\mathcal{N}$ computed in the language of coherent structures) and $d_{\mathcal{N}}$ (the Dodd-parameter of $\mathcal{N}$) in place of $\tilde{h}_{\mathcal{N}}^{n+1}$ and $p_{\mathcal{N}}$, respectively. It turns out, see [38, Section 2], that the notion of divisor/strong divisor does not depend on which language one uses to compute it.

As in [38], if $\mathcal{N}$ has strong divisors, *the canonical strong divisor* $(\mu_{\mathcal{N}}, q_{\mathcal{N}})$ of $\mathcal{N}$ is chosen as follows: $q_{\mathcal{N}}$ is the shortest initial segment of $p_{\mathcal{N}}$ such that for some ν^*, $(\nu^*, q_{\mathcal{N}})$ is a strong divisor of $\mathcal{N}$ and $\mu_{\mathcal{N}}$ is the largest ν^* such that $(\nu^*, q_{\mathcal{N}})$ is a strong divisor of $\mathcal{N}$. Now we define our collapsing structure $\mathcal{N}_\tau$ for $\tau \in \mathcal{S}$.

DEFINITION 11.3.2. Fix $\tau \in \mathcal{S}$. Suppose there is a pointclass $\Omega \subsetneq \Gamma$ such that there is a hod pair $(\mathcal{R}, \Sigma_{\mathcal{R}})$ such that

(i) $\mathcal{N}_\tau^* | \gamma_\tau \lhd \mathcal{R}$,
(ii) $\rho_{\mathcal{R}}^{\omega} = \kappa$,
(iii) γ_τ is a cutpoint of $\mathcal{R}$ and $\Sigma_{\mathcal{R}|\gamma_\tau} = \Sigma_{\mathcal{P}|\gamma_\tau}$,
(iv) $\mathcal{R}$ is γ_τ-sound,
(v) the order type of $\mathcal{R}$'s layer Woodin cardinals above γ_τ is a limit ordinal,
(vi) $\mathcal{R}$ has a strong divisor of the form (μ, q) where $p_{\mathcal{R}} = q \cup r$ for r above the supremum λ of the layer Woodin cardinals of $\mathcal{R}$ and $\max(q)$ is below $(\gamma_\tau^+)^{\mathcal{R}}$ and $(\gamma_\tau^+)^{\mathcal{R}} < \lambda$,
(vii) $\Sigma_{\mathcal{R}}$ has branch condensation, is Ω-fullness preserving, and $(\mathcal{R}, \Sigma_{\mathcal{R}})$ generates Ω; that is $\Gamma(\mathcal{R}, \Sigma_{\mathcal{R}}) = \Omega$.

We call $(\mathcal{R},\Sigma_{\mathcal{R}})$ with the above properties the pointclass generator of Ω. Let Γ_τ be the Wadge-minimal such pointclass and $\mathcal{N}_\tau$ be the pointclass generator of Γ_τ, $(\mu_\tau,q_\tau,\lambda_\tau)$ be the (μ,q,λ) associated with $\mathcal{N}_\tau$ as above (note that $\mathcal{N}_\tau$ must be distinct from $\mathcal{N}_\tau^*$ in this case). If $(\Gamma_\tau,\mathcal{R},\mu,q,\lambda)$ does not exist, we let $\mathcal{N}_\tau=\mathcal{N}_\tau^*$.

The properties of pointclass generators seem technical; these properties are abstracted from various situations in interpolation arguments. It seems hard to do with much less. Here is a very rough explanation for why we would consider such objects before going into details: suppose $(\Gamma_\tau,\mathcal{R},\mu,q,\lambda)$ exists, then letting $\pi:\mathcal{N}_\tau^*(\mu,q)\to\mathcal{R}$ be the uncollapse map and F be the (μ,γ_τ)-extender derived from π. Letting $\mathcal{R}^*\lhd\mathcal{P}|\mu^+$ be the largest such that $\rho_\omega^{\mathcal{S}}=\mu$ and $\wp(\mu)\cap\mathcal{R}^*=\wp(\mu)\cap\mathcal{N}_\tau^*(\mu,q)$., then we can show that $\mathcal{R}=\mathrm{Ult}(\mathcal{R}^*,F)$. Since $\wp(\mu)\cap\mathcal{R}^*\subsetneq\wp(\mu)\cap\mathcal{P}$, $\mathcal{R}$ belongs to Γ and will generate a pointclass (Γ_τ) strictly smaller than Γ. It turns out that there is a unique such $(\mathcal{R},\Sigma_{\mathcal{R}})$ that generates Γ_τ. This canonicity is important in the $\square$-construction.

The following proposition justifies the uniqueness of pointclass generators.

PROPOSITION 11.3.3. *Let $\mathcal{P},\tau,\Omega$ be as in Definition 11.3.2. Let $(\mathcal{R}_0,\Sigma_0)$ and $(\mathcal{R}_1,\Sigma_1)$ be pointclass generators of Ω with the properties described in 11.3.2. Then $(\mathcal{R}_0,\Sigma_0)=(\mathcal{R}_1,\Sigma_1)$.*

PROOF. We compare the pair $(\mathcal{R}_0,\Sigma_0)$ against $(\mathcal{R}_1,\Sigma_1)$, lining up the models and the strategies (as done in Section 4.13). The comparison is possible by the assumption, namely $\Gamma(\mathcal{R}_0,\Sigma_0)=\Gamma(\mathcal{R}_1,\Sigma_1)=\Omega$, and is above γ_τ (this is because γ_τ is a cutpoint, in fact a strong cutpoint, of both models and $\mathcal{R}_0|\gamma_\tau=\mathcal{R}_1|\gamma_\tau$). The end model is, say, $\mathcal{S}$ and the tail strategies of Σ_0 and Σ_1 on $\mathcal{S}$ are the same. The usual proof using the fact that $\mathcal{R}_0$ and $\mathcal{R}_1$ are γ_τ-sound and the comparison is above γ_τ shows that $\mathcal{S}=\mathcal{R}_0=\mathcal{R}_1$ (the comparison is trivial) and $\Sigma_0=\Sigma_1$. The equality of models (i.e., $\mathcal{R}_0=\mathcal{R}_1$) follows from the fact that $(\mathcal{R}_0,\Sigma_0)$ and $(\mathcal{R}_1,\Sigma_1)$ generate the same pointclass. Otherwise, say $\mathcal{R}_0=\mathcal{R}_1(\alpha)\lhd\mathcal{R}_1$ for some α and $\Sigma_0=(\Sigma_1)_{\mathcal{R}_0}$. It is easy to see that from (ii) and (v), there is a layer Woodin cardinal β of $\mathcal{R}_1$ such that $\mathcal{R}_0\lhd\mathcal{R}_1(\beta)$. But this means that $\Omega=\Gamma(\mathcal{R}_0,\Sigma_0)\subsetneq\Gamma(\mathcal{R}_1(\beta),(\Sigma_1)_{\mathcal{R}_1(\beta)})\subseteq\Gamma(\mathcal{R}_1,\Sigma_1)=\Omega$. Contradiction. $\dashv$

We simply use the notations from [38, page 49] in the definition of our square sequence below. For instance, (μ_τ,q_τ) denotes the canonical strong divisor of $\mathcal{N}_\tau$ (if exists) in the case $\mathcal{N}_\tau=\mathcal{N}_\tau^*$ and denotes the (μ_τ,q_τ) in Definition 11.3.2 in the case $\mathcal{N}_\tau\neq\mathcal{N}_\tau^*$ (note that (μ_τ,q_τ) is the unique strong divisor of $\mathcal{N}_\tau$ with the properties as in Definition 11.3.2). If $\mathcal{N}_\tau=\mathcal{N}_\tau^*$ is a pluripotent level that has no strong divisors, then (μ_τ,q_τ) denotes $(\mathrm{cr}(E^{\mathrm{top}}_{\mathcal{N}_\tau}),d(\mathcal{N}_\tau))$, where $d(\mathcal{N}_\tau)$ is the Dodd-parameter of $\mathcal{N}_\tau$.

Suppose (ν,q) is a divisor of $\mathcal{N}_\tau$; let r,λ,n be as in the definition of divisor. Let $\pi:\mathcal{N}_\tau^*(\nu,q)\to\tilde{h}_{\mathcal{N}}^{n+1}(\nu\cup\{r\})$ be the uncollapse map. We let the associated protomouse $\mathcal{N}_\tau(\nu,q)$ be the coherent structure $(\mathcal{N}_\tau|\xi,F)$ where $\xi=\pi((\nu^+)^{M^*})$ and $F=E_\pi\restriction(\wp(\nu)\cap\mathcal{N}_\tau^*(\nu,q))$. We denote the λ associated to ν,q in the definition of divisor $\lambda_{\mathcal{N}_\tau}(\nu,q)$. We let $\mathcal{M}_\tau = \mathcal{N}_\tau(\mu_\tau,q_\tau)$ be the protomouse associated with (μ_τ,q_τ). If $\mathcal{N}_\tau$ is pluripotent, we let $\mathcal{M}_\tau = \mathcal{N}_\tau$.

The following proposition is easy to see and justifies that the structure $\mathcal{N}_\tau(\nu,q)$ are protomice (and not hod premice). See [38, Section 2.1] for a detailed discussion and proof.

PROPOSITION 11.3.4. *Suppose (ν,q) be a divisor of $\mathcal{N}_\tau$ and $\pi : \mathcal{N}^*_\tau(\nu,q) \to \tilde{h}^{n+1}_{\mathcal{N}_\tau}(\nu \cup \{r\})$ be the uncollapse map (and in the case $\mathcal{N}_\tau \neq \mathcal{N}^*_\tau$, assume $\nu = \mu$). Then $\wp(\nu) \cap \mathcal{N}^*_\tau(\nu,q) \subsetneq \wp(\nu) \cap \mathcal{N}_\tau$. Furthermore, ν is an (inaccessible) cardinal of $\mathcal{N}^*_\tau(\nu,q)$ and a limit cardinal of $\mathcal{N}_\tau$, and $\lambda_{\mathcal{N}_\tau}(\nu,q)$ is an (inaccessible) cardinal of $\mathcal{N}_\tau$.*

DEFINITION 11.3.5. Let $\mathcal{S}^1 \subset \mathcal{S}$ be the set of τ such that (μ_τ,q_τ) is defined and $\mathcal{S}^0 = \mathcal{S} - \mathcal{S}^1$.

Suppose $\mathcal{N}_\tau = \mathcal{N}^*_\tau$, then no divisors of $\mathcal{N}_\tau$ are of the form (μ,q). This is because otherwise, $\lambda = \lambda_{\mathcal{N}_\tau}(\mu,q)$ is a limit of Woodin cardinals because it is the image of μ under the uncollapse map and μ itself is a limit of Woodin cardinals. Let $\gamma_0 < \gamma_1$ be consecutive Woodin cardinals in the interval (μ,λ); then by definition of $\mathcal{P}$, $\mathcal{P}|\gamma_1$ is a Λ^{sts}-mouse where Λ is the strategy of $M^+(\mathcal{P}|\gamma_0)$. On the other hand, by elementarity, $\mathcal{P}|\gamma_1$ is a Λ-mouse. Contradiction.[15]

A similar argument applies to show that no divisors for $\mathcal{N}_\tau$ are of the form (ξ,q) for $\xi < \mu$; though we do not need this fact in our construction as no divisors (ν,q) in this proof will have the property that $\nu < \mu$. So if (ν,q) is a divisor of $\mathcal{N}_\tau$, then $\nu > \mu$. This allows us to simply quote results of [38, Section 2] in this case (in light of Theorem 11.1.5). In the case that $\mu_\tau = \mu$ (so $\mathcal{N}_\tau \neq \mathcal{N}^*_\tau$), more care needs to be taken since it is not obvious that all results in [38, Section 2.4] can be generalized to this case.

Using the remarks above, it is easy to see that if $\mathcal{N}_\tau \neq \mathcal{N}^*_\tau$, then $\tau \in \mathcal{S}^1$ and in fact $\mathcal{N}_\tau$ is not an initial segment of $\mathcal{P}$ (though $\mathcal{N}_\tau \in \mathcal{P}$ by Proposition 11.3.6); also, if $\mathcal{N}_\tau = \mathcal{N}^*_\tau$ is pluripotent, then $\tau \in \mathcal{S}^1$. For $\tau \in \mathcal{S}^0$, $\mathcal{N}_\tau = \mathcal{N}^*_\tau$ is not pluripotent and does not admit a strong divisor.

The following lemma allows us to define our $\square_{\kappa,2}$-sequence in a uniform manner.

PROPOSITION 11.3.6. *Suppose $\mathcal{N}_\tau \neq \mathcal{N}^*_\tau$. Then $\mathcal{N}_\tau$ is definable over $\mathcal{P}$ (in fact, over any $\mathcal{N}^*_\xi$ or $\mathcal{N}_\xi$ with $\gamma_\xi \geq \xi$ for $\xi > \tau$ in $\mathcal{S}$) unformly from $\{\tau,\gamma_\tau\}$.*

PROOF. Fix $\xi > \tau$ in $\mathcal{S}$ with $\gamma_\xi \geq \xi$. We first claim that $\gamma_\xi > \gamma_\tau$. To see this, note that $\tau \leq \gamma_\tau < o(\mathcal{N}^*_\tau) < \xi$. This is because ξ is a cardinal (successor of κ) in $\mathcal{N}_\xi$ while there is a surjection from κ onto γ_τ in $\mathcal{N}_\xi$. Since $\xi \leq \gamma_\xi$, the claim follows.

---

[15]Another argument is as follows. Note that each Woodin cardinal in the interval (μ,λ) is $> (\mu^+)^{\mathcal{P}}$, and hence μ is strong to λ (in $\mathcal{P}$) by the initial segment condition. This contradicts the definition and smallness assumption on $\mathcal{P}$ since one can easily find an active ω Woodin lsa mouse in $\mathcal{P}$ (as defined in Definition 8.2.2).

Now let E be the extender on the $\mathcal{N}_\xi$-sequence such that $\mathrm{cr}(E) = \mu$, $\mathrm{lh}(E) > \gamma_\tau$, and is the least such.[16] Let $\mathcal{S} = \mathrm{Ult}(\mathcal{N}_\xi, E)$ (this is a Σ_0-ultrapower). Let $i : \mathcal{S} \to \mathcal{S}_\infty$ be an $\mathbb{R}$-genericity iteration (above γ_τ). Now it is easy to see that in the derived model of $\mathcal{S}_\infty$ (at the sup of its Woodin cardinals), the pointclass Ω in the definition of $\mathcal{N}_\tau$ is a strict Wadge initial segment of $\wp(\mathbb{R})$ and is definable there from $\{\tau, \gamma_\tau\}$ (by Lemma 11.3.3). Then $\mathcal{N}_\tau \in \mathcal{S}_\infty$ and in fact is definable there from parameters $\{\tau, \gamma_\tau\}$. The same holds in $\mathcal{S}$ by elementarity and the fact that $\mathrm{cr}(i) > \gamma_\tau$. Finally, $\mathcal{N}_\tau \in \mathcal{N}_\xi$ and is definable there from parameters $\{\tau, \gamma_\tau, E\}$. But E is definable in $\mathcal{N}_\xi$ from $\{\tau, \gamma_\tau\}$. So $\mathcal{N}_\tau$ is definable in $\mathcal{N}_\xi$ from $\{\tau, \gamma_\tau\}$. ⊣

REMARK 11.3.7. By our smallness assumption on $\mathcal{P}$, the set $\mathfrak{A} = \{\xi \mid \kappa < \xi < \kappa^+ \wedge \mathcal{P}|\xi \text{ is } E\text{-active}\}$ is non-stationary in $\mathcal{P}$. The reason is $\mathfrak{A} = \mathfrak{A}_0 \cup \mathfrak{A}_1$. Here $\mathfrak{A}_0$ consists of ξ's such that the top extender of $\mathcal{P}|\xi$ has critical point μ and $\mathfrak{A}_1 = \mathfrak{A} - \mathfrak{A}_0$. $\mathfrak{A}_0$ in nonstationary by Footnote 16. $\mathfrak{A}_1$ is nonstationary because otherwise, κ is subcompact by [38]. As in [38], the fact that $\mathfrak{A}$ is nonstationary is crucial in our construction. We use this fact in various arguments to follow.

11.3.2. Approximation of a $\square_{\kappa,2}$ sequence. We use the notation established in the previous section. Below, as in [38], n_τ is the unique n such that $\rho^{n+1}_{\mathcal{N}_\tau} = \kappa < \rho^n_{\mathcal{N}_\tau}$ and p_τ is the standard parameter of $\mathcal{N}_\tau$. Let also p^*_τ be the standard parameter of $\mathcal{N}^*_\tau$.

DEFINITION 11.3.8. Suppose $\tau \in \mathcal{S}^0$, let $\vec{B}_\tau = \{B^0_\tau\}$ be the set of $\bar{\tau} \in \mathcal{S} \cap \tau$ satisfying:

1. $\mathcal{N}_{\bar{\tau}}$ is a hod premouse of the same type as $\mathcal{N}_\tau$.[17]
2. $n_\tau = n_{\bar{\tau}}$.
3. There is a map $\sigma^0_{\bar{\tau}\tau} : \mathcal{N}^*_{\bar{\tau}} \to \mathcal{N}_\tau$ that is $\Sigma_0^{(n_\tau)}$-preserving with respect to the language of hod premice such that
   (a) $\bar{\tau} = \mathrm{cr}(\sigma^0_{\bar{\tau}\tau})$ and $\sigma^0_{\bar{\tau}\tau}(\bar{\tau}) = \tau$.
   (b) $\sigma^0_{\bar{\tau}\tau}(p^*_{\bar{\tau}}) = p_\tau$.
   (c) for each $\alpha \in p_\tau$, there is a generalized witness for α with respect to $\mathcal{N}_\tau$ and p_τ in the range of $\sigma_{\bar{\tau}\tau}$.

Note that if $\tau \in \mathcal{S}^0$, then $\mathcal{N}^*_\tau = \mathcal{N}_\tau$ and either $\mathrm{crt}(E^{top}_{\mathcal{N}_\tau}) \geq \kappa$ or $\rho_1^{\mathcal{N}_\tau} > \kappa$. Recall the definition of (μ_τ, q_τ), $p_\tau, d_\tau, \mathcal{M}_\tau$ for $\tau \in \mathcal{S}^1$ in Section 11.3.1. Below, m_τ is $|q_\tau|$. We also let $r_\tau = d_\tau - q_\tau$ be the top part of d_τ.

DEFINITION 11.3.9. Suppose $\tau \in \mathcal{S}^1$. Let B^1_τ be the set of $\bar{\tau} \in \mathcal{S}^1 \cap \tau$ satisfying:

[16] We note that the set of indices for extenders with critical point μ is nowhere continuous.

[17] In this case, it simply means: $\mathcal{N}_\tau$ is E (B)-active if and only if $\mathcal{N}_{\bar{\tau}}$ is E (B)-active. If $\mathcal{N}_\tau$ is E-active (equivalently, $\mathcal{N}_{\bar{\tau}}$ is E-active), then $E^{\mathrm{top}}_{\mathcal{N}_\tau}$ is indexed according to the cutpoint (Jensen) indexing scheme if and only if $E^{\mathrm{top}}_{\mathcal{N}_{\bar{\tau}}}$ is indexed according to the cutpoint (Jensen, respectively) indexing scheme. Recall that all E-active hod mice, where E is indexed according to the Jensen indexing scheme, in this proof will be of type A, i.e., the set of cutpoints is empty.

1. $(\mu_{\bar{\tau}}, m_{\bar{\tau}}) = (\mu_\tau, m_\tau)$.
2. There is a map $\sigma^1_{\bar{\tau}\tau} : \mathcal{M}_{\bar{\tau}} \to \mathcal{M}_\tau$ that is Σ_0-preserving with respect to the language of coherent structures such that
   (a) $\bar{\tau} = \mathrm{cr}(\sigma^1_{\bar{\tau}\tau})$ and $\sigma^1_{\bar{\tau}\tau}(\bar{\tau}) = \tau$.
   (b) $\sigma^1_{\bar{\tau}\tau}(q_{\bar{\tau}}) = q_\tau$.
   (c) for each $\alpha \in q_\tau$, there is a generalized witness for α with respect to $\mathcal{M}_\tau$ and q_τ in the range of $\sigma^1_{\bar{\tau}\tau}$.

Suppose in addition that either $\mathrm{crt}(E^{top}_{\mathcal{N}^*_\tau}) \geq \kappa$ or $\rho_1^{\mathcal{N}^*_\tau} > \kappa$, let B^0_τ be the set of $\bar{\tau} \in \mathcal{S} \cap \tau$ satisfying:

1. $\mathcal{N}^*_{\bar{\tau}}$ is a hod premouse of the same type as $\mathcal{N}^*_\tau$.
2. $n_\tau = n_{\bar{\tau}}$.
3. There is a map $\sigma^0_{\bar{\tau}\tau} : \mathcal{N}^*_{\bar{\tau}} \to \mathcal{N}^*_\tau$ that is $\Sigma_0^{(n_\tau)}$-preserving with respect to the language of hod premice such that
   (a) $\bar{\tau} = \mathrm{cr}(\sigma^0_{\bar{\tau}\tau})$ and $\sigma^0_{\bar{\tau}\tau}(\bar{\tau}) = \tau$.
   (b) $\sigma^0_{\bar{\tau}\tau}(p^*_{\bar{\tau}}) = p_\tau$.
   (c) for each $\alpha \in p_\tau$, there is a generalized witness for α with respect to $\mathcal{N}^*_\tau$ and p_τ in the range of $\sigma^0_{\bar{\tau}\tau}$.

Finally, if B^0_τ exists, let $\vec{B}_\tau = \{B^0_\tau, B^1_\tau\}$. Otherwise, let $\vec{B}_\tau = \{B^1_\tau\}$.

As in [38], it is easy to see that in both cases $\sigma_{\bar{\tau}\tau}, \sigma^0_{\bar{\tau}\tau}, \sigma^1_{\bar{\tau}\tau}$ (if exist) are uniquely determined, Σ_0 (and not Σ_1), and non-cofinal. By [38, Lemma 3.3], for each $\tau \in \mathcal{S}$ such that B^0_τ is defined, and $\bar{\tau} \in B^0_\tau$,

$$B^0_\tau \cap \bar{\tau} = B^0_{\bar{\tau}} - \min B^0_\tau. \tag{5}$$

And similarly, if B^1_τ is defined, then for all $\bar{\tau} \in B^1_\tau$,

$$B^1_\tau \cap \bar{\tau} = B^1_{\bar{\tau}} - \min B^1_\tau. \tag{6}$$

The following is the key lemma (cf. [38, Lemma 3.5]).

LEMMA 11.3.10. *For each $\tau \in \mathcal{S}$ of uncountable cofinality, for $i \in \{0,1\}$, if B^i_τ is defined, then B^i_τ is a club subset of τ on a tail end. That is, there is a $\bar{\tau} < \tau$ such that $B^i_\tau - \bar{\tau}$ is closed and unbounded in τ. If $i = 0$, we can take $\bar{\tau} = 0$.*

Using the lemma and (5), (6), by the argument on [38, pg 52-55] , we can construct a $\square'_{\kappa,2}$-sequence on $\mathcal{S}$. We summarize the construction next. First for $\tau \in \mathcal{S}$, for i such that B^i_τ is defined, let

1. $\tau^i(0) = \tau$;
2. $\tau^i(j+1) = \min(B^i_{\tau(j+1)})$;
3. $l^i_\tau =$ the least j such that $B^i_{\tau(j)} = \emptyset$.

Now let

1. $B^{i,*} = B^i_{\tau^i(0)} \cup \cdots \cup B^i_{\tau^i(l^i_\tau - 1)}$;

2. $\sigma^{i,*}_{\bar{\tau}\tau} = \sigma^i_{\tau^i(1)\tau^i(0)} \circ \cdots \circ \sigma^i_{\tau^i(j)\tau^i(j-1)} \circ \sigma^i_{\bar{\tau}\tau^i(j)}$ whenever $\bar{\tau} \in B^{i,*}_\tau$ and j is such that $\bar{\tau} \in B^i_{\tau(j)}$.

By the exact same proof as in [38, Lemma 3.4], we get the coherency of the $B^{i,*}_\tau$ sets.

LEMMA 11.3.11. *For $\tau \in \mathcal{S}$, for i such that B^i_τ is defined, suppose $\bar{\tau} \in B^{i,*}_\tau$. Then $B^i_{\bar{\tau}}$ is defined and $B^{i,*}_{\bar{\tau}} = B^{i,*}_\tau \cap \bar{\tau}$.*

For each $\tau \in \mathcal{S}$, for i such that B^i_τ is defined, let β^i_τ be the least β in $B^{i,*}_\tau \cup \{\tau\}$ such that $B^{i,*}_\tau - \beta$ is closed in τ. Using Lemmas 11.3.10 and 11.3.11, we easily get that letting

$$C^{i,*}_\tau = B^{i,*}_\tau - \beta^i_\tau, \tag{7}$$

then for $\bar{\tau} \in \beta^{i,*}_\tau$, $\bar{\tau} \geq \beta_\tau$,

$$\beta^i_\tau = \beta^i_{\bar{\tau}} \quad \text{and} \quad C^{i,*}_\tau \cap \bar{\tau} = C^{i,*}_{\bar{\tau}}. \tag{8}$$

Now note that if $C^{0,*}_\tau$ is defined, then o.t.$(C^{0,*}_\tau)$ may not be $\leq \kappa$, while if $C^{1,*}_\tau$ is defined then o.t.$(C^{0,*}_\tau) \leq \kappa$. As in [38, pg 54-55], we can shrink $C^{0,*}_\tau$ to a set $C^{0,'}_\tau \subseteq C^{0,*}_\tau$ such that

1. o.t.$(C^{0,'}_\tau) \leq \kappa$;
2. $C^{0,'}_\tau$ is a closed subset of $\mathcal{S} \cap \tau$ and if cof$(\alpha) > \omega$, then $C^{0,'}_\tau$ is also unbounded in τ;
3. $C^{0,'}_\tau \cap \bar{\tau} = C^{0,'}_{\bar{\tau}}$.

So letting $\vec{C}'_\tau = \{C^{i,'}_\tau \mid i \in \{0,1\} \wedge C^{i,'}_\tau \text{ is defined}\}$, we get that the sequence $\langle \vec{C}'_\tau \mid \tau < \kappa^+ \rangle$ is a $\square'_{\kappa,2}$-sequence on $\mathcal{S}$. Since $\mathcal{S}$ is club subset of κ^+, by a standard combinatorial argument (cf. [6]), the $\square'_{\kappa,2}$-sequence on $\mathcal{S}$ can be turned into a $\square_{\kappa,2}$-sequence. Our main task is to prove Lemma 11.3.10. This will take up the rest of the section.

REMARK 11.3.12. It is clear from [38, pg 54-55], Definitions 11.3.8 and 11.3.9 and Proposition 11.3.6 that the square sequence $\square_{\kappa,2}$ is definable from κ in $\mathcal{P}$ and the definition is uniform in κ.

11.3.3. When $\tau \in \mathcal{S}^0$. Fix $\tau \in \mathcal{S}^0$. Assume τ is a limit point of $\mathcal{S}$ uncountable cofinality. Recall B^0_τ is defined to be the set of $\bar{\tau} \in \mathcal{S}$ such that

1. $n_\tau = n_{\bar{\tau}}$.
2. $\mathcal{N}^*_{\bar{\tau}}$ is a hod premouse of the same type as $\mathcal{N}_\tau$.
3. There is an embedding $\sigma^0_{\bar{\tau}\tau} : \mathcal{N}^*_{\bar{\tau}} \to \mathcal{N}_\tau$ such that $\sigma^0_{\bar{\tau}\tau}$ is $\Sigma^{(n_\tau)}_0$-preserving (in the language of hod premice) and
   (a) $\bar{\tau} = \text{cr}(\sigma^0_{\bar{\tau}\tau})$ and $\sigma^0_{\bar{\tau}\tau}(\bar{\tau}) = \tau$.
   (b) $\sigma^0_{\bar{\tau}\tau}(p^*_{\bar{\tau}}) = p_\tau$, where recall $p^*_{\bar{\tau}}$ is the standard parameter of $\mathcal{N}^*_{\bar{\tau}}$.

(c) for each $\alpha \in p_\tau$, there is a generalized witness for α with respect to $\mathcal{N}_\tau$ and p_τ in the range of $\sigma_{\bar{\tau}\tau}$.

To simplify the notation, let D denote B^0_τ and $\sigma_{\bar{\tau},\tau}$ denote $\sigma^0_{\bar{\tau},\tau}$.

LEMMA 11.3.13. *D is unbounded in τ.*

PROOF. Given $\tau' < \tau$, we find $\tilde{\tau} \geq \tau'$ in D. In $\mathcal{P}$, form an countable elementary hull of $\{\mathcal{N}_\tau, \tau', \mathcal{S}\}$ in $H_{(\kappa^{++})}$ (in which everything relevant is present). Let H be the transitive collapse of the hull and $\sigma_0 : H \to H_{\kappa^{++}}$ be the uncollapse map. Set:

1. $\bar{x} = \sigma_0^{-1}(x)$ for any x in range of σ_0,
2. $\sigma = \sigma_0 \restriction \bar{\mathcal{N}_\tau} : \bar{\mathcal{N}_\tau} \to \mathcal{N}_\tau$,
3. $\tilde{\tau} = \sup(\sigma''\bar{\tau})$.

Note that since $\tau \in \mathcal{S}^0$, $\mathcal{N}_\tau = \mathcal{N}^*_\tau$ and either $\mathrm{cr}(E^{\mathrm{top}}_{\mathcal{N}_\tau}) \geq \kappa$ or $\omega\rho^1_{\mathcal{N}_\tau} > \kappa$. Set $n = n_\tau$. Using the interpolation lemma (Lemma [38, Lemma 1.2]), we can find a map $\tilde{\sigma} : \bar{\mathcal{N}_\tau} \to \tilde{\mathcal{N}}$ which is $\Sigma_0^{(n)}$-preserving and cofinal (the map $\tilde{\sigma}$ is the ultrapower map via the $(\mathrm{cr}(\sigma), \tilde{\tau})$-extender derived from σ). Note that $\tilde{\tau} = (\kappa^+)^{\tilde{\mathcal{N}}}$. Also, by the interpolation lemma, there is a map $\sigma' : \tilde{\mathcal{N}} \to \mathcal{N}_\tau$ satisfying $\sigma' \restriction \tilde{\tau} = \mathrm{id}$, $\sigma'(\tilde{\tau}) = \tau$, and $\sigma' \circ \tilde{\sigma} = \sigma$.

We have that

1. $\tilde{\mathcal{N}}$ is a hod premouse of the same type as $\mathcal{N}_\tau$.
2. $\tilde{\mathcal{N}}$ is sound.
3. $\omega\rho^\omega_{\tilde{\mathcal{N}}} = \omega\rho^{n+1}_{\tilde{\mathcal{N}}} \leq \kappa$.

The above follow from the proof of [38, Lemma 3.7] for the most part, except for the first item in the case when $\mathcal{N}_\tau$ is B-active. In this case, the first item follows from Lemma 11.1.3 (or see [50, Lemma 2.36]) and hull condensation of Σ.[18]

It remains to see that $\tilde{\mathcal{N}}$ is indeed $\mathcal{N}^*_{\tilde{\tau}}$. We apply Theorem 11.1.5. (a) cannot hold since the map σ' is a $\Sigma_0^{(n)}$, non-cofinal map and hence cannot be a core map. (c) cannot hold because $\tilde{\mathcal{N}}$ is sound. So (b) holds; in fact, $\tilde{\mathcal{N}}$ is a strict segment of $\mathcal{N}_\tau$ because $\tilde{\tau} = \mathrm{cr}(\sigma') = (\kappa^+)^{\tilde{\mathcal{N}}} < \tau = (\kappa^+)^{\mathcal{N}_\tau}$. This easily implies $\tilde{\mathcal{N}} = \mathcal{N}^*_{\tilde{\tau}}$. ⊣

LEMMA 11.3.14. *D is a closed subset of τ.*

PROOF. Let $\tilde{\tau}$ be a limit point of D. We show that $\tilde{\tau} \in D$. Form the direct limit $\langle \tilde{\mathcal{N}}, \sigma_{\bar{\tau}\tilde{\tau}} \mid \bar{\tau} \in D \cap \tilde{\tau} \rangle$ of the system $\langle \mathcal{N}^*_{\bar{\tau}}, \sigma_{\tau^*\bar{\tau}} \mid \tau^* \leq \bar{\tau} \wedge \tau^*, \bar{\tau} \in D \cap \tilde{\tau} \rangle$. The direct limit is well-founded and there is a Σ_0 embedding $\sigma : \tilde{\mathcal{N}} \to \mathcal{N}_\tau$ (defined by $\sigma(\sigma_{\bar{\tau}\tilde{\tau}}(x)) = \sigma_{\bar{\tau},\tau}(x)$). It is easy to check that:

(a) $\sigma \circ \sigma_{\bar{\tau}\tilde{\tau}} = \sigma_{\bar{\tau}\tau}$.
(b) $\tilde{\tau} = \sigma_{\bar{\tau}\tilde{\tau}}(\bar{\tau})$, $\sigma_{\tilde{\tau}\tau}(\tilde{\tau}) = \tau$, and $\tilde{\tau} = \mathrm{cr}(\sigma)$.
(c) σ is $\Sigma_0^{(n)}$ preserving where $n = n_\tau$ (with respect to the language of coherent structures).

[18]Indexing branches using the $\mathfrak{B}$-operator allows the proof of Lemma 11.1.3 and [50, Lemma 2.36] to go through in this situation. The traditional approach to indexing branches does not seem to imply that $\tilde{N}$ is a hod premouse.

We need to see that $\tilde{\mathcal{N}} = \mathcal{N}^*_{\tilde{\tau}}$. First, we show that $\tilde{\mathcal{N}}$ is a hod premouse of the same type as $\mathcal{N}_\tau$. Note that Π_2-properties which hold on a tail end are upward preserved under direct limit maps (cf. [38, pg 8-9]). Furthermore, $\mathcal{N}^*_{\bar{\tau}}$ is of the same type as $\mathcal{N}_\tau$ for each $\bar{\tau} \in D \cap \tau$. So $\tilde{\mathcal{N}}$ is of the same type as $\mathcal{N}_\tau$ (as either a passive hod premouse, or a B-active hod premouse, or an E-active hod premouse with $\mathrm{cr}(E^{\mathrm{top}}_{\tilde{\mathcal{N}}}) > \mu$, in which case $\tilde{\mathcal{N}}$ is of type A, or else an E-active hod premouse with $\mathrm{cr}(E^{\mathrm{top}}_{\tilde{\mathcal{N}}}) = \mu$, in which case $\omega\rho^1_{\tilde{\mathcal{N}}} > \kappa$; these statements can be expressed in a Π_2-fashion).

Recall that for $\bar{\tau} \in D$, we use $\tilde{h}_{\bar{\tau}}$ to denote $\tilde{h}^{(n_{\bar{\tau}}+1)}_{\mathcal{N}^*_{\bar{\tau}}}$, the $\Sigma_1^{(n_{\bar{\tau}})}$-Skolem function of $\mathcal{N}^*_{\bar{\tau}}$. Here note that $n_{\bar{\tau}} = n_\tau = n$. Let $\tilde{p} = \sigma_{\bar{\tau}\tilde{\tau}}(p^*_{\bar{\tau}})$ for $\bar{\tau} \in D \cap \tilde{\tau}$. Given any $x \in \tilde{\mathcal{N}}$, there is $\bar{\tau} \in D \cap \tilde{\tau}$ and $\bar{x} \in \mathcal{N}^*_{\bar{\tau}}$ such that $x = \sigma_{\bar{\tau}\tau}(\bar{x})$. There is $\xi < \kappa$ such that $\bar{x} = \tilde{h}_{\bar{\tau}}(\xi, p_{\bar{\tau}})$. This $\Sigma_1^{(n)}$-statement is preserved by $\sigma_{\bar{\tau}\tilde{\tau}}$, so $x = \tilde{h}^{n+1}_{\tilde{\mathcal{N}}}(\xi, \tilde{p})$. So $\tilde{\mathcal{N}} = \tilde{h}^{n+1}_{\tilde{\mathcal{N}}}(\kappa \cup \{\tilde{p}\})$.

This gives $\omega\rho^{n+1}_{\tilde{\mathcal{N}}} = \omega\rho^{\omega}_{\tilde{\mathcal{N}}} \leq \kappa$. But κ is a cardinal in $\mathcal{P}$, so we indeed have equality. For each $\alpha \in p_\tau$, there is a generalized witness for α with respect to $(\mathcal{N}_\tau, p_\tau)$ in range of σ. This is because $\mathrm{rng}(\sigma)$ contains $\mathrm{rng}(\sigma_{\bar{\tau},\tau})$ for any $\bar{\tau} \in D \cap \tilde{\tau}$ and $\mathrm{rng}(\sigma_{\bar{\tau},\tau})$ contains such a witness. This takes care of (c) in the definition of D. This easily implies that $\tilde{\mathcal{N}}$ is sound and $\tilde{p}$ is the standard parameter of $\tilde{\mathcal{N}}$. We can now apply Theorem 11.1.5 as in the proof of Lemma 11.3.13 to conclude that $\tilde{\mathcal{N}} = \mathcal{N}^*_{\tilde{\tau}}$. ⊣

Lemmas 11.3.13 and 11.3.14 together complete the proof of Lemma 11.3.10 in the case $\tau \in \mathcal{S}^0$.

11.3.4. When $\tau \in \mathcal{S}^1$. Fix $\tau \in \mathcal{S}^1$ a limit point of $\mathcal{S}$ of uncountable cofinality. If B^0_τ is defined, then as in the previous section, using the fact that $\mathrm{crt}(E^{top}_{\mathcal{N}^*_\tau}) \geq \kappa$ or $\rho_1^{\mathcal{N}^*_\tau} > \kappa$, we can show that B^0_τ is closed and unbounded in τ. So let us now focus on the case B^1_τ is defined. Define $D \subset \tau$ to be the set of $\bar{\tau} \in \mathcal{S}$ such that

1. $(\mu_\tau, q^*_{\bar{\tau}})$ is a strong divisor of $\mathcal{N}_{\bar{\tau}}$ where $q^*_{\bar{\tau}}$ is the bottom segment of $p_{\bar{\tau}}$ of length m_τ (recall m_τ is the length of q_τ).
2. Letting $\mathcal{M}^*_{\bar{\tau}}$ be the protomouse of $\mathcal{N}_{\bar{\tau}}$ associated with $(\mu_\tau, q^*_{\bar{\tau}})$, there is a map $\sigma_{\bar{\tau}\tau} : \mathcal{M}^*_{\bar{\tau}} \to \mathcal{M}_\tau$ that is Σ_0-preserving (with respect to the language of coherent structures) such that
   (a) $\bar{\tau} = \mathrm{cr}(\sigma_{\bar{\tau}\tau})$ and $\sigma_{\bar{\tau}\tau}(\bar{\tau}) = \tau$.
   (b) $\sigma_{\bar{\tau}\tau}(q^*_{\bar{\tau}}) = q_\tau$.
   (c) for each $\alpha \in q_\tau$, there is a generalized witness for α with respect to $\mathcal{M}_\tau$ and q_τ in the range of $\sigma_{\bar{\tau}\tau}$ (in the language of coherent structures).

We will show that there is some $\bar{\tau} < \tau$ such that $B^1_\tau - \bar{\tau} = D - \bar{\tau}$. Part of this is to show that for all sufficiently large $\bar{\tau} \in D$, $(\mu_\tau, q^*_{\bar{\tau}}) = (\mu_{\bar{\tau}}, q_{\bar{\tau}})$.

LEMMA 11.3.15. *D is unbounded in τ.*

PROOF. Let $\tau' < \tau$. As before, we find $\tilde{\tau} \in D$ above τ'. Since protomice are present, we carry out the argument in the language of coherent structures.

We let σ_0, H be defined as in Lemma 11.3.13. Again, we denote $\bar{x}$ for $\sigma_0^{-1}(x)$. We let $\sigma : \bar{\mathcal{M}}_\tau \to \mathcal{M}_\tau$ and $\tilde{\tau} = \sup \sigma''\bar{\tau}$. As before, $\tau' \leq \tilde{\tau} < \tau$. Let $\tilde{\sigma} : \bar{\mathcal{M}}_\tau \to \tilde{\mathcal{M}}$ be the $(\mathrm{cr}(\sigma), \tilde{\tau})$-ultrapower map derived from σ and $\sigma' : \tilde{\mathcal{M}} \to \mathcal{M}_\tau$ be the factor map. As in [38, Lemma 3.10], we have:

1. $\tilde{\sigma}(\bar{\kappa}, \bar{\tau}) = (\kappa, \tilde{\tau})$.
2. $\mathrm{cr}(\sigma') = \tilde{\tau}$ and $\sigma(\tilde{\tau}) = \tau$.
3. $h_{\tilde{\mathcal{M}}}(\kappa \cup \{\tilde{q}\}) = \tilde{\mathcal{M}}$ where $\tilde{q} = \tilde{\sigma}(\bar{q}_\tau)$; in other words, $\tilde{\mathcal{M}}$ is Σ_1-generated by $\kappa \cup \{\tilde{q}\}$.
4. $\omega\rho^\omega_{\tilde{\mathcal{M}}} = \omega\rho^1_{\tilde{\mathcal{M}}} = \kappa$ and $\tilde{q} \in R_{\tilde{\mathcal{M}}}$, the set of very good parameters for $\tilde{\mathcal{M}}$.
5. The range of $\tilde{\sigma}$ contains a generalized solidity witness for α with respect to $(\mathcal{M}_\tau, q_\tau)$ for each $\alpha \in q_\tau$.
6. $\tilde{q} = p_{\tilde{\mathcal{M}}}$ and $\tilde{\mathcal{M}}$ is solid and sound.

Note that as in Lemma 11.3.13, $\tilde{\sigma}$ is Σ_0 (but not Σ_1) and is not cofinal. This implies that $\tilde{\mathcal{M}}$ is a protomouse, even if $\mathcal{M}_\tau$ is a hod premouse (in which case, $\mathcal{M}_\tau = \mathcal{N}_\tau = \mathcal{N}^*_\tau$ is pluripotent).

We show $\tilde{\mathcal{M}} = \mathcal{N}_{\tilde{\tau}}(\mu_\tau, \tilde{q})$. Let $\mathcal{R}_0, \mathcal{R}_1$ be the hod premice associated with $\mathcal{M}_\tau, \tilde{\mathcal{M}}$, respectively. We have that $\mathcal{R}_0 = \mathrm{Ult}_n(\mathcal{N}^*_0, F)$, where F is the top extender (fragment) of $\mathcal{M}_\tau$ and $\mathcal{N}^*_0$ is largest (strict) segment of $\mathcal{M}_\tau$ such that $\omega\rho^\omega_{\mathcal{N}^*_0} \leq \mathrm{cr}(F) = \mu$ and F measures all sets in $\mathcal{N}^*_0$ if exists, otherwise, $\mathcal{N}^*_0 = \mathcal{M}_\tau$;[19] in the other case, $\mathcal{R}_1 = \mathrm{Ult}(\mathcal{N}^*_1, \tilde{F})$, where $\tilde{F}$ is the top extender (fragment) of $\tilde{\mathcal{M}}$ and $\mathcal{N}^*_1$ is the largest (strict) segment of $\tilde{\mathcal{M}}$ (equivalently, of $\mathcal{M}_\tau$) such that $\omega\rho^\omega_{\mathcal{N}^*_1} \leq \mathrm{cr}(\tilde{F}) = \mu$ and $\tilde{F}$ measures all sets in $\mathcal{N}^*_1$. Let $\pi_i : \mathcal{N}^*_i \to \mathcal{R}_i$ be the ultrapower maps and $\pi_2 : \mathcal{R}_1 \to \pi_0(\mathcal{N}^*_1)$ be the factor map

$$\pi_2(\pi_1(f)(a)) = \pi_0(f)(\tilde{\sigma}(a)).$$

Note that $\pi_2 \restriction \lambda^{\mathcal{R}_1} = \tilde{\sigma} \restriction \lambda^{\mathcal{R}_1}$ and therefore $\mathrm{crt}(\pi_2) = \tilde{\tau}$. Furthermore, if $\omega\rho^\omega_{\mathcal{N}^*_i} = \mu$, then $\mathcal{N}^*_i \lhd \mathcal{P}^b$ and therefore if $\mathcal{N}^*_i$ is E-active, $E^{\mathrm{top}}_{\mathcal{N}^*_i}$ has critical point $> \mu$ because μ is a strong cutpoint of $\mathcal{P}^b$. This easily gives that $\mathcal{R}_i$ is of the same type as $\mathcal{N}^*_i$ (as potential premice).

Note that $p_{\mathcal{R}_1} = \pi_1(p_{\mathcal{N}^*_1}) \cup p_{\tilde{\mathcal{M}}}$ (cf. [38, Lemma 2.16, 2.19]). In the case $\mathcal{N}^*_\tau \neq \mathcal{N}_\tau$, and hence $\mu_\tau = \mu$, $\pi_1(p_{\mathcal{N}^*_1})$ is the part of $p_{\mathcal{R}_1}$ above $\pi_1(\mu)$, the supremum of $\mathcal{R}_1$'s layer Woodin cardinals, and $p_{\tilde{\mathcal{M}}}$ is the part below $\pi_1(\mu)$.

The argument in [38, Lemma 3.10] then shows that $(\mu_\tau, \tilde{q})$ is a strong divisor of $\mathcal{R}_1$.[20] To show $\tilde{\mathcal{M}} = \mathcal{N}_{\tilde{\tau}}(\mu_\tau, \tilde{q})$, we show $\mathcal{R}_1 = \mathcal{N}_{\tilde{\tau}}$. This then will show $\tilde{\tau} \in D$ as desired. There are two cases to consider.

---

[19]The first case is the case $\mathcal{M}_\tau$ is a protomouse and the second case is when $\mathcal{M}_\tau$ is a pluripotent level.

[20]The proof of this fact does not depend on whether $\mu_\tau > \mu$.

Case 1. $\mathcal{N}_\tau = \mathcal{N}^*_\tau$.

If $\mathrm{cr}(F) = \mathrm{cr}(\tilde{F}) > \mu$, then it is easy to see that $\mathcal{P}^b \lhd \mathcal{R}_0, \mathcal{R}_1$. Note that in this case, $\mathcal{R}_0 = \mathcal{N}_\tau = \mathcal{N}^*_\tau$ (see [38, Section 2]). So we can apply Theorem 11.1.5 as in the proof of Lemma 11.3.13 and conclude that $\mathcal{R}_1 = \mathcal{N}_{\tilde{\tau}} = \mathcal{N}^*_{\tilde{\tau}}$. Now suppose $\mathrm{cr}(F) = \mu$ (so $\mu_\tau = \mu$). Recall from the discussion above that we know $(\mu_\tau, \tilde{q})$ is a strong divisor of $\mathcal{R}_1$ and $\tilde{q}$ is the bottom part of the standard parameter of $\mathcal{R}_1$ below $\pi_1(\mathrm{cr}(\tilde{F}))$. We show that $\mathcal{R}_1 = \mathcal{N}_{\tilde{\tau}} \neq \mathcal{N}^*_{\tilde{\tau}}$ by the following claims. We also will get then that $(\mu_\tau, \tilde{q}) = (\mu_{\tilde{\tau}}, q_{\tilde{\tau}})$ in this case.

Let γ_τ be defined as in the paragraphs before 11.3.1 for $\mathcal{N}_\tau$; let $\gamma_{\tilde{\tau}}, \tilde{\gamma}$ be defined similarly for $\mathcal{N}^*_{\tilde{\tau}}, \mathcal{R}_1$, respectively. Let Λ be $\mathcal{R}_0$'s iteration strategy.

CLAIM 11.3.16. $\tilde{\gamma} = \gamma_{\tilde{\tau}}$.

PROOF. Suppose not. Assume $\tilde{\gamma} < \gamma_{\tilde{\tau}}$ (the other case is similar). Let E be least on the extender sequence of $\mathcal{N}_{\tilde{\tau}}$ (equivalently of $\mathcal{N}^*_{\tilde{\tau}}$) such that

1. $\mathrm{cr}(E) = \mu$,
2. $\mathrm{lh}(E) \geq \tilde{\gamma}$.

Let $\mathcal{S} = \mathrm{Ult}(\mathcal{R}_0, E)$. Note that $\tilde{\gamma}$ is a cutpoint of $\mathcal{S}$ and $i_E(\mu)$ is a limit of Γ-full Woodin cardinals above $\tilde{\gamma}$. By SMC in Γ, we can conclude that $\mathcal{R}' \in \mathcal{S}$, where $\mathcal{R}'$ is a sound hod mouse extending $\mathcal{R}_1|\tilde{\gamma}$, $\tilde{\tau} = (\kappa^+)^{\mathcal{R}'}$, $\tilde{\gamma}$ as a cutpoint of $\mathcal{R}'$, and $\mathcal{R}'$ projects to κ.[21]

Fix $\mathcal{R}' \in \mathcal{S}$ as above. $\mathcal{R}'$ defines a surjection f from κ onto $\tilde{\tau}$. Since $\mathcal{R}' \in \mathcal{S}$, $f \in \mathcal{S}$. This contradicts the fact that $\mathcal{S} \models \tilde{\tau} = \kappa^+$. ⊣

CLAIM 11.3.17. *There is a pointclass Ω with pointclass generator a sound hod mouse that projects to κ, extends $\mathcal{P}|\tilde{\gamma}$, having $\tilde{\tau} = \kappa^+$, $\tilde{\gamma}$ as a cutpoint, and the set of layer Woodin cardinals above $\tilde{\gamma}$ has limit order type. $\mathcal{R}_1$ is the generator for the Wadge minimal such pointclass.*

PROOF. Clearly, such Ω exists since the pointclass generated by $\mathcal{R}_1$ is such. Let Ω_0 be the pointclass $\mathcal{R}_1$ generates and Ω_1 be the minimal pointclass satisfying the hypothesis of the claim. Let $(\mathcal{N}, \Psi)$ generate Ω_1 with the properties in the statement of the claim. Note that at this point, we know $\mathcal{R}_1$ and $\mathcal{N}$ are sound, projects to κ, extends $\mathcal{P}|\gamma_{\tilde{\tau}}$, satisfies $\kappa^+ = \tilde{\tau}$, have $\gamma_{\tilde{\tau}}$ as cutpoint, and the set of layer Woodin cardinals above $\gamma_{\tilde{\tau}}$ of both models is of limit order type.

We claim that $\Omega_0 = \Omega_1$. Suppose for contradiction that $\Omega_0 \subsetneq \Omega_1$ (the other case is similar). Then, using $\mathbb{R}$-genericity iteration above $\gamma_{\tilde{\tau}}$ and elementarity, in the derived model of $\mathcal{N}$ (at the supremum of its Woodin cardinals) there is a pointclass

[21] By genericity iterations, without loss of generality, we may assume that a real witnessing the Wadge reduction of Λ^{π_2} to Λ is generic over $\mathcal{S}$. In $\mathcal{S}$'s derived model at $i_E(\mu)$, we can find $\mathcal{R}_1$. This means, in the derived model of $\mathcal{S}$, there is some hod mouse $\mathcal{R}$ extending $\mathcal{R}_1|\tilde{\gamma}$, having $\tilde{\tau} = \kappa^+$, $\tilde{\gamma}$ as a cutpoint, and projects to κ; furthermore, we can demand that $(\mu_\tau, \tilde{q})$ is a strong divisor of $\mathcal{R}$ and $\tilde{q}$ is the bottom part of the standard parameter of $\mathcal{R}$ below the supremum of $\mathcal{R}$'s layer Woodin cardinals. Let Ω be the Wadge-minimal pointclass that has a pointclass generator with these properties. Note that this determines the unique pointclass generator $\mathcal{S}_\Omega$ for Ω. This implies that $\mathcal{S}_\Omega \in \mathcal{S}$.

with a generator $\mathcal{S}$ that is sound, projects to κ, extends $\mathcal{P}|\gamma_{\tilde{\tau}}$, satisfies $\kappa^+ = \tilde{\tau}$, and have $\gamma_{\tilde{\tau}}$ as cutpoint. Some such $\mathcal{S}$ is in $\mathcal{N}$ by a similar argument as in Footnote 21. This implies as in Claim 11.3.16 that $\tilde{\tau}$ is not a cardinal in $\mathcal{N}$. Contradiction. We have shown $\Omega_0 = \Omega_1$.

Now we can compare $\mathcal{R}_1$ against $\mathcal{N}$. Since $\Omega_0 = \Omega_1$, both models are κ-sound, projects to κ, and $\kappa < \gamma_{\tilde{\tau}}$, just as Lemma 11.3.3, we conclude that $(\mathcal{N},\Psi) = (\mathcal{R}_1,\Lambda)$. ⊣

Using the claims and the fact that $(\mu_\tau,\tilde{q})$ is a strong divisor of $\mathcal{R}_1$ (note that $\max(\tilde{q}) < (\gamma_{\tilde{\tau}}^+)^{\mathcal{R}_1}$ and $\tilde{q}$ is the bottom part below $\pi_1(\mathrm{cr}(\tilde{F}))$ of the standard parameter of $\mathcal{R}_1$) we easily verify that $\mathcal{R}_1 = \mathcal{N}_{\tilde{\tau}}$ and hence $\tilde{\mathcal{M}} = \mathcal{N}_{\tilde{\tau}}(\mu_\tau,\tilde{q})$. Hence $\tilde{\tau} \in D$ as desired.

Case 2. $\mathcal{N}_\tau \neq \mathcal{N}_\tau^*$.

In this case, $\mu_\tau = \mu$. As above, $\mathcal{R}_0 = \mathcal{N}_\tau$ and (μ_τ,q_τ) is a strong divisor of $\mathcal{N}_\tau$. We aim to show that $\mathcal{R}_1 = \mathcal{N}_{\tilde{\tau}}$. As above, $(\mu,\tilde{q})$ is a strong divisor of $\mathcal{R}_1$ by the proof of [38, Lemma 3.10]. Furthermore, $\max(\tilde{q}) < (\tilde{\gamma}^+)^{\mathcal{R}_1} = (\gamma_{\tilde{\tau}}^+)^{\mathcal{R}_1}$ and $\tilde{q}$ is the bottom part of the standard parameter of $\mathcal{R}_1$ below $\mathcal{R}_1$'s limit of layer Woodin cardinals $\pi_1(\mathrm{cr}(\tilde{F}))$. This easily implies, using Claim 11.3.17, that $\mathcal{N}_{\tilde{\tau}} \neq \mathcal{N}_{\tilde{\tau}}^*$, $\mathcal{R}_1 = \mathcal{N}_{\tilde{\tau}}$, $\mu = \mu_{\tilde{\tau}}$, $\tilde{q} = q_{\tilde{\tau}}$ and hence $\tilde{\mathcal{M}} = \mathcal{N}_{\tilde{\tau}}(\mu_{\tilde{\tau}},q_{\tilde{\tau}})$. So $\tilde{\tau} \in D$ as desired. ⊣

LEMMA 11.3.18. *D is a closed subset of τ.*

PROOF. Let $\tilde{\tau}$ be a limit point of D. We show that $\tilde{\tau} \in D$. As in Lemma 11.3.14, form the direct limit $\langle \tilde{\mathcal{M}}, \sigma^1_{\bar{\tau}\tilde{\tau}} \mid \bar{\tau} \in D \cap \tilde{\tau}\rangle$ of the system $\langle \mathcal{M}^*_{\bar{\tau}}, \sigma^1_{\tau^*\bar{\tau}} \mid \tau^* \leq \bar{\tau} \wedge \{\tau^*,\bar{\tau}\} \subset D \cap \tilde{\tau}\rangle$. The direct limit is well-founded (so we identify $\tilde{\mathcal{M}}$ with its transitive collapse) and there is a Σ_0 embedding $\sigma : \tilde{\mathcal{M}} \to \mathcal{M}_\tau$ (defined by $\sigma(\sigma^1_{\bar{\tau}\tilde{\tau}}(x)) = \sigma^1_{\bar{\tau},\tau}(x)$). It is easy to check that (cf. [38, Lemma 3.11]):

(a) $\tilde{\mathcal{M}}$ is a coherent structure.
(b) $\sigma \circ \sigma^1_{\bar{\tau}\tilde{\tau}} = \sigma^1_{\bar{\tau}\tau}$.
(c) $\tilde{\tau} = \sigma^1_{\bar{\tau}\tilde{\tau}}(\bar{\tau})$, $\sigma^1_{\bar{\tau}\tau}(\bar{\tau}) = \tau$, and $\tilde{\tau} = \mathrm{cr}(\sigma)$.
(d) $h_{\tilde{\mathcal{M}}}(\kappa \cup \{\tilde{q}\}) = \tilde{\mathcal{M}}$ where $\tilde{q} = \sigma^1_{\bar{\tau}\tilde{\tau}}(q^*_{\bar{\tau}})$, so $\omega\rho^\omega_{\tilde{\mathcal{M}}} = \omega\rho^1_{\tilde{\mathcal{M}}} = \kappa$ and $\tilde{q} \in R_{\tilde{\mathcal{M}}}$.
(e) For every $\alpha \in q_\tau$, there is a generalized witness for α with respect to $(\mathcal{M}_\tau,q_\tau)$ in the range of σ. Hence $\tilde{q} = p_{\tilde{\mathcal{M}}} = \sigma^{-1}(q_\tau)$ and $\tilde{\mathcal{M}}$ is sound and solid.

The first four clauses are clear. The last follows from the fact that $\mathrm{rng}(\sigma)$ contains $\mathrm{rng}(\sigma^1_{\bar{\tau},\tau})$ for sufficiently large $\bar{\tau} < \tau$ and $\mathrm{rng}(\sigma^1_{\bar{\tau},\tau})$ has all relevant generalized witnesses.

Note that $\tilde{\mathcal{M}}$ is always a protomouse (this is because σ is not cofinal). If $\mu_\tau > \mu$ (or equivalently $\mathcal{N}_\tau = \mathcal{N}^*_\tau$), we can appeal to the proof of [38, Lemma 3.11] to get that $\tilde{\mathcal{M}} = \mathcal{N}_{\tilde{\tau}}(\mu_\tau,\tilde{q})$ and $(\mu_\tau,\tilde{q})$ is a strong divisor of $\tilde{\mathcal{M}}$. Otherwise, the same conclusion follows from the proof of Claim 11.3.17.

The previous paragraph gives $\tilde{\tau} \in D$ as desired. ⊣

LEMMA 11.3.19. *There is a $\bar{\tau} < \tau$ such that for all $\tau' \in D - \bar{\tau}$, $(\mu_\tau, q^*_{\tau'}) = (\mu_{\tau'}, q_{\tau'})$. Consequently, $B^1_\tau - \bar{\tau} = D - \bar{\tau}$.*

PROOF. We need to prove that there is $\bar{\tau} < \tau$ such that for every $\tau' \in D - \bar{\tau}$, $(\mu_\tau, q^*_{\bar{\tau}}) = (\mu_{\tau'}, q_{\tau'})$. Assume for contradiction that there is a sequence $\langle \tau_i \mid i < \delta \rangle$ that is increasing, cofinal in τ such that $(\mu_{\tau_i}, q_{\tau_i}) \neq (\mu_\tau, q^*_{\tau_i})$. We may assume without loss of generality that the sequence $\langle \mu_{\tau_i} \mid i < \delta \rangle$ is monotonic and all q_{τ_i}'s have the same length, say m.

If $\mu_\tau = \mu$, then we claim that for each $i < \delta$, $(\mu_{\tau_i}, q_{\tau_i}) = (\mu_\tau, q^*_{\tau_i})$. This follows from the proof of Lemma 11.3.15, where we prove that in this case, for each $i < \delta$, $\mathcal{N}_{\tau_i} \neq \mathcal{N}^*_{\tau_i}$ and $\mu_{\tau_i} = \mu = \mu_\tau$ and $q_{\tau_i} = q^*_{\tau_i}$. This contradicts the assumption that $(\mu_{\tau_i}, q_{\tau_i}) \neq (\mu_\tau, q^*_{\tau_i})$. So we must have that $\mu_\tau > \mu$, so $\mathcal{N}_\tau = \mathcal{N}^*_\tau$. This implies that for each $i < \delta$, $\mathcal{N}_{\tau_i} = \mathcal{N}^*_{\tau_i}$ (again, by remarks in Section 11.3.1 and the argument in Lemma 11.3.15). So it must be the case then that $\mu_{\tau_i} > \mu$ (note that since $\mathcal{N}^*_{\tau_i} = \mathcal{N}_{\tau_i}$, $\mathcal{N}^*_{\tau_i}$ cannot have strong divisors of the form (μ, q) for some q) and so $(\mu_{\tau_i}, q_{\tau_i})$, by definition, is the canonical strong divisor of $\mathcal{N}_{\tau_i}$.

By the definition of $(\mu_{\tau_i}, q_{\tau_i})$, each q_{τ_i} is a bottom part of $q^*_{\tau_i}$, say $q^*_{\tau_i} = q_{\tau_i} \cup s_{\tau_i}$ (s_{τ_i} may be empty). Recall we have shown $\mu_\tau, \mu_{\tau_i} > \mu$ (so we can freely quote results of [38, Section 2.4 and Lemma 3.12] in the arguments that follow). Now we observe that $\mu_{\tau_i} > \mu_\tau$ for all $i < \delta$. This is because the argument in [38, Lemma 3.12] shows: if $q_{\tau_i} = q^*_{\tau_i}$, then μ_{τ_i} must be $> \mu_\tau$ by maximality of μ_{τ_i} for $\mathcal{N}_{\tau_i}$ and the assumption that $(\mu_\tau, q^*_{\tau_i}) \neq (\mu_{\tau_i}, q_{\tau_i})$; otherwise, q_{τ_i} is a strict bottom segment of $q^*_{\tau_i}$, and hence [38, Lemma 2.26] shows that no $\nu \leq \mu_\tau$ is such that (ν, q_{τ_i}) is a strong divisor of $\mathcal{N}_{\tau_i}$.

Set for some (equivalently for all sufficiently large) $i < \delta$, $q = \sigma_{\tau_i \tau}(q_{\tau_i})$, $s = \sigma_{\tau_i \tau}(s_{\tau_i})$, $r = r_\tau$, $\nu = \sup_{i<\delta} \mu_{\tau_i}$. Now (ν, q) is a divisor of $\mathcal{N}_\tau$ (see [38, Lemma 3.12]). Since $\nu > \mu_\tau > \mu$, (ν, q) cannot be a strong divisor of $\mathcal{N}_\tau$. Then a calculation as in [38, Lemma 3.12] shows that for some $i < \delta$, $(\mu_{\tau_i}, q_{\tau_i})$ is not a strong divisor of $\mathcal{N}_{\tau_i}$. Contradiction. ⊣

Lemmas 11.3.15, 11.3.18, 11.3.19 together complete the proof of Lemma 11.3.10 in the case $\tau \in \mathcal{S}^1$. This finishes the construction of our $\square_{\kappa,2}$ sequence.

Chapter 12

LSA FROM PFA

For a cardinal κ, let $\wp_0(\kappa) = \kappa$; $\wp_{n+1}(\kappa) = 2^{\wp_n(\kappa)}$ for all $n < \omega$.

DEFINITION 12.0.20. A sequence $\langle \vec{C}_\alpha \mid \alpha \in \lambda \rangle$ is a $\square(\kappa, \lambda)$ sequence if it satisfies the following properties.

(i) $0 < |\vec{C}_\alpha| < \kappa$ for all limit $\alpha \in \lambda$.

(ii) $C \subseteq \alpha$ is club in α for all limit $\alpha \in \lambda$ and $C \in \vec{C}_\alpha$.

(iii) $C \cap \beta \in \vec{C}_\beta$ for all limit $\alpha \in \lambda$, $C \in \vec{C}_\alpha$ and $\beta \in \mathrm{Lim}(C)$.

(iv) There is no club $D \subseteq \lambda$ such that $D \cap \alpha \in \vec{C}_\alpha$ for all $\alpha \in \mathrm{Lim}(D)$.

We say that $\square(\kappa, \lambda)$ holds if a $\square(\kappa, \lambda)$-sequence exists.

Clearly, $\square_{\lambda, <\kappa}$ implies $\square(\kappa, \lambda^+)$ and if $\kappa \leq \kappa'$, then $\square(\kappa, \lambda)$ implies $\square(\kappa', \lambda)$. $\square(2, \lambda)$ is $\square(\lambda)$. The following is the main result of this chapter.

THEOREM 12.0.21. *Suppose κ is a cardinal such that $\kappa^\omega = \kappa$ and $2^{2^{\aleph_0}} \leq \kappa$. Suppose there is a regular cardinal $\gamma \in [\omega_2, \kappa)$ such that for all $\alpha \in [\gamma, (\wp_4(\kappa))^+]$, $\neg\square(3, \alpha)$. Then there is a model M containing* $\mathrm{OR} \cup \mathbb{R}$ *such that $M \vDash$* LSA.

We immediately have the following corollary.

COROLLARY 12.0.22. *Assume one of the following theories.*

1. PFA.

2. There is a strongly compact cardinal.

Then there is a model M containing $\mathrm{OR} \cup \mathbb{R}$ *such that $M \vDash$* LSA.

PROOF. It is well-known that (1) implies the hypothesis of Theorem 12.0.21 (cf. [66]); this is because PFA implies $\neg\square(3, \gamma)$ for all $\gamma \geq \omega_2$. For (2), let κ be a cardinal above a strongly compact cardinal γ such that $\kappa^\omega = \kappa$. The hypothesis for Theorem 12.0.21 holds at κ by the construction in [51]. $\dashv$

In the previous chapter, we show LSA is consistent assuming the existence of a Woodin cardinal which is a limit of Woodin cardinals by analyzing the HOD of the minimal model of LSA, here we use the core model induction method to construct some model of determinacy that satisfies LSA. The proof of Theorem 12.0.21 is built on that of [65], which in turns is inspired by [55] and [30].

The rest of the chapter is dedicated to proving Theorem 12.0.21. We assume the hypothesis of Theorem 12.0.21 along with the following simplifying assumption:

$$\kappa \text{ is measurable and } \forall \xi \in [\kappa, \kappa^{++}]\ 2^{\xi} = \xi^{+}. \tag{9}$$

From the theorem's assumption, we have that

$$\forall \alpha \in [\gamma, \kappa^{+4}],\ \neg\square(3,\alpha).$$

Later, we show how to get rid of assumption (9). Our smallness assumption throughout this section is:

(†) in $V[G]$, where $G \subseteq Col(\omega, <\kappa)$ is V-generic, there is no model M containing $\mathbb{R} \cup \mathrm{OR}$ such that $M \vDash$ "$\mathsf{ZF} + \mathsf{AD}^{+} + \Theta = \vartheta_{\alpha+1} + \vartheta_{\alpha}$ is the largest Suslin cardinal below $\vartheta_{\alpha+1}$".

Before plunging into the details, we give a very rough outline of the proof of Theorem 12.0.21. Fix κ as in the hypothesis of Theorem 12.0.21. We operate under assumptions (†) and (9). Let $\mathbb{P} = Col(\omega, <\kappa)$. In $V^{\mathbb{P}}$, let Ω be the "maximal pointclass of determinacy" (to be defined in the next section). Let $\mathcal{P}^{-}$ be the direct limit of hod pairs $(\mathcal{M}, \Sigma)$ such that $\Sigma \restriction \mathrm{HC} \in \Omega$ and Σ is Ω-fullness preserving and has branch condensation. Let $\mathcal{P}$ be the appropriate "Lp"-closure of $\mathcal{P}^{-}$ (defined in Section 12.1). So $\mathcal{P}^{-}$ is an initial segment of $\mathcal{P}$. [65] and the results of Chapter 9 show that $\mathcal{P} \vDash$ "$o(\mathcal{P}^{-})$ is a regular limit of Woodin cardinals." In $V^{\mathbb{P}}$, we carry out a mixture of validated sts constructions and hybrid K^{c}-constructions over some transitive set W containing $\mathcal{P}$ (to be explained in Section 12.6). Either the constructions stop prematurely (before stage κ^{+++} for various reasons to be specified in Section 12.6), in which case we show that we can obtain a model of LSA; otherwise, we reach a model $\mathcal{P}^{+}$ (extending $\mathcal{P}$) of height κ^{+++}. Then we consider the stack $\mathcal{S}$ of (appropriately defined) hod mice over $\mathcal{P}^{+}$. Using the proof of [12, Theorem 3.4], we show that $\mathrm{cof}(o(\mathcal{S})) \geq \kappa^{+++}$. Using the fact that $\mathcal{S} \in V$ and our hypothesis $\forall \alpha \in [\gamma, \kappa^{+4}],\ \neg\square(3,\alpha)$, we show that $\mathrm{cof}(o(\mathcal{S})) < \kappa^{+++}$. This contradiction shows that the second case cannot occur; therefore, we must reach a model of LSA.

12.1. Some core model induction backgrounds

We continue to assume (†) and (9) in this section. We recall some notions and results from [65]. In $V[G]$, where $\mathbb{P} = Col(\omega, <\kappa)$ and $G \subseteq \mathbb{P}$ is V-generic, let

$$\Omega = \bigcup\{\wp(\mathbb{R}) \cap M \mid \mathbb{R} \cup \mathrm{OR} \subset M \wedge M \vDash \mathsf{AD}^{+}\}.$$

[65] shows that, under (†),[1] the Solovay sequence $\langle \vartheta^{\Omega}_{\alpha} \mid \alpha \leq \gamma^* \rangle$ of Ω is of limit length. Furthermore, we get that if $A \in \Omega$, then there is a hod pair (or sts hod pair) $(\mathcal{P},\Sigma) \in \Omega$ such that $A \in \Gamma^b(\mathcal{P},\Sigma)$.

Let $\mathcal{P}^-$ be the direct limit of all hod pairs $(\mathcal{M},\Sigma)$ such that $\mathcal{M}$ is countable in $V^{\mathbb{P}}$ and Σ is an (ω_1,ω_1+1)-strategy of $\mathcal{M}$ that is strongly Ω-fullness preserving, has strong branch condensation, and $\Sigma \restriction \mathrm{HC} \in \Omega$. We will say that a pair $(\mathcal{M},\Sigma)$ with these properties is *nice* and let $\mathcal{F}$ be the direct limit system of all nice hod pairs. [65] shows that if $(\mathcal{M},\Sigma \restriction V) \in V$, then Σ can be uniquely extended to a $(\kappa^{+4},\kappa^{+4})$-strategy Σ^+ (and hence $\Sigma^+ \restriction V \in V$) (see Remark 12.7.11 and the discussion before it for a somewhat more general argument). Say $\mathcal{M}$ iterates (via Σ^+) to a complete layer $\mathcal{P}^-(\alpha)$ of $\mathcal{P}^-$, where $\alpha < \gamma^*$, we let Σ_α be the Σ^+-tail of Σ^+.[2] Σ_α only depends on α and does not depend on any particular choice of $(\mathcal{M},\Sigma^+)$ as long as Σ^+ is nice. Let

$$\Sigma = \oplus_{\alpha<\gamma^*}\Sigma_\alpha,$$

and

$$\mathcal{P} = \mathrm{Lp}^{\Omega,\Sigma}(\mathcal{P}^-),$$

be defined as in Section 9.1. So for every countable (in $V[G]$), transitive $\mathcal{M}^*$ embeddable into a level $\mathcal{M} \lhd \mathcal{P}$ via π, $\mathcal{M}^*$ is (ω_1,ω_1+1)-iterable as an (anomalous) hod mouse with strategy Λ such that Λ has a unique iteration strategy Λ such that $\Lambda \restriction \mathrm{HC} \in \Omega$; furthermore, if $\pi \in V$, then $\Lambda \restriction V \in V$ and Λ can be uniquely extended to a $(\kappa^{+4},\kappa^{+4})$-strategy in $V[G]$.

REMARK 12.1.1. As in Chapter 9, we let $\varphi(U,W)$ be the formula that expresses the fact that U is a mousefull pointclass with all the properties that Ω has and W is a hod pair $(\mathcal{Q},\Lambda)$ such that $Code(\Lambda) \in U$ and Λ is strongly U-fullness preserving and has strong branch condensation. Then the $\mathcal{F}$ above is $\mathcal{F}_{\varphi,\Omega}$ etc. From this point on, we will often suppress φ,Ω from our notations; this should not be confusing since all the notations that come into the following constructions are relative to (φ,Ω).

The hypothesis of Theorem 12.0.21 implies that "(φ,Ω) is full, maximal, homogenous and lower part (φ,Ω)-covering fails". In particular, the conclusions of Theorem 9.2.2 hold for $V[G]$.

LEMMA 12.1.2. *Let λ be the ordinal height of Ω, so $\lambda = \mathsf{ord}(\mathcal{P}^-) = \delta^{\mathcal{P}}$.*

1. *No levels $\mathcal{M} \lhd \mathcal{P}$ is such that $\rho_\omega(\mathcal{M}) < \lambda$. Hence $\rho_\omega(\mathcal{P}) = \mathsf{ord}(\mathcal{P})$ and $\mathcal{P} \models \mathsf{ZFC}^-$.*
2. *$\mathcal{P} \models \delta^{\mathcal{P}}$ is a regular limit of Woodin cardinals*

[1] [65] uses a stronger assumption, namely no models of "$\mathsf{AD}_{\mathbb{R}} + \Theta$ is regular" exist. But the proof there combined with the results in Section 9, particularly Theorem 9.2.2, work using (†); the main point is that the HOD analysis now can be carried out up to models of LSA.

[2] The complete layers of $\mathcal{P}^-$ are $(\mathcal{P}^-(\beta) : \beta < \gamma^*)$ and for each β, $\vartheta^{\Omega}_{\beta}$ is either the largest Woodin cardinal or the limit of Woodin cardinals in $\mathcal{P}^-(\alpha)$.

3. $\lambda < \kappa^+$.
4. *In* V, $\mathrm{ord}(\mathcal{P}) < \kappa^+$ *and* $\mathrm{cof}(\mathrm{ord}(\mathcal{P})) < \gamma$.

PROOF. (1) follows from [65, Theorem 3.78]. (2) follows by adapting the core model induction argument in [65] to get that γ^* is a limit ordinal and using Theorem 9.2.2 to show the existence of condensing sets in $V[G]$, which in turns will give us that γ^* is a regular cardinal in $\mathcal{P}$ (see for example the paragraph above [65, Remark 3.86]). For (3) first note that $2^{<\kappa} = \kappa$ and $\omega_1 = \kappa$ in $V[G]$. For any $\mathcal{Q} \lhd^c_{hod} \mathcal{P}$, note that there is a hod pair $(\mathcal{R}, \Psi) \in \Omega$ such that $\mathcal{R} \in H_\kappa[G \restriction \alpha]$, $\Psi \restriction V[G \restriction \alpha] \in V[G \restriction \alpha]$ for some $\alpha < \kappa$. This easily implies (3).

For (4), first note that $\mathcal{P} \in V$. Let $\vec{C}$ be the canonical $\square_\lambda$-sequence built in $\mathcal{P}$ (using the construction in [38]), where $\lambda = \mathrm{ord}(\mathcal{P}^-)$ is the ordinal height of Ω as defined above. $\vec{C}$ is not threadable (by the maximality of $\mathcal{P}$). So if $\mathrm{ord}(\mathcal{P}) = \kappa^+$ or $\mathrm{cof}(\mathrm{ord}(\mathcal{P})) \geq \gamma$, then using our hypothesis $\forall \alpha \in [\gamma, \kappa^{+4}], \neg\square(3, \alpha)$, we can find a thread for $\vec{C}$ by standard arguments. Contradiction. ⊣

12.2. Condensing sets

In $V[G]$, as done in Chapter 9, for each $X \in \wp_{\omega_1}(\mathcal{P})$, we let $\mathcal{Q}_X$ be the transitive collapse of $Hull_1^{\mathcal{P}}(X)$, $\delta_X = \delta^{\mathcal{Q}_X}$, and $\tau_X : \mathcal{Q}_X \to \mathcal{P}$ be the uncollapse map. Let Σ_X be the τ_X-pullback strategy for $\mathcal{Q}_X$.[3] For $X \subseteq Y \in \wp_{\omega_1}(\mathcal{P})$, let $\tau_{X,Y} = \tau_Y^{-1} \circ \tau_X$.

If $Y \in \wp_{\omega_1}(\mathcal{P}^-)$ and $X \cap \mathcal{P}^- \subseteq Y$, we let $\mathcal{Q}_Y^X$ be the transitive collapse of $Hull_1^{\mathcal{P}}(X \cup Y)$, Σ_Y^X be the $\tau_{X \oplus Y}$-pullback of Σ, and $\sigma_Y^X : \mathcal{Q}_Y^X \to \mathcal{P}$ be given by

$$\sigma_Y^X(q) = \tau_X(f)(\pi^{\Sigma_Y^X}_{\mathcal{Q}_Y^X, \infty}(a))$$

where $a \in (\mathcal{Q}_Y | \delta^{\mathcal{Q}_Y})^{<\omega}$ and $q = \tau_{X, X \oplus Y}(f)(a)$. We also write τ_Y^X for $\tau_{X \oplus Y}$ and π_Y^X for $\tau_{X, X \oplus Y}$.

Recall the notions of $((\varphi, \Omega))$-condensing sets, extensions, and honest extensions discussed in Chapter 9. Suppose $Y \subset Z$ are extensions of X and $X \in Cnd(\mathcal{P})$. We write $\pi_{Y,Z}^X$ for the natural, uncollapse map from $\mathcal{Q}_Y^X$ to $\mathcal{Q}_Z^X$. By Lemma 12.2.1, $\pi_{Y,Z}^X \restriction \delta^{\mathcal{Q}_Y^X}$ agrees with the iteration map $\pi^{\Sigma_Y}_{\mathcal{Q}_Y^X, \mathcal{Q}_Z^X}$. We note that $\mathcal{Q}_X^X = \mathcal{Q}_X$ and if Y is an extension of X, then $\pi_{X,Y}^X$ is just $\tau_{X,Y}$. When X is a fixed condensing set and Y extends X, we sometimes write $\mathcal{Q}_Y$ for $\mathcal{Q}_Y^X$ when no confusion arises.

Let $\mathfrak{S} \in V$ be the set of $X \prec H_{\kappa^{+4}}$ such that

1. $\kappa \cap X \in \kappa$,
2. $\gamma < |X| < \kappa$,
3. $\mathcal{P} \in X$, $X \cap \mathcal{P}$ is cofinal in $o(\mathcal{P})$, and
4. $X^\xi \subset X$ for any $\xi < |X|$.

[3] Typically, $X = X^* \cap \mathcal{P}$ for some countable $X^* \prec H_{\kappa^{+4}}$. And Σ_X is the τ_{X^*}-realization map, where τ_{X^*} is the uncollapse map of X^*.

Note that $\mathfrak{S}$ is stationary. We say that $\mathfrak{S}$ is the collection of *good* hulls. For $X \in \mathfrak{S}$, we let $\pi_X : M_X \to H_{\kappa^{+4}}$ be the uncollapse map. Note that letting κ_X be the critical point of π_X, π_X extends to an elementary map from $M_X[G \restriction \kappa_X] \to H_{\kappa^{+4}}[G]$, where $G \restriction \kappa_X = G \cap Coll(\omega, <\kappa_X)$. We also call this map π_X.

The following facts follow easily from [65] and Chapter 9. The point is $j[\mathcal{P}]$ is a (strongly) (φ,Ω)-condensing set in $M[H]$ where $H \subseteq Coll(\omega, < j(\kappa))$ is V-generic and $G = H \restriction \kappa$. Furthermore, because $\mathcal{P} \in V$ and $\mathrm{ord}(\mathcal{P}) < \kappa^+$ by Lemma 12.1.2, $j[\mathcal{P}] \in M$.

LEMMA 12.2.1. *(i) (φ,Ω) is full, maximal, homogenous and lower part (φ,Ω)-covering fails.*

(ii) $\forall^* X' \in \mathfrak{S}_{\varphi,\Omega}$, $X = X' \cap \mathcal{P}$ is a (strongly) condensing set.

(iii) Suppose Y is an honest extension of a (strongly) condensing set X and there are elementary maps $i : \mathcal{Q}_Y \to \mathcal{R}$ and $\sigma : \mathcal{R} \to \mathcal{P}$ such that $\sigma \circ i = \tau_Y \restriction \mathcal{Q}_Y$ and every $x \in \mathcal{R}$ has the form $i(f)(a)$ for $f \in \mathcal{Q}_Y$ and $a \in [\delta^{\mathcal{R}}]^{<\omega}$. Then letting Λ be the σ-pullback strategy of $\mathcal{R}$, and $\tau(i(f)(a)) = \tau_Y(f)(\pi^{\Lambda}_{\mathcal{R}|\delta^{\mathcal{R}},\infty}(a))$, then τ is well-defined, elementary, and $\tau \restriction \mathcal{R}|\delta^{\mathcal{R}} = \pi^{\Lambda}_{\mathcal{R}|\delta^{\mathcal{R}},\infty} \restriction \mathcal{R}|\delta^{\mathcal{R}}$.

(iv) Suppose X is (strongly) condensing and Y, Z are honest extensions of X such that $\mathcal{Q}^X_Y = \mathcal{Q}^X_Z$, then $\Sigma^X_Y = \Sigma^X_Z$.

REMARK 12.2.2. Let X be as in (ii) of the lemma. Then it is easy to se that any $Y = Y^* \cap \mathcal{P}$ where $Y^* \prec H_{\kappa^{+4}}$ is such that Y^* is countable (in $V[G]$) is an honest extension of X.

From now on, by "condensing set", we mean "strongly condensing set" and we will omit (φ,Ω) from our terminology.

We let $Cnd(\mathcal{P})$ be the collection of condensing sets $X \in \wp_{\omega_1}(\mathcal{P})$ in $V[G]$. If X' is a good hull, and $X = X' \cap \mathcal{P}$ is a condensing set, then we say that X' is a X-good hull. Similarly, any good hull Y such that $X' \subset Y$ and $\{\mathcal{P}, X\} \in Y$ is called an X-*good* hull.

We also get the following easy consequences.

PROPOSITION 12.2.3. *Suppose X is a condensing set and $\mathcal{Q}$ is such that for some extension Y of X, $\mathcal{Q} = \mathcal{Q}^X_Y$. Then there is a unique honest extension W of X such that $\mathcal{Q} = \mathcal{Q}^X_W$.*

PROPOSITION 12.2.4. *Suppose X is a condensing set. Suppose Y and W are extensions of X such that there is a Σ_1-embedding $i : \mathcal{Q}^X_Y \to \mathcal{Q}^X_W$ and $\tau^X_W \circ i[\mathcal{Q}^X_Y]$ is an extension of X. Then $(\Sigma^X_W)^i = \Sigma^X_Y$.*

PROOF. Letting $Y^* = \tau^X_W \circ i[\mathcal{Q}^X_Y]$, then $\mathcal{Q}^X_Y = \mathcal{Q}^X_{Y^*}$. Moreover, $\Sigma^X_Y = \Sigma^X_{Y^*}$ by Lemma 12.2.1, and hence $\Sigma^X_Y = (\Sigma^X_W)^i$. ⊣

An easy modification of the proof of Theorem 9.2.2 also gives us the following useful fact. A proof of this can also be found in [61, Lemma 2]. Note also that Corollary 9.1.15 is a consequence of Lemma 12.2.5. A standard argument, using the Vopenka algebra and the fact that $\mathcal{P} \models$ "$\delta^{\mathcal{P}}$ is a regular limit of Woodin cardinals", gives us that $L(\Omega, \mathcal{P}) \models$ "$\mathsf{AD}_{\mathbb{R}} + \Theta$ is regular." See [61] for a proof.

LEMMA 12.2.5. *Suppose X is a condensing set. Suppose $\mathcal{Q}, \mathcal{R}$ are countable, transitive in $V[G]$ with the property that there are elementary maps $i : \mathcal{P}_X \to \mathcal{Q}$, $k : \mathcal{Q} \to \mathcal{R}$, $\tau : \mathcal{Q} \to \mathcal{P}$, $\sigma : \mathcal{R} \to \mathcal{P}$ such that $\tau \circ i = \pi_X \restriction \mathcal{P}_X$, and $\tau = \sigma \circ k$. Letting $\Sigma_{\mathcal{Q}} = \Sigma^{\tau}$ and $\Sigma_{\mathcal{R}} = \Sigma^{\sigma}$, then for any $A \subseteq \delta_X$ in $\mathcal{P}_X$ and any formula φ,*

$$L(\Omega, \mathcal{P}) \models \text{"}\forall s \in \mathcal{Q}(\varphi[\mathcal{Q}, s, \Sigma_{\mathcal{Q}}, \mathcal{P}, \pi_X(A)] \iff \varphi[\mathcal{R}, k(s), \Sigma_{\mathcal{R}}, \mathcal{P}, \pi_X(A)])\text{"}.$$

12.3. X-suitable hod premouse and X-validated iterations

We summarize some key notions developed in [36]. The reader can read [36] for more details. We continue using the notations from the previous section.

DEFINITION 12.3.1. Let X be a condensing set. $\mathcal{Q}$ *nicely extends* $\mathcal{Q}_X$ if $\mathcal{Q}$ is non-meek and $\mathcal{Q}^b = \mathcal{Q}_X$. We also say that $\mathcal{Q}$ is a *nice extension* of $\mathcal{Q}_X$. Similarly, we can define what it means for $\mathcal{Q}$ to nicely extend $\mathcal{Q}_Y^X$ for an extension Y of X and $\mathcal{Q}$ to nicely extend $\mathcal{P}$.

Suppose X is a condensing set and Y is an extension of X. Suppose further that $\mathcal{Q}$ nicely extends $\mathcal{Q}_Y^X$. A stack (of normal iteration trees) $\mathcal{T}$ on $\mathcal{Q}$ has the form

$$\vec{\mathcal{T}} = ((\mathcal{M}_\alpha)_{\alpha<\eta}, (E_\alpha)_{\alpha<\eta-1}, D, R, (\beta_\alpha, m_\alpha)_{\alpha\in R}, T),$$

where the displayed objects are introduced in 2.4.1. The above notation is quite standard. D is the set of drops, R is the set of stages where player I starts a new round of the iteration game, (β_α, m_α) is the place player I drops at the beginning of the αth round, and T is the tree order. Sometimes to make clear these objects are associated with $\vec{\mathcal{T}}$, we write $D^{\vec{\mathcal{T}}}$ etc. Recall the convention in Chapter 2, we assume that all our stacks are proper (see Remark 2.7.27). One of the key aspects of being proper is that if $\beta < lh(\vec{\mathcal{T}})$ is such that $\vec{\mathcal{T}}_{\geq\beta}$ is a stack on $\mathcal{M}_\beta^{\vec{\mathcal{T}}}$ then $\beta \in R$[4]. We will also use the notation introduced in 2.4.4. In particular, for $\alpha \in R^{\vec{\mathcal{T}}}$, $\mathsf{next}^{\vec{\mathcal{T}}}(\alpha) = min(R^{\vec{\mathcal{T}}} - (\alpha+1))$ if this minimum exists and otherwise $\mathsf{next}^{\vec{\mathcal{T}}}(\alpha) = lh(\vec{\mathcal{T}})$. For $\alpha \in R^{\vec{\mathcal{T}}}$, we also set $\mathsf{nc}_\alpha^{\vec{\mathcal{T}}} = \vec{\mathcal{T}}_{[\alpha,\alpha')}$ where $\alpha' = \mathsf{next}^{\vec{\mathcal{T}}}(\alpha)$.

DEFINITION 12.3.2. Suppose $X \in Cnd(\mathcal{P})$ and Y is an extension of X. Suppose further that $\mathcal{Q}$ nicely extends $\mathcal{Q}_Y^X$. Given $E \in \vec{E}^{\mathcal{Q}}$ such that $\mathrm{crit}(E) = \delta^{\mathcal{Q}_Y^X}$, we say E is (X,Y)-*realizable* if there is W, an extension of $X \oplus Y$ such that $E = E_{Y,W}^X$, where $E_{Y,W}^X$ is the extender defined by

$$(a, A) \in E_{Y,W}^X \iff \tau_W^X(a) = \pi^{\Sigma_W^X}_{\mathcal{Q}_W^X,\infty}(a) \in \tau_Y^X(A), \tag{10}$$

[4]Thus, no normal component of $\vec{\mathcal{T}}$ can be split into two normal components.

for any $a \in [\mathrm{lh}(E)]^{<\omega}$ and $A \in \wp(\mathrm{crit}(E))^{|a|} \cap \mathcal{Q}$.

We are continuing with the notation of Definition 12.3.2. Suppose $\vec{\mathcal{T}}$ is a stack on $\mathcal{Q}$. We say $\vec{\mathcal{T}}$ is a (X,Y)-*realizable* iteration if there is a sequence $(W_\alpha : \alpha \in R^{\vec{\mathcal{T}}})$ such that

1. $W_0 = Y$,
2. if $\alpha, \beta \in R^{\vec{\mathcal{T}}}$ and $\alpha < \beta$ then W_β is an extension of $X \oplus W_\alpha$,
3. if $\alpha, \beta \in R^{\vec{\mathcal{T}}}$, $\alpha < \beta$ and $\pi^{\vec{\mathcal{T}},b}_{\alpha,\beta}$ is defined then $\pi^{\vec{\mathcal{T}},b}_{\alpha,\beta} = \pi^X_{W_\alpha,W_\beta}$[5], and
4. if $\alpha \in R^{\vec{\mathcal{T}}}$ and $\vec{\mathcal{U}}$ is the largest fragment of $\vec{\mathcal{T}}_{\geq\alpha}$ that is based on $\mathcal{M}^b_\alpha$ then $\vec{\mathcal{U}}$ is according to $\Psi^X_{W_\alpha}$.

We say $\vec{\mathcal{T}}$ is X-*realizable* if Y is an honest extension of X and $\vec{\mathcal{T}}$ is (X,Y)-realizable.

The following lemma shows that (X,Y)-realizability is equivalent to X-realizability. The proof can be found in [36, Section 6].

LEMMA 12.3.3. *Suppose Y is an extension of X and $\mathcal{Q}$ nicely extends $\mathcal{Q}^X_Y$. Suppose $\vec{\mathcal{T}}$ is a (X,Y)-realizable iteration as witnessed by $(W'_\alpha : \alpha \in R^{\vec{\mathcal{T}}})$. For $\alpha \in R^{\vec{\mathcal{T}}}$ let W_α be the unique honest extension of X with the property that $(\mathcal{M}^{\vec{\mathcal{T}}}_\alpha)^b = \mathcal{Q}^X_{W_\alpha}$. Then $(W_\alpha : \alpha \in R^{\vec{\mathcal{T}}})$ witnesses that $\mathcal{Q}$ is X-realizable.*

We fix a condensing set X throughout this section. Suppose $\mathcal{Q}$ nicely extends $\mathcal{Q}^X_Y$ and $\vec{\mathcal{T}}$ is a X-realizable iteration of $\mathcal{Q}$. We cannot in general prove that $\vec{\mathcal{T}}$ picks unique branches mainly because we say nothing about $\mathcal{Q}$-structures that appear in $\vec{\mathcal{T}}$ when we iterate above $\delta^{\mathcal{S}^b}$ for some $\mathcal{S} = \mathcal{M}^{\vec{\mathcal{T}}}_\beta$ and $\beta \in R^{\vec{\mathcal{T}}}$. The next definition introduces a notion of a premouse that resolves this issue.

DEFINITION 12.3.4. We say $\mathcal{R}$ is *weakly X-suitable* if $\mathcal{R}$ is a hod premouse of lsa type such that $\mathcal{R} = (\mathcal{R}|\delta^{\mathcal{R}})^\#$, $\mathcal{R}$ has no Woodin cardinals in the interval $(\delta^{\mathcal{R}^b}, \delta^{\mathcal{R}})$ and for some extension Y of X, $\mathcal{R}$ nicely extends $\mathcal{R}^b = \mathcal{Q}^X_Y$. We say $\mathcal{R}$ is *weakly suitable* if $\mathcal{R}$ is a hod premouse of lsa type such that $\mathcal{R} = (\mathcal{R}|\delta^{\mathcal{R}})^\#$, $\mathcal{R}$ has no Woodin cardinals in the interval $(\delta^{\mathcal{R}^b}, \delta^{\mathcal{R}})$, and $\mathcal{R}^b = \mathcal{P}$.

We now define the notion of *X-approved sts premouse of depth n* by induction on n. The induction ranges over all weakly X-suitable hod premice. Suppose $\mathcal{R}$ is weakly X-suitable hod premouse and Y is an extension of X such that $\mathcal{R}$ nicely extends $\mathcal{R}^b = \mathcal{Q}^X_Y$.

(1) We say that $\mathcal{M}$ is a *X-approved sts premouse over $\mathcal{R}$ of depth* 0 if $\mathcal{M}$ is an sts premouse over $\mathcal{R}$[6] such that if $\mathcal{T} \in \mathcal{M}$ is according to $S^{\mathcal{M}}$ then $\mathcal{T}$ is (X,Y)-realizable.

[5] The embedding $\pi^{\vec{\mathcal{T}}}_{\alpha,\beta}$ is defined similarly to $\pi^{\vec{\mathcal{T}},b}$, it is essentially the embedding $\pi^{\vec{\mathcal{T}}}_{\alpha,\beta} \restriction \mathcal{M}^b_\alpha$. See Section 2.8.

[6] This in particular means that the strategy indexed on the sequence of $\mathcal{M}$ is a strategy for $\mathcal{R}$.

(2) We say that $\mathcal{M}$ is a *X-approved sts premouse over $\mathcal{R}$ of depth $n+1$* if $\mathcal{M}$ is a X-approved sts premouse over $\mathcal{R}$ of depth n such that if $\mathcal{T} \in \mathcal{M}$ is a nuvs and $S^{\mathcal{M}}(\mathcal{T})$ is defined then letting $b = S^{\mathcal{M}}(\mathcal{T})$, $\mathcal{Q}(b,\mathcal{T})$ is a X-approved sts premouse over $\mathrm{m}^+(\mathcal{T}) = (\mathcal{M}(\mathcal{T}))^\#$ of depth n.

DEFINITION 12.3.5. We say $\mathcal{M}$ is a *X-approved sts premouse* over $\mathcal{R}$ if for each $n < \omega$, $\mathcal{M}$ is an X-approved sts premouse over $\mathcal{R}$ of depth n. We say $\mathcal{M}$ as above is a *X-approved sts mouse* (over $\mathcal{R}$) if $\mathcal{M}$ has a μ-strategy Σ such that whenever $\mathcal{N}$ is a Σ-iterate of $\mathcal{M}$, $\mathcal{N}$ is a X-approved sts premouse over $\mathcal{R}$.

We say $\mathcal{M}$ is an *X-approved hod premouse* if whenever $\mathcal{T} \in \mathcal{M}$ is according to $S^{\mathcal{M}}$, then $\mathcal{T}$ is X-realizable. We say $(\mathcal{M},\Sigma)$ is an *X-approved hod mouse* if whenever $\mathcal{U}$ is according to Σ with last model $\mathcal{N}$, then $\mathcal{N}$ is an X-approved hod premouse and $\Sigma_{\mathcal{U},\mathcal{N}} \restriction \mathcal{N} = S^{\mathcal{N}}$.

We let $Lp^{Xa,sts}(\mathcal{R})$ be the union of all X-approved sound sts mice over $\mathcal{R}$ that project to $\leq \mathsf{ord}(\mathcal{R})$. Finally, we can define the *correctly guided X-realizable iterations*.

DEFINITION 12.3.6. Suppose $\mathcal{R}$ is a weakly X-suitable hod premouse and $\vec{\mathcal{T}}$ is a X-realizable iteration of $\mathcal{R}$. We say $\vec{\mathcal{T}}$ is *correctly guided* if whenever $\alpha \in R^{\vec{\mathcal{T}}}$, $\mathcal{U} = \mathsf{nc}_\alpha^{\vec{\mathcal{T}}}$ is above $\delta^{\mathcal{M}^b_\alpha}$, and $\alpha < lh(\mathcal{U})$ is a limit ordinal such that $\mathrm{m}^+(\mathcal{M}(\mathcal{U} \restriction \alpha)) \models$ "$\delta(\mathcal{U} \restriction \alpha)$ is a Woodin cardinal", then letting $b = [0,\alpha]_{\mathcal{U}}$, $\mathcal{Q}(b,\mathcal{U} \restriction \alpha)$ is an X-approved sts mouse over $\mathrm{m}^+(\mathcal{M}(\mathcal{U} \restriction \alpha))$.

The following facts follow straightforwardly from the definitions above (see [36, Section 6] for proofs).

PROPOSITION 12.3.7. *(i) Suppose $\mathcal{R}$ and $\mathcal{S}$ are weakly X-suitable hod premice. Suppose further that $\vec{\mathcal{T}}$ is a X-realizable iteration of $\mathcal{S}$ and $\vec{\mathcal{U}}$ is an iteration of $\mathcal{R}$ such that $(\mathcal{R},\vec{\mathcal{U}})$ is a hull of $(\mathcal{S},\vec{\mathcal{T}})$. Then $\vec{\mathcal{U}}$ is also X-realizable.*

(ii) Suppose $\mathcal{R}$ and $\mathcal{S}$ are weakly X-suitable, $\mathcal{N}$ is an sts premouse over $\mathcal{R}$ and $\mathcal{M}$ is a X-approved premouse (mouse) over $\mathcal{S}$. Suppose $\pi : \mathcal{N} \to_{\Sigma_1} \mathcal{M}$. Then $\mathcal{N}$ is also a X-approved premouse (mouse).

(iii) Suppose $\mathcal{R}$ and $\mathcal{S}$ are weakly X-suitable hod premice. Suppose further that $\vec{\mathcal{T}}$ is a correctly guided X-realizable iteration of $\mathcal{S}$ and $\vec{\mathcal{U}}$ is an iteration of $\mathcal{R}$ such that $(\mathcal{R},\vec{\mathcal{U}})$ is a hull of $(\mathcal{S},\vec{\mathcal{T}})$. Then $\vec{\mathcal{U}}$ is also correctly guided X-realizable iteration.

Our uniqueness theorem applies to $\mathcal{R}$ that are not *infinitely descending*.

DEFINITION 12.3.8. We say that a weakly X-suitable hod premouse $\mathcal{R}$ is *infinitely descending* if there is a sequence $(p_i, \mathcal{R}_i, Y_i : i < \omega)$ such that

1. $\mathcal{R}_0 = \mathcal{R}$,
2. for every $i < \omega$, $\mathcal{R}_i$ is weakly X-suitable and nicely extends $\mathcal{R}_i^b = \mathcal{Q}^X_{Y_i}$,
3. for every $i < \omega$, p_i is a correctly guided X-realizable iteration of $\mathcal{R}_i$,

4. for every $i < \omega$, p_i has a last normal component $\mathcal{T}_i$ of successor length such that $\alpha_i =_{\text{def}} \text{lh}(\mathcal{T}_i) - 1$ is a limit ordinal and $\mathcal{R}_{i+1} = \text{m}^+(\mathcal{M}(\mathcal{T}_i \restriction \alpha_i))$,
5. for every $i < \omega$, setting $b_i =_{\text{def}} [0, \alpha_i)_{\mathcal{T}_i}$, b_i is a cofinal branch of $\mathcal{T}_i$ such that $\mathcal{Q}(b_i, \mathcal{T}_i)$ exists and is X-approved.

Note that in the above definition, for each i, $\mathcal{R}_{i+1}$ is a strict initial segment of $\mathcal{Q}(b_i, \mathcal{T}_i)$. The following is the uniqueness result we need.

PROPOSITION 12.3.9. *Suppose $\mathcal{R}$ is a weakly X-suitable hod premouse that is not infinitely descending and $\vec{\mathcal{T}}$ is a correctly guided X-realizable iteration of limit length on $\mathcal{R}$. There is then a unique branch b of $\vec{\mathcal{T}}$ such that $\vec{\mathcal{T}}^\frown\{b\}$ is correctly guided and X-realizable.*

PROOF. The proof easily follows from results of Section 4, particularly Lemma 4.7.2, and Section 9.2. Suppose first that

(a) if $\vec{\mathcal{T}}$ does not have a last component or
(b) if there is $\alpha \in R^{\vec{\mathcal{T}}}$ such that $\vec{\mathcal{T}}_{\geq\alpha}$ is based on $\mathcal{M}_\alpha^b$.

In case (a), there is nothing to prove. In case (b), let $\mathcal{S} = \mathcal{M}_\alpha^{\vec{\mathcal{T}}}$, and let $W_{\mathcal{S}}$ be as in Definition 12.3.2, $\Psi_{W_{\mathcal{S}}}^X$ only depends on $\mathcal{S}^b$ (by Lemma 12.2.1)[7]. Suppose we are not in case (a) or (b). Let now $\mathcal{T} = \mathsf{nc}_\alpha^{\vec{\mathcal{T}}}$ be the last normal component of $\vec{\mathcal{T}}$. If b, c are two different branches of $\mathcal{T}$ such that $\vec{\mathcal{T}}^\frown\{b\}$ and $\vec{\mathcal{T}}^\frown\{c\}$ are correctly guided X-realizable iterations then $\mathcal{Q}(b, \mathcal{T}) \neq \mathcal{Q}(c, \mathcal{T})$ and both are X-approved sts mice over $\text{m}^+(\mathcal{T})$. It now follows from Lemma 4.7.2 and the fact that $\mathcal{R}$ is not infinitely descending that we can reduce the disagreement of $\mathcal{Q}(b, \mathcal{T})$ and $\mathcal{Q}(c, \mathcal{T})$ to a disagreement between Ψ_Y^X and Ψ_Z^X for some extensions Y, Z of X with $\mathcal{Q}_Y^X = \mathcal{Q}_Z^X$. However, this cannot happen by Lemma 12.2.1. ⊣

DEFINITION 12.3.10. Suppose X is a condensing set. Suppose $\mathcal{R}_0$ extends $\mathcal{P}$, p is an iteration of $\mathcal{R}_0$ such that if p is nuvs, then setting $\mathcal{R} = \text{m}^+(p)$, $\mathcal{M}$ is an sts premouse over $\mathcal{R}$. Suppose $(\mathcal{R}, \mathcal{M}, p) \in H_{\kappa^{+4}}$.

1. We say $\mathcal{R}$ is not infinitely descending if whenever U is an X-good hull such that $\mathcal{R} \in U$, $\pi_U^{-1}(\mathcal{R})$ is not infinitely descending.
2. We say *p is X-validated* if whenever U is an X-good hull such that $\{\mathcal{R}, p\} \subseteq U$, $\pi_U^{-1}(p)$ is a correctly guided X-realizable iteration of $\pi_U^{-1}(\mathcal{R})$.
3. Suppose $\mathcal{R}$ is weakly suitable. We say *$\mathcal{M}$ is a X-validated sts premouse* over $\mathcal{R}$ if for every X-good hull U such that $\{\mathcal{R}, \mathcal{M}\} \subseteq U$, letting $\mathcal{N} = \pi_U^{-1}(\mathcal{M})$, $\mathcal{N}$ is a X-approved sts premouse over $\pi_U^{-1}(\mathcal{R})$.
4. Suppose $\mathcal{R}$ is weakly suitable. We say *$\mathcal{M}$ is a X-validated sts mouse* over $\mathcal{R}$ if whenever U is an X-good hull such that $\{\mathcal{R}, \mathcal{M}\} \subseteq U$, letting $\mathcal{N} = \pi_U^{-1}(\mathcal{M})$, $\mathcal{N}$ is a X-approved sts mouse over $\pi_U^{-1}(\mathcal{R})$.

[7]Notice that in this case there is a branch b such that $\vec{\mathcal{T}}^\frown\{b\}$ is correctly guided and X-realizable.

5. Suppose $\mathcal{M}$ is a X-validated sts mouse over $\mathcal{R}$ and ξ is an ordinal. We say $\mathcal{M}$ has an X-validated ξ-strategy Σ if whenever $\mathcal{N}$ is an iterate of $\mathcal{M}$ via Σ, $\mathcal{N}$ is a X-validated sts mouse over $\mathcal{R}$.
6. Suppose q is an iteration of $\mathcal{R}$. We say q is X-*validated* if $p^\frown q$ is X-validated.
7. We say a hod premouse $\mathcal{M}$ such that $\mathcal{P} \trianglelefteq \mathcal{M}$ is an X-*validated* hod premouse (mouse) if for every X-good hull U such that $\{\mathcal{P}, \mathcal{M}\} \subseteq U$, letting $\mathcal{N} = \pi_U^{-1}(\mathcal{M})$, $\mathcal{N}$ is a X-approved hod premouse (mouse, respectively).

The following proposition is very useful and is an immediate consequence of Proposition 12.3.7. When U is a good hull we will use it as a subscript to denote the π_U-preimages of objects that are in U.

PROPOSITION 12.3.11. *Suppose $\mathcal{R}_0, p, \mathcal{R}, \mathcal{M}$ are as in Definition 12.3.10. Suppose U is an X-good hull such that $\{\mathcal{R}, \mathcal{M}\} \subseteq U$ and $\mathcal{M}_U$ is not X-approved. Then whenever U^* is an X-good hull such that $U \cup \{U\} \subseteq U^*$, $\mathcal{M}_{U^*}$ is not X-approved. Hence, $\mathcal{M}$ is not X-validated.*

A similar result holds for iterations.

PROPOSITION 12.3.12. *Suppose $\mathcal{R}_0$ is as in Proposition 12.3.11. Suppose $p \in H_{\kappa^{+4}}$ is an iteration of $\mathcal{R}_0$. Suppose U is an X-good hull such that $\{\mathcal{R}_0, p\} \subseteq U$ and p_U is not X-realizable. Then whenever U^* is an X-good hull such that $U \cup \{U\} \subseteq U^*$, p_{U^*} is not X-realizable. Hence, p is not X-validated.*

DEFINITION 12.3.13. We say $\mathcal{R}$ is X-*suitable* if it is weakly X-suitable and whenever $\mathcal{M}$ is an X-approved sts mouse over $\mathcal{R}$ then $\mathcal{M} \models$ "$\delta^{\mathcal{R}}$ is a Woodin cardinal". We say $\mathcal{R}$ is *suitable* if it is weakly suitable and whenever $\mathcal{M}$ is an X-validated sts mouse over $\mathcal{R}$, then $\mathcal{M} \models$ "$\delta^{\mathcal{R}}$ is a Woodin cardinal".

We let $Lp^{Xv,sts}(\mathcal{R})$ be the union of all X-validated sound sts mice over $\mathcal{R}$ that project to $\leq \mathrm{ord}(\mathcal{R})$. The following proposition is a consequence of Proposition 12.3.9.

PROPOSITION 12.3.14. *Suppose $(\mathcal{R}_0, p, \mathcal{R})$ are as in Definition 12.3.10 and $\mathcal{R}$ is not infinitely descending. Suppose $\vec{\mathcal{T}}$ is an X-validated iteration of $\mathcal{R}$ of limit length. Then there is at most one branch b of $\vec{\mathcal{T}}$ such that $\vec{\mathcal{T}}^\frown\{b\}$ is X-validated.*

In the next two sections, we describe two kinds of constructions: *the hybrid K^c-construction* over $\mathcal{P}$ or some $\mathcal{P}'$ extending $\mathcal{P}$,[8] and the *X-validated sts constructions* over some weakly suitable $\mathcal{R}$. We use the notations and definitions from the previous section. We fix a condensing set $X \in V$ (X exists by the previous section); and we assume that $X = X' \cap \mathcal{P}$ where $X' \prec H_{\kappa^{+4}}$ is of size κ in V. The hybrid K^c-construction proceeds more or less according to the usual procedure for building hod pairs (as described many times in this book) except that extenders F we put on the sequence of the models are correctly backgrounded (described

[8]This is a variation of the mixed hod pair construction in Definition 10.2.41.

below) instead of being fully backgrounded. This construction will reach a weakly suitable stage $\mathcal{R}$. We then continue with the X-validated sts construction over $\mathcal{R}$. If this construction produces a $\mathcal{Q}$-structure $\mathcal{Q}$ extending $\mathcal{R}$, then we attempt to construct an X-validated strategy for $\mathcal{Q}$. If this is successful, we then continue with the hybrid K^c-construction over $\mathcal{Q}$. If not, we show that an honest suitable $\mathcal{R}'$ must exist (see below). Producing such an $\mathcal{R}'$ means that the X-validated sts construction over $\mathcal{R}'$ will no longer produce $\mathcal{Q}$-structures for $\mathcal{R}'$ and will either reach a model that produces $\mathcal{N}^{\sharp}_{\omega.2,lsa}$ and hence a model of LSA or go on for κ^{+++} many steps. We will rule out the latter by an argument using the technique of stacking mice developed in [12] (see the next section). This argument also rules out the case that the two aforementioned constructions alternate for κ^{+++} many times.

The two constructions described above will produce a sequence of models $(\mathcal{M}_\xi, \mathcal{N}_\xi : \xi \leq \Upsilon)$. Before defining the sequence, we discuss the kind of background extenders being used in this construction. Suppose $\mathcal{M}_\xi$ has been constructed as part of one of such constructions and is in V, is passive, $\wp(\delta^{\mathcal{P}})^{\mathcal{M}_\xi} = \wp(\delta^{\mathcal{P}})^{\mathcal{P}}$. Suppose F is a $(\mathrm{crit}(F), \mathsf{ord}(\mathcal{N}_\xi))$-extender that coheres the sequence of $\mathcal{N}_\xi = \mathcal{C}(\mathcal{M}_\xi)$.[9] Suppose $Y \prec H_{\kappa^{+4}}$ (or sometimes, we'll let $Y \prec H_\gamma$ for $\gamma \geq \kappa^{+4}$) is in V and is a good X-hull and Y contains all relevant objects. Let π_Y be the corresponding uncollapse map. Suppose $\mathcal{N}^Y_\xi =_{\mathrm{def}} \pi_Y^{-1}(\mathcal{N}_\xi)$ has a unique X-realizable strategy Σ^Y_ξ such that $\Sigma^Y_\xi \restriction \mathrm{HC} \in \Omega$ (these properties will be maintained during the course of our construction). Alternatively, we sometimes write $\Sigma_{Y,\xi}$ for Σ^Y_ξ.

DEFINITION 12.3.15. We say that an extender F is *correctly backgrounded* if one of the following holds:

1. if $\mathrm{crit}(F) = \delta^{\mathcal{P}}$ and the least cutpoint above $\delta^{\mathcal{P}}$ is the largest cardinal in $\mathcal{M}_\xi$, then $(a, A) \in F$ if and only if $\forall^* Y$, Y is X-good, letting $a_Y = \pi_Y^{-1}(a)$, $\pi^{\Sigma^Y_\xi}_{\mathcal{N}^Y_\xi, \infty}(a_Y) \in A$. We say that F is *X-certified.*
2. if $\mathrm{crit}(F) > \delta^{\mathcal{P}}$, then say, $\lambda = F(\mathrm{crit}(F))$, F is *certified by a collapse* in the sense of [12], that is, there is $Z \prec H^V_{\kappa^{+4}}$ (in V) such that $|Z| < \kappa^{+++}$, where $o(\mathcal{P}) + 1 \subset Z$, $Z \cap \kappa^{+++} \in \kappa^{+++}$, $Z^{<\kappa} \subseteq Z$,[10] and letting $\pi_Z : M_Z \to Z$ be the uncollapse, we have: $\mathcal{M}_\xi | (\mathrm{crit}(F))^{+, \mathcal{M}_\xi} \in M_Z$, $\mathrm{crit}(\pi_Z) = \mathrm{crit}(F)$, and
$$F \text{ is the Jensen completion of } (\pi \restriction \wp(\mathrm{crit}(\pi_Z)) \cap \mathcal{N}_\xi) \restriction \lambda.$$

In either case, we will sometimes say "F is certified".

---

[9] $\mathcal{N}_\xi$ is the appropriate fine-structural core of $\mathcal{M}_\xi$, as dictated by the construction. See the next section.

[10] Note that $|o(\mathcal{P})| = |\Omega| = \kappa$. Futhermore, $Z[G]^{<\omega_1} \subseteq Z[G]$ in $V[G]$.

Suppose F is X-certified and Y is X-good. Let $F_Y = \pi_Y^{-1}(F)$. Then it is easy to check that $(a,A) \in F_Y$ iff $\pi^{\Sigma^Y_\xi}_{\mathcal{N}^Y_\xi,\infty}(a) \in \pi_Y(A)$. We say that F_Y is *π_Y-certified over* $(\mathcal{N}^Y_\xi, \Sigma^Y_\xi)$.

12.4. The X-validated sts constructions

We first describe the X-validated sts construction for a fixed condensing set X. Suppose $\mathcal{R}$ is weakly suitable and $Y \in H_{\kappa^{+++}}$ is transitive such that either $\mathcal{R} = Y$ or $\mathcal{R} \in Y$. Recall the conventions for hod premice introduced earlier in the book. A (hod) premouse has the form $(\mathcal{M}, k)$, where $\mathcal{M}$ is a k-sound, acceptable J-structure. $k(M) = k$ is the degree of soundness of $\mathcal{M}$. We write the core $\mathcal{C}(\mathcal{M})$ for the $(k(\mathcal{M})+1$-)core of $\mathcal{M}$ (if this makes sense, i.e., when $\mathcal{M}$ is $k(\mathcal{M})+1$-solid, or just solid; the same abbreviation will be applied when we say $\mathcal{M}$ is universal, meaning $\mathcal{M}$ is $k(\mathcal{M})+1$-universal). Similarly, we write $\rho(\mathcal{M})$ for the $k(\mathcal{M})+1$-projectum and $p(\mathcal{M})$ for the $k(\mathcal{M})+1$-standard parameters of $\mathcal{M}$. When $\mathcal{C}(\mathcal{M})$ exists, $k(\mathcal{C}(\mathcal{M})) = k(\mathcal{M})+1$. $\mathcal{M}$ is sound iff $\mathcal{M} = \mathcal{C}(\mathcal{M})$. For brevity, we suppress the degree of soundness of the models constructed below. For instance, if $k(\mathcal{M}_\xi) = k$, then we write $\mathcal{M}_\xi$ for $(\mathcal{M}_\xi, k)$. Before, describing the next construction, we advise the reader to consult [36, Section 8] for a similar construction.

DEFINITION 12.4.1 (X-validated sts construction). We say $(\mathcal{M}_\xi, \mathcal{N}_\xi : \xi \leq \Upsilon)$ are the models of the *X-validated sts construction over* Y if the following conditions hold:

1. $\Upsilon \leq \kappa^{+++}$, and for all $\xi < \Upsilon$ if $\mathcal{M}_\xi, \mathcal{N}_\xi$ are defined then $\mathcal{M}_\xi, \mathcal{N}_\xi \in H_{\kappa^{+++}}$.
2. For every $\xi \leq \Upsilon$, $\mathcal{M}_\xi$ and $\mathcal{N}_\xi$ are X-validated sts hod premice over $\mathcal{R}$ or are X-validated sts hod premice over Y based on $\mathcal{R}$.
3. Suppose the sequence $(\mathcal{M}_\xi, \mathcal{N}_\xi : \xi < \eta)$ has been constructed. Suppose further that there is a total (κ, ν)-extender F such that F is certified by a collapse and letting G be the Jensen completion of $\mathcal{N}_{\eta-1} \cap F$, $(\mathcal{N}_{\eta-1}, G)$ is a X-validated sts hod premouse over $\mathcal{R}$ or over Y based on $\mathcal{R}$. Let then $\mathcal{M}_\eta = (\mathcal{N}_{\eta-1}, F)$ and $\mathcal{N}_\eta = \mathcal{C}(\mathcal{M}_\eta)$.
4. Suppose the sequence $(\mathcal{M}_\xi, \mathcal{N}_\xi : \xi < \eta)$ has been constructed, and $\mathcal{T} \in \mathcal{N}_{\eta-1}$ is the $<_{\mathcal{N}_{\eta-1}}$-least uvs without an indexed branch. Suppose further that there is a branch b of $\mathcal{T}$ such that $(\mathcal{N}_{\eta-1}, B_b)$[11] is a X-validated sts hod

[11] B_b is a code for b as done in [50, Section 2] and outlined in Chapter 11. We only note that this amenable coding ensures condensation under very weak hull embeddings, cf. [50, Lemma 3.10] and this fact is in turns used to show that $\square_{\kappa,2}$ holds in hod mice. From now on, we may confuse the structure $(\mathcal{N}_{\eta-1}, B_b)$ with $(\mathcal{N}_{\eta-1}, b)$.

premouse[12] over $\mathcal{R}$ or over Y based on $\mathcal{R}$. Let then $\mathcal{M}_\eta = (\mathcal{N}_{\eta-1}, B_b)$ and $\mathcal{N}_\eta = \mathcal{C}(\mathcal{M}_\eta)$.

5. Suppose the sequence $(\mathcal{M}_\xi, \mathcal{N}_\xi : \xi < \eta)$, and for some nuvs tree $\mathcal{T} \in \mathcal{N}_{\eta-1}$ there is a branch b such that $(\mathcal{N}_{\eta-1}, B_b)$ is an X-validated sts hod premouse over $\mathcal{R}$ or over Y based on $\mathcal{R}$. Let $\mathcal{T}$ be the $\mathcal{N}_{\eta-1}$-least such tree and b be such a branch for $\mathcal{T}$. Then $\mathcal{M}_\eta = (\mathcal{N}_{\eta-1}, B_b)$ and $\mathcal{N}_\eta = \mathcal{C}(\mathcal{M}_\eta)$.
6. Suppose the sequence $(\mathcal{M}_\xi, \mathcal{N}_\xi : \xi < \eta)$ has been constructed and all of the above cases fail. In this case we let $\mathcal{M}_\eta = \mathcal{J}_1(\mathcal{N}_{\eta-1})$ and provided $\mathcal{M}_\eta$ is a X-validated sts hod premouse over $\mathcal{R}$ or over Y based on $\mathcal{R}$, $\mathcal{N}_\eta = \mathcal{C}(\mathcal{M}_\eta)$.
7. Suppose the sequence $(\mathcal{M}_\xi, \mathcal{N}_\xi : \xi < \eta)$ has been constructed and η is a limit ordinal. Then $\mathcal{M}_\eta = \liminf_{\xi \to \eta} \mathcal{M}_\xi$.

The construction fails at η if one of the following holds.

$(\mathrm{i})_\eta$ $\mathcal{M}_\eta$ is not solid or universal.

$(\mathrm{ii})_\eta$ $\mathcal{M}_\eta$ is not X-validated.

$(\mathrm{iii})_\eta$ There is a uvs $\mathcal{T} \in \mathcal{N}_{\eta-1}$ such that the indexing scheme demands that a branch of $\mathcal{T}$ must be indexed yet $\mathcal{T}$ has no branch b such that $(\mathcal{N}_{\eta-1}, B_b)$ is a X-validated sts premouse over $\mathcal{R}$ or over Y based on $\mathcal{R}$.

$(\mathrm{iv})_\eta$ There is a nuvs $\mathcal{T} \in \mathcal{N}_{\eta-1}$ such that the indexing scheme demands that a branch of $\mathcal{T}$ must be indexed yet $\mathcal{T}$ has no branch b such that $(\mathcal{N}_{\eta-1}, B_b)$ is a X-validated sts premouse over $\mathcal{R}$ or over Y based on $\mathcal{R}$.

$(\mathrm{v})_\eta$ $\rho(\mathcal{M}_\eta) \leq \delta^{\mathcal{H}}$.

We note that in (3) above, $\mathrm{crit}(F) > \delta^{\mathcal{P}}$. If the construction holds at η (i.e., $(1)_\eta$-$(5)_\eta$ all fail) then $\mathcal{M}_\eta, \mathcal{N}_\eta$ are hp-indexed lses. Indeed, we inductively maintain that the models in our construction are hp-indexed lses. Verifying $(\mathrm{iii})_\eta$, $(\mathrm{iv})_\eta$ fail for all η in the X-validated sts construction roughly corresponds to showing that the important anomaly does not occur in the construction in 10.2.41.[13]

One shows $(\mathrm{i})_\eta$ fails by the usual arguments, namely showing that countable substructures of $\mathcal{M}_\eta$ are iterable. That $(\mathrm{v})_\eta$ fails will be shown when Y is specified. The proof that $(\mathrm{v})_\eta$ fails for certain precisely defined Y is to show that $\mathcal{M}_\eta$ is iterable via a fullness preserving strategy. The proof that $(\mathrm{v})_\eta$ fails will subsume the proof that $(\mathrm{i})_\eta$ fails and will be given later (see Lemma 12.7.18). [36, Section 8] shows that $(\mathrm{ii})_\eta$, $(\mathrm{iii})_\eta$, $(\mathrm{iv})_\eta$ cannot fail if $\mathcal{R}$ is honest as witnessed by $(\vec{\mathcal{V}}, p) \in H_{\kappa^{+4}}$, where $\vec{\mathcal{V}} = (\mathcal{V}_\alpha : \alpha \leq \xi)$ is an array with the X-realizability property (defined in [36, Definitions 8.1 and 8.2]) and either $\mathcal{R} = \mathcal{V}_\xi$ and $p = \emptyset$ or p is an X-validated iteration of $\mathcal{V}_\xi$ of limit length such that $\pi^{p,b}$ exists and $\mathcal{R} = \mathrm{m}^+(p)$. We summarize some main points of the arguments in [36] below and outline the proof that $(\mathrm{iii})_\eta$ fails and $(\mathrm{iv})_\eta$ fails. First we define the objects $(\vec{\mathcal{V}}, p)$ more precisely.

---

[12] This in particular implies that $b \in \mathcal{N}_{\eta-1}$. For brevity, we suppress the other predicates that are part of the hod premouse, like $\varepsilon, \vec{E}$ etc.

[13] We note that $\mathcal{T}$ as in $(\mathrm{iii})_\eta$, $(\mathrm{iv})_\eta$ are of the form $(\mathcal{R}, \mathcal{T}_0, \mathcal{S}, \mathcal{T}_1)$ but we will suppress this notation for brevity.

DEFINITION 12.4.2. We say $\vec{\mathcal{V}} = (\mathcal{V}_\alpha : \alpha \leq \eta)$ is an *array* of length η if the following conditions hold.

1. For every $\alpha < \eta$, $(\mathcal{V}_\beta : \beta \leq \alpha)$ is an array of length α at μ.
2. $\mathcal{V}_\eta$ nicely extends $\mathcal{P}$ and is a X-validated hod premouse.
3. For all $\alpha < \eta$, if $\mathcal{V}_\alpha$ is weakly X-suitable then there is $\beta \leq \eta$ such that $\mathcal{V}_\beta$ is a X-validated sts mouse over $\mathcal{V}_\alpha$ and $\mathcal{J}_1(\mathcal{V}_\beta) \models$ "there are no Woodin cardinals $> \delta^{\mathcal{P}}$".
4. For all $\alpha < \eta$, if $\mathcal{J}_1(\mathcal{V}_\alpha) \models$ "there are no Woodin cardinals $> \delta^{\mathcal{P}}$" then $\mathcal{V}_\alpha$ has a X-validated iteration strategy.

We say $\vec{\mathcal{V}}$ is *small* if $rud(\mathcal{V}_\eta) \models$ "there are no Woodin cardinals $> \delta^{\mathcal{V}_\eta^b}$". We let $\eta = \mathrm{lh}(\vec{\mathcal{V}})$ and for $\alpha \leq \eta$, we let $\vec{\mathcal{V}} \restriction \alpha = (\mathcal{V}_\beta : \beta \leq \alpha)$.

Recall the notions of k-maximal iteration trees in [59, Definition 3.4], weak k-embeddings [59, Definition 4.1]. For an iteration tree $\mathcal{T}$ on $\mathcal{M}$, letting $\mathcal{M}_\alpha^\mathcal{T}$ be the α-th model in the tree; for $\alpha + 1 < \mathrm{lh}(\mathcal{T})$, recall the notion of degree $deg^\mathcal{T}(\alpha+1)$ [59, Definition 3.7]. Recall the definition of $D^\mathcal{T}$: if $\alpha + 1 \in D$, then the extender $E_{\alpha+1}^\mathcal{T}$ is applied to a strict initial segment of $\mathcal{M}_\beta^\mathcal{T}$ where $\beta = T - pred(\alpha+1)$. For λ limit, $deg^\mathcal{T}(\lambda)$ is the eventual values of $deg^\mathcal{T}(\alpha+1)$ for $\alpha + 1 \in [0, \lambda]_\mathcal{T}$. For a cofinal branch b of $\mathcal{T}$, $deg^\mathcal{T}(b)$ is defined to be the eventual value of $deg^\mathcal{T}(\alpha+1)$ for $\alpha + 1 \in b$. We write $\mathcal{C}_k(\mathcal{M})$ for the k-th core of $\mathcal{M}$. Sometimes, we confuse $\mathcal{C}_0(\mathcal{M})$ with $\mathcal{M}$ itself.

DEFINITION 12.4.3. Suppose $\vec{\mathcal{V}}$ is an array. We say $\vec{\mathcal{V}}$ has the *X-realizability property* if for all $\alpha < \mathrm{lh}(\mathcal{V})$, $\vec{\mathcal{V}} \restriction \alpha$ has the X-realizability property and whenever $g \subseteq Coll(\omega, <\kappa)$ is generic, in $V[g]$, whenever $\pi : \mathcal{W} \to \mathcal{C}_k(\mathcal{V}_\eta)$ is a weak k-embedding and $\mathcal{T}$ are such that

1. $X \subseteq \mathrm{rng}(\pi)$
2. $\mathcal{W}, \mathcal{T} \in HC$,
3. $\mathcal{T}$ is X-approved normal, k-maximal iteration of $\mathcal{W}$ that is above $\delta^{\mathcal{W}^b}$,

then one of the following holds (in $V[g]$).

1. $\mathcal{T}$ is of limit length and there is a cofinal well-founded branch c such that c has no drops in model (i.e., $D^\mathcal{T} \cap b = \emptyset$); letting $l = deg^\mathcal{T}(b)$, there is a weak l-embedding $\tau : \mathcal{M}_c^\mathcal{T} \to \mathcal{C}_l(\mathcal{V}_\eta)$ such that $\pi \restriction \mathcal{W} = \tau \circ \pi_c^\mathcal{T}$.
2. $\mathcal{T}$ is of limit length and there is a cofinal well-founded branch c such that c has a drop in model, and there is $\beta < \eta$ and a weak l-embedding $\tau : \mathcal{M}_c^\mathcal{T} \to \mathcal{C}_l(\mathcal{V}_\beta)$ such that $\tau \restriction (\mathcal{M}_c^\mathcal{T})^b = \pi \restriction (\mathcal{M}_c^\mathcal{T})^b$, where $l = deg^\mathcal{T}(c)$.
3. $\mathcal{T}$ has a last model and letting $\gamma = \mathrm{lh}(\mathcal{T}) - 1$, $[0, \gamma]_\mathcal{T} \cap \mathcal{D}^\mathcal{T} = \emptyset$ and there is a weak l-embedding $\tau : \mathcal{M}_\gamma^\mathcal{T} \to \mathcal{C}_l(\mathcal{V}_\eta)$ such that $\pi \restriction \mathcal{W} = \tau \circ \pi^\mathcal{T}$, where $l = deg^\mathcal{T}(\gamma)$.
4. $\mathcal{T}$ has a last model and letting $\gamma = \mathrm{lh}(\mathcal{T}) - 1$, $[0, \gamma]_\mathcal{T} \cap \mathcal{D}^\mathcal{T} \neq \emptyset$ and for some $\beta < \eta$ there is a weak l-embedding $\tau : \mathcal{M}_\gamma^\mathcal{T} \to \mathcal{C}_l(\mathcal{V}_\beta)$ such that $\tau \restriction (\mathcal{M}_\gamma^\mathcal{T})^b = \pi \restriction (\mathcal{M}_\gamma^\mathcal{T})^b$, where $l = deg^\mathcal{T}(\gamma)$.

When the above 4 clauses hold we say that $\mathcal{T}$ is $(\pi,\vec{\mathcal{V}})$-realizable. In cases where $\vec{\mathcal{V}}$ is clear from the context, we omit it from our notation.

DEFINITION 12.4.4. Suppose $\mathcal{R}$ is a weakly X-suitable hod premouse. We say $\mathcal{R}$ is *honest* if there is an array $\vec{\mathcal{V}} = (\mathcal{V}_\alpha : \alpha \leq \eta)$ at μ with the X-realizability property such that $\mathcal{R},\vec{\mathcal{V}} \in H_{\kappa^{+4}}$, the following conditions hold.

1. Either $\mathcal{V}_\eta = \mathcal{R}$ or there is a X-validated iteration p of $\mathcal{V}_\eta$ of limit length such that $\pi^{p,b}$ exists and $\mathcal{R} = \mathrm{m}^+(p)$.
2. $\vec{\mathcal{V}}$ is small if and only if $\mathcal{V}_\eta \neq \mathcal{R}$.

If $\mathcal{R}$ is honest and $\vec{\mathcal{V}}$ is as above then we say that $\vec{\mathcal{V}}$ is an honesty certificate for $\mathcal{R}$.

We fix $\mathcal{R},\vec{\mathcal{V}},p$ as in Definition 12.4.4 and the models $(\mathcal{M}_\eta,\mathcal{N}_\eta : \eta \leq \Upsilon)$ are constructed by the above X-validated sts construction over $\mathcal{R}$. We proceed to verify the failure of (ii)$_\eta$, (iii)$_\eta$, (iv)$_\eta$ for $\eta \leq \Upsilon$.

12.4.0.1. *Failure of (ii)$_\eta$.* Let us first verify (ii)$_\eta$ fails. Towards contradiction assume that there is some model $\mathcal{W}^*$ that appears in the X-validated sts construction such that $\mathcal{W}^*$ is not X-validated. Let $\mathcal{W}$ be the least such model.

Suppose first that $\mathcal{W}$ is $\mathcal{W} = \mathcal{M}_\alpha$ for some α. Suppose first α is a limit ordinal. Let U be an X-good hull such that $\{\mathcal{R},\mathcal{W}\} \subseteq U$ and $(\mathcal{M}_\beta,\mathcal{N}_\beta : \beta < \alpha) \in U$. We have that $\mathcal{M}_\beta,\mathcal{M}_\beta$ are X-validated for every $\beta < \alpha$. Let $(\mathcal{K}_\xi : \xi \leq \alpha_U) = \pi_U^{-1}(\mathcal{M}_\beta : \beta < \alpha)$. Fix $\mathcal{T} \in \mathcal{K}_{\alpha_U}$ according to $S^{\mathcal{K}_{\alpha_U}}$. We need to see that $\mathcal{T}$ is X-approved. Fix $\xi < \alpha_U$ such that $\mathcal{T} \in \mathcal{K}_\xi$ and is according to $S^{\mathcal{K}_\xi}$. Then $\pi_U(\mathcal{T}) \in \mathcal{M}_{\pi_U(\xi)}$ and is according to $S^{\mathcal{M}_{\pi_U(\xi)}}$. Therefore, $\mathcal{T}$ is X-approved. In this case, it is also easy to see that $\mathcal{N}_\alpha = \mathcal{C}(\mathcal{M}_\alpha)$ is X-validated.

Suppose next that $\alpha = \beta + 1$. Suppose the least model that is not X-validated is $\mathcal{N}_\alpha$. We must have that $\mathcal{M}_\alpha$ is X-validated and that all models $(\mathcal{M}_\xi,\mathcal{N}_\xi : \xi \leq \beta)$ are X-validated. Let now U be an X-good hull such that $\{\mathcal{R},\mathcal{W}\} \subseteq U$. But then $\pi_U^{-1}(\mathcal{N}_\alpha) = \mathcal{C}(\pi_U^{-1}(\mathcal{M}_\alpha))$. It then follows that $\pi_U^{-1}(\mathcal{N}_\alpha)$ is X-approved (see Proposition 12.3.7).

We now assume that $\mathcal{W} = \mathcal{M}_\alpha$ for some α. Suppose first that $\mathcal{M}_\alpha = (\mathcal{N}_{\alpha-1}, b)$ where b is a branch. Then it follows from the definition of X-validated sts constructions that $\mathcal{M}_\alpha$ is X-validated. The case that $\mathcal{M}_\alpha = (\mathcal{N}_{\alpha-1}, E)$ for some extender E is trivial as no new iterations of $\mathcal{R}$ have been introduced.

Finally suppose $\mathcal{M}_\alpha = \mathcal{J}_1(\mathcal{N}_{\alpha-1})$. If $\mathcal{M}_\alpha$ is not X-validated then it is because there is $\mathcal{T} \in \mathcal{M}_\alpha - \mathcal{N}_{\alpha-1}$ such that $\mathcal{T}$ is according to $S^{\mathcal{M}_\alpha}$ yet $\mathcal{T}$ is not X-validated. Let $\xi = \sup\{\zeta : \mathcal{T} \restriction \zeta \in \mathcal{N}_{\alpha-1}\}$. Then all proper initial segments of $\mathcal{T} \restriction \xi$ is in $\mathcal{N}_{\alpha-1}$ and hence, all of the proper initial segments of $\mathcal{T} \restriction \xi$ are X-validated. Because $\mathcal{T}$ is not X-validated, $\xi + 1 \leq \mathrm{lh}(\mathcal{T})$.

The following is the key point. There is no limit ordinal $\beta \in (\xi, \mathrm{lh}(\mathcal{T}))$. This is because to define $[0,\beta]_\mathcal{T}$ we need to "leave behind" a level that at the minimum is a model of ZFC, while there is no such level between $\mathcal{M}_\alpha$ and $\mathcal{N}_{\alpha-1}$. Thus, it must be that $\mathrm{lh}(\mathcal{T}) = \xi + n$ for some $n \in [1, \omega)$. Let then m be least such that $\mathcal{T} \restriction \xi + m$ is X-validated but $\mathcal{T} \restriction \xi + m + 1$ is not. We then have three cases.

The first case is the following.

1. $\pi^{\mathcal{T} \restriction \xi, b}$ is defined,
2. $\mathcal{T}_{\geq \xi}$ is above $(\mathcal{M}_{\xi}^{\mathcal{T}})^b$, and
3. letting $\mathcal{Q} = \mathcal{M}_{\xi+m}^{\mathcal{T}}$, $\mathrm{crit}(E_{\xi+m}^{\mathcal{T}}) = \delta^{\mathcal{Q}^b}$.

Let now U be an X-good hull such that $\{\mathcal{R}, \mathcal{W}\} \subseteq U$. Let $E = \pi^{-1}(E_{\xi+m}^{\mathcal{T}})$ and let $\mathcal{Y}$ be the least node of $\mathcal{T}_U$ to which E must be applied. Using [36, Proposition 8.8], fix β, an $l < \omega$, and a weak l-embedding $k : \mathcal{Y} \to \mathcal{C}_l(\mathcal{V}_\beta)$ such that $X \subseteq \mathrm{rng}(k)$. Using the fact that $(\mathcal{T}_U)_{\geq \mathcal{Y}}$ is $(k, \vec{\mathcal{V}} \restriction \beta)$-realizable, we can find $\gamma \leq \beta$ and a weak n-embedding $\tau : \mathcal{Q}_U \to \mathcal{C}_n(\mathcal{V}_\gamma)$ such that $X \subseteq \mathrm{rng}(\tau)$. Therefore, as $\tau(E)$ is X-validated, letting $Z = k[\mathcal{Y}^b] \cap \mathcal{P}^-$, there is Y an extension of $Z \oplus X$ such that $\mathrm{Ult}(\mathcal{Y}^b, E) = \mathcal{Q}_Y^X$. Hence, $\mathcal{T}$ is X-validated.

The next possibility is when $\mathcal{T}_{\geq \xi}$ is either a tree of finite length based on $\pi^{\mathcal{T}_{\geq \xi}, b}(\mathcal{R}^b)$ or it only uses extenders with critical points $> \pi^{\mathcal{T}_{\geq \xi}, b}(\delta^b)$. The second case is easy because the tree $\mathcal{T}_{\geq \xi}$ is finite. The first case follows from an argument similar to the one given above.

Finally we could have that $\xi + 1 = \mathrm{lh}(\mathcal{T})$, where ξ is a limit ordinal, and for a cofinal set of $\beta < \xi$, letting $\gamma_\beta = \mathit{pred}_T(\beta + 1)$, $\mathrm{crit}(E_\beta) = \delta^{\pi_{0,\gamma_\beta}^{\mathcal{T}}(\delta^{\mathcal{R}^b})}$. This case, however, easily follows from the direct limit construction. This completes our argument that (ii)$_\eta$ fails.

12.4.0.2. *Failure of (iii)$_\eta$.* We now consider (iii)$_\eta$. Suppose $\mathcal{R}$ is honest as witnessed by $(\vec{\mathcal{V}}, p)$. Then we say $\mathcal{T}$ is a X-validated iteration of $\mathcal{R}$ if $p^\frown \mathcal{T}$ is a Z-validated iteration of $\mathcal{V}_\eta$ where $\eta + 1 = \mathrm{lh}(\vec{\mathcal{V}})$. The array $\vec{\mathcal{V}}$ typically comes from an X-validated sts construction or a hybrid K^c-construction (described later).

Let $\mathcal{R}$ be honest as witnessed by $(\vec{\mathcal{V}}, p)$. Suppose $\mathcal{T}$ is a normal tree on $\mathcal{R}$ such that $\pi^{\mathcal{T}, b}$ exists and δ is a Woodin cardinal of $\pi^{\mathcal{T}, b}(\mathcal{P})$ and δ^* is a Woodin cardinal of $\mathcal{P}$. Then we have:

1. $\mathrm{cof}(\delta^*) < \kappa$ because there is a hod pair $(\mathcal{Q}, \Lambda) \in \mathcal{F}$ and γ such that $\mathcal{Q} \models$ "ξ is Woodin" and $\delta^* = \pi_{\mathcal{Q}, \infty}^{\Lambda}(\xi)$.
2. if $\delta > \sup(\pi^{\mathcal{T}, b}[\delta^{\mathcal{H}}])$, then by [36, Lemma 8.11], $\mathrm{cof}(\delta) = \mathrm{cof}(\mathsf{ord}(\mathcal{P}))$. By our assumption and Lemma 12.1.2, $\mathrm{cof}(\mathsf{ord}(\mathcal{P})) < \gamma$, so $\mathrm{cof}(\delta) < \gamma$.

Suppose $\mathcal{M}$ is the η-th model appearing in the X-validated sts construction over Y based on $\mathcal{R}$ ($\mathcal{M} = \mathcal{N}_{\eta-1}$) and $\mathcal{T}^* \in \mathcal{M}$ is a uvs iteration of $\mathcal{R}$ such that the indexing scheme requires that we index a branch of $\mathcal{T}^*$ at $\mathrm{ord}(\mathcal{M})$. We need to show that there is a branch b of $\mathcal{T}^*$ such that $(\mathcal{M}, b)$ is X-validated. Because of Proposition 12.3.14 and [36, Proposition 8.8], there can be at most one such branch. The proof of this is given in [36, Proposition 8.4]. We outline the main points of the proof in the following. Suppose c is a certified branch; we need to see that c

is $(X,\vec{\mathcal{V}})$-embeddable. If U is an X-good hull such that $(\mathcal{R},\vec{\mathcal{V}},p,\mathcal{T}^*,c)\in U$. Let $\mathcal{V}'=\pi_U^{-1}(C_n(\mathcal{V}_\eta))$ and $k:\mathcal{R}_U\to C_n(\mathcal{V}_\eta)$ be such that $\pi_U\restriction\mathcal{V}'=k\circ\pi^{p_U}$. We now suppose that there is a cofinal branch d of $\mathcal{T}_U^*$ such that for some $\beta\le\eta$ there is $m:\mathcal{M}_d^{\mathcal{T}_U^*}\to\mathcal{V}_\beta$ and $\mathcal{Q}(d,\mathcal{T}_U^*)$-exists. Let $\mathcal{M}=\mathcal{Q}(d,\mathcal{T}_U^*)$ and $\mathcal{N}=\mathcal{Q}(c_U,\mathcal{T}_U^*)$. Both $\mathcal{M}$ and $\mathcal{N}$ are X-approved. Let $\mathcal{S}_0=\mathrm{m}^+(\mathcal{T}_U^*)$. If we could conclude that $\mathcal{M}=\mathcal{N}$ then we would get that $c_U=d$, and that would finish the proof. To conclude that $\mathcal{M}=\mathcal{N}$, [36, Proposition 8.4] argues that $\mathcal{S}_0$ is not infinitely descending. Otherwise, there is a sequence $(p_i,\mathcal{S}_i:i<\omega)$ witnessing that $\mathcal{S}_0$ is infinitely descending such that for some $\beta<\eta$ and for some $i_0<\omega$ for every $i<j\in(i_0,\omega)$ there are weak n_i-embeddings $m_i:\mathcal{S}_i\to C_{n_i}(\mathcal{V}_\beta)$ such that $m_i=m_j\circ\pi^{p_i}$. The existence of a sequence as in the claim above gives us a contradiction, as the sequence must have a well-founded branch. The uniqueness proof is similar.

Because $\mathcal{T}^*$ is uvs, we have a normal iteration $\mathcal{T}\in\mathcal{M}$ with last model $\mathcal{S}$ such that $\pi^{\mathcal{T}}$ is defined and a normal iteration $\mathcal{U}$ based on $\mathcal{S}^b$ such that $\mathcal{T}^\frown\mathcal{U}=\mathcal{T}^*$. At this point, we assume that (ii)$_\eta$ fails, so we have that $\mathcal{M}$ is X-validated and therefore, $\mathcal{T}$ is X-validated. Also, we can assume that $\mathcal{U}$ is not based on $\mathcal{S}|\xi$ where $\xi=\sup(\pi^{\mathcal{T}}[\delta^{\mathcal{H}}])$, as otherwise the desired branch of $\mathcal{U}$ is given by Σ.

We now show that $\mathcal{U}$ has a branch b such that $(\mathcal{M},b)$ is X-validated. Given an X-good hull U such that $\{\mathcal{M},\mathcal{T},\mathcal{S},\mathcal{U}\}\subseteq U$, let $b_U=\Sigma_W(\pi_U^{-1}(\mathcal{U}))$ where W is any extension of X such that $\pi_U^{-1}(\mathcal{S}^b)=\mathcal{Q}_W^X$. First we claim that for all U as above,

CLAIM 12.4.5. $b_U\in M_U$.

PROOF. Fix then a U as above. Suppose first that $\mathcal{Q}(b_U,\mathcal{U}_U)$ does not exist. As we are assuming $\mathcal{U}$ is not based on $\mathcal{S}|\xi$, the remark above gives that $\mathrm{cf}(\delta(\mathcal{U}))<\gamma$. Because M_U is γ-closed it follows that $b_U\in M_U$.

Suppose next that $\mathcal{Q}(b_U,\mathcal{U}_U)$ exists. It is easy to see that letting Σ_U be the π_U-pullback of Σ, then $\mathrm{Lp}^{\Omega,\Sigma_U}(a)\in M_U$[14], where $a=\pi_U^{-1}(A)$ and A is a transitive set that codes $\{\mathcal{M},\mathcal{T},\mathcal{S},\mathcal{U}\}$.

Let $Y=U\cap\mathcal{H}$. Clearly Y is an extension of X and because $\mathcal{M}$ is X-validated, we must have W^* an extension of $X\cup Y$ such that $\mathcal{S}_U^b=\mathcal{Q}_{W^*}^X$. Notice that because Σ_{W^*} is computable from Σ_U and because $Lp^{\Omega,\Sigma_U}(a)\in M_U$, we must have that $\mathcal{Q}(b_U,\mathcal{U}_U)\in M_U$. Hence, $b_U\in M_U$. ⊣

Suppose first that $\mathrm{cf}(\mathrm{lh}(\mathcal{U}))>\omega$. In this case, let U be as above and set $c=\pi_U(b_U)$. Then c is the unique well-founded branch of $\mathcal{U}$ and hence, for any X-good hull Z such that $U\cup\{(\mathcal{M},c),U\}\in Z$, $c_Z=b_Z$. Hence, $(\mathcal{M},c)$ is X-validated (see Proposition 12.3.11).

Suppose then $\mathrm{lh}(\mathcal{U})=\omega$. We now claim that there is an X-good hull Z such that for all X-good hull Y such that $Z\cup\{\mathcal{M},Z\}\in Y$, $\pi_{X,Y}(b_X)=b_Y$. Assuming not we get a continuous chain $(X_\alpha:\alpha<\kappa)$ such that

[14]This is true because $\mathrm{Lp}^{j(\Omega),\Sigma}(A)\in V$ and $\Sigma=j(\Sigma)^j$.

1. $\mathcal{M}, \mathcal{U} \in X_0$,
2. for all $\alpha < \kappa$, $X_{\alpha+1}$ is an X-good hull,
3. for all $\alpha < \kappa$, $X_\alpha \cup \{X_\alpha\} \in X_{\alpha+1}$,
4. for all $\alpha < \kappa$, $\pi_{X_{\alpha+1}, X_{\alpha+2}}(b_{X_{\alpha+1}}) \neq b_{X_{\alpha+1}}$.

Let $\nu \in (\gamma, \kappa)$ be an inaccessible cardinal such that $X_\nu \cap \kappa = \nu$. Fix $\alpha < \nu$ such that

$$\sup(b_{X_\nu} \cap \mathrm{rng}(\pi_{X_\alpha, X_\nu})) = \delta(\mathcal{U}_{X_\nu}).$$

As $\mathrm{cf}(\mathrm{lh}(\mathcal{U}_{X_\nu})) = \omega$ this is easy to achieve. For $\beta \in [\alpha, \nu)$ let $c_\beta = \pi^{-1}_{X_\alpha, X_\nu}[b_{X_\nu}]$. Let for $\beta \in [\alpha, \nu]$, W_β be such that $\mathcal{S}^b_{X_\beta} = \mathcal{Q}^X_{W_\beta}$. It follows that c_β is according to π_{X_β, X_ν}-pullback of Σ_{W_ν}. Because Σ_{W_β} depends only $\mathcal{S}^b_{X_\beta}$, we have that $c_\beta = b_{X_\beta}$[15] It follows that for all $\beta < \gamma \in [\alpha, \nu)$, $\pi_{X_\beta, X_\gamma}(b_{X_\beta}) = b_{X_\gamma}$. This contradiction proves the claim.

Fix now a Z as above. Set $c = \pi_Z(b_Z)$. The the above property of Z guarantees that $(\mathcal{M}, c)$ is X-validated. Indeed, fix an X-good hull U such that $\mathcal{M}, c \in U$. Let Y be an X-good hull such that $Z \cup U \cup \{Z, U\} \in Y$. Then $\pi_{U,Y}(c_U) = \pi_{Z,Y}(b_Z) = b_Y$. It follows that $c_U = \Sigma_W^{\pi_{U,Y}}(\pi_U^{-1}(\mathcal{U}))$ where W is such that $\mathcal{S}^b_Y = \mathcal{Q}^X_W$. Hence, $c_U = b_U$.

This completes the outline of the proof that the X-validated sts construction over an honest $\mathcal{R}$ cannot break down because $(\mathrm{iii})_\eta$ holds for some η.

12.4.0.3. *Failure of $(iv)_\eta$.* Now we sketch the proof that $(\mathrm{iv})_\eta$ fails. Suppose that the first time the X-validated sts construction over X breaks down because $(\mathrm{iv})_\eta$ holds, where η is the least such. This means that setting $\mathcal{W} = \mathcal{N}_{\eta-1}$

1. $\mathcal{W}$ is X-validated,
2. there is a nuvs $\mathcal{T} \in \mathcal{W}$ that is according to $S^{\mathcal{W}}$ and is such that one of the following holds:
   (a) there is a cofinal branch $b \in \mathcal{W}$ such that $\mathcal{Q}(b, \mathcal{T})$ exists and is authenticated in $\mathcal{W}$ but $(\mathcal{W}, b)$ is not X-validated, or
   (b) there is a $\mathcal{Q}$-structure $\mathcal{Q} \in \mathcal{W}$ that is authenticated but there is no branch $b \in \mathcal{W}$ such that $\mathcal{Q}(b, \mathcal{T}) = \mathcal{Q}$.

We first show that case 2(a) holds as it illustrates the main arguments and utilizes most of the concepts introduced above. Let β be such that $\mathcal{W}|\beta$ authenticates b. Thus $\mathcal{W}|\beta$ is a model of ZFC in which there is a limit of Woodin cardinals ν and the derived model of $\mathcal{W}|\beta$ at ν has a strategy for $\mathcal{Q}(b, \mathcal{T})$ that is $\mathcal{W}|\beta$-authenticated.

Fix now a X-good hull U such that $(\mathcal{R}, \mathcal{W}, \mathcal{T}) \in U$ and $\mathcal{T}_U^\frown\{b_U\}$ is not a correctly guided X-realizable iteration of $\mathcal{R}_U$. Because $\mathcal{W}$ is X-validated, we can assume that $\mathcal{T}_U$ is correctly guided X-realizable iteration. It must then be that $\mathcal{Q}(b_U, \mathcal{T}_U)$ is not X-approved.

[15] The π_{X_β, X_ν}-pullback of Σ_{W_ν} is a strategy of the form Σ_Y where $\mathcal{Q}^X_Y = \mathcal{S}^b_{X_\beta}$ by Lemma 12.2.1.

We show that $\mathcal{N} =_{\text{def}} \mathcal{Q}(b_U, \mathcal{T}_U)$ is X-approved of depth 1. The proof of depth n is the same, we will leave the rest to the reader. To start with, notice that since $\mathcal{T}_U$ itself is correctly guided X-realizable, we have that $\mathcal{S} = \mathrm{m}^+(\mathcal{T}_U)$ is weakly X-suitable. To prove that $\mathcal{N}$ is X-approved of depth 1 we need to show that if $\mathcal{U} \in \mathcal{N}$ is according to $S^{\mathcal{N}}$ then $\mathcal{U}$ is X-realizable.

Fix then $\alpha \in R^{\mathcal{U}}$ and $\mathcal{X} = \mathcal{M}^{\mathcal{U}}_{\alpha}$. First we show that there is Z, an extension of X such that $\mathcal{Q}^X_Z = \mathcal{X}^b$. Because $\mathcal{T}_U^\frown\{b_U\}$ is authenticated inside $\mathcal{W}_U|\beta_U$, we must have an iteration $\mathcal{Y}$ of $\mathcal{R}_U$ according to $S^{\mathcal{W}_U}$ with last model $\mathcal{R}_1$ such that there is an embedding $k : \mathcal{X}^b \to \mathcal{R}^b_1$ with the property that $\pi^{\mathcal{Y},b} = k \circ \pi^{\mathcal{U}_{\leq \mathcal{X}},b} \circ \pi^{\mathcal{T}_U,b}_{b_U}$. Because $\mathcal{Y}$ is X-realizable, we must have Y an extension of X such that $\mathcal{R}^b_1 = \mathcal{Q}^X_Y$. Composing k with τ^X_Y we have that $\mathcal{X}^b = \mathcal{Q}^X_Z$ for some Z.

The rest is similar. If $\mathcal{U}^*$ is the longest initial segment of $\mathcal{U}_{\geq \mathcal{X}}$ that is based on $\mathcal{X}^b$ then there are $\mathcal{Y}$ and k as above such that $\mathcal{U}^*$ is according to k-pullback of $S^{\mathcal{W}_U}_{\mathcal{R}^b_1}$. But because $\mathcal{W}_U$ is X-approved, $S^{\mathcal{W}_U}_{\mathcal{R}^b_1}$ is a fragment of Σ_Y where Y is as above. Hence, $\mathcal{U}^*$ is according to Σ_Z for Z extending X as above (see 12.2.1). This finishes proof of 2(a).

Next we assume that case 2(b) holds. Let then U be a X-good hull such that $(\mathcal{R}, \mathcal{W}, \mathcal{T}, \mathcal{Q}) \in U$. Because $\mathcal{T}$ is X-validated, we have that the π_U-realizable branch of d of $\mathcal{T}_U$ is cofinal. Suppose then $\mathcal{Q}(d, \mathcal{T}_U)$ exists. Then because it is X-approved, we must have that $\mathcal{Q}(d, \mathcal{T}_U) = \mathcal{Q}_U$ (see [36, Proposition 8.4]). It follows that $d \in \mathcal{W}_U$, and so $\pi_U(d)$ is our desired branch.

We claim that $\mathcal{Q}(d, \mathcal{T}_U)$ exists. Suppose not. We regard $\mathcal{T}_U$ as a tree on $\mathcal{W}_U$ based on $\mathcal{R}_U$. First set $\mathcal{N} = \mathcal{M}^{\mathcal{T}_U}_d$ and $j = \pi^{\mathcal{T}_U}_d$. Note that $d \cap D^{\mathcal{T}_U} = \emptyset$ and $\pi^{\mathcal{T}_U}_d(\delta^{\mathcal{R}_U}) = \delta(\mathcal{T}_U)$; so $\mathcal{N} \models$ "$\delta(\mathcal{T}_U)$ is a Woodin cardinal". We have that $j(\mathcal{Q}_U) \in \mathcal{N}$ and is authenticated in $\mathcal{N}$. Let $\gamma = j(\beta_U)$. Then $\mathcal{N}|\gamma$ has Woodin cardinals bigger than $\delta(j(\mathcal{T}_U))$. Let δ be the least one that is $> \delta(j(\mathcal{T}_U))$. We can now iterate $\mathcal{N}$ below δ and above $j(\mathcal{Q}(b_U, \mathcal{T}_U))$ to make $\mathcal{Q}_U$ generic for the extender algebra at the image of δ. This iteration produced $i : \mathcal{N} \to \mathcal{N}_1$ such that $\mathrm{crit}(i) > \delta(j(\mathcal{T}_U))$. Letting $h \subseteq Coll(\omega, i(\delta))$ be $\mathcal{N}_1$-generic such that $\mathcal{Q}_U \in \mathcal{N}_1[h]$, we can find $l : \mathcal{Q}_U \to i(j(\mathcal{Q}_U)) = j(\mathcal{Q}_U)$ such that

1. $l \in \mathcal{N}_1[h]$, and
2. $l \restriction (\mathrm{m}^+(\mathcal{T}_U))^b = \pi^{j(\mathcal{T}_U),b}$.

As $\mathcal{N}_1[h] \models$ "$j(\mathcal{Q}_U)$ is authenticated and has an authenticated strategy", $\mathcal{N}_1[h] \models$ "$\mathcal{Q}_U$ has an authenticated iteration strategy", and hence $\mathcal{Q}_U$ is definable in $\mathcal{N}_1[h]$ from objects in $\mathcal{N}_1$. It follows that $\mathcal{Q}_U \in \mathcal{N}_1$, implying that $\mathcal{N}_1 \models$ "$\delta(\mathcal{T}_U)$ is not a Woodin cardinal". Hence, $\mathcal{N} \models$ "$\delta(\mathcal{T}_U)$ is not a Woodin cardinal". Therefore, $\mathcal{Q}(d, \mathcal{T}_U)$ exists. This completes the proof of case 2(b), and the proof that $(\mathrm{iv})_\eta$ fails.

The proofs above give us the following corollary.

COROLLARY 12.4.6. *Suppose the X-validated sts construction above breaks down because of $(iv)_\alpha$ for some α, then*

1. $\vec{\mathcal{V}}$ *is not small (so* $\mathcal{V}_\eta = \mathcal{R}$*), and*
2. *letting* $(\mathcal{T}, b) \in \mathcal{N}_{\alpha-1}$ *witnessing the construction fails because of* $(iv)_\alpha$*, then* $\mathcal{N}_{\alpha-1} \models$ *"*$\delta^{\mathcal{R}}$ *is not a Woodin cardinal".*

The following corollary is also useful. We will use it in later sections. See [36, Proposition 8.15] for the corresponding version of the following corollary.

COROLLARY 12.4.7. *Suppose* $\vec{\mathcal{V}}$ *is a small array with the* X*-realizability property. Then either*

1. $\mathcal{V}_\eta$ *has a* X*-validated iteration strategy*

or

2. *there is a* X*-validated* nuvs *iteration* p *of* $\mathcal{V}_\eta$ *such that* $\mathrm{m}^+(p)$ *is* Z*-suitable*[16].

PROOF. We outline the argument here: the argument earlier in this section shows that if p is an X-validated uvs of $\mathcal{M}_\xi$ of limit length then there is a unique branch b of p such that $p^\frown\{b\}$ is X-validated. Therefore, since picking X-validated branches is not defining an iteration strategy for $\mathcal{M}_\xi$, we must have an nuvs X-validated iteration p of $\mathcal{M}_\xi$ which does not have a X-validated branch.[17] We now claim that $\mathrm{m}^+(p)$ is a X-suitable hod premouse. Indeed, suppose there is some X-validated sts premouse $\mathcal{Q}$ extending $\mathcal{R} =_{\mathrm{def}} \mathrm{m}^+(p)$ such that $\mathcal{Q}$ is a $\mathcal{Q}$-structure for p. Let then U be an X-good hull such that $\{\vec{\mathcal{V}}, p, \mathcal{Q}\} \in U$. It is not hard to see, using the fact that $\vec{\mathcal{V}}$ is a small array with the X-realizability property, that there is $\beta \leq \xi$, a branch b of p_U such that $\mathcal{Q}(b, p_U)$ exists and a weak l-embedding $k : \mathcal{M}_b^{p_U} \to \mathcal{C}_l(\mathcal{M}'_\beta)$ for an appropriate l. It follows that $\mathcal{Q}(b, p_U)$ is X-approved and hence, $\mathcal{Q}(b, p_U) = \mathcal{Q}_U$. Because $\mathcal{Q}_U \in M_U$, we have that $b \in M_U$. Then $c =_{\mathrm{def}} \pi_U(b)$ is a (cofinal) branch of p such that $p^\frown\{c\}$ is X-validated. ⊣

We end this section by defining the following.

DEFINITION 12.4.8. We say that the X-validated sts construction over Y based on $\mathcal{R}$ *stops prematurely* if Υ is the least such that the following hold for $\mathcal{M}_\Upsilon$:

(i) There is an increasing sequence $(\delta_n : n < \omega.2)$ of Woodin cardinals above $\delta^{\mathcal{P}}$ such that $\delta^{\mathcal{P}}$ is the least $< \delta_0$-strong and $(\delta_n : n < \omega.2)$ are the only Woodin cardinals above δ_0.
(ii) There are no extenders E on the $\mathcal{M}_\Upsilon$-sequence such that there is some n such that $\mathrm{cr}(E) \leq \delta_i < \mathrm{lh}(E)$.
(iii) $\mathcal{M}_\Upsilon$ is an X-validated sts hod premouse over Y based on $\mathcal{R}$.
(iv) $\mathcal{M}_\Upsilon$ is E-active with top extender F such that $\mathrm{cr}(F) > \delta_n$ for all $n < \omega.2$.
(v) $\rho_\omega(\mathcal{M}_\Upsilon) \geq \mathsf{ord}(\mathcal{R})$.

We note that (ii) easily follows from (i), but for clarity, we make it explicit. In the later sections, we will obtain a contradiction (using stacking mice techniques) from

[16] See Definition 12.3.13.

[17] Note that there may not be any $\mathcal{Q}$-structure for p.

the assumption that the construction does not stop prematurely and $\Upsilon = \kappa^{+++}$. The subsequent several sections will show that the construction does not fail and stop prematurely. From this, we then show that there must be a model of LSA.

12.5. Hybrid K^c-constructions

We continue with the notations of the previous section. Now we describe the hybrid K^c-construction. We first describe the "bottom structure" that the hybrid K^c-construction is built on top of. We say $\mathcal{S}$ is *K^c-appropriate* if $\mathcal{S} = \mathcal{P}$ or $\mathcal{S} = \mathcal{R}$ where the following hold for $\mathcal{R}$:

1. $\mathcal{P} \lhd \mathcal{R}$ and $o(\mathcal{R})$ is Woodin in $\mathcal{R}$, where $o(\mathcal{R}) = \sup\{\xi : E_\xi^{\mathcal{R}} \neq \emptyset \wedge \mathrm{crit}(E_\xi^{\mathcal{R}}) = \delta^{\mathcal{P}}\}$.
2. $\mathcal{R}$ is a sound, X-validated sts premouse over $(\mathcal{R}|o(\mathcal{R}))^\sharp$ such that $\mathcal{J}_1[\mathcal{R}] \models$ "$o(\mathcal{R})$ is not Woodin".
3. There exists a (unique) X-validated strategy for $\mathcal{R}$.

We build in $V[G]$ a sequence $(\mathcal{M}_\xi, \mathcal{N}_\xi : \xi \leq \Upsilon)$ of levels of our K^c-construction such that $\mathcal{N}_0 = \mathcal{M}_0 = \mathcal{R}$ for a K^c-appropriate $\mathcal{R}$, $\mathcal{N}_\xi = \mathfrak{C}(\mathcal{M}_\xi)$ for all $\xi \leq \Upsilon$ and $\Upsilon \leq \kappa^{+++}$. It will be clear from the construction that $\mathcal{N}_\xi, \mathcal{M}_\xi \in V$ for all ξ.

We maintain during the construction that $\mathcal{M}_\xi$ (and hence $\mathcal{M}_\xi^Y$) is *small*, i.e., either $\mathcal{M}_\xi$ has no Woodin cardinal $> \delta^{\mathcal{P}}$ or else letting δ be the least such Woodin, then $\rho(\mathcal{M}_\xi) \leq \delta$ and $\mathcal{M}_\xi$ defines a failure of Woodinness of δ. Note that this implies that

$$\mathcal{J}_1[\mathcal{N}_\xi] \models \text{"there are no Woodin cardinals} > \delta^{\mathcal{P}}\text{"}.$$

If $\mathcal{M}_\xi$ is not small or that we fail to construct an X-validated strategy for $\mathcal{M}_\xi$, then we let $\Upsilon = \xi$ and stop the construction.

Suppose we have constructed $\mathcal{M}_\xi, \mathcal{N}_\xi$ for some ξ. Let $\gamma_\xi = o(\mathcal{N}_\xi)$ be the supremum of indices of extenders on the $\mathcal{N}_\xi$-sequence with critical point $\delta^{\mathcal{P}}$ if there are such extenders; otherwise, let $\gamma_\xi = \mathrm{ord}(\mathcal{P})$. Suppose $\gamma_\xi < \mathrm{ord}(\mathcal{N}_\xi)$ and let $\gamma_\xi \leq \lambda_\xi \leq \mathrm{ord}(\mathcal{N}_\xi)$ be such that $\rho_\omega(\mathcal{N}_\xi) \geq \lambda_\xi$. Suppose there is a stack $\vec{\mathcal{T}} \in \mathcal{N}_\xi$ based on $\mathcal{N}_\xi|\lambda_\xi$ according to the internal strategy $S^{\mathcal{N}_\xi}$ such that $S^{\mathcal{N}_\xi}(\vec{\mathcal{T}})$ is undefined.[18] Suppose also $\vec{\mathcal{T}}$ is such that the theory developed above (Chapter 3) dictates that a cofinal branch b for $\vec{\mathcal{T}}$ needs to be added to $\mathcal{N}_\xi$ and $\mathcal{N}_\xi$ is so that $(\mathcal{N}_\xi, B_b)$ is amenable. We call such a tuple $(\mathcal{N}_\xi, \lambda_\xi, \vec{\mathcal{T}})$ *branch-ready*.

For a branch-ready tuple $(\mathcal{N}_\xi, \lambda_\xi, \vec{\mathcal{T}})$ in our hybrid K^c-construction, we need to see that $(\vec{\mathcal{T}}, b)$ is X-validated and that $(\mathcal{N}_\xi, B_b)$ is an X-validated hod premouse; this is accomplished by constructing an external X-validated strategy Λ for $\mathcal{N}_\xi|\lambda_\xi$ and let $b = \Lambda(\mathcal{T})$, see below for a more detailed discussion. Furthermore, we need

[18] By the notations earlier in the book, $\mathcal{N}_\xi|\lambda_\xi$ is a complete hod initial segment of $\mathcal{N}_\xi$.

to construct an external X-validated strategy for $(\mathcal{N}_\xi, B_b)$. So we maintain that $S^{\mathcal{N}_\xi}$[19] above λ_ξ is according to:

(a) either the strategy Σ_ξ of $\mathcal{N}_\xi|\lambda_\xi$, where Σ_ξ is the canonical Q-structure guided, X-validated strategy of $\mathcal{N}_\xi|\lambda_\xi$ if $\mathrm{ord}(\mathcal{N}_\xi) < \mathrm{ord}(\mathcal{M}_2^{\Sigma_\xi,\sharp})$;

(b) or else the canonical Σ_ξ-strategy Λ_ξ of $\mathcal{N}_\xi|\lambda_\xi = \mathcal{M}_2^{\Sigma_\xi,\sharp}(\mathcal{N}_\xi|\varepsilon)$, where Σ_ξ is the canonical Q-structure guided, X-validated strategy of $\mathcal{N}_\xi|\varepsilon$.

We let Ψ_ξ denote Σ_ξ in case (a), and Λ_ξ in case (b).

We will discuss the construction of the strategy in (a), or (b) in the next section. At this point, we assume it exists and just want to extend the internal strategy of $\mathcal{N}_\xi$ one more step. The key thing we want to maintain here is that the indexed branch b for $\mathcal{N}_\xi$ is according to the strategies Ψ_ξ. Roughly, what we need to do to construct such a strategy is as follows.

In the following, we write $\forall^* Y$ to mean "for some club $\mathfrak{C}$, $Y \in \mathfrak{C} \cap \mathfrak{S}_{\varphi,\Omega}$". What we show in the next sections is that $\forall^* Y$ such that Y is X-good, we can construct an X-realizable strategy $\Psi_{Y,\xi}$ as in case (a), (b). Ψ_ξ is then determined from the strategies $\Psi_{Y,\xi}$ by the procedure described in the previous section. $b = \Psi_\xi(\vec{\mathcal{T}})$ if there is an X-good Z such that letting $b_Z = \Psi_{Z,\xi}(\vec{\mathcal{T}}_Z)$, then for any X-good hull Y such that $\mathcal{N}_\xi, Z \in Y$,

$$\pi_{Z,Y}[b_Z] \subseteq b_Y \quad \text{and} \quad \tau_Z^X[b_Z] \subseteq \tau_Y^X[b_Y] \subseteq b.$$

We say that b described above is *suitable* for $(\mathcal{N}_\xi, \lambda_\xi, \vec{\mathcal{T}})$. We define the above notions in a similar manner for $\mathcal{M}_\xi$; for brevity, we also use the symbols γ_ξ, λ_ξ for $\mathcal{M}_\xi$ when no confusion arises from the context.

REMARK 12.5.1. If such a strategy does not exist, we will stop the hybrid K^c-construction over $\mathcal{R}$ at stage ξ. In this case, there will be an X-validated iteration p witnessing this. We will then switch to the sts X-validated construction over $\mathcal{M}(p)^\sharp$.

The argument in the previous section also shows that if such a strategy exists and is constructed according to the aforementioned procedure, then it must agree with $S^{\mathcal{N}_\xi}$ (similarly for $\mathcal{M}_\xi$).

REMARK 12.5.2. The reason we have case (b) is because we want our hod mice to be g-organized in the sense of [50]. g-organization ensures that S-constructions go through as discussed in Chapter 6.

The procedure above allows us to define the object $\mathrm{Lp}^{\Psi_\xi}(\mathcal{N}_\xi)$ in the case $\gamma_\xi < \lambda_\xi = \mathrm{ord}(\mathcal{N}_\xi)$ as follows. In the definition below, following [65], we let $\mathrm{Lp}^{\Psi_{Y,\xi},\Omega}(\mathcal{N}_\xi^Y)$ be the union of sound, $\Psi_{Y,\xi}$-mouse $\mathcal{R}$ over $\mathcal{N}_\xi^Y$ such that $\rho(\mathcal{R}) \leq \mathrm{ord}(\mathcal{N}_\xi^Y)$ with unique X-realizable iteration strategy (above $\mathrm{ord}(\mathcal{N}_\xi^Y)$) in Ω.

[19] Recall, this is the sequence of branch predicates that codes up some internal strategy of $\mathcal{N}_\xi$.

DEFINITION 12.5.3. Suppose $\gamma_\xi < \lambda_\xi = \text{ord}(\mathcal{N}_\xi)$. We let $\text{Lp}^{\Psi_\xi}(\mathcal{N}_\xi)$ be the union of $\mathcal{N}_\xi \lhd \mathcal{M}$ such that $\rho_\omega(\mathcal{M}) \leq \text{ord}(\mathcal{N}_\xi)$, $\mathcal{M}$ is $\text{ord}(\mathcal{N}_\xi)$-sound, and $\forall^* Y$, Y is X-good and contains all relevant objects, $\pi_Y^{-1}(\mathcal{M}) \lhd \text{Lp}^{\Psi_{Y,\xi},\Omega}(\mathcal{N}_\xi^Y)$. Sometimes, we write $\text{Lp}^{\Sigma_{\mathcal{N}_\xi}}(\mathcal{N}_\xi)$ for $\text{Lp}^{\Psi_\xi}(\mathcal{N}_\xi)$.

We also define $\Sigma_{\mathcal{M}_\xi}$ and $\text{Lp}^{\Sigma_{\mathcal{M}_\xi}}(\mathcal{M}_\xi)$ in a similar manner.

REMARK 12.5.4. By [65, Lemma 3.78] and Lemma 12.2.1,

$$\forall^* Y \pi_Y^{-1}(\text{Lp}^{\Sigma_{\mathcal{N}_\xi}}(\mathcal{N}_\xi)) = \text{Lp}^{\Psi_{Y,\xi},\Omega}(\mathcal{N}_\xi^Y).$$

We also remind the reader, for reasons mentioned before, levels of $\text{Lp}^{\Psi_{Y,\xi},\Omega}(\mathcal{N}_\xi^Y)$, etc are g-organized in the sense of [50].

DEFINITION 12.5.5 (Relevant extender). Suppose F is on the $\mathcal{N}_\xi$-extender sequence for some $\xi \leq \Upsilon$. We say that F is *relevant* if $F = G \cap \mathcal{N}_\xi$[20] for G a correctly backgrounded extender.

We let $\mathcal{N}_\xi^+ = \mathcal{J}_\gamma[\mathcal{N}_\xi]$ for γ being least such that

1. either $\mathcal{J}_\gamma[\mathcal{N}_\xi]$ is not sound or $\rho(\mathcal{J}_\gamma[\mathcal{N}_\xi]) < \rho(\mathcal{N}_\xi)$,
2. or else $\mathcal{J}_\gamma[\mathcal{N}_\xi]$ satisfies ZFC^- and there is a correctly backgrounded extender F that coheres the $\mathcal{J}_\gamma[\mathcal{N}_\xi]$-sequence,
3. or else there are $\lambda_\xi \geq \gamma_\xi$, $\vec{\mathcal{T}} \in \mathcal{J}_\gamma[\mathcal{N}_\xi]$ such that $(\mathcal{J}_\gamma[\mathcal{N}_\xi], \lambda_\xi, \vec{\mathcal{T}})$ is branch-ready.

DEFINITION 12.5.6 (Hybrid K^c-construction). The models $\mathcal{M}_\xi$, $\mathcal{N}_\xi$ are defined as follows: for all $\xi \leq \Upsilon$,

(a) if $\xi = 0$, then $\mathcal{N}_\xi = \mathcal{M}_\xi = \mathcal{R}$ for a K^c-appropriate $\mathcal{R}$;
(b) if ξ is limit, let $\mathcal{M}_\xi$ be $\liminf_{\xi^* < \xi} \mathcal{N}_{\xi^*}$;
(c) if $\xi = \xi^* + 1$, the following hold:
  (i) if $\mathcal{N}_{\xi^*}$ is passive and there is a correctly backgrounded extender F that coheres the $\mathcal{N}_{\xi^*}$-sequence, then let $\mathcal{M}_\xi = (\mathcal{N}_{\xi^*}, F)$.[21]
  (ii) if $\mathcal{N}_{\xi^*}$ is passive and there is some $\vec{\mathcal{T}} \in \mathcal{N}_{\xi^*}$ and some λ such that $\vec{\mathcal{T}}$ is based on $\mathcal{N}_{\xi^*}|\lambda$ and $(\mathcal{N}_{\xi^*}, \lambda, \vec{\mathcal{T}})$ is branch-ready, then letting b be given by the procedure above, we set $\mathcal{M}_\xi = (\mathcal{N}_{\xi^*}, B_b)$.
  (iii) if $\mathcal{M}_{\xi^*}$ is passive and cases (i) and (ii) do not hold, then we set $\mathcal{M}_\xi = \mathcal{N}_{\xi^*}^+$.
(d) $\mathcal{N}_\xi = \mathcal{C}(\mathcal{M}_\xi)$.

[20] If $\text{crit}(F) = \delta^{\mathcal{P}}$, we confuse F with its amenable code for $G \cap \mathcal{N}_\xi$ and in the case $\text{crit}(F) > \delta^{\mathcal{P}}$, we think of F as a "map" as in [70].

[21] See Definition 12.3.15. The uniqueness of the extender F with $\text{crit}(F) > \delta^{\mathcal{P}}$ follows from a standard bicephalous argument, cf. [23, Section 7], and tools developed in the previous Chapters. If $\text{crit}(F) = \delta^{\mathcal{P}}$, the uniqueness of F follows from the proof of Lemma 12.7.14.

REMARK 12.5.7. If $\mathcal{N}_{\xi^*}$ is as in (c)(i), we say that $\mathcal{N}_{\xi^*}$ is *extender-ready*. We give priority to adding extenders, that is, if $\mathcal{N}_\xi$ is both an extender-ready level and a branch-ready level, then we are in case (c)(i).

If $\mathcal{N}_\xi$ is weakly suitable or that $\Sigma_{\mathcal{M}_\xi}$ is not defined and p witnesses this, then as mentioned above, we let $\Upsilon = \xi$ and stop the construction. We then start a new X-validated sts construction over $\mathcal{N}_\xi$ in the first case and over $\mathcal{M}(p)^\#$ in the second case.

Let $Y \in \mathfrak{S}_{\varphi,\Omega}$, so $Y \cap \mathcal{P}$ is an honest extension of X. Let π_Y be the uncollapse map and $\mathcal{N}_\xi^Y = \pi_Y^{-1}(\mathcal{N}_\xi)$. We recall that Λ is the *X-realizable strategy of $\mathcal{N}_\xi^Y$* if whenever $\vec{\mathcal{T}}$ is according to Λ, $i : \mathcal{N}_\xi^Y \to \mathcal{Q}$ is the iteration map according to Λ, where $\mathcal{Q} = \mathcal{M}_\alpha^{\vec{\mathcal{T}}}$ and $\alpha \in R^{\vec{\mathcal{T}}}$, then there is some Z, honest extension of X, such that $\mathcal{Q}^b = \mathcal{Q}_Z^X$, and the map $k : \mathcal{Q} \to \mathcal{N}_\xi$ defined as: for $f \in \mathcal{N}_\xi^Y$, $a \in (\delta^{\mathcal{Q}})^{<\omega}$,

$$k(i(f)(a)) = \pi_Y(f)(\pi_{\mathcal{Q},\infty}^{\Lambda_{\vec{\mathcal{T}},\mathcal{Q}}}(a))$$

is well-defined, elementary[22], $k \circ i = \pi_Y$, and $k \restriction \delta^{\mathcal{Q}} = \pi_{\mathcal{Q},\infty}^{\Lambda_{\vec{\mathcal{T}},\mathcal{Q}}} \restriction \delta^{\mathcal{Q}}$. Similar definitions are given for $\mathcal{M}_\xi, \mathcal{M}_\xi^Y$.

We maintain as part of the construction the following for $\xi < \Upsilon$:

$(1)_\xi$ (a) $\mathcal{M}_\xi$ is X-validated.

(b) $\forall^* Y$, there is an X-realizable strategy for $\mathcal{M}_\xi^Y$, called $\Sigma_{\mathcal{M}_\xi^Y}$.[23] So $\Sigma_{\mathcal{M}_\xi}$ is an X-validated strategy for $\mathcal{M}_\xi$.

(c) $\forall^* Y$, $\Sigma_{\mathcal{M}_\xi^Y}$ has (locally) strong branch condensation and is (locally) strongly Ω-fullness preserving.

$(2)_\xi$ $\rho_\omega(\mathcal{M}_\xi) \geq \mathsf{ord}(\mathcal{P})$. In other words, $\mathsf{ord}(\mathcal{P})$ is $(\delta^{\mathcal{P}})^+$ in $\mathcal{M}_\xi$ and in $\mathcal{N}_\xi$.

$(3)_\xi$ $\mathcal{M}_\xi$ is solid and universal. So $\mathcal{N}_\xi$ is sound.

See $(\dagger\dagger)_\xi$ in Section 12.7 for how these statements are precisely maintained. Υ is the least such that one of the conditions fail at Υ or that the hybrid K^c-construction stops prematurely. By Section 12.7, if one of $(1)_\Upsilon, (2)_\Upsilon, (3)_\Upsilon$ fails, then in fact $(1)_\Upsilon(b)$ fails.

In the next section, we will obtain a contradiction (using stacking mice techniques) from the assumption that $\Upsilon = \kappa^{+++}$.

REMARK 12.5.8. The extender sequence of $\mathcal{N}_\xi$ utilizes two indexing schemes: the cutpoint indexing scheme (for extenders with critical point $\delta^{\mathcal{P}}$) and the Jensen indexing scheme (for extenders with critical point $> \delta^{\mathcal{P}}$). This follows from the definition of correctly backgrounded extenders for relevant extenders. The Jensen

[22] By this, we mean if $\vec{\mathcal{T}}$ is a k-maximal stack then k is a weak k-embedding in the sense of [23].

[23] It will be clear that $\Sigma_{\mathcal{M}_\xi^Y}$ is unique. In essence, $(1)_\xi(b)$ is equivalent to the statement that the natural X-realizable strategy defined in Section 12.7 is total; in particular, all iterates according to this strategy are X-approved.

indexing scheme could be replaced by the Mitchell-Steel indexing scheme, but we choose not to do so out of convenience; we want to quote direct results from [12] and [3] as well as using results of Chapter 11.

12.6. Stacking mice

Suppose the constructions described above do not stop prematurely and therefore result in a model $\mathcal{M}_\Upsilon$ such that $\mathrm{ord}(\mathcal{M}_\Upsilon) = \kappa^{+++}$ (see Lemma 12.8.2). Let $\mathcal{N} = \mathcal{N}_\Upsilon = \mathcal{C}(\mathcal{M}_\Upsilon)$. It is clear that $\mathcal{N} = \mathcal{M}_\Upsilon$.[24] So $\mathcal{N}$ is an X-validated hod premouse or an X-validated sts premouse. There are two possible cases on how we reach such an $\mathcal{N}$. In the first case, we must have alternated the two constructions (the X-validated sts construction and the hybrid K^c-construction) until we reach a K^c-appropriate $\mathcal{R}$ and the rest of the construction is the hybrid K^c-construction with $\mathcal{N}_0 = \mathcal{R}$ and $\mathcal{N}_\Upsilon = \mathcal{N}$[25]; in the second case, we reached a suitable $\mathcal{R}$ and $\mathcal{N}$ is obtained by the X-validated sts construction over $\mathcal{R}$.

Let $\delta^{\mathcal{N}} > \delta^{\mathcal{P}}$ be the unique γ such that $\mathcal{N} \vDash$ "$\delta^{\mathcal{P}}$ is strong to γ and γ is Woodin" if it exists; otherwise we let $\delta^{\mathcal{N}} = 0$. Note that by the remarks above, which is a consequence of our smallness assumption (†), $\delta^{\mathcal{N}}$ is a strong cutpoint of $\mathcal{N}$. Following [12], we define the following *stack of hod mice above* $\mathcal{N}$. The following definition takes place in $V[G]$ but it is easily seen that $S(\mathcal{N}) \in V$ (see Lemma 12.6.2).

DEFINITION 12.6.1. Let δ denote $\delta^{\mathcal{N}}$. Let $S(\mathcal{N})$ be the stack of sound X-validated hod premice $\mathcal{M}$ if $\delta = 0$ or else X-validated sts-premice $\mathcal{M}$ extending $\mathcal{N}$ such that $\rho_\omega(\mathcal{M}) = \mathrm{ord}(\mathcal{N}) = \kappa^{+++}$ and for every $\mathcal{M}^*$ embeddable into $\mathcal{M}$ via $\pi_{\mathcal{M}^*}$ such that $|\mathcal{M}^*| < \kappa$, $X \cup \{X, \mathcal{P}^-, \mathcal{P}, \mathcal{N}, \delta\} \subset \mathrm{rng}(\pi_{\mathcal{M}^*})$, $\mathrm{rng}(\pi_{\mathcal{M}^*}) \cap \mathcal{P}$ is an honest extension of X, $\mathcal{M}^*$ is (ω_1+1)-iterable above $\pi^{-1}(\delta)$. Furthermore, in the case $\delta = 0$, the strategy for $\mathcal{M}^*$ is X-realizable and in the case $\delta > 0$, the strategy witnesses $\mathcal{M}^*$ is an X-approved sts mouse.

$\mathcal{N}$ is the κ^{+++}-th model in the hybrid K^c-construction or the X-validated sts construction, and hence is passive. In the first case, one can show easily that items (1)-(3) hold for $\mathcal{N}$. In the above, if $\delta = 0$, then $\mathcal{M}^*$ has an X-realizable strategy $\Lambda_{\mathcal{M}^*}$ such that $\Lambda_{\mathcal{M}^*} \restriction \mathrm{HC} \in \Omega$, and $\Lambda_{\mathcal{M}^*}$ is locally Ω-fullness preserving and has local strong branch condensation (see next section). Furthermore, by the fact that X is a condensing set and Section 12.4, $\Lambda_{\mathcal{M}^*}$ witnesses $\mathcal{M}^*$ is an X-approved hod mouse. In particular, if E is on the $\mathcal{M}$-sequence such that $\mathrm{cr}(E) = \delta^{\mathcal{P}}$ and $\mathrm{lh}(E) \geq o(\mathcal{N})$, then for every such $\mathcal{M}^*$ as above such that $E \in \mathrm{rng}(\pi_{\mathcal{M}^*})$, letting ν be the length of $\pi^{-1}_{\mathcal{M}^*}(E)$, then for any $a \in [\nu]^{<\omega}$, $A \in \wp(\delta^{\mathcal{P}})^{|a|} \cap \mathcal{P}$ such that

[24]One way to see that is to recall that $\mathcal{M}_\Upsilon = \liminf_{\xi\to\Upsilon} \mathcal{M}_\xi$. Since $\Upsilon = \kappa^{+++}$, and PFA holds, $\mathrm{ord}(\mathcal{M}_\Upsilon) = \kappa^{+++}$ is a limit cardinal in $\mathcal{M}_\Upsilon$ and $\mathcal{M}_\Upsilon \vDash \mathsf{ZFC}$. So $\rho(\mathcal{M}_\Upsilon) = \kappa^{+++}$.

[25]It could be that $\mathcal{R} = \mathcal{N}_\Upsilon$.

$(a,A) \in E \cap \mathrm{rng}(\pi_{\mathcal{M}^*})$, then $\pi^{\Lambda_{\mathcal{M}^*}}_{\mathcal{M}^*||\nu,\infty}(\pi^{-1}_{\mathcal{M}^*}(a)) \in A$. In the case $\delta > 0$, we demand as part of Definition 12.6.1 that $\mathcal{M}^*$ is iterable above $\pi^{-1}(\delta)$ as an X-approved sts mouse; note also that δ is a strong cutpoint of $\mathcal{M}$. The following facts about $S(\mathcal{N})$ more or less follow immediately from results in [12].

LEMMA 12.6.2. *Suppose* $\Upsilon = \kappa^{+++}$ *and* $\mathcal{N} = \mathcal{N}_\Upsilon$.

(i) For $\mathcal{M}_0, \mathcal{M}_1 \in S(\mathcal{N})$, *either* $\mathcal{M}_0 \trianglelefteq \mathcal{M}_1$ *or* $\mathcal{M}_1 \trianglelefteq \mathcal{M}_0$.

(ii) For all $\mathcal{M} \trianglelefteq S(\mathcal{N})$, *there is some* $\mathcal{R} \lhd S(\mathcal{N})$ *such that* $\mathcal{M} \lhd \mathcal{R}$. *In particular,* $S(\mathcal{N}) \models \mathsf{ZFC}^-$.

(iii) $\mathrm{cof}(\mathrm{ord}(S(\mathcal{N}))) \geq \kappa^{+++}$.

PROOF. (i) and (ii) are analogs of [12, Lemma 3.1] and [12, Lemma 3.3] respectively and follow straightforwardly from the condensation lemma, Theorem 11.1.5. The point is that if $\delta^{\mathcal{N}} = 0$, then the theory developed above allows us to perform comparisons (and shows that no strategy disagreement can occur); otherwise, the construction above δ is an X-validated sts construction over a fixed suitable $\mathcal{R}$, so the comparison is again an extender comparison (and is above $\mathrm{ord}(\mathcal{R}) = \delta$).[26] Therefore, by an easy application of Theorem 11.1.5 (see also [12, Lemma 1.3]), letting $\pi : H \to H_{\kappa^{+4}}$ be elementary and such that $\{\mathcal{M}_1, \mathcal{M}_2, X, \mathcal{R}\} \in H$, $H \cap \kappa^{+++} \in \kappa^{+++}$, then $\pi^{-1}(\mathcal{M}_0) \lhd \mathcal{N}$ and $\pi^{-1}(\mathcal{M}_1) \lhd \mathcal{N}$. By elementarity of π, (i) holds for $\mathcal{M}_0, \mathcal{M}_1$. The proof of (ii) follows from [12, Lemma 3.3] and the discussion above.

(iii) follows from the proof of [12, Theorem 3.4] with obvious modifications, noting that by our assumption, κ^+, κ^{++} are κ-closed, and $2^{\kappa^{++}} = \kappa^{+++}$ in $V[G]$. We note that κ plays the role of ω, κ^{+++} plays the role of κ in that proof and all hulls taken are closed under κ-sequences in this case. ⊣

12.7. Iterability of lsa-small, non-lsa type levels

First, we verify that (1)-(3) holds for $\xi = 0$ and $\mathcal{M}_0 = \mathcal{N}_0 = \mathcal{P}$. By Lemma 12.1.2, no $\mathcal{P}^- \lhd \mathcal{M} \lhd \mathcal{P}$ projects across $\delta^{\mathcal{P}}$; also $\mathcal{P} \models \mathsf{ZFC}^-$, and hence $\rho_\omega(\mathcal{P}) = \mathrm{ord}(\mathcal{P})$.

LEMMA 12.7.1. *(1)-(3) hold for* $\xi = 0$.

PROOF. Fix Y as in the statement of (1); so $\mathcal{M}_0^Y = \mathcal{N}_0^Y$. Let $\delta_Y = \delta^{\mathcal{N}_0^Y}$ and $\Sigma_0^Y = \Sigma_{\mathcal{N}_0^Y}$ be Σ_Y. By definition, Σ_0^Y has branch condensation as it is the join of strategies with those properties. Furthermore, note that Σ_0^Y acts on $\mathcal{N}_0^Y$ in the following way. Let $(\mathcal{Q}, \vec{\mathcal{T}}) \in I(\mathcal{N}_0^Y, \Sigma_0^Y)$ and let $i : \mathcal{N}_0^Y \to \mathcal{Q}$ be the iteration map and $\Sigma_{\mathcal{Q},\vec{\mathcal{T}}}$ be the $\vec{\mathcal{T}}$-tail of Σ_0^Y.

[26] See Corollary 12.7.18 for a similar argument with more details. The point is that the comparisons involve two X-approved sts mice, so no strategy disagreements appear.

Suppose $x \in \mathcal{Q}$, then there is some $f \in \mathcal{N}_0^Y$ and $a \in i(\delta_Y)^{<\omega}$ such that

$$x = i(f)(a).$$

Let $k : \mathcal{Q} \to \mathcal{N}_0$ be defined as follows:

$$k(i(f)(a) = \pi_Y(f)(\pi^{\Sigma_{\mathcal{Q},\mathcal{T}}}_{\mathcal{Q}(i(\delta_Y)),\infty}(a)),$$

for any $f \in \mathcal{N}_0^Y$ and any $a \in i(\delta_Y)^{<\omega}$. Note that since X is a condensing set and $i \circ \pi_{X,Y} \restriction \delta_X$ is according to Σ_0^X, $\operatorname{rng}(k)$ is an honest extension of X. By Lemma 12.2.1, k is well-defined, Σ_1-elementary (and cofinal), $k \circ i = \pi_Y$, and $k \restriction \delta^{\mathcal{Q}} = \pi^{\Sigma_{\mathcal{T},\mathcal{Q}}}_{\mathcal{Q}|\delta^{\mathcal{Q}},\infty} \restriction \delta^{\mathcal{Q}}$. It is clear that this is the only way to define k; the uniqueness of Σ_0^Y also follows.

We remark that local strong branch condensation is just branch condensation in this case. Now to see that Σ_0^Y is Ω-fullness preserving, it suffices to show $\mathcal{Q}$ is Ω-full. But this follows from the definition of condensing sets and the fact that Y and $\operatorname{rng}(k)$ are honest extensions of X. Also, we get local strong Ω-fullness preservation.

We have shown (1). (2) holds by the remark immediately before the lemma and (3) follows from (2) and (1) by Remark 12.5.8. The usual proof of universality and solidity goes through with the iterability proved in (1). ⊣

Now we inductively in ξ prove the following:

$(\dagger\dagger)_\xi$ suppose $(1)_{\xi^*}$-$(3)_{\xi^*}$ hold for all $\xi^* < \xi$, then $(1)_\xi$(a) holds and suppose $(1)_\xi$(b) holds, then $(1)_\xi$(c), $(2)_\xi$, $(3)_\xi$ hold.

We continue with proving $(\dagger\dagger)_\xi$. So suppose $(\dagger\dagger)_{\xi^*}$ holds for all $\xi^* < \xi$. Now, we verify $(\dagger\dagger)_\xi$. Let $Y \in V$ be an honest extension of X; as before, we assume also $Y = Y^* \cap \mathcal{M}_\xi$ for some $Y^* \prec H^V_{\kappa^{+4}}$. We recall that $\mathcal{M}_\xi$ (and hence $\mathcal{M}_\xi^Y$) is *small*, i.e., either $\mathcal{M}_\xi$ has no Woodin cardinal $> \delta^{\mathcal{P}}$ or else letting δ be the least such Woodin, then $\rho(\mathcal{M}_\xi) \leq \delta$ and $\mathcal{M}_\xi$ defines a failure of Woodinness of δ. Note that this implies that

$$\mathcal{J}_1[\mathcal{N}_\xi] \models \text{“there are no Woodin cardinals} > \delta^{\mathcal{P}}\text{”}.$$

If $\mathcal{M}_\xi$ is not small and that $\mathcal{M}_\xi$ is of lsa type, i.e., $\mathcal{M}_\xi = \mathrm{m}^+(\mathcal{M}_\xi|\delta) \models$ “δ is Woodin and $\delta^{\mathcal{P}}$ is $< \delta$-strong”, then we stop the hybrid K^c-construction, set $\Upsilon = \xi$, and switch to the X-validated sts construction over $\mathcal{M}_\xi$.

If $(1)_\xi(b)$ fails, we stop the hybrid K^c-construction and let $\Upsilon = \xi$. In this case Corollary 12.4.7 shows that there is an X-validated nuvs p of $\mathcal{M}_\Upsilon$ such that $\mathcal{R} = \mathrm{m}^+(p)$ is X-suitable. We then continue with our X-validated sts construction over $\mathcal{R}$ or over some transitive W containing $\mathcal{R}$.

So we assume $\mathcal{M}_\xi$ is small and $(1)_\xi(b)$. We verify the other clauses. We now define the strategy $\Sigma_{\mathcal{M}_\xi^Y}$ for $\mathcal{M}_\xi^Y$; we sometimes write Σ_ξ^Y for $\Sigma_{\mathcal{M}_\xi^Y}$.[27] We write x^Y for $\pi_Y^{-1}(x)$ for $x \in \mathcal{M}_\xi \cap \mathrm{rng}(\pi_Y)$.

DEFINITION 12.7.2 (Normal form). An iteration $((\mathcal{P}_\alpha, \vec{\mathcal{T}}_\alpha) \mid \alpha < \eta)$ on $\mathcal{P}_0 = \mathcal{M}_\xi^Y$ is said to be in *normal form* if the following hold:

(i) $\vec{\mathcal{T}}_\alpha$ is a stack of normal trees with base model $\mathcal{P}_\alpha$ and last model $\mathcal{P}_{\alpha+1}$.
(ii) If $\lambda \leq \eta$ is limit, $\mathcal{P}_\lambda = \lim_{\alpha<\lambda} \mathcal{P}_\alpha$.
(iii) Either $\vec{\mathcal{T}}_\alpha$ uses no extenders in the top block of $\mathcal{P}_\alpha$ or its images or $\mathcal{P}_{\alpha+1} = \mathrm{Ult}(\mathcal{P}_\alpha, E)$ for some extender E on the $\mathcal{P}_\alpha$-sequence with $\mathrm{cr}(E) = \delta^{\mathcal{P}_\alpha}$ or else $\vec{\mathcal{T}}_\alpha$ is completely above $\delta^{\mathcal{P}_\alpha}$.
(iv) If $\eta = \alpha + 1$ for some α, then for all $\beta < \alpha$, $\vec{\mathcal{T}}_\beta$ does not drop.

We define Σ_ξ^Y for stacks in normal form. We say that a stack $((\mathcal{P}_\alpha, \vec{\mathcal{T}}_\alpha) \mid \alpha < \eta)$ in normal form, where $\mathcal{P}_0 = \mathcal{M}_\xi^Y$, is according to Σ_ξ^Y if: letting $\tau_0 = \pi_Y \restriction \mathcal{P}_0$, $i_{\gamma,\tau} : \mathcal{P}_\gamma \to \mathcal{P}_\tau$ be iteration maps, and $\kappa^{\mathcal{P}_\gamma} = i_{0,\gamma}(\kappa^{\mathcal{P}_0})$, where $\kappa^{\mathcal{P}_0} = \delta^{\mathcal{P}_0^b} = \delta^{\mathcal{P}}$, then

(A) $\mathcal{P}_\alpha$ is X-approved and there are maps $\tau_\alpha : \mathcal{P}_\alpha \to \mathcal{M}_\xi$ for all $\alpha < \eta$;
(B) for all $\gamma \leq \alpha < \eta$, if $i_{\gamma,\alpha}$ exists then $\tau_\gamma = \tau_\alpha \circ i_{\gamma,\alpha}$;
(C) for all $\alpha < \eta$, letting Λ_α be the τ_α-pullback strategy and $\pi^{\Lambda_\alpha}_{\mathcal{P}_\alpha|\kappa^{\mathcal{P}_\alpha},\infty} = \tau_\alpha \restriction \mathcal{P}_\alpha|\kappa^{\mathcal{P}_\alpha}$; furthermore, $\mathcal{P}_\alpha^b = \mathcal{Q}_Z^X$ for some Z extending X;
(D) if $\eta = \alpha + 1$ and $\vec{\mathcal{T}}_\alpha$ drops, then there is a (unique) branch b of $\vec{\mathcal{T}}_\alpha$, some $\xi' < \xi$, and a weak-$deg(b)$-embedding[28] $\tau_\eta : \mathcal{M}_b^{\vec{\mathcal{T}}_\alpha} \to \mathcal{M}_{\xi'}$. Otherwise, there is a (unique) branch b and map $\tau_\eta : \mathcal{M}_b^{\vec{\mathcal{T}}_\alpha} \to \mathcal{M}_\xi$ such that $\tau_\alpha = \tau_\eta \circ i_{\alpha,\eta}$.

It is clear how to extend Σ_ξ^Y to all stacks of normal trees. This is because all stacks of normal trees on $\mathcal{N}_\xi^Y$ can be decomposed into stacks in normal form. We will need to define maps τ_α in the definition of Σ_ξ^Y in such a way that makes Σ_ξ^Y an X-realizable strategy.

LEMMA 12.7.3. *Suppose $Y \prec H_{\kappa^{+4}}$ is a countable such that $Y \cap \mathcal{P}$ is an extension of X. Suppose $i : \mathcal{M}_\xi^Y \to \mathcal{R}$ and $\sigma : \mathcal{R} \to \mathcal{M}_\xi$ are such that $\pi_Y = \sigma \circ i$, and* $\mathrm{rge}(\sigma) \cap \mathcal{P}$ *is an honest extension of $Y \cap \mathcal{P}$. Let Λ be the σ-pullback strategy on $\mathcal{R}$.*

---

[27]Technically, we construct $\Sigma_{\mathcal{M}_\xi^Y}$ in $V[G]$ but $\Sigma_{\mathcal{M}_\xi^Y} \cap V \in V$ and $\Sigma_{\mathcal{M}_\xi^Y}$ does not depend on the choice of G. This will be clear from the construction of $\Sigma_{\mathcal{M}_\xi^Y}$. So in effect, we are constructing an invariant name $\dot{\Sigma}$ in V whose interpretation in $V[G]$ is $\Sigma_{\mathcal{M}_\xi^Y}$ for any G. This justifies our notation Σ_ξ^Y.

[28]$deg(b)$ is the degree of soundness of model corresponding to the last drop along b.

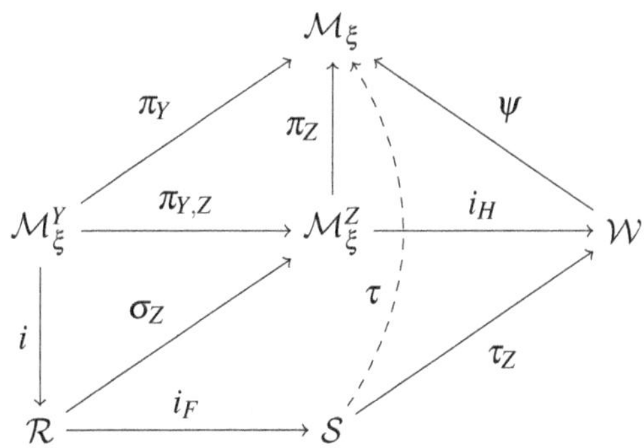

FIGURE 6. Hypothesis of Lemma 12.7.3.

Then:

(a) *If $j : \mathcal{R} \to \mathcal{S}$ is a Λ-iteration based on $\mathcal{R}^b$ and suppose $\kappa^{\mathcal{S}} = \sup j[\kappa^{\mathcal{R}}] = j(\kappa^{\mathcal{R}})$, then letting $\tau : \mathcal{S} \to \mathcal{M}_\xi$ be the map: $\tau(j(f)(a)) = \sigma(f)(\pi^{\Lambda_{\mathcal{S}}}_{\mathcal{S}|\kappa^{\mathcal{S}},\infty}(a))$, where $f \in \mathcal{R}$, $a \in [\kappa^{\mathcal{S}}]^{<\omega}$, and $\Lambda_{\mathcal{S}}$ is the tail of Λ. Then τ is well-defined, elementary, and $\pi^{\Lambda_{\mathcal{S}}}_{\mathcal{S}|\kappa^{\mathcal{S}},\infty} = \tau \restriction (\mathcal{S}|\kappa^{\mathcal{S}})$. As before (and later on in this chapter), $\kappa^{\mathcal{R}} = \delta^{\mathcal{R}_b}$ etc. is the cutpoint cardinal that begins the top block of $\mathcal{R}$.*

(b) *Suppose F is an extender on the $\mathcal{R}$-sequence with $\mathrm{cr}(F) = \kappa^{\mathcal{R}} = i(\delta^{\mathcal{P}})$. Then F is σ-certified over $(\mathcal{R}', \Lambda_{\mathcal{R}'})$, where $\mathcal{R}' = \mathcal{R}||\mathrm{lh}(F)$. This means for $a \in \mathrm{lh}(F)^{<\omega}$, $A \subseteq \kappa^{\mathcal{R}}$ in $\mathcal{R}$, $(a,A) \in F$ if and only if $\pi^{\Lambda_{\mathcal{R}'}}_{\mathcal{R}',\infty}(a) \in \sigma(A)$.*

REMARK 12.7.4. The extender F in the lemma is said to be *certified* for short instead of "σ-certified over $(\mathcal{R}||\mathrm{lh}(F), \Lambda_{\mathcal{R}||\mathrm{lh}(F)})$".

PROOF. (a) follows from Lemma 12.2.1 and the fact that the iteration map j is continuous at $\delta^{\mathcal{R}}$.

For (b), first, note that i is continuous at $(\delta^+)^{\mathcal{N}_\xi^Y}$ and is cofinal in $((\kappa^{\mathcal{R}})^+)^{\mathcal{R}}$. This is because π_Y is continuous at $(\delta^+)^{\mathcal{N}_\xi^Y}$ and is cofinal in $(\delta^+)^{\mathcal{P}}$. Finally, F is total over $\mathcal{R}$; this follows from the continuity of i.

Now, let $\mathcal{S} = \mathrm{Ult}(\mathcal{R}, F)$, i_F be the ultrapower map. Let $Y \prec Z$ be countable such that $Z \cap \mathcal{P}$ is an honest extension of X and such that $\mathrm{rng}(\sigma) \subseteq \mathrm{rng}(\pi_Z)$. Let $\sigma_Z = \pi_Z^{-1} \circ \sigma$. Let $H = \sigma_Z(F)$[29] and $i_H : \mathcal{M}_\xi^Z \to \mathrm{Ult}(\mathcal{M}_\xi^Z, H) =_{\mathrm{def}} \mathcal{W}$ be the ultrapower map. Let $\tau_Z : \mathcal{S} \to \mathcal{W}$ be the copy map and $\psi : \mathrm{Ult}(\mathcal{M}_\xi^Z, H) \to \mathcal{M}_\xi$ be the map

$$\psi(i_H(f)(a)) = \pi_Z(f)(\pi^{\Psi}_{\mathcal{N}_\xi^Z||\mathrm{lh}(H),\infty}(a)),$$

[29] If F is the top extender of $\mathcal{R}$, then by $\sigma_Z(F)$, we mean $\sigma_Z[F]$.

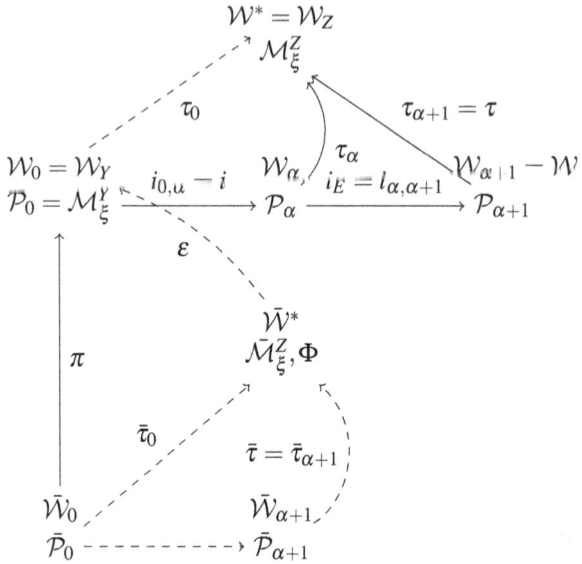

FIGURE 7. Sketch of Remark 12.7.5(ii).

where Ψ is $(\Sigma_\xi^Z)_{\mathcal{M}_\xi^Z || \mathrm{lh}(H)}$ Since H is π_Z-certified over $(\mathcal{N}_\xi^Z || \mathrm{lh}(H), \Psi)$, ψ is well-defined, elementary, and $\psi \circ i_H = \pi_Z$. Now,

$$\sigma = \psi \circ \tau_Z \circ i_F,$$

so letting $\Lambda_{\mathcal{S}}$ be the $\psi \circ \tau_Z$-pullback strategy for $\mathcal{S}$, then by strategy coherence for hod mice, $\Lambda_{\mathcal{S}}$ agrees with $\Lambda_{\mathcal{R}}$ on $\mathcal{R} || \mathrm{lh}(F)$ (see Remark 12.7.5(ii)). Now let $\tau : \mathcal{S} \to \mathcal{M}_\xi$ be defined as follows: for all $a \in [\mathrm{lh}(F)]^{<\omega}$ and $f \in \mathcal{R}$,

$$\tau(i_F^{\mathcal{R}}(f)(a)) = \sigma(f)(\pi^{\Lambda_{\mathcal{R}}}_{\mathcal{R}||\mathrm{lh}(F),\infty}(a)).$$

By Lemma 12.2.1, τ is well-defined, elementary, and agrees with $\pi^{\Lambda_{\mathcal{S}}}_{\mathcal{S},\infty}$ up to $\delta^{\mathcal{S}}$ and with $\pi^{\Lambda_{\mathcal{R}}}_{\mathcal{R}||\mathrm{lh}(F),\infty}$ up to $\mathcal{R} || \mathrm{lh}(F)$. This proves part (b). ⊣

The following remarks summarize how we can inductively define maps τ_α and hence define Σ_ξ^Y on stacks in normal form.

REMARK 12.7.5. (i) If $\vec{\mathcal{T}}_\alpha = \langle E \rangle$ for $\mathrm{cr}(E) = \kappa^{\mathcal{P}_\alpha}$, then

$$\tau_{\alpha+1}(i_E^{\mathcal{P}_\alpha}(f)(a)) = \tau_\alpha(f)(\pi^{\Lambda_\alpha}_{\mathcal{P}_\alpha||\mathrm{lh}(E),\infty}(a)).$$

Lemma 12.7.3(b) shows that $\tau_{\alpha+1}$ is well-defined, elementary,[30], agrees with τ_α up to $\kappa^{\mathcal{P}_\alpha}$ and with $\pi^{\Lambda_\alpha}_{\mathcal{P}_\alpha||\mathrm{lh}(E),\infty}$ up to $\mathrm{lh}(E)$.

[30] If τ_α is a weak k-embedding for some k, as is typical of realization maps, then so is $\tau_{\alpha+1}$.

(ii) With the exact same situation as (i) and suppose $\mathrm{cof}^V(\mathrm{ord}(\mathcal{M}_\xi)) < \kappa$,[31] we claim that the $\mathcal{S} =_{\mathrm{def}} \mathcal{P}_{\alpha+1}$-tail of $\Psi =_{\mathrm{def}} \Sigma^Y_{\xi-1}$ agrees with $\Psi_\mathcal{S}$, the $\tau_{\alpha+1}$-pullback strategy of $\mathcal{S}$. This is *strategy coherence at* $\alpha+1$. Suppose not. Write τ for $\tau_{\alpha+1}$ and i for $i_{0,\alpha}$. This is basically combining the proof of Theorem 2.7.6 in [29] and Lemma 12.2.1 (see Figure 7). We briefly sketch it here. Let $Y \prec Z$ and $Z \in V$ be countable (in $V[G]$), such that $Z \cap \mathcal{P}$ is an honest extension of X. Furthermore, we assume $Y \cap \kappa \in \kappa$, $Y^{<|Y|} \subset Y$, and letting $\iota = \mathrm{cof}^V(\mathrm{ord}(\mathcal{M}_\xi))$, then $\iota < |Y|$.[32] Let $\mathcal{W}_Y$ be a Ψ-hod mouse with $\mathrm{cof}(\lambda^{\mathcal{W}_Y}) = \omega$ and $\mathcal{W}_Z = \mathrm{Ult}(\mathcal{W}_Y, F)$, where F is the $(\mathrm{crit}(\pi_Y), \mathrm{ord}(\mathcal{M}^Z_\xi))$-extender derived from $\pi_{Y,Z}$. So letting $j = i_{\alpha,\alpha+1} \circ i$, j extends to $j^+ : \mathcal{W}_Y \to \mathcal{W}$ and τ extends to $\tau^+ : \mathcal{W} \to \mathcal{W}_Z$, where $\mathcal{M}^Z_\xi \lhd \mathcal{W}_Z$ (this is because $\mathrm{ord}(\mathcal{M}^Y_\xi)$ is a cutpoint in $\mathcal{W}_Y$ and $\pi_{Y,Z}$ is cofinal in $\mathrm{ord}(\mathcal{M}^Z_\xi)$). Let $\pi : M \to H^V_{\kappa^{+4}}$ be the inverse of the transitive collapse of some elementary substructure of $H^V_{\kappa^{+4}}$ in V containing all relevant objects such that $X \subset \mathrm{ran}(\pi)$ and $|M| < |Y|$. For any $a \in H^V_{\kappa^{+4}} \cap \mathrm{ran}(\pi)$, let $\bar{a} = \pi^{-1}(a)$. Let $\bar{g} \subseteq Col(\omega, \bar{\kappa})$ be M-generic with $g \in V$ and $\bar{\mathcal{S}}$, $\bar{j}$, $\bar{\tau}$ be the objects in $M[\bar{g}]$ witnessing the failure of the claim in $M[\bar{g}]$. Since $|M| < |Y|$ and $Y^{<|Y|} \subseteq Y$, there is a map $\varepsilon : \bar{\mathcal{W}}_Z \to \mathcal{W}_Y$ such that $\pi \restriction \bar{\mathcal{W}}_Y = \varepsilon \circ \bar{\pi}_Y \restriction \bar{\mathcal{W}}_Y$. Let Φ be the ε-pullback of Ψ. By the proof of Theorem 2.7.6 in [29], working in $M[\bar{g}]$, the uB code for $\bar{\Psi}$ gets moved to the uB code for its $\bar{\mathcal{S}}$-tail and also to the uB code for the $\bar{\tau}$-pullback of Φ; by Lemma 12.2.1, this is also the $\tau \circ \pi = \bar{\tau} \circ \pi$-pullback of $\Sigma^Z_{\xi-1}$. This is a contradiction.

(iii) If $\vec{\mathcal{T}}_\alpha$ is below $\delta^{\mathcal{P}_\alpha}$ then it is according to Λ_α and so $\tau_{\alpha+1}$ is given by the inductive assumption on Λ_α. Strategy coherence at $\alpha+1$ is maintained here. See Lemma 12.7.3(a).

(iv) If $\vec{\mathcal{T}}_\alpha$ is above $\delta^{\mathcal{P}_\alpha}$ then $\vec{\mathcal{T}}_\alpha$ is correctly guided.[33] The map $\tau_{\alpha+1}$ is given by the K^c-construction theorem (cf. [3, Theorem 3.2]) and our smallness assumption on the hod mice that we are constructing; in fact, using the argument in Section 12.4 and $(1)_\xi(b)$, we get that the branch giving rise to $\tau_{\alpha+1}$ is the unique branch b such that $\mathcal{Q}(b, \vec{\mathcal{T}}_\alpha)$ exists and is X-validated.

(v) Suppose $\lambda < \eta$ is limit. Let for $\alpha < \lambda$

$$\tau_\lambda(i_{\alpha,\lambda}(x)) = \tau_\alpha(x).$$

So we get $\tau_\lambda : \mathcal{P}_\lambda \to \mathcal{M}^Y_\xi$ is such that for all $\alpha < \lambda$, $\tau_\alpha = \tau_\lambda \circ i_{\alpha,\lambda}$. Using the above argument, we get strategy coherence at λ. Finally, we verify that letting $\pi : \mathcal{P}_\lambda | \delta^{\mathcal{P}_\lambda} \to \mathcal{M}_\xi$ be the iteration maps by the τ_λ-pullback strategy Λ_λ, $\pi = \tau_\lambda \restriction \delta^{\mathcal{P}_\lambda}$. Let $\nu < \delta^{\mathcal{P}_\lambda}$. We note that Λ_λ is the Λ_α-tail by strategy coherence

[31] Cofinally many ξ' has the property that $\mathrm{cof}^V(o(\mathcal{M}_{\xi'})) < \kappa$. In our case, $\xi = \xi^* + 1$, this holds because $\mathcal{N}_{\xi^*} \models \forall \xi \Box_{\xi,2}$ by Chapter 11 and the hypothesis of Theorem 12.0.21.

[32] This is possible because κ is strongly inaccessible.

[33] Recall this means that for $\beta < \mathrm{lh}(\vec{\mathcal{T}}_\alpha)$, letting $c = [0,\beta]_{\vec{\mathcal{T}}_\alpha}$, then $\mathcal{Q}(c, \vec{\mathcal{T}}_\alpha \restriction \beta)$ exists and is X-validated.

at λ. Let $i_{\alpha,\lambda}(\nu^*) = \nu$ for some $\alpha < \lambda$ and $\nu^* < \delta^{\mathcal{P}_\alpha}$. Then

$$\tau_\lambda(\nu) = \tau_\alpha(i_{\alpha,\lambda}(\nu^*)) = \pi_{\mathcal{P}_\alpha|\kappa^{\mathcal{P}_\alpha},\infty}(\nu^*) = \pi(i_{\alpha,\lambda}(\nu^*)) = \pi(\nu).$$

The following lemma gives some useful consequences regarding uniqueness of strategies.

LEMMA 12.7.6. *(i) Suppose $\pi : \mathcal{Q} \to \mathcal{P}$ is elementary such that* $\mathrm{rng}(\pi)$ *is an extension of X. Suppose $i : \mathcal{Q} \to \mathcal{R}$ is such that $i \restriction \delta^{\mathcal{P}}$ is according to the π-pullback strategy Σ^π and $\tau_0, \tau_1 : \mathcal{R} \to \mathcal{P}$ are such that $\tau_0 \circ i = \tau_1 \circ i = \pi$. Then the τ_0-pullback strategy Σ^{τ_0} is the same as the τ_1-pullback strategy Σ^{τ_1}.*

(ii) Suppose Y is countable, elementary in $\mathcal{M}_\xi$ and $Y \cap \mathcal{P}$ is an (honest) extension of X. Suppose n is such that $\omega\rho^{n+1}_{\mathcal{M}^Y_\xi} < \omega\rho^n_{\mathcal{M}^Y_\xi}$. Let $\Psi = \Sigma^Y_\xi$.

(a) Suppose $(\vec{\mathcal{T}}, \mathcal{R}) \in I(\mathcal{M}^Y_\xi, \Psi)$ is such that $\pi^{\vec{\mathcal{T}},b}$ exists and $\tau : \mathcal{M}^Y_\xi \to \mathcal{S}$ is $\Sigma^{(n)}_0$ and cardinal preserving and $\mathcal{S} \trianglelefteq \mathcal{R}$. Suppose $(\vec{\mathcal{U}}, \mathcal{Q}) \in I(\mathcal{M}^Y_\xi, \Psi)$ is such that $\pi^{\vec{\mathcal{U}},b}$ exists, $\mathcal{Q}^b = \mathcal{S}^b$, and $\tau \restriction \mathcal{P}_Y = \pi^{\vec{\mathcal{U}},b} \restriction \mathcal{P}_Y$, then $\Psi^\tau_{\vec{\mathcal{T}},\mathcal{S}} = \Psi$.

(b) Suppose $(\vec{\mathcal{T}}, \mathcal{R})$ is such that $\pi^{\vec{\mathcal{T}},b}$ exists and is according to Ψ. Suppose $\mathcal{U}$ is a normal tree of limit length on $\mathcal{R}(\beta)$ according to $\Psi_{\vec{\mathcal{T}},\mathcal{R}}$, where $\mathcal{R}(\beta) \lhd^{hod}_c \mathcal{R}$.[34] Suppose c is a cofinal branch of $\mathcal{U}$ (considered as a tree on $\mathcal{R}$) and there is a map $\tau_c : \mathcal{M}^{\mathcal{U}}_c \to \mathcal{M}_\xi$ such that $\pi_Y \restriction \mathcal{P}_Y = \tau_c \circ \pi^{\mathcal{U}}_c \circ \pi^{\vec{\mathcal{T}},b}$. Then $c = \Psi_{\vec{\mathcal{T}},\mathcal{R}}(\mathcal{U})$.

PROOF. (i) follows straightforwardly from Lemma 12.2.1 (iv). The main point is that, letting Λ_i be the τ_i-pullback strategy Σ^{τ_i} (for $i = 0, 1$), then letting $\sigma_i : \mathcal{R} \to \mathcal{P}$ be

$$\sigma(i(f)(a)) = \pi(f)(\pi^{\Lambda_i}_{\mathcal{R},\infty}(a))$$

for $f \in \mathcal{Q}$ and $a \in \delta^{\mathcal{R}}$. Then $\sigma_i[\mathcal{R}]$ is an honest extension of X.

(ii)(b) follows easily from (i) and Remark 12.7.5(ii). For (ii)(a) (see Figure 8), suppose $\Psi^\tau_{\vec{\mathcal{T}},\mathcal{S}} \neq \Psi$, then by results of Section 4.7, there is a (minimal) low-level disagreement, i.e., there is $(\vec{\mathcal{W}}, \mathcal{R}_0, \mathcal{W}^*)$ such that:

1. $\vec{\mathcal{W}}$ is according to both strategies.
2. $\mathcal{R}_0$ is the last model of $\vec{\mathcal{W}}$.
3. $\mathcal{W}^*$ is a tree of limit length on $\mathcal{R}_0(\beta)$ for some $\mathcal{R}_0(\beta) \lhd^{hod}_c \mathcal{R}_0$.

Let $b = \Psi(\vec{\mathcal{W}}^\frown \mathcal{W}^*)$ and $c = \Psi^\tau_{\vec{\mathcal{T}},\mathcal{S}}(\vec{\mathcal{W}}^\frown \mathcal{W}^*)$. Let $\sigma : \mathcal{Q} \to \mathcal{M}_\xi$ be the realization map; hence

$$\pi_Y \restriction \mathcal{Q}_Y = \sigma \circ \pi^{\vec{\mathcal{U}},b} = \sigma \circ \tau \restriction \mathcal{Q}_Y.^{35} \tag{11}$$

[34]Recall this means that $\mathcal{R}(\beta)$ is a complete layer of $\mathcal{R}$ and $\mathcal{R}(\beta) \neq \mathcal{R}$.

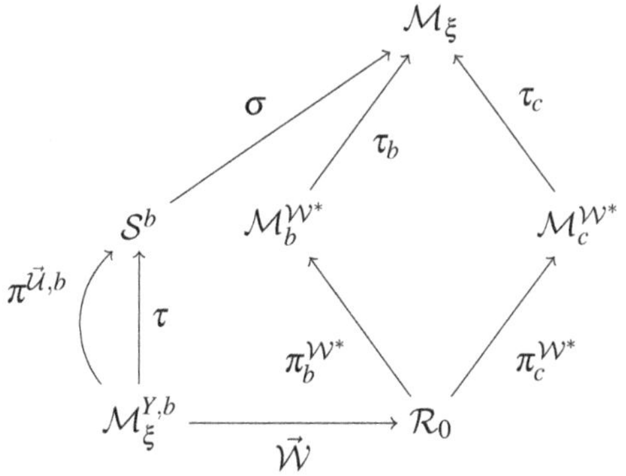

FIGURE 8. Lemma 12.7.6(ii)(a).

Let $\tau_b : \mathcal{M}_b^{\mathcal{W}^*,b} \to \mathcal{P}$ and $\tau_c : \mathcal{M}_c^{\mathcal{W}^*,b} \to \mathcal{P}$ be the natural realization maps. We have:

$$\sigma \circ \tau \restriction \mathcal{Q}_Y = \tau_c \circ \pi_c^{\mathcal{W}^*} \circ \pi^{\vec{\mathcal{W}}} \tag{12}$$

and

$$\pi_Y \restriction \mathcal{Q}_Y = \tau_b \circ \pi_b^{\mathcal{W}^*} \circ \pi^{\vec{\mathcal{W}}}. \tag{13}$$

By (i) and Equations (11), (12), (13), $b = c$. Contradiction. ⊣

LEMMA 12.7.7. *Suppose Y is countable, elementary in $\mathcal{M}_\xi$ and $Y \cap \mathcal{P}$ is an (honest) extension of X. Then Σ_ξ^Y has locally strong branch condensation, and is Ω-fullness preserving.*

PROOF. Ω-fullness preservation follows from the construction of Σ_ξ^Y and the fact that X is a condensing set (see Lemma 12.2.1). We first prove branch condensation (see Figure 9). Suppose not. Let $\mathcal{N} = \mathcal{M}_\xi^Y$ and $\Psi = \Sigma_\xi^Y$ and suppose the following hold: there are stacks $\vec{\mathcal{T}}^\frown\mathcal{U}$ and $\vec{\mathcal{W}}$ on $\mathcal{N}$ such that

1. $\vec{\mathcal{T}}$ is via Ψ with end model $\mathcal{R}$.
2. $\mathcal{U}$ is according to $\Psi_\mathcal{R}$, $\vec{\mathcal{W}}$ is according to Ψ, and $i = \pi^{\vec{\mathcal{W}}} : \mathcal{Q}_Y \to \mathcal{Q}$ is the iteration map.
3. There are cofinal branches b, c of $\mathcal{U}$ and $\pi : \mathcal{M}_b^\mathcal{U} \to \mathcal{Q}$ such that
   (a) $i = \pi \circ \pi_b^\mathcal{U} \circ i^{\vec{\mathcal{T}}}$.
   (b) $c = \Psi(\vec{\mathcal{T}}^\frown\mathcal{U})$.
   (c) $b \neq c$.

[35]Recall $\mathcal{Q}_Y = \pi_Y^{-1}(\mathcal{P})$.

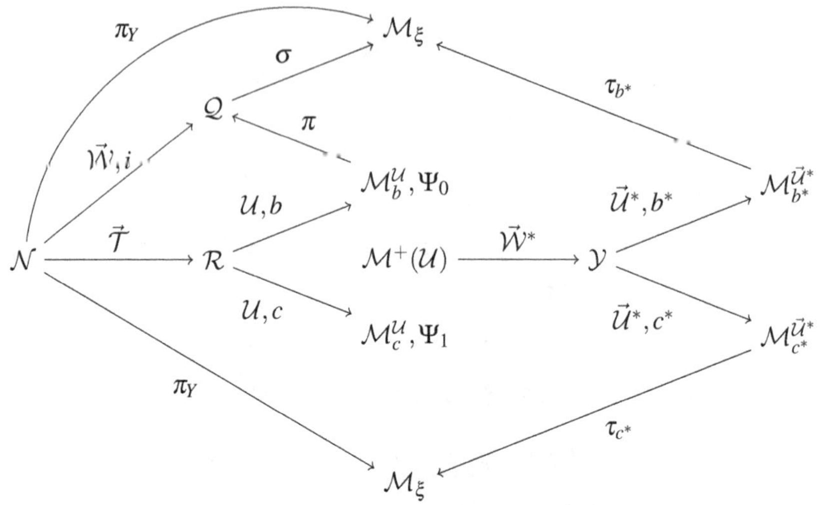

FIGURE 9. Branch condensation.

Let Ψ_0 be the π-pullback strategy of $\Psi_{\vec{\mathcal{W}},\mathcal{Q}}$ and Ψ_1 be $\Psi_{\vec{\mathcal{T}}^\frown\mathcal{U}^\frown c}$. Recall $\mathrm{m}^+(\mathcal{U}) = \mathcal{M}(\mathcal{U})^\sharp$. We first show:

$$\Lambda_0 =_{\mathrm{def}} (\Psi_0)^{sts}_{\mathrm{m}^+(\mathcal{U})} = (\Psi_1)^{sts}_{\mathrm{m}^+(\mathcal{U})} =_{\mathrm{def}} \Lambda_1. \tag{14}$$

In the case there is $\mathcal{Q} \trianglelefteq \mathrm{m}^+(\mathcal{U})$ which is a Q-structure for $\delta(\mathcal{U})$ then

$$(\Psi_0)^{sts}_{\mathrm{m}^+(\mathcal{U})} = (\Psi_0)_{\mathrm{m}^+(\mathcal{U})}$$

and similarly for Ψ_1. We assume this is not the case; otherwise, the argument is similar and simpler.

Let $\sigma : \mathcal{Q}^b \to \mathcal{P}$ be the π_Y-realization map, so that

$$\pi_Y \restriction \mathcal{Q}_Y = \sigma \circ \pi \circ \pi^{\vec{\mathcal{T}}^\frown\mathcal{U},b}.$$

In the above, we note that $\pi^{\vec{\mathcal{T}}^\frown\mathcal{U},b}$ exists and is the same as $\pi_c^{\vec{\mathcal{T}}^\frown\mathcal{U},b}$ and this map does not depend on the choice of the cofinal branch; so $\pi^{\vec{\mathcal{T}}^\frown\mathcal{U},b}$ is also $\pi_b^{\vec{\mathcal{T}}^\frown\mathcal{U},b} = \pi^{\mathcal{U},b} \circ \pi^{\vec{\mathcal{T}}}$.

By results of Section 4.7, if (14) fails, then there is a minimal disagreement $(\vec{\mathcal{W}}^*, \mathcal{Y}) \in B(\mathrm{m}^+(\mathcal{U}), \Lambda_0) \cap B(\mathrm{m}^+(\mathcal{U}), \Lambda_1)$. Note that $\mathcal{Y}$ is of successor type and $(\Lambda_0)_{\vec{\mathcal{W}}^*,\mathcal{Y}(\alpha)} = (\Lambda_1)_{\vec{\mathcal{W}}^*,\mathcal{Y}(\alpha)}$ for all $\mathcal{Y}(\alpha) \triangleleft_c^{hod} \mathcal{Y}$. Furthermore, there is a stack $\vec{\mathcal{U}}^*$ on $\mathcal{Y}$ such that there are distinct branches $b^* = (\Lambda_0)_{\vec{\mathcal{W}}^*,\mathcal{Y}} \neq c^* = (\Lambda_1)_{\vec{\mathcal{W}}^*,\mathcal{Y}}$. Note that

$$\pi_b^{\mathcal{U},b} \circ \pi^{\vec{\mathcal{T}}} \restriction \mathcal{Q}_Y = \pi_c^{\mathcal{U}} \circ \pi^{\vec{\mathcal{T}}} \restriction \mathcal{Q}_Y.$$

Note further that there are $\tau_{b^*} : \mathcal{M}_{b^*}^{\vec{\mathcal{U}}^*,b} \to \mathcal{P}$ and $\tau_{c^*} : \mathcal{M}_{c^*}^{\vec{\mathcal{U}}^*,b} \to \mathcal{P}$ such that

$$\pi_Y \restriction \mathcal{Q}_Y = \tau_{b^*} \circ \pi_{b^*}^{\vec{\mathcal{U}}^*} \circ \pi^{\vec{\mathcal{W}}^*,b} \circ \pi_b^{\mathcal{U},b} \circ \pi^{\vec{\mathcal{T}}} \restriction \mathcal{Q}_Y, \tag{15}$$

and

$$\pi_Y \restriction \mathcal{P}_Y = \sigma \circ i = \tau_{c^*} \circ \pi_{c^*}^{\vec{\mathcal{U}}^*} \circ \pi^{\vec{\mathcal{W}}^*,b} \circ \pi_c^{\mathcal{U},b} \circ \pi^{\vec{\mathcal{T}}} \restriction \mathcal{P}_Y. \tag{16}$$

This is because

$$\pi_{b^*}^{\vec{\mathcal{U}}^*} \circ \pi^{\vec{\mathcal{W}}^*,b} = \pi_{c^*}^{\vec{\mathcal{U}}^*} \circ \pi^{\vec{\mathcal{W}}^*,b}.$$

Equations (15), (16) give us

$$\tau_{b^*} \circ \pi_{b^*}^{\vec{\mathcal{U}}^*} \circ \pi^{\vec{\mathcal{W}}^*,b} \restriction (\mathcal{M}^+(\mathcal{U}))^b = \tau_{c^*} \circ \pi_{c^*}^{\vec{\mathcal{U}}^*} \circ \pi^{\vec{\mathcal{W}}^*,b} \restriction (\mathcal{M}^+(\mathcal{U}))^b.$$

Lemma 12.7.6 then implies that $b^* = c^*$. This is a contradiction.

So (14) holds. By our assumption, $\mathcal{Q}(c,\mathcal{U}) \trianglelefteq \mathcal{M}_c^{\mathcal{U}}$ and is a X-validated Λ_1-mouse and $\mathcal{Q}(b,\mathcal{U}) \trianglelefteq \mathcal{M}_b^{\mathcal{U}}$ and is a X-validated Λ_0-mouse. Results of Chapter 6 and earlier sections of this chapter imply that $\mathcal{Q}(b,\mathcal{U}) = \mathcal{Q}(c,\mathcal{U})$ and hence $b = c$. Contradiction.

The argument above shows branch condensation. The other clause of strong branch condensation follows from a very similar argument, so we leave it to the reader. ⊣

LEMMA 12.7.8. *Σ_ξ^Y is locally strongly Ω-fullness preserving.*

PROOF. Ω-fullness preservation follows from the previous lemma. We now prove the other clause of locally strongly Ω-fullness preservation (see Figure 10). Let $\mathcal{N} = \mathcal{M}_\xi^Y$ and $\Psi = \Sigma_\xi^Y$. Suppose $(\vec{\mathcal{T}},\mathcal{S}) \in I(\mathcal{N},\Psi)$ (so $\pi^{\vec{\mathcal{T}},b}$ exists). Suppose $\mathcal{S}^b \triangleleft \mathcal{W} \trianglelefteq \mathcal{S}$ is such that for some n, $\mathcal{W}$ is n-sound and,

$$o(\mathcal{S}^b) \leq \omega\rho_{\mathcal{W}}^{n+1} < \omega\rho_{\mathcal{W}}^n.$$

Suppose $\tau : \mathcal{R} \to \mathcal{W}$ is cardinal preserving, is $\Sigma_0^{(n)}$, and $\omega\rho_{\mathcal{R}}^n > \mathrm{cr}(\tau) \geq \omega\rho_{\mathcal{R}}^{n+1} = \omega\rho_{\mathcal{W}}^{n+1}$. We want to show the τ-pullback of the strategy $\Sigma_{\vec{\mathcal{T}},\mathcal{W}}$ is Ω-fullness preserving.

Note that $\tau \restriction \mathcal{R}^b = \mathrm{id}$ and $\mathcal{R}^b = \mathcal{W}^b$. This implies $\mathrm{rng}(\pi^{\vec{\mathcal{T}},b}) \subseteq \mathrm{rng}(\tau)$. Let $\sigma : \mathcal{W}^b = \mathcal{S}^b \to \mathcal{P}$ be the π_Y-realization map, so that $\pi_Y \restriction \mathcal{N}^b = \sigma \circ \pi^{\vec{\mathcal{T}},b}$. Note that $\mathrm{rng}(\sigma)$ is an honest extension of X.

We now show $\Sigma_{\vec{\mathcal{T}},\mathcal{W}}^\tau$ is Ω-fullness preserving. To see this, let $(\mathcal{W}^*,\vec{\mathcal{U}}) \in I(\mathcal{W},\Sigma_{\vec{\mathcal{T}},\mathcal{W}}^\tau)$ be such that $\pi^{\vec{\mathcal{U}},b} : \mathcal{R}^b \to (\mathcal{R}^*)^b$ exists and let $\tau\vec{\mathcal{U}}$ be the copy tree on $\mathcal{W}$ with last model $\mathcal{W}^*$. So $\pi^{\tau\vec{\mathcal{U}},b} : \mathcal{W}^b \to (\mathcal{W}^*)^b$ exists. Let $\psi : (\mathcal{R}^*)^b \to (\mathcal{W}^*)^b$ be the copy map and $\sigma^* : (\mathcal{W}^*)^b \to \mathcal{P}$ be given by the construction of Ψ, so $\sigma^* \circ \pi^{\tau\vec{\mathcal{U}},b} = \sigma$ and $\sigma^* \circ \pi^{\tau\vec{U},b} \circ \sigma = \pi_Y \restriction \mathcal{N}^b$.

Note that $\psi = \mathrm{id}$ and $\mathrm{rng}(\sigma^*)$ is an honest extension of X. So $(\mathcal{W}^*)^b$ is Ω-full. This is our desired conclusion. ⊣

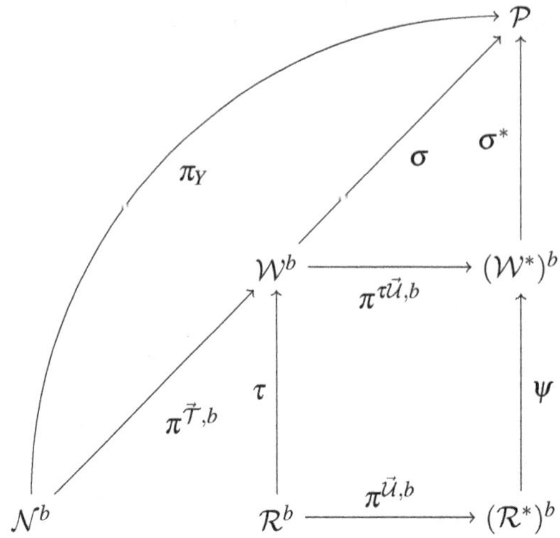

FIGURE 10. Strong Ω-fullness preservation.

An easy corollary of the above Lemmas is the following.

COROLLARY 12.7.9. *Suppose $Y \prec Z \prec \mathcal{M}_\xi$ are countable (in $V[G]$), and such that $Y \cap \mathcal{P}, Z \cap \mathcal{P}$ are honest extensions of X, $Y, Z \in V$, and $Y = Y^* \cap \mathcal{M}_\xi$, $Z = Z^* \cap \mathcal{M}_\xi$ for some $Y^* \prec Z^* \prec H^V_{\kappa+4}$. Let $\pi_{Y,Z} = \pi_Z^{-1} \circ \pi_Y$. Then $\Sigma^Y_\xi = (\Sigma^Z_\xi)^{\pi_{Y,Z}}$.*

PROOF. Let $\delta_Y = \pi_Y^{-1}(\delta^{\mathcal{P}})$ and $\delta_Z = \pi_Z^{-1}(\delta^{\mathcal{P}})$. By our assumption on Y and Z, we have:

$$\pi_Z \restriction \delta_Z = \pi^{\Sigma^Z_\xi}_{\mathcal{N}^Z_\xi,\infty} \restriction \delta_Z,$$

and

$$\pi_Y \restriction \delta_Y = \pi^{\Sigma^Y_\xi}_{\mathcal{N}^Y_\xi,\infty} \restriction \delta_Y = \pi_Z \circ \pi_{Y,Z} \restriction \delta_Y.$$

Using the above equations and the proof of Lemma 12.7.6 (especially the idea that if two strategies disagree, then there is a lower-level disagreement), we obtain the desired conclusion. ⊣

COROLLARY 12.7.10. *Σ^Y_ξ is positional and commuting.*

PROOF. This follows from Lemmas 12.7.7, 12.7.8, and results of Section 4.7. ⊣

We have verified $(1)_\xi(c)$ holds, assuming $(1)_\xi(b)$. Let Y be as above, i.e., $Y \prec H^V_{\kappa+4}$ is such that $|Y| < \kappa$, $Y \cap \mathcal{P}$ is an honest extension of X. We discuss how to lift $\Psi = \Sigma^Y_\xi$ to a (necessarily unique) (κ^+,κ^+)-strategy Ψ^+ with branch condensation and show $Code(\Psi) \in \Omega$.

Recall Ψ is an (ω_1,ω_1)-strategy for $\mathcal{M}^Y_\xi$ with branch condensation, is positional and Ω-fullness preserving. Furthermore, $\Psi \cap V \in V$ and is independent of the choice of generic G. $\Psi \cap V$ can be uniquely extended to an (κ^+,κ^+) strategy with branch condensation and is positional in V. We also call this extension Ψ. We outline how this extension works. We define $\Psi(\mathcal{T})$ for $\mathcal{T}$, a normal tree of length $<\kappa^+$. Suppose $\mathrm{cof}(\mathrm{lh}(\mathcal{T})) \geq \omega_2$, then letting $\xi = \mathrm{cof}(\mathrm{lh}(\mathcal{T}))$, we can construe $\vec{C} = ([0,\alpha]_T : \alpha < \mathrm{lh}(\mathcal{T}) \wedge \alpha$ is a limit ordinal) as a coherent sequence. Applying $\neg\Box(cof(\xi))$ to $\vec{C}$, we get a club $D \subset \mathrm{lh}(\mathcal{T})$ that threads the sequence $\vec{C}$. D gives a cofinal branch b through $\mathcal{T}$. This branch is necessarily the unique well-founded branch of $\mathcal{T}$. We define $\Psi(\mathcal{T}) = b$. Suppose $\mathrm{cof}(\mathrm{lh}(\mathcal{T})) < \omega_2$, the arguments in [55] or [65, Lemma 3.62] show that there is $W \prec H_{\kappa^{+4}}$ with the properties:

(a) $|W| < \kappa$, $W \cap \kappa \in \kappa$;
(b) $W^{<|W|} \subset W$;
(c) $\{\mathcal{M}^Y_\xi, \mathcal{P}, \mathcal{T}, \Psi\} \in W$;
(d) for any $W \prec W_0 \prec W_1$ with properties (a)-(c), letting $\pi_{W_i} : W_i \to H_{\kappa^{+4}}$, $\mathcal{T}_{W_i} = \pi_{W_i}^{-1}(\mathcal{T})$, and $b_i = \Psi(\mathcal{T}_{W_i})$ for $i \in \{0,1\}$, then $\pi_{W_0}[b_0]\downarrow \subseteq \pi_{W_1}[b_1]\downarrow$, where $\pi_{W_i}[b_i]\downarrow$ is the downward closure of $\pi_{W_i}[b_i]$ in $\mathcal{T}$.

The W as above is called $\mathcal{T}$-stable and we define $\Psi(\mathcal{T}) = b$ where b is the downward closure of $\pi_W[\Psi(\mathcal{T}_W)]$ in $\mathcal{T}$. It is clear that the definition of $\Psi(\mathcal{T})$ does not depend on the choice of $\mathcal{T}$-stable W.

We briefly give a sketch as to how to obtain a (κ^+,κ^+)-strategy Ψ^+ extending Ψ with branch condensation and is positional in $V[G]$. In $V[G]$, suppose $\mathcal{T}$ is of limit length $<\kappa^+$ and is according to Ψ^+. We show how to define $\Psi^+(\mathcal{T})$ (stacks of normal trees can be handled similarly). In V, let $A \subseteq \kappa$ code H_κ and a (nice) $Col(\omega,<\kappa)$-name $\dot{\mathcal{T}} \in H_{\kappa^+}$ for $\mathcal{T}$. Let

$$M_A = L^\Lambda_{\kappa^+}[A, \mathcal{M}_2^{\Psi,\sharp}]$$

where Λ is the unique (κ^+,κ^+)-strategy for $\mathfrak{M} =_{\mathrm{def}} \mathcal{M}_2^{\Psi,\sharp}$, the minimal E-active Ψ-mouse with two Woodin cardinals. We note that the existence of $\mathcal{M}_2^{\Psi,\sharp}$ follows from [65, Section 3.2]. By $\neg\Box(\kappa^+)$,

$$M_A \models \text{ there are no largest cardinals.}$$

In particular $(\kappa^+)^{M_A} < \kappa^+$, so in M_A, which is closed under Λ, we can use Λ to perform a generic genericity iteration to make A-generically generic (see [29] or [50] for more on generic genericity iterations). Let $\mathcal{Q} \in M_A$ be the result of such an iteration. There is a $\mathcal{Q}$-generic $h \subseteq Col(\omega,\delta_0^{\mathcal{Q}})$ such that $H_\kappa, G, \dot{\mathcal{T}} \in \mathcal{Q}[h]$, where $\delta_0^{\mathcal{Q}}$ is the first Woodin cardinal of $\mathcal{Q}$. Since $\mathcal{Q}$ is closed under Ψ; we can generically interpret Ψ on any generic extensions of $\mathcal{Q}$ (as done in [50] or in Chapter 6).[36] This allows us to define $\Psi^+(\mathcal{T})$ as the branch chosen by the interpretation of Ψ applied to $\mathcal{T}$ in $\mathcal{Q}[h]$. The well-definition and uniqueness of Ψ^+ follow from hull arguments in [65, Section 3.2].[37]

Using Ψ^+ and assuming Ψ is a strategy, we can define the stack of Θ-g-organized mice over $\mathbb{R}$, $\mathrm{Lp}^{G_{\Psi^+}}(\mathbb{R}, Code(\Psi))$, in $V[G]$ (cf. [50, Definition 4.23]),[38] and show that there is a maximal initial segment $\mathcal{M} \trianglelefteq \mathrm{Lp}^{G_{\Psi^+}}(\mathbb{R}, Code(\Psi))$ such that $\mathcal{M}$ is constructibly closed and $\mathcal{M} \vDash \mathsf{AD}^+ + \mathsf{SMC} + \Theta = \vartheta_\Psi$. This implies $Code(\Psi) \in \Omega$.

REMARK 12.7.11. The arguments given above show that we can further extend Ψ^+ to a $(\kappa^{+4}, \kappa^{+4})$-strategy.

If Ψ is not total (so $(1)_\xi(b)$ fails) and that $\mathcal{M}_\xi$ is of $\sharp$-lsa-type, then we stop the hybrid K^c-construction and continue the X-validated sts construction above $\mathcal{C}(\mathcal{M}_\xi) = \mathcal{N}_\xi$. The idea is that we'll wait until we reach a level $\mathcal{M}'_\gamma$ (if exists) of the X-validated sts construction extending $\mathcal{N}_\xi$ such that some $\mathcal{R} \trianglelefteq \mathcal{M}'_\gamma$ is a $\mathcal{Q}$-structure for $\delta^{\mathcal{N}_\xi}$ and then $\forall^* Y$, we can construction the canonical X-realizable strategy (Σ^Y_ξ) of $\mathcal{R}^Y$ and show that it is in Ω. If we reach a level $\mathcal{M}'_\gamma$ such that there is no X-validated strategy for $\mathcal{M}'_\gamma$ as witnessed by p, then we need to continue with an X-validated sts construction over $\mathcal{M}(p)^\sharp$.

DEFINITION 12.7.12 (Certified-extender-ready levels). For $\xi < \Upsilon$, $\mathcal{N}_\xi$ is *certified-extender-ready* if for a V-club $\mathcal{C}_\xi$ of $Y \prec \mathcal{N}_\xi$ such that $Y \cap \mathcal{P}$ is an honest extension of X and $Y \in V$ is countable in $V[G]$, letting $\mathcal{N}^Y_\xi = \pi_Y^{-1}(\mathcal{N}_\xi)$, $\Psi = \Sigma^Y_\xi$, and γ^Y_ξ be the supremum of the indices of extenders on the $\mathcal{N}^Y_\xi$-sequence with critical point $\delta_Y =_{\text{def}} \pi_Y^{-1}(\delta^{\mathcal{P}})$ (we let $\gamma^Y_\xi = ((\delta_Y)^+)^{\mathcal{N}^Y_\xi}$ if $\mathcal{N}^Y_\xi$ has no such extenders on its sequence), we have that Ψ is a strategy[39] and no $\mathcal{M} \trianglelefteq \mathrm{Lp}^{\Psi,\Omega}(\mathcal{N}^Y_\xi)$ projects across γ^Y_ξ. For $Y \in \mathcal{C}_\xi$, we also say $\mathcal{N}^Y_\xi$ is *π_Y-certified extender-ready*.

REMARK 12.7.13. Extender-ready levels are those $\mathcal{N}_\xi$'s that are eligible to be extended to a hod premouse $(\mathcal{N}_\xi, F)$ where F has critical point $\delta^{\mathcal{P}}$. Let $Y, \mathcal{M}$ be as in the above definition, it is easy to see that $\mathcal{M}$ also does not project across $\mathrm{ord}(\mathcal{N}^Y_\xi)$.

---

[36] If Ψ is a strategy, we could have simply let $\mathfrak{M} = \mathcal{M}_1^{\Psi,\sharp}$; but if Ψ is a short-tree strategy, then one seems to need $\mathcal{M}_2^{\Psi,\sharp}$ to apply results in Chapter 6. Relevant results in [50] can be applied to $\mathcal{M}_2^{\Psi,\sharp}$ as well.

[37] Let M, M^* be such that $\mathcal{T} \in M \cap M^*$; let τ, τ^* be nice $Col(\omega, <\kappa)$-terms for M, M^* respectively. In $V[G]$, let $W[G]$ contain all relevant objects and $W \prec H_{\kappa^{+4}}$ is good. Let $\bar{a} = \pi_W^{-1}(a)$ for all $a \in W[G]$. Then letting b_0, b_1 be the branches of $\bar{\mathcal{U}}$ given by applying [50, Lemma 4.8] in $L^\Lambda[trc(\bar{\tau}), <_1, \mathfrak{M}], L^\Lambda[trc(\bar{\tau}^*), <_2, \mathfrak{M}]$ (built inside $M_W[G]$), where $<_1$ is a well-ordering of $\bar{\tau}$ and $<_2$ is a well-ordering of $\bar{\tau}^*$. Then $b_0 = b_1$ as both are according to Ψ, since $(\mathfrak{M}, \Lambda)$ generically interprets Ψ in $V[G]$.

[38] [65] shows that $Code(\Psi)$ is self-scaled in the sense of [50, Definition 4.22] if Ψ is a strategy.

[39] This means $(1)_\xi(b)$ holds and hence $(\mathcal{N}^Y_\xi, \Psi)$ is not a sts hod pair.

The lemma below shows that the collection of correctly-backgrounded extenders with critical point $\delta^{\mathcal{P}}$ is sufficiently rich. For instance, if $\mathcal{P}_Y = \pi_Y^{-1}(\mathcal{P})$, and $\mathcal{N}_\xi^Y = \mathrm{Lp}^{\Sigma_{\mathcal{P}_Y},\Omega}(\mathcal{P}_Y)$, then $\mathcal{N}_\xi^Y$ is extender-ready (Corollary 12.7.18 shows that no level of $\mathcal{N}_\xi^Y$ projects below $\mathrm{ord}(\mathcal{P}_Y)$ and Theorem 11.1.5 and Corollary 12.7.18 show that every level of $\mathrm{Lp}^{\Sigma_{\mathcal{P}_Y},\Omega}(\mathcal{N}_\xi^Y)$ is sound). Lemma 12.7.14 shows that if $\mathcal{N}_\xi$ is extender-ready then for every $Y \in \mathcal{C}_\xi$, there is a correctly backgrounded extender E with critical point δ_Y such that $(\mathcal{N}_\xi^Y, E)$ is a hod premouse.

LEMMA 12.7.14. *Suppose $\mathcal{N}_{\xi^*}$ is extender-ready where $\xi = \xi^* + 1$ and suppose $(1)_{\xi^*} - (3)_{\xi^*}$ hold. Fix $Y \prec \mathcal{N}_{\xi^*}$ in $\mathcal{C}_{\xi^*}$. Let $\mathcal{N} = \mathcal{N}_{\xi^*}^Y$, $\delta^Y = \pi_Y^{-1}(\delta^{\mathcal{P}})$, and $\Psi = \Sigma_{\xi^*}^Y$ be the X-realizable strategy for $\mathcal{N}$. Then there is an extender E_Y with $\mathrm{crit}(E_Y) = \delta^Y$ such that E_Y is π_Y-certified over $(\mathcal{N}, \Psi)$.*

PROOF. Let $\gamma = \mathrm{ord}(\mathcal{N})$. Let $E = E_Y$ be the following extender over $\mathcal{N}$: for $a \in [\gamma]^{<\omega}$ and $A \in \wp(\delta^Y)^{|a|} \cap \mathcal{N}$,

$$(a, A) \in E \iff \pi_{\mathcal{N},\infty}^{\Psi}(a) \in \pi_Y(A).$$

Fix a $Y \prec Z \in \mathcal{C}_{\xi^*}$ such that $Z = Z' \cap H_{\kappa+4}^V$ and $Z' \prec H_{\kappa+4}^V$ contains all relevant objects. Let $\pi_{Z'} : M_{Z'} \to Z'$ be the uncollapse map and $\iota = \mathrm{crit}(\pi_{Z'}) = Z' \cap \kappa$. Naturally, $\pi_{Z'}$ extends to act on all of $M_{Z'}[G \restriction \iota]$ and induces an elementary embedding from $M_{Z'}[G \restriction \iota]$ into $H_{\kappa+4}[G]$; we also denote the extension map $\pi_{Z'}$. Let $\pi = \pi_{\mathcal{N},\infty}^{\Psi}$ and $\pi' = (\pi_{\mathcal{N},\infty}^{\Psi})^{M_{Z'}}$. By our assumption, Ψ is Ω-fullness preserving, commuting, and has branch condensation; furthermore, $\pi \restriction \mathcal{N}|\delta^Y = \pi_Y \restriction \mathcal{N}|\delta^Y$ and $\pi' \restriction \mathcal{N}|\delta^Y = \pi_{Y,Z} \restriction \mathcal{N}|\delta^Y$.

It is easy to see that E is the extender E' defined as follows: for $a \in \gamma^{<\omega}$ and $A \in \wp(\delta^Y)^{|a|} \cap \mathcal{N}$,

$$(a, A) \in E' \iff \pi'(a) \in \pi_{Y,Z}(A).$$

We need to see that $(\mathcal{N}, E)$ is a hod premouse.

Amenability. Let $\eta < \gamma$ and $\xi < (\delta^{Y,+})^{\mathcal{N}}$, we show: $E \cap (\eta^{<\omega} \times \mathcal{N}|\xi) \in \mathcal{N}$. Let $\mathcal{A} = (A_\alpha \mid \alpha < \delta^Y)$ enumerate $\mathcal{N}|\xi \cap \wp(\delta^{Y,<\omega})$. Let

$$B = \pi_Y(\mathcal{A}) \cap (\pi(\eta) \times \pi(\eta)).$$

Then $B \in \mathcal{N}_\xi|\delta^{\mathcal{P}}$ and so is OD^{Ω}. Now for all $a \in \eta^{<\omega}$, for all $\alpha < \delta^Y$,

$$(a, A_\alpha) \in E \iff \pi(a) \in B_{\pi(\alpha)}.$$

This shows $E \cap (\eta^{<\omega} \times \mathcal{N}|\xi)$ is OD_{Ψ}^{Ω}. By SMC and the fact that $\mathcal{N}$ is π_Y-certified extender-ready, $E \cap (\eta^{<\omega} \times \mathcal{N}|\xi) \in \mathcal{N}$.

Normality. Let $c \in \gamma^{<\omega}$, $f : [\delta_Y]^{|c|} \to \delta_Y$ be such that $f \in \mathcal{N}^b$ and $\forall^*_{E_c} u\ f(u) < \max(u)$ or equivalently $\pi_Y(f)(\pi(c)) < \max(\pi(c))$. We want to find a $\xi < \max(c)$ such that

$$\pi_Y(f)(\pi(c)) = \pi(\xi) = c_{\pi(\xi)}(\pi(c)),$$

where c_ξ is the constant function with range $\{\xi\}$

Let $\mathcal{M}$ be a Ψ-iterate of $\mathcal{N}$ such that $\pi^{\Psi}_{\mathcal{N},\mathcal{M}} = \pi_{\mathcal{N},\mathcal{M}}$ exists, $\tau_{\mathcal{M}} : \mathcal{M} \to \mathcal{N}_\xi$ be the π_Y-realization map given by the definition of Ψ, and let $\Psi_{\mathcal{M}}$ be the $\tau_{\mathcal{M}}$-pullback of $\Sigma_{\mathcal{N}^*_\xi}$. Let $E_{\mathcal{M}}$ be the extender that is $\tau_{\mathcal{M}}$-certified over $(\mathcal{M}, \Psi_{\mathcal{M}})$, that is:

$$(a, A) \in E_{\mathcal{M}} \iff \pi^{\Psi_{\mathcal{M}}}_{\mathcal{M},\infty}(a) \in \tau_{\mathcal{M}}(A).$$

It is easy to see, using Lemma 12.2.5 that $\pi_{\mathcal{N},\mathcal{M}}[E_{\mathcal{N}}] \subseteq E_{\mathcal{M}}$ and $\Psi_{\mathcal{M}}$ extends the tail strategy induced by Ψ and $\pi_{\mathcal{N},\mathcal{M}}$.

We can find $\mathcal{M}$ such that $\pi_Y(f)(\pi(c)) \in \mathrm{rng}(\tau_{\mathcal{M}})$. Let $\mathcal{M}^* = \mathrm{Ult}(\mathcal{M}, E_{\mathcal{M}})$ and we note that $\Psi_{\mathcal{M}^*||\mathrm{lh}(E_{\mathcal{M}})} = \Psi_{\mathcal{M}||\mathrm{lh}(E_{\mathcal{M}})}$; call this strategy Λ. Note that $\pi(c) = \pi^{\Lambda}_{\mathcal{M}||\mathrm{lh}(E_{\mathcal{M}}),\infty} \circ \pi^{\Psi}_{\mathcal{N}||\mathrm{lh}(E),\mathcal{M}}(c)$. We have then that

$$L(\Omega, \mathcal{P}) \models \text{“}\pi_Y(f)(\pi(c)) \in \mathrm{rng}(\pi^{\Lambda}_{\mathcal{M}||\mathrm{lh}(E_{\mathcal{M}}),\infty})\text{”}.$$

By Lemma 12.2.5,

$$L(\Omega, \mathcal{P}) \models \text{“}\pi_Y(f)(c) \in \mathrm{rng}(\pi^{\Psi}_{\mathcal{N}||\mathrm{lh}(E),\infty})\text{”}.$$

This is what we want.

Coherence. We now show:

1. $\mathrm{Ult}_0(\mathcal{N}, E)|\gamma = \mathcal{N}$.
2. Let $\nu = \max\{(\delta_Y^+)^{\mathcal{N}}, \gamma^Y_{\xi^*}\}$.[40] Then ν is a cutpoint of $\mathrm{Ult}_0(\mathcal{N}, E)$ and $\gamma = ((\nu)^+)^{\mathrm{Ult}_0(\mathcal{N},E)}$.

For 1), let $\tilde{\tau} : \mathrm{Ult}_0(\mathcal{N}, E) \to \mathcal{N}_{\xi^*}$ be the natural map:

$$\tilde{\tau}(i_E(f)(a)) = \pi_Y(f)(\pi^{\Psi}_{\mathcal{N},\infty}(a))$$

for all $f \in \mathcal{P}_X$ and $a \in \gamma^{<\omega}$. It is clear from the fact that Ψ is X-realizable and $\mathrm{rge}(\pi_Y)$ is an honest extension of X that $\tilde{\tau} \restriction \gamma = \pi^{\Psi}_{\mathcal{N},\infty} \restriction \gamma$. This implies $\mathrm{Ult}_0(\mathcal{N}, E)|\gamma$ is isomorphic to $\pi_Y[\mathcal{N}]$ and hence isomorphic to $\mathcal{N}$.

For 2), suppose not. Using the fact that $\mathcal{N}$ is extender-ready, we first observe that,

$$\mathcal{N} \models \forall \nu \leq \alpha < \gamma\, (|\alpha| \leq \nu). \tag{17}$$

Let F be on the sequence of $\mathrm{Ult}_0(\mathcal{N}, E)$ such that

(i) $\mathrm{crit}(F) = \delta_Y$.
(ii) $\mathrm{lh}(F) \geq \nu$.

[40] See Definition 12.7.12.

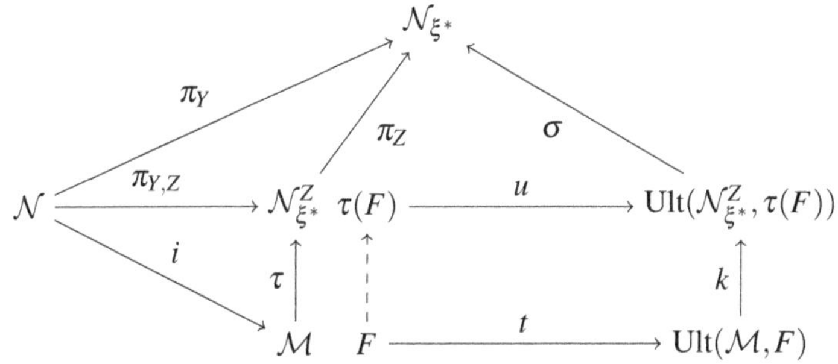

FIGURE 11. Coherence.

(iii) $\mathrm{lh}(F)$ is the least such that (i) and (ii) hold.

We have then that $\mathrm{lh}(F) \geq \gamma$ by the definition of ν and the fact that $\mathrm{Ult}_0(\mathcal{N}, E)|\gamma = \mathcal{N}$.[41]

Let $\tilde{\tau}$ and Z be defined as above. Recall $Y \prec Z \in \mathcal{C}_\xi$. Let $\mathcal{M} = \mathrm{Ult}_0(\mathcal{N}, E)$, i be the corresponding ultrapower map. Let $\tau : \mathcal{M} \to \mathcal{N}^Z_{\xi^*}$ be the natural map so that $\tilde{\tau} = \pi_Z \circ \tau$. Let $t : \mathcal{M} \to \mathrm{Ult}(\mathcal{M}, F)$ be the ultrapower map by F and $u : \mathcal{N}^Z_\xi \to \mathrm{Ult}(\mathcal{N}^Z_\xi, \tau(F))$ be the ultrapower map by $\tau(F)$. Let $k : \mathrm{Ult}(\mathcal{M}, F) \to \mathrm{Ult}(\mathcal{N}^Z_\xi, \tau(F))$ be the natural map and $\sigma : \mathrm{Ult}(\mathcal{N}^Z_\xi, \tau(F)) \to \mathcal{N}_\xi$ be the realization map. The existence of σ comes from the fact that $\tau(F)$ is π_Z-certified over $(\mathcal{N}^Z_{\xi^*} || \mathrm{lh}(\tau(F)), (\Sigma^Z_{\xi^*})_{\mathcal{N}^Z_{\xi^*} || \mathrm{lh}(\tau(F))})$.

CLAIM 12.7.15. $\mathrm{lh}(F) = \gamma$.

PROOF. Note that ν is a cutpoint in $\mathrm{Ult}(\mathcal{M}, F)$ and is the least such $> \delta_Y$. So by (17),

$$\mathrm{lh}(F) = (\nu^+)^{\mathrm{Ult}(\mathcal{M}, F)}.$$

Suppose $\mathrm{lh}(F) > \gamma$. Let $\mathcal{Q} \lhd \mathcal{M} || \mathrm{lh}(F)$ be least such that

$$\mathcal{N} \lhd \mathcal{Q} \wedge \mathcal{Q} \vDash |\gamma| = \nu.$$

Note that $\mathcal{Q}$ is a level of $Lp^{\Psi,\Omega}(\mathcal{N})$. This is by SMC and the fact that

$$\pi_Z \circ \tau \restriction \mathcal{N} = \tilde{\tau} \restriction \mathcal{N} = \pi^\Psi_{\mathcal{N},\infty}.$$

This contradicts the assumption that $(\mathcal{N}, \Psi)$ is extender-ready. ⊣

Now we show F is π_Y-certified over $(\mathcal{N}, \Psi)$. This would give $E = F \in \mathrm{Ult}(\mathcal{N}, E)$. Contradiction.

Let $\Lambda = \Sigma^Z_{\xi^*}$. First note that

---

[41] If ν is not a cutpoint of $\mathrm{Ult}_0(\mathcal{N}, E)$, then there is some extender H on the sequence of $\mathrm{Ult}_0(\mathcal{N}, E)$ such that $\mathrm{cr}(H) \leq \nu < \mathrm{lh}(H)$. This easily implies that there is some extender F on the sequence of $\mathrm{Ult}_0(\mathcal{N}, E)$ such that $\mathrm{crit}(F) = \delta_Y$ and $\mathrm{lh}(E) \geq \nu$.

(a) $\pi^{\Psi}_{\mathcal{N},\infty} = \pi^{\Lambda_{\tau(\mathcal{N})}}_{\tau(\mathcal{N}),\infty} \circ \pi^{\Psi}_{\mathcal{N},\mathcal{N}^Z_{\xi^*}}$.[42]

(b) $\tau \restriction \mathcal{N} = \pi^{\Psi}_{\mathcal{N},\mathcal{N}^Z_{\xi^*}}$.

(c) $\sigma \restriction \tau(\mathcal{N}) = \pi^{\Lambda_{\tau(\mathcal{N})}}_{\tau(\mathcal{N}),\infty}$.

Let $c \in [\mathrm{ord}(\mathcal{N})]^{<\omega}$, $A \in \mathcal{P}_Y$, we have:

$$\begin{aligned} A \in F_c &\iff c \in t(A) \\ &\iff c \in t(i(A)) \\ &\iff k(c) \in k(t(i(A))) \\ &\iff \tau(c) \in u \circ \tau(i(A)) \\ &\iff \tau(c) \in u(\pi_{Y,Z}(A)) \\ &\iff \sigma(\tau(c)) \in \pi_Y(A) \\ &\iff \pi^{\Psi}_{\mathcal{N},\infty}(c) \in \pi_Y(A). \end{aligned}$$

The second equivalence holds because $i(A) \cap \delta_Y = A$. The third equivalence uses Corollary 9.1.15, noting that $\mathrm{rng}(\sigma)$ is an honest extension of $\mathrm{rng}(\sigma \circ k)$. The fourth equivalence uses the fact that $\tau(c) = k(c)$ and $k \circ t = u \circ \tau$. The fifth equivalence is true because $\pi_{Y,Z}(A) = \tau(i(A))$. The sixth equivalence is true because $\pi_Y(A) = \sigma(u(\pi_{Y,Z}(A)))$. The last equivalence follows from equations (a)–(c). This finishes the proof of the lemma. $\dashv$

Lemma 12.7.14 implies that if $\mathcal{N}_\xi$ is extender-ready then $\mathcal{M}_{\xi^*+1} = (\mathcal{N}_{\xi^*}, E)$ where using the notation of Lemma 12.7.14

$$(a,A) \in E \iff \forall^* Y \in \mathcal{C}_\xi((a,A) \in Y \to \pi_Y^{-1}(a,A) \in E_Y).$$

It also follows from Lemma 12.7.14 that $(1)_\xi(a)$ holds if $(1)_{\xi^*} - (3)_{\xi^*}$ hold. We continue by proving another condensation lemma for relevant extenders with critical point $\delta^{\mathcal{P}}$. This condensation property does not seem to follow from Theorem 11.1.5.

LEMMA 12.7.16. *(a) Suppose $\mathcal{M}_\xi$ is of the form $(\mathcal{M}_\xi^-, F_\xi)$, where* $\mathrm{crit}(F_\xi) = \delta^{\mathcal{P}}$. *Suppose $\pi : \mathcal{M} = (\mathcal{M}^-, \tilde{F}) \to \mathcal{M}_\xi$ is Σ_0 and cofinal, or Σ_2, with* $\mathrm{crit}(\pi) > \mathrm{ord}(\mathcal{P})$ *and suppose further that $\mathcal{M}^- \trianglelefteq \mathcal{M}_\xi^-$. Furthermore, let Y be a good hull that contains all relevant objects, let $\pi_Y : M_Y[G] \to H_{\kappa^{+4}}[G]$ be the uncollapse map, and let $\mathcal{M}^Y = \pi_Y^{-1}(\mathcal{M})$. Let Ψ be the π_Y-pullback strategy for $\mathcal{M}^Y$ and suppose that $\Psi_{\mathcal{M}^{Y,-}} = (\Sigma^Y_\xi)_{\mathcal{M}^{Y,-}} = (\Sigma^Y_\xi)^{\pi^Y}_{\mathcal{M}^{Y,-}}$. Then $\tilde{F}$ is on the sequence of $\mathcal{M}_\xi$ and $\Psi = (\Sigma^Y_\xi)_{\mathcal{M}^Y} = (\Sigma^Y_\xi)^{\pi^Y}_{\mathcal{M}^Y}$, where Σ^Y_ξ is the strategy for $\mathcal{N}^Y_\xi$ defined above.*

[42] $\mathcal{N}^Z_{\xi^*}$ is not literally a Ψ-iterate of $\mathcal{N}$, but $\mathcal{N}$ iterates into a hod initial segment of $\mathcal{N}^Z_{\xi^*}$. By $\pi^{\Psi}_{\mathcal{N},\mathcal{N}^Z_{\xi^*}}$, we mean $(\pi^{\Psi}_{\mathcal{N},\infty})^{M_Z}$.

(b) More generally, suppose $\mathcal{M}_\xi$ is as above and $\pi : \mathcal{M} = (\mathcal{M}^-, \tilde{F}) \to \mathcal{M}_\xi$ is Σ_0 and cofinal, or Σ_2, with $\mathrm{crit}(\pi) > \mathrm{ord}(\mathcal{P})$. *Suppose Y, π_Y, Ψ are as above and $\tilde{F}^Y = \pi_Y^{-1}(\tilde{F})$, then $\tilde{F}^Y$ is π_Y-certified over $(\mathcal{M}^{Y,-}, \Psi_{\mathcal{M}^{Y,-}})$.*

PROOF. The preservation of π guarantees that $\mathcal{M}$ is a hod premouse. Recall that $\mathrm{ord}(\mathcal{P})$ is the cardinal successor of $\delta^{\mathcal{P}}$ in both $\mathcal{M}_\xi$ and $\mathcal{M}$ and the models agree up to $\mathcal{P}$.

We first prove (a). Let Y be as in the hypothesis. Let $\tilde{F}^Y = \pi_Y^{-1}(\tilde{F})$,

$$(\mathcal{P}^Y, \delta^Y, F_\xi^Y) = \pi_Y^{-1}((\mathcal{P}, \delta^{\mathcal{P}}, F_\xi)),$$

and $\pi^Y = \pi_Y^{-1}(\pi)$. We work with $\mathcal{M}^Y$ and $\mathcal{N}_\xi^Y$ and first show that $\tilde{F}^Y$ is on the sequence of $\mathcal{N}_\xi^Y$. Let $\Lambda = \Psi_{\mathcal{M}^{Y,-}} = (\Sigma_\xi^Y)_{\mathcal{M}^{Y,-}}$.

CLAIM 12.7.17. *For $a \in [\mathrm{ord}(\mathcal{M}^Y)]^{<\omega}$ and $A \subset [\delta^Y]^{|a|}$ in $\mathcal{P}^Y$, $(a, A) \in \tilde{F}^Y$ if and only if $\pi^\Lambda_{\mathcal{M}^{Y,-},\infty}(a) \in \pi_Y(A)$.*

PROOF. First, note that $\tilde{F}^Y$ is total over $\mathcal{N}_\xi^Y$ and hence it makes sense to apply $\tilde{F}^Y$ to $\mathcal{N}_\xi^Y$. Also, $\mathrm{Ult}(\mathcal{N}_\xi^Y, \tilde{F}^Y)$ embeds into $\mathrm{Ult}(\mathcal{N}_\xi^Y, F_\xi^Y)$ via the natural map τ:

$$\tau(i_{\tilde{F}^Y}(f)(b)) = i_{F_\xi^Y}(f)(\pi^Y(b)).$$

Note that

$$\tau \restriction \mathcal{M}^Y || \mathrm{lh}(\tilde{F}^Y) = \pi^Y \restriction \mathcal{M}^Y || \mathrm{lh}(\tilde{F}^Y).$$

Now,

$$\begin{aligned}(a, A) \in \tilde{F}^Y &\iff (\tau(a) = \pi^Y(a), A) \in F_\xi^Y \\ &\iff \pi^\Lambda_{\mathcal{M}^{Y,-},\infty}(\pi^Y(a)) \in \pi_Y(A) \\ &\iff \pi^\Lambda_{\mathcal{M}^{Y,-},\infty}(a) \in \pi_Y(A).\end{aligned}$$

The first equivalence holds because $\pi^Y(A) = \tau(A) = A$. The second equivalence holds by the definition of F_ξ^Y and our assumption on Λ. The last equivalence follows from Lemma 12.2.5. This finishes the proof of the claim. ⊣

The claim and Lemma 12.7.14 imply that $\tilde{F}^Y$ is on the $\mathcal{N}_\xi^Y$-sequence. By elementarity, $\tilde{F}$ is on the $\mathcal{N}_\xi$-sequence.

$\Psi = (\Sigma_\xi^Y)_{\mathcal{M}^Y} = (\Sigma_\xi^Y)^{\pi^Y}_{\mathcal{M}^Y}$ follows from Lemma 12.7.6 and the proof of Lemma 12.7.7 (the main points are $\pi_Y \circ \pi^Y \restriction \mathcal{P}^Y = \pi_Y \restriction \mathcal{P}^Y$, and the fact that if the strategies disagree then we can find a lower-level disagreement just as in the proof of Lemma 12.7.7). This proves (a).

The proof of (b) is very similar, note that we have the following equivalences

$$\begin{aligned}(a,A)\in\tilde{F}^Y &\iff (\pi^Y(a),A)\in F^Y_\xi\\ &\iff \pi^{\Sigma^Y_\xi}_{\mathcal{M}^{Y,-}_\xi,\infty}(\pi^Y(a))\in\pi_Y(A)\\ &\iff \pi^{\Psi}_{\mathcal{M}^{Y,-},\infty}(a)\in\pi_Y(A).\end{aligned}$$

The last equivalence easily follows from Lemma 12.2.5 and shows that $\tilde{F}^Y$ is π_Y-certified over $(\mathcal{M}^{Y,-},\Psi_{\mathcal{M}^{Y,-}})$. ⊣

COROLLARY 12.7.18. *1. Let $\mathcal{N}=\mathcal{M}_\xi$ be the ξ-th model in the hybrid K^c-construction. Suppose $(1)_\xi$ hold. Then $\rho(\mathcal{N})\geq\mathsf{ord}(\mathcal{N}^b)=\mathsf{ord}(\mathcal{P})$ and $\mathcal{N}$ is $k(\mathcal{N})+1$-solid and $k(\mathcal{N})+1$-universal.*

2. Suppose $\mathcal{M}_\xi$ is the ξ-th model in the X-validated sts construction over a weakly suitable $\mathcal{R}$ that is the result of a hybrid K^c-construction. Suppose $\rho(\mathcal{M}_\xi)<\delta$, where δ is the lsa Woodin cardinal of $\mathcal{R}$. Let $\gamma=max(\rho(\mathcal{M}_\xi),\delta^{\mathcal{P}})$ and let $\mathcal{N}$ be the transitive collapse of $Hull^{\mathcal{M}_\xi}(\gamma\cup\{p(\mathcal{M}_\xi)\})$.[43] If there is an X-validated iteration strategy Ψ of $\mathcal{N}$ (i.e., if $(1)_\xi(b)$ holds), then $\rho(\mathcal{N})>\delta^{\mathcal{P}}$.

PROOF. We prove (1) first. We prove $\rho(\mathcal{N})\geq\mathsf{ord}(\mathcal{N}^b)$ and $\mathcal{N}$ is n-solid and n-universal, where $n=k(\mathcal{N})+1$. Without loss of generality we assume $n=1$. The case $n>1$ is similar (one just has to work with the $n-1$-reduct).

CLAIM 12.7.19. $\rho_1(\mathcal{N})\geq\mathsf{ord}(\mathcal{N}^b)$.

PROOF. Suppose not. Let $Y\prec H^V_{\kappa+4}$ be X-good such that $\mathcal{N}\in Y$. Let $\delta_Y=\pi_Y^{-1}(\delta^{\mathcal{P}})$, $\mathcal{N}_Y=\pi_Y^{-1}(\mathcal{N})$, and $\Psi=\Sigma^Y_\xi$. Let $\mathcal{Q}=\mathrm{Ult}_0(\mathcal{N}_Y,\nu)$ where ν is the order 0 total measure with critical point δ_Y and $i_\nu:\mathcal{N}_Y\to\mathcal{Q}$ be the canonical embedding. Let $q=i_\nu(p)$ where $p=p_1(\mathcal{N}_Y)$. Hence

(i) $\mathcal{N}^b_Y$ is a cutpoint initial segment of $\mathcal{Q}$ and $\mathsf{ord}(\mathcal{N}^b_Y)$ is the cardinal successor of δ_Y in $\mathcal{Q}$.
(ii) We can regard $\mathcal{Q}$ as a hod premouse over $(\mathcal{N}^b,\Psi_{\mathcal{N}^b_Y})$ with strategy $\Sigma_{\mathcal{Q}}\in\Omega$ that is commuting and is Ω-fullness preserving.[44]
(iii) There is some $A\subseteq\delta_Y$ such that A is Σ_1-definable over $\mathcal{Q}$ from q and $A\notin\mathcal{N}^b_Y$.

We say that a triple $(\mathcal{Q},\Sigma_{\mathcal{Q}},q)$ satisfying (i)-(iii) is *minimal* if there is no iteration $\vec{\mathcal{T}}$ according to $\Sigma_{\mathcal{Q}}$ with iteration map $i:\mathcal{Q}\to\mathcal{R}$ and some $r<i(q)$ (in the reverse lexicographic order) such that $(\mathcal{R},\Sigma_{\mathcal{R},\vec{\mathcal{T}}},i(\mathcal{N}^b),r)$ satisfies (i)-(iii).

Fix two minimal triples $(\mathcal{R},\Sigma_{\mathcal{R}},r)$ and $(\mathcal{S},\Sigma_{\mathcal{S}},s)$. We can then compare them above $\mathcal{N}^b_Y$. Letting $i:\mathcal{R}\to\mathcal{W}$ and $j:\mathcal{S}\to\mathcal{W}$ be iteration maps. Note that

[43] $\mathcal{N}$ is obtained by decoding the Σ_1-hull of the $k(\mathcal{M}_\xi)$-reduct with parameters in $\gamma\cup\{p(\mathcal{M}_\xi)\}$.

[44] We can take $\Sigma_{\mathcal{Q}}$ be the $\mathcal{Q}$-tail of Ψ. By Lemma 12.7.8, $\Sigma_{\mathcal{Q}}$ is Ω-fullness preserving. By Corollary 12.7.10 and results of Section 4.7, $\Sigma_{\mathcal{Q}}$ is positional and commuting.

$i(r) = j(s)$ and so

$$\mathrm{Th}^{\mathcal{R}}_{\Sigma_1}(\delta_Y \cup \{r\}) = \mathrm{Th}^{\mathcal{S}}_{\Sigma_1}(\delta_Y \cup \{s\}).$$

This means $\mathrm{Th}^{\mathcal{R}}_{\Sigma_1}(\delta_Y \cup \{r\})$ is $OD^{\Omega}_{\mathcal{N}^b_Y, \Psi_{\mathcal{N}^b_Y}}$ for any minimal $(\mathcal{R}, \Sigma_{\mathcal{R}}, r)$.

By $\mathsf{MC}(\Psi_{\mathcal{N}^b_Y})$,

$$\mathrm{Th}^{\mathcal{R}}_{\Sigma_1}(\delta_Y \cup \{r\}) \in \mathcal{N}^b_Y.^{45}$$

This contradicts (iii). ⊣

The claim and Theorem 11.1.5 (which is built on the results of Section 4.9) imply that $\mathcal{N}_Y$ and hence $\mathcal{N}$ is 1-solid and 1-universal. The point is that relevant phalanx comparisons of the form $(H, \mathcal{N}_Y, \alpha)$ where H is the transitive collapse of $Hull_1^{\mathcal{N}_Y}(\alpha \cup \{p - (\alpha+1)\})$ for $\alpha \in p$ are such that $\alpha > \mathsf{ord}((\mathcal{N}_Y)^b)$ (by the claim), no strategy disagreements can occur (see Lemma 11.1.4) and do not use extenders with critical point δ_Y (by Lemma 12.7.16). The last item holds because Lemma 12.7.16 shows that extenders with critical point δ_Y on the H-sequence are certified and the proof of Lemma 12.7.14 shows that extenders with critical point δ_Y and its images on the sequence of iterates of $H, \mathcal{N}_Y$ are certified.[46] These comparisons terminate successfully by the usual arguments. By similar arguments, we get the conclusion for all $n \in \omega$.

For (2), letting Y be X-good and using the notations as above, we describe the X-realizable iteration strategy Ψ_Y for $\mathcal{N}_Y$ witnessing $\mathcal{N}_Y$ is an X-approved sts mouse; then the X-validated strategy Ψ for $\mathcal{N}$ is defined from the Ψ_Y's as before. First, note that by [3], for such a Y, $\mathcal{M}^Y_\xi$ has a τ-realizable strategy above γ for some map $\tau : \mathcal{M}^Y_\xi \to \mathcal{M}_\xi$.[47] The usual proof of solidity/universality then shows that $\mathcal{N}$ is γ-sound and if $\gamma = \rho(\mathcal{M}_\xi)$, then $\mathcal{N}$ is sound. In the case $\mathcal{N}$ is sound, then it is just $\mathcal{N}_\xi$.

Let $(\mathcal{S}, \delta')$ be the image of $(\mathcal{R}, \delta)$ under the collapse map π^{-1} and let $(\mathcal{S}_Y, \delta'_Y)$ be the image of $(\mathcal{S}, \delta')$ under π_Y^{-1}. Note that $\rho(\mathcal{N}) < \delta'$. Now we outline the description of Ψ_Y. Ψ_Y on stacks based on $\mathcal{S}^b_Y$ has been defined in great detail before (using the fact that X is a condensing set),[48] so we focus on stacks $\vec{\mathcal{T}}$ above $\mathcal{S}^b_Y$. Suppose $\mathcal{T}$ is above $\mathcal{S}^b_Y$, based on $\mathcal{S}_Y$, and is correctly guided. Then by [3], there is a maximal branch b and a realizing map $\sigma : \mathcal{M}^{\mathcal{T}}_b \to \mathcal{M}_\xi$ such that $\sigma \circ i^{\mathcal{T}}_b = \pi \circ \pi_Y \restriction \mathcal{N}_Y$. But note that there is a $\mathcal{Q} \trianglelefteq \mathcal{M}^{\mathcal{T}}_b$ such that $\mathcal{Q} = \mathcal{Q}(\mathcal{T}, b)$. This comes from the fact that $\rho(\mathcal{N}) < \delta'$ and that $\mathcal{N}$ is γ-sound; so since $\gamma < \delta'$, $\mathcal{J}_1[\mathcal{N}] \vDash$ "there are no Woodin cardinals $> \gamma$", and hence $\mathcal{Q}$ exists. Therefore, the branch b is the canonical $\mathcal{Q}$-structure guided branch for $\mathcal{T}$. The case where $\mathcal{T}$ is

---

[45]Note that we take Y so that $\mathcal{N}^b_Y = \mathrm{Lp}^{\Psi_{\mathcal{N}_Y|\delta_Y}, \Omega}(\mathcal{N}_Y|\delta_Y)$.

[46]See the proof of Claim 12.8.4 for a similar argument.

[47]τ is a minimal map relative to some enumeration $\vec{e}$ of $\mathcal{M}^Y_\xi$ in order type ω.

[48]Extenders on these stacks have critical point $\leq \delta^{\mathcal{S}^b_Y}$ and their images. Note also that $\mathcal{R}^b = \mathcal{S}^b$.

above $\mathcal{S}$ is similar. We can then easily define Ψ_Y on arbitrary stacks on $\mathcal{N}_Y$. An argument similar to the proof of Claim 12.7.19 then shows that $\rho(\mathcal{N}) > \delta^{\mathcal{P}}$. $\dashv$

Corollary 12.7.18 verifies $(2)_\xi, (3)_\xi$ hold, given that $(1)_\xi$ holds.

Now suppose for some ξ, $\mathcal{M}_\xi$ and $\mathcal{N}$ are as in Corollary 12.7.18(2). By Corollary 12.7.18(2), $\mathcal{N} = \mathcal{N}_\xi$. Then as in 12.7.18, $\forall^* Y$, Y is X-good, $\mathcal{N}_\xi^Y$ is iterable via the X-realizable strategy. This induces an X-validated strategy Λ for $\mathcal{N}_\xi$[49]; so $\mathcal{S} = \mathcal{N}_\xi$ is K^c-appropriate. We can then start a hybrid K^c-construction over $\mathcal{S}$, producing models $(\mathcal{M}'_\xi, \mathcal{N}'_\xi : \xi \leq \Upsilon')$, maintaining $(1)-(3)$ along the way. At some $\xi \leq \Upsilon'$, if $(1)_\xi(b)$ fails then this implies that there is a nuvs p witnessing such a failure; so $(\vec{\mathcal{V}} =_{\text{def}} \{\mathcal{M}'_\alpha, \mathcal{N}'_\alpha : \alpha < \xi\} \cup \{\mathcal{M}'_\xi\}, p)$ witnesses that $\mathrm{m}^+(p)$ is honest weakly X-suitable. Corollary 12.4.7 then shows that $\mathrm{m}^+(p)$ is suitable. At this point, we will continue with the X-validated sts construction over $\mathrm{m}^+(p)$ until it stops prematurely or it produces $\mathcal{M}_{\kappa^{+++}}$.

Suppose $\mathcal{M}_\xi$ is the ξ-th model in an X-validated sts construction over some $\mathcal{R}$ and does not define a failure of Woodinness of δ, the lsa Woodin of $\mathcal{R}$, then we continue with the X-validated sts construction over $\mathcal{R} = \mathrm{m}^+(\mathcal{M}_\xi|\delta)$. There are several cases. The first case occurs when we reach an X-validated model $\mathcal{M}_\varepsilon$ for some $\varepsilon > \xi$ such that ε is the least such that $\rho(\mathcal{M}_\varepsilon) < \delta$ and $\rho(\mathcal{M}_\varepsilon) \geq \mathsf{ord}(\mathcal{P})$. So letting $\mathcal{N}_\varepsilon = \mathcal{C}(\mathcal{M}_\varepsilon)$, then as in Section 12.4 and Corollary 12.7.18, we can show $\mathcal{N}_\varepsilon$ is X-validated and if it has an X-validated iteration strategy, then $\mathcal{N}_\varepsilon$ is sound and by the assumption on $\mathcal{M}_\varepsilon$, we have that $\mathcal{J}_1[\mathcal{N}_\varepsilon] \vDash$ "there are no Woodin cardinals $> \delta^{\mathcal{P}}$". In this case, letting $\mathcal{S} = \mathcal{N}_\varepsilon$, then $\mathcal{S}$ is K^c-appropriate as before and we continue with our hybrid K^c-construction over $\mathcal{S}$ until we reach a level $\mathcal{M}'_\xi$ with no X-validated iteration strategy, so letting p witness this, we then continue with the sts X-validated construction over $\mathrm{m}^+(p)$, which is suitable, as above. If $\mathcal{M}_\varepsilon$ does not have an X-validated strategy, then letting p witness this, we continue with the X-validated sts construction above $\mathrm{m}^+(p)$ as before. The cases where the construction reaches $\mathcal{M}_{\kappa^{+++}}$ or stops prematurely, are handled in the next section.

12.8. K^c stops prematurely and a model of LSA

Suppose the construction lasts κ^{+++} steps; as in the previous subsection, let $\mathcal{N} = \mathcal{N}_{\kappa^{+++}}$ and $\mathcal{S} = \mathcal{S}(\mathcal{N})$.

LEMMA 12.8.1. $\mathrm{cof}(\mathcal{S}) < \kappa^{+++}$.

PROOF. Let $\lambda = \kappa^{+++}$. Note that $\mathcal{S} \in V$. If $\mathcal{N}$ is lsa-small,[50] then as shown in Chapter 11, $\mathcal{S} \vDash \square_{\lambda,2}$. Suppose $\mathcal{N}$ is the result of the X-validated sts construction

[49] We assumed this strategy exists.

[50] In this case, $\mathcal{N}$ is the result of a hybrid K^c-construction over some K^c-appropriate $\mathcal{R}$ or the result of alternating the hybrid K^c-constructions and the X-validated sts construction in a manner described in the previous section.

over some suitable $\mathcal{R}$, then $\mathcal{N}$ is not lsa-small but it is an sts mouse over $\mathcal{R}$ and so is $\mathcal{S}$. In this case, the standard proof ([38]) shows $\mathcal{S} \models \square_{\lambda,2}$.[51] Working in V, $\neg\square(3,\kappa^{+4})$ implies then that $o(\mathcal{S}) < \kappa^{+4}$ and $\neg\square(3,\lambda)$ now implies that $\mathrm{cof}(\mathrm{ord}(\mathcal{S})) < \lambda$ since otherwise, the canonical $\square_{\lambda,2}$-sequence $\vec{C}$ of $\mathcal{S}$ (as defined in Chapter 11) has a thread D. The thread D will produce a sound hod mouse (or sts mouse) $\mathcal{M}$ such that $\mathrm{ord}(\mathcal{M}) \geq \mathrm{ord}(\mathcal{S})$ and $\rho_\omega(\mathcal{M}) \leq \lambda$. This contradicts (ii) of Lemma 12.6.2. $\dashv$

Lemma 12.8.1 contradicts (iii) of Lemma 12.6.2. Now we assume the construction stops prematurely. Without loss of generality, we assume the X-validated sts construction over some $\mathcal{Q}$, where $\mathcal{Q}$ is produced by a hybrid K^c-construction,[52] reaches a model $\mathcal{N}_\Upsilon$ which is a sts hod premouse that satisfies:

(i) There is a unique Woodin cardinal $\delta_0 > \delta^{\mathcal{P}}$ such that $\delta^{\mathcal{P}}$ is the least $< \delta_0$-strong.
(ii) There are $\omega.2$ many Woodin cardinals above δ_0, say these Woodin cardinals are $(\delta_n : 1 \leq n < \omega.2)$.
(iii) There is an extender F with $\mathrm{crt}(F) > \sup_n \delta_n$ such that $\mathcal{N}_\Upsilon = (\mathcal{N}_\Upsilon^-, F)$ for some extender F. In fact, $\mathcal{N}_\Upsilon = (\mathcal{N}_\Upsilon^-)^\sharp$.
(iv) $\mathcal{N}_\Upsilon$ is a sts hod premouse over $\mathcal{Q} =_{\mathrm{def}} (\mathcal{N}_\Upsilon|\delta_0)^\sharp$, $\mathcal{Q}$ is of lsa type with $\delta^{\mathcal{P}}$ is $< \delta_0$-strong in $\mathcal{Q}$.
(v) $\rho_\omega(\mathcal{N}_\Upsilon) \geq \mathrm{ord}(\mathcal{Q})$.

Let $\lambda = \sup_n \delta_n$ and for every $\mathcal{Q} \lhd \mathcal{M} \unlhd \mathcal{N}_\Upsilon$, let $\Sigma^{\mathcal{M}}$ be the internal sts strategy of $\mathcal{Q}$ as defined in $\mathcal{M}$.

LEMMA 12.8.2. *Suppose the construction stops prematurely. Then $\Upsilon < \kappa^{+++}$.*

PROOF. If the construction stops prematurely, then $\mathcal{N}_\Upsilon$ is E-active. This clearly implies that $\Upsilon < \kappa^{+++}$ because if $\Upsilon = \kappa^{+++}$, then $\mathcal{N}_\Upsilon$ is the lim inf of $\mathcal{N}_\alpha$ for $\alpha < \Upsilon$ and hence is passive. $\dashv$

Now suppose $\mathcal{N}_\Upsilon$ satisfies Definition 8.2.2, then the results of Section 8.2 show that the derived model of $\mathcal{N}_\Upsilon$ (at the supremum of its Woodin cardinals) satisfies LSA. Suppose this is not the case. We would like to produce an active $\omega.2$ Woodin lsa mouse as in Definition 8.2.2 from $\mathcal{N}_\Upsilon$.

LEMMA 12.8.3. *Then there is a countable substructure of some $\mathcal{M}^* \unlhd \mathcal{N}_\Upsilon$ satisfying Definition 8.2.2.*

PROOF. Recall that we have $\rho_\omega(\mathcal{N}_\Upsilon) \geq \mathrm{ord}(\mathcal{Q})$. Let $W \prec \mathcal{N}_\Upsilon$ be such that $\mathcal{P} \cup \{\mathcal{P}\} \subset W$ and $W \cap \delta_0 \in \delta_0$.[53] Let $\pi^* : \mathcal{M}^* \to W$ be the uncollapse map.

[51] In fact $\mathcal{S} \models \square_\lambda$. This is because $\mathrm{ord}(\mathcal{R})$ is a strong cut point of $\mathcal{S}$ and all relevant comparisons are above $\mathcal{R}$ and in fact extender comparisons.

[52] The other case where $\mathcal{Q} = \mathrm{m}^+(p)$ for some p is similar.

[53] Such a W can be found easily because $\rho_\omega(\mathcal{N}_\Upsilon) \geq \delta_0$ and δ_0 is definably inaccessible over $\mathcal{N}_\Upsilon$.

Let $\mathcal{M}$ be the transitive collapse of $Hull^{\mathcal{M}^*}(\mathcal{P} \cup p(\mathcal{M}^*))$ and $\pi' : \mathcal{M} \to \mathcal{M}^*$ be the uncollapse map. Let $\pi = \pi^* \circ \pi'$. First, note that we have the following:

$$\rho_\omega(\mathcal{N}_\Upsilon) \geq \mathrm{ord}(\mathcal{P}) \quad \text{and} \quad \rho_\omega(\mathcal{M}) \leq \mathrm{ord}(\mathcal{P}) \text{ (so in fact, } \rho_\omega(\mathcal{M}) = \mathrm{ord}(\mathcal{P})).$$

Now we claim that

CLAIM 12.8.4. *(i)* $\mathcal{M}^* \lhd \mathcal{N}_\Upsilon$.

(ii) If Y is X-good such that $\{\mathcal{M}, \mathcal{M}^*, \mathcal{N}_\Upsilon\} \in Y$, letting π_Y be the uncollapse map and $x^Y = \pi^{-1}(x)$ for x in range π_Y or $x = \mathcal{M}$, then $\mathcal{M}^Y$ is iterable via the X-realizable strategy.

(iii) Suppose Y is as in (ii) and $\tau : \mathcal{N} \to \mathcal{M}^Y$ is either Σ_0 cofinal or Σ_2 elementary and $\mathrm{cr}(\tau) > o(\mathcal{P}^Y)$, then the comparison $(\mathcal{M}^Y, \mathcal{N}, \mathrm{crit}(\tau))$ against $\mathcal{M}^Y$ does not use extenders with critical point $(\delta^{\mathcal{P}^Y})$.

(iv) $\mathcal{M}$ is ω-sound.

PROOF. For (i), first note that letting $\xi = \mathrm{crit}(\pi^*)$ and $\mathcal{Q}^* = (\pi^*)^{-1}(\mathcal{Q})$, then $\xi = \delta_0^{\mathcal{M}^*}$ and $\pi^*(\xi) = \delta_0^{\mathcal{N}_\Upsilon}$. Now, note that $\mathcal{Q}^* \lhd \mathcal{Q}$. Now let $\mathcal{R} \lhd \mathcal{Q}$ be the largest such that $\mathcal{R} \models$ "ξ is Woodin". Let Y be X-good such that $\{\mathcal{M}^*, \mathcal{Q}^*, \mathcal{R}, \mathcal{N}_\Upsilon, \xi\} \in Y$. Let $((\mathcal{M}^*)^Y, \mathcal{R}^Y, (\mathcal{Q}^*)^Y, \xi^Y) = \pi_Y^{-1}(\mathcal{M}^*, \mathcal{R}, \mathcal{Q}^*, \xi)$. Note that ξ is a strong cut point of both $\mathcal{M}^*, \mathcal{R}$. Since $(\mathcal{M}^*)^Y$ is ξ^Y-sound and projects to ξ^Y, by [3], $(\mathcal{M}^*)^Y$ has an iteration strategy Ψ above ξ^Y and Ψ can be taken to be a τ-realizable strategy for some $\tau : (\mathcal{M}^*)^Y \to \mathcal{M}^*$ such that $X \subset \mathrm{rng}(\tau)$.[54] We can then compare $(\mathcal{M}^*)^Y, \mathcal{R}^Y$ above ξ^Y.[55] Since both models are ξ-sound and $\mathcal{R}^Y$ is a $\mathcal{Q}$-structure for ξ^Y, $(\mathcal{M}^*)^Y \trianglelefteq \mathcal{R}^Y$. This implies $\mathcal{M}^* \lhd \mathcal{N}_\Upsilon$.

(ii) follows from the fact that $\mathcal{M}^*$ is an initial segment of a model in a hybrid K^c-construction that produces $\mathcal{Q}$. Furthermore, since there is a $\mathcal{Q}$-structure $\mathcal{R} \lhd \mathcal{Q}$ for ξ extending $\mathcal{M}^*$, the previous section indeed produces a unique X-realizable iteration strategy for $\mathcal{M}^Y$. This strategy is $\pi' \circ \pi_Y$-realizable since it is the $(\pi')^Y$-pullback of the unique X-realizable strategy Λ of $(\mathcal{M}^*)^Y$, and by [3] and the fact that $(\mathcal{M}^*)^Y \trianglelefteq \mathcal{R}^Y$, Λ is π-realizable.

(iii) follows from an argument as in Corollary 12.7.18; the main point is that by Lemmas 12.7.16 and 12.7.14, letting $\mathcal{R}$ be $\mathcal{M}^Y$ or $\mathcal{N}$ or its iterate, extenders on the sequence of $\mathcal{R}$ with critical point $\delta^{\mathcal{R}^b}$ are certified. More precisely, let $\mathcal{S}$ be the tree on the phalanx side and $\mathcal{T}$ be the tree on the $\mathcal{M}^Y$-side that participate in the comparison. Suppose $\mathcal{R} = \mathcal{M}^{\mathcal{S}}_\alpha$ and $\mathcal{R}' = \mathcal{M}^{\mathcal{T}}_\alpha$ and that $\mathcal{S} \restriction \alpha, \mathcal{T} \restriction \alpha$ have not used extenders with critical point $\delta^{\mathcal{P}^Y}$. Suppose ξ is the largest such that $\mathcal{R}||\xi = \mathcal{S}||\xi$

[54] We fix an enumeration $\vec{e}$ of $(\mathcal{M}^*)^Y$, in type ω in $V[g]$. Let $\tau : (\mathcal{M}^*)^Y \to \mathcal{M}^*$ be a $\vec{e}$-minimal embedding τ^* with the property that $\tau^* \restriction (\mathcal{Q}^*)^Y = \pi_Y \restriction (\mathcal{Q}^*)^Y$. Such an embedding can be obtained as the left-most branch of the tree that builds approximations to embeddings from $(\mathcal{M}^*)^Y$ into $\mathcal{M}^*$ that agrees with π_Y on $(\mathcal{Q}^*)^Y$. By results of [3], we get a τ-realizable strategy for $(\mathcal{M}^*)^Y$ for stacks above $(\mathcal{Q}^*)^Y$.

[55] By what has been shown, the comparison does not encounter strategy disagreements. The point is that since Ψ is a τ-realizable strategy and $X \subset \mathrm{rng}(\tau)$, it witnesses that Ψ-iterates of $(\mathcal{M}^*)^Y$ are X-approved. A similar comment applies to the canonical strategy of $\mathcal{R}^Y$.

and that $\mathcal{R}|\xi \neq \mathcal{S}|\xi$. Since no strategy disagreement occurs by Lemma 11.1.4 and the fact that $\mathcal{R},\mathcal{S}$ are X-approved, letting Ψ be the common strategy for $\mathcal{R}||\xi, \mathcal{S}||\xi$, then if ξ indices an extender E with $\text{crit}(E) = \delta^{\mathcal{P}^Y}$ on the $\mathcal{R}$-sequence, then E is π^Y-certified over $(\mathcal{R}||\xi,\Psi) = (\mathcal{S}||\xi,\Psi)$, this implies E is the extender indexed at ξ on the $\mathcal{S}$-sequence. Contradiction.

For (iv), the point is that in the relevant phalanx comparisons in the proof of solidity and universality, no extenders with critical point $(\delta^{\mathcal{P}})^Y$ are used by (iii), no strategy disagreements occur, and hence these comparisons are successful. $\dashv$

Suppose without loss of generality, no countable substructures of any $\mathcal{N} \lhd \mathcal{M}$ satisfies Definition 8.2.2. We claim that for Y as in (ii) of the above claim, $\mathcal{M}_Y$ does. Again, let Y be as above and it suffices to show $\mathcal{M}^Y$ satisfies Definition 8.2.2. Everything is clear except, perhaps, for (1). So let Λ be the $\pi' \circ \pi_Y$-realizable strategy for $\mathcal{M}^Y$ and $\mathcal{Q} = (\mathcal{M}^Y|\delta_0^Y)^\sharp$. By the argument as in Claim 12.8.4 and Lemma 12.7.7, $\Lambda_{\mathcal{Q}}^{stc}$ has (locally) strong branch condensation. Similarly to 12.7.8, $\Lambda_{\mathcal{Q}}^{stc}$ is also (locally) strongly Ω-fullness preserving and hence is (locallly) strongly $\Gamma(\mathcal{Q},\Lambda_{\mathcal{Q}}^{stc})$-fullness preserving. Lastly, the arguments in Section 12.4 (particularly the proof of case 2(b)) show that $\Lambda_{\mathcal{Q}}^{stc} \restriction \mathcal{M}^Y \supseteq \Sigma^{\mathcal{M}_Y}$. $\dashv$

Again, Lemma 12.8.3 and results in Section 8.2 show that the new derived model of $\mathcal{N}$ as in the conclusion of Lemma 12.8.3 (at the sup of its Woodin cardinals) satisfies LSA.

Now by boolean comparisons, there is some $(\mathcal{M},\Sigma) \in V$ satisfying Definition 8.2.2. By taking a countable hull of $\mathcal{M}$ if necessary, we may assume $\mathcal{M}$ is countable (in V). Let $\mathcal{M}^-$ be the class model obtained by iterating the top extender of $\mathcal{M}$ OR many times and $\mathcal{M}_\infty$ be the result of an $\mathbb{R}$-genericity iteration of $\mathcal{M}^-$ via Σ. Then (new) derived model N of $\mathcal{M}_\infty$ satisfies LSA as shown by Section 8.2. By homogeneity of $Col(\omega,<\kappa)$, there is in V a model M containing $\mathbb{R} \cup \text{OR}$ such that $M \models \text{LSA}$.

PROOF OF THEOREM 12.0.21. The arguments above prove the consistency of LSA from the hypothesis of Theorem 12.0.21 plus the simplifying assumption (9). For this argument, since $2^{<\kappa} = \kappa$ and the core model induction is carried out in $V^{Coll(\omega,<\kappa)}$, $\text{ord}(\mathcal{P}) < \kappa^+$ as shown in Lemma 12.1.2. The constructions on top of $\mathcal{P}$ go on for at most κ^{+++} many steps and the hypothesis we need to carry out the arguments in the previous sections is $\forall\alpha \in [\gamma,\kappa^{+4}]\ \neg\square(3,\alpha)$.

To eliminate (9), we carry out the core model induction in $V^{Coll(\omega,\kappa)}$ much like that in [65] to obtain objects like $\Gamma,\mathcal{P}$. We note that in this case, using the argument in 12.1.2, we can show $\text{ord}(\mathcal{P}^-) \leq (2^\kappa)^+$ and $\text{ord}(\mathcal{P}) < (2^\kappa)^{++}$. Now similar to the proof of [12, Theorem 4.1], letting $\xi = 2^\kappa$, $\mathbb{P} = Coll(\omega,\kappa) \star Coll(\xi^+,\xi^+) \star Coll(\xi^{++},\xi^{++}) \star Coll(\xi^{+++},\xi^{+++})$, we carry out the hybrid K^c-constructions and the X-validated sts constructions over $\mathcal{P}$ in $V^{\mathbb{P}}$. By homogeneity, the objects constructed are in V. We use Theorem 9.2.15, which in turns was built on 9.2.14, 9.2.13, and 9.2.11, to obtain condensing sets and adapt the constructions in the previous sections of this chapter to obtain a model of LSA in a straightforward way. Note that in $V^{\mathbb{P}}$, ξ^{+++} is countably closed and $2^{<\xi^{+++}} = \xi^{+++}$; this allows the proof of Lemma 12.6.2 to go through in this case. The constructions above $\mathcal{P}$ in the previous sections go on for at most ξ^{+++} many steps and we need the full hypothesis $\forall\alpha \in [\gamma, (\wp_4(\kappa)^+] \neg\Box(3,\alpha)$ to carry out the arguments.[56] We leave the details to the kind reader.[57] $\dashv$

[56]In $V^{\mathbb{P}}$, $\wp_1^V(\xi)$ is collapsed to ξ^+, $\wp_2^V(\xi)$ is collapsed to ξ^{++}, and $\wp_3^V(\xi)$ is collapsed to ξ^{+++}. It seems very plausible that Lemma 12.6.2 can be proven with less than what we assumed, but we have not checked this thoroughly.

[57]Some definitions are modified in an obvious way. For instance, a good hull will now have size κ and be countably closed; in Definition 12.3.15, we demand that $|Z| \geq 2^{\kappa^+}$ and $Z^\kappa \subset Z$.

REFERENCES

[1] DOMINIK ADOLF and GRIGOR SARGSYAN, *Derived models of mice below the least fixpoint of the Solovay sequence*, ***The Journal of Symbolic Logic***, vol. 84 (2019), no. 1, pp. 27–53.

[2] DOMINIK ADOLF, GRIGOR SARGSYAN, NAM TRANG, TREVOR WILSON, and MARTIN ZEMAN, *Ideals and strong axioms of determinacy*, available at https://arxiv.org/abs/2111.06220, 2021.

[3] ALESSANDRO ANDRETTA, ITAY NEEMAN, and JOHN STEEL, *The domestic levels of K^c are iterable*, ***Israel Journal of Mathematics***, vol. 125 (2001), pp. 157–201.

[4] ANDRÉS EDUARDO CAICEDO, PAUL LARSON, GRIGOR SARGSYAN, RALF SCHINDLER, JOHN STEEL, and MARTIN ZEMAN, *Square principles in $\mathbb{P}_{\max}$ extensions*, ***Israel Journal of Mathematics***, vol. 217 (2017), pp. 231–261.

[5] SCOTT CRAMER, *Implications of very large cardinals*, ***Essays in honor of W. Hugh Woodin's 60th birthday. Proceedings of the Logic at Harvard Conference held at Harvard University, Cambridge, MA, March 27–29, 2015*** (A. E. Caicedo, J. Cummings, P. Koellner, and P. B. Larson, editors), American Mathematical Society, Providence, RI, 2017, pp. 225–257.

[6] KEITH J DEVLIN, ***Constructibility***, Springer, 1984.

[7] ILIJAS FARAH, *The extender algebra and Σ^2_1 absoluteness*, ***Large Cardinals, Determinacy, and Other Topics. The Cabal Seminar, Volume IV*** (A. S. Kechris, B. Löwe, and J. R. Steel, editors), Lecture Notes in Logic, vol. 43, Cambridge University Press and Association for Symbolic Logic, 2020, pp. 155–192.

[8] GUNTER FUCHS, *λ-structures and s-structures: translating the iteration strategies*, ***Annals of Pure and Applied Logic***, vol. 162 (2011), no. 9, pp. 710–751.

[9] ———, *λ-structures and s-structures: translating the models*, ***Annals of Pure and Applied Logic***, vol. 162 (2011), no. 4, pp. 257–317.

[10] RONALD JENSEN, *The fine structure of the constructible hierarchy*, ***Annals of Mathematical Logic***, vol. 4 (1972), pp. 229–308; erratum, ibid. 4 (1972), 443, with a section by Jack Silver.

[11] ———, *Manuscript on fine structure, inner model theory, and the core model below one Woodin cardinal*, available at https://www.mathematik.hu-berlin.de/~raesch/org/jensen.html, 2020.

[12] RONALD JENSEN, ERNEST SCHIMMERLING, RALF SCHINDLER, and JOHN STEEL, *Stacking mice*, ***The Journal of Symbolic Logic***, vol. 74 (2009), no. 1, pp. 315–335.

[13] ALEXANDER S. KECHRIS, EUGENE M. KLEINBERG, YIANNIS N. MOSCHOVAKIS, and W. HUGH WOODIN, *The axiom of determinacy, strong partition properties and nonsingular measures*, ***Cabal Seminar 77 – 79. Proceedings Caltech-UCLA Logic Seminar, 1977–79)***, Lecture Notes in Mathematics, vol. 839, Springer, Berlin, 1981, pp. 75–99.

[14] ALEXANDER S. KECHRIS, BENEDIKT LÖWE, and JOHN R. STEEL, ***Games, Scales and Suslin Cardinals – The Cabal Seminar, Volume I***, Lecture Notes in Logic, vol. 31, Cambridge University Press and Association for Symbolic Logic, 2008.

[15] ———, ***Wadge Degrees and Projecive Ordinals – The Cabal Seminar, Volume II***, Lecture Notes in Logic, vol. 37, Cambridge University Press and Association for Symbolic Logic, 2012.

[16] ———, ***Ordinal Definability and Recursion Theory – The Cabal Seminar, Volume III***, Lecture Notes in Logic, vol. 43, Cambridge University Press and Association for Symbolic Logic, 2016.

[17] ———, ***Large Cardinals, Determinacy and Other Topics – The Cabal Seminar, Volume IV***, Lecture Notes in Logic, vol. 49, Cambridge University Press and Association for Symbolic Logic, 2021.

[18] PETER KOELLNER and W. HUGH WOODIN, *Large cardinals from determinacy*, ***Handbook of Set Theory. Vols. 1, 2, 3***, Springer, Dordrecht, 2010, pp. 1951–2119.

[19] PAUL B. LARSON, ***The Stationary Tower: Notes on a Course by W. Hugh Woodin***, University Lecture Series, vol. 32, American Mathematical Society, Providence, RI, 2004.

[20] ———, *An introduction to AD^+*, unpublished, available at https://paulblarson.github.io/, 2022.

[21] PAUL B. LARSON and GRIGOR SARGSYAN, *Failures of square in Pmax extensions of Chang models*, available at https://arxiv.org/abs/2105.00322, 2021.

[22] DONALD A. MARTIN and JOHN R. STEEL, *Iteration trees*, ***Journal of the American Mathematical Society***, vol. 7 (1994), no. 1, pp. 1–73.

[23] WILLIAM J. MITCHELL and JOHN R. STEEL, ***Fine Structure and Iteration Trees***, Lecture Notes in Logic, vol. 3, Springer-Verlag, Berlin, 1994.

[24] YIANNIS N. MOSCHOVAKIS, ***Descriptive Set Theory***, Studies in Logic and the Foundations of Mathematics, vol. 100, North-Holland, Amsterdam, 1980.

[25] ITAY NEEMAN, *Optimal proofs of determinacy*, ***The Bulletin of Symbolic Logic***, vol. 1 (1995), no. 3, pp. 327–339.

[26] ———, *Inner models in the region of a Woodin limit of Woodin cardinals*, ***Annals of Pure and Applied Logic***, vol. 116 (2002), no. 1-3, pp. 67–155.

[27] ITAY NEEMAN and JOHN STEEL, *Equiconsistencies at subcompact cardinals*, ***Archive for Mathematical Logic***, vol. 55 (2016), no. 1-2, pp. 207–238.

[28] GRIGOR SARGSYAN, *Descriptive inner model theory*, ***The Bulletin of Symbolic Logic***, vol. 19 (2013), no. 1, pp. 1–55.

[29] ———, ***Hod Mice and the Mouse Set Conjecture***, vol. 236, Memoirs of the American Mathematical Society, no. 1111, American Mathematical Society, 2014.

[30] ———, *Nontame mouse from the failure of square at a singular strong limit cardinal*, ***Journal of Mathematical Logic***, vol. 14 (2014), no. 01, p. 1450003.

[31] ———, *Covering with universally Baire operators*, ***Advances in Mathematics***, vol. 268 (2015), pp. 603–665.

[32] ———, *$AD_{\mathbb{R}}$ implies that all sets of reals are Θ universally Baire*, ***Archive for Mathematical Logic***, vol. 60 (2021), no. 1-2, pp. 1–15.

[33] ———, *Negative results on precipitous ideals on ω_1*, ***The Journal of Symbolic Logic***, vol. 88 (2023), pp. 490–509.

[34] GRIGOR SARGSYAN and JOHN STEEL, *The mouse set conjecture for sets of reals*, ***The Journal of Symbolic Logic***, vol. 80 (2015), no. 2, pp. 671–683.

[35] GRIGOR SARGSYAN and NAM TRANG, *Non-tame mice from tame failures of the unique branch hypothesis*, ***Canadian Journal of Mathematics***, vol. 66 (2014), no. 4, pp. 903–923.

[36] ———, *The exact strength of generic absoluteness for the universally Baire sets*, available at arxiv.org/abs/2110.02725, 2021.

[37] ERNEST SCHIMMERLING, *Coherent sequences and threads*, ***Advances in Mathematics***, vol. 216 (2007), no. 1, pp. 89–117.

[38] ERNEST SCHIMMERLING and MARTIN ZEMAN, *Characterization of $\square_\kappa$ in core models*, ***Journal of Mathematical Logic***, vol. 4 (2004), no. 01, pp. 1–72.

[39] RALF SCHINDLER and JOHN R. STEEL, *The self-iterability of $L[E]$*, ***The Journal of Symbolic Logic***, vol. 74 (2009), no. 3, pp. 751–779.

[40] ———, *The core model induction*, available at math.berkeley.edu/~steel, 2014.

[41] RALF SCHINDLER and MARTIN ZEMAN, *Fine structure*, ***Handbook of Set Theory. Vols. 1, 2, 3***, Springer, Dordrecht, 2010, pp. 605–656.

[42] RALF-DIETER SCHINDLER, JOHN STEEL, and MARTIN ZEMAN, *Deconstructing inner model theory*, ***The Journal of Symbolic Logic***, vol. 67 (2002), no. 2, pp. 721–736.

[43] ———, *Deconstructing inner model theory*, ***The Journal of Symbolic Logic***, vol. 67 (2002), no. 2, pp. 721–736.

[44] FARMER SCHLUTZENBERG, ***Measures in Mice***, ProQuest LLC, Ann Arbor, MI, 2007, Ph.D. thesis, University of California, Berkeley.

[45] ———, *Reconstructing resurrection*, to appear, available at https://arxiv.org/abs/1811.04236, 2018.

[46] ———, *A premouse inheriting strong cardinals from V*, ***Annals of Pure and Applied Logic***, vol. 171 (2020), no. 9, p. 66, Id/No 102826.

[47] ———, *Background construction for λ-indexed mice*, available at https://arxiv.org/abs/2101.00889, 2021.

[48] ———, *The definability of* $\mathbb{E}$ *in self-iterable mice*, ***Annals of Pure and Applied Logic***, vol. 174 (2023), no. 2, p. 59, Id/No 103208.

[49] FARMER SCHLUTZENBERG and JOHN R. STEEL, *Comparison of fine structural mice via coarse iteration*, ***Archive for Mathematical Logic***, vol. 53 (2014), no. 5-6, pp. 539–559.

[50] FARMER SCHLUTZENBERG and NAM TRANG, *Scales in hybrid mice over* $\mathbb{R}$, available at https://arxiv.org/abs/1210.7258, 2014.

[51] ROBERT M. SOLOVAY, *Strongly compact cardinals and the GCH*, ***Proceedings of the Tarski Symposium. An international symposium held to honor Alfred Tarski on the occasion of his seventieth birthday*** (L. Henkin, J. Addison, C. C. Chang, W. Craig, D. Scott, and R. Vaught, editors), Proceedings of Symposia in Pure Mathematics, vol. 25, American Mathematical Society, Providence, RI, 1974, pp. 365–372.

[52] JOHN R. STEEL, ***The Core Model Iterability Problem***, Lecture Notes in Logic, vol. 8, Springer-Verlag, Berlin, 1996.

[53] ———, *Core models with more Woodin cardinals*, ***The Journal of Symbolic Logic***, vol. 67 (2002), no. 3, pp. 1197–1226.

[54] ———, *A theorem of Woodin on mouse sets*, unpublished notes, available at http://math.berkeley.edu/~steel/, 2004.

[55] ———, PFA *implies* $\mathsf{AD}^{L(\mathbb{R})}$, ***The Journal of Symbolic Logic***, vol. 70 (2005), no. 4, pp. 1255–1296.

[56] ———, *A stationary-tower-free proof of the derived model theorem*, ***Advances in logic***, Contemp. Math., vol. 425, Amer. Math. Soc., Providence, RI, 2007, pp. 1–8.

[57] ———, *Derived models associated to mice*, ***Computational Prospects of Infinity. Part I. Tutorials*** (C. Chong, Q. Feng, T. A. Slaman, W. H. Woodin, and Y. Yang, editors), Lecture Notes Series, Institute for Mathematical Sciences, National University of Singapore, vol. 14, World Scientific, 2008, pp. 105–193.

[58] ———, *The derived model theorem*, ***Logic Colloquium 2006*** (S. B. Cooper, H. Geuvers, A. Pillay, and J. Väänänen, editors), Lecture Notes in Logic, vol. 32, Cambridge University Press and Association for Symbolic Logic, 2009, pp. 280–327.

[59] ———, *An outline of inner model theory*, ***Handbook of Set Theory. Vols. 1, 2, 3***, Springer, Dordrecht, 2010, pp. 1595–1684.

[60] ———, *Hod mice below* LST^-, *part I*, handwritten notes, available at https://math.berkeley.edu/~steel/, 2013.

[61] ———, *Remarks on a paper by Sargsyan*, handwritten notes, available at ttps://math.berkeley.edu/~steel/papers/remarks.sargsyan.pdf, 2015.

[62] ———, *Mouse pairs and Suslin cardinals*, to appear, available at https://math.berkeley.edu/~steel/, 2016.

[63] ———, ***A Comparison Process for Mouse Pairs***, Lecture Notes in Logic, vol. 51, Cambridge University Press, 2023.

[64] NAM TRANG, ***Generalized Solovay Measures, the HOD Analysis, and the Core Model Induction***, Ph.D. thesis, University of California, Berkeley, 2013.

[65] ———, PFA *and guessing models*, ***Israel Journal of Mathematics***, vol. 215 (2016), no. 2, pp. 607–667.

[66] MATTEO VIALE and CHRISTOPH WEISS, *On the consistency strength of the proper forcing axiom*, ***Advances in Mathematics***, vol. 228 (2011), no. 5, pp. 2672–2687.

[67] TREVOR M. WILSON, *The envelope of a pointclass under a local determinacy hypothesis*, ***Annals of Pure and Applied Logic***, vol. 166 (2015), no. 10, pp. 991–1018.

[68] W. HUGH WOODIN, ***The Axiom of Determinacy, Forcing Axioms, and the Nonstationary Ideal***, 2nd revised ed., Walter de Gruyter, Berlin, 2010.

[69] ———, *In search of ultimate-L. The 19th Midrasha mathematicae lectures*, ***The Bulletin of Symbolic Logic***, vol. 23 (2017), no. 1, pp. 1–109.

[70] MARTIN ZEMAN, ***Inner Models and Large Cardinals***, de Gruyter Series in Logic and its Applications, vol. 5, Walter de Gruyter & Co., Berlin, 2002.

INDEX

For EU product safety concerns, contact us at Calle de José Abascal, 56–1°,
28003 Madrid, Spain or eugpsr@cambridge.org.

www.ingramcontent.com/pod-product-compliance
Lightning Source LLC
Chambersburg PA
CBHW[illegible]1125
35866CB00079B/7
9781009520713